ALUMINUM CASTING TECHNOLOGY

2nd Edition

ALUMINUM CASTING TECHNOLOGY

2nd Edition

Reviewed and Revised by

John L. Jorstad, Reynolds Metals Co.

Wayne M. Rasmussen, AFS

Edited by

Donna L. Zalensas, AFS

AMERICAN FOUNDRYMEN'S SOCIETY, INC.

DES PLAINES, ILLINOIS

Published and distributed

by the

American Foundry Society
Des Plaines, Illinois 60016-8399

www.afsinc.org

Reprinted with corrections 1997
Reprinted 2001

ISBN 0-87433-157-9

Printed in the United States of America

CONTENTS

FOREWORD i

Chapter 1
CASTABILITY OF ALUMINUM ALLOYS

Characteristics 1
Choosing a Casting Alloy 2

Chapter 2
MELTING PRACTICES & FURNACES

Melting Practices 5
Tools and Their Maintenance 6
Effects of Temperature & Holding Time 6
Automatic Ladling Devices 8
Types of Furnaces 8
Selecting Melting Equipment 9
Crucible Furnaces 9
Reverberatory Furnaces 10
Electric Induction Furnaces 14
Electric Resistance Furnaces 15
Filter Systems for Melting Equipment 16
Pumping Molten Metal 17

Chapter 3
CHEMICAL & PHYSICAL CHARACTERISTICS OF MOLTEN ALUMINUM

General Characteristics 19
Dross Formation 19
Gas Porosity in Aluminum Castings 20
Fluxing the Melt 26
Determining Which Flux to Use 27
Grain Refinement 28
Structural Modification of Hypoeutectic Al-Si alloys 32
Structural Modification of Hypereutectic Al-Si Alloys 35
Thermal Analysis 37

Chapter 4
ALUMINUM CASTING ALLOYS

Introduction 45
Properties and Characteristics 46
Selecting Alloys and Casting Methods 47
Aluminum Ingot 48
Chemical Composition Limits, Properties and Treatments 50
Definitions of Terms Used with Alloys 52
Hypereutectic Al-Si Alloys 53
General Properties of Hypereutectic Al-Si Alloys 53
General Foundry Characteristics 53
Properties and Characteristics of Principal Casting Alloys 56
Chapter 4 Appendix
Registration Record of Aluminum Association Alloy Designations and Chemical Composition Limits for Aluminum Alloys in the Form of Castings and Ingot 105

Chapter 5
DESIGNING ALUMINUM CASTINGS

Introduction to Casting Design 113
Predesign Considerations 113
Design for Minimum Casting Stresses 113
Design for Strength 114
Calculating Service Stresses 114
Design for Directional Solidification 114
Design for Metal Flow 115
Composite Design 116
Safety and Allowable Design Stresses 116
Design for Low Pattern and Mold Cost 116
Modelmaking as a Design Aid 117
Summary of Basic Casting Design 117
Design Considerations 117
General Design Suggestions 119
Designing Die Castings 127
Die and Permanent Mold Casting Guides 129
Designing Permanent Mold Castings 130
Designing Sand Castings 133
Designing Shell Mold Castings 135
Designing Plaster Mold Castings 136
Designing Centrifugal Castings 137
Designing Investment Castings 139
Designing Expendable Pattern Castings 141

Chapter 6
DIECASTING

The Process 145
Economics of Diecasting 145
Factors to Consider about Diecasting 147
Comparisons with Other Manufacturing Processes 148
Aluminum Diecasting Alloys 150
Machines for Diecasting Aluminum 150
Dies for Aluminum Casting 153
Designing and Building Dies 154
Runners and Gating 157
Venting 158
Ejector Pins 158
Cores 158
Die Release Agents and Lubricants 159
Trimming and Finishing 159
Automating the Diecasting Process 160
Designing Better Diecasting Machines 161
Variations to Traditional Diecasting 163
Correcting Defects 164
Diecasting Pure Aluminum 166

Chapter 7
PERMANENT MOLD CASTING

The Process 167
Advantages and Disadvantages 168
Applications for Permanent Molding 169
Alloys Used for Permanent Molding 170
Effect of Cooling Rate 170
Design Factors That Affect Soundness 170
Preventing Hot Cracking 172
Determining Mold Cavity Dimensions 173
Mold Materials 173
Cores in Permanent Molds 174
Mold Operations 176
Melting Practices 177
Pouring into Permanent Molds 178
Variations to the Process 181
Quality Control 181
Defects in Permanent Mold Castings 181

Chapter 8
SAND CASTING

Introduction to Sand Casting 185
Alloys Typically Used for Sand Casting 186
Automated Foundry Systems 186

Green Sand Molding

Introduction 187
Description of the Process 187
Sand for Green Sand Molding 187
Sand Segregation 191
Clay Characteristics 191
Types of Clays Used in Green Sand 192
Water in Green Sand Mixtures 193
Oil-Bonded Molding Sands 193
Natural and Synthetic Green Sands 193
Special Additives for Green Sands 195
Mulling Green Sands 196
Effect of Heat on System Components 197
Maintaining the System 197
Sand Testing for Clay-Bonded Systems 197
Calibration of Testing Equipment 202

Chemically Bonded Sand Molding and Coremaking

Resin Binder Materials and Practices 203
Binder System Development 203
Binder Identification & Classification 203
Heat-Activated Systems 206
Coldbox Systems 207
Nobake Systems 210
Core and Mold Testing for Resin-Bonded Sands 215
Sand for Producing Molds and Cores 217
Additives 217
Reclamation and Reuse of Spent Sand 217
Refractory Coatings 219
Venting 220
Effect of Relative Humidity on Binders 220
Shakeout 221
Conclusions 221
Bibliography 221

Chapter 9
THE SHELL PROCESS

Introduction 223
Coreboxes and Patterns 224
Shell Process Equipment 225
Materials 226
Producing Shell Cores 227
Producing Shell Molds 229
Testing and Process Control 232
Casting Defects 232

Chapter 10
PLASTER MOLD CASTING

Introduction 235
Prototype and Simulated Die Castings 236
Designing Plaster Mold Castings 236
The Process 236
Pattern Material 238
Metalcasting Plaster Process 238
Variations to the Metalcasting Plaster Method 241
Permeable Plaster Methods 241
Plaster Cores 243

Chapter 11
CENTRIFUGAL CASTING

Introduction 245
Methods for Centrifugal Casting 245
Centrifugal Casting Machines 246
Designing Centrifugal Castings 248
Aluminum Alloys & Foundry Practices 248
The Centrifuge Process 248
Defects in Centrifugal Castings 249

Chapter 12
INVESTMENT CASTING

Background 251
The Process 251
Patternmaking 252
Materials and Mold Manufacturing 255
Solid Mold Production 256
Ceramic Shell Production 257
Dewaxing Solid Molds 259
Dewaxing Ceramic Shell Molds 259
Mold Firing 260
Metal Melting and Casting 260
Postcasting Operations 261
Mechanization of Investment Casting 261

Chapter 13
EXPENDABLE PATTERN CASTING

History 263
Expandable Polystyrene Patterns 264
Coatings 267
Sand 267
Flasks 267
Gates and Risers 268
Pouring 268
Casting Quality 269

Chapter 14
GATING & RISERING

Velocity, Flow Rate and Choke 271
Fluid Flow 271
Requirements of Gating Systems 272
Horizontal Gating 272
Vertical Gating 276
Filters 277
Gating for Aluminum Composites 279
Solidification 280
Risers 281
Chills 284

Chapter 15
HEAT TREATMENT & STRENGTHENING OF ALUMINUM CASTINGS

Heat Treatment Effects on Properties 287
Heat Treatment Practices 290
Metallurgical Aspects 291
Strengthening Aluminum 293
A Practical Example 295
Summary 296
Chapter 15 Appendix A
    Vacancies 297
    Physical Metallurgy of Deformation 297
Chapter 15 Appendix B
    Effects of Aging on Properties
    of Alloys 355 and 356 299

Chapter 16
METALLOGRAPHY

Introduction 303
Sampling 303
Macroscopic Examination 303
Microscopic Examination 305
Simple, As-Cast Microstructures 311
Heat Treatment 313
Modification of Al-Si Alloys 318
Refinement of Primary Si
    in Hypereutectic Al-Si Alloys 318
Other Al-Si-Mg and
    Al-Si-Cu-Mg Alloys 318
Die Castings 320
Discontinuities 321
Surface Coatings 322
Service-Related Effects 322
Typical Microstructures of
    Aluminum Casting Alloys 323

Chapter 17
POSTCASTING OPERATIONS

Introduction 325
Cleaning and Trimming 325
Inspection of Castings 330
Machining Aluminum Castings 332
Repairing Castings 334
Finishing Aluminum Castings 338
Chapter 17 Appendix
    Machining Data Tables 343

INDEX 349

FOREWORD

The American Foundrymen's Society presents *Aluminum Casting Technology, 2nd Edition,* the most complete text and reference source that we know of for the producers and users of aluminum castings. Those of us who remember *Casting Kaiser Aluminum,* in its day, as the authority on aluminum casting will find this volume the authority for today.

This textbook has been updated extensively from the previous version to reflect current technology. Chapters have been added on expendable pattern casting (Chapter 13) and on gating and risering (Chapter 14). The chapters on sand casting (Chapter 8) and heat treatment and strengthening (Chapter 15) were completely rewritten.

The format of the chapter on aluminum casting alloys (Chapter 4) was changed to make it easier for the reader to find information about specific alloys. The remaining chapters were reviewed and revised to provide the user with updated information in an easy-to-read and easy-to-understand format.

The American Foundrymen's Society has acknowledged contributing authors and companies in each of the chapters. The Society also wishes to acknowledge the authors and committee members from the previous edition and from *Casting Kaiser Aluminum,* where this book started.

John L. Jorstad

Wayne M. Rasmussen

Chapter 1

CASTABILITY OF ALUMINUM ALLOYS

CHARACTERISTICS

Aluminum casting alloys are the most versatile of all common foundry alloys and generally have the highest castability ratings. Aluminum is adaptable to many of the commonly used casting methods and can be readily cast in metal molds or in dies mounted on automated machines for high volume, cost-effective casting. Aluminum also is castable by the sand, shell mold, centrifugal, investment, expendable pattern casting and plaster processes. Single or multiple units may be cast in one mold.

As casting materials, aluminum alloys have the following favorable characteristics:

- good fluidity—important for filling thin sections (Fig. 1-1);
- low melting point—lower melting and casting temperatures are needed relative to those required for many other metals;
- lower density—the lighter weight of aluminum alloys allows easier handling in the foundry, compared to ferrous casting alloys;
- fast heat exchange—heat transfer from the molten aluminum to the mold is relatively fast, providing faster casting cycles in metal molds;
- good chemical reproduction—chemical stability is relatively good;
- good as-cast surface finish—surfaces are lustrous and have no blemishes (Fig. 1-2).

Aluminum casting alloys have a considerably lower operating temperature range (1200–1400F, 650–760C) than do most other common structural metals. During sand casting, lower pouring temperatures reduce sand problems because there is a minimum amount of burn-in during casting. The molding sand used has a longer life and can be used over and over again—new sand additions are required only to make up for mechanical and thermal losses.

Low operating temperatures enable greater flexibility in handling the molten metal, and the melting equipment requires less maintenance. The lower pouring temperatures also allow the use of gypsum-based plaster molds for casting aluminum.

One of the most useful casting properties of aluminum is its low specific gravity. Its light weight permits small castings and many medium-sized castings to be poured efficiently with hand ladles. Only the largest castings must be poured from mechanically transported ladles or crucibles.

Fig. 1-1. Good fluidity and excellent mechanical properties allow for the casting of thin blades for a compressor wheel casting.

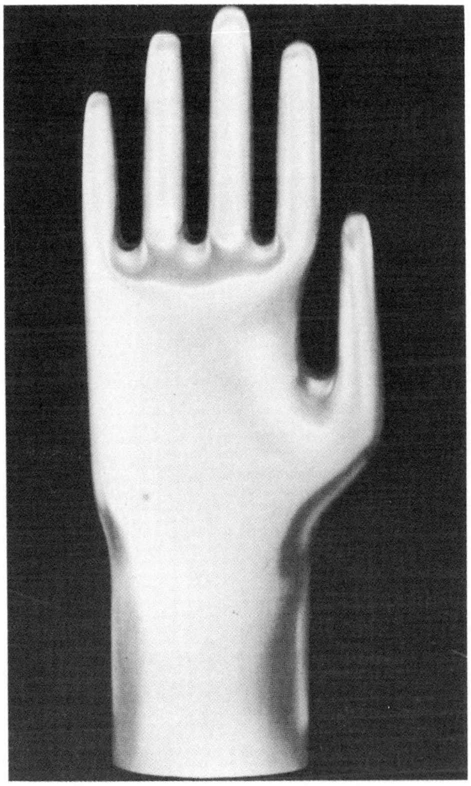

Fig. 1-2. Extraordinary surface smoothness was specified for this cast aluminum surgical glove mold. Plaster was chosen for the mold material, and a CO_2 core was used.

With the lower density and lower casting temperatures of aluminum, the problem of sand washes—which are common in the pouring of heavy metals—is practically eliminated. Also, lower mold pressures are obtained, permitting the use of more lightly rammed and more permeable molds and lighter molding equipment. Small- and medium-sized castings can be poured without placing weights on the mold to hold the cope and drag together. It is fairly common aluminum foundry practice to make molds using snap flasks and to pour the molds without placing them in jackets for support.

A less favorable characteristic that accompanies the lower specific gravity of aluminum is that nonmetallic impurities (such as oxides) that may be present are close to the specific gravity of aluminum itself. Therefore, greater attention is required to separate these impurities from the metal during melting and casting operations than is usually the case with heavier metals.

Aluminum alloys contract considerably during cooling. They undergo liquid contraction as they cool from the pouring temperature to the solidification temperature, solidification contraction as they freeze, and solid contraction as the solid casting cools to room temperature. The total range of contraction is approximately 6.5–8.5%. When allowance is made for this contraction in the design of patterns and equipment, castings can be produced within relatively close dimensional tolerances.

The quality and cost of castings are directly affected by the type of material that is melted, the melting equipment that is used and the casting methods that are employed. Because molten aluminum alloys absorb hydrogen readily and oxidize easily, good foundry practices must be followed to obtain optimum results.

Hydrogen gas, soluble in liquid aluminum and essentially insoluble in solid aluminum, is responsible for much of the porosity in aluminum castings. When molten aluminum containing dissolved hydrogen solidifies, the hydrogen gas is rejected from the molten metal and forms porosity in the casting.

Gas absorption in aluminum increases with increased temperature, long holding times in the molten state and proximity of the metal to products of combustion. High humidity in the atmosphere surrounding the melt and in the fuel mixtures of the melting furnace also will increase gas absorption. Some gas absorption also can be caused if there is excessive moisture in molding sand. Likewise, moisture or other hydrogen compounds (oil, for example) on ingot or other metal to be added to the melt can contribute hydrogen to the melt. Preheating the metal to be charged to about 900F (480C) before adding it to the melt will greatly reduce this source of hydrogen.

Molten aluminum surrounds itself completely with a thin envelope of oxide. As long as this oxide layer remains unbroken, the rate at which gas is absorbed by the melt is quite low, and further oxidation is retarded.

A furnace that is used for melting aluminum should permit minimal hydrogen gas absorption and oxidation, and it should be of a size that provides adequate melting capacity with overall economy. Good temperature controls are also an essential part of the furnace operation because excessively high melting and processing temperatures tend to yield defective castings with coarse grain, gas porosity and dross defects.

A fairly recent process that has evolved is the *expendable pattern casting* (EPC) process. In the EPC process, expandable polystyrene patterns in loose, dry sand are used to make large quantities of castings. The process also has enhanced aluminum casting by allowing intricate shapes to be cast without cores.

The favorable characteristics of molten aluminum make it highly desirable for casting in permanent metal molds or dies, which are made of iron or steel. In the *gravity permanent mold* method, the molten aluminum is poured into a gate at the top of the mold, and the mold is filled by force of the weight of the metal itself (gravity).

A modification of the gravity permanent mold method is the *low-pressure permanent mold* method whereby the mold is positioned above a hermetically sealed holding furnace. A tube extends from below the surface of the molten metal in the furnace to the bottom of the mold cavity. The mold is filled by applying a small amount of gas pressure above the melt in the sealed furnace, which is sufficient to force the metal up the tube into the mold cavity. In a variation of this low-pressure method, an electromagnetic pump is used to bottom feed a mold instead of using the hermetically sealed and pressurized furnace.

In *pressure diecasting,* molten aluminum is injected into a mold (die) under high pressure by a hydraulically actuated piston (Fig. 1-3).

These methods, when sufficiently automated, offer important economic advantages in high-volume production of aluminum castings. Permanent molds and dies can be used to make thousands of castings.

CHOOSING A CASTING ALLOY

Many casting alloys are available to the foundry industry, with combinations of properties suitable for almost limitless applications. Because of the individual nature of each alloy and each casting process, the selection of the most suitable and economical casting alloy for a specific application depends on a multitude of factors.

Its relatively low strength and limited castability limits the use of pure aluminum largely to the production of rotor castings for electrical motors and other applications in which high electrical conductivity is required. By alloying aluminum, mechanical properties as well as casting characteristics are improved appreciably. Therefore, all practical aluminum casting alloys contain varying percentages of alloying elements, and each alloy has been developed to provide specific properties for various applications.

The selection of the alloy and the most economical casting process to use must take into account the capabilities of the foundry, the mechanical and physical properties of the alloy, and the specific service for which the part is designed. The alloy that has the most desirable balance of properties necessary for good performance of the casting is the one that should be chosen.

After all requirements have been considered, the alloy chosen should have:

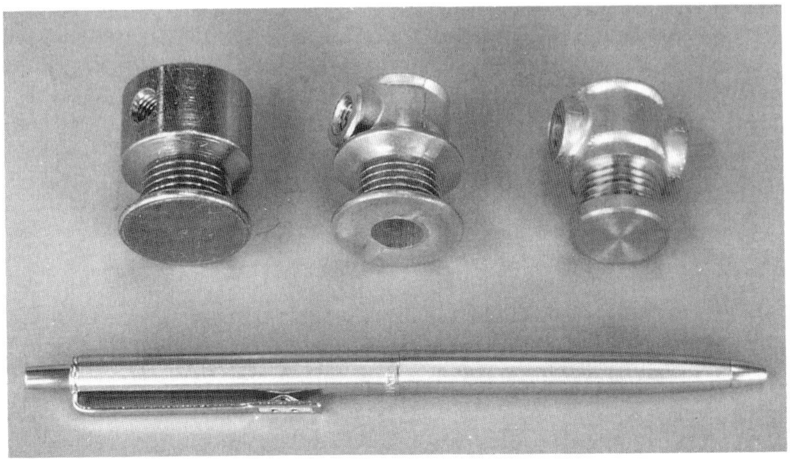

Fig. 1-3. Diecast parts often can be designed with thinner walls, reduced draft and closer tolerances than parts made using other methods. The pulley for a home appliance (left) was produced from 1 in. (25 mm) diameter bar stock using automatic screw machines. The diameter of the part was controlled by the need for a tapped hole that received a set screw. The diecast version (center) uses a smaller-diameter hub with a raised boss to accommodate the screw. The pulley on the right is a prototype of the diecast version. [Photo courtesy of the Diecasting Development Council.]

- good foundry characteristics for the casting process that will be used;
- the required balance of mechanical and other properties for the service intended;
- stable mechanical properties over the required service life;
- adequate corrosion resistance for the service application;
- satisfactory machinability, good pressure tightness and good finishability, if these are of importance.

The cost of the finished part is directly affected by the following:

- the alloy chosen and the inherent technical requirements of casting that alloy;
- the quantity of castings to be produced;
- the casting process used;
- the complexity of the shape of the part;
- the dimensional tolerances required;
- the inspection and quality control standards required.

Chapter 2

MELTING PRACTICES & FURNACES

MELTING PRACTICES

To consistently produce castings of good quality, foundries must control two major variables of the casting process: melting practices and mold conditions. Design, fabrication, use and maintenance of molds are discussed later in this book.

Melting must be regarded as much more than merely the transformation of solid metal into a liquid state. The melting operation is important to the soundness and strength of all cast products. Good melting equipment and the use of the correct techniques are primary requisites for achieving an end product that retains, within practical limits, certain desired metallurgical properties of the original alloy from which it is cast.

For pure aluminum, the melting point is 1221F (660C); the heat of fusion is 169 Btu/lb; and the specific heat at 212F (100C) is 0.226 cal/g. Aluminum, like other metals, comes to the melting temperature quite readily, then it must take on more heat before it actually changes to a liquid. Figure 2-1, a graph of the heat content of aluminum at various temperatures, illustrates this important physical characteristic.

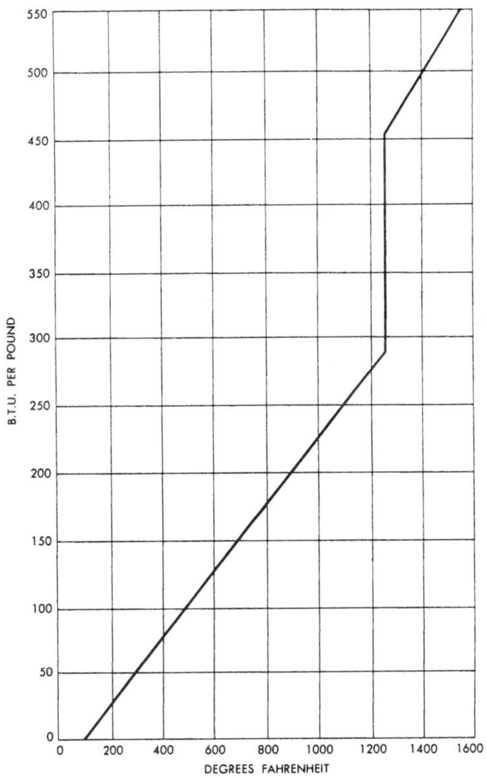

Fig. 2-1. Thermal capacity of pure aluminum. Note that aluminum doesn't just get hot enough to melt. After reaching the melting temperature, it requires more than 160 additional Btu/lb before the aluminum actually takes the liquid form.

Aluminum is seldom used in its pure form and is generally alloyed with other metals or elements. Consequently, the melting temperatures and heats of fusion of various aluminum alloys vary quite a bit.

Problems and Factors When Melting Aluminum

Aluminum does not behave well in all respects in the melting process. It has certain characteristics that cause reactions that must be controlled to produce a satisfactory end product.

At elevated temperatures, aluminum combines rapidly with oxygen to form oxides, and during melting, the quantity of oxides increases with the temperature. Because the specific gravity of molten aluminum alloys is close to that of the oxides, the latter are easily mixed into the melt if proper precautions are not taken. Therefore, agitation of molten metal should be avoided or at least held to a minimum.

Another problem confronting the aluminum foundry is the fact that the metal will absorb certain gases, of which hydrogen is the most troublesome. In an open-flame melter, there is a risk of gas absorption because water vapor is produced in the combustion process. Hydrogen-forming materials also can be carried into the molten bath on the surface of cold or dirty chargings.

Furnace atmosphere and characteristics of the burner flame are important factors to consider. It has been found that the envelope of oxide that usually covers the metal offers considerable protection against gas absorption. As long as this envelope remains unbroken, the absorption of gas is relatively low. Generally, a slightly oxidizing flame is best to maintain the oxide film without excessive oxidation.

A melting problem that is usually present when large reverberatory furnaces are being used is a tendency for hard oxide formations to develop at the metal-refractory interface. This buildup can reduce both furnace capacity and refractory life. There are other detrimental effects: oxide buildup causes metal contamination and increases fuel cost. One extreme case revealed that the capacity of a 50,000 lb furnace was reduced to 35,000 lb because of excessive oxide buildup alone.

Special furnace-cleaning compounds can be used to remove oxide buildup and to prevent future oxide formation on furnace walls, but the most economical preventative measure (and for that matter, one of the best ways to assure quality melting) is faithful adherence to a good housekeeping policy. Good housekeeping should include rigid supervision of methods and routines (including special training for employees who operate furnaces and handle molten aluminum); frequent drossing or cleaning of the bath; proper use of fluxes or flushing gases; automatic control of furnace and metal temperatures; and periodic maintenance of refractories and burner equipment.

There are no substitutes for good equipment and good melting practices when melting aluminum. Good furnace equipment is a primary requisite. Even though degassing agents have their place in aluminum melting, they should not be considered a remedy or cover-up for poor melting and handling practices.

Melting Aluminum in the Foundry

After new furnace linings or refractory crucibles have been carefully and slowly preheated, the furnace may be charged. Furnaces that have been in use can be charged cold. Crucible furnaces should not be packed tightly because the charge expands and will tend to crack the crucible. Bridging, which causes overheating of the metal in the bottom of the crucible, also may occur if the charge is overpacked.

After the furnace has been charged and melting has started, more charge metal should be added on top as the solid material in the crucible settles into the molten metal. This will serve to preheat the ingot and scrap before it sinks below the melt surface. Preheating is important because it helps to eliminate the adsorbed and chemically combined gases and the moisture that is present on the ingot surface.

The charge should consist of alloy ingot and clean foundry scrap of the same alloy. The alloy ingot can be prepared in foundry reverberatory furnaces or purchased from aluminum producers who manufacture refined standard alloyed foundry ingot (see Chapter 4).

Some foundries prefer to melt foundry scrap separately. In large-volume operations, scrap is sometimes melted in a reverberatory furnace. The melt is analyzed and brought to the correct composition, if necessary, and then cast into ingots or transferred as liquid for subsequent use in the foundry. A quick analysis of the chemical composition of the melt can be made with a spectrometer, which can complete an electronic analysis of aluminum melt samples in about three minutes.

In general, it is not desirable to prepare foundry melts from mixed charges that include pure aluminum unless extreme care is exercised in obtaining the correct weight and composition of all the materials in the charge.

When a foundry is producing its own casting alloys, the most economical method is to use rich alloy ingot (often called master alloys or hardeners). The foundry adds standard pure ingot and alloying elements in proper proportions to the rich alloy ingot to obtain the specified composition for the casting alloy.

When adding alloying elements to foundry melts of aluminum, low-melting elements, such as zinc and magnesium, can be added in their pure forms. However, alloying elements that have high melting points (such as manganese, nickel, silicon, titanium, chromium and copper) are best added in the form of a master alloy of known composition. Silicon and copper may be added in their pure forms, but their solution in the melt is relatively slow—considerable care must be exercised if this is done. Silicon tends to float on the surface of the melt and become oxidized, thereby greatly reducing its rate of solution. Magnesium also floats and must be submerged into the melt immediately to prevent magnesium burnout and the formation of dross and spinel (a form of magnesium oxide).

Foundry alloying is commonly done in reverberatory furnaces, which may be operated as batch furnaces or as continuous or semicontinuous melting units. When one is used as a batch furnace, all or most of the metal is drained from the hearth after each melting operation. In the semicontinuous method, the melt is tapped in batches, and additional charges are added and melted between taps.

In the continuous method, molten metal is removed continuously from the hearth at one end while charges are added at the other. Continuous melting is best done in a twin-chamber furnace in which the charge is melted in one chamber and the metal is tapped continuously from the other.

TOOLS AND THEIR MAINTENANCE

Foundry tools such as stirrers, plungers (for adding volatile materials), skimmers and hand ladles are usually made of steel. These tools should be kept clean and properly coated with a refractory wash. Before inserting them into the melt, they should be thoroughly preheated to a red glow, or nearly so. Otherwise, they may be a source of hydrogen absorption and a hazard to workers because of the danger of explosion from moisture on the tool surfaces.

Graphite tools are sometimes used, especially graphite fluxing tubes for degassing. Though they are more fragile than iron or steel, these tools are not sources of metallic contamination. As with steel tools, these also must be carefully preheated to a red glow before they are inserted into the melt. This is especially true of carbon tubes and other devices used to introduce chlorine or halide fluxing compounds into the melt, for such devices invariably become impregnated with hygroscopic compounds. These compounds, present from previous use, must be thoroughly dehydrated at a high temperature.

EFFECTS OF TEMPERATURE & HOLDING TIME

During the preparation of melts in the foundry, it is important to avoid unreasonably high temperatures and excessive holding times. In general, gating practices should be designed to permit pouring at the lowest temperatures possible. This is an important phase of good foundry practice. Excessive temperatures and long holding periods can result in coarse-grained castings with low mechanical properties. They also increase the chances of gas absorption and dross formation. Melts should never be overheated, and unless required for the casting, melt temperature should not exceed 1400F (760C).

Because overheating is so harmful to the quality of the melt, it should be emphasized that the temperature of the aluminum tends to increase rapidly after the charge has completely melted. Great amounts of heat are required to change the metal from the solid to the liquid state. Reliable furnace controls will prevent overheating.

Adequate precautions should be taken to prevent residual heat in the furnace from raising the temperature of the melt above the desired setting after the burners are turned off. It is possible to equip the melting room with automatic signals that will warn the operator when the charge has been melted. Automatic temperature control equipment also will minimize this problem.

Temperature Control and Measurement

Temperatures of molten aluminum are usually measured by chromel-alumel thermocouples. Figure 2-2 shows the thermocouple wire separated by double-hole porcelain insulators in a closed-end, 1 in. ID x 1½ in. OD cast iron protective tube for continuous measurement of the melt.

Silicon carbide or porcelain-enameled, cast iron protective tubes also are used. No. 8 B&S gauge thermocouple wire is normally used, but for fast, intermittent spot checking, No. 20 B&S gauge is recommended. To prevent the melt from picking up iron from the protective tube and also to prolong the tube's life, the tube should be kept clean and coated with a refractory wash similar to that used on tools or pots. Fibrous glass sleeves also are used.

Measuring melt temperature in pouring ladles requires a sensitive thermocouple that reacts rapidly. Either open-end or closed-end thermocouples can be used. Numbers 8 and 14 B&S gauge chromel-alumel wires wrapped tightly with insulation to within ½ in. of the lower, open end are submerged into the melt. This is a rapid-acting couple, but it is not as reliable as a closed-end couple.

Closed-end couples for rapid reading should be of light gauge. They may be the stationary type for measuring temperature at a fixed location in the foundry or the portable type for use on the pouring floor (Fig. 2-3). Thin-wall or silicon carbide protection tubes are sometimes used on couples of this type, but use of a protection tube makes the action of the couples slow.

Both open and closed unprotected couples require a great deal of maintenance because they are attacked rapidly by the molten aluminum. On open-end couples, the burned insulation is removed, and the ends of the wires that were attacked by the aluminum are cut off. The damaged part of closed-end chromel-alumel couples is cut off and the ends rewelded. Some closed-tip thermocouples are made with replaceable tips (Fig. 2-4), thereby simplifying maintenance.

Leads from the thermocouple are connected to a direct-reading millivoltmeter or potentiometer. Millivoltmeters are usually preferred in smaller shops because they do not require a power source for operation and can be moved easily. Potentiometers are preferred by many foundries because they give truer temperature readings, are easier to read and can make permanent heat record charts. Both potentiometers and millivoltmeters must be carefully maintained and serviced.

Stationary, direct-reading potentiometers or recording pyrometers with selector switches are used in some foundries. Such equipment is placed at strategic locations in the melting room and on the pouring floor. The pyrometer serves several thermocouples, which are connected to a selector switch.

A thermocouple can be checked readily by placing it in a closed-end, thin-walled carbon tube and immersing it in a small pot of pure aluminum that is allowed to solidify. The temperature will drop continuously to the solidification point where it will remain constant for a considerable period of time. This stable temperature is the melting point of the aluminum and corresponds to about 1220F (660C). Frequent checking of thermocouples, potentiometers and millivoltmeters is necessary to avoid serious errors when measuring pouring temperatures. Portable instruments, especially, require frequent checking.

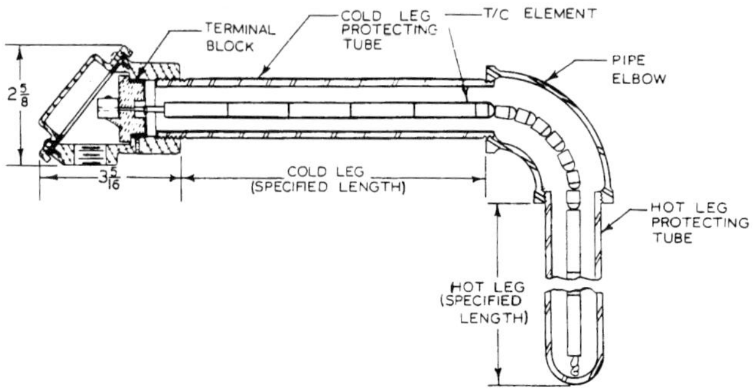

Fig. 2-2. Cross-sectional drawing of a closed-end thermocouple.

Fig. 2-3. Checking melt temperature with a portable thermocouple. The melt temperature is read on a compact, lightweight pyrometer.

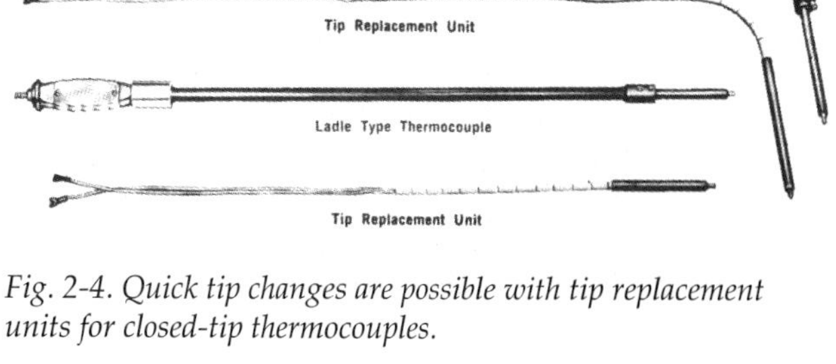

Fig. 2-4. Quick tip changes are possible with tip replacement units for closed-tip thermocouples.

AUTOMATIC LADLING DEVICES

Reliability of process control in aluminum casting production is enhanced by proper application of automatic ladling devices. This equipment helps assure quality and minimizes wasted production time and scrap generation from misruns. In volume production, the savings from reducing cycle time can be significant. Automatic pouring or ladling equipment is available for use in all types of high-production casting.

For the expendable pattern casting (EPC) process, automatic pouring is used to prevent variations from pour to pour and to prevent interruptions in pouring, which can cause the unbonded sand used in the process to collapse into the casting cavity.[1] In the system shown in Fig. 2-5, EPC canisters rotate on an indexed turntable. A hydraulically operated robot arm submerges a ladle into molten aluminum inside an electric resistance furnace. The ladle (which is covered with a nonwetting coating) fills with metal through a slot on its back side. When the ladle is filled, the robot arm raises, allowing excess metal to drain off, then rotates to the pour position over an EPC canister where it ladles molten metal into a pouring cup positioned above the sprue.[2]

One type of automatic ladling system for green sand molding is shown in Fig. 2-6. The two pouring chambers are filled pneumatically using a vacuum. Green sand molds move by the pouring area on a conveyor at a controlled rate. When the chambers have filled to the desired level and the molds are in the correct position, the vacuum is released, and pressure is applied. This pressure pushes the molten aluminum into launders (troughs), which channel it into the pouring basins of the molds. After the molds are filled, they are transported by the conveyor to cooling and shakeout areas.[3]

Other automatic ladling systems use electromagnetic pumps to move molten metal at exact rates and to release it in precise doses. The pumps used for aluminum alloys are rectangular channels made of ceramic materials. Two magnets create an electromagnetic force, which moves the molten aluminum. This method for moving molten aluminum has the advantage of having no moving parts.

Fig. 2-5. This automatic ladling device provides a consistent pouring rate when pouring expendable pattern castings.

TYPES OF FURNACES

Fuel conservation and improved aluminum casting processes have brought about many new developments in melting furnaces. Both the need for larger-capacity equipment to melt and hold aluminum and the search for faster, more economical melting methods have increased due to greater demands for cast aluminum products. Moreover, today's quality specifications for aluminum castings are more rigid. As a result, more attention is being given to the types of furnaces used and to their construction and operation to ensure proper melting and pouring conditions.

While foundries can, in some instances, use or adapt existing aluminum melting equipment to meet their stringent specifications, they will usually find it to their advantage to check their melting equipment and practices from time to time against a list of factors that directly influence the quality of the castings produced. Two of these factors are the type of furnace and its condition. In refractory-lined furnaces, the lining must be cleaned periodically and kept in good repair. Exhaust gases should be controlled adequately. Combustion gases must be sealed off to help prevent gas absorption by the molten aluminum.

Furnaces for melting and/or holding aluminum may be classed or grouped into many types. Some of these types can be divided and then subdivided according to size, construction material, shape, methods of charging, mobility, melting rates and various other characteristics. Aluminum melting furnaces for foundry use are divided by heat source into two main groups: fuel-fired and electric.

Fig. 2-6. A green sand molding line with pressure-pour auto ladling. [Photo courtesy of CMI-Wabash Cast.]

Fuel-fired furnaces generally burn either gas or oil. Some fuel-fired furnaces are direct-fired furnaces, in which products of combustion pass directly over the metal charge. Others are indirect-fired furnaces, in which products of combustion do not contact the metal charge directly.

There are two main types of electric melting furnaces: the induction type and the resistance type for melting and holding. Individual furnaces of the same type or of common nomenclature may vary with the manufacturer's design, but their operating principles remain the same.

SELECTING MELTING EQUIPMENT

When selecting melting and holding furnaces, the quantities and types of alloys to be melted are of primary importance. If large quantities of molten metal are required for production runs, a single-chamber, direct-charging reverberatory furnace would be the best selection. Where only a few hundred pounds of metal are being melted to produce small castings, a small, indirect-fired crucible furnace should suffice. If only a single alloy is to be cast, the selection probably would be a reverberatory or induction furnace.

The cost of energy to melt aluminum is an important factor to consider when selecting melting and holding furnaces. The ideal heating method is the least expensive one for the particular locality, but foundry efficiency and casting quality should not be compromised for slightly lower heating costs.

Other important considerations are fuel consumption, initial cost of the equipment, installation costs, maintenance of the equipment, and floor space required for charging and handling molten metal from furnace to pouring floor or casting station. Working conditions in the melting room or furnace area should also be considered, as should temperature control, quantity of metal required, size of individual melts and the quality of each melt.

Different types of furnaces can and do affect the chemical and physical characteristics of the aluminum as it is melted. The changes in the composition and physical qualities are discussed in detail later in this chapter and in Chapter 3. Foundries should consider these when selecting furnaces for their specific needs.

CRUCIBLE FURNACES

The crucible melting furnace has been the workhorse in many foundries. This indirect-fired melting unit usually consists of a circular steel outer shell with 3–5 in. of refractory into which a crucible is placed. Burners located inside the refractory lining apply heat that is transferred by conduction through the crucible into the metal to be melted.

The crucible or pot-type melting furnace is available in the following commercial designs:

- the bowl type, from which molten metal is removed by ladling;
- the tilting type, from which the molten metal is poured;
- the lift-out crucible type where the crucible is removed and used as a ladle.

These systems are fairly inexpensive and practical for the smaller caster and job shops that melt small volumes of a number of alloys (Figs. 2-7, 2-8 and 2-9). Fuel consumption in

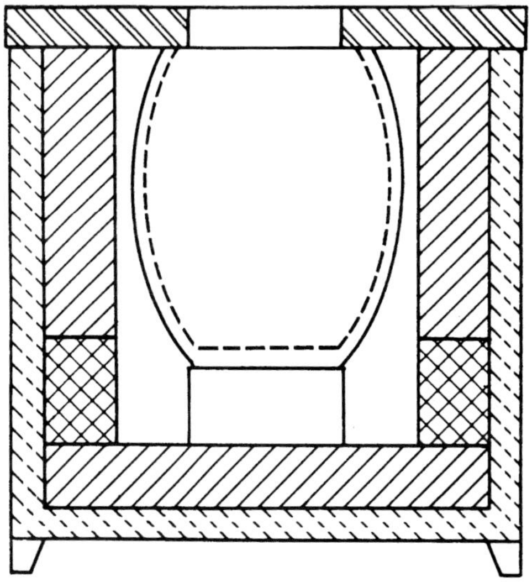

Fig. 2-7. Cross section of a sealed crucible furnace. The casing is a welded steel shell; the top (which is removable on lift-out types) is a refractory cover with a steel top plate. The lining consists of 4½ in. of insulating firebrick backed by 2½ in. of high-temperature insulation.

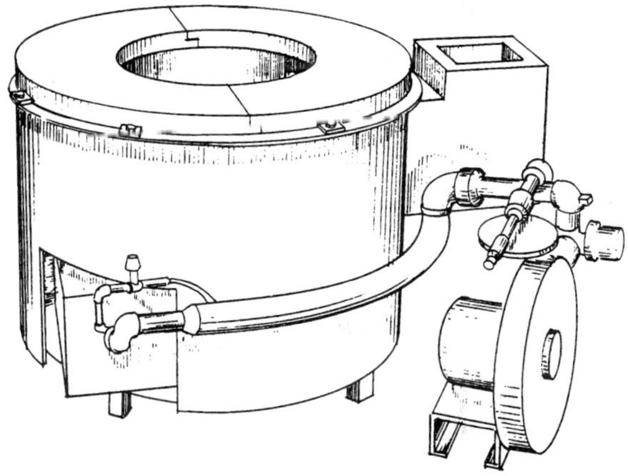

Fig. 2-8. Stationary crucible (bowl-type) furnace.

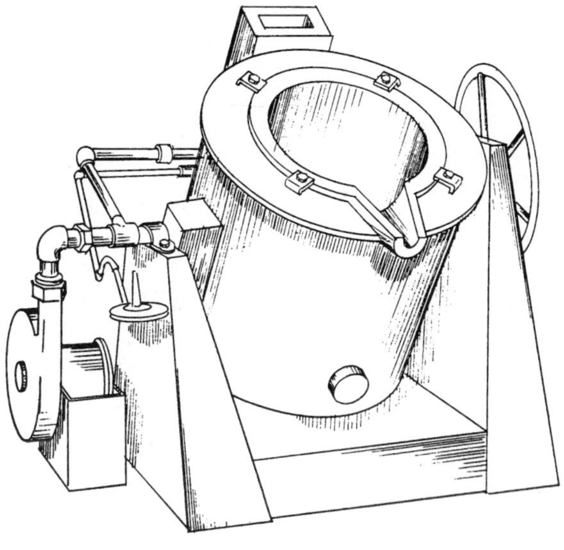

Fig. 2-9. Tilting crucible furnace.

this type of furnace is relatively high: about 3300 Btu per pound of metal melted. The efficiency range is in the area of 7–17%. The metal loss is about 3%.

Crucible Selection and Maintenance

The open-top crucible is the most expensive of the two types, but it has a longer life at high temperatures. Because the open-top crucible is made of either a clay graphite base or a silicon carbide material, iron pickup will not occur as it does with cast iron pots. This makes the open-top crucible a necessity in pure aluminum applications, such as rotor casting. Today, it is being used more and more for all aluminum alloys.

An important complaint about indirect-fired crucible bowl furnaces is the short life of the crucible. Flame impingement on the crucible is a common cause of short crucible life. It occurs in furnaces that do not have a large enough heating chamber, enough distance between the burners and the vessel, or properly positioned burners. Longer life can be obtained by allowing maximum travel of the hot gases around the crucible, thus eliminating local hot spots that are a major reason for failure.

Short crucible life is also caused when a molten heel is not left in the vessel at all times. No more than two-thirds of the molten metal should be removed from the melting vessel at any time. This leaves at least 33% of the molten metal to receive the next cold metal charge. Thus, less excessive temperatures are needed to push heat through the vessel wall in order to keep up with the melt rate. More conservative use of the stored Btus already in the molten bath is accomplished. A 30 or 40% molten heel left in the furnace will cause less of a heat load on the crucible when melting the next cold charge, and longer life of the crucible will result.

The most common crucible materials used for aluminum melting are clay graphite and silicon carbide (usually bonded with carbon). If refractory crucibles for stationary and tilt furnaces are to provide long service, they must be installed carefully with proper support (the flame should not impinge directly on them). Typical crucibles are shown in Fig. 2-10.

Fig. 2-10. Refractory crucibles for melting aluminum. Crucibles made of graphite or carbon-bonded silicon carbide eliminate the possibility of iron pickup.

Cast iron crucibles are rarely used in aluminum foundries because molten aluminum tends to erode ferrous containers and because the mechanical properties of certain alloys are affected by iron pickup, making it necessary to apply and maintain a protective wash or coating on iron pots.

Silicon carbide is preferred over clay graphite for crucibles because its thermal conductivity is higher and heat transfer is, therefore, more rapid. When using silicon carbide crucibles, chips and pieces broken from them must be removed from the melt and not be allowed to become trapped in the metal to form hard spots in the castings. It is especially important that a highly oxidizing flame does not impinge upon or even be used to heat silicon carbide crucibles. If this is done, the crucible will have a relatively short life because the silicon carbide will oxidize at high temperatures.

After the melting cycle has been completed and while crucibles are still hot enough to keep any adhering metal molten, they should be scraped down with a blunt, rounded steel blade. Care must be taken to avoid damaging refractory crucibles in this operation. If this cleaning is done when the crucibles are still red hot, the dross is more readily removed, and the crucibles are less likely to be damaged. Clay graphite crucibles are the easiest to clean after the melting operation.

REVERBERATORY FURNACES

Reverberatory or open hearth furnaces may be of many designs ranging in capacity from about 2000 to more than 200,000 lb. These direct-fired furnaces are usually stationary, although there are some tilting types in use. They may have one or two chambers (see Figs. 2-12 and 2-15). The hearths may be divided. Metal may be charged in front or side door openings (see Fig. 2-11) or through a door at the base of a flue stack (see Fig. 2-17).

The modern aluminum reverberatory furnace has the potential of many years of service life if proper operating practices are followed. For lining the bath area, the refractory industry has developed a brick that is high in alumina and specially treated for high resistance to wetting by and absorption of aluminum. This lining generally may be restricted to the molten bath area and to the first 6–12 in. of the side wall above the metal line.

Any refractory will last longer if it is not exposed to the thermal shock of severe temperature changes. High firing rates on the heating side and door openings on the cooling side are frequent sources that can lead to early damage or failure of the lining. Temperature increases occur most often when weekend shutdowns require high firing rates to bring the furnace temperature back up to the operating point. If the furnace is full of aluminum, the floor and walls will be protected by this excessive change in temperatures. A drained furnace will expose the side walls to the attack of the aluminum coated thereon, and this aluminum will be superheated to a highly active state that is damaging to the brick.

Logical safety practices require some means of ensuring that no runout will occur during weekend shutdown periods when full manpower is not available to cope with such an emergency. Under these circumstances, it is desirable to lower the furnace temperature to about 1100F (593C), where the aluminum will remain in the "mushy" stage between solid and liquid and cannot run out.

Sludges can form in certain alloys when the temperature is reduced. Temperature reduction below the liquidus point also endangers the furnace lining because of the contraction of the solidifying aluminum and subsequent expansion of the melting metal when the furnace is brought back to heat. No brick construction can withstand these destructive forces for long.

Experience has shown that optimum furnace lining life is obtained from those furnaces that:

- always contain a substantial heel of metal close to capacity loading;
- are maintained at a temperature close to the operating point;
- are protected from rapid chilling caused by frequent opening of large doors;
- are not subjected to extreme firing rates.

When it becomes necessary to shut down the furnace for maintenance purposes, the cooling rate should not exceed 50°F (28°C) per hour until the temperature reaches 400F (204C). During this period, all doors should remain closed to maintain as even a temperature reduction as possible. Below 400F (204C), doors may be opened and cooling air applied to facilitate cooling to a habitable temperature. When returning the cold furnace to service, the firing rate should be such that temperature does not increase more than 50°F (28°C) per hour.

Direct-Fired Reverberatory Furnace

With the increase in demand for cast aluminum products came the need for increased volume and melting speeds in furnaces. As a result, the direct-fired (open-flame) reverberatory furnace (Fig. 2-11) has become popular. Melting in this type of furnace is accomplished by direct contact of the flame and combustion products with the solid and molten metal.

Because conditions in the open hearth are likely to produce rather large quantities of dross and a high rate of gas absorption, it is important that melting practices be controlled carefully. Direct flame on the surface of the bath can cause turbulence, gas absorption and excessive dross formation. Controlling charging procedures and combustion furnace atmospheres can protect against these hazards. There are several reverberatory furnace designs.

Sloping Dry-Hearth Reverberatory Furnace

The sloping dry-hearth furnace (Fig. 2-12) consists of a sloping refractory hearth on which scrap and ingot are placed until melted. As the metal melts, it runs down the hearth into a secondary holding chamber.

Melt loss in this type of furnace, compared to that of other furnace types, is one of the highest. It ranges between 2 and 12% for every pound of aluminum melted. The fuel efficiency ranges from about 28% down to a poor 10%, and the fuel consumption runs about 1500–5000 Btu per pound of aluminum melted.

To help reduce the melt loss, flame impingement should be minimized, and the designed melting rate of the system should not be exceeded. When the furnace capacity is exceeded, the result is excessive thermal temperatures in the melt zone, which increases melt loss. The best way to reduce melt loss is to minimize charging light bulky scrap in this type of furnace.

Stack Melting Reverberatory Furnace

Stack or tower-type reverberatory furnaces are high-volume, high-efficiency melters. They have melting capacities of 1000–4000 lb/hr. Improved efficiency of 40–60% or 1250 to 850 Btu/lb is achieved by using the flue gases to preheat the metal charge before it enters the molten bath. Stack discharge temperatures are reduced significantly, and melt losses usually average about 2–3% of the charge.

In stack melters (Figs. 2-13 and 2-14), the charge is loaded in the top of the stack and is heated to the mushy point as it moves down the stack before reaching the dry hearth. There, final melting takes place, and the metal flows into the holding bath.

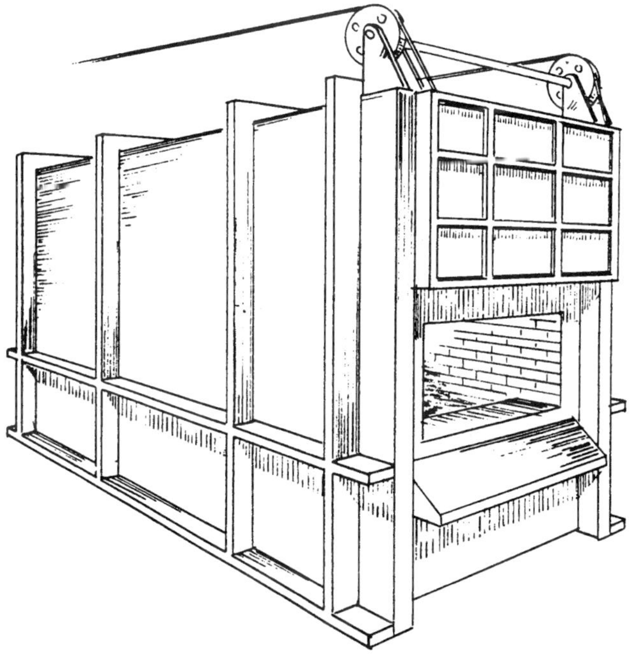

Fig. 2-11. Front-charging, single-chamber reverberatory furnace.

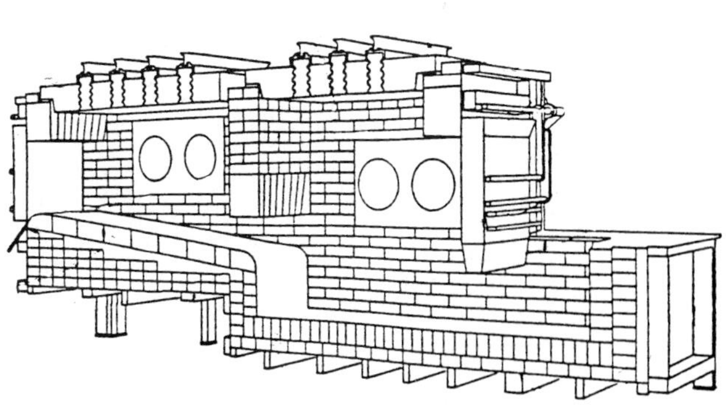

Fig. 2-12. Sloping dry-hearth reverberatory furnace.

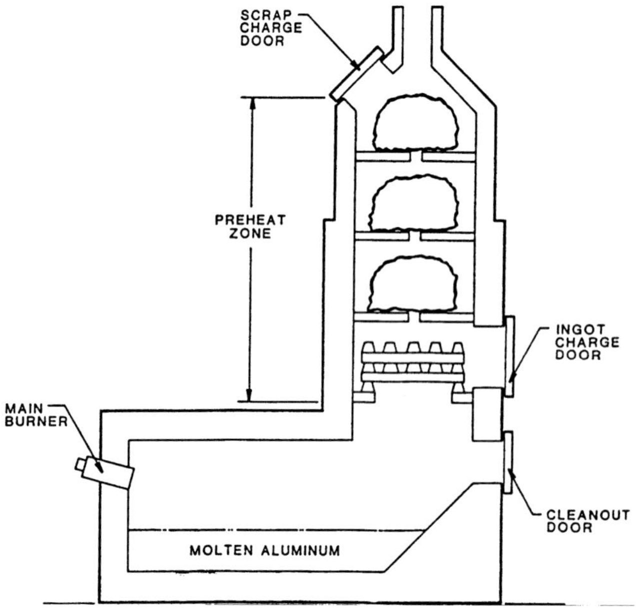

Fig. 2-13. A side-charging stack melting furnace. [Drawing courtesy of K. Boeckenhauer, Seco/Warwick Corp.]

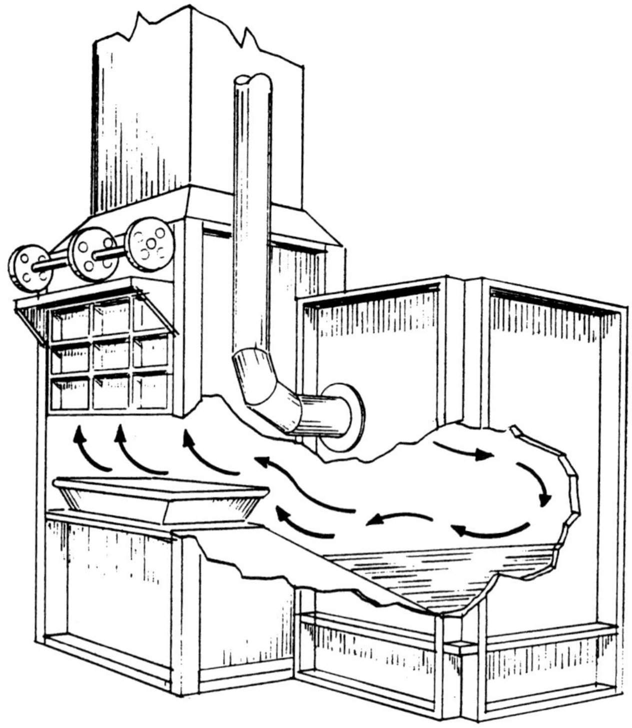

Fig. 2-15. Front-door, stack preheat reverberatory furnace.

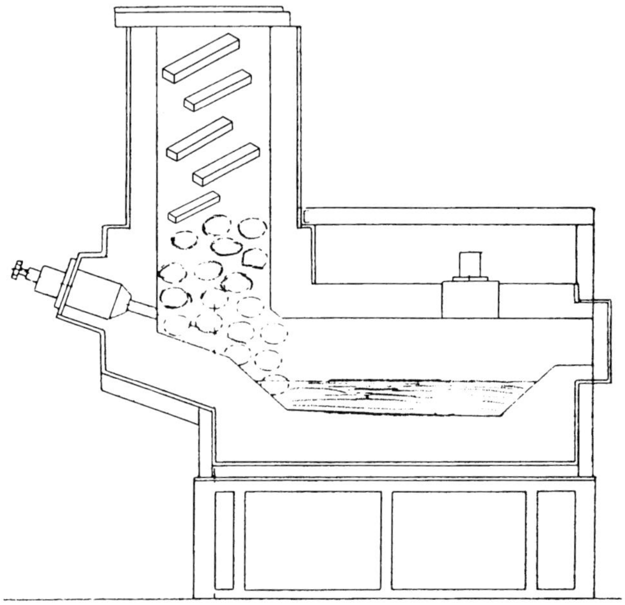

Fig. 2-14. A tower-type "jet melter" furnace.

A stack charging furnace is illustrated in Fig. 2-15. The solid charge is made through a door onto a dry hearth under a flue, dried by the initial flue gases, and then melted. The metal flows into a holding bath as rapidly as the metal becomes molten.

Melters of this type should be operated continuously at nearly the maximum rated capacity to realize lowest fuel costs. Some furnace manufacturers state that, when operated continuously at rated capacity, these large melters will melt aluminum using as little as 1500 Btu per pound. These large reverberatory furnaces have a designed melting capacity of 40 lb of aluminum per square foot of bath per hour.

Wet Bath Reverberatory Furnace

The wet bath reverberatory furnace (Fig. 2-16) has become one of the most-used systems for melting aluminum. It essentially consists of an exterior charging well that is separated from the main heat chamber by a submerged arch. It may have a ladling well at the other end or a bottom outlet for pouring the metal. Recent designs have greatly reduced the metal-to-roof distance, and the furnaces now use roof-mounted gas radiant burners. These innovations have increased fuel efficiency while appreciably reducing oxide and metal loss problems.

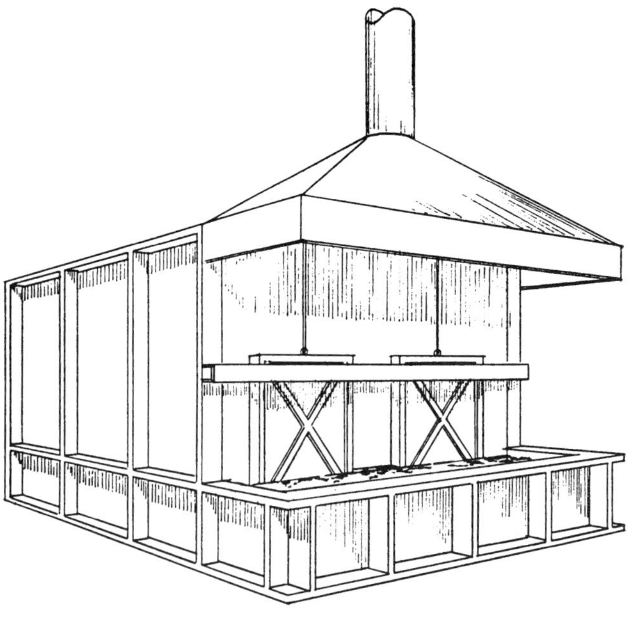

Fig. 2-16. Direct-fired, well-charged reverberatory furnace.

The wet bath reverberatory furnace normally is not used for batch charging. It is more adaptable to continuous loading at a well-defined, prescribed rate with careful planning for correct volume of charge per hour. With gas-fired radiant burners, the furnace uses about 3000 Btu per pound of metal melted. This furnace has a low melt loss relative to some other fossil fuel melting units. Melt loss per pound of metal melted ranges from 3–5% in normal practice. These figures are based on cold charging.

Electric Radiant Reverberatory Furnace

The melting of aluminum by radiant heat is a new concept in furnaces. The electric radiant reverberatory furnace (Fig. 2-17) is similar to oil- or gas-fired reverberatory furnaces. It has a remote charging well separated by a submerged arch (or door) from an isolated heat chamber incorporating silicon carbide heating elements over the molten bath.

This furnace uses about 820 Btu per pound of metal melted. Based on a cold charging operation, the melting efficiency is approximately 70–75%. The furnace will melt from a cold start. The melt loss is about 1% per pound of aluminum melted. Because there is no combustion process, no flues are required, and the thermal head runs much lower than it does in a fossil fuel-fired furnace.

In the heat chamber, the metal is tranquil with no agitation to cause dross formation. The short distance between the radiant burners and the bath provides a high transfer of Btu into the bath. Oxide formation and gas absorption are low. Less cleaning of the furnace is required. There is no noise or excess heat load.

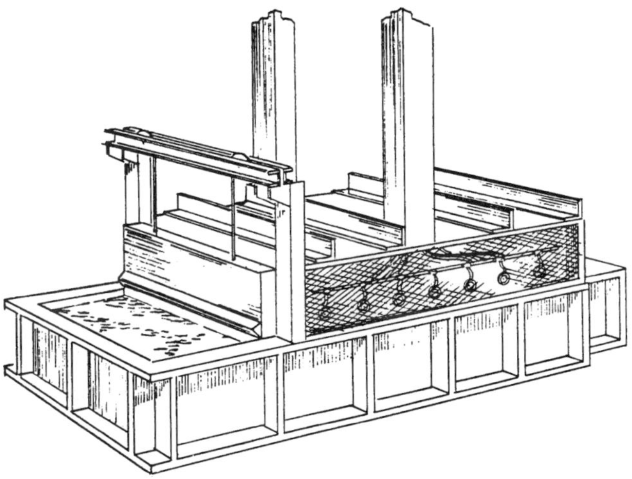

Fig. 2-17. Single-chamber electric reverberatory furnace with well.

Front Charging Reverberatory Furnace

Smaller reverberatory furnaces with one or more dipping wells are frequently used to prepare foundry melts. Reverberatory furnaces equipped with dipping wells have a charging door at one end where charging, fluxing, degassing and skimming are done before ladling and pouring the melt directly into the molds. Alloyed ingots and scrap are charged into the furnace more or less continuously for the casting operations.

The single-well, dip-out type furnace is one of the most frequently used open-flame aluminum furnaces (Fig. 2-18). Many large-volume foundries use a number of these as holding furnaces for die- and permanent mold casting operations (molten metal is usually supplied to these units by larger reverberatory furnaces). When these units are used as holding furnaces, the burner equipment need only be sufficient to handle the normal heat loss of the furnace itself.

Large, stationary reverberatory furnaces for volume melting are used in melting and refining plants where metal is alloyed and cast into ingots. Diecasting plants with considerable volume also use large stationary reverberatory furnaces for breaking down and alloying scrap or returns.

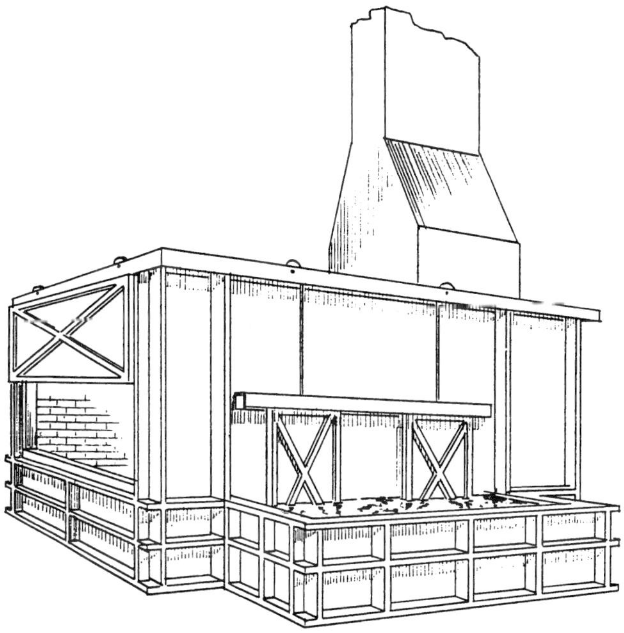

Fig. 2-18. A single-chamber, gas-fired reverberatory furnace with well.

Dual-Energy Reverberatory Furnace

Unpredictable energy conditions led to the development of a furnace that could be the ultimate in melting flexibility. The dual-energy reverberatory furnace (Fig. 2-19) has become quite practical because it can melt with fossil fuels and electricity in the same chassis. When operating in electric mode, the furnace consists of an upper chassis with electric radiant elements and plugged flue and burner ports. If electric service is interrupted, the following procedures are followed:

1) The electric control panel is shut off, and the electric heating elements are removed.
2) Insulating fiber plugs are installed in the electric element holds.
3) The fossil fuel burner plugs are removed and the burner nozzles reconnected.

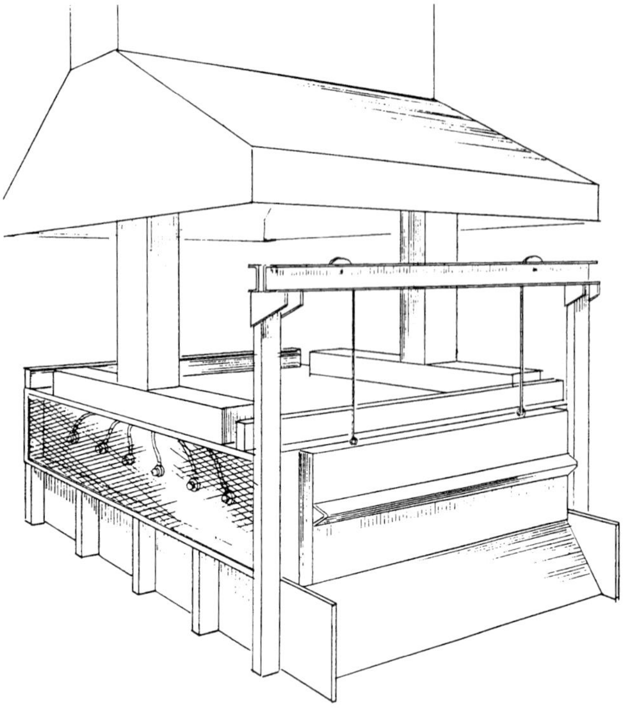

Fig. 2-19. *Dual-energy reverberatory furnace.*

4) The combustion panel is energized, and burners are lit.

The furnace then operates by burning oil, gas or propane. When converting back to using electricity, the preceding procedures are used in reverse. Depending on the size of the furnace, the changeover takes from 1–4 hr. The aluminum bath remains molten throughout the procedure. This is a new concept that appears to have promising possibilities.

Regenerative and Recuperative Burner Systems

Heat reclamation systems improve the efficiency of reverberatory furnaces by preheating combustion air using heat from flue gases. There are two basic heat exchanger systems for waste heat recovery: recuperators and regenerators.

Recuperators are heat exchangers that separate the flue gas from the combustion air by a ceramic or metallic wall through which the heat transfer occurs. The combustion air is preheated as it passes through the recuperator on its way to the burner.

Regenerators have two completely separate chambers that contain a media capable of storing heat as the flue gas passes through it (Fig. 2-20). The stored heat is then used to heat

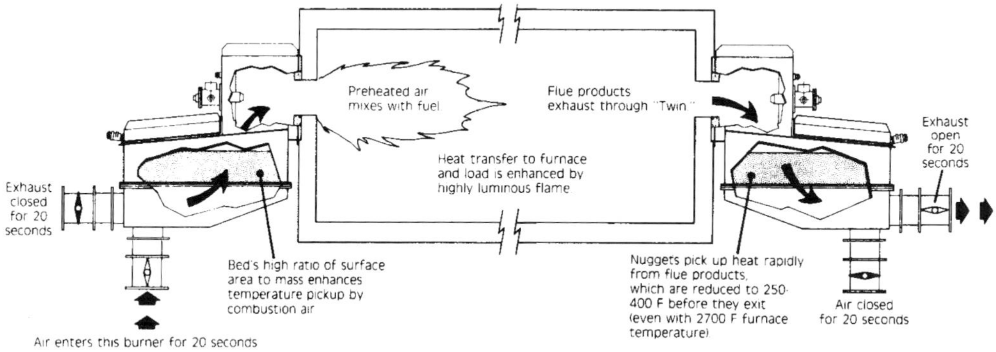

Fig. 2-20. A schematic drawing of a regenerator. During a cycle, burners operate in this configuration for 20 seconds then reverse for the next 20 seconds. The burner on the right would be firing, the one on the left exhausting. [Drawing courtesy of James Bowers, North American Manufacturing Co.]

combustion air as it is passed through during alternate combustion/exhaust cycles. When two burner/recuperator chambers are fired alternately, a constant heat source is obtained.

ELECTRIC INDUCTION FURNACES

Induction furnaces range in size from small, stationary, single-chamber holding units with capacities of a few hundred pounds to large units with capacities of about 100,000 lb. There are two basic types of electric induction furnaces. The channel-type (Fig. 2-21) is basically a transformer with a shorted secondary loop of molten metal. The heat is generated in channels, forming the secondary loop. This type of furnace is more efficient electrically, melting about 5 lb per kilowatt hour input. When used with aluminum, this type of furnace builds up nonmetallics in the channels, and these must be removed periodically.

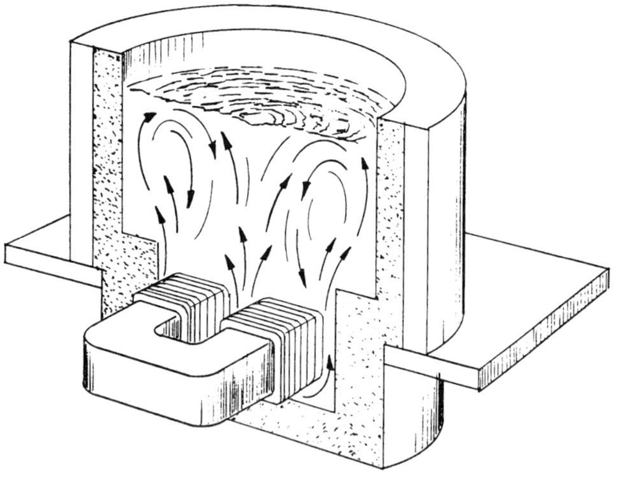

Fig. 2-21. Channel induction furnace.

The channel induction furnace is used in foundries to prepare melts of both ingot and scrap. Most often, it serves as a holding furnace for aluminum melted in other furnaces.

The second type of induction furnace is the coreless induction furnace (Fig. 2-22). In this type of furnace, eddy currents, generated by the primary coil, heat the metal by resistance to flow. These furnaces have been used for years operating on what is known as high frequency (above 60 cycles per second). In recent years, furnaces operating on 60 cycles have been developed. These furnaces operate at an electrical efficiency of from 3.5–4.0 lb of aluminum per kilowatt hour.

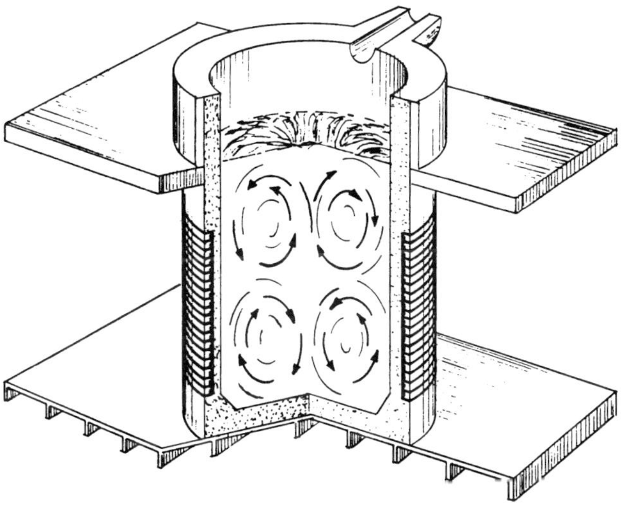

Fig. 2-22. Coreless induction furnace.

The coreless furnace has many operating advantages that overcome its poor electrical utilization. The main advantage is the absence of channels and their problems. Unlike channel furnaces, which need molten metal for starting, coreless induction furnaces can be completely drained and restarted with solid metal. (In a power outage of any duration, channel furnaces must be drained quickly or a complete reline could be necessary due to the metal freezing.) The coreless induction furnace also requires less maintenance.

Two-chamber induction furnaces for continuous operation are frequently used in foundries casting aluminum. Melting is accomplished in one chamber and pouring from the other. Good temperature control may be obtained on the pouring side if care is used when charging the metal.

ELECTRIC RESISTANCE FURNACES

Electric resistance furnaces (Fig. 2-23) are used mainly as holding units and are available in many types and sizes. The lining is composed of lightweight insulating brick placed in front of insulation slabs. Rugged nickel-chrome elements are mounted on the sidewalls of the lining. In this type of holding furnace, heat created by the nickel-chrome elements is transferred through a crucible into the molten bath. Standard sizes of this furnace range from 1000–4000 lb.

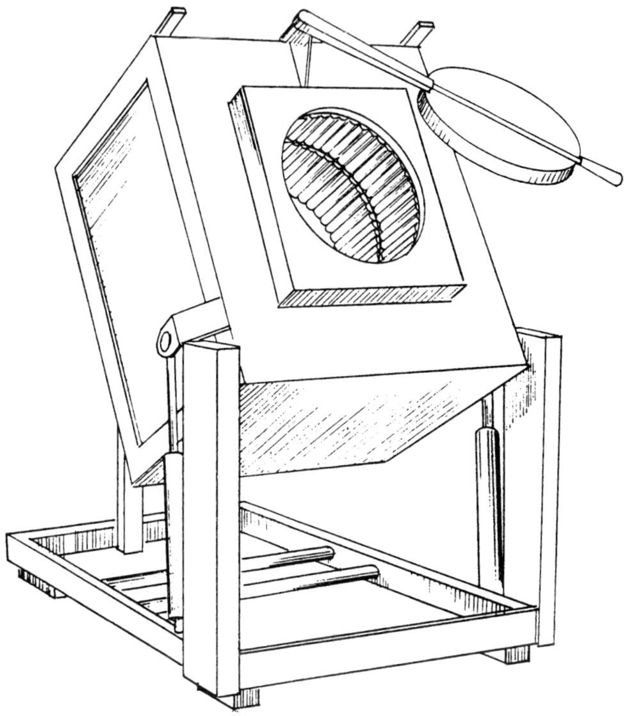

Fig. 2-23. Electrical resistance furnace.

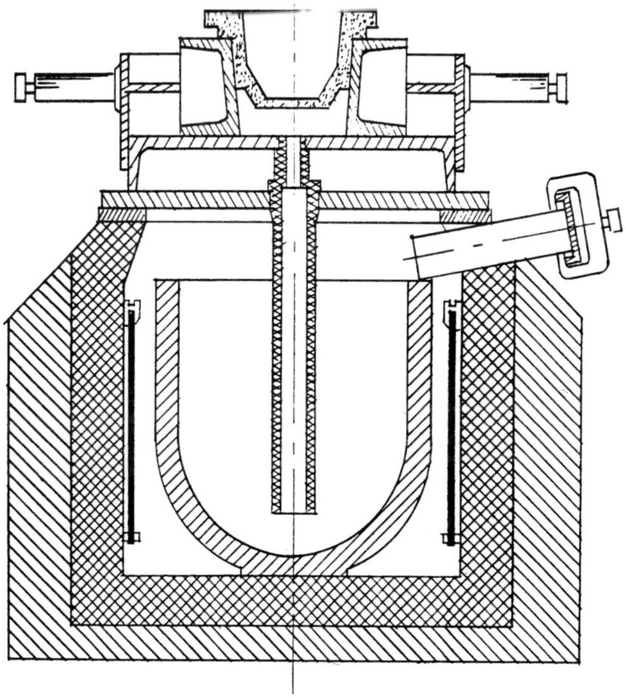

Fig. 2-24. Electrical resistance furnace for low-pressure processes.

The furnace shown in Fig. 2-24 is a pressurized electric resistance holding furnace used in the low-pressure permanent mold casting process. The mold is located over the sealed furnace cover, and metal is forced through a tube into the melt by gas pressure applied to the surface of the metal in the crucible.

FILTER SYSTEMS FOR MELTING EQUIPMENT

The technology of filtering molten aluminum alloys has advanced to the point where filter systems are now practical for most aluminum foundry operations.

Basically, filter systems employ a porous refractory material formed in such a manner as to allow only clean molten metal to pass through. Contaminants (such as oxides, nonmetallics and sludge particles) remain behind. The filters are replaceable and contain pore openings of 20, 30, 45 and 60 pores per inch.

Reverberatory Well Filter System

The reverberatory furnace system consists of a swinger hoist that is used to manipulate a special filtering vessel into and out of the pouring well (Fig. 2-25). The vessel is attached to a mounting frame that, in turn, is fastened to the hoist by a quick disconnect. Thus, the system can be removed easily from the well for either service of the filtering mechanism or for routine maintenance of the melting furnace.

During operation, the vessel is immersed in the pouring well, and metal enters through bottom and side filter elements. Oxides and contaminants in the melt are isolated from the pouring vessel, and clean metal is available for ladling on an as-needed basis. Any extra metal is normally returned to the pouring well outside of the vessel; thus, the melt in the vessel is not contaminated with oxides from the agitation caused by the metal transfer.

In addition to providing a ready means of manipulating the vessel, the swinger hoist permits the vessel filters to be backflushed during normal production. This is accomplished by raising the vessel several inches above the melt and allowing the clean metal inside the vessel to drain back out and wash the accumulated oxide cake from the filters. When performed on a routine basis, the backflushing operations will increase the life of the filter elements by approximately 10 times.

Through careful selection of the filter element area, pore opening and filter placement, the temperature difference between the vessel and the pouring well can be greatly reduced. This difference changes with the amount of metal processed by the system but normally will be approximately 30°F (17°C).

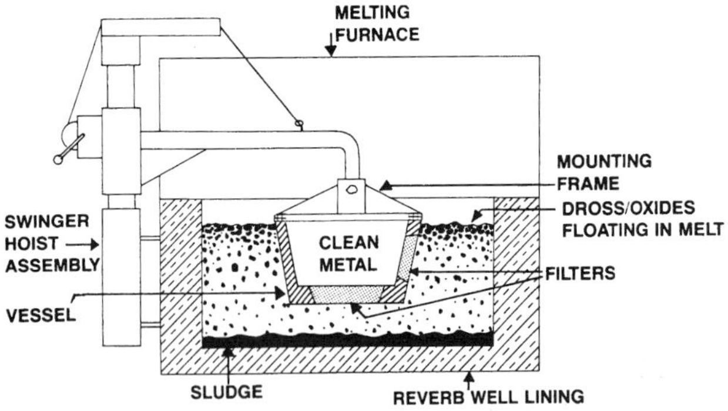

Fig. 2-25. Reverberatory well filter system.

The filter elements, while cemented in place to facilitate backflushing, are replaceable upon a schedule that is determined by the melt cleanliness. Several replacements are normally possible during the course of the vessel life if backflushing does not provide sufficient cleansing.

Filter Melting/Holding Crucible

Melting/holding crucibles (Fig. 2-26) for the system are constructed with a permanent baffle and are available in either silicon carbide or cast iron vessels. A filter paddle containing a replaceable filter element is cammed and clamped in place against the baffle, which cleanses the metal as it passes from the charging side of the crucible. The baffle is positioned to provide a ladling area that is equivalent to approximately 60% of the total diameter of the vessel. On a 600 lb bowl melter, the ladling area measures approximately 14x23 in.

The crucibles can be furnished for either stationary or tilting melting furnaces. The tilting furnace type requires design modifications that are not apparent in the figure.

The filter paddle may be left in place overnight in a continuous melting operation, or it may be removed and the vessel drained and cleaned at the user's option. In the latter case, it is not necessary to replace the filter element as long as some modified basic procedures are followed. The life of the replaceable filter element is only limited by the quality of the charge materials and the rate at which the filter pores become obstructed by oxides. It is only necessary that the seal gasket be replaced and the system restarted with essentially standard crucible melt operating procedures.

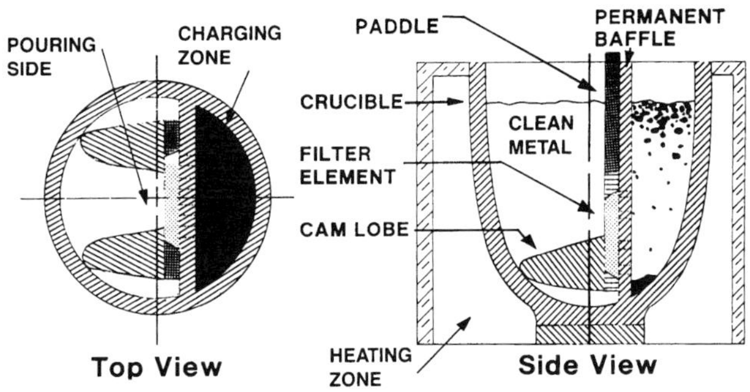

Fig. 2-26. Melting/holding crucible.

Filter Pouring Ladles

Special pouring vessels are available in several basic designs, depending upon the size and pouring rate requirements of the application. Normally, the vessels are constructed with a replaceable filter element mounted in a filter paddle, which is positioned against a permanent baffle (Fig. 2-27). The filter area and pore openings are varied to meet individual requirements for pouring rate and economy of operation.

On the initial pour, a new filter element is placed in the paddle, and the assembly is preheated in the usual fashion, before it is filled with molten aluminum. The filling operation may be conducted in the normal manner and the ladle taken directly to the pouring or transfer station to discharge

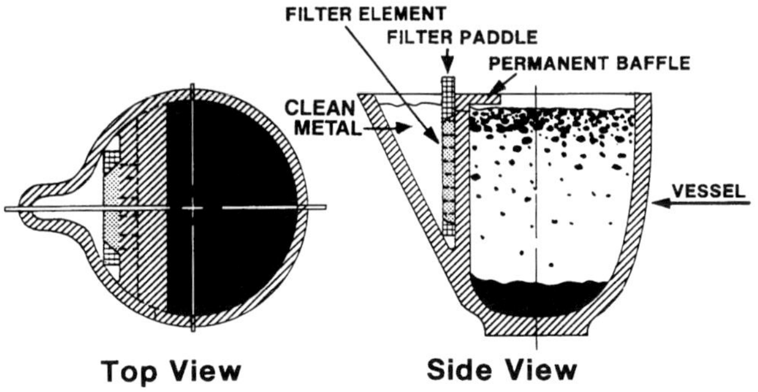

Fig. 2-27. Filter pouring ladles.

into a gating system or an auxiliary holding furnace. No special priming or pretreatment of the filter element is required if the vessel received an adequate preheat before filling. Any heels or residual aluminum left at the conclusion of the transfer operation are normally returned to the pouring furnace through the back edge of the ladle. The ladle is then returned to a preheat torch, if desired, for continued pouring and application in the near term. No special practices are required other than to make sure that most of the metal was drained out of the filter element. This, in itself, is a back-flushing operation that adds to the life of the filter element.

The filter element is not limited to a single pour or even to a single day of operation when proper procedures are followed. Thus, an economic metal cleansing treatment can be conducted directly in the final transfer/pouring operation.

The filter element and, to a lesser extent, the filter paddle are expendable supplies and will require replacement. Their life spans are affected by a number of operating variables, including the level of contamination in the metal.

Filter Pressure-Pour Ladle

This system is used in automatic-pour installations serving diecasting, permanent mold or automatic sand molding lines. The pressure-pour vessel (Fig. 2-28) typically is immersed in the well of a reverberatory furnace, and it delivers molten metal to the pouring location via a launder system.

Metal entering the pressure-pour vessel passes through a filter mounted in a carrier paddle, thus cleansing the metal on the entry cycle. The rate of transfer through the filter is controlled by the application of a vacuum to the internal chamber and is regulated to a constant volume by the return of the vessel to atmospheric pressure. Pouring is accomplished by pressurizing the vessel with either air or an inert gas to discharge the proper amount of metal to the launder system. The volume of metal poured is controlled by the discharge pressure and the length of application time, and it can be controlled within precise limits.

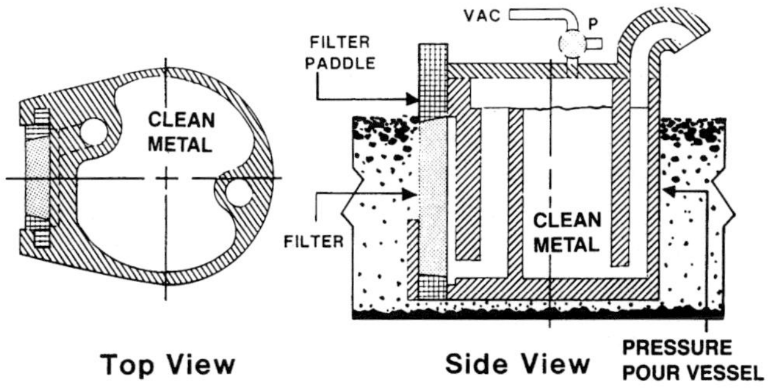

Fig. 2-28. Filter pressure-pour ladle.

PUMPING MOLTEN METAL

Because many casters of aluminum are recycling large tonnages of scrap, more efficient methods of furnace remelting are required. One such development is the improvement of metal transfer from one furnace chamber to another for uniting scrap metal with the bath.

Molten aluminum is a liquid and should be capable of being handled by pumps, as with other liquids. The main problem with pumping liquid metal has centered around finding materials that are capable of handling the metal but that are not too fragile or that are attacked by the aluminum. Pumping of molten metal represents new technology that is being put to practical economical operation. Considerable progress has been made in this area, and today metal can be carried through pipelines by using pumps (Fig. 2-29).

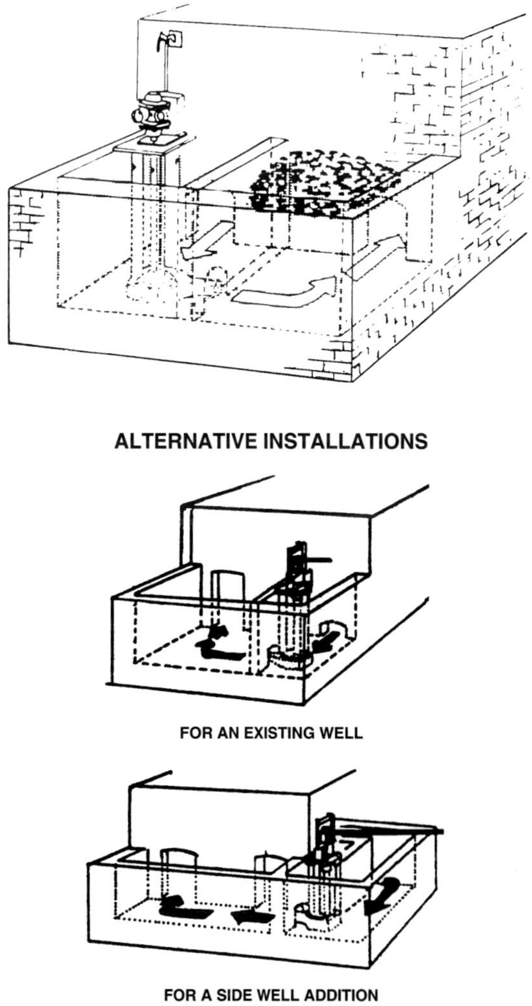

Fig. 2-29. Submerged discharge pump in a reverberatory furnace.

Several types of pumps have been built and operated successfully. They generally may be classed as being the high-volume, low-pressure type. Extensive experience has been accumulated on the use of mechanical pumps. They have been used for metal transfer, furnace stirring and uniting scrap metal with the bath. In each case, the operation was improved by using the pump. The pumps are made of graphite with special refractories for bearings. In some cases, flux gas may also be introduced through the pump for demagging and cleaning the charge.

Pumping action may also be generated electrically. The three-channel inductor by one manufacturer can be used as an electrical pump with no moving parts. Electrical pumps also can be built using three-phase alternating current supplied to coils so that an unidirectional pulse is generated. When small-sized scrap is being melted, metal pumps can be used to introduce it efficiently into a moving molten stream before excessive oxidation occurs.

REFERENCES

1. R.W. Monroe; *Expendable Pattern Casting,* © 1992, American Foundrymen's Society, Inc., Des Plaines, Illinois.
2. K.G. Hudson; "Automatic Ladling," *Innovations and Advancements in Aluminum Casting Technology,* proceedings of an AFS special conference; City of Industry, California (Feb 22–23, 1988).
3. P.R. Gouwens; "Liquid Metal Handling for Permanent Mold Foundries," from proceedings of AFS 2nd International Conference on Permanent Mold Casting of Aluminum; Nashville, Tennessee (Nov 4–5, 1991).

Chapter 3

CHEMICAL & PHYSICAL CHARACTERISTICS OF MOLTEN ALUMINUM

GENERAL CHARACTERISTICS

The degree to which the principles of good foundry practice are followed directly affects the quality of the casting, regardless of the metal. However, some aluminum alloys are more sensitive in this respect than are others.

As in quality casting of any metal, it is advisable to be familiar with the principal chemical and physical characteristics of the various aluminum melts and to know how they influence the founding operation.

The general physical and chemical characteristics of molten aluminum include:

- the ease with which aluminum melts combine with oxygen, moisture or other oxidizing materials to form dross and the tendency and ease with which this dross can be entrained in the melt;
- the readiness with which the melt will absorb nascent hydrogen and aluminum's evolution of hydrogen during solidification of the casting to form porosity (the principal source of hydrogen is moisture);
- the 6.5–8.5% contraction in volume that occurs when the melt solidifies and cools;
- the low density of molten aluminum, which results in a low hydrostatic pressure in the mold. This, combined with the contraction during solidification, necessitates careful risering practice to avoid shrinkage defects in castings;
- the high degree of protection against gas absorption and dross formation given by the oxide film on the aluminum alloy melt, provided this film is not broken by excessive agitation of the melt or turbulence during ladling and pouring;
- hot shortness or the fracture or tearing of a metal during and soon after solidification (Fig. 3-1). This can be due almost entirely to improper casting design, but it is more pronounced in certain alloys and can be made more severe by various impurities in the melt that segregate at dendrite or grain boundaries during solidification;
- the solution by molten aluminum of many common metallic materials. Unprotected steel is rapidly attacked; cast iron less rapidly so. High-chromium steel, because it has a heavily oxidized surface, is quite resistant to attack;
- aluminum is reduced from alumina. Many furnaces are lined with high-alumina refractory brick and materials. Any appreciable amount of cryolite or related fluorides may accelerate a harmful reaction between the aluminum and these refractories.

DROSS FORMATION

When aluminum melts react with the atmosphere or moisture, they form a dross of aluminum oxide and nitride, which contains some mechanically entrained gas and metal. Because the dross is wetted by the aluminum melt and has about the same density, it often becomes entrained in the melt during melting, handling or casting and does not readily separate at the surface of the melt.

The quantity of dross formed during melting increases with: 1) the use of fine, badly weathered or corroded scrap; 2) the presence of magnesium in the alloys in the charge; 3) an increase in turbulence, which breaks the protective oxide surface of the melt in the furnace; and 4) an increase in the temperature of gases in contact with the surface.

The oxide on the melt surface contains a considerable amount of liquid metal, causing the dross layer to be "wet." This type of dross formation is the principal cause of metal loss during melting. Dross defects in castings usually are a result of unsatisfactory melting, handling and gating practices. They can be reduced by:

- producing melts substantially free of suspended dross particles or dross films;
- avoiding turbulence, which breaks the oxide film on the melt during the melting and pouring operations. Turbulence is the least-appreciated cause of dross and

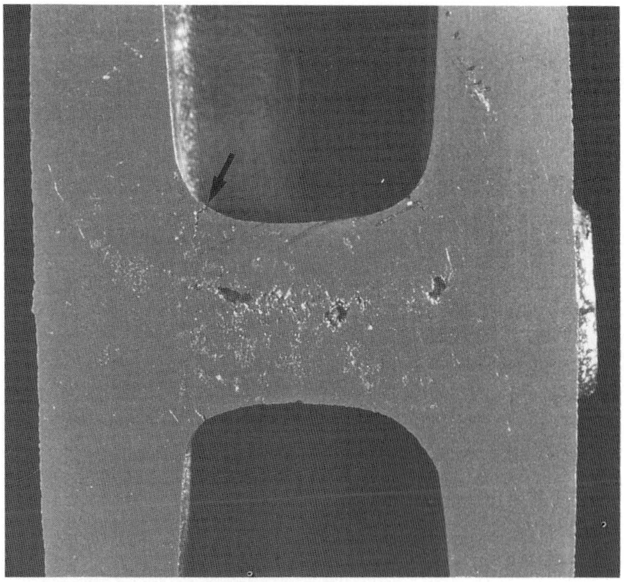

Fig. 3-1. Hot tearing in an A356.0 casting.

oxides and is the most difficult to minimize in processing and handling aluminum;

- using good gating (rigging) practices.

GAS POROSITY IN ALUMINUM CASTINGS

Liquids, including liquid metals, can absorb considerable amounts of gas under certain conditions. During melting, the metal is surrounded by objectionable gases, including hydrogen. These gases may come from the burning fuel and can even penetrate the walls of the crucible. When the metal is liquid, hydrogen gas dissolves in it easily and rapidly—the gas is actually in solution in the metal.

When molten metal is poured into a mold, it cools rapidly. The metal then can no longer hold appreciable quantities of gas in solution, and bubbles are produced. The evolution of the absorbed hydrogen during solidification of a casting is shown in Fig. 3-2. Figure 3-3 shows a machined section of an aluminum alloy casting with widespread porosity due entirely to trapped hydrogen gas.

Hydrogen in Aluminum Alloys

It is generally recognized that only gases in the atomic state will dissolve in molten metal. When two atoms combine to form a molecule, the gas escapes but still may become trapped as bubbles. As has been pointed out, the principal gas that causes unsoundness in aluminum alloys is hydrogen.

Little hydrogen is absorbed from the negligible amount of free hydrogen in the atmosphere. The main sources of hydrogen are water vapor and steam. Water vapor is the most

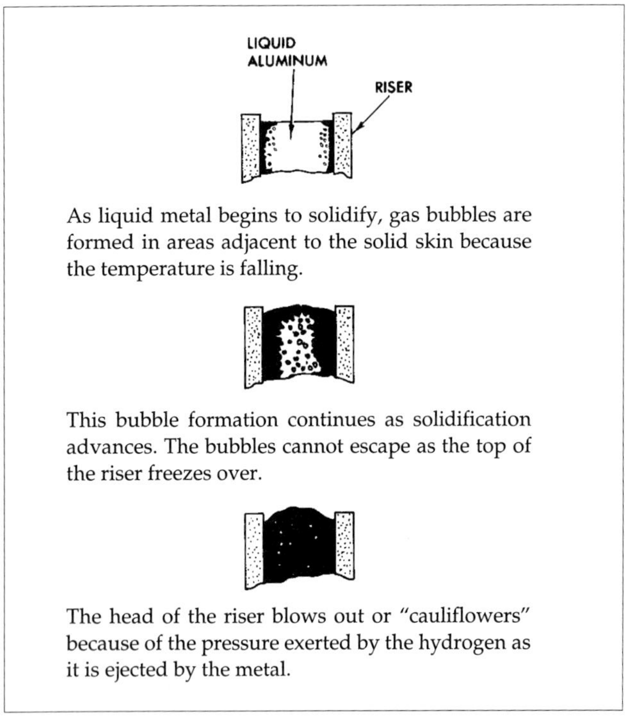

Fig. 3-2. Hydrogen gas dissolves in liquid aluminum. When aluminum cools, bubbles are produced, and gas porosity may result in the casting.

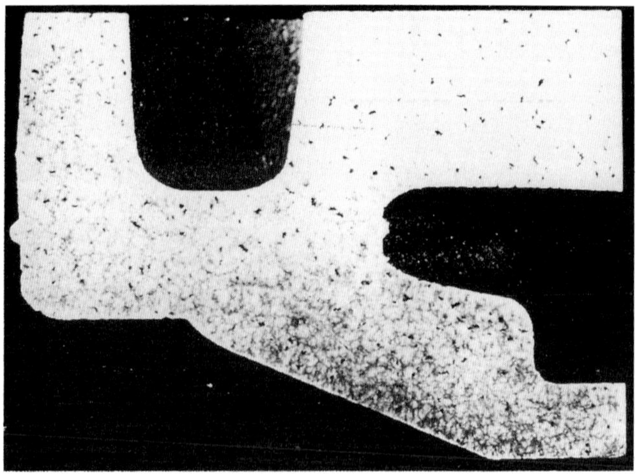

Fig. 3-3. The severe and widespread porosity in this section of an aluminum casting was caused by gas in the metal.

common source and is 9% hydrogen by weight. One of the most likely sources of water vapor is the furnace combustion atmosphere, which contains about 18% water vapor. Nascent or atomic hydrogen, which is very active and quickly absorbed by molten aluminum, is produced by the chemical reaction:

$3H_2O$ +	$2Al$	→	$6H$	+	Al_2O_3
steam	aluminum		hydrogen		aluminum oxide

Sources of hydrogen porosity are:

- the furnace atmosphere, which contains appreciable amounts of water vapor in addition to some hydrogen;
- moisture from refractories, dirty skimmers and furnace tools;
- hydrated corrosion products that form part of the charge, such as weathered ingot and scrap;
- oil-contaminated turnings, chips or scrap;
- damp fluxes;
- oil and hydroxide coating on metallic sodium used for modifying aluminum-silicon alloys.

It is important to appreciate the small amount of hydrogen necessary to gas a melt.

The water vapor that may be contained in one cubic foot of air is capable of ruining a ton of metal. The temperature of the molten metal determines the amount of hydrogen absorbed. As the temperature rises, the volume of hydrogen taken into solution increases rapidly. Solubility in the solid metal is practically nil; however, it jumps from a very small measurable quantity at just below the melting point to over 0.7 cc/100 g when the metal is molten at a little above the melting point (Table 3-1). Alloys with a hydrogen content of less than about 0.01 cc/100 g remain practically free of gas pores and blisters.

In Fig. 3-4, the graph (based on the figures given in the Table 3-1) shows how rapidly the hydrogen content rises as the temperature of the molten aluminum increases. Taken in reverse, it shows equally well the decline in solubility as the temperature drops, and also shows that the rapid expulsion of gas causes pinhole porosity.

Table 3-1.
Hydrogen Solubility in Aluminum Based on Temperature

Temperature °C	Temperature °F	Hydrogen Solubility (cc/100 g)*
0	32	0.0000001
400	752	0.005
660 Solid	1220	0.036
660 Liquid	1220	0.69
700	1292	0.92
750	1382	1.23
800	1472	1.67
850	1562	2.15

*Measured at a pressure of one atmosphere and a temperature of 68F (20C).

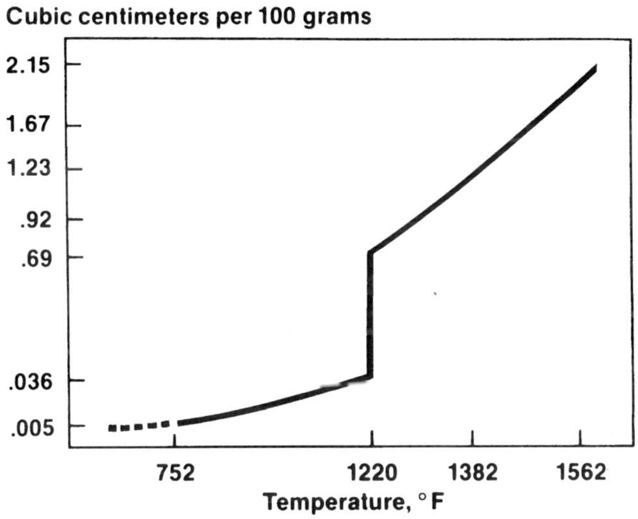

Fig. 3-4. Effect of temperature on the solubility of hydrogen in aluminum.

A skin of aluminum oxide forms on the surface of molten aluminum in contact with air. Hydrogen trapped in this skin will produce porosity in castings if mixed in with the melt during turbulent pouring.

A vast amount of information shows that there is a relationship between gas in aluminum and suspended nonmetallic (oxide) particles. It is quite probable that gas can be present in a metal bath, both on suspended particles, which act as sites for precipitation of hydrogen gas, and in the liquid solution.

Measuring Hydrogen Concentration

The measurement of hydrogen in molten alloys allows the foundry to assess the melt quality before the casting operation. There are several methods available to the foundryman to assess the hydrogen concentration. These methods are:

- Reduced Pressure Test (Straube-Pfeiffer Test);
- Quantitative Reduced Pressure Test (Severn Science);
- Density measurements;
- Initial Bubble Test (Hycon Test);
- Recirculating Gas Test (Telgas, Alscan).

Each of these methods will be discussed briefly in the following sections.

Reduced Pressure Test (Straube-Pfeiffer)

The reduced pressure test (RPT) is commonly used to assess hydrogen content of a melt. The test consists of solidifying a sample under reduced pressure (usually 1–100 mm Hg).[1] Gas pores that form are molecular and obey the gas law; therefore, they expand because of the reduced pressure and produce larger pores than if the sample were allowed to solidify under atmospheric pressure.

Reduced pressure testers, such as that shown in Fig. 3-5, are available at relatively low cost. The tester consists of a vacuum pump connected to a bell jar with a vacuum seal. Molten aluminum is placed in a small, preheated crucible on an insulated pad and covered with the bell. A vacuum is then applied to the bell.

Relatively small quantities of gas can be detected. If the melt contains a large quantity of gas, the sample will swell; if gas-free, it normally will sink. The solid slug can then be sectioned to check for relative gas content (Fig. 3-6). This test gives semi-qualitative results.

Fig. 3-5. A reduced pressure tester for detecting gas in molten aluminum alloys.

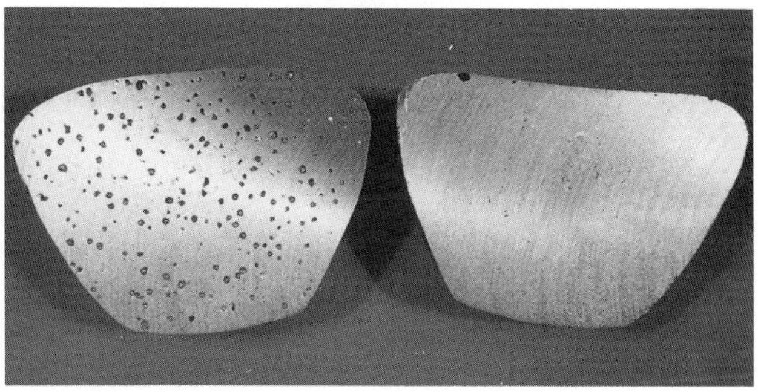

Fig. 3-6. Metal samples solidified at reduced pressure using an instrument similar to that shown in Fig. 3-5. At right, the cross section of a degassed specimen reveals sound metal with no porosity. Non-degassed metal in the specimen on the left shows porosity caused by entrapped gases that were not removed from the melt.

RPT is most effective if the sampling is done in a repeatable sequence that eliminates or reduces variables. Care should be taken to keep testing conditions (the pressure and the solidification rate/temperature, for example) the same from test to test to avoid inconsistent results.

- *Reduced pressure*—The pressure applied to the melt in this test has a major impact on gas bubble formation.[1] The vacuum must be drawn in the shortest possible time and be constant from sample to sample. A higher vacuum is not always best. For most applications, the higher the vacuum, the greater the RPT sensitivity. Too high a vacuum could degas the sample. However, the same is true for too low a vacuum and a sample allowed to solidify over a long period.
- *Solidification rate*—Solidification rate is important in terms of hydrogen distribution and the period the alloy spends between liquidus and solidus temperatures.[1] Isolating the sample crucible on an insulated base prevents it from cooling too quickly (however, cooling it too slowly can cause degassing and lead to false results).
- *Inclusions*—The number and size of hydrogen pores depends on the presence of nonmetallic microinclusions.[2] Inclusions and hydrogen contamination in the sample can change test parameters and introduce problems with test procedures. Without finely dispersed microinclusions to initiate bubble formation, relatively high concentrations of hydrogen are required in the melt to overcome molten aluminum's surface tension and form voids of significant size in the casting. This can lead to deceiving results when filtered metal is used. A sample that appears pore-free may mean the melt is relatively free of microinclusions not that it is relatively free of hydrogen. Because of the widespread use of filtered metal, foundries need to understand the relationship between hydrogen and finely dispersed nonmetallic inclusions with regard to the formation of pinhole porosity.

Quantitative Reduced Pressure Test

A constant amount of metal is allowed to solidify in a vacuum chamber. Hydrogen gas is evolved as the solidification proceeds and raises the pressure in the chamber. The amount of hydrogen is calculated from the pressure rise and displayed on a digital readout. Sensitivity is good at high hydrogen concentrations, but difficulties are experienced with concentrations of less than 0.1 cc/100 g Al.

Density Measurements

This is a simple way to indirectly measure the hydrogen content in the melt. A sample is cast and the density measured by weighing the sample in air and water (Archimedes Principle), and then calculating the density:

$$D_S = W_A/(W_A - W_W)$$

where D_S = density of sample, W_A = weight in air and W_W = weight in water. The volume of hydrogen can then be determined by the following calculation:

$$H_2 = 1/(D_S - D_T)$$

where D_T = theoretical density of the alloy, which can be determined by alloy chemistry or by measuring the density of a sample that has been thoroughly degassed and chilled.

This technique makes the assumption that internal porosity is all due to hydrogen. If the porosity is due to gas and shrinkage, the amount of hydrogen will be overestimated. Because pores will form on solid inclusions, filtered metal is more dense than unfiltered metal with the same hydrogen content. Differences between D_S and D_T are extremely small, so detecting these differences requires precise instrumentation. Hydrogen concentration can, therefore, be calculated more easily for samples solidified under a vacuum using a suitable correction factor.

Initial Bubble Test (Hycon Test)

In this test, the pressure over a molten metal sample in a closed chamber is slowly reduced. When the first bubble of hydrogen breaks the surface, the pressure and temperature are recorded. This pressure equals the partial pressure of hydrogen in the melt. Knowing the temperature, hydrogen content can be determined based on Sievert's Law:

$$[H] = k\,(P_{H_2})^{1/2}$$

where [H] is hydrogen concentration, P_{H_2} is the partial pressure of hydrogen in the surrounding atmosphere and k is a temperature-dependent constant. Nomographs that relate pressure, temperature and hydrogen solubility of the alloy being tested also can be used to determine the hydrogen content.

Recirculating Gas Test (Telgas, Alscan)

These methods measure for hydrogen content directly. A small volume of nitrogen is recirculated through the molten bath. As the nitrogen circulates through the bath, it picks up hydrogen, and eventually the hydrogen in the nitrogen and in the bath reach equilibrium. The partial pressure of the hydrogen in the nitrogen equals the partial pressure of the hydrogen in the molten bath. The hydrogen content is determined by measuring the thermal conductivity of the hydrogen-nitrogen mixture.

These methods are precise and useful for the research lab or for companies that produce a high volume of castings. However, they have had limited foundry use because the probes used are fragile and costly. Detailed additional information on these methods can be found in the literature. Foundries will need to evaluate individual circumstances to determine the proper method of analysis.

Degassing

Degassing is typically carried out by introducing a sparging gas into the molten aluminum.[3,4] As gas bubbles move up through the bath, hydrogen diffuses into the bubbles because of the difference between its partial pressure in the bath and its partial pressure in the bubbles. The methods used for degassing generally differ in two aspects: the type of sparging gas used to purge the melt and the method used to introduce the gas.

Methods for Introducing Sparging Gas

A lance, porous refractory (plug or tile) or rotary impeller can be used to inject sparging gases. Tablets or powders that form gas bubbles also can be added to the melt.

Fig. 3-7. A lance is inserted into molten aluminum for degassing.

- A *graphite lance* with an outside diameter of 2.5–5 cm (1–2 in.) is the simplest but least effective way to introduce the treatment gas (Fig. 3-7). Lances generally produce large gas bubbles that rise quickly through the metal and create turbulence at the surface (Fig. 3-8).

  Because the gas bubbles tend to rise in the vicinity of the lance, in order to degas a large melt effectively, several lances must be used. If only one is used, it must be moved constantly.
- By attaching a *porous graphite plug* to the end of the lance, a finer dispersion of bubbles can be achieved. However, bubbles again tend to rise only in the area of the plug, limiting the amount of metal that comes in contact with the plug. Thorough degassing requires a fairly large amount of time and gas.
- Using a *rotary impeller* provides the most effective means of degassing. This method introduces treatment gas into the melt through a special impeller head that rotates rapidly, producing fine bubbles and dispersing them throughout the liquid.

  This method produces small bubbles that rise slowly to the surface, permitting a longer time for hydrogen to react. Rotary degassing does not generate the turbulence typically associated with lancing (Fig. 3-9).
- Holding *chlorine-based tablets* below the surface of the melt (flux degassing) is the least controllable method because the tablets decompose rapidly. Several treat-

Fig. 3-8. Large bubbles are obtained with a lance in a water model study. Note the turbulence at the surface.

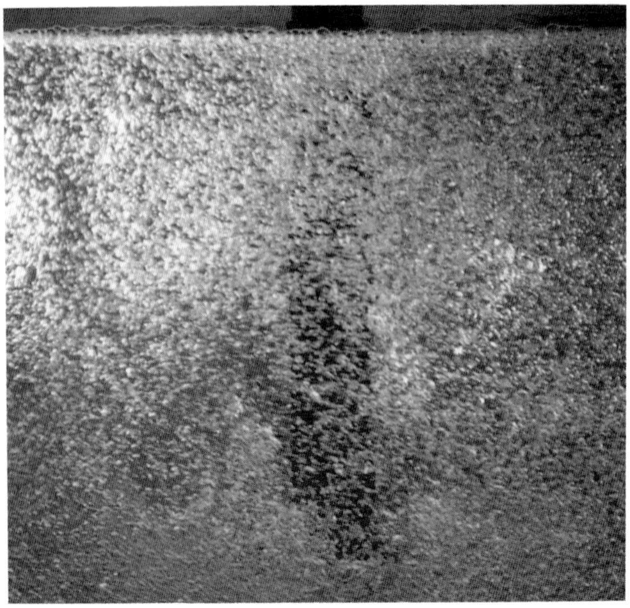

Fig. 3-9. Small, well-dispersed bubbles are obtained in a water model study when using a rotary impeller. Compare the relatively undisturbed surface with the agitated surface in Fig. 3-8.

ments may be necessary to provide enough reaction time. This method promotes dross formation and generates obnoxious fumes. Despite these disadvantages, tablet degassing is often suitable for treating small melts and is used where precise control of degassing is not required. For more information, see the section on degassing fluxes later in this chapter.

- A *powder flux* can be injected into the melt using a system like the one shown in Fig. 3-10. The flux is picked up by a low-pressure stream of an inert carrier gas (such as argon or nitrogen) and blown into the bottom of the metal bath through a submerged lance. As the powder floats to the surface, it reacts with the hydrogen throughout the depth of the bath.[5]

Types of Gases Used

The gases used for degassing aluminum may be inert (such as argon or nitrogen), meaning they show little or no reaction to the liquid aluminum, or reactive (such as chlorine, freon, fluorine or a combination of these with an inert gas), meaning they undergo a reaction with liquid aluminum or the alloying elements or additives in the melt. (Using pure chlorine has been largely discontinued because of the toxicity of the fumes generated and the corrosive effect it has on metal surfaces.)

One of the benefits to using a reactive gas or a mixture of reactive and insert gases is the cleaning effect reactive gases have on the melt. Chlorides that are formed change the surface tension of the aluminum so that oxide inclusions stick to the bubbles and are floated out of the melt. Inert gas also has a cleaning effect, but this effect is enhanced by using some reactive gas.

The advantages to using a reactive gas for hydrogen removal are not always clear. Sometimes, using a reactive gas produces better degassing; other times it appears to impair the process. If bubbles are larger and degassing efficiency is low (as with lance degassing), an improvement in hydrogen removal occurs. When bubbles are small and gas removal is highly efficient (as in rotary degassing), there is evidence that a reactive gas can slow down hydrogen removal.

A reactive gas should not be used in a melt modified with strontium or sodium. Degassing should be performed prior to modification if a reactive gas is to be used (it should be performed only briefly if it is done afterward).

Various reactive and some not-so-reactive gases and mixtures of gases have been used to remove hydrogen and suspended oxides in molten aluminum: chlorine, nitrogen, argon, 90% nitrogen-10% chloride, 95% nitrogen-5% Freon 12, 95% argon-5% Freon 12, and a tri-gas 80% nitrogen-10% carbon monoxide-10% chlorine. The most reactive single gas is chlorine, and it rates as the best flux cleaner of molten aluminum. The Freon 12 mixtures and the tri-gas mixtures are proprietary. Sulfa hexafluoride (SF_6) has recently been introduced as a reactive gas for degassing aluminum.

For many years, chlorine and nitrogen have been the most common degassers for molten aluminum. The gas is bubbled

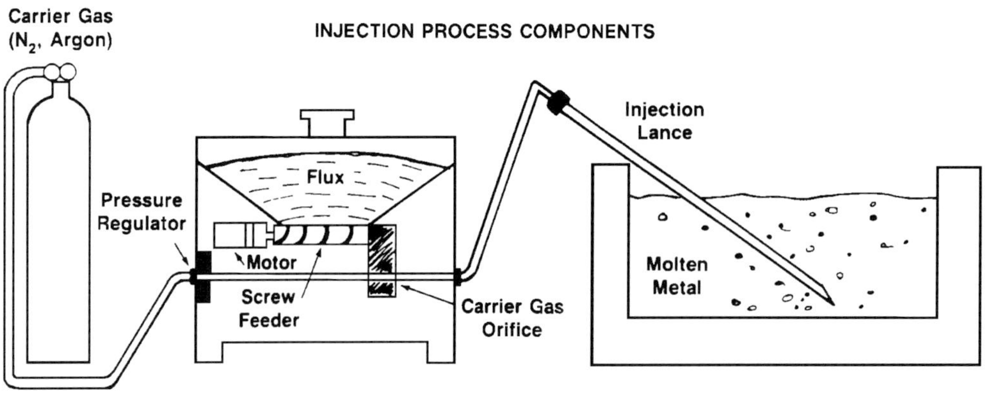

Fig. 3-10. This drawing illustrates the flux injection process.

through molten aluminum to remove absorbed hydrogen. The bubbling action also helps to float oxide particles to the surface. In this sense, chlorine and nitrogen can be classed as cleaning fluxes.

Chlorine has two advantages. Mechanically, it creates low-pressured bubbles that carry hydrogen and other impurities. Chemically, it forms aluminum chloride, which will wet nonmetallic inclusions and bring them to the surface—nitrogen is unable to remove nonmetallic particles by wetting. The relatively nonreactive nitrogen performs a mechanical fluxing action. Experience has shown that argon acts in a similar way to nitrogen. Chlorine is a more efficient cleaner of molten aluminum than nitrogen or argon.

However, chlorine is a toxic gas and, without tight control, pollutes the atmosphere with solid and gaseous effluents. Chlorine also readily reacts with magnesium by removing it—the reaction is pronounced in alloys of high magnesium content. Depending upon the fluxing period, the loss of magnesium is small in alloys of low magnesium content.

With the OSHA and EPA Clean Air Acts, the use of straight chlorine has been greatly reduced and replaced by use of mixtures of 10% chlorine-90% nitrogen or 10% chlorine-90% argon. Also, fluxing with 100% nitrogen or 100% argon has become more popular.

To ensure that the molten metal entering the mold cavity is clean of oxide films and other nonmetallics, many foundries use fiberglass cloth screens and strainer cores of various designs in the gating systems. These items have been successful in removing large inclusions but less so in removing smaller, less dense oxides. To remove most of the nonmetallics that may enter with the metal stream, a continuous open pore ceramic filter has been introduced for use in gating systems.

Comparison standards for the relative soundness of aluminum cast from a given melt are provided by the samples (taken from melts degassed to varying degrees) shown in Fig. 3-11. (See also the description of the reduced pressure test for detecting gas porosity in this chapter.)

Fig. 3-11. Comparison standards of cast A356.0 poured under 100 mm pressure, which can be used for porosity quality control. For example, No. 1 above may be specified as the standard for high quality castings, and No. 3 for general commercial quality. The center number for each indicates percent surface area porosity, the bottom number the density. [Photo courtesy of Stahl Specialty Co.]

FLUXING THE MELT

Foundries are relying more and more on fluxing to assure the best physical and mechanical properties in metals. There are so many acceptable fluxes commercially available that foundries have ample opportunity to improve product quality and to cut casting cost by using the right fluxes for foundry operations.

Common fluxes fall into four broad classes:

1) *covering fluxes,* which are used to prevent gas pickup and to reduce dross formation by oxidation (and thereby reduce metal loss);
2) *cleaning fluxes,* which are used to remove solid nonmetallic inclusions;
3) *degassing fluxes,* which are added to the melt to remove the entrapped gases (especially hydrogen) that can contribute to casting porosity;
4) *drossing-off fluxes,* which are used to recover metal from drosses.

Specialized fluxes include impurity removers, alloying fluxes and grain refiners. Using fluxes to refine grain structures of aluminum alloys is discussed later in this chapter.

There is no universal flux because there are many variables to consider when selecting a flux. The purpose for which the metal is being treated and the composition of the alloy should determine the ingredients for the flux. A flux that is suitable for one aluminum alloy may not be good for another because of differences in the oxidation rates. Melting and casting temperatures also can affect flux composition.

The end use of the alloy is an important consideration when determining proper fluxing. The type of melting unit—induction furnace, reverberatory furnace or other—may require the use of a specific type of fluxing compound.

Covering Fluxes

Simple covering fluxes, which form a liquid cover or "blanket," are used chiefly to reduce oxidation and to prevent gas absorption. Mixtures of chlorides are effective covering fluxes available at reasonable cost.

While covering fluxes provide some incidental cleansing properties, they are not efficient cleaners. They have no capacity for wetting and coalescing the nonmetallic impurities that are suspended in the molten alloy. Only considerable time or the use of a cleansing flux will bring these impurities together so that they can be removed easily.

Some people think covering fluxes cause high melting losses. This need not be true. The amount of metal lost in liquid covering fluxes depends on the furnace atmosphere and on the viscosity and melting point of the flux. If conditions are right and the flux is of the proper composition, metal entrapment can be negligible.

Sometimes, however, when liquid fluxes have been on the bath for some time and are in contact with dirty metal (as may occur in a reverberatory aluminum melting operation), thickening occurs. The dross formed tends to hold globules of aluminum. Ultimately, an intimate mixture of dross and metal forms, but even more dross will form under the same conditions without a covering flux.

A common remedy for this condition is the use of a second type of flux that, by exothermic action, will break down the rather solid material and release any entrapped molten metal back into the melt. At the same time, a fine powdery residue that can be readily skimmed off the surface of the melt is produced.

Nonhygroscopic salt fluxes are often used during and after meltdown to reduce metal loss in the dross. Salt fluxes usually consist of halide mixtures. A typical salt flux contains equal parts of potassium chloride and sodium chloride with 5–10% additions of cryolite (Na_3AlF_6) or sodium fluoride. Such fluxes are called drossing-off fluxes and will be covered later in this section.

Cleaning Fluxes

When cleaning action is important as well as a protective cover, fluorides are often added to the covering flux. Doing this produces a liquid flux that has excellent cleaning powers. The fluorides wet the suspended particles or the skins of aluminum oxide and float them to the surface.

By using fluorides, it is possible to produce excellent metal from a crucible or reverberatory furnace. If the drosses become thick and sticky, entrapped metal can be recovered by using a drossing-off flux.

Degassing Fluxes

In tablet (or flux) degassing, a sparging gas is introduced into the melt in solid pill form. The pills, submerged in a perforated bell, are composed of chemicals that thermally decompose to release gas bubbles. This is the least controllable method of degassing because the tablets decompose rapidly, and several treatments may be necessary to provide a long enough reaction time.

Most chemical degassers depend on chlorine-generating compounds. A popular solid degasser is hexachloroethane (C_2Cl_6). Like all solid fluxes, these materials are hygroscopic and must be stored in a dry atmosphere, otherwise their use will add hydrogen to the melt rather than remove it.

Degassing fluxes in powder form can be introduced into a melt using an inert carrier gas (nitrogen or argon). These powder fluxes are added through a lance, making it possible to blow the flux into the bottom of the bath (see Fig. 3-10). A flux powder that is dispersed in this manner will float to the surface of the metal, reacting chemically throughout the depth of the bath. Powder fluxes that can degas, remove nonmetallic inclusions, and perform modification and/or grain refinement during one injection process also are available.

Drossing-Off Fluxes

Quite often, the alloy composition, melting furnace and product specifications are such that the use of covering or cleaning fluxes is not necessary. Application of a drossing-off flux before tapping the furnace is generally necessary, however, to avoid excessive metal loss.

Drossing-off fluxes are exothermic fluxes that give off heat when they contact metal-laden drosses. These fluxes react with caked, semisolid dross that forms on the melt surface. In

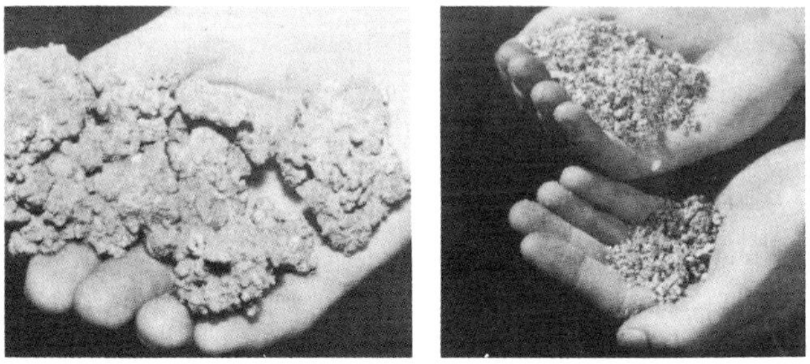

Fig. 3-12. (Left) The lumpy, metallic dross resulting from an unfluxed melt is shown. (Right) Powdery, metal-free dross results when an effective drossing-off flux is used.

doing this, drossing-off fluxes release the entrapped metal particles to the melt and reduce the sticky dross to a fine powder that is skimmed off (Fig. 3-12). The proper use of a drossing-off flux may reduce the free metal content of aluminum drosses from a high of 85% to a low of around 15%.

The reaction of drossing-off fluxes occurs by oxidation of a small percentage of the –200 mesh free aluminum contained in the drosses. The aluminum burns, releasing considerable heat that actually melts the +200 mesh aluminum particles in the dross. These remelted particles run together through the porous dross residue and back into the melt. About 5% of the total metallics in the dross are burned to recover up to 85% of the entrapped metal without further handling, processing, special equipment or treatment outside the melting unit.

DETERMINING WHICH FLUX TO USE

Based on Alloy Composition

The chemical properties of alloys determine, to a great extent, the fluxing treatment that will be used. Many of the popular aluminum alloys—319.0, 355.0 and 356.0, for example—are basically similar in that they oxidize only moderately. All may be treated with the same covering and/or drossing-off fluxes and, finally, with a degassing and/or grain refining flux.

However, some alloys containing high percentages of magnesium or silicon may require specialized treatment. Such high-aluminum/magnesium alloys as 535.0, 518.0 and 520.0 are treated much the same as magnesium alloys. In this case, the fluid cleaning flux brings the nonmetallic skins and inclusions to the surface where they may be readily drossed off and removed. Fluxes with sodium should not be used.

In the hypoeutectic aluminum-silicon alloys (those containing less than 11.7% Si), modification of the coarse silicon in the eutectic may be accomplished by adding a small amount of strontium-rich master alloy or by means of metallic sodium or sodium-bearing fluxes.

However, in the case of the hypereutectic aluminum-silicon alloys (those containing more than 13% Si), the primary silicon refinement is accomplished by adding a small amount of phosphorus. There are several refining agents that may be used, but if they are salts, they must be free of sodium and calcium, as both are detrimental to the refining action of phosphorus.

Based on Casting Method

Casting method also determines proper fluxing treatment. Except for removal of nonmetallic inclusions, treatment of an aluminum alloy is usually different for permanent mold casting than for high-pressure diecasting or sand casting.

In permanent mold operations for commercial low-stressed castings, it may be advantageous to maintain in the melt a gas content level higher than in melts used for sand casting. This is true because many permanent mold castings are not adequately fed in the mold. As a result, hot tears and/or surface sinks often appear on relatively large, uniformly thick walls of the castings. To overcome this condition, it is common practice to apply a flux in either tablet or granular form to produce a comparatively small, controlled amount of hydrogen, which forms minute, round bubbles during freezing. These bubbles, distributed uniformly throughout the metal, compensate for the shrinkage (Fig. 3-13).

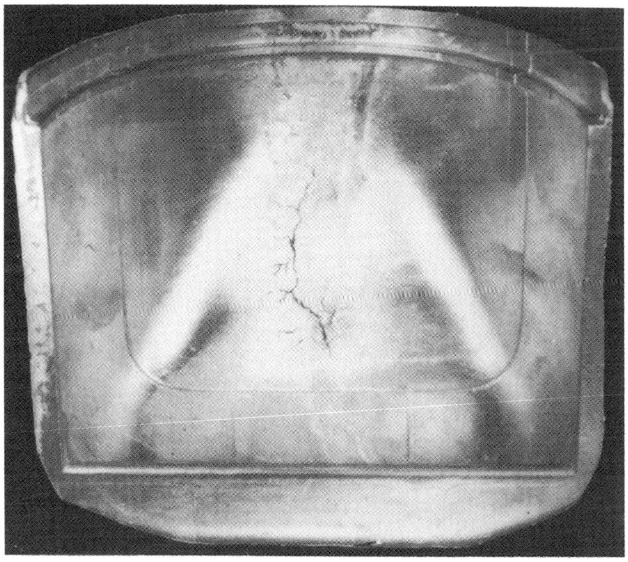

(3-13a)

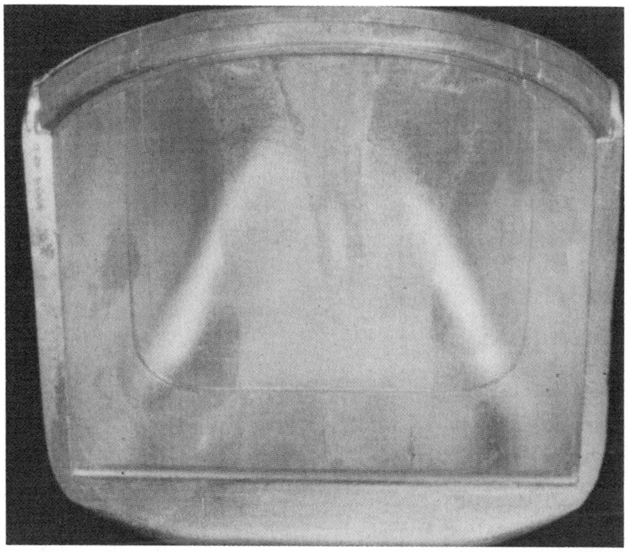

(3-13b)

Fig. 3-13. (a) Hot tears in this aluminum permanent mold casting are the result of inadequate feeding. (b) When a boiling flux was used, finely dispersed gas filled out the casting to prevent hot tears.

Admittedly, this is inducing artificial and intentional porosity. In some applications, however, controlled porosity has little effect on mechanical properties, and gas induction is simpler and often far less expensive than correcting a feeding problem that might require extensive mold work.

In diecasting, hydrogen gas is not a problem because of the high pressures under which metal solidifies and the fast chill (preventing pores of significant size). A more serious problem is air entrapment, which can be avoided by proper die venting, overflows, metal temperature control and vacuum casting. Major requirements in diecasting are to maintain a smooth surface and to meet specified mechanical properties.

In a vacuum diecasting operation, the melt should be cleaned thoroughly, then covered to prevent metal oxidation during the holding period while making the castings. With the ordinary cold chamber process, the cleaning treatment should be repeated frequently to remove oxide that has accumulated on the surface of the bath.

Based on End Use

Aluminum piston alloys are an example of how the end use of a casting determines the nature of the fluxing treatment, for these alloys must be free from even microscopic inclusions. To ensure thorough cleaning, these alloys require application of a cleaning flux that contains fluorides. The fluxes form a liquid with good cleaning powers—the best method for producing extra-clean metal. Subsequent use of a drossing-off flux will help ensure more complete removal of drosses.

GRAIN REFINEMENT

Aluminum alloys, like alloys of other light metals, will normally form coarse, equiaxed and columnar crystals (grains) during solidification (Fig. 3-14). The degree of coarseness or the length of the columnar crystals depends on the pouring temperature of the metal, thermal gradients set up within the mold, and the presence of naturally occurring grain nuclei (such as intermetallics, nonmetallics or mold-surface features). It should be noted that all common alloying elements added to aluminum decrease grain size somewhat—the more soluble the element, the greater the decrease in grain size.

Grain refinement is one of the inoculations made late in the melting operation to alter the structure of cast metal. It is done to produce a finer, equiaxed grain during solidification. For example, less than 0.2% titanium or 0.02% boron will reduce cast grain size of an aluminum alloy from as much as 0.10 in. in diameter to as little as 0.005 in. in diameter.

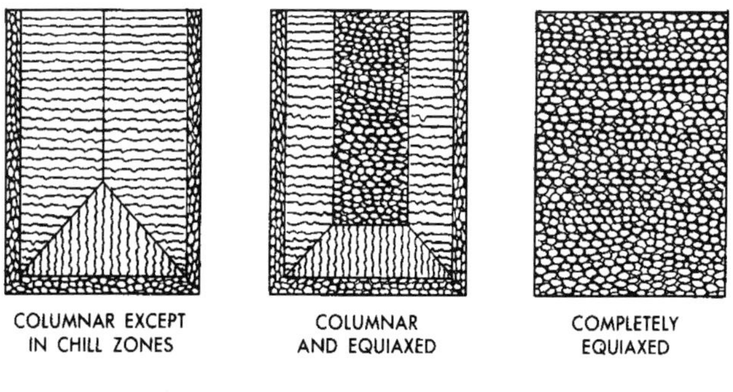

Fig. 3-14. Grain structures in aluminum castings.

Alloys containing considerable amounts of elements that have high solid solubility in aluminum (copper, magnesium and zinc) naturally develop a compact, equiaxed grain of smaller size than do the alloys with appreciable amounts of silicon (i.e., silicon-magnesium, silicon-copper and silicon-copper-magnesium alloys. Copper alloys are more easily grain refined than those containing large amounts of silicon.

A thorough presentation of grain formation, grain refinement, dendrites, solid solubility and other specific metallurgical terms applicable to the structure of aluminum alloys is beyond the scope of this book. Readers who wish to gain detailed knowledge of the grain structure of metals can find information in a number of books on physical metallurgy.

Effects of Grain Refinement on Casting Alloys

Refining grains in aluminum alloys has a variety of effects on castability. Some of the recognized advantages include:

- *Less hot tearing.* Coarse-grained metals are more prone to hot tearing during solidification because they develop more localized strain areas (Fig. 3-15). Smaller grains also tend to prevent the hairline contraction cracks in solid solution alloys.

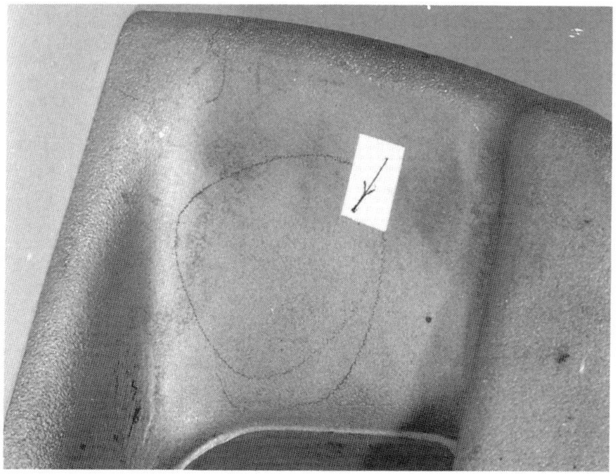

(3-15a)

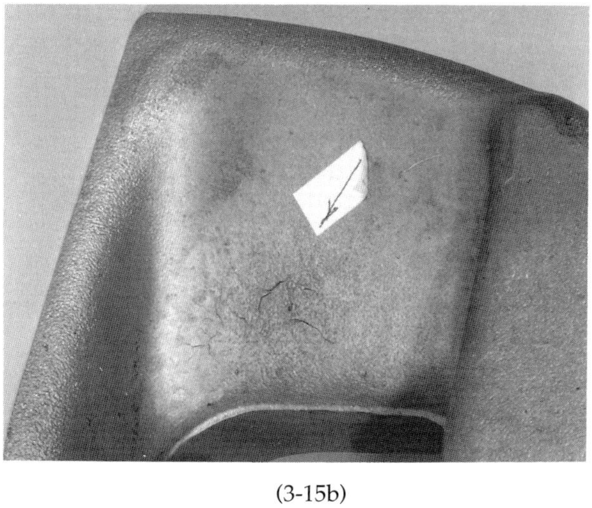

(3-15b)

Fig. 3-15. Castings that are grain refined (a) show finer grains and less tendency for hot cracking than castings that are not grain refined (b).

- *Less shrinkage porosity.* Shrinkage that occurs in sprue and risers can form pipes, which can be indicators of grain size. These pipes, which frequently form when pouring coarse-grained metal, become shallow cavities when the grains are refined. The reduction in porosity is directly related to the decrease in average pore size. With smaller pore sizes, the contribution of interfacial energies is great enough to reduce pore formation.
- *Increased pressure tightness.* Porosity is smaller and more evenly distributed in fine-grained castings. This helps make the casting more pressure-tight because there is less tendency for shrinkage in pipes.
- *Improved mechanical properties in heavier sections.* Mechanical properties of grain refined, unchilled heavier sections are 10–20% higher. When hydrogen is present and the grain unrefined, small amounts of gas porosity can decrease tensile strength by 20% in castings that solidify under low thermal gradients. Since insoluble components of an alloy help determine mechanical properties, properties are improved considerably less in high-purity alloys than in commercial-quality alloys.

The effect of grain refinement on mechanical properties is shown in Table 3-2, which compares the results of test samples cast using the Aluminum Association's standard permanent mold casting with A356.0

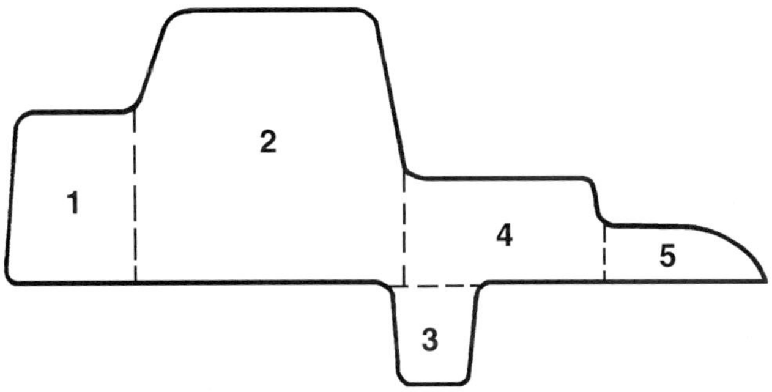

Fig. 3-16. A cross-sectional drawing of the standard Aluminum Association permanent mold casting. The sections labeled 1–5 show how the casting was sectioned for metallographic examination.

(3-17a)

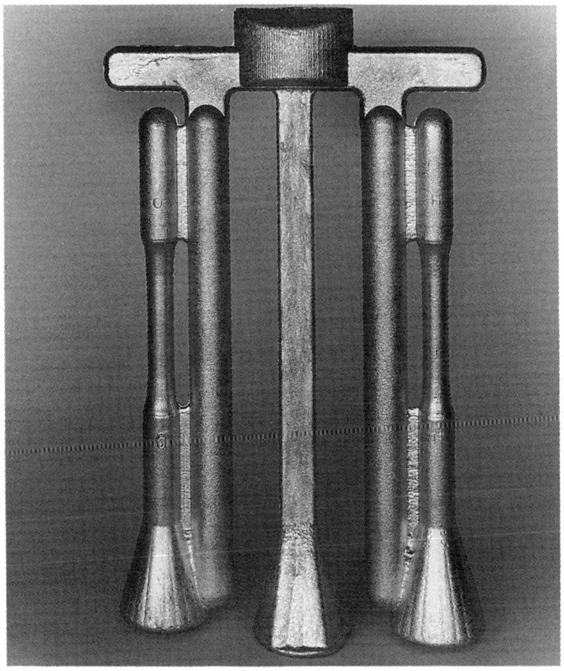

(3-17b)

Fig. 3-17. (a) A permanent mold that produces test bars per ASTM B108 specifications. (b) An ASTM B108 permanent mold test bar with gating.

Table 3-2.
Effect of Grain Refinement, Solidification Rate and Heat Treatment on Mechanical Properties of A356.0 Alloy †

	Section 2 (slow cooling rate) *				**Section 5 (high cooling rate) ***				**ASTM B108 Perm. Mold Test Bars**			
	As-Cast		**–T6 Heat Treat**		**As-Cast**		**–T6 Heat Treat**		**As-Cast**		**–T6 Heat Treat**	
Mechanical Property	**A**	**B**	**A**	**B**	**A**	**B**	**A**	**B**	**A**	**B**	**A**	**B**
Ultimate Strength (ksi)	20.815	22.648	39.520	41.674	24.363	24.150	43.342	43.370	24.949 S 1.109	26.965 0.590	46.385 1.694	47.603 0.899
Yield Strength (ksi)	13.043	13.572	35.765	36.452	12.150	13.580	34.043	35.452	12.700 S 0.540	12.956 0.424	36.361 0.615	36.650 0.580
Elongation (%)	3.0	3.0	3.5	4.0	5.4	4.6	6.0	6.5	4.4 S 1.0	5.4 0.8	5.0 2.0	6.0 2.0

A—chemistry: 0.39 Mg; 0.2 Cu; 0.19 Fe; 6.79 Si; 0.08 Ti; 0.0000B; 0.0000 Sr; 0.0027 Ph. Not grain refined.
B—chemistry: 0.39 Mg; 0.2 Cu; 0.19 Fe; 6.85 Si; 0.14 Ti; 0.0065 B; 0.0000 Sr; 0.0028 Ph. Grain refined.
★ See schematic shown in Fig. 3-16.
† According to ASTM B108 and cast per the Aluminum Association permanent mold casting.
S Signifies one standard deviation.

- *Consistent properties after heat treatment.* Because grain refinement results in a finer distribution of the alloying constituents throughout the casting, a faster response to and more consistent mechanical properties after heat treatment can be expected.
- *Improved feeding characteristics.* Grain refinement leads to improved feeding characteristics. Coarse grains of intermetallics (alloying elements and impurities) solidify at temperatures higher than the desired casting temperature and can interfere with feeding by nucleating the primary solid solution and bridging across dendrites. When the alloy is grain refined, dendrites are broken down into small, equiaxed grains that are transported easily within the casting, aiding in feeding.
- *Improved appearance in anodized coatings.* Because large grains show up as areas of different reflectivity, a fine grain size will improve the appearance of anodized castings.
- *Improved machinability.* As the grain size is reduced, secondary phases and porosity are distributed on a finer scale, thereby improving machinability.

Other Effects

Additions of titanium, a widely used grain refiner, decrease the electrical conductivity of aluminum alloys. Fluidity also is lowered slightly by grain refinement. For example, addition of approximately 0.17% titanium by weight to alloy 295.0 will reduce fluidity by about 10%.

The effect of the grain refining addition is lost in varying degrees by subsequent remelting, prolonged holding, excessive temperatures, and degassing and cleaning treatments.

Mechanical Action of Grain Refining

Grain-refined castings solidify in finely grained, completely equiaxed structures. To obtain this desirable condition, solidification must start from numerous locations, and excessive growth of grains must be prevented (Fig. 3-18). Methods for achieving grain refinement include: rapid cooling rate, mechanical vibration of the melt during cooling and addition of grain-refining elements.

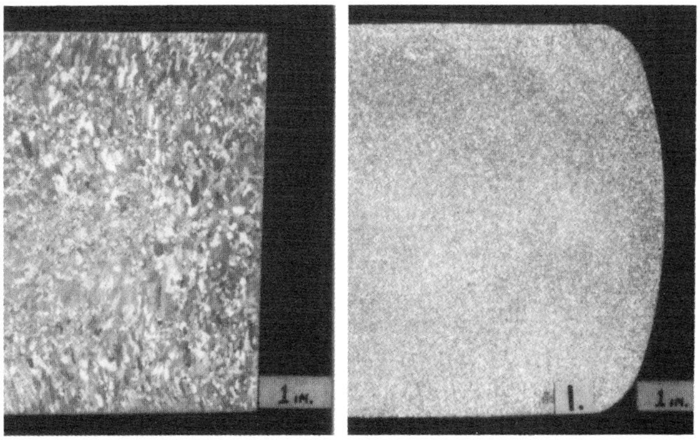

Fig. 3-18. Sections of chill cast 319.0 alloy poured under identical conditions. These two polished casting sections were chemically etched to show their grain structure. The casting on the left, made without grain refining the melt, shows large, columnar crystals at the edges. Grain refining the melt produced the equiaxed grain structure shown on the right.

Rapid cooling can retard the growth of grains in a casting, but rapid cooling alone will not change the grain structure of pure aluminum from columnar to finely equiaxed throughout the casting. As stated, adding the common alloying elements to pure aluminum results in some decrease in grain size because the added elements promote constitutional supercooling.

Mechanical vibration has proved to be effective in grain refining; however, this method is not widely used because of size limitations and the expensive equipment required.

Grain Refiners

A grain refiner, when added to a melt in a small amount, does not change the composition of the base material to a substantial degree. The most common and practical means of controlling grain size in aluminum casting alloys is by the deliberate introduction of grain-refining elements, such as titanium, boron or zirconium, immediately before the casting operation.

Titanium and boron are highly efficient grain refiners. They form nuclei or seeds of titanium aluminide, titanium diboride or more complex intermetallic compounds throughout the mass of molten metal. These countless "seeds" of insoluble material (at normal melt temperatures) nucleate fine, equiaxed grains as the casting solidifies.

Commercial Hardeners

Difficulties were experienced in the 1950s with hardeners that were made in electric arc furnaces. Contamination from oxides and nitrides resulted in unreliable recoveries and discoloration of castings. Today, titanium hardeners are produced from scrap in coreless induction furnaces.

Hardeners most used for grain refinement are aluminum-titanium (Al-Ti), aluminum-boron (Al-B) and aluminum-titanium-boron (Al-Ti-B). The most widely used hardeners are the Al-Ti and the Al-Ti-B types. The Ti or Ti-B mixtures are added in the form of master alloys available in ingot or waffle form or as salt mixtures.[4]

The straight aluminum-titanium hardeners contain from 5–10% titanium. Even at these low concentrations, gravity segregation is difficult to control, and, because of rapidly rising melting points, higher contents are not practical. The same problem occurs with aluminum-titanium-boron hardeners, which usually contain 5% titanium. Their boron content is kept low (0.1–2.5%) because titanium borides have an even greater tendency to segregate.

Grain-Refining Fluxes

A number of flux materials (grain-refining salts) compounded especially for grain refinement of aluminum alloys are available. Some foundries prefer to use fluxes instead of hardeners, claiming the fluxes give greater uniformity of grain refinement; require less labor to add to the melt; and can be added at lower temperatures, thereby lessening the danger of oxidation and gas pickup. However, poor recovery of boron usually results when grain-refining fluxes are used in alloys containing a large percentage of magnesium.

Grain-refining fluxes commonly used for treatment of most aluminum alloys contain titanium, boron, zirconium and carbon. Titanium and boron, or a combination of the two, are considered most effective. A combination of salts, such as potassium fluorotitanate and potassium fluoroborate, are used to introduce titanium and boron into the melt. These materials form intermetallic compounds—titanium carbide or titanium-boron carbide, for example.

Metallurgists differ as to the precise nature of these compounds, but they agree that the compounds serve as nuclei for crystal formation at many points rather than just a few. The result is a much smaller and more nearly uniform grain size with the attendant advantages of improved mechanical properties and better appearance (Fig. 3-19).

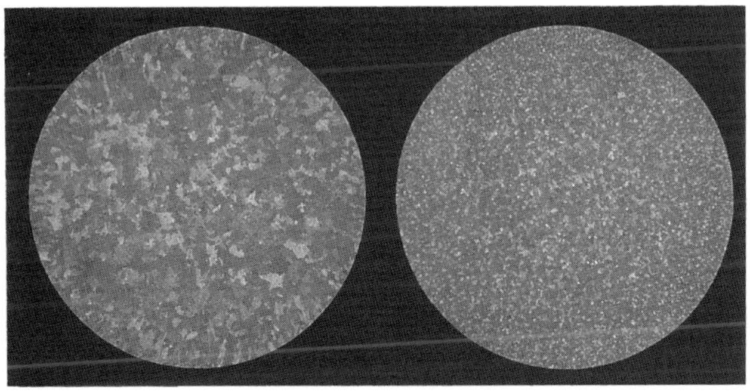

Fig. 3-19. Fluxes and master alloys containing boron and titanium are used to refine grains in aluminum castings. The casting on the left was cast without grain refinement; the casting on the right was cast after treatment with a master alloy grain refiner. [Photo courtesy of KB Alloys, Inc.]

(3-20a)

(3-20b)

Fig. 3-20. (a) Test samples for evaluating aluminum grain refiner additions are formed by pouring molten metal into iron rings set on a block of fused silica. (b) The three surfaces of each test sample show grain structures produced by the most widely used aluminum casting methods. The sample's bottom (lower right) resembles a die casting, the sample's top (lower left) a sand casting, and the sample's side (top) a permanent mold casting. [Photos courtesy of KB Alloys, Inc.]

Amounts of Grain Refiner

Considerable quantities of foundry ingot are prerefined, even though it is commonly recognized that the grain refiner additions are most effective when added immediately before pouring. With normal operating temperatures, the usual amounts of refiners added are approximately 0.05–0.15% titanium, 0.04% boron, or 0.01–0.08% titanium plus 0.003% boron. An addition of 0.003–0.008% boron is preferred for high-silicon aluminum alloys, such as piston alloys. At higher temperatures, 0.03% titanium plus 0.01% boron is recommended.

Larger additions of refiners frequently result in gravity segregation of intermetallic complexes, especially in the more complex alloys. Excessive boron additions can cause undesirable mold reactions and also can result in borides settling out in the bottom of the furnace.

Simple Grain Size Test Procedure

A quick and simple test procedure for evaluating grain size has been developed for users of foundry and wrought alloys. It consists of one or more iron rings with 3 in. (76 mm) ID and 1 in. (25 mm) high (Fig. 3-20).

These open-ended rings are set on a block of fused silica. Fused silica is a light, porous, soft material that serves as an excellent insulator for the base. The iron rings are not preheated. The molten metal is poured into the top of the rings. The fused silica prevents fast heat conduction away from the base of the rings; thus, the molten metal heats the ring molds, simulating a heated mold in production.

When cool, the samples are removed from the molds. In these samples (which require no polishing or cutting because the grain structure is sharp and clear), the finest grains appear on the sides next to the iron mold, as the mold provides the most chill. The bottom freezes next, and the top surface with the largest grains freezes last. Based on the history of the melt, a comparison of how high processing temperatures, long holding times, grain refiner additions and such, affect grain size can be made.

STRUCTURAL MODIFICATION OF HYPOEUTECTIC Al-Si ALLOYS

Background and Theory

There has always been considerable interest in the structural modification of the hypoeutectic and eutectic aluminum-silicon foundry alloys. These alloys generally fall in the 5–13% silicon range and include much-used alloys such as A332.0, 319.0, 356.0 and 357.0.

Over the years, studies have been carried out to better understand the structural modification of these alloys. Much effort has gone into finding better ways to implement modification and into finding elements other than sodium that can be used to modify these alloys. From this effort came the desire to achieve a means of more permanent modification.

As early as 1920, it was found that a very small addition of sodium modified the structure of these alloys, causing an increase in strength and an appreciable improvement in ductility. As time passed, this generally became the established modification treatment for Al-Si alloys.

The term "modification" describes a condition of marked refinement of the silicon crystal structure in the aluminum-silicon eutectic of this type of alloy. As mentioned earlier, modifiers such as sodium and strontium in hypoeutectic silicon alloys (those containing less than 12% silicon) tend to reduce interfacial tension between the eutectic phases. The reduction in interfacial tension increases the contact angle between aluminum and silicon, allowing the aluminum to envelope and arrest the growth of the silicon crystal.

Calcium also is used as a modifying agent, but it is considerably less effective than sodium.

Sodium Modification

During solidification of hypoeutectic aluminum-silicon alloys, the freezing process occurs over a range of temperatures. At the end of the freezing process, the liquid phase that surrounds the dendritic array undergoes a transformation, resulting in the formation of eutectic phases.

An unmodified alloy contains large, brittle flakes of silicon that cause the casting to exhibit poor ductility. Unmodified alloys typically have elongations of no more than a few percent, and the fracture surfaces are seen to be largely brittle.

With a successful modification treatment, achieved by adding either strontium or sodium, the silicon flakes change morphology and assume a fine fibrous structure (Fig. 3-21). These fibers appear to be small individual particles on a conventionally polished surface, but, in fact, they are connected in a coral- or seaweed-like structure. The fracture mode in the eutectic portion of the modified casting is ductile, and the elongation and tensile strengths have increased substantially. The effect of modification on mechanical properties is shown in Table 3-3.

There have been some problems with sodium modification. Adding sodium in the foundry by immersing it in the metal bath creates considerable surface agitation and objectionable fume generation. Depending on how it is added, sometimes a questionable degree of modification is achieved. After adding the sodium, the modification effect decreases and is lost in an hour or so.

Almost all producers of Al-Si alloys modify them before they are shipped to foundries. This modification has been found to vary in effectiveness, depending on how foundries

Table 3-3.
Typical Mechanical Properties of Modified and Unmodified Cast Aluminum Alloys

Alloy & Temper	Product	Modification Treatment	Yield Strength (0.2%) ksi	UTS ksi	Elongation (%)
13%Si	Sand Cast Test Bars	None	—	18.0	2.0
		Na-modified	—	28.0	13.0
13%Si	Permanent Mold Test Bars	None	—	28.0	3.6
		Na-modified	—	32.0	8.0
359.0	Permanent Mold Test Bars	None	—	26.1	5.5
		0.07% Sr	—	30.5	12.0
356.0—T6	Sand Cast Test Bars	None	30.1	41.9	2.0
		0.07% Sr	34.5	42.5	3.0
356.0—T6	Bars Cut from Chilled Sand Casting	None	30.9	41.2	4.4
		0.07% Sr	31.6	42.2	7.2
A356.0—T6	Sand Cast Test Bars	None	26.0	40.0	4.8
		0.01% Sr	30.0	43.0	8.0
A444.0—T4	Permanent Mold Test Bars	None	—	21.9	24.0
		0.07% Sr	—	21.6	30.0
A413.2	Sand Cast Test Bars	None	16.3	19.8	1.8
		0.005–0.05% Sr	15.6	23.0	8.4
A413.2	Permanent Mold Test Bars	None	18.1	24.4	6.0
		0.005–0.08% Sr	18.1	27.7	12.0
A413.2	Test Bar Cut from Auto Wheel	0.05% Sr	17.5	28.0	10.6
		0.06% Sr	18.2	28.0	12.8

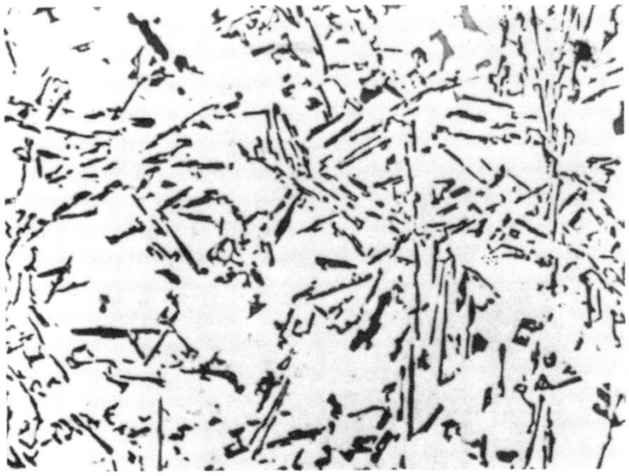

(3-21a)

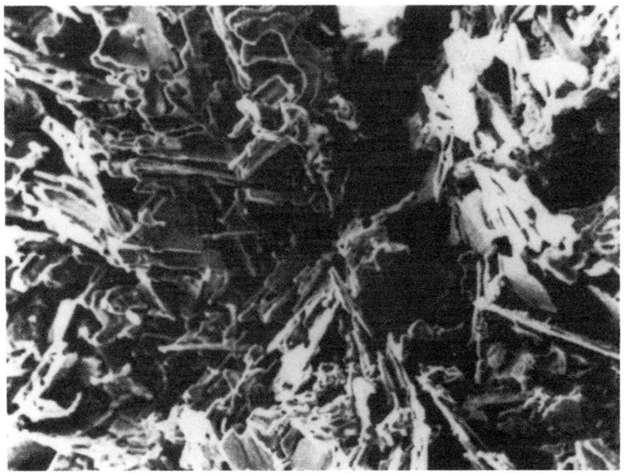

(3-21b)

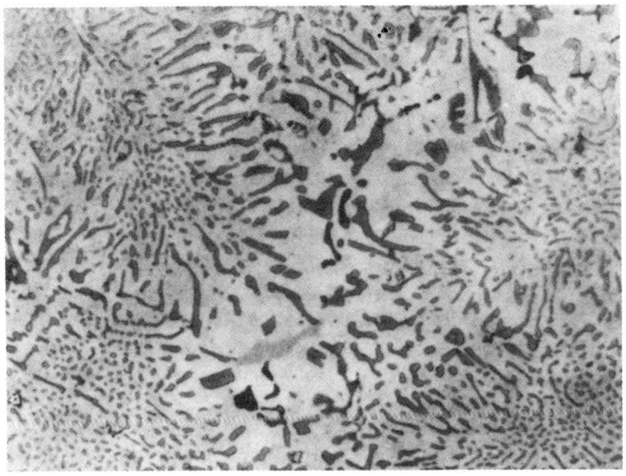

(3-21c)

(3-21d)

Fig. 3-21. Effect of modification on the morphology of the eutectic phase in 356.0. The SEM samples (b and d) were deep etched to bring out a 3-D perspective of the eutectic phases. (a) Unmodified, 0.5% HF, 300X. (b) Unmodified, deep etched, 300X. (c) Modified, 0.5% HF, 800X. (d) Modified, deep etched, 2000X. [Photos courtesy of G.K. Sigworth.]

process the metal. Production experience has proved that, to fully control modification, the modifier is best added in the foundry just before the metal is cast.

One producer approached the problem of the rather elusive characteristic of sodium modification in a different manner. Instead of alloying the silicon and subsequently modifying it in the holding furnace, this producer alloys the silicon in the reduction cell where it receives a long (24 hour) soak in a natural sodium environment—the electrolyte of the cell. Alloys that have been produced by this type of cell action possess good modification and retention of modification when melted a number of times by the foundry.

If a foundry wants to add sodium (assuming the metal has not been pretreated), good modification can be obtained by adding not less than 0.015% by weight. This usually results in a retained level of about 0.002% in the casting.

Figure 3-22 is a photomicrograph of the area of a typical Al-Si alloy casting showing structure in the absence of modification. The background or matrix of the structure consists

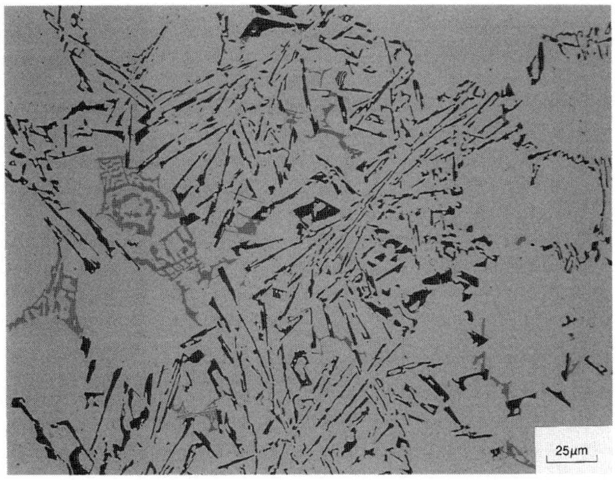

Fig. 3-22. As-cast structure of A356.0 cast without a modifier. Rectangular and needle-like dark shapes are plate-like particles of silicon. Lighter background area is aluminum. [Photo courtesy of KB Alloys, Inc.]

predominantly of aluminum and is soft and ductile. Plate-like hard particles of silicon look like needles. The general mixture of aluminum and these coarse silicon particles is known as the aluminum-silicon eutectic. This is the typical structure of a casting having low tensile and elongation properties. The hard silicon particles represent discontinuities in the soft, ductile matrix of aluminum.

When a small amount of sodium is added to a melt, the eutectic mixture with the coarse silicon platelets is converted into a finely dispersed mixture, the same as for strontium additions (Fig. 3-23). Sodium modifies the eutectic into a uniform dispersion of very fine particles. This fine eutectic is more ductile than a structure with coarse plates of silicon. Also, tensile strength increases, and elongation increases substantially. Shrinkage behavior during solidification is dispersed instead of being centered in large cavities. Machinability is greatly improved.

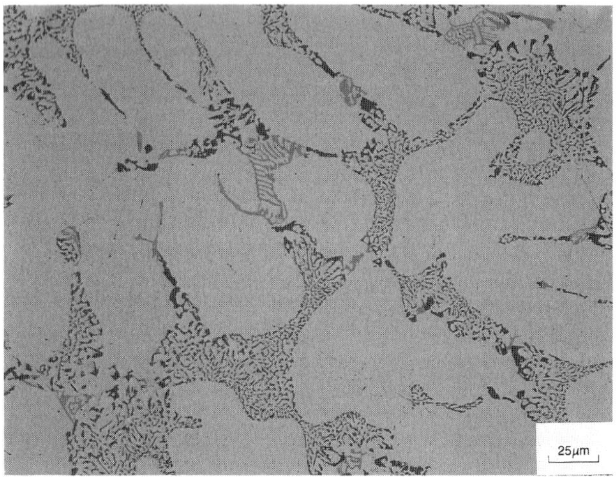

Fig. 3-23. As-cast structure of A356.0 modified with strontium. Coarse silicon particles (as shown in Fig. 3-22) have been changed to very fine particles, making the casting more ductile. [Photo courtesy of KB Alloys, Inc.]

Strontium Modification

In 1975, an aluminum master alloy containing 10% strontium and 14% silicon was introduced for modifying aluminum-silicon alloys. It offered a convenient way to add a modifier while eliminating the uncertain modification associated with sodium modification.

Early work indicated that, when comparing strontium modification to sodium modification, a somewhat greater amount of strontium needed to be added. Investigators in Europe reported adding as much as 0.07–0.08% strontium to the higher-silicon hypoeutectic silicon alloys for the best results. However, much smaller amounts have subsequently proved to be adequate. Strontium appeared to retain modification longer than sodium. There were initial reports that small amounts of sodium and calcium in the melt interfered with strontium refinement.

Strontium is an effective modifying agent for hypoeutectic aluminum-silicon alloys. In the foundry, modification can be accomplished effectively, safely and more conveniently. Modification will be retained longer than with some sodium treatments. However, investigations show that reduction cell sodium modification by producers of the silicon alloys is of equal effectiveness and has the same retained modification as strontium.

When strontium is initially added to a furnace melt, there is about a 50% loss of the strontium. It is believed that this initial loss is to the furnace linings. In subsequent melts, the strontium loss is appreciably less. Generally, in repeated melts, additions of 0.01–0.02% strontium by weight is sufficient to cause adequate modification in the final castings. Strontium has greater "staying power," and if a minimum of 0.008% is retained in the metal, then modification is retained in subsequent melts.

Investigations indicate that sodium and strontium are, in fact, compatible. The presence of both will not destroy the modification effect of the other. Microstructural transformations caused by strontium additions are the same as for sodium additions (see Figs. 3-22 and 3-23).

Straight chlorine fluxing will tend to destroy the modifying effects of both sodium and strontium. It is recommended that metal pretreated by strontium be fluxed by dry nitrogen or argon gas.

Strontium has proved to be easy to handle in the foundry and effective for modification. However, there has been a tendency to add too much strontium to the melts, and in slow-cooled sand casting, there have been experiences of gassing the metal when strontium is used.

Strontium can be added to the melt in several different forms. Table 3-4 shows the main forms and compositions in which it is added. The master alloys are added directly to the molten metal. It is possible to add Al-90% Sr alloy directly to the metal in the foundry without going through the master alloy state.

The low-strontium master alloys in the table can be added to the surface of the melt because of their low reactivity in air. Master alloys D and E will dissolve gradually in a normal melt temperature range (1290–1380F, 700–750C) because of their higher liquidus temperature. While master alloys D and E will sink to the bottom of the melt, F will float on the surface because of its lower density. A gentle stirring should be applied to obtain good dissolution rates and high strontium recoveries.

Table 3-4.
Form and Composition of Strontium Alloys and Master Alloys[4]

Alloy	Composition	Sr Source	Form
A	Al-90% Sr	Pure	Rod, Can
B	Al-3.5% Sr	Pure	Rod
C	Al-5% Sr	Pure	Rod, Waffle
D	Al-10% Sr	Pure	Waffle, Buttons
E	Al-10% Sr-1% Ti	Pure	Rod
F	Al-10% Sr-14% Si	Sr_2Si	Waffle, Buttons

Master alloys B and C melt readily if the metal is kept in the 1290–1380F (700–750C) range. In rod form, master alloys B, C and E can be added into the launder by an automatic wire feeding machine.

The dissolution characteristics of the Al-90% alloy are different from the low-strontium master alloys. Dissolution rates are higher at low melt temperatures (1200–1290F, 650–700C) than at higher temperatures. This alloy can be added by one of several techniques: surface addition, cage addition and ladle addition.

By using the appropriate product form and addition technique, dissolution of strontium modifier is achieved in less than 10 minutes. High strontium recovery levels (usually in the 90–100% range) are obtained.[4]

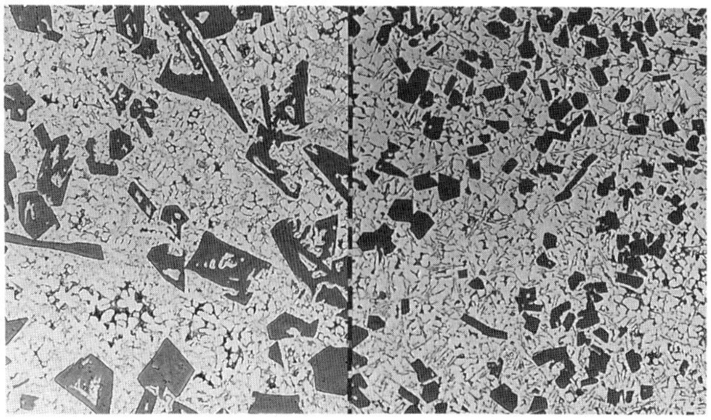

Fig. 3-24. Photomicrographs of Alloy 390.0 (17% Si) show the effect phosphorus has on primary silicon crystal size. The untreated metal on the left contains coarse plates of silicon. The metal on the right (0.03% phosphorus) contains fine crystals of silicon that are dispersed through the matrix. Unetched, original magnification 100X, reduced by ⅓ for publication.

Modification by Antimony Treatment

In Europe, a small addition of antimony has been used to obtain structural modification of aluminum-silicon alloys. It is reported that about 0.12% antimony is added to refine the eutectic in alloys such as 356.0. The antimony is a permanent constituent of the alloy and, thus, performs as a *permanent* modifying agent. The antimony is reported to be unaffected by holding time, remelting or degassing.

Alloys with antimony are distinguished by their low susceptibility to gassing, as compared to alloys treated with sodium or strontium. They are intended for low-pressure and permanent mold castings.

Antimony is not compatible with strontium and sodium, and the eutectic modifying properties of these elements can be adversely affected if cross contamination occurs. Antimony-modified alloys are not recommended for food contact applications.

The use of antimony as a modifier in North America has been limited because concerns about potentially toxic antimony compounds have not been fully resolved.

STRUCTURAL MODIFICATION OF HYPEREUTECTIC Al-Si ALLOYS

Background and Theory

The mechanism for refining the primary silicon structure by phosphorus is different than that for sodium or strontium in eutectic silicon modification.

Modifiers, such as sodium and strontium in hypoeutectic silicon alloys (alloys with less than 12% silicon), tend to reduce the interfacial tension between eutectic phases, which consist of lamellar structures of eutectic aluminum and eutectic silicon. The reduction in interfacial tension increases the contact angle between aluminum and silicon, allowing the aluminum to periodically envelope and arrest the growth of the silicon crystals.

On the other hand, phosphorus inoculation (termed "refinement") is, theoretically, a nucleation phenomenon that influences the size and shape of only the primary silicon (Fig. 3-24). Phosphorus reportedly combines with aluminum in a melt to form tiny insoluble aluminum-phosphide (AlP_3) particles that, due to their crystallographic similarity to silicon, act as suitable nuclei on which primary silicon crystals grow during solidification. Both aluminum-phosphide and silicon have a diamond cubic crystal habit and a similar lattice constant (Si, a_o = 5.43Å; AlP_3, a_o = 5.45Å) (Fig. 3-25).

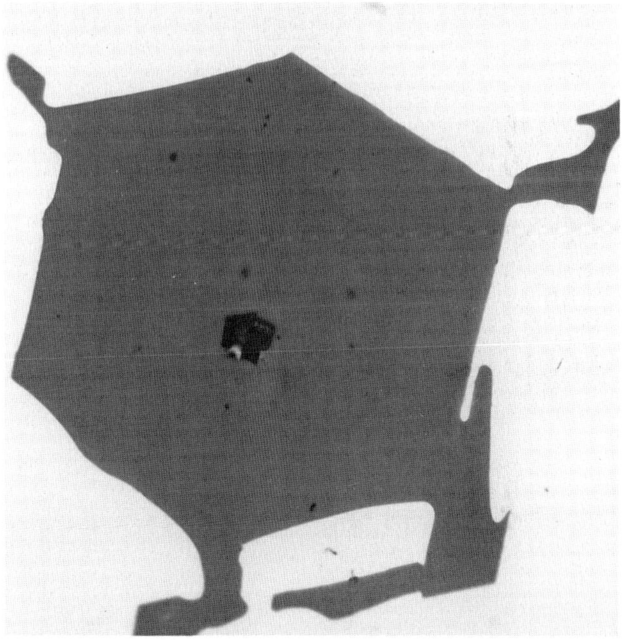

Fig. 3-25. Primary silicon particle with aluminum-phosphide seed crystals (1000X).

Refining Agents

Many agents other than phosphorus have been investigated for refining hypereutectic aluminum-silicon alloys and are reported or proposed in the literature. Although most of these agents have been tried in numerous domestic laboratories, none has proven as consistently effective as phosphorus (Fig. 3-26).

Phosphorus is (or has been) available for refinement in several forms including: 1) phos-copper alloy; 2) aluminum-phosphide, silicon-phosphide and magnesium-phosphide master alloys; 3) the phosphides of nickel, magnesium, iron and/or manganese; 4) the compounds PCl_3 and PCl_5; and 5) various proprietary salt mixtures containing phosphorus.

Refining Procedures

The refining agent should be added to the aluminum-silicon alloy after the melt has been degassed, the analysis of the melt is correct, and the temperature is in the range of 1300–1450F (705–788C). The melt temperature at the time of addition should generally be the approximate desired pouring temperature—excess superheat is not required unless pouring is intended to be below 1300F.

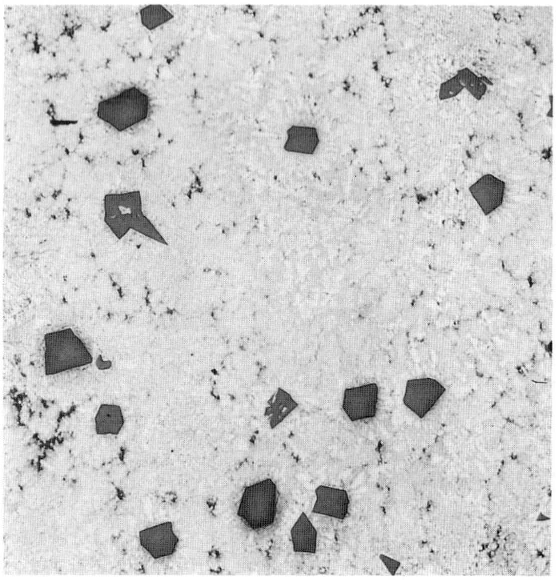

(3-26a)

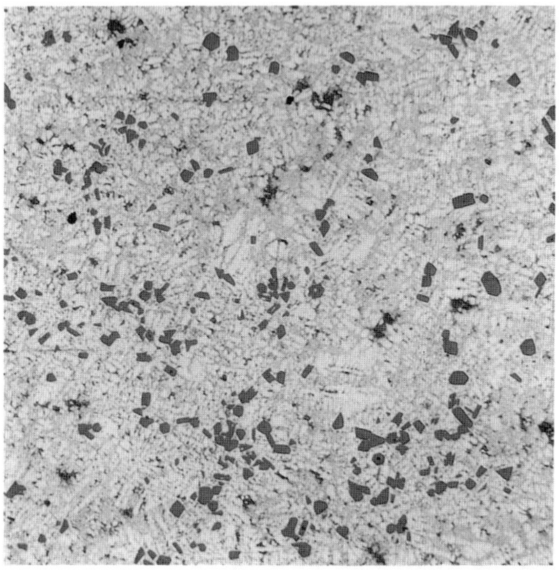

(3-26b)

Fig. 3-26. Comparison of primary silicon sizes in an unrefined casting (a) and a casting refined with phosphorus (b). Unetched, 100X.

Eutectic phos-copper is stirred into the melt, and after a few minutes, the melt is again gently stirred to assure that all the phos-copper is dissolved.

Proprietary salts (in tubes or tablets) should be added per the manufacturer's instructions. These salts are generally plunged beneath the melt surface with an appropriate tool (inverted bell or phosphorizer) and held there until active fuming ceases. Caution: any salt-type preparation must be kept dry. The tools used to plunge it must be thoroughly dried by preheating to guard against an explosion hazard and to avoid gassing the melt, which will react with moisture to generate oxides and hydrogen.

PCl_5 can be added in the same manner as the salt preparation; or, because it sublimes at only 324F (162C), it can be boiled and bubbled through the melt on a carrier gas such as chlorine or nitrogen.

Effect of Variables

The precise microstructure achieved in any hypereutectic aluminum-silicon alloy casting will be governed by a number of factors other than the addition of a phosphorus-bearing inoculant. Among these are: 1) solidification rate (and hence the casting process being used); 2) melt temperature and pouring rate; 3) length of time the melt is held after inoculation; 4) degassing and stirring after inoculation; 5) the presence of certain alkaline metals or alkaline earths; and 6) excess refinement.

Solidification Rate

Even if a melt is refined with phosphorus, the ultimate primary silicon size is still controlled by the rate at which the alloy cools from the liquidus temperature to the aluminum-silicon eutectic temperature. Casting processes characterized by high solidification rates or thin sections of a casting that naturally solidify at higher rates will generally yield finer primary silicon crystals.

For this reason and because higher solidification rates also improve the strength of the matrix (by improving the size, shape and distribution of elements other than primary silicon), it is desirable to chill areas of any hypereutectic aluminum-silicon alloy casting that are critical to performance of the part, that are to be highly stressed or that are to undergo critical machining operations.

Melt Temperature

Primary crystals that initiate and grow during flow conditions (that is, while the metal is still in movement in the mold gating system or cavity) do not exhibit the same characteristics as those crystals that initiate and form in quiescent metal entirely under the influence of the surface heat-transfer rate and conductivity of the mold. Crystals formed prematurely typically are larger and often are segregated into clusters.

Pouring Rate

Within the limits common to gravity and low-pressure casting, pouring rate does not have a significant effect on primary silicon size, except relative to its combined effect with factors such as pouring temperature on the prevention of premature primary crystal formation. Only at the extremely quick fill rates of conventional diecasting does this rate alone play a significant role.

Melt Holding Time

Inoculation of a melt with phosphorus is not permanent. The useful life of a refinement treatment is dependent on melt size and the amount of stirring or agitation that the melt undergoes.

Loss of refinement is gradual. Although phosphorus is indeed lost from a melt in time, the loss of refinement seems to be due to the agglomeration of aluminum-phosphide particles (which reduces the number of effective silicon nucleation sites). Effective refinement is generally lost well before any appreciable decrease in phosphorus content.

High-magnification inspection of melt samples taken after complete loss of refinement often reveals clusters of aluminum phosphide. Chemical analysis of the same samples shows that the phosphorus content of the melt was nearly the same as during periods of effective refinement.

Agitation

Any agitation of a melt seems to hasten the loss of refinement. Large melts that were held at relatively low temperatures (1225–1275F, 663–691C) and were allowed to remain undisturbed—except for occasional sampling—have been observed to retain refinement for 10 or more days. Similar melts that were held at more normal operating temperatures (1350–1400F, 732–760C) and with more normal stirring and ladling retained their refinement for little more than one day.

Fluxing or Degassing

Fluxing with chlorine, nitrogen or mixed gases represents extreme agitation of a melt and, therefore, hastens loss of refinement. In addition to the pure agitation effect on the rate of agglomeration of aluminum-phosphide particles, flux gas bubbles also seem to float aluminum-phosphide particles to or near the melt surface where they are entrapped with dross and removed during skimming.

Remodification

Solidification and remelting per se do not destroy refinement, except as they contribute to the total holding time and to melt agitation. Experimental melts have been solidified and remelted as many as five times with no significant loss of refinement.

Sodium and Strontium

Various investigators have attempted to promote both primary refinement and eutectic modification by using phosphorus together with sodium or strontium. This has not been successful because sodium and strontium interact with phosphorus to form phosphides, compounds that are ineffective in altering either primary or eutectic silicon and that seem to pass quickly from the melt.

Only if excess phosphorus remains after its reaction with sodium or strontium is refinement likely to occur. Similarly, only when excess sodium or strontium remains is eutectic modification possible. Therefore, refinement with phosphorus and modification with sodium or strontium cannot be achieved simultaneously.

THERMAL ANALYSIS

It is clear that both the extent of grain refinement and the level of modification achieved are important because of their consequent effects on mechanical and other properties. Spectrographic analysis will confirm the presence of grain refining and modification elements, but the elements may have lost their functional role if the melt was held at bath temperatures longer than the elements' critical active time. Metallography enables the observation of both the relative grain size and modification of the eutectic phase, and if the melt is held for this quality check, the time lag may result in decreased effectiveness of the elements.

In some castings, test coupons have been added to the design of the pattern to be sectioned and metallographically evaluated to check the grain size and level of modification. A negative evaluation obviously results in scrap.

To have a high degree of quality assurance in the product, a control situation that assesses the melt quality prior to casting is required. If the melt quality index is unacceptable, changes in the melt can be made. This is accomplished through microprocessor-assisted thermal analysis of the melt.

Temperature readings (T) from samples of solidifying molten aluminum are collected using a sand cup mold containing the temperature sensor. The data comprising a cooling curve for Al-Si foundry alloys is shown in Fig. 3-27. The curve is divided into three regions: A, B and C. The liquidus temperature is denoted by T_L; at time t_1, the solidification process starts.

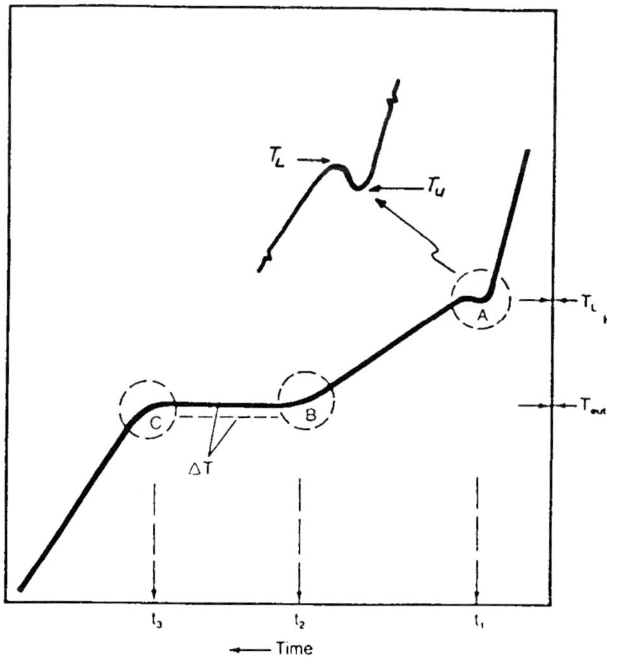

Fig. 3-27. Cooling curve obtained during thermal analysis.

In region A, "fingerprints" left by the thermal analysis curve indicate the extent of grain refinement that has taken place in the melt. For example, if no grain refiners are present, the melt will remain molten at a temperature below the liquidus point, T_L, because the nucleation of the solid phase has not commenced. (In effect, the absence of grain refiners can be viewed as a nucleation barrier.) Thus, the nature of the cooling curve recorded in region A yields information about the extent of grain refinement in the melt.

Between regions A and B, in the time interval between t_1 and t_2, the sample is in the two-phase region of the cooling curve. In this region, the solid and liquid phases coexist. As the temperature drops, more liquid freezes, and the solid phase fraction increases. At time t_2, the liquid phase surrounding the solid dendritic network is enriched in composition by nearly all of the Si (solute element) and undergoes eutectic transformation. The eutectic is invariant, and the liquid phase transforms into two solid phases at a constant temperature.

The eutectic transformation occurs between t_2 and t_3. If the melt is not modified (by adding Na or Sr), then the resulting eutectic phases are coarse. If, however, the melt is modified, then the resulting phases surrounding the solid dendrites are fine and fibrous in nature (see Fig. 3-21).

The cooling curve also indicates the modification level in the melt. If the melt is not modified, the solid line between times t_2 and t_3 on the cooling curve will be obtained.

If the melt is modified, then the eutectic reaction occurs at a lower temperature with a certain differential, ΔT. The dashed line between the time interval t_2 and t_3 indicates a modified melt. In region C, the solidification process is complete. When appreciable amounts of other impurities are also present, the formation of other phases (e.g., MgSi and $CuAl_2$) after point C are observed. The impurity elements also lower the temperatures T_e and T_{eut}, but it is still possible to assess grain refinement and modification.

A commonly used thermal analysis instrument is the unit shown in Fig. 3-28. It features a digital display of the grain size, the eutectic arrest and the ΔT of the eutectic arrest when compared to the unmodified alloy. A printer provides a hard copy of the pertinent data displayed, along with the time, date, casting reference code, and the temperature difference and time duration between T_U and T_L.

Figure 3-29 is a correlation of the grain size number digitally displayed by the unit in Fig. 3-28 and the grain size found in the microstructure of aluminum-silicon alloys. The system used is similar to the ASTM grain size numbering system, where the grain size number increases as a finer grain size is obtained in the microstructure.

Experiments have shown[6] that if undercooling (T_L–T_U) is large, a relatively large grain size is indicated. When undercooling is less than 0.5°F (0.3°C), there is virtually no barrier to nucleation, and the grain size is relatively fine. In addition to the grain size number that is displayed during a thermal analysis test, the thermal analysis unit also may display the following digits that each have a particular significance:

- 1111—Formula is negative; a very large grain size (600–1000μm);
- 3333—$T_U - T_L$ < 0.5°F (0.3°C); a very fine grain size (approximately 500μm);
- 6666—No sinusoidal is observed ($T_U = T_L$); grain size is extremely fine (approximately 200–300μm);
- 8888—Maximum temperature of poured sample is less than 1100F (600C).

Figure 3-30 shows the average grain size achieved in samples cast from A356.0 and Al-7% Si melts having different amounts of grain refiner (Al-5Ti-1B master alloy) additions. The corresponding undercooling (T_L–T_U) obtained from thermal analysis of each of these melts also is shown.

As previously noted, a large undercooling from thermal analysis indicates that a cast structure has a large grain size (>2000 μm). The lower dashed curve represents the measured grain size at the bottom surface of the sample, which cools quicker than the rest of the casting. The upper dashed curve represents the measured grain size at the upper surface of the sample. Data obtained from the upper surface of the sample is typical for what would be expected to be

Fig. 3-28. A thermal analysis instrument used in assessing the grain refinement and modification levels in Al-Si foundry alloys.

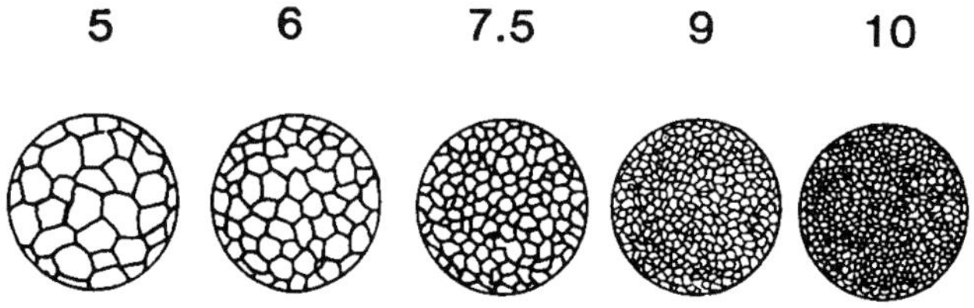

Fig. 3-29. Schematic representation of grain sizes in a casting as evaluated and displayed by the instrument shown in Fig. 3-28.

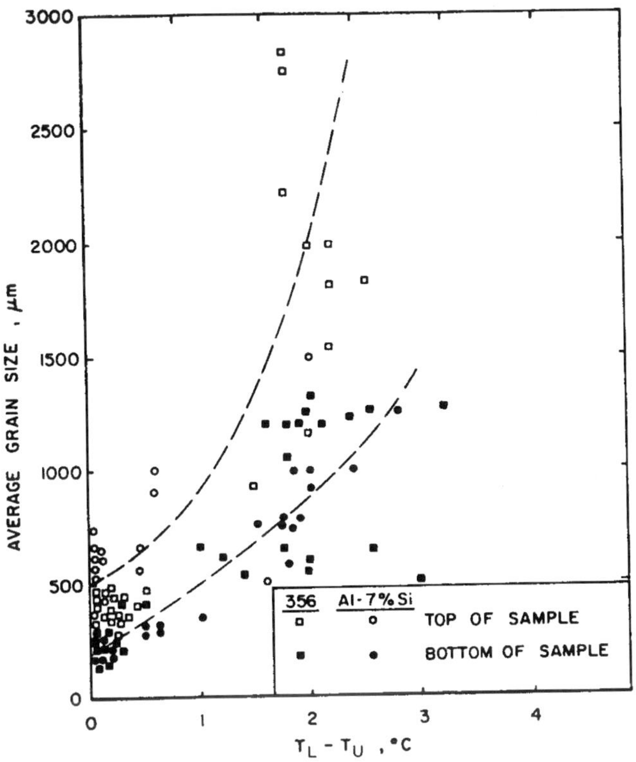

Fig. 3-30. Average grain size as a function of the temperature change ($T_L - T_U$) during the first arrest in thermal analysis cooling curves of A356.0 and Al-7% Si alloys. T_L is the liquidus temperature and T_U the undercooled temperature.

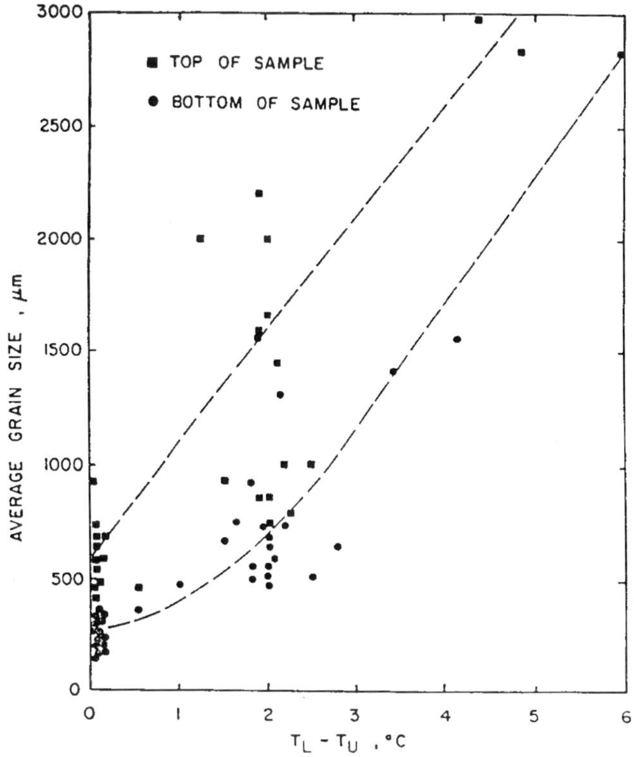

Fig. 3-31. Average grain size as a function of the temperature change ($T_L - T_U$) during thermal analysis of A319.0 alloy. T_L is the liquidus temperature, and T_U the undercooled temperature.

obtained in sand castings. Data obtained from the bottom surface would be representative of permanent mold castings.

Figure 3-31 shows the average grain size as a function of undercooling and the grain size number obtained from the thermal analysis of A319.0 alloy melts having different amounts of grain refinement. Figure 3-32 is a graph of the grain size numbers obtained from the thermal analysis of A319.0 alloy melts having different amounts of grain refinement. Similar trends as those found in the A356.0 alloy melts also were observed in the A319.0 melts. Again, the presence of Cu, Zn, Fe and Mn in this alloy does not interfere with the thermal analysis for grain size.

When reasonable amounts of a grain refiner are present, many nuclei are available. These control the grain structure, such that (T_L–T_U) becomes small, and numerous fine grains are produced in the test casting.

At lower grain refiner addition levels, there are insufficient nuclei, and the grain structures are larger. In this case, other factors become important. For example, vibrating the casting or stirring the metal during solidification provides some degree of grain refinement. Variation in these other features must be responsible for the observed scatter in large-grained castings, because the thermal analysis test was found to be consistent.

Thermal analysis is used in Europe to detect primary silicon refinement, although interpretation of the cooling curve for this treatment is somewhat different than for primary aluminum grain refinement or eutectic silicon modification.

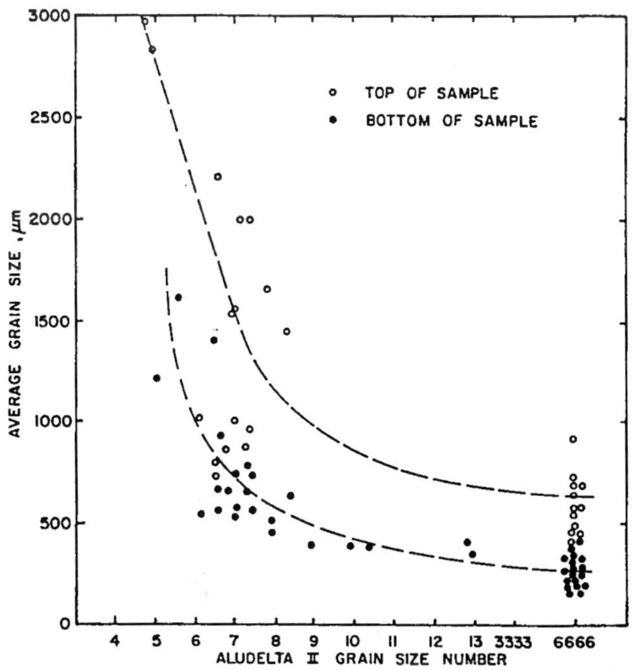

Fig. 3-32. Correlation between the average grain size of A319.0 alloy determined metallographically and the grain size number evaluated and displayed by the instrument shown in Fig. 3-28.

Representative Thermal Analysis Curves

Representative thermal analysis curves for two different melts are shown in Figs. 3-33 and 3-34. Figure 3-33 shows the thermal analysis for an unrefined melt, and Fig. 3-34 shows the analysis of a grain-refined melt.

Comparing the two cases, the disappearance of the sinusoidal temperature undercooling after grain refinement of the melt is evident in Fig. 3-34. On-line interpretation of the sinusoidal behavior (or the lack of it) of the cooling curve at the onset of freezing is accomplished by a quick test by a programmed microprocessor.

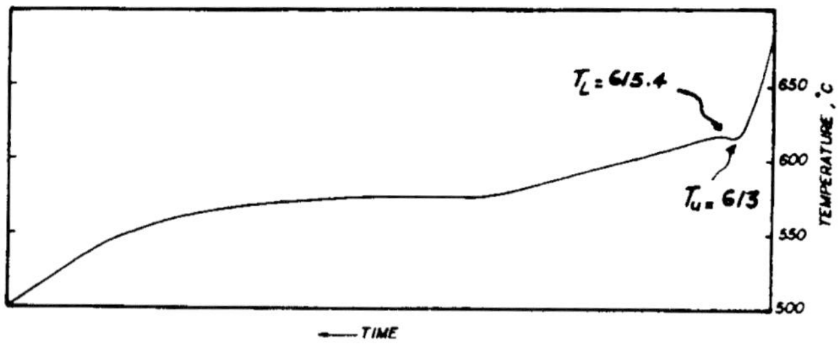

Fig. 3-33. Thermal analysis cooling curve of A356.0 alloy that is not grain refined. Note the sinusoidal arrest at ~1140F (~615C).

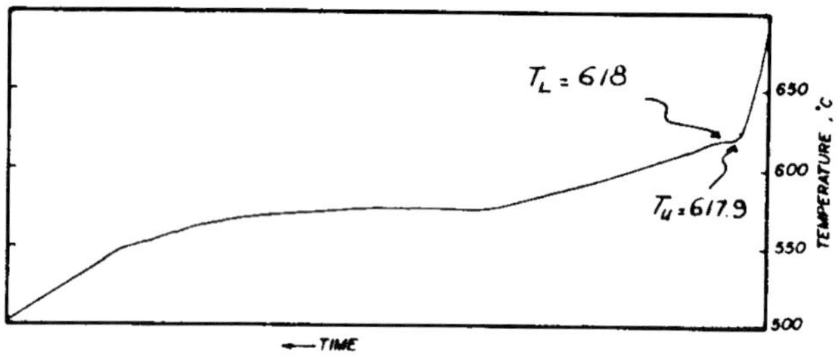

Fig. 3-34. Thermal analysis cooling curve of grain refined A356.0 alloy. Note the lack of undercooling during the first arrest.

Sodium and Strontium Additions

The range of modified structures that have been observed in modified Al-Si alloys are rated on a scale of 1 to 6. A description of each rating follows:

1) *Fully Unmodified Structure.* Silicon is present as large plates and also in an acicular form (Fig. 3-35);
2) *Lamellar Structure.* At low strontium levels, a linear lamellar structure is obtained. Some acicular silicon remains, but the large plates are not present (Fig. 3-36);
3) *Partial Modification.* The lamellar structure starts to break up into smaller pieces (Fig. 3-37);
4) *Absence of Lamellar Structure.* The lamellar structure continues to break up until it has disappeared. A small amount of acicular phase remains (Fig. 3-38);
5) *Fibrous Silicon Eutectic.* The acicular phase is completely absent. Only a fibrous silicon is present. This structure is generally referred to as "fully modified" (Fig. 3-39);
6) *Very Fine Structure.* The fibrous silicon becomes so small that individual particles can no longer be distinguished under an optical microscope. It is in a way a "super-modified" structure that has been observed in several alloys after an incubation time of 1–3 hours (Fig. 3-40).

Thermal analysis has shown that the eutectic arrest temperature is depressed by the presence of a modifier. Moreover, the amount of the depression (ΔT) as shown in Fig. 3-27 correlates well with the eutectic structure of the casting.

Experimental results for A356 alloys with varying levels of strontium are shown in Fig. 3-41. In this alloy, the unmodified eutectic (structure No. 1) occurs at about 1063F (573C). The well-modified structures (types 4, 5 and 6) occur at eutectic temperatures equal to or less than about 1054F (568C). In other words, when $\Delta T \geq 9.9°F$ (5.5°C), the casting has a fine eutectic silicon that is 90% or more a fine fibrous structure.

In some cases (for example, when permanently modified metal is used), it is not practical to measure the temperature of the unmodified eutectic. The arrest temperature may be calculated, in some cases, from the composition of the alloy.

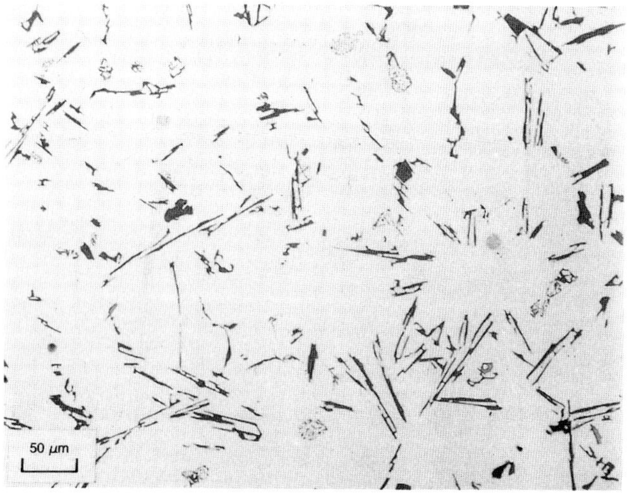

Fig. 3-35. Fully unmodified structure of Al-Si alloy. #1 on scale. [Photos for Figs. 3-35 to 3-40 courtesy of G.K. Sigworth.]

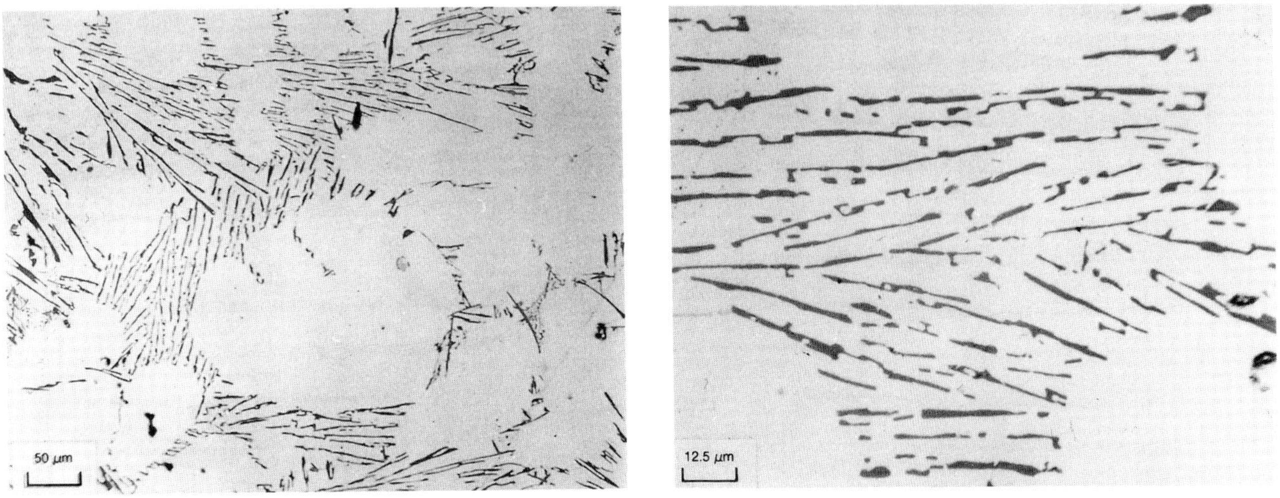

Fig. 3-36. Lamellar structure of Al-Si alloy. #2 on scale.

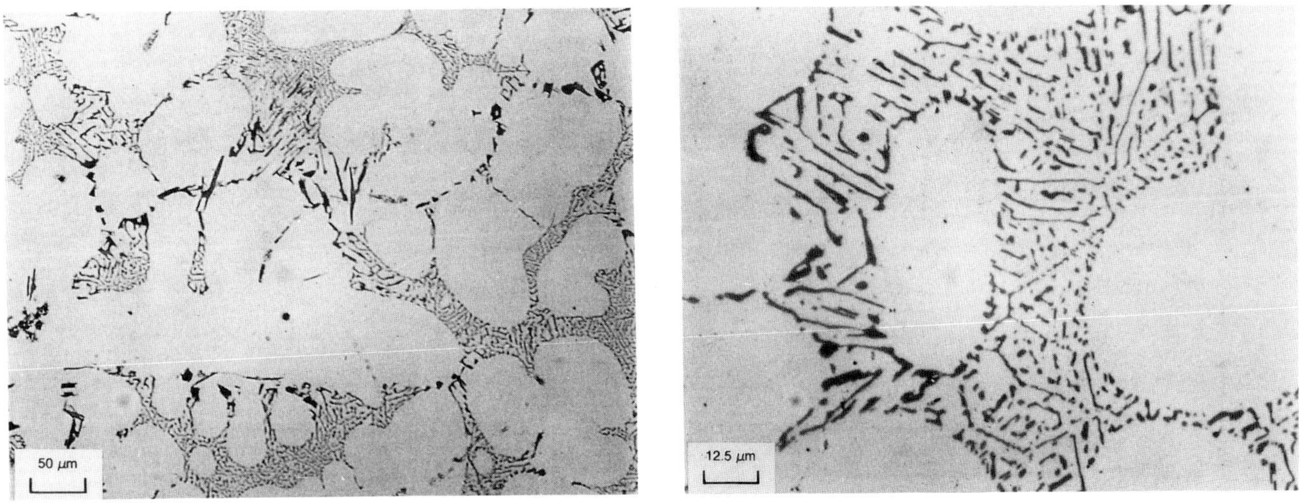

Fig. 3-37. Partial modification of Al-Si alloy. #3 on scale.

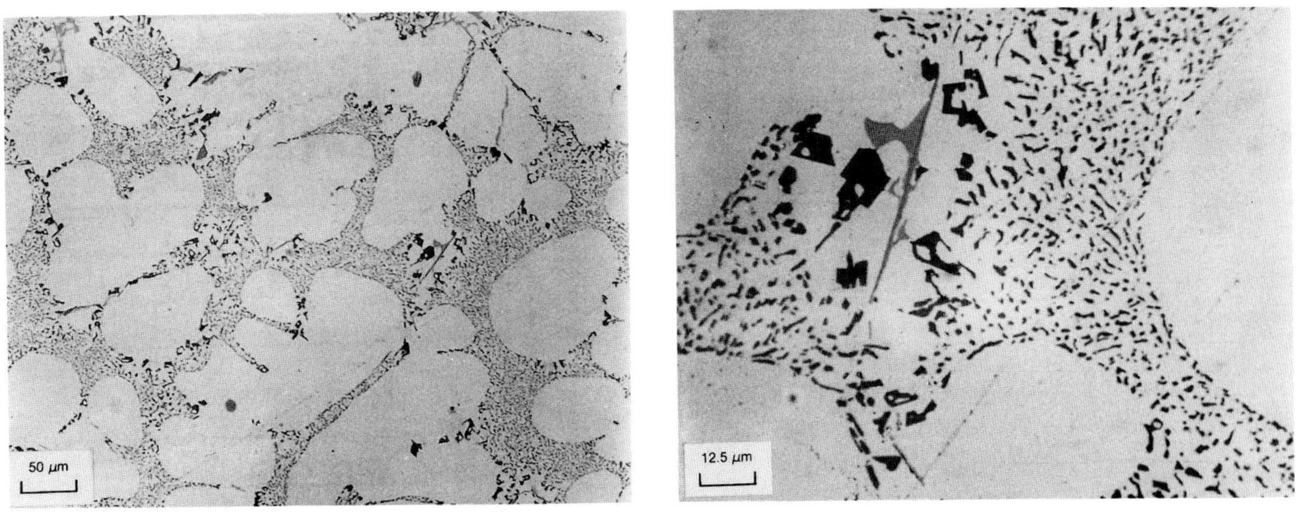

Fig. 3-38. Absence of lamellar structure in Al-Si alloy. #4 on scale.

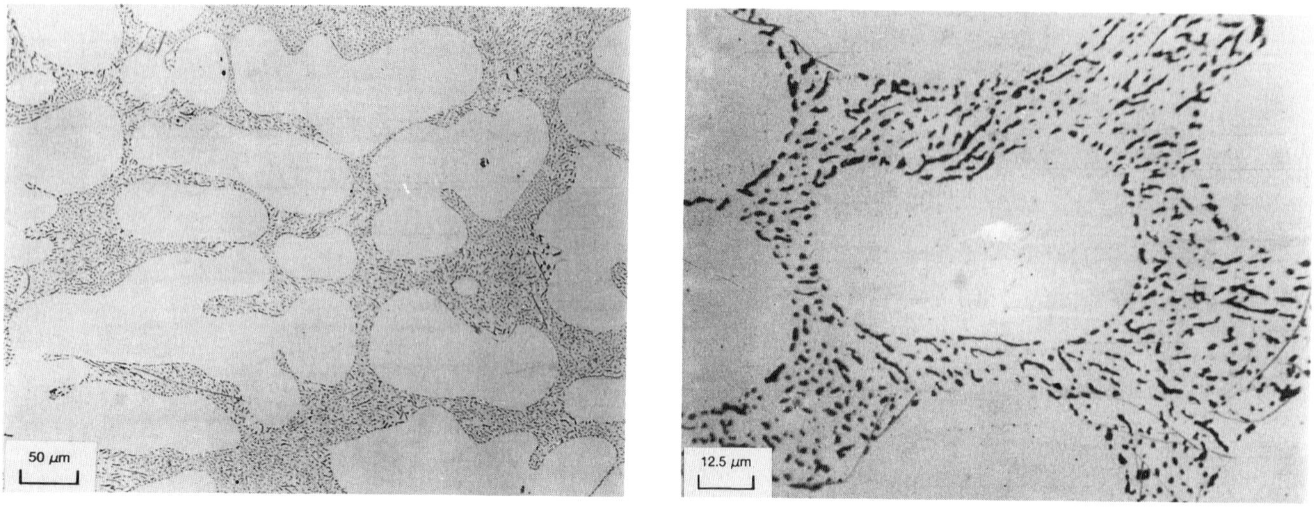

Fig. 3-39. Fibrous silicon eutectic in Al-Si alloy. #5 on scale.

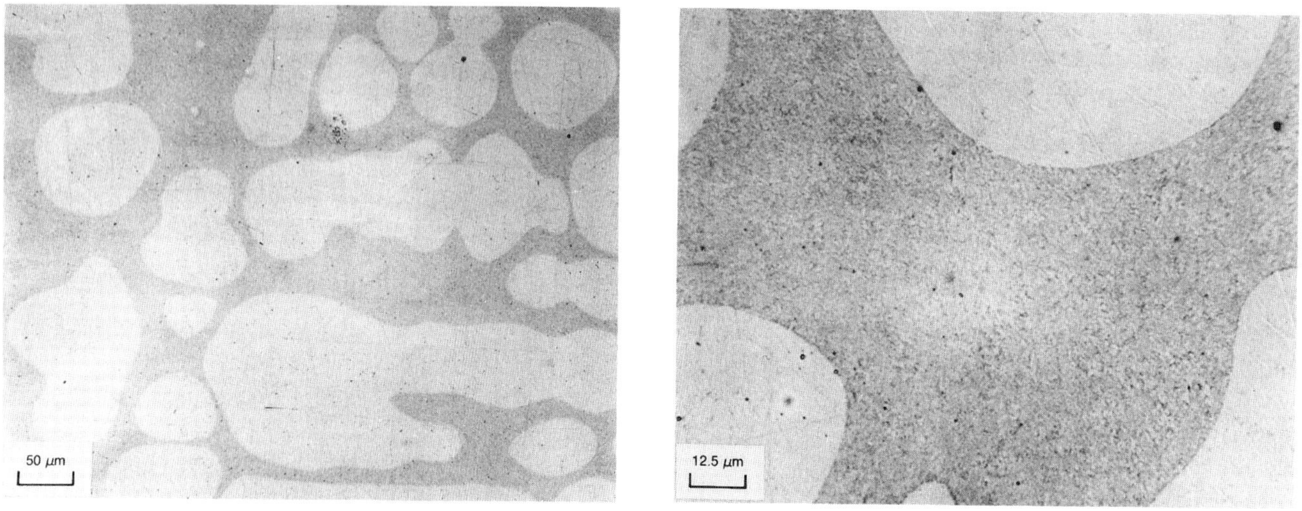

Fig. 3-40. Very fine structure of Al-Si alloy. #6 on scale.

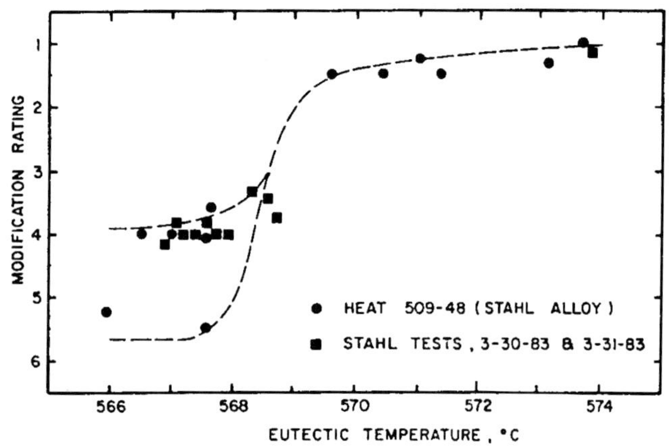

Fig. 3-41. Correlation between level of modification (on scale of 1 to 6) and the eutectic arrest temperature found in A356.0 alloys.

In unmodified Al-Si alloys of high purity, the eutectic arrest temperature has been reported to be 1071 ± 0.5°F (577 ± 0.3°C). These results are confirmed by experimental findings. In commercial alloys, however, this eutectic temperature will be lowered by the presence of other impurities, such as Mg, Fe, Cu, Zn, Mn and Ni.

The effect of these elements has been estimated from the ternary phase diagrams, Al-Si-X, given by Mondolfo,[7] by assuming that the elements accumulate in the liquid in the same proportion as the silicon. The result is:

$$T_e = 577 - (12.5/\%\,Si)\,[4.43\,(\%\,Mg) + 1.43\,(\%\,Fe) + 1.93\,(\%\,Cu) + 1.7\,(\%Zn) + 3.0\,(\%Mn) + 4.0\,(\%Ni)]$$

This equation was confirmed experimentally to be accurate to better than 1.8°F (1°C) in many A356.0 melts. The equation probably applies reasonably well to other alloys where the total of %Al + %Si is high, near 99% (for example, in alloys 324, 355, 357, 359 and 360). However, the equation is not valid for A319.0 and A308.0 alloys or for similar alloys that contain higher amounts of Cu, Fe, Mg or Zn.

The cooling curve obtained from thermal analysis of a modified A356.0 alloy is shown in Fig. 3-42. The eutectic temperature arrest depression is noted. The temperature is reduced by 18.2°F (10.1°C), down to 1052.4F (566.9C) from 1070.6F (577C).

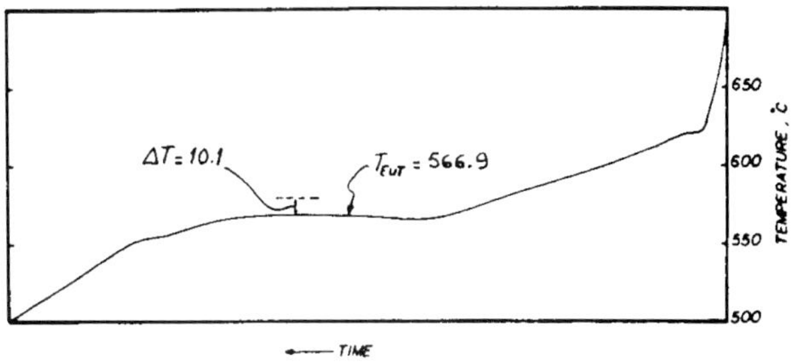

Fig. 3-42. Thermal analysis cooling curve of A356.0 alloy that is modified with strontium. A eutectic temperature arrest depression of 18.2°F (10.1°C) is noted.

REFERENCES

1. W.M. Rasmussen, C.E. Eckert; "Reduced Pressure Testing: Part 1 of 2—RPT Gauges Aluminum Porosity," *Modern Casting*, pp 29–31 (Mar 1992).
2. *International Molten Aluminum Processing*, p 294 (Feb 1986).
3. D.P. Kanicki, W.M. Rasmussen; "Processing Molten Aluminum—Part 2: Cleaning Up Your Metal," *Modern Casting*, pp 55–58 (Feb 1990).
4. J.E. Gruzleski, B.M. Closset; *The Treatment of Liquid Aluminum-Silicon Alloys*, AFS, Des Plaines, Illinois (1990).
5. J.M. Fuqua; "An Advanced Injection Treatment System for Aluminum-Silicon Alloys," *AFS Transactions*, pp 635–642 (1987).
6. D. Apelian, G.K. Sigworth, K.R. Whaler; "Assessment of Grain Refinement and Modification of Al-Si Foundry Alloys by Thermal Analysis," *AFS Transactions*, pp 297–307 (1984).
7. L.F. Mondolfo; *Aluminum Alloys. Structures and Properties*, Butterworths, London, pp 213–624 (1979).

Chapter 4

ALUMINUM CASTING ALLOYS

INTRODUCTION

Traditional materials are being replaced increasingly by improved aluminum casting alloys. In volume and variety, the potential for aluminum is apparently unlimited. In the U.S. automobile industry, for example, the average per-car use of aluminum products (both wrought and cast) has increased substantially in the last 10 years, to well over 170 lb per car.

Availability and Production

Aluminum is the most abundant metallic element in the earth's crust, forming nearly 8% of it. Aluminum always occurs in a compound, such as the minerals bauxite, cryolite, corundum, alunite, diaspore, turquoise and spinel and the silicates kaolin, feldspar and mica. Bauxite, a ferruginous aluminum hydroxide, is the basic raw material from which the metal aluminum is produced.

Impurities are removed from bauxite by chemical processing to make alumina (aluminum oxide). The bauxite is crushed, mixed in a caustic soda (sodium hydroxide) solution, and then heated by steam under pressure to dissolve the alumina. Impurities settle and are filtered out. After it cools, the solution is agitated, and crystals of hydrated alumina form. The crystals are washed and then dried in long rotating kilns. The final product, alumina, is a hard, crystalline substance with the appearance of refined sugar. Four pounds of bauxite will yield two pounds of alumina.

Aluminum is made by electrolyzing a solution of alumina in molten fluorides. Electrolysis takes place in reduction pots or cells at reduction plants, which produce pure, high-quality primary aluminum. Reduction removes the oxygen from alumina, which consists of almost equal parts of aluminum and oxygen, leaving pure aluminum. Two pounds of alumina yields one pound of aluminum. A typical reduction pot (electric smelting furnace) is a shallow, carbon-lined steel tank designed for continuous operation (Figs. 4-1 and 4-2).

Fig. 4-1. This photograph shows a typical line of reduction cells.

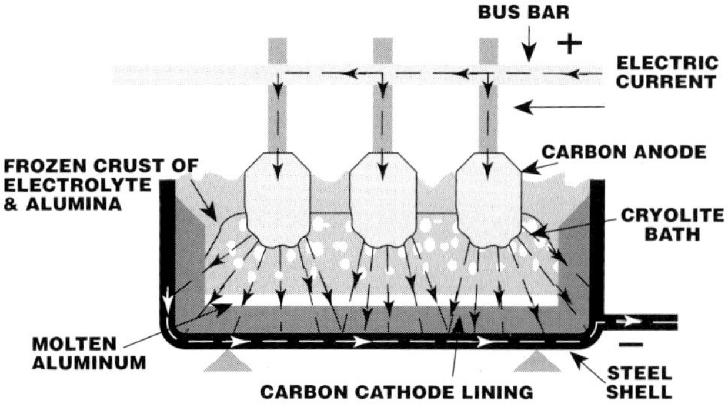

Fig. 4-2. A reduction cell using prebaked carbon anodes.

The pot is partially filled with molten cryolite (sodium aluminum fluoride), which dissolves the alumina and serves as the electrolyte. An electric current is introduced through a carbon anode in the solution. The oxygen in the alumina combines with the carbon, reducing the alumina to pure aluminum, which settles to the bottom of the pot. The aluminum, at a temperature of approximately 1700F (927C), is siphoned off periodically and poured into ingot molds or charged directly into remelt furnaces for the next stage of processing.

Figure 4-3 shows the method for transferring molten aluminum from an alloy smelter to a casting facility.

Unique Characteristics of Aluminum

There are many reasons for aluminum's continued expansion into new and wider fields of application. Light weight, excellent strength, high thermal and electrical conductivity, high reflectivity, good corrosion resistance, excellent workability, and attractive appearance are just some of aluminum's most appealing properties. Aluminum can be given almost any surface finish. It is nonmagnetic, nontoxic and nonsparking. The characteristics and properties obtainable in fabricated aluminum components are of major significance to both manufacturer and consumer.

Fig. 4-3. Truck transfer of molten aluminum using insulated crucibles containing approximately 15,000 lb each.

Aluminum weighs about 0.1 lb/in.^{3}, approximately one-third the density of steel, copper and brass. Some of the stronger aluminum alloys exceed the strength of mild steel. This favorable strength-to-weight ratio is spurring aluminum's use for many new applications.

The high thermal conductivity of aluminum is a marked advantage in any application where it is desirable to conduct or dissipate heat quickly and uniformly. On a weight basis, aluminum is the most efficient heat conductor of the common metals.

Pound for pound, aluminum conducts electricity better than most metals. Aluminum conductivity is 61% of the International Annealed Copper Standard, but copper is more than three times as heavy. An aluminum conductor with equivalent current-carrying capacity weighs only about half as much as a copper conductor.

PROPERTIES AND CHARACTERISTICS

Specific data for the individual casting alloys are given in this chapter. The alloys are arranged in numerical order using the Aluminum Association's alloy designation system for castings and ingot, which is described in detail later in this chapter.

Principal casting alloys are produced either by primary aluminum (produced from ore) producers or by secondary (recycled aluminum) smelters. Chemical composition, typical mechanical properties in different tempers, common applications, castability ratings and fabricating characteristics, recommended thermal treatments, and, for many, typical tensile properties at elevated temperatures are given.

Mechanical Properties

The mechanical properties given as tabular data for aluminum casting alloys are obtainable only if the chemical and heat treating specifications are followed carefully.

It should be noted that the properties obtained from one particular combination of casting alloy, foundry practice and thermal treatment may not necessarily be identical to those achieved with the same alloy in a different foundry or with a different thermal treating source. In all aluminum casting alloys, primary or secondary, the percentages of alloying elements and impurities must be controlled carefully. If they are not, characteristics such as soundness, machinability, corrosion resistance and conductivity are affected adversely.

Elements such as calcium and sodium, in ranges of a few thousandths of a percent, can have either a beneficial or a detrimental effect upon both soundness and machinability of alloys containing a high percentage of silicon. Similarly, the use of phosphorus in controlled amounts has a beneficial effect on the hypereutectic aluminum-silicon alloys containing silicon in excess of 12%. (See Chapter 3 for discussion of modification of aluminum-silicon alloys.)

Special mechanical properties are obtainable in many alloys through accurate control of chemistry and of heat-treating processes. Reliable ingot producers can supply foundries with special ingot and with heat treating specifications designed to obtain special mechanical properties. Within practical limits, certain mechanical properties can be improved at the expense of others. For example, tensile and yield strengths can be increased, but this results in lower elongation and higher hardness. Higher elongation and lower hardness usually mean slight losses in tensile and yield strengths.

Thermal Treatment

Sand and permanent mold castings may be heat treated to improve mechanical and physical properties. Die castings can be stress relieved only (and not solution heat treated) because they generally have a porous internal structure containing some included gas, and they may blister at high temperatures. Some alloys, such as 443.0 (which is cast in sand or permanent molds but contains little or no copper, zinc or magnesium), do not respond to heat treatment to increase mechanical properties. Heat treating processes and the effects they have on the metallurgical structure of aluminum castings are discussed in detail in Chapter 15.

Thermal treatment designations (tempers) and what they specify are listed below. For aluminum castings, –T designates thermal treatment and is always followed by one or more digits that indicate specific sequences of basic treatments. Specifications for the thermal treatment of individual alloys are given in the alloy descriptions later in this chapter.

- O Stress Relieve or Anneal.
- –T4 Solution Heat Treat and Quench: This is an unstable heat treatment. Although it improves mechanical properties (such properties increase to some degree through aging at room temperature over a period of weeks), it is usual practice to artificially age to attain maximum mechanical values.
- –T5, –T51, –T551 Artificially Age: This type of heat treatment is done at a comparatively low temperature and serves to eliminate growth on thermal cycle and to stabilize castings dimensionally (improving mechanical properties somewhat), to improve machinability and to relieve stress.
- –T6, –T61, –T62, –T65 Solution Heat Treat, Quench and Artificially Age: Such heat treatment usually results in maximum tensile and yield strengths with adequate elongation. Aging stabilizes the properties.
- –T7, –T71, –T75, –T77 Solution Heat Treat, Quench and Artificially Overage: This heat treatment improves mechanical properties to a large degree, completely degrowths and stabilizes the castings, and usually results in a slightly lower tensile and yield strength but an increased elongation value compared to the –T6 series of heat treatments.

Characteristics of Aluminum

Aluminum is highly reflective to the common forms of radiant energy: light, radiant heat, and radio or radar waves. Polished aluminum has a light reflectivity of over 80% and will reflect as much as 96% of both short-wave light or long-wave heat radiation.

Aluminum's resistance to commonly corrosive conditions is well-known. Corrosive attack, which may be classed as chemical deterioration, is a rather rare occurrence under average conditions and should be clearly distinguished from normal surface oxidation of aluminum on exposure to the elements. The latter is a natural and actually protective phenomenon and is an important reason for the excellent service given by aluminum under ordinary weathering conditions.

The natural appearance of aluminum is attractive for many applications. Aluminum ordinarily will not stain other materials. It responds well to mechanical finishes, such as polishing and buffing. Paint coatings are easily applied to it. A number of chemical conversion coatings and finishes may be used on it, and electrochemical finishes, chiefly anodizing, are employed extensively on it.

Aluminum's nonsparking quality is important in containers and other equipment where explosive or flammable materials or atmospheres may be present. Because it is nonmagnetic, aluminum is of vital importance to the electrical and electronics industries.

Aluminum is used widely in food processing and handling equipment due chiefly to its nontoxicity.

Tensile Properties at Low Temperatures

Tests conducted to date have shown that aluminum alloy castings subjected to freezing and subzero temperatures generally show a rise in tensile property values at these low temperatures.

Results of a series of tests on Alloys 355.0–T6 and 355.0 –T61 appear in the section for Alloy 355.0.

SELECTING ALLOYS AND CASTING METHODS

Choosing the alloy, casting process and thermal treatment requires a complete knowledge of the service conditions of the part under consideration. A good knowledge of the alloys, heat treatments and various casting methods in common use is necessary.

Nearly 60 casting alloys are described in this chapter, with as many as five different thermal treatment possibilities for some. This results in a large number of alternatives to choose from to satisfy individual requirements. Because of these many alloy and thermal treatment combinations, the possible range of typical mechanical properties varies widely.

Frequently, commercial castings do not have critical service requirements. In such cases, the foundry should be consulted to ascertain the most economical alloy and production method for the job.

In most instances, aluminum castings are designed for maximum efficiency. In such cases, the alloy and heat treatment must be selected carefully from those available. In doing this, designers, with their greater knowledge of the service requirements for the casting, must confine the choice of alloys to those that provide the necessary properties. They must then be guided by the foundry in the final choice.

Sometimes, the alloy that shows the best properties in the tables may have some production characteristics that would make it less desirable on an overall basis than other eligible alloys. The foundry is in the best position to advise on such factors as availability, relative ingot costs, production costs and reproducibility of results. When this is coordinated with the designer's knowledge of service requirements, such as strength, hardness, corrosion resistance, impact strength and machinability, the best possible selection will result.

Because of this coordination, changes from initial design may be indicated to improve design efficiency, to lower production costs, or both. For instance, a casting with sound design from other standpoints may have a size or shape conducive to distortion in heat treating, which could be minimized through design changes.

The production and service requirements have a large bearing on the casting method, as do the size and shape of the part. For example, castings required in fairly large numbers should be made either by the permanent mold or diecasting processes, provided the size and design features of the casting and available alloys are suitable.

Sand casting is usually confined to producing parts with hollow cavities and a complex arrangement of ribs, pockets, etc., and to sizes that make parts unsuitable for casting in metal molds. In some cases, it is advantageous to redesign a casting to make it suitable for either permanent mold or diecasting methods. Sand casting usually requires minimum tooling charge, but the unit price of the castings and the finished part can be high. Permanent molding involves higher tooling charge, but the unit price is lower, particularly for longer runs. Die casting usually involves the highest tooling charge and lowest piece price on large quantities.

Once the casting method is determined, the alloy choice is narrowed down appreciably. Next considered are the service requirements. If high strength is required, heat treatable alloys should be used. Alloy choice can be narrowed down further when remaining requirements, such as pressure tightness, corrosion resistance and machinability, are considered.

In some instances, it may be required to maximize one certain property—for example, highest possible yield strength. This immediately limits the alloy and heat treatment choices, as well as the casting method, to a possible one or two. Greater compromises will have to be made on the other requirements, particularly ductility.

In other cases, the large size and shape of a casting may not permit the use of a solution heat treatment but will limit it to a straight aging treatment taking place at approximately 300–500F (149–260C) or to no treatment at all, thereby sacrificing, to a degree, some mechanical properties and machinability. This limitation exists because the first half of a solution heat treatment takes place at a temperatures of 930–1000F (499–538C), followed by rapid quench in water. At those temperatures, the mechanical strength of aluminum is extremely low; therefore, large unsupported areas of the casting (or perhaps the whole casting) may sag or distort from its own weight. Sometimes, the size and shape of the part are such that it can withstand the effects of the high temperatures but cannot be quenched without danger of warping or cracking.

Because die castings are usually used in the as-cast condition, thermal treatment is generally of little significance, and choice of a diecasting alloy is based primarily on properties and castability of the alloy. Sand casting, permanent mold casting and diecasting are the more widely used production methods. However, in special cases, a part may be produced more efficiently by other methods, such as shell molding, plaster molding, investment casting or centrifugal casting.

ALUMINUM INGOT

Standard Pure Ingot

Standard pure aluminum ingot is produced for casting in several grades of purity. A pure ingot designated as EC (electrical conductor), which is boron-treated and 99.50% minimum aluminum in composition, also is produced. These grades of pure aluminum are cast from reduction pots into special ingots (sows) or into conventional-sized ingots.

Rotor Ingot

Rotor ingot designates those special aluminum casting alloys developed to provide the high degree of electrical conductivity required for electric motor rotor applications and to minimize casting problems. The rotor alloys have a low, carefully balanced amount of alloying elements.

Aluminum Hardeners

Aluminum hardeners, rich alloys made following standard designations, are manufactured in a reduction cell, reverberatory furnace or induction furnace and cast into ingots of various sizes. Rich alloy ingot is used principally by foundries desiring to produce their own casting alloys. Normal foundry practice is to melt measured amounts of rich ingot and pure ingot, adding such pure alloy elements required to bring the melt up to the specified composition.

Standard Alloyed Foundry Ingot

Standard alloyed foundry ingot is the ingot material used most frequently to produce aluminum castings. It is produced in all the common casting alloys and also is manufactured to special customer specifications. Metallic impurities are kept at low levels because of the careful metallurgical controls practiced in the reduction plant. Metal treatment and transfer and ingot casting techniques all are designed to minimize the formation of nonmetallic impurities, thus producing ingot for aluminum castings of the highest quality.

Ingot Quality Assurance

In Fig. 4-4, freshly poured foundry ingot is being skimmed of dross (aluminum oxide skin). The continuous-cast foundry ingot illustrated in Fig. 4-5 produces a product of superior quality. Ingot bundles are compact and easily handled and stored. Continuous-cast ingot has won wide acceptance in aluminum casting plants due to its uniform, controlled grain refinement, improved mechanical properties, casting characteristics, and economics in handling and storage.

Quality has come to mean much more, however, than for an ingot simply to be well-skimmed and within the alloy specification. Consistency of composition from heat to heat, as well as reduced variation within a given heat; absence of nonmetallic inclusions; silicon modification; grain refinement; and even reduced hydrogen gas content are quality characteristics of ingot that are being specified by foundries.

To meet ever-increasing quality requirements, alloyed ingot suppliers have implemented quality assurance systems covering every phase of the manufacturing process. Defect prevention and continuous quality improvement are achieved with the use of statistical process control (SPC) techniques.

SPC is the application of mathematical (statistical) techniques to control process variation. Every employee is given training in SPC theory and practice, making each person an integral part of the quality system. Raw material suppliers to ingot producers are encouraged to use SPC and to develop quality systems. Periodically, these suppliers are audited to monitor their progress.

Fig. 4-4. A worker skims freshly poured aluminum ingot. Ingot molds are filled from a holding furnace where the aluminum is fluxed to eliminate impurities and alloying elements are added.

Fig. 4-5. Typical compact skid of continuous-cast aluminum foundry ingot. Storage areas can be reduced by as much as 50%.

Each heat of ingot is manufactured to a specific foundry's requirements, including targets or desired ranges for alloying elements well within the usual alloy specification limits. Throughout the building of every heat of ingot, analytical samples are taken on a statistical basis. These spectrographic samples are analyzed according to ASTM standards.

An optical emission spectrograph is the instrument that is used to determine an alloy's chemical composition. The spectrograph must be calibrated based on National Bureau of Standards specifications. The accuracy and repeatability of analyses is verified by statistically charting the results of calibration analysis and by conducting periodic Gage R&R (repeatability and reproducibility) testing.

Furnaces used to produce alloys are equipped with automatic temperature controls and molten metal pumps. These pumps continuously stir the liquid metal bath, ensuring a homogeneous composition throughout the heat. All heats are cleansed of oxides and nonmetallic inclusions with chlorine gas or fluoride salts. Just prior to casting the alloy into ingot, any remaining oxides or nonmetallic inclusions are removed by filtering the metal with ceramic foam filters.

Silicon modification and/or grain refinement (when required) is accomplished by adding the appropriate modifying and refining elements with hardeners. Spectrographic analysis for the proper concentration of modifying and refining elements is one form of control. Thermal analysis of the cooling curves from different heats also verifies that proper silicon modification and grain refinement have taken place.

If ingot with a low hydrogen gas content is required, the heat is degassed with nitrogen. Verification of proper degassing is done by periodically performing a reduced gas pressure test during ingot casting.

Once cast, the ingot is stacked, marked, strapped and color-coded according to each customer's requirements. Plastic bags are placed over each bundle as protection against moisture and dirt (Fig. 4-6). A certified chemical analysis accompanies each delivery of alloy. SPC charts on significant characteristics of the alloy may be requested by the foundry.

Fig. 4-6. Stacked ingots protected by sturdy plastic are securely bundled and ready for shipment. Alloy and heat treatment numbers are stamped on each ingot.

CHEMICAL COMPOSITION LIMITS, PROPERTIES AND TREATMENTS

Aluminum casting alloys in the U.S. are numbered according to a three-digit (plus decimal) system adopted by the Aluminum Association (AA) in 1954 and approved by the American National Standards Institute in 1957 (ANSI H35.1). The American Society for Testing and Materials (ASTM), the Society of Automotive Engineers (SAE), and the Federal and Military specifications for aluminum castings now conform to the AA designation system.

In this system, major alloying elements and certain combinations of elements are indicated by the following number series:

Series	Alloy Type
1XX.X	99.0% minimum aluminum
2XX.X	Al + Cu
3XX.X	Al + (Si-Mg), (Si-Cu) or (Si-Mg-Cu)
4XX.X	Al + Si
5XX.X	Al + Mg
7XX.X	Al + Zn
8XX.X	Al + Sn

The 6XX.X and 9XX.X series are not currently in use, but they are being held open for possible use in the future.

The digit following the decimal in each alloy number indicates the form of product:

- a "0" (zero) following the decimal indicates the chemistry limits applied to an alloy casting;
- a "1" (one) following the decimal indicates the chemistry limits for ingot used to make the alloy casting;
- a "2" (two) following the decimal also indicates ingot but with somewhat different chemical limits (typically tighter, but still within the limits for ingot).

Generally, the XXX.1 ingot version can be supplied as a secondary product (remelted from scrap, etc.), whereas the XXX.2 ingot version is made from primary aluminum (reduction cell). The current version of the Aluminum Association's "Pink Sheets," giving the chemistry limits for all casting alloys, is located in the appendix to this chapter.

Some alloy names include a letter. Such letters, which precede an alloy number, distinguish between alloys that differ only slightly in percentages of impurities or relatively minor alloying elements (for example, 356.0, A356.0, B356.0 and F356.0).

Former alloy designations (if any) are shown in parentheses in the headings of the alloy tables. The properties shown for the alloys were obtained from ASTM, SAE and

Table 4-1.
Density and Solidification Range of Selected Aluminum Alloys

Alloy	Density (lb/cu in.)	Approximate Solidification Range (°F)
208.0	0.10	1160–970
213.0	0.11	1160–970
222.0	0.11	1155–965
238.0	0.11	1110–945
240.0	0.10	1110–950
242.0	0.10	1175–990
295.0	0.10	1190–970
296.0	0.10	1170–970
308.0	0.10	1135–970
319.0	0.10	1220–960
A319.0	0.10	1120–960
332.0	0.10	1080–970
333.0	0.10	1085–960
336.0	0.10	1050–1000
354.0	0.10	1105–1000
355.0	0.10	1150–1015
C355.0	0.10	1150–1015
356.0	0.10	1135–1035
A356.0	0.10	1135–1035
357.0	0.10	1135–1035
A357.0	0.10	1135–1035
359.0	0.09	1115–1045
360.0	0.10	1105–1035
A360.0	0.10	1105–1035
364.0	0.10	1115–1045
380.0	0.10	1100–1000
A380.0	0.10	1100–1000
B380.0	0.10	1090–970
383.0	0.10	1080–960
384.0	0.10	1080–900
390.0	0.10	1200–945
B390.0	0.10	1200–950
413.0	0.10	1080–1065
A413.0	0.10	1085–1065
443.0	0.10	1170–1065
A444.0	0.10	1145–1065
511.0	0.10	1185–1090
512.0	0.10	1170–1090
513.0	0.10	1180–1075
514.0	0.10	1185–1110
518.0	0.09	1150–995
535.0	0.10	1165–1020
A535.0	0.09	1150–1020
B535.0	0.10	1165–1020
520.0	0.09	1120–840
705.0	0.10	1180–1105
707.0	0.10	1165–1085
710.0	0.10	1195–1105
711.0	0.10	1190–1120
712.0	0.10	1200–1135
713.0	0.10	1180–1100
771.0	0.10	1185–1120
850.0	0.10	1200–435
852.0	0.10	1175–410

federal specifications as applied (sand, permanent mold and diecasting). Chemical composition limits are for standard foundry alloy ingot, percent maximum, unless shown as a range; the remainder is aluminum.

Data on density, solidification range, specific gravity, thermal expansion coefficients, electrical and thermal conductivity, and Young's modulus for select aluminum casting alloys appear in Tables 4-1 and 4-2.

Table 4-2.
Physical Properties of Select Aluminum Casting Alloys

Alloy	Specific Gravity	Thermal Expansion Coefficient per °F x 10^{-6} (68–212F)	Electrical Conductivity %IACS	Thermal Conductivity Btu·in./hr·ft²·°F	Young's Modulus 10^6 psi
100.0	2.70	12.8	54	1165	9.6
201.0–T6	2.80	10.7	30	840	10.3
A206.0–T6	2.80	10.7	—	840	9.6
222.0–F	2.95	12.3	34	925	10.7
224.0–T62	2.81	—	30	810	10.5
242.0–T571	2.81	12.5	34	925	10.3
295.0–T62	2.81	12.7	35	980	10.0
296.0–T6	2.80	12.2	33	925	10.0
319.0–F	2.79	12.0	27	780	10.7
332.0–T5	2.76	11.5	26	720	11.2
336.0–T551	2.72	10.5	29	810	—
355.0–T61	2.71	12.4	37	1010	10.2
C355.0–T61	2.71	12.4	39	1010	10.2
356.0–T6	2.69	11.9	39	1040	10.5
A356.0–T6	2.69	11.9	40	1040	10.5
A357.0–T6	2.69	11.9	40	1100	—
360.0–F	2.68	11.6	37	1010	10.3
A360.0–F	2.68	11.6	37	1010	—
364.0–F	2.63	11.6	30	840	—
A380.0–F	2.76	11.7	27	750	10.3
B380.0–F	2.76	11.7	25	750	10.3
383.0–F	2.71	11.8	27	750	10.3
384.0–F	2.71	11.3	23	750	10.3
A390.0–T	2.73	10.3	25	925	11.9
B390.0–F	2.73	10.3	25	925	11.9
413.0–F	2.66	11.4	39	1070	10.3
A413.0–F	2.66	11.4	39	1070	—
A443.0–F	2.69	12.3	37	1010	10.3
C443.0–F	2.69	12.3	37	1010	10.3
A444.0–T4	2.68	12.1	41	1100	—
511.0–F	2.66	13.1	36	980	—
512.0–F	2.65	13.3	38	1010	—
513.0–F	2.68	13.3	34	925	—
514.0–F	2.65	13.3	35	954	10.3
518.0–F	2.53	13.4	24	695	—
520.0–T4	2.57	14.0	21	605	9.5
A535.0–F	2.54	13.4	23	695	—
705.0–F	2.76	13.1	25	720	—
707.0–F	2.77	13.2	25	720	—
710.0–F	2.81	13.4	35	955	9.7
711.0–F	2.84	13.1	40	1100	11.0
712.0–F	2.82	13.1	40	1100	—
713.0–F	2.84	13.3	37	1070	9.7
850.0–T5	2.87	—	47	1290	10.3
851.0–T5	2.83	12.6	43	1155	10.3
852.0–T5	2.88	12.9	45	1215	10.3

DEFINITIONS OF TERMS USED WITH ALLOYS

Even though various terms used on the alloy description pages of this chapter and throughout this book are probably familiar to the reader, they are defined in the following sections to enhance the usefulness of the information that will be presented.

Fabricating Characteristics

Castability ratings are relative. They are used to indicate the suitability of a particular alloy for casting by a specific casting process (economically and under normal foundry conditions) to produce a sound, finished part that requires little machining. Characteristics such as fluidity, resistance to hot cracking, pressure tightness and shrinkage during solidification affect a metal's castability rating.

It is difficult to measure a material's fitness for fabrication (castability in this sense) in truly quantitative terms. It should be noted that castability does not depend solely on the metal; foundry techniques and casting design can facilitate or impede the production of a sound part from a readily castable metal.

Fluidity is the term used to define the molten alloy's capability, at normal pouring temperature, to flow readily in the cavity and fill thin sections. Some alloys will fill an intricately shaped casting at low casting temperatures. Others, under similar conditions, will perform less well and, in some cases, produce unsound castings with misruns and cold shuts.

Resistance to hot cracking is an alloy's capacity to withstand contraction stresses while cooling through the hot-short or brittle temperature range. This characteristic is very important, and the alloy must have a high degree of resistance when sections of castings solidify around metal cores in the permanent mold and diecasting processes.

Pressure tightness is the resistance to leakage through the casting under pressure. Casting leakage usually is tested by closing all openings of the casting, pressurizing the interior, and immersing it in water. Air bubbles denote leaks. Such a pressure test reveals the relative soundness of the casting.

Solidification shrinkage tendency (not applicable for die castings): During solidification (freezing), the volume of the metal decreases. This contraction causes shrink voids that must be filled with molten metal from risers or feeders; otherwise, unsound castings will result. Ratings of this characteristic indicate the relative amount of compensating feed metal that must be provided by the risers.

Corrosion resistance ratings indicate an alloy's resistance in standard salt spray tests. Corrosion is an electrochemical process that affects surface finish and reduces quality and strength of the metal. Fresh water, soil and the atmosphere contain compounds that usually make weak electrolytes. Salts, acids and alkalis create strong corrosive environments.

Machinability rating of the alloy is a combination of cutting ease, chip characteristics, finish smoothness and tool life. For heat treatable alloys, the machinability ratings are given for the –T6 tempers. With the current popularity of polycrystalline diamond cutting tools, tool wear has become less and less of an issue with hypereutectic alloys. It continues to be an important consideration, however, where high-speed steel (HSS) or carbide tools are still used.

Polishing ratings are based on the speed and ease of polishing and the appearance and quality of the finish resulting from various standard polishing techniques.

Electroplating ratings indicate the degree of receptivity of the alloy to take and its capacity to hold an electroplate as applied by standard methods.

Anodizing is rated on lightness of color, brightness and uniformity of coating formed in sulfuric acid electrolyte.

Chemical conversion coating ratings indicate the combined effectiveness of the coating and the base to resist corrosion.

Weldability ratings of casting alloys are based on how readily the parts can be joined (fused)—with or without filler material—and the strength and quality of the weld.

Brazing cannot be performed satisfactorily on most aluminum alloys, but aluminum-zinc-magnesium alloys, such as the 712 and 713 series, can withstand brazing temperatures. After aging at room temperature for a period of time, aluminum-zinc-magnesium alloys will regain their initial strength.

Mechanical Properties

Ultimate or *tensile strength* is the maximum tensile stress that a material is capable of developing under a gradual and uniformly applied load. Tensile strength is calculated from the maximum load carried during a tension test and the original cross-sectional area of the specimen.

Yield strength is the stress at which a material exhibits a specified permanent set. The value of the set used for aluminum and its alloys is 0.002 in./in. or 0.2%. For many aluminum alloys, the yield strength in tension and compression are approximately equal.

Elongation is an increase in distance between two gauge marks from stressing the specimen in tension to fracture.

Shearing strength is the maximum shearing stress that a material is capable of developing. In practice, it is considered to be the maximum average stress computed by dividing the ultimate load in the plane of shear by the original area subject to shear. Shearing strength is usually determined by inserting a cylindrical specimen through round holes in three hardened steel blocks, the center of which is pulled (or pushed) between the other two so as to shear the specimen on two planes. The maximum load divided by the combined cross-sectional area of the two planes is the shearing strength.

Brinell hardness is a measure of resistance to indentation. It is obtained by applying a load through a ball indenter and measuring the diameter of the impression in the material. The hardness value is obtained by dividing the applied load in kilograms by the spherical area of the impression in square millimeters. In testing aluminum alloys, a load of 500 kg is applied to a ball 10 mm in diameter for 30 sec.

Endurance limit is the limiting stress below which a material will withstand an infinitely large number of cycles of stress. In the case of aluminum alloys, endurance limits are based on 500,000,000 cycles of completely reversed stress, using the rotating beam-type of machine and specimen.

Compressive yield strength is the maximum stress that the metal can withstand under compression without a predefined amount of deformation.

HYPEREUTECTIC Al-Si ALLOYS

Hypereutectic aluminum-silicon alloys contain silicon concentrations greater than 11.7% (Fig. 4-7). There are several commercial alloys of this type, such as A- and B390.0, 392.0, and 393.0. They contain nominal silicon contents of 17%, 19% and 22%, respectively, along with other alloying elements such as copper, magnesium, manganese and nickel. All of these alloys possess great hardness and wear resistance as compared to other aluminum casting alloys.

In general, as the silicon content of hypereutectic alloys increases, wear resistance and scuff resistance increase, and the coefficient of thermal expansion decreases. All of the hypereutectic Al-Si alloys offer excellent castability, low coefficients of thermal expansion, good wear resistance and mechanical properties. Ductility, however, is low.

The following information on hypereutectic silicon technology primarily concerns Alloys A390.0 and B390.0 because considerable work has been done in establishing their technology for automotive engine cylinder blocks, pistons, compressors, pumps, pulleys and brake-system components. These alloys represent the development of one hypereutectic Al-Si alloy system with enough hard primary silicon particles to achieve the required wear and scuff resistance—but not too many to impede commercial casting and machining requirements.

During the development of A390.0 and B390.0 alloys, much consideration was given to cylinder bore finish and its effect on wear and compatibility. Out of this came the technology of chemical and mechanical finishing of the surface to leave the silicon particles in relief above the aluminum matrix (Fig. 4-8). Many suitable applications for these alloys do not require the sophisticated surface finish applicable to the automotive engine cylinder walls.

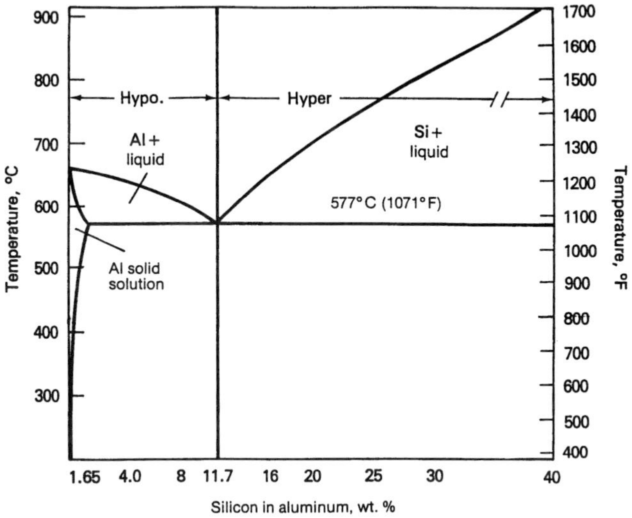

Fig. 4-7. Aluminum end of the Al-Si equilibrium diagram.

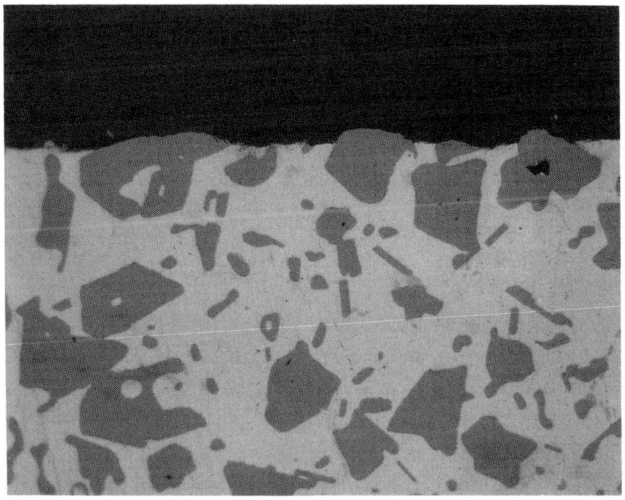

Fig. 4-8. A cross section through properly prepared 390.0 alloy engine bore shows silicon particles standing proud of the surface.

GENERAL PROPERTIES OF HYPEREUTECTIC Al-Si ALLOYS

General Microstructure

The general microstructure of a hypereutectic aluminum-silicon alloy is illustrated in Fig. 4-9. The most important single feature of the microstructure is, of course, the primary silicon. It is this phase that gives aluminum-silicon alloys their desirable wear characteristics.

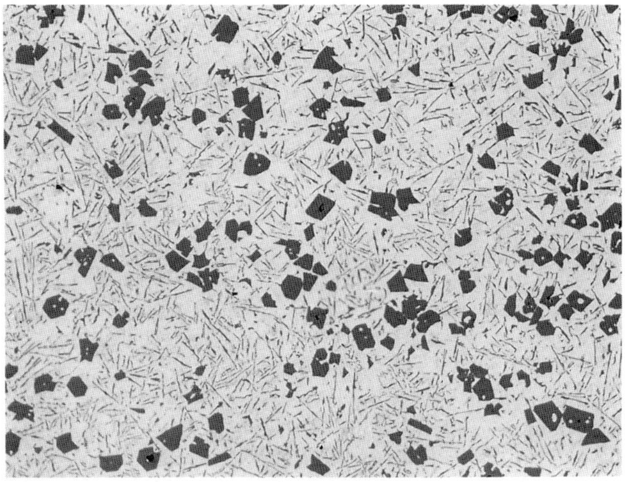

Fig. 4-9. General microstructure of Alloy 390.0 (100X).

Thermal Expansion

The coefficient of thermal expansion of the hypereutectic aluminum-silicon alloys, while above that of cast iron, is about 15% less than that of the more conventional aluminum casting alloys.

GENERAL FOUNDRY CHARACTERISTICS

Certain foundry characteristics of hypereutectic Al-Si alloys differ somewhat from those of many conventional aluminum casting alloys:

- For a given mold cavity volume and casting temperature, there is a greater release of heat during solidification. This heat release affects fluidity and feeding and also can influence gating, risering, core venting and mold chilling.

- There is a more critical need to control the casting's microstructure in order to achieve desired operating characteristics and acceptable machinability.
- The cast parts are harder and more abrasive, sometimes requiring modified degating, trimming and machining techniques.

However, regardless of these differences and contrary to the beliefs of many foundries, hypereutectic aluminum-silicon alloys are not difficult to cast if their important foundry characteristics are understood and dealt with appropriately. Casting of these alloys worldwide in the early 1990s exceeds 100,000 metric tons annually.

Isothermal Solidification

Pure aluminum and pure eutectic composition alloys (such as Al-12% Si, for example) solidify isothermally, meaning that they solidify progressively from the mold surface toward the center of the casting's cross section with a narrow plane of demarcation between the solid formed and the remaining liquid.

Such a solidification pattern provides the least tendency for an alloy to hot tear or hot crack during casting. This type of solidification pattern also tends to provide castings that have excellent pressure tightness because whatever shrinkage porosity is apt to occur does so at the thermal center of the cross section—that is, well below and not connected to the casting's surface.

Fluidity

Figure 4-10 shows the effect that silicon has on an aluminum casting alloy's fluidity as determined in a cast-spiral test. Fluidity, in the foundry sense, is not the reciprocal of viscosity. Instead, fluidity defines how well a particular alloy is likely to fill out a mold during casting—that is, how far the molten metal will flow before it is too solid to flow further. Therefore, the fluidity of an alloy tends to relate most directly to its latent heat of crystallization. Generally, the more heat a mold must remove as an alloy solidifies, the further that alloy can flow.

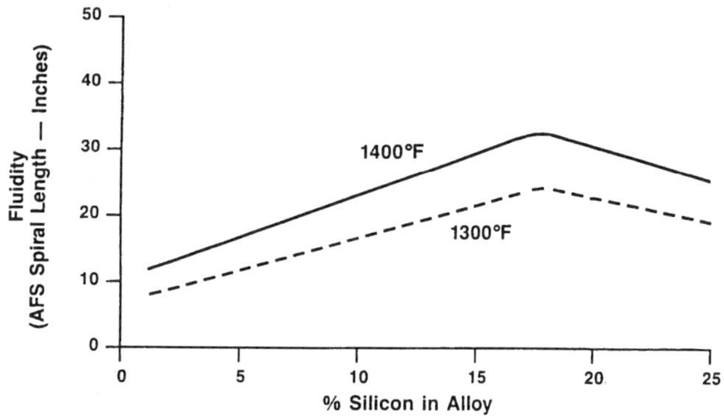

Fig. 4-10. Fluidity of an aluminum-silicon binary alloy as a function of percent silicon content.

Latent Heat

Table 4-3 lists the approximate heat values for aluminum and several important alloying elements, including silicon. The relative latent heat of crystallization of an alloy can be estimated roughly by summing the weighted latent heat values for each of its alloying elements, according to a rule of mixtures.

Aluminum itself has a relatively high latent heat content—higher than iron, copper, magnesium or zinc. But silicon has the highest latent heat of all elements alloyed with aluminum. [Note: Beryllium is the only alloying element with a latent heat higher than that of aluminum, but it is used in such small quantities that its contribution to the total heat of the system is relatively insignificant.]

Figure 4-11 depicts, in very rough form, the incremental release of heat to the mold from several alloys as they solidify. Two alloys stand out on that chart: 713.0 because it gives up the least heat (it contains no silicon), and B390.0 because it gives up the greatest amount of heat (it contains 17% silicon).

An important feature relative to the B390.0 alloy curve concerns when heat is released. Because the first phase to solidify in B390.0 is primary silicon (relatively pure silicon), an exceptionally large amount of heat is released during that early formation of only a small volume fraction of solid. In fact, nearly three times as much heat is released from B390.0 as from 356.0 alloy, for example, during formation of the same volume fraction of solid during solidification. That difference contributes to B390.0 alloy's having the greatest fluidity of all aluminum alloys commonly cast.

Table 4-3.
Latent Heat of Crystallization (Btu/lb) Released by Aluminum and Various Alloying Elements

Aluminum	170
Silicon	778
Copper	91
Magnesium	158
Zinc	43

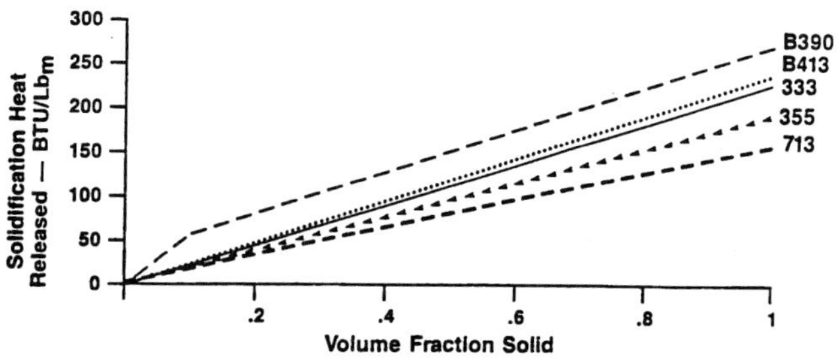

Fig. 4-11. Approximate incremental release of heat during solidification of several permanent mold casting alloys.

The solidification cooling curve for a eutectic Al-Si alloy (B413.0) is shown in Fig. 4-12 compared to the curves for several other alloys. [Note: The shapes of the curves shown in Fig. 4-12 are approximations. They are "ideal" cases intended to depict the differing solidification patterns characteristic of several aluminum casting alloys. More detailed and exacting information on solidification patterns is available in the literature.[1]]

Note especially the shape of the curve for 713.0 alloy in Fig. 4-12. As alloys with this characteristic pattern solidify, each incremental increase in volume fraction solid requires an incremental decrease in temperature. Once a contiguous solid network of dendrites has formed (requiring as little as 30–40% solid), the whole mass then "contracts" as cooling and solidification continue. Especially in large and/or complex shapes, the strength of the solid fraction often is insufficient to overcome the resistance to contraction imposed by the mold, and the casting is apt to tear or crack during the last stages of solidification. Alloys that share this solidification pattern are said to be "hot short."

Alloys that have this solidification pattern are primarily responsible for aluminum foundries often labeling broad freezing range alloys as "difficult to cast." It is not, however, their solidification temperature range that makes them difficult, but rather their characteristic cooling curve shapes.

Contrast, for example, the 713.0 alloy curve in Fig. 4-12 with those of 355.0, 333.0 and the hypereutectic alloy B390.0 (all three of these alloys, incidentally, have much broader solidification temperature ranges than 713.0). However, those three alloys all contain silicon. A substantial volume fraction of virtually all 3XX and 4XX alloys solidifies isothermally at the major Al-Si eutectic arrest. By the time cooling resumes below the arrest temperature, the bulk of the solid has already formed, and only the lowest-melting-temperature phases remain liquid (generally, the copper and/or magnesium-bearing phases).

Such alloys already have, at that point, sufficient integrity and strength to overcome whatever restrictions the mold might impose during the small contraction that occurs as the casting cools from the eutectic arrest to the solidus temperature. For example, alloys such as A390.0 or B390.0 have little or no tendency for hot tearing or hot cracking (unless, of course, severe "hot spots" are allowed to develop during solidification). The isothermal solidification pattern also improves the alloys' shrinkage feeding characteristics, resulting in sounder castings.

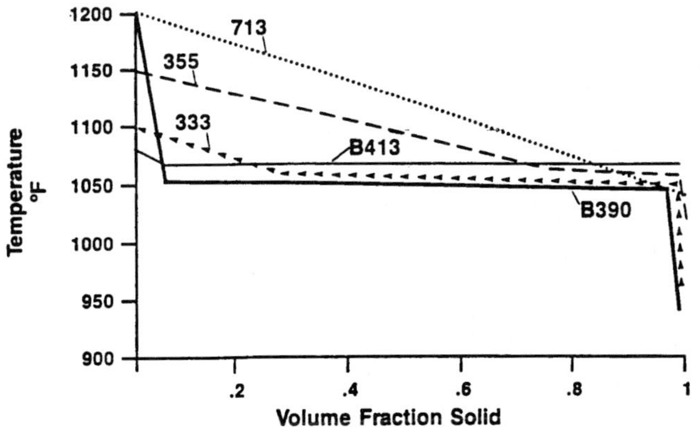

Fig. 4-12. Idealized solidification cooling curves for several permanent mold casting alloys.

Microstructure Control

As was shown in Fig. 4-9, the outstanding feature of the hypereutectic aluminum-silicon alloy microstructure is the primary silicon phase. It is this phase that gives these alloys their desirable wear characteristics. Therefore, every aspect of the processing and preparation of the hypereutectic aluminum-silicon alloys must be aimed at providing a desirable microstructure.

Optimum machinability and performance of such alloys require that the foundry produce a microstructure in which the primary silicon 1) is of reasonably uniform size and spatial distribution, and 2) is no larger than need be for the part's intended application.

Control of the primary silicon microstructures rests in two areas: solidification rate and artificial nucleation (refinement). Without the benefit of an artificial nucleation treatment, primary silicon crystals grow quite large and irregular in shape, generally forming in a somewhat segregated spatial pattern. In this case, the size of the crystal is governed solely by the rate at which the alloy is cooled from the liquidus to the aluminum-silicon eutectic temperature, with faster rates yielding smaller sizes.

Except for during conventional diecasting, control of primary silicon size cannot be left to the solidification rate alone; instead, it must be assisted by artificial nucleation. Refinement is accomplished by adding phosphorus to the melt (a section on refinement practices can be found in Chapter 3). Typically, refined primary silicon particles will be only 8–10% of the size of unrefined particles for any given solidification rate.

Surface Depletion of Primary Silicon

A characteristic of hypereutectic Al-Si alloy die castings (not of sand or permanent mold castings) is a surface skin that contains little or no primary silicon of significant size. This condition is known as surface depletion. The depth of this depleted area (or the thickness of the skin) is a function of several casting variables and, therefore, can be controlled (if not eliminated) in critical wear areas of a casting.

The normal acceptable depth for surface depletion is up to 0.015 in. (0.38 mm). Hence, sufficient machining stock (0.020 in./0.51 mm minimum) must be allowed on wear surfaces to assure that the depleted area will be removed and that sufficient quantities of primary silicon will exist on the wear surface for proper operating characteristics.

Shakeout and Trimming

Because the hypereutectic Al-Si alloys are harder, more abrasive and less ductile than many conventional alloys, modified handling is sometimes required during core shakeout, degating and trimming.

Primary silicon is extremely hard and (especially if large in size) can render the hypereutectic Al-Si alloys quite abrasive to saw blades and other cutting tools. Slowly solidified castings and particularly the riser areas of those castings (because they require the longest solidification time) can be difficult or costly to degate by sawing. In such cases, abrasive cutoff wheels are a better choice of degating tools. Risers can be broken off easily if pad-neck restrictions, such as knockoffs or Washburn cores, are used during casting.

ACKNOWLEDGMENT

AFS would like to thank B. Cochran, Wabash Alloys, for updating the section on ingot quality assurance.

REFERENCES

1. L. Backerud et al; *Solidification Characteristics of Aluminum Alloys—Volume 2, Casting Alloys;* AFS, Des Plaines, IL (1990.)

Properties and Characteristics of Principal Casting Alloys

NOTES TO MECHANICAL PROPERTIES IN THE ALLOY TABLES

The following notes apply to most of the alloys in the alloy tables in this chapter:

- Tensile properties and hardness values of sand and permanent mold castings are determined by tests on standard ½-in. diameter test specimens without surface finishing.
- Tensile properties of die castings are determined for ASTM standard round ¼-in. diameter test specimens cast on a cold-chamber (high-pressure) machine.
- Compression yield test results are from specimens with length:radius ratio of 12.
- Endurance limits are based on 500 million cycles of completely reversed stresses using rotating beam-type machine and specimen.

Alloy 100.0

Alloy 100.0 was designed especially for casting larger electric motor rotors. It is also used in other applications where high electrical conductivity is of prime importance. The alloy is recommended for rotors larger than 6 in. in diameter.

Alloy 100.0 has a guaranteed minimum aluminum purity of 99.0% and a controlled iron-to-silicon ratio. The iron content of this alloy is a nominal 0.6%, which is generally six times the silicon content of the alloy. Increasing the iron content in Alloy 100.0 greatly improves the feeding characteristics of the metal.

Alloy 100.0*

Used in: Die, Sand and Permanent Mold Casting

Chemical Composition Limits

Si	Fe	Cu	Mn	Mg	Cr	Ni	Zn	Ti	Sn	Al	Other Elements Each	Other Elements Total
0.15	0.6–0.8	0.10	**	–	**	–	0.05	**	–	99.0	0.03**	0.10

Typical Mechanical Properties

Casting Process and Temper	Tension: Ultimate Strength (ksi)	Tension: Yield Strength (set 0.2%) (ksi)	Tension: Elongation (% in 2 in.)	Brinell Hardness (500 kg load on 10 mm ball)
P.M.—F	13.4	3.9	39	25
Sand—F	12.5	4.5	38	22

* 54% minimum electrical conductivity (ingot) based on the International Annealed Copper Standard.
** Manganese, chromium, titanium & vanadium 0.025 max.

Alloy 170.0

Alloy 170.0 is designed principally for casting small-diameter electric motor rotors. It also is used in connector castings for joining transmission cables and in other electrical applications. As a general rule, Alloy 170.0 is recommended for use in rotors having diameters up to 6 in. The alloy has a guaranteed minimum aluminum content of 99.7% and a carefully controlled iron-to-silicon ratio (the iron content nominally is twice the silicon content). The combination of iron and silicon results in a eutectic material that improves the feeding characteristics of the alloy.

Finishing

The anodized appearance of this alloy is excellent.

Alloy 170.0 (formerly 100 or RP-S)*

Used in: Die, Sand and Permanent Mold Casting

Chemical Composition Limits

Si	Fe	Cu	Mn	Mg	Cr	Ni	Zn	Ti	Sn	Al	Other Elements Each	Other Elements Total
***	***	–	**	–	**	–	0.05	**	–	99.70	0.03**	0.10

Typical Mechanical Properties

Casting Process and Temper	Tension: Ultimate Strength (ksi)	Tension: Yield Strength (set 0.2%) (ksi)	Tension: Elongation (% in 2 in.)	Brinell Hardness 500 kg load on 10 mm ball)
P.M.—F	10.5	2.6	55	20
Sand—F	9.8	3.0	47	17

* 59% minimum electrical conductivity (ingot) based on the International Annealed Copper Standard.
** Manganese, chromium, titanium & vanadium 0.025 max
*** Iron : Silicon ratio = 1.5 min.

Alloys 201.0 & A201.0

Alloys 201.0 and A201.0, which contain silver, provide exceptionally high mechanical strength after heat treatment. Silver is added primarily to improve resistance to stress corrosion cracking and also to increase the response of these alloys to the aging heat treatment.

These alloys are used where highest tensile and yield strengths and moderate elongation are required. Elevated temperature and impact strengths are very good. Applications for Alloys 201.0 and A201.0 include structural casting members, truck and trailer castings, cylinder heads, pistons, gear and pump housings, aerospace housings, and aircraft landing gear components.

Castability

Resistance to hot cracking and solidification shrinkage is rated fair for these alloys. Pressure tightness is good. Successful production of high-quality castings requires close control of alloy composition, grain refining, melt temperature, fluxing and heat treat practices. Melt temperature should not exceed 1450F (788C).

Alloy 201.0 Typical Tensile Properties at Elevated Temperatures*

Test Temp (°F)	Ultimate Strength (ksi)	Yield Strength (Set 0.2%) (ksi)	Elongation (% in 2 in.)
201.0—T6			
300	55	52	6
400	36	33	9
500	22	19	19
600	10	9	39
201.0—T7			
300	60	54	9
400	39	33	16
500	16	13	25
600	9	8	48

* Lowest strengths during 1000 hours of heating at test temperatures.

Machinability

Machinability rating is excellent; best results are obtained with sharp, polished tools with adequate rake and clearance angles. Moderate to fast speeds and feeds are recommended.

Weldability

Good results can be obtained when joining castings of these alloys to components of similar composition using arc or resistance welding methods. Weld repair methods require close control of the casting's temperature during welding to prevent hot tearing and solidification cracking.

Finishing

Electroplating imparts an excellent finish to these alloys. Mechanical polishing also gives excellent results. Anodizing produces very good appearance and good protection.

Corrosion Resistance

Alloys 201.0 and A201.0 exhibit only fair resistance to common corrosive environments.

Sand Cast Thermal Treatment

*T4, T6, T7 Solution Heat Treatment**
Soak at 960F (516C) for 2 hr, then soak at 985F (529C) for 14–20 hr. Quench in water at 150–212F (66–100C).

*T4 Precipitation Treatment***
Age at room temperature for 5 days.

T43 Precipitation Treatment
Age at room temperature for 12–24 hr, then age at 320F (160C) for ½–1 hr.

T6 Precipitation Treatment
Age at room temperature for 12–24 hr, then age at 310F (154C) for 20 hr.

T7 Precipitation Treatment
Age at room temperature for 12–24 hr, then age at 370F (188C) for 5 hr.

Permanent Mold Thermal Treatment

*T4, T6, T7 Solution Heat Treatment**
Soak at 955F (513C) for 2 hr, then soak at 978F (525C) for 14 hr. Quench in water at 150–212F (66–100C).

*T4 Precipitation Treatment***
Age at room temperature for 5 days.

T43 Precipitation Treatment
Age at room temperature for 12–24 hr, then age at 320F (160C) for ½–1 hr.

T6 Precipitation Treatment
Age at room temperature for 12–24 hr, then age at 310F (154C) for 20 hr.

T7 Precipitation Treatment
Age at room temperature for 12–24 hr, then age at 370F (188C) for 5 hr.

Notes to text: * Solution heat treatment is critical. Careful composition and temperature control must be maintained both to attain adequate solution and to prevent incipient melting.

** To expedite testing, test pieces may be precipitation hardened after 12–24 hr at room temperature by holding at 320F (160C) for ½–1 hr.

Alloys 201.0 (formerly K01) & A201.0

Both alloys used in: Sand and Permanent Mold Casting

Alloy 201.0—Chemical Composition Limits

Si	Fe	Cu	Mn	Mg	Cr	Ni	Zn	Ag	Ti	Sn	Other Elements Each	Other Elements Total
0.10	0.15	4.0–5.2	0.20–0.50	0.15–0.55	–	–	–	0.40–1.0	0.15–0.35	–	0.05	0.10

Alloy A201.0—Chemical Composition Limits

Si	Fe	Cu	Mn	Mg	Cr	Ni	Zn	Ag	Ti	Sn	Other Elements Each	Other Elements Total
0.05	0.10	4.0–5.0	0.20–0.40	0.15–0.35	–	–	–	0.40–1.0	0.15–0.35	–	0.03	0.10

Typical Mechanical Properties

Casting Process and Temper	Tension: Ultimate Strength (ksi)	Tension: Yield Strength (Set 0.2%) (ksi)	Tension: Elongation (% in 2 in.)	Shearing Strength (ksi)	Compressive Yield Strength (Set 0.2%) (ksi)	Brinell Hardness (500 kg load on 10 mm ball)	Endurance Limit (ksi)
Sand & P.M.—T43	60	37	17.0	–	–	–	–
Sand & P.M.—T6	65	55	8.0	42	56	130	–
Sand & P.M.—T7	68	60	5.5	40	62	146	14

Alloy A206.0

Alloy A206.0 is a high-strength aluminum alloy often used where damage-tolerant castings are required. This alloy is used for structural castings in the automotive and aerospace industries where high tensile and yield strength and moderate elongation are needed.

The yield strength and fracture toughness values of this alloy are approximately two times those of 356- and 357-type alloys. Other applications where high fracture toughness is required are gear housings and truck spring hangers. High-temperature applications include cylinder heads and also turbine and supercharger applications.

Castability

This alloy exhibits only fair resistance to hot cracking and solidification shrinkage. It requires rigging techniques that promote steep thermal gradients.

Molten metal treatment techniques that thoroughly degas and remove oxides are recommended. Grain refinement will aid heat treatment response and casting soundness. The alloy is sensitive to contamination, especially from silicon.

Machinability

Machinability for A206.0 is rated excellent. Sharp, polished tools with adequate rake and clearance angles obtain the best results when using moderate to fast speeds and feeds.

Weldability

This alloy shows good welding characteristics using arc or resistance processes. The casting's temperature requires close control during welding to prevent cracking from induced stresses.

Finishing

The alloy may be mechanically polished, electroplated or anodized with very good results.

Corrosion Resistance

This alloy exhibits fair corrosion resistance and is comparable to other Al-Cu alloys.

Thermal Treatment

T4, T6, T7 Solution Heat Treatment

For thin-walled, rapidly solidified castings, hold for 2 hr at 920–940F (493–504C); then gradually increase to 980–990F (527–532C) and hold for 8 hr. Quench in 150–212F (66–100C) water.

For slowly solidified or heavy-walled ($^3/_4$–2 in.) castings, hold for 2 hr at 900–920F (468–493C); increase to 940–960F (504–516C) and hold for 2 hr; increase to 980–990F (527–532C) and hold for 12 hr. Quench in 150–212F (66–100C) water.

T6 Precipitation Treatment

Hold at room temperature for 12–24 hr; then at 305–315F (152–157C) for 20 hr.

T7 Precipitation Treatment

Hold at room temperature 12–24 hr; then at 385–395F (196–202C) for 4 hr.

Alloy A206.0

Used in: Sand and Permanent Mold Casting

Chemical Composition Limits

Si	Fe	Cu	Mn	Mg	Cr	Ni	Zn	Ag	Ti	Sn	Other Elements Each	Other Elements Total
0.05	0.010	4.2–5.0	0.20–0.50	0.15–0.35	–	0.05	0.10	–	0.15–0.30	0.05	0.05	0.15

Typical Mechanical Properties

Casting Process and Temper	Tension: Ultimate Strength (ksi)	Tension: Yield Strength (Set 0.2%) (ksi)	Tension: Elongation (% in 2 in.)	Shearing Strength (ksi)	Compressive Yield Strength (Set 0.2%) (ksi)	Brinell Hardness (500 kg load on 10 mm ball)	Endurance Limit (ksi)
P.M.—T4	62	38	17.0	42	41	–	–
Sand—T4	51	36	7.0	40	38	–	–
P.M.—T7	63	50	11.7	37	54	–	–

Alloy 208.0

Alloy 208.0 is widely used where good castability, weldability, pressure tightness and moderate strength are required.

Typical applications involve valve bodies, manifolds and other parts that must be pressure tight. Alloy 208.0 is also a popular alloy for general-purpose castings. This alloy was developed for use in the as-cast condition.

Castability

Very good castability is exhibited by 208.0. Pressure tightness, fluidity and resistance to solidification shrinkage are all above average.

Machinability

Alloy 208.0 machines readily when tools have high rake and clearance angles and sharp, polished cutting edges. Fast speeds and feeds and adequate coolant are recommended.

Weldability

The weldability of 208.0 is good for resistance and arc methods. Brazing is not recommended.

Finishing

Electroplating leaves a very good finish on this alloy. Anodizing and mechanical finishing characteristics are good.

Corrosion Resistance

Alloy 208.0 exhibits fair resistance to most corrosive environments and takes oxidized finishes readily for further protection as well as appearance.

Thermal Treatment

O Annealing

To relieve stresses and reduce growth, heat castings of Alloy 208.0 at 650F (343C) for 2 hr. Air cool to room temperature.

Alloy 208.0 (formerly 108)

Used in: Sand Casting

Chemical Composition Limits

Si	Fe	Cu	Mn	Mg	Cr	Ni	Zn	Ag	Ti	Sn	Other Elements Each	Other Elements Total
2.5–3.5	1.2	3.5–4.5	0.50	0.10	–	0.35	1.0	–	0.25	–	–	0.50

Typical Mechanical Properties

Casting Process & Temper	Tension: Ultimate Strength (ksi)	Tension: Yield Strength (Set 0.2%) (ksi)	Tension: Elongation (% in 2 in.)	Shearing Strength (ksi)	Compressive Yield Strength (Set 0.2%) (ksi)	Brinell Hardness (500 kg load on 10 mm ball)	Endurance Limit (ksi)
Sand—F	21	14	2.5	17	15	55	11.0

Alloy 222.0

Alloy 222.0 retains its strength at elevated temperatures and is used extensively for bushings, meter parts, bearings, bearing caps, pistons, air-cooled cylinder heads and tappet guides.

Castability

Castability of Alloy 222.0 is considered fair to good. Fluidity and resistance to hot cracking are good. The alloy has fair pressure tightness and resistance to solidification shrinkage.

Machinability

The machinability of the alloy is excellent, approaching that of Alloys 520.0 and 512.0.

Weldability

Alloy 222.0 joins very well when it is resistance welded and fairly well when it is welded by the arc and gas methods.

Brazing is not recommended for this alloy.

Finishing

Electroplated coatings are excellent for this alloy. Polishing results in a very good finish. Anodizing is good in appearance and gives fair protection.

Corrosion Resistance

Alloy 222.0 shows poor resistance to most corrosive environments. However, chemical conversion coatings may be used quite successfully for added protection.

Thermal Treatment

Sand —T2 Artificial Aging Treatment
Age at 600F (316C) for 3 hr.

Sand —T61 Solution Heat Treatment & Precipitation
Soak at 950F (510C) for 12 hr; quench in hot water (150–212F, 66–100C). Age at 310F (154C) for 11 hr.

Perm. Mold —T52 Precipitation Treatment
Age at 310F (154C) for 5–7 hr.

Perm. Mold —T551 Precipitation Treatment
Age at 340F (171C) for 16–22 hr.

Perm. Mold —T65 Solution Heat Treatment & Precipitation
Soak at 950F (510C) for 4–12 hr; quench in hot water (150–212F, 66–100C). Age at 340F (171C) for 7–9 hr.

Alloy 222.0 Typical Tensile Properties at Elevated Temperatures*

Test Temp (°F)	Ultimate Strength (ksi)	Yield Strength (Set 0.2%) (ksi)	Elongation (% in 2 in.)
Sand 222.0 —O			
75	27	20	1.0
212	26	19	1.0
300	25	17	1.0
400	22	14	1.5
500	17	11	3.0
600	8	4.5	14.0
Sand 222.2 —T6			
75	41	40	< 0.5
212	39	38	0.5
300	36	35	1.0
400	24	17	2.0
500	17	11	6.0
600	8.5	5	14.0
700	5	2.5	30.0
Perm. Mold 222.0 —T55			
75	37	35	< 0.5
212	34	33	< 0.5
300	30	27	< 0.5
400	25	20	1.0
500	18	12	3.0
600	8.5	5	10.0
700	5	2.5	25.0

* Lowest strengths during 10,000 hours of heating at test temperatures.

Alloy 222.0 (formerly 122)

Used in: Sand and Permanent Mold Casting

Chemical Composition Limits

Si	Fe	Cu	Mn	Mg	Cr	Ni	Zn	Ag	Ti	Sn	Other Elements Each	Other Elements Total
2.0	1.5	9.2–10.7	0.50	0.15–0.35	–	0.50	0.8	–	0.25	–	–	0.35

Typical Mechanical Properties

Casting Process & Temper	Tension: Ultimate Strength (ksi)	Tension: Yield Strength (Set 0.2%) (ksi)	Tension: Elongation (% in 2 in.)	Shearing Strength (ksi)	Compressive Yield Strength (Set 0.2%) (ksi)	Brinell Hardness (500 kg load on 10 mm ball)	Endurance Limit (ksi)
Sand—O	27	20	1.0	21	20	80	9.5
Sand—T61	41	40	< 0.5	32	43	115	8.5
P.M.—T52	35	31	1.0	25	31	100	–
P.M.—T551	37	35	< 0.5	30	40	115	8.5
P.M.—T65	48	36	< 0.5	36	36	140	9.0

Alloys 242.0 & A242.0

Alloys 242.0 and A242.0 are used extensively for applications where excellent strength and hardness at high temperatures are required.

Heavy-duty pistons; motorcycle, diesel and aircraft pistons; aircraft generator housings; and air-cooled cylinder heads are typical applications.

Castability

These alloys have good fluidity, are fair for pressure tightness, and show fair resistance to hot cracking and solidification shrinkage.

Machinability

Machining characteristics of 242.0 and A242.0 are very good. The usual precautions for machining aluminum will give the best results.

Weldability

Arc and resistance methods are very good for welding these alloys. Gas welding is satisfactory, but brazing is not recommended.

Finishing

Electroplated finishes are excellent on both Alloy 242.0 and Alloy A242.0.

Mechanical finishes are very good, and anodized appearance is good.

Corrosion Resistance

The resistance of Alloys 242.0 and A242.0 to most forms of common corrosion is fair.

Some additional protection can be gained by using chemical conversion coatings.

Alloys 242.0 (formerly 142) & A242.0 (formerly A142)

Used in: Sand and Permanent Mold Casting

Alloy 242.0—Chemical Composition Limits

Si	Fe	Cu	Mn	Mg	Cr	Ni	Zn	Ag	Ti	Sn	Other Elements Each	Other Elements Total
0.7	1.0	3.5–4.5	0.35	1.2–1.8	0.25	1.7–2.3	0.35	–	0.25	–	0.05	0.15

Used in: Sand Casting

Alloy A242.0—Chemical Composition Limits

Si	Fe	Cu	Mn	Mg	Cr	Ni	Zn	Ag	Ti	Sn	Other Elements Each	Other Elements Total
0.6	0.8	3.7–4.5	0.10	1.2–1.7	0.15–0.25	1.8–2.3	0.10	–	0.07–0.20	–	0.05	0.15

Typical Mechanical Properties

Casting Process and Temper	Tension: Ultimate Strength (ksi)	Tension: Yield Strength (Set 0.2%) (ksi)	Tension: Elongation (% in 2 in.)	Shearing Strength (ksi)	Compressive Yield Strength (Set 0.2%) (ksi)	Brinell Hardness (500 kg load on 10 mm ball)	Endurance Limit (ksi)
Sand 242.0—F	31	30	< 0.5	–	–	–	–
Sand 242.0—O	27	18	1.0	21	18	70	8.0
Sand 242.0—T571	32	30	0.5	26	34	85	11.0
Sand 242.0—T77	30	23	2.0	24	24	75	10.5
P.M. 242.0—T571	40	34	1.0	30	34	105	10.5
P.M. 242.0—T61	47	42	0.5	35	44	110	9.5
Sand A242.0—T75	31	–	2.0	–	–	75	–

Thermal Treatment

Sand —O Annealing
To relieve stresses and reduce growth, heat at 650F (343C) for 2 hr; air cool to room temperature.

Sand —O Annealing
Age at 650F (343C) for 3 hr.

Sand —T571 Precipitation Treatment
Age at 400F (204C) for 8 hr.

Sand —T77 Solution Heat Treatment & Stabilization
Soak at 960F (516C) for 6 hr; quench in air blast. Age at 650F (343C) for 2 hr.

Perm. Mold —T571 Precipitation Treatment
Age at 330–340F (166–171C) for 22–26 hr.

Perm. Mold —T61 Solution Heat Treatment & Precipitation
Soak castings at 960F (516C) for 4–12 hr, then quench in hot water (150–212F, 66–100C). Age at 400–450F (204–232C) for 3–5 hr.

Alloy 242.0 Typical Tensile Properties at Elevated Temperatures*

Test Temp (˚F)	Ultimate Strength (ksi)	Yield Strength (Set 0.2%) (ksi)	Elongation (% in 2 in.)	Test Temp (˚F)	Ultimate Strength (ksi)	Yield Strength (Set 0.2%) (ksi)	Elongation (% in 2 in.)
242.0 —F				Sand 242.0 —T571			
75	31	30	< 0.5	75	32	30	0.5
212	31	30	< 0.5	212	32	30	0.5
300	30	27	0.5	300	30	28	0.5
400	24	20	1.5	400	26	21	1.0
500	12	6	9.0	500	13	8	8.0
600	8	4.5	14.0	600	8	4	20.0
700	5.5	3	36.0	700	5	3	40.0
Sand 242.0 —T77				Perm. Mold 242.0 —T571			
75	30	23	0.5	75	40	34	1.0
212	30	23	2.0	212	40	34	1.0
300	27	21	2.0	300	37	33	1.0
400	20	15	3.0	400	28	22	2.0
500	13	8	6.0	500	28	22	2.0
600	8	4	20.0	600	8	4	35.0
700	5	3	40.0	700	5	3	60.0

Sand 242.0 —O

Test Temp (˚F)	Ultimate Strength (ksi)	Yield Strength (Set 0.2%) (ksi)	Elongation (% in 2 in.)
75	27	18	1.0
300	23	15	1.0
400	18	11	3.0
500	12	5	8.0
600	7	3	20.0

* Lowest strengths during 10,000 hours of heating at test temperatures.

Alloy 295.0

Although used extensively prior to the mid-1930s, Alloy 295.0 has largely been replaced by 355.0 and 356.0 because of their superior casting characteristics and better service performance.

Castability

This alloy shows good fluidity and resistance to solidification shrinkage. It has fair resistance to hot cracking and is moderately pressure-tight.

Machinability

The machinability rating of 295.0 is very good. Best results are obtained with sharp, polished tools with sufficient rakes and clearances. Moderate to fast speeds and feeds are recommended.

Weldability

Very good results can be obtained using arc or resistance welding methods. Gas welding is generally satisfactory. Brazing is not used.

Finishing

Electroplating imparts an excellent finish to this alloy. Mechanical finishes from polishing are also very good. Anodizing shows up well and gives a good protective finish.

Corrosion Resistance

Fair corrosion resistance to common corrosive environments is exhibited by Alloy 295.0.

Thermal Treatment

T4, T6, T62 Solution Heat Treatment
Soak at 960F (516C) for 12 hr; quench in hot water (150–212F, 66–100C).

T6 Precipitation Treatment
Age at 310F (154C) for 3–6 hr.

T62 Precipitation Treatment
Age at 310F (154C) for 12–24 hr.

Alloy 295.0 Typical Tensile Properties at Elevated Temperatures*

Test Temp (°F)	Ultimate Strength (ksi)	Yield Strength (Set 0.2%) (ksi)	Elongation (% in 2 in.)
295.0 —T4			
75	32	16	8.5
212	30	15	5.0
300	28	15	5.0
400	15	9	15.0
500	9	6	25.0
600	4	3	75.0
700	2.5	1.5	100.0
295.0 —T6			
75	36	24	5.0
212	34	23	5.0
300	28	20	5.0
400	15	9	15.0
500	9	6	25.0
600	4	3	75.0
700	2.5	1.5	100.0

* Lowest strengths during 10,000 hours of heating at test temperatures.

Alloy 295.0 (formerly 195)

Used in: Sand Casting

Chemical Composition Limits

Si	Fe	Cu	Mn	Mg	Cr	Ni	Zn	Ag	Ti	Sn	Other Elements Each	Other Elements Total
0.7–1.5	1.0	4.0–5.0	0.35	0.03	–	–	0.35	–	0.25	–	0.05	0.15

Typical Mechanical Properties

Casting Process and Temper	Tension: Ultimate Strength (ksi)	Tension: Yield Strength (Set 0.2%) (ksi)	Tension: Elongation (% in 2 in.)	Shearing Strength (ksi)	Compressive Yield Strength (Set 0.2%) (ksi)	Brinell Hardness (500 kg load on 10 mm ball)	Endurance Limit (ksi)
Sand—T4*	32	16	8.5	26	17	60	7.0
Sand—T6	36	24	5.0	30	25	75	7.5
Sand—T62	41	32	2.0	33	34	90	8.0

* Properties of —T4 will approach those of —T6 after several weeks at room temperature.

Alloy 296.0

Alloy 296.0 has a higher silicon content than the 295.0 sand casting version, which reduces shrinkage and increases fluidity in the permanent mold. High tensile properties and good machinability are characteristics of this alloy. It is used for gun control parts, gear housings, fittings for aircraft, compressor connecting rods and railway car seat frames. Alloy 296.0 has largely been replaced by 355.0 and 356.0.

Castability

The fluidity, pressure tightness and resistance to solidification shrinkage are all rated good for this alloy. Resistance to hot cracking is fair.

Machinability

This alloy has good machinability, especially when moderate to high speeds and feeds and smooth, sharp tools with high rakes and clearance angles are used.

Weldability

Arc and resistance methods are very satisfactory, and gas welding is considered good. Brazing is not recommended.

Finishing

Electroplated finishes are excellent on this alloy. Mechanical finishes are very good. Anodized appearance is good.

Corrosion Resistance

Although Alloy 296.0 has only fair resistance to most corrosive environments, chemical conversion coatings give very good protection.

Thermal Treatment

T4, T6, T7 Solution Heat Treatment
Soak at 960F (516C) for 12 hr; quench in hot water (150–212F, 66–100C).

T6 Precipitation Treatment
Age at 310F (154C) for 3–6 hr.

T62 Precipitation Treatment
Age at 310F (154C) for 12–24 hours.

Alloy 296.0 Typical Tensile Properties at Elevated Temperatures*

Test Temp (°F)	Ultimate Strength (ksi)	Yield Strength (Set 0.2%) (ksi)	Elongation (% in 2 in.)
296.0—T4			
75	36	22	7.5
212	33	22	7.5
300	30	20	8.0
400	15	9	12.0
500	7.5	5	25.0
600	4	2.5	65.0
296.0—T6			
75	40	26	5.0
212	35	23	5.0
300	29	23	5.0
400	17	11	15.0
500	7	4	25.0
600	3.5	2.5	75.0
700	2.5	1.5	100.0

* Lowest strengths during 10,000 hours of heating at test temperatures.

Alloy 296.0 (formerly B295.0 & B195)

Used in: Permanent Mold Casting

Chemical Composition Limits

Si	Fe	Cu	Mn	Mg	Cr	Ni	Zn	Ag	Ti	Sn	Other Elements Each	Other Elements Total
2.0–3.0	1.2	4.0–5.0	0.35	0.05	–	0.35	0.50	–	0.25	–	–	0.35

Typical Mechanical Properties

Casting Process and Temper	Tension: Ultimate Strength (ksi)	Tension: Yield Strength (Set 0.2%) (ksi)	Tension: Elongation (% in 2 in.)	Shearing Strength (ksi)	Compressive Yield Strength (Set 0.2%) (ksi)	Brinell Hardness (500 kg load on 10 mm ball)	Endurance Limit (ksi)
P.M.—T4*	37	19	9.0	30	20	75	9.5
P.M.—T6	40	26	5.0	32	26	90	10.0
P.M.—T7	39	20	4.5	30	20	80	9.0

*After standing for several weeks at room temperature, the tension properties of —T4 will approach those of —T6.

Alloy 308.0

Typical applications include ornamental grills, reflectors, and other permanent mold castings requiring good castability, weldability, pressure tightness and moderate strength.

Castability

The alloy is good in fluidity, pressure tightness and resistance to hot cracking. Solidification shrinkage is comparatively low.

Machinability

Machinability is good. Carbide-tipped tools are recommended, for the alloy tends to be abrasive. Other general precautions for machining aluminum should be followed for best results.

Finishing

Electroplating gives an excellent finish to 308.0. Anodizing is good in appearance and adds protection against corrosion. Mechanical finishes are good.

Weldability

The weldability of 308.0 is rated good when gas, arc or resistance methods are used. Brazing is not recommended.

Corrosion Resistance

This alloy shows fair resistance to most common forms of corrosive environments. Anodizing and chemical finishing give additional protection.

Thermal Treatment

O Annealing
To relieve stresses and reduce growth, heat at 650F (343C) for 2 hr. Air cool to room temperature.

Alloy 308.0 (formerly A108)

Used in: Sand and Permanent Mold Casting

Chemical Composition Limits

Si	Fe	Cu	Mn	Mg	Cr	Ni	Zn	Ag	Ti	Sn	Other Elements	
											Each	Total
5.0–6.0	1.0	4.0–5.0	0.50	0.10	–	–	1.0	–	0.25	–	–	0.50

Typical Mechanical Properties

Casting Process and Temper	Tension: Ultimate Strength (ksi)	Tension: Yield Strength (Set 0.2%) (ksi)	Tension: Elongation (% in 2 in.)	Shearing Strength (ksi)	Compressive Yield Strength (Set 0.2%) (ksi)	Brinell Hardness (500 kg load on 10 mm ball)	Endurance Limit (ksi)
P.M.—F	28	16	2.0	22	17	70	13

Alloys 319.0, A319.0, B319.0 & 320.0

Alloys 319.0 and A319.0 exhibit very good castability, weldability and pressure tightness and moderate strength. They are very stable alloys (i.e., their good casting and mechanical properties are not affected seriously by fluctuations in the impurity content).

Alloys B319.0 and 320.0 show higher strength and hardness than 319.0 and A319.0 and are generally used in the permanent mold casting process. Characteristics other than strength and hardness are similar to those of 319.0 and A319.0.

Typical applications for sand castings of these alloys are internal combustion and diesel engine crankcases, gasoline and oil tanks, and oil pans. Permanent mold cast parts include water-cooled cylinder heads, typewriter frames, rear axle housings and engine parts.

Castability

Pressure tightness, fluidity, resistance to hot cracking and solidification shrinkage tendencies are all classed as very good for these alloys.

Machinability

Machining characteristics are good. To preclude possible adverse effects of abrasiveness and inclusions, carbide-tipped tools are recommended.

Finishing

Electroplated finishes are very good on these alloys. Mechanical and anodized finishes are fair.

Weldability

Arc welding results in the most satisfactory welds. Resistance and gas methods are both good. Brazing is not recommended.

Corrosion Resistance

The alloys exhibit good resistance to most common forms of corrosion and react well to chemical conversion coatings for additional protection.

Thermal Treatment

Sand & Perm. Mold —O Annealing
To relieve stresses and reduce growth, heat at 650F (343C) for 2 hr. Air cool to room temperature.

Sand —T5 Precipitation Treatment
Age at 400F (204C) for 7–9 hr.

Sand —T6 Solution Heat Treatment & Precipitation
Soak at 940F (504C) for 12 hr; quench in hot water (150–212F, 66–100C). Age at 310F (154C) for 2–5 hr.

Perm. Mold —T6 Solution Heat Treatment & Precipitation
Soak at 940F (504C) for 8 hr; quench in hot water (150–212F, 66–100C). Age at 310F (154C) for 2–5 hr.

Alloys 319.0 (formerly AllCast) & A319.0

Both alloys used in: Sand and Permanent Mold Casting

Alloy 319.0—Chemical Composition Limits

Si	Fe	Cu	Mn	Mg	Cr	Ni	Zn	Ag	Ti	Sn	Other Elements Each	Other Elements Total
5.5–6.5	1.0	3.0–4.0	0.50	0.10	–	0.35	1.0	–	0.25	–	–	0.50

Alloy A319.0—Chemical Composition Limits

Si	Fe	Cu	Mn	Mg	Cr	Ni	Zn	Ag	Ti	Sn	Other Elements Each	Other Elements Total
5.5–6.5	1.0	3.0–4.0	0.50	0.10	–	0.35	3.0	–	0.25	–	–	0.50

Typical Mechanical Properties

Casting Process & Temper	Tension: Ultimate Strength (ksi)	Tension: Yield Strength (Set 0.2%) (ksi)	Tension: Elongation (% in 2 in.)	Shearing Strength (ksi)	Compressive Yield Strength (Set 0.2%) (ksi)	Brinell Hardness (500 kg load on 10 mm ball)	Endurance Limit (ksi)
Sand—F	27	18	2.0	22	19	70	10
Sand—T5	30	26	1.5	24	27	80	11
Sand—T6	36	24	2.0	29	25	80	11
P.M.—F	34	19	2.5	24	19	85	–
P.M.—T6	40	27	3.0	–	27	95	–

Alloys B319.0 & 320.0

Both alloys used in: Permanent Mold Casting

Alloy B319.0—Chemical Composition Limits

Si	Fe	Cu	Mn	Mg	Cr	Ni	Zn	Ag	Ti	Sn	Other Elements Each	Other Elements Total
5.5–6.5	1.2	3.0–4.0	0.8	0.10–0.50	–	0.50	1.0	–	0.25	–	–	0.50

Alloy 320.0—Chemical Composition Limits

Si	Fe	Cu	Mn	Mg	Cr	Ni	Zn	Ag	Ti	Sn	Other Elements Each	Other Elements Total
5.0–8.0	1.2	2.0–4.0	0.8	0.05–0.60	–	0.35	3.0	–	0.25	–	–	0.50

Typical Mechanical Properties

Casting Process and Temper	Tension: Ultimate Strength (ksi)	Tension: Yield Strength (Set 0.2%) (ksi)	Tension: Elongation (% in 2 in.)	Compressive Yield Strength (Set 0.2%) (ksi)	Brinell Hardness (500 kg load on 10 mm ball)
P.M.—F	35	21	1.0	19	90
P.M.—T6	42	30	1.0	27	105

Alloy 332.0

As with Alloy 336.0, Alloy 332.0 is used in applications where properties such as good strength at high temperatures, low coefficient of thermal expansion and good resistance to wear are required. This alloy has many characteristics similar to 336.0 and, in addition, offers improved castability and machinability.

Typical applications include automotive and diesel pistons, pulleys, and sheaves.

Castability

This alloy shows excellent fluidity and resistance to hot cracking. Pressure tightness is very good. There is also high resistance to solidification shrinkage tendencies.

Machinability

The machinability of the alloy is fair. Abrasiveness due to high silicon content and inclusions can be troublesome if carbide-tipped tools (or better) are not used.

Weldability

Good results are obtained by the arc or resistance welding methods. Brazing is not recommended.

Finishing

Alloy 332.0 responds very well to chemical conversion coatings for protection but is considered poor in appearance when anodized. Electroplating is rated fair for this alloy, but mechanical finishes are poor.

Corrosion Resistance

The alloy's resistance to most common forms of corrosion is good.

Thermal Treatment

T5 Precipitation Treatment
Age at 400F (204C) for 7–9 hr.

Alloy 332.0 Typical Tensile Properties at Elevated Temperatures*

Test Temp (°F)	Ultimate Strength (ksi)	Yield Strength (Set 0.2%) (ksi)	Elongation (% in 2 in.)
332.0—T5			
75	36	28	1.0
212	33	27	1.0
300	31	24	2.0
400	25	16	3.0
500	19	12	6.0
600	12	8	15.0
700	8	6	25.0

* Lowest strengths during 10,000 hours of heating at test temperatures.

Alloy 332.0 (formerly F332.0 & F132)

Used in: Permanent Mold Casting

Chemical Composition Limits

Si	Fe	Cu	Mn	Mg	Cr	Ni	Zn	Ag	Ti	Sn	Other Elements Each	Other Elements Total
8.5–10.5	1.2	2.0–4.0	0.50	0.5–1.5	–	0.50	1.0	–	0.25	–	–	0.50

Typical Mechanical Properties

Casting Process and Temper	Tension: Ultimate Strength (ksi)	Tension: Yield Strength (Set 0.2%) (ksi)	Tension: Elongation (% in 2 in.)	Brinell Hardness (500 kg load on 10 mm ball)
P.M.—T5	36	28	1.0	105

Alloys 333.0 & A333.0

The combination of a low coefficient of expansion and good mechanical properties make Alloys 333.0 and A333.0 very suitable for applications such as gas meter and regulator parts, gear blocks, pistons and general automotive, and electric iron sole plates. The alloys have very good casting characteristics.

Castability

Fluidity, pressure tightness and resistance to hot cracking are excellent. The solidification shrinkage tendency is rated as good.

Machinability

Machinability is rated good for Alloys 333.0 and A333.0 when they are machined with wear-resistant cutting tools.

Weldability

Good welds are obtained by the gas, arc or resistance welding methods. These alloys are not brazed.

Finishing

Good finishes are obtained by polishing. The appearances of anodized finishes are rated as only fair and are grayish brown. Electroplate finishes are very good.

Corrosion Resistance

In their as-cast state, Alloys 333.0 and A333.0 have good corrosion resistance. Chemical conversion coatings increase protection.

Thermal Treatment

O Annealing
To relieve stresses and reduce growth, heat at 650F (343C) for 2 hr; air cool to room temperature.

T5 Precipitation Treatment
Age at 400F (204C) for 7–9 hr.

T6 Solution Heat Treatment & Precipitation
Soak at 940F (504C) for 6–12 hr; quench in hot water (150–212F, 66–100C). Age at 310F (154C) for 2–5 hr.

T7 Solution Heat Treatment & Stabilization
Soak at 940F (504C) for 6–12 hr; quench in hot water (150–212F, 66–100C). Age at 500F (260C) for 4–6 hr.

Alloys 333.0 & A333.0

Both alloys used in: Permanent Mold Casting

Alloy 333.0—Chemical Composition Limits

Si	Fe	Cu	Mn	Mg	Cr	Ni	Zn	Ag	Ti	Sn	Other Elements Each	Other Elements Total
8.0–10.0	1.0	3.0–4.0	0.50	0.05–0.50	–	0.50	1.0	–	0.25	–	–	0.50

Alloy A333.0—Chemical Composition Limits

Si	Fe	Cu	Mn	Mg	Cr	Ni	Zn	Ag	Ti	Sn	Other Elements Each	Other Elements Total
8.0–10.0	1.0	3.0–4.0	0.50	0.05–0.50	–	0.50	3.0	–	0.25	–	–	0.50

Typical Mechanical Properties

Casting Process and Temper	Tension: Ultimate Strength (ksi)	Tension: Yield Strength (Set 0.2%) (ksi)	Tension: Elongation (% in 2 in.)	Shearing Strength (ksi)	Compressive Yield Strength (Set 0.2%) (ksi)	Brinell Hardness (500 kg load on 10 mm ball)	Endurance Limit (ksi)
P.M.—F	34	19	2.0	27	19	90	14.5
P.M.—T5	34	25	1.0	27	25	100	12.0
P.M.—T533	32	25	1.0	–	–	100	–
P.M.—T6	42	30	1.5	33	30	105	15.0
P.M.—T7	37	28	2.0	28	28	90	12.0

Alloy 336.0

Good resistance to wear, a low coefficient of thermal expansion and strength at elevated temperatures are some main attributes of this alloy.

Typical uses include automotive and heavy-duty diesel pistons, pulleys, sheaves, and similar applications.

Castability

Alloy 336.0 shows excellent fluidity and resistance to hot cracking. Pressure tightness is very good. There is also high resistance to solidification shrinkage tendencies.

Machinability

The machinability of the alloy is fair. Abrasiveness due to high silicon content and inclusions can be troublesome if carbide-tipped tools (or better) are not used.

Weldability

Good results are obtained by the arc or resistance welding methods. Brazing is not recommended.

Finishing

Alloy 336.0 responds very well to chemical conversion coatings for protection but is considered poor in appearance when anodized.

Electroplating is rated fair for this alloy, but mechanical finishes are poor.

Corrosion Resistance

The alloy's resistance to most common forms of corrosion is good.

Thermal Treatment

T551 Precipitation Treatment
Age at 400F (204C) for 7–9 hr.

T65 Solution Heat Treatment & Precipitation
Soak at 960F (516C) for 8 hr; quench in hot water (150–212F, 66–100C). Age at 400F (204C) for 7–9 hr.

Alloy 336.0 Typical Tensile Properties at Elevated Temperatures*

Test Temp (°F)	Ultimate Strength (ksi)	Yield Strength (Set 0.2%) (ksi)	Elongation (% in 2 in.)
336.0 —T551			
75	36	28	0.5
212	35	25	1.0
300	31	22	1.0
400	26	15	2.0
500	18	10	5.0
600	10	4	10.0
700	5	3	45.0

* Lowest strengths during 10,000 hours of heating at test temperatures.

Alloy 336.0 (formerly A332.0 & A132)

Used in: Permanent Mold Casting

Chemical Composition Limits

Si	Fe	Cu	Mn	Mg	Cr	Ni	Zn	Ag	Ti	Sn	Other Elements Each	Other Elements Total
11.0– 13.0	1.2	0.50–1.50	0.35	0.7–1.3	–	2.0–3.0	0.35	–	0.25	–	0.05	–

Typical Mechanical Properties

Casting Process and Temper	Tension: Ultimate Strength (ksi)	Tension: Yield Strength (Set 0.2%) (ksi)	Tension: Elongation (% in 2 in.)	Shearing Strength (ksi)	Compressive Yield Strength (Set 0.2%) (ksi)	Brinell Hardness (500 kg load on 10 mm ball)	Endurance Limit (ksi)
P.M.—T65	47	43	0.5	36	43	125	–
P.M.—T551	36	28	0.5	28	28	105	13.5

Alloy 354.0

Alloy 354.0 is basically similar to C355.0, but its increased silicon and copper content give it greatly improved mechanical properties. Its tensile and elongation properties show almost no change up to temperatures of at least 300F (149C). The outstanding strength characteristics of this alloy make it especially useful in the aerospace industries in applications where premium-strength castings are required.

Castability

The alloy has excellent castability. Its excellent fluidity and resistance to hot cracking make it possible to produce castings larger in size and more intricate in design. When cast by the permanent mold process, the alloy shows less tendency to surface shrinkage than C355.0.

Machinability

The machining characteristics of this alloy are comparable to those of C355.0.

Weldability

Welds produced by all the common welding processes are excellent. Brazing is not recommended.

Finishing

Polished finishes are relatively good. Anodized appearances are fair. Chemical conversion and electroplated coatings are very good.

Corrosion Resistance

The alloy exhibits high resistance to corrosion, but it is susceptible to stress corrosion cracking in its highest-strength condition.

Thermal Treatment

T6 Solution Heat Treatment & Precipitation

Soak at 980F (527C) for 10–12 hr; quench in hot water (140–180F, 60–82C). Allow castings to remain at room temperature for 8–16 hr and then age for 3–5 hr at 310F (154C).

Perm. Mold —T61 Solution Heat Treatment & Precipitation

Soak at 980F (527C) for 10–12 hr; quench in hot water (140–180F, 60–82C). Allow castings to remain at room temperature for 8+ hr and then age for 10–12 hr at 310F (154C).

Sand & Perm. Mold —T62 Solution Heat Treatment & Precipitation

Soak at 980F (527C) for 10–12 hr; quench in hot water (140–180F, 60–82C). Allow castings to remain at room temperature for 8+ hr and then age for 6–10 hr at 340F (171C).

Alloy 354.0 Typical Tensile Properties at Elevated Temperatures

Test Temp (°F)	Ultimate Strength (ksi)	Yield Strength (ksi)	Elongation (% in 4D)
Perm. Mold 354.0 —T61			
75*	55	41	6.0
212*	55	49	6.0
300*	41	35	7.0
400*	16	11	43.0
500**	27	24	16.0
600**	11	10	30.0

* Properties after 10,000 hours at test temperatures.
** Properties after 1/2 hour at test temperatures.

Alloy 354.0

Used in: Permanent Mold Casting

Chemical Composition Limits

Si	Fe	Cu	Mn	Mg	Cr	Ni	Zn	Ag	Ti	Sn	Other Elements Each	Other Elements Total
8.6–9.4	0.20	1.6–2.0	0.10	0.40–0.60	–	–	0.10	–	0.20	–	0.05	0.15

Typical Mechanical Properties*

Casting Process and Temper	Tension: Ultimate Strength (ksi)	Tension: Yield Strength (Set 0.2%) (ksi)	Tension: Elongation (% in 2 in.)	Endurance Limit (ksi)
P.M.—T61	52	39	6.0	19.5
Sand & P.M.—T62	55	45	3.0	–

* Mechanical properties of specimens machined from castings.

Alloy 355.0

Alloy 355.0 is a general-purpose, high-strength material with many desirable characteristics that make it popular for a wide variety of products. This alloy is less ductile than 356.0.

Some of this alloy's applications in sand casting are air compressor pistons, printing press bedplates, water jackets, crankcases, electric motor fans and blowers, and accessory and gear housings.

Some of its applications in permanent mold casting include impellers, aircraft fittings, timing gears, aircraft supercharger covers, liquid-cooled cylinder heads, accessory housings for internal combustion engines, compressors, jet engine compressor cases, passenger car wheels, engine blocks, cylinder heads, spring brackets, transmission cases, and miscellaneous castings for trucks and trailers. Marine usage includes those applications where high strength is needed along with pressure tightness, corrosion resistance and high strength at elevated temperatures.

Castability

Excellent castability is shown by 355.0. Resistance to hot cracking, pressure tightness and solidification shrinkage tendencies are all classified as excellent.

Machinability

This alloy is considered to have good machinability. Tungsten carbide-tipped cutting tools may be required, however, because of the alloy's abrasiveness. This can often add to machining costs.

Weldability

Arc, resistance and gas welding methods all give excellent welds on Alloy 355.0. Brazing is not desirable.

Finishing

Electroplate and chemical conversion coatings are classed as very good for this alloy. Polishing characteristics and other mechanical finishes are good, but anodizing for appearance is only fair because the color is gray.

Corrosion Resistance

Alloy 355.0 exhibits good corrosion resistance but somewhat less than that of 333.0 or 356.0. Added protection through chemical conversion coatings is very satisfactory.

Alloy 355.0

Used in: Sand and Permanent Mold Casting

Chemical Composition Limits

Si	Fe	Cu	Mn	Mg	Cr	Ni	Zn	Ag	Ti	Sn	Other Elements Each	Other Elements Total
4.5–5.5	0.6*	1.0–1.5	0.5*	0.4–0.6	0.25	–	0.35	–	0.25	–	0.05	0.15

Typical Mechanical Properties

Casting Process and Temper	Tension: Ultimate Strength (ksi)	Tension: Yield Strength (Set 0.2%) (ksi)	Tension: Elongation (% in 2 in.)	Shearing Strength (ksi)	Compressive Yield Strength (Set 0.2%) (ksi)	Brinell Hardness (500 kg load on 10 mm ball)	Endurance Limit (ksi)
Sand—F	23	12	3.0	–	–	–	–
Sand—T51	28	23	1.5	22	24	65	8.0
Sand—T6	35	25	3.0	28	26	80	9.0
Sand—T61	39	35	1.0	31	37	90	9.5
Sand—T7	38	36	0.5	28	38	85	10.0
Sand—T71	35	29	1.5	26	30	75	10.0
P.M.—F	27	15	4.0	–	–	–	–
P.M.—T51	30	24	2.0	24	24	75	–
P.M.—T6	42	27	4.0	34	27	90	10.0
P.M.—T62	45	40	1.5	36	40	105	10.0
P.M.—T7	40	30	2.0	30	30	85	10.0
P.M.—T71	36	31	3.0	27	31	85	10.0

* If Fe exceeds 0.45, Mn level shall not be less than half the Fe content.

Thermal Treatment

Sand & Perm. Mold O Annealing
To relieve stresses and reduce growth, heat at 650F (343C) for 2 hr; air cool to room temperature.

Sand & Perm. Mold —T51 Precipitation Treatment
Age at 440F (227C) for 7–9 hr.

Sand —T6 Solution Heat Treatment & Precipitation
Soak at 980F (527C) for 12 hr; quench in hot water (150–212F, 66–100C). Age at 310F (154C) for 3–5 hr.

Sand —T61 Solution Heat Treatment & Precipitation
Soak at 980F (527C) for 12 hr; quench in hot water (150–212F, 66–100C). Age at 310F (154C) for 8–10 hr.

Sand —T7 Solution Heat Treatment & Stabilization
Soak at 980F (527C) for 12 hr; quench in hot water (150–212F, 66–100C). Age for stress relief at 440F (227C) for 3–5 hr.

Sand —T71 Solution Heat Treatment & Stabilization
Soak at 980F (527C) for 12 hr; quench in hot water (150–212F, 66–100C). Age for stress relief at 475F (246C) for 4–6 hr.

Perm. Mold —T6 Solution Heat Treatment & Precipitation
Soak at 980F (527C) for 4–12 hr; quench in hot water (150–212F, 66–100C). Age at 310F (154C) for 2–5 hr.

Perm. Mold —T62 Solution Heat Treatment & Precipitation
Soak at 980F (527C) for 4–12 hr; quench in hot water (150–212F, 66–100C). Age at 340F (171C) for 14–18 hr.

Perm. Mold —T7 Solution Heat Treatment & Stabilization
Soak at 980F (527C) for 4–12 hr; quench in hot water (150–212F, 66–100C). Age 7–9 hr at 440F (227C) to relieve stresses.

Perm. Mold —T71 Solution Heat Treatment & Stabilization
Soak at 980F (527C) for 4–12 hr; quench in hot water (150–212F, 66–100C). Age for stress relief at 475F (246C) for 3–6 hr.

Alloy 355.0 Tensile Properties at Low Temperatures

Test Temp (°F)	Tensile Strength (ksi)	Yield Strength (ksi)	Elongation (%)	Reduction in Area (%)
Sand 355.0 —T6				
78	32.7	26.3	4.0	4.0
–104	52.0	41.2	4.0	6.0
–320	59.2	46.5	2.0	4.0
–423	63.6	56.8	2.0	2.0
Perm. Mold 355.0 —T6				
78	49.9	43.6	3.0	5.0
–104	50.3	45.5	4.0	4.0
–320	59.7	53.2	3.0	3.0
–423	70.8	60.8	2.0	2.0
Perm. Mold 355.0 —T61				
78	47.4	34.8	5.0	6.0
–104	49.8	36.7	6.0	7.0
–320	55.6	49.0	3.0	6.0
–423	69.9	53.2	2.0	3.0

Alloy 355.0 Typical Tensile Properties at Elevated Temperatures*

Test Temp (°F)	Ultimate Strength (ksi)	Yield Strength (Set 0.2%) (ksi)	Elongation (% in 2 in.)
Sand 355.0 —F			
75	28	23	1.5
212	28	22	2.0
300	24	19	3.0
400	14	10	8.0
500	9.5	5	16.0
600	6.0	3	36.0
700	3.5	2	50.0
Sand 355.0 —T6			
75	35	25	3.0
212	35	25	2.0
300	33	25	1.5
400	17	13	8.0
500	9.5	5	16.0
600	6	3	36.0
700	3.5	2	50.0
Sand 355.0 —T71			
75	35	29	1.5
212	34	28	2.0
300	30	26	3.0
400	17	13	8.0
500	9.5	5	16.0
600	6.0	3	36.0
700	3.5	2	50.0
Perm. Mold 355.0 —T51			
75	30	24	2.0
212	28	24	3.0
300	23	20	4.0
400	15	10	19.0
500	9.5	5	33.0
600	6.0	3	38.0
700	3.5	2	60.0
Perm. Mold 355.0 —T6			
75	42	27	4.0
212	40	27	5.0
300	32	25	10.0
400	19	13	20.0
500	9.5	5	40.0
600	6.0	3	50.0
700	3.5	2	60.0
Perm. Mold 355.0 —T71			
75	36	31	3.0
212	33	29	4.0
300	29	26	8.0
400	19	13	20.0
500	9.5	5	40.0
600	6.0	3	50.0
700	3.5	2	50.0

* Lowest strengths during 10,000 hours of heating at test temperatures.

Alloy C355.0

Although C355.0 is a modification of 355.0, its characteristics are much the same (elongation and strengths are higher). Parts are usually heat treated. Strengths at elevated temperatures are relatively high.

C355.0 sand castings include printing press bedplates, air compressor pistons, water jackets, crankcases, electric motor fans and blowers. Permanent mold castings are used for aircraft, missile, and other components requiring high strength, and for fittings, impellers, timing gears and aircraft supercharger covers. Automotive uses include passenger car wheels, engine blocks, cylinder heads, spring brackets, transmission cases, and miscellaneous truck and trailer castings.

Castability

Excellent castability is exhibited by C355.0. Pressure tightness, resistance to hot cracking, fluidity and solidification shrinkage tendencies are all rated excellent.

Machinability

Machining characteristics are considered good, being somewhat better than those shown by 356.0, but not as good as those shown by alloys that have lower silicon content, such as 514.0. Metal removal should be light and at high speeds.

Weldability

Welding characteristics of Alloy C355.0 are excellent for standard welding procedures. Brazing is considered impractical.

Finishing

Polished finishes are relatively good. Anodized appearance is fair. Chemical conversion and electroplated coatings are very good.

Corrosion Resistance

The resistance to most atmospheric conditions is good, only slightly less than that of Alloys 333.0 or 356.0.

Thermal Treatment

Sand & Perm. Mold —T6 Solution Heat Treatment & Precipitation
Soak at 980F (527C) for 12 hr; quench in hot water (150–212F, 66–100C). Age at room temperature for 8+ hr, then at 310F (154C) for 3–5 hr.

Perm. Mold —T61 Solution Heat Treatment & Precipitation
Soak at 980F (527C) for 6–12 hr; quench in hot water (150–212F, 66–100C). Age at room temperature for 8+ hr, then at 310F (154C) for 10–12 hr.

Perm. Mold —T62 Solution Heat Treatment & Precipitation
Soak at 980F (527C) for 8–10 hr; quench in hot water (150–212F, 66–100C). Age at 340F (171C) for 14–18 hr.

Alloy C355.0 Typical Tensile Properties at Elevated Temperatures*

Test Temp (°F)	Ultimate Strength (ksi)	Yield Strength (Set 0.2%) (ksi)	Elongation (% in 2 in.)
Perm. Mold C355.0 —T61			
75	46	34	6.0
212	43	34	6.0
300	38	34	10.0
400	14	10	40.0
500	6.5	5.5	60.0
600	4.0	3.0	70.0
700	3.5	2.5	90.0

* Lowest strengths during 10,000 hours of heating at test temperatures.

Alloy C355.0

Used in: Sand and Permanent Mold Casting

Chemical Composition Limits

Si	Fe	Cu	Mn	Mg	Cr	Ni	Zn	Ag	Ti	Sn	Other Elements Each	Other Elements Total
4.5–5.5	0.20	1.0–1.5	1.10	0.40–0.60	–	–	0.10	–	0.20	–	0.05	0.15

Typical Mechanical Properties

Casting Process and Temper	Tension: Ultimate Strength (ksi)	Tension: Yield Strength (Set 0.2%) (ksi)	Tension: Elongation (% in 2 in.)	Shearing Strength (ksi)	Compressive Yield Strength (Set 0.2%) (ksi)	Brinell Hardness (500 kg load on 10 mm ball)	Endurance Limit (ksi)
Sand—T6	39	29	5.0	–	–	85	–
P.M.—T6	48	28	8.0	–	–	90	–
P.M.—T61	46	34	6.0	34	36	100	14
P.M.—T62	48	37	5.0	–	–	100	–

Alloy 356.0

Applications for 356.0 are similar to those for 355.0. Alloy 356.0 has excellent casting characteristics and has largely replaced Alloy 295.0.

Permanent mold castings of this alloy are used for machine tool parts, aircraft wheels and handwheels, pump parts, tank car fittings, marine hardware, valve bodies, bridge railing parts, and for such parts as aileron control sectors, rudder control supports, fuselage fittings, and fuel tank elbows for airplanes and missiles. Automotive applications include miscellaneous castings for trucks and trailers, spring brackets, cylinder heads, engine blocks, passenger car wheels, and transmission cases.

Uses for sand castings of Alloy 356.0 include flywheel housings, automotive transmission cases, oil pans, rear axle housings, brackets, water-cooled cylinder blocks, various fittings and pump bodies. This alloy is used in various marine applications in the T6 condition where pressure tightness and/or corrosion resistance are major requirements.

Castability

Fluidity, resistance to hot cracking and resistance to solidification shrinkage are all excellent for Alloy 356.0.

Machinability

After heat treatment (notably in the T6 condition), Alloy 356.0 has good machinability. Because of the high silicon content, savings can be realized by using carbide-tipped cutting tools (or better). Rakes should be positive and high. Speeds should approach the maximum, and light cuts should be taken.

Weldability

Very good welding characteristics are shown by Alloy 356.0 for all standard welding methods. The alloy is not brazed.

Finishing

Alloy 356.0 has good polishing characteristics. It is excellent for electroplating and very good for chemical conversion coatings. Anodizing appearance is very good; the color is gray.

Corrosion Resistance

Excellent resistance to corrosion is shown by Alloy 356.0. Its corrosion resistance is higher than that of 355.0. Chemical conversion coatings give further protection.

Thermal Treatment

Sand & Perm. Mold —O Annealing
To relieve stresses and reduce growth, heat at 650F (343C) for 2 hr. Air cool to room temperature.

Sand & Perm. Mold —T51 Precipitation Treatment
Heat at 440F (227C) for 7–9 hr.

Sand —T6 Solution Heat Treatment & Precipitation
Soak at 1000F (538C) for 12 hr; quench in hot water (150–212F, 66–100C). Age at 310F (154C) for 3–5 hr.

Alloy 356.0

Used in: Sand and Permanent Mold Casting

Chemical Composition Limits

Si	Fe	Cu	Mn	Mg	Cr	Ni	Zn	Ag	Ti	Sn	Other Elements Each	Other Elements Total
6.5–7.5	0.60*	0.25	0.35*	0.20–0.45	–	–	0.35	–	0.25	–	0.05	0.15

Typical Mechanical Properties

Casting Process and Temper	Tension: Ultimate Strength (ksi)	Tension: Yield Strength (Set 0.2%) (ksi)	Tension: Elongation (% in 2 in.)	Shearing Strength (ksi)	Compressive Yield Strength (Set 0.2%) (ksi)	Brinell Hardness (500 kg load on 10 mm ball)	Endurance Limit (ksi)
Sand—F	24	18	6.0	–	–	–	–
Sand—T51	25	20	2.0	20	21	60	8.0
Sand—T6	33	24	3.5	26	25	70	8.5
Sand—T7	34	30	2.0	24	31	75	9.0
Sand—T71	28	21	3.5	20	22	60	8.5
P.M.—F	26	18	5.0	–	–	–	–
P.M.—T51	27	20	2.0	–	–	–	–
P.M.—T6	38	27	5.0	30	27	80	13
P.M.—T7	32	24	6.0	25	24	70	11

*If Fe exceeds 0.45, Mn level shall not be less than half the Fe content.

Sand —T7 Solution Heat Treatment & Stabilization
Soak at 1000F (538C) for 12 hr; quench in hot water (150–212F, 66–100C). Age to relieve stresses at 400F (204C) for 3–5 hr.

Sand —T71 Solution Heat Treatment & Stabilization
Soak at 1000F (538C) for 12 hr; quench in hot water (150–212F, 66–100C). Age to relieve stresses for 2–4 hr at 475F (246C).

Perm. Mold —T6 Solution Heat Treatment & Precipitation
Soak at 1000F (538C) for 4–12 hr; quench in hot water (150–212F, 66–100C). Age at 310F (154C) for 2–5 hr.

Perm. Mold —T7 Solution Heat Treatment & Stabilization
Soak at 1000F (538C) for 4–12 hr; quench in hot water (150–212F, 66–100C). Age to relieve stresses for 7–9 hr at 440F (227C).

Alloy 356.0 Typical Tensile Properties at Elevated Temperatures*

Test Temp (°F)	Ultimate Strength (ksi)	Yield Strength (Set 0.2%) (ksi)	Elongation (% in 2 in.)	Test Temp (°F)	Ultimate Strength (ksi)	Yield Strength (Set 0.2%) (ksi)	Elongation (% in 2 in.)
Sand 356.0 —T6				Perm. Mold 356.0 —T6			
75	33	24	3.5	75	38	27	5.0
212	32	24	4.0	212	30	25	6.0
300	23	20	6.0	300	21	17	10.0
400	12	8.5	18.0	400	12	8.5	30.0
500	7.5	5.0	35.0	500	7.5	5.0	55.0
600	4.0	3.0	60.0	600	4.0	3.0	70.0
700	2.5	2.0	80.0	700	2.5	2.0	80.0
Sand 356.0 —T7				Perm Mold 356.0 —T7			
75	34	30	2.0	75	32	24	6.0
212	30	28	2.0	212	27	23	10.0
300	23	20	6.0	300	21	17	20.0
400	12	8.5	18.0	400	12	8.5	40.0
500	7.5	5.0	35.0	500	7	5.0	55.0
600	4.0	3.0	60.0	600	4.0	3.0	70.0
700	2.5	2.0	80.0	700	2.5	2.0	80.0

* Lowest strengths during 10,000 hours of heating at test temperatures.

Alloy A356.0

Alloy A356.0 has greater elongation, higher strength and considerably higher ductility than 356.0. It has these improved mechanical properties because impurities are lower in A356.0 than in 356.0. Typical applications are airframe castings, machine parts, truck chassis parts, aircraft and missile components, and structural parts requiring high strength.

Castability

All casting characteristics are excellent for this alloy.

Machinability

Alloy A356.0 has good machinability. Abrasiveness can be overcome and high tool wear can be minimized by using sharp, carbide-tipped (or better) tools with high rakes and clearances. Moderate to fast speeds are recommended.

Finishing

Electroplated finishes are very good. Chemical conversion coatings give very good protection, but anodized appearance is only fair. Mechanical finishes on A356.0 are good.

Weldability

All common welding methods are excellent for joining this alloy. Brazing is not performed.

Corrosion Resistance

This alloy has very good resistance to most common forms of corrosion.

Thermal Treatment

Sand & Perm. Mold —O Annealing
To relieve stresses and reduce growth, heat at 650F (343C) for 2 hr; air cool to room temperature.

Sand —T6 Solution Heat Treatment & Precipitation
Soak at 1000F (538C) for 12 hr; quench in hot water (150–212F, 66–100C). Age at room temperature for 8+ hr, then at 310F (154C) for 2–5 hr.

Perm. Mold —T6 Solution Heat Treatment & Precipitation
Soak at 1000F (538C) for 12 hr; quench in hot water (150–212F, 66–100C). Age at room temperature for 8+ hr, then at 310F (154C) for 3–5 hr.

Perm. Mold —T61 Solution Heat Treatment & Precipitation
Soak at 1000F (538C) for 6–12 hr; quench in hot water (150–212F, 66–100C). Age at room temperature for 6–12 hr, then at 310F (154C) for 10–12 hr.

Sand & Perm. Mold —T7 Solution Heat Treatment & Stabilization
Soak at 1000F (538C) for 12 hr; quench in hot water (150–212F, 66–100C). Age at room temperature for 8+ hr, then at 440F (227C) for 8 hr.

Sand —T71 Solution Heat Treatment & Stabilization
Soak at 1000F (538C) for 12 hr; quench in hot water (150–212F, 66–100C). Age at room temperature for 8+ hr, then at 475F (246C) for 3 hr.

Alloy A356.0 Typical Tensile Properties at Elevated Temperatures*

Test Temp (°F)	Ultimate Strength (ksi)	Yield Strength (Set 0.2%) (ksi)	Elongation (% in 2 in.)
Perm. Mold A356.0 —T61			
75	41	30	10.0
300	21	17	20.0
400	12	8.5	40.0
500	7.5	5.0	55.0
600	4.0	3.0	70.0
700	2.5	2.0	80.0

* Lowest strengths during 10,000 hours of heating at test temperatures.

Alloy A356.0

Used in: Sand and Permanent Mold Casting

Chemical Composition Limits

Si	Fe	Cu	Mn	Mg	Cr	Ni	Zn	Ag	Ti	Sn	Other Elements Each	Other Elements Total
6.5–7.5	0.20	0.20	0.10	0.25–0.45	–	–	0.10	–	0.20	–	0.05	0.15

Typical Mechanical Properties

Casting Process and Temper	Tension: Ultimate Strength (ksi)	Tension: Yield Strength (Set 0.2%) (ksi)	Tension: Elongation (% in 2 in.)	Shearing Strength (ksi)	Compressive Yield Strength (Set 0.2%) (ksi)	Brinell Hardness (500 kg load on 10 mm ball)	Endurance Limit (ksi)
Sand—F	23	12	6.0	–	–	–	–
Sand—T51	26	18	3.0	–	–	–	–
Sand—T6	40	30	6.0	–	–	75	–
Sand—T71	30	20	3.0	–	–	–	–
P.M.—F	27	13	8.0	–	–	–	–
P.M.—T51	29	20	5.0	–	–	–	–
P.M.—T6	41	30	12.0	–	–	80	–
P.M.—T61	41	30	10.0	28	32	90	13

Alloy 357.0

The characteristics and uses of this alloy are similar to those of 356.0. However, 357.0 exhibits greater ultimate and yield strengths. This alloy shows high strength and good ductility after heat treatment.

Applications for Alloy 357.0 include highly stressed aircraft and missile structures and high velocity blowers and impellers.

Castability

This alloy has excellent pressure tightness, fluidity and resistance to hot cracking. Solidification shrinkage tendency is slight.

Machinability

Machining of 357.0 is good. However, like all high-silicon alloys, there is appreciable abrasive action on the cutting tools. Carbide-tipped tools (or better) give best results.

Weldability

The weldability rating for 357.0 is considered excellent for all standard methods including gas, arc and resistance. Brazing is not recommended.

Finishing

Polished finishes on 357.0 are considered good. Anodized appearance is fair. Electroplated finishes are excellent.

Corrosion Resistance

The alloy has very good corrosion resistance. Good additional protection is afforded by chemical coatings.

Thermal Treatment

Sand & Perm. Mold O Annealing
To relieve stresses and reduce growth, heat at 650F (343C) for 2 hr; air cool to room temperature.

Sand & Perm. Mold —T51 Precipitation Treatment
Age at 440F (227C) for 7–9 hr.

Sand —T6 Solution Heat Treatment & Precipitation
Soak at 1000F (538C) for 12 hr; quench in hot water (150–212F, 66–100C). Age at room temperature for 8+ hr, then at 310F (154C) for 2–5 hr.

Perm. Mold —T6 Solution Heat Treatment & Precipitation
Soak at 1000F (538C) for 8 hr; quench in hot water (150–212F, 66–100C). Age at room temperature for 8+ hr, then at 310F (154C) for 3–5 hr.

Sand & Perm. Mold —T7 Solution Heat Treatment & Stabilization
Soak at 1000F (538C) for 8–12 hr (sand castings require the longer soaking time); quench in hot water. Age to relieve stresses for 5–7 hr at 440F (227C).

Alloy 357.0

Used in: Sand and Permanent Mold Casting

Chemical Composition Limits

Si	Fe	Cu	Mn	Mg	Cr	Ni	Zn	Ag	Ti	Sn	Other Elements Each	Other Elements Total
6.5–7.5	0.15	0.05	0.03	0.45–0.6	–	–	0.05	–	0.20	–	0.05	0.15

Typical Mechanical Properties

Casting Process and Temper	Tension: Ultimate Strength (ksi)	Tension: Yield Strength (Set 0.2%) (ksi)	Tension: Elongation (% in 2 in.)	Brinell Hardness (500 kg load on 10 mm ball)
Sand—F	25	13	5.0	–
Sand—T51	26	17	3.0	–
Sand—T6	50	43	2.0	90
Sand—T7	40	34	3.0	60
P.M.—F	28	15	6.0	–
P.M.—T51	29	21	4.0	–
P.M.—T6	52	43	5.0	100
P.M.—T7	38	30	5.0	70

Alloy A357.0

Alloy A357.0 provides very high mechanical properties and has excellent castability. Beryllium is one of its alloying elements.

The alloy is used for aircraft, missile and other applications that require high tensile, yield and ultimate strengths. Strengths at elevated temperatures are very good.

Castability

Fluidity, pressure tightness, solidification shrinkage and resistance to hot cracking of this alloy are all rated as excellent.

Machinability

The machining characteristics of this alloy are fair.

Weldability

The gas, arc and resistance welding methods give excellent results. Brazing is not recommended.

Finishing

Good mechanical finishes are produced by polishing. Anodized appearance is gray in color.

Corrosion Resistance

The corrosion resistance classification of this aluminum alloy is very good. Additional protection is provided by chemical conversion coatings.

Thermal Treatment

Sand —T6 Solution Heat Treatment & Precipitation
Soak at 1005F (540C) for 12 hr; quench in hot water (150–212F, 66–100C). Age at room temperature for 8+ hr, then at 315F (157C) for 3–5 hr.

Perm. Mold —T6 Solution Heat Treatment & Precipitation
Soak at 1000F (538C) for 10 hr; quench in hot water (150–212F, 66–100C). Age at room temperature for 8+ hr, then at 340F (171C) for 3–5 hr.

Alloy A357.0 Typical Tensile Properties at Elevated Temperatures*

Test Temp (°F)	Ultimate Strength (ksi)	Yield Strength (ksi)	Elongation (% in 2 in.)
Sand A357.0 —T6			
75	48	36	3.0
300	30	28	3.0
400	13	10	24.0
500	7	6	30.0
Perm. Mold A357.0 —T6			
75	50	40	10.0
300	31	29	11.0
400	12	9	29.0
500	7	6	—

* Lowest properties during 1000 hours at test temperatures.

Alloy A357.0

Used in: Sand and Permanent Mold Casting

Chemical Composition Limits

Si	Fe	Cu	Mn	Mg	Cr	Ni	Zn	Be	Ti	Sn	Other Elements Each	Other Elements Total
6.5–7.5	0.20	0.20	0.10	0.40–0.70	–	–	0.10	0.04–0.07	0.04–0.20	–	0.05	0.15

Typical Mechanical Properties

Casting Process and Temper	Tension: Ultimate Strength (ksi)	Tension: Yield Strength (Set 0.2%) (ksi)	Tension: Elongation (% in 2 in.)	Shearing Strength (ksi)	Compressive Yield Strength (Set 0.2%) (ksi)	Brinell Hardness (500 kg load on 10 mm ball)	Endurance Limit (ksi)
Sand—T6	46	36	3.0	40	35	85	12
P.M.—T6	50	40	10.0	43	40	85	16

Alloy 359.0

Alloy 359.0, like 354.0, exhibits outstanding strength characteristics of particular usefulness to the aerospace industries. The alloy has excellent castability and exhibits little evidence of corrosion cracking when stressed at high levels.

Castability

The alloy has excellent castability. Its resistance to hot cracking is rated as excellent, and its fluidity in permanent molds is better than that of A356.0.

Machinability

Alloy 359.0 has good machining characteristics when wear-resistant tools with liberal rakes and clearance angles are used at moderate to fast speeds.

Weldability

All common welding methods are excellent for welding this alloy. Brazing is not recommended.

Finishing

Polished finishes are good. Anodized appearances are only fair. Very good finishes are achieved with electroplated and chemical conversion coatings.

Corrosion Resistance

Alloy 359.0 has very high resistance to corrosion and to stress corrosion cracking.

Thermal Treatment

T6 Solution Heat Treatment & Precipitation
Soak at 1000F (538C) for 10–14 hr; quench in hot water (140–180F, 60–82C). Allow castings to remain at room temperature for 8+ hr, then age for 10–12 hr at 310F (154C).

T61 Solution Heat Treatment & Precipitation
Soak at 1000F (538C) for 10–14 hr; quench in hot water (140–180F, 60–82C). Allow castings to remain at room temperature for 8+ hr, then age for 10–12 hr at 310F (154C).

T62 Solution Heat Treatment & Precipitation
Soak at 1000F (538C) for 10–14 hr; quench in hot water (140–180F, 60–82C). Allow castings to remain at room temperature for 8+ hr, then age for 6–10 hr at 340F (171C).

Alloy 359.0 Typical Tensile Properties at Elevated Temperatures

Test Temp (°F)	Ultimate Strength (ksi)	Yield Strength (ksi)	Elongation (% in 4D)
359.0—T61			
75*	47	37	7.0
212*	45	38	6.0
300*	18	14	30.0
350**	33	28	15.0

* Properties after 10,000 hours at test temperatures.
** Properties after 100 hours at test temperatures.

Alloy 359.0

Used in: Sand and Permanent Mold Casting

Chemical Composition Limits

Si	Fe	Cu	Mn	Mg	Cr	Ni	Zn	Ag	Ti	Sn	Other Elements Each	Other Elements Total
8.5–9.5	0.20	0.20	0.10	0.50–0.70	–	–	0.10	–	0.20	–	0.05	0.15

Typical Mechanical Properties*

Casting Process and Temper	Tension: Ultimate Strength (ksi)	Tension: Yield Strength (Set 0.2%) (ksi)	Tension: Elongation (% in 2 in.)	Endurance Limit (ksi)
P.M.—T61	48	37	6.0	–
Sand** & P.M.—T62	50	42	5.5	16

* Mechanical properties of specimens cut from premium strength castings.
** Chilled dry sand.

Alloys 360.0 & A360.0

Where excellent castability, mechanical properties and corrosion resistance are required, miscellaneous thin-walled and intricate parts are typical applications for Alloys 360.0 and A360.0. Strength at elevated temperatures is excellent.

Some of the uses for these versatile alloys are outboard motor parts, instrument cases, cover plates, a wide variety of general-purpose castings, and aircraft and marine items of many types.

Castability

The excellent castability of Alloys 360.0 and A360.0 make them favorite die alloys for high-strength, thin-walled, complex castings.

They have excellent fluidity, pressure tightness and resistance to hot cracking. It should be noted, however, that these alloys do tend to stick in the die.

Machinability

Machinability of 360.0 and A360.0 is only fair, due partially to the abrasiveness caused by the high silicon content. Tungsten carbide wear-resistant tools (or better) are recommended.

Moderate to high machine speeds and feeds, tools ground with high rakes and clearances, and plenty of lubricant-coolant will also assist in satisfactorily machining these alloys.

Weldability

Like many of the other diecasting alloys, welding by standard methods is limited because of outgassing of porosity entrapped in conventional die castings. Brazing is not recommended.

Finishing

Electroplated finishes are excellent on Alloys 360.0 and A360.0. Chemical conversion coatings give good protection, and anodized appearance and mechanical finishes are rated good. The anodized color is gray.

Corrosion Resistance

The corrosion resistance of 360.0 and A360.0 is very good, being only slightly below that of 518.0. The value of chemical conversion coatings for protection is rated good.

Thermal Treatment

Stress Relieving
To reduce internal stress, heat at 350–500F (177–260C) for 4–6 hr. Cool in still air to room temperature.

O Annealing
To increase ductility, heat at 500–700F (260–371C) for 4–6 hr. Cool to room temperature in the furnace or in still air.

Alloys 360.0 & A360.0 Typical Tensile Properties at Elevated Temperatures*

Test Temp (°F)	Ultimate Strength (ksi)	Yield Strength (Set 0.2%) (ksi)	Elongation (% in 2 in.)
360.0 & A360.0—F			
75	47	25	3.0
212	44	25	2.0
300	35	24	4.0
400	22	14	8.0
500	12	7.5	20.0
600	7	4.5	35.0
700	4.5	3	40.0

* Lowest strengths during 10,000 hours of heating at test temperatures.

Alloys 360.0 & A360.0

Used in: Diecasting

Alloy 360.0—Chemical Composition Limits

Si	Fe*	Cu	Mn	Mg	Cr	Ni	Zn	Ag	Ti	Sn	Other Elements Each	Other Elements Total
9.0–10.0	2.0	0.6	0.35	0.40–0.60	–	0.50	0.50	–	–	0.15	–	0.25

Alloy A360.0—Chemical Composition Limits

Si	Fe*	Cu	Mn	Mg	Cr	Ni	Zn	Ag	Ti	Sn	Other Elements Each	Other Elements Total
9.0–10.0	1.3	0.6	0.35	0.40–0.60	–	0.50	0.50	–	–	0.15	–	0.25

Typical Mechanical Properties

Casting Process and Temper	Tension: Ultimate Strength (ksi)	Tension: Yield Strength (Set 0.2%) (ksi)	Tension: Elongation (% in 2 in.)	Shearing Strength (ksi)	Compressive Yield Strength (Set 0.2%) (ksi)	Brinell Hardness (500 kg load on 10 mm ball)	Endurance Limit (ksi)
Die 360.0—F	44	25	2.5	28	20	75	18
Die A360.0—F	46	24	3.5	26	20	75	20

* When produced to Fed Spec. QQ-A-371d, maximum Fe content will be 0.8%.

Alloys A380.0 & B380.0

These alloys are widely used for making general-purpose die castings.

They have good mechanical properties and are used to make housings for lawn mowers and radio transmitters, air brake castings, gear cases, and air-cooled cylinder heads.

Castability

Fluidity, pressure tightness and resistance to hot cracking are all very good.

Machinability

Although the machinability of A380.0 and B380.0 is considered good, carbide-tipped (or better) tooling is recommended because of the alloys' tendency to be abrasive.

Tools should be sharp with high rake and clearance angles, and machines should be operated at moderate to fast speeds and feeds.

Weldability

Welding of A380.0 and B380.0 is fair for all common types of welding processes. Brazing is not employed.

Finishing

Electroplating provides an excellent finish on these alloys. Mechanical and anodized finishes are good.

Corrosion Resistance

These alloys have only fair resistance to most corrosive atmospheres, and the protective value of chemical conversion coatings is poor.

Thermal Treatment

Stress Relieving
To reduce internal stresses, heat at 350–500F (177–260C) for 4–6 hr. Cool to room temperature in still air.

O Annealing
To increase ductility, heat at 500–700F (260–371C) for 4–6 hr. Cool to room temperature in furnace or in still air.

Alloys A380.0 & B380.0 Typical Tensile Properties at Elevated Temperatures*

Test Temp (°F)	Ultimate Strength (ksi)	Yield Strength (Set 0.2%) (ksi)	Elongation (% in 2 in.)
A380.0 & B380.0 —F			
75	48	24	3.0
212	45	24	4.0
300	34	22	5.0
400	24	16	8.0
500	13	8	20.0
600	7	4	30.0
700	4	2.5	35.0

* Lowest strengths during 10,000 hours of heating at test temperatures.

Alloys A380.0 & B380.0

Both alloys used in: Diecasting

Alloy A380.0—Chemical Composition Limits

Si	Fe	Cu	Mn	Mg	Cr	Ni	Zn	Ag	Ti	Sn	Other Elements Each	Other Elements Total
7.5–9.5	1.3	3.0–4.0	0.50	0.10	–	0.50	3.0	–	–	0.35	–	0.50

Alloy B380.0—Chemical Composition Limits

Si	Fe	Cu	Mn	Mg	Cr	Ni	Zn	Ag	Ti	Sn	Other Elements Each	Other Elements Total
7.5–9.5	1.3	3.0–4.0	0.50	0.10	–	0.50	1.0	–	–	0.35	–	0.50

Typical Mechanical Properties

Casting Process and Temper	Tension Ultimate Strength (ksi)	Tension Yield Strength (Set 0.2%) (ksi)	Tension Elongation (% in 2 in.)	Shearing Strength (ksi)	Brinell Hardness (500 kg load on 10 mm ball)	Endurance Limit (ksi)
Die—F	48	24	3.0	31	80	21

Alloys 383.0 & 384.0

These are high-strength alloys used for pistons and other components designed for severe service conditions.

They are excellent for castings with thin walls and large areas. Other uses include wringer frames, pulleys, automatic transmissions, air-cooled cylinder heads, and body and end shields for electric motors and toys.

Castability

Fluidity is excellent. Pressure tightness is very good. Resistance to hot cracking is good.

Machinability

This alloy has good machinability. Use of carbide-tipped (or better) tools is recommended to minimize tool wear.

Weldability

Welding by most of the common methods is fair. Brazing is not recommended.

Finishing

Polished finishes are good. Electroplate finishes are very good, but anodized finishes are only fair.

Corrosion Resistance

The corrosion resistance of Alloys 383.0 and 384.0 is good. In addition, chemical conversion coatings give very good added protection against corrosion.

Thermal Treatment

Stress Relieving

To reduce internal stresses, heat at 350–500F (177–260C) for 4–6 hr. Cool to room temperature in still air.

O Annealing

To increase ductility, heat at 500–700F (260–371C) for 4–6 hr. Cool to room temperature in the furnace or in still air.

Alloys 383.0 & 384.0 Typical Tensile Properties at Elevated Temperatures*

Test Temp (°F)	Ultimate Strength (ksi)	Yield Strength (Set 0.2%)(ksi)	Elongation (% in 2 in.)
384.0 —F			
75	48	24	2.5
383.0 —F			
75	45	22	3.5
212	46	25	1.0
300	38	24	2.0
400	26	18	6.0
500	14	9	25.0
600	7	4	45.0
700	4.5	2.5	45.0

* Lowest strengths during 10,000 hours of heating at test temperatures.

Alloys 383.0 & 384.0

Both alloys used in: Diecasting

Alloy 383.0—Chemical Composition Limits

Si	Fe	Cu	Mn	Mg	Cr	Ni	Zn	Ag	Ti	Sn	Other Elements Each	Other Elements Total
9.5–11.5	1.3	2.0–3.0	0.50	0.10	–	0.30	3.0	–	–	0.15	–	0.50

Alloy 384.0—Chemical Composition Limits

Si	Fe	Cu	Mn	Mg	Cr	Ni	Zn	Ag	Ti	Sn	Other Elements Each	Other Elements Total
10.5–12.0	1.3	3.0–4.5	0.50	0.10	–	0.50	3.0	–	–	0.35	–	0.50

Typical Mechanical Properties

Casting Process and Temper	Tension: Ultimate Strength (ksi)	Tension: Yield Strength (Set 0.2%) (ksi)	Tension: Elongation (% in 2 in.)	Shearing Strength (ksi)	Endurance Limit (ksi)
Die 383.0—F	45	22	3.5	29	20
Die 384.0—F	48	24	2.5	29	20

Alloys A390.0 & B390.0

These companion alloys are hypereutectic aluminum-silicon alloys. The optimum structure of either alloy should consist of fine, uniformly distributed primary silicon crystals in a eutectic matrix. Structural modification of primary silicon is assisted by refinement—adding phosphorus to the melt in the form of phosphorus-copper, nickel-phosphorus, phosphorus pentachloride or other phosphorus-bearing salts.

The low coefficient of thermal expansion, high hardness and good wear resistance of these alloys make them suitable for internal combustion engine pistons and blocks and cylinder bodies for compressors, pumps and brakes.

Castability

For Alloy B390.0, the die castability rating is very good, and relatively thin and intricate sections can be produced. Pressure tightness and resistance to hot cracking are very good. Resistance to die soldering is excellent.

For Alloy A390.0, permanent mold castability is very good. Sand castability is only good because the slower cooling rates adversely affect casting microstructure. Pressure tightness and resistance to hot cracking are very good. Gating designs for proper directional solidification and feeding are essential for sound castings. Pressure tightness is rated good.

Machinability

Tool life when machining A390.0 or B390.0 is at its best with adequate refinement of primary silicon crystals. Cutting tools of cast iron cutting grade carbide and M-7 tool steel may be acceptable, but the recommended cutting material is polycrystalline diamond.

Alloys A390.0 and B390.0 generally require a cutting fluid in most operations. Many of the commercial water-emulsion cutting fluids are satisfactory. Aside from tool wearing, A390.0 and B390.0 alloys have excellent machining characteristics. Chips are short and easy to remove, and high-quality surface finishes are easily generated.

Weldability

Weldability is fair—the best processes are inert gas arc or oxyacetylene welding. Brazing is not recommended.

Corrosion Resistance

Resistance to corrosion is good to very good, depending on the alloys' intended environments. Chemical conversion coatings increase corrosion resistance.

Thermal Treatment

T5 Precipitation Treatment
Soak at 450F (232C) for 8 hr; degrowth.

T6 Solution Heat Treatment (A390.0)
Soak at 925F (496C) for 8 hr; quench in boiling water; age at 350F (177C) for 8 hr.

T7 Solution Heat Treatment (A390.0)
Soak at 925F (496C) for 8 hr; quench in boiling water; age at 450F (232C) for 8 hr.

Alloys A390.0 & B390.0

Used in: Sand and Permanent Mold Casting

Alloy A390.0—Chemical Composition Limits

Si	Fe	Cu	Mn	Mg	Cr	Ni	Zn	Ag	Ti	Sn	Other Elements Each	Other Elements Total
16.0–18.0	0.50	4.0–5.0	0.10	0.45–0.65	–	–	0.10	–	0.20	–	0.10	0.20

Used in: Diecasting

Alloy B390.0—Chemical Composition Limits

Si	Fe	Cu	Mn	Mg	Cr	Ni	Zn	Ag	Ti	Sn	Other Elements Each	Other Elements Total
16.0–18.0	1.3	4.0–5.0	0.50	0.45–0.65	–	0.10	1.5	–	0.20	–	0.10	0.20

Typical Mechanical Properties

Casting Process and Temper	Tension: Ultimate Strength (ksi)	Tension: Yield Strength (Set 0.2%) (ksi)	Tension: Elongation (% in 2 in.)	Compressive Yield Strength (Set 0.2%) (ksi)	Brinell Hardness (500 kg load on 10 mm ball)	Endurance Limit (ksi)
Die B390—F	40.5	35.0	1	–	120	20.0
Die B390—T5	43.0	38.5	1	–	–	–
P.M. A390—F & T5	29.0	29.0	< 1	–	110	–
P.M. A390—T6	45.0	45.0	< 1	60.0	145	17.0
P.M. A390—T7	38.0	38.0	< 1	51.7	120	14.5
Sand A390—F & T5	26.0	26.0	< 1	–	100	–
Sand A390—T6	40.0	40.0	< 1	–	140	13.0
Sand A390—T7	36.0	36.0	< 1	–	115	–

Alloys 413.0 & A413.0

Alloys 413.0 and A413.0 are general-purpose alloys with good characteristics for casting large, intricate parts with thin sections. Mechanical properties are good.

The alloys are recommended for architectural, ornamental, marine, and food and dairy equipment applications. Other uses are for pressure-tight parts, large instrument cases, street lamp housings, typewriter frames, dental equipment and outboard motor pistons.

Castability

The fluidity, resistance to hot cracking and pressure tightness of Alloys 413.0 and A413.0 are all excellent.

Machinability

Because of the abrasiveness caused by the high silicon content of these alloys, tungsten carbide (or better) tooling is recommended. Otherwise, the alloys have good machinability.

Weldability

These alloys have good welding characteristics in arc, resistance or gas methods. Brazing is not recommended but is sometimes performed on a limited basis.

Finishing

Mechanical finishes, including polishing, are rated good. Anodizing appearance is poor; however, the chemical conversion coatings for protection are considered good. Electroplating can be accomplished readily.

Corrosion Resistance

Corrosion resistance of Alloys 413.0 and A413.0 are very good. Chemical conversion coatings are often used to further increase resistance.

Thermal Treatment

Stress Relieving
To reduce internal stresses, heat at 350–500F (177–260C) for 4–6 hr. Cool in still air.

O Annealing
To increase ductility, heat at 500–700F (260–371C) for 4–6 hr. Cool in the furnace or in still air.

Alloy 413.0 Typical Tensile Properties at Elevated Temperatures*

Test Temp (°F)	Ultimate Strength (ksi)	Yield Strength (Set 0.2%) (ksi)	Elongation (% in 2 in.)
413.0			
75	43	21	2.5
212	37	20	5.0
300	32	19	8.0
400	24	15	15.0
500	13	9	30.0
600	7	4.5	35.0
700	4.5	2.5	40.0

* Lowest strengths during 10,000 hours of heating at test temperatures.

Alloy 413.0 (formerly 13) & A413.0 (formerly A13)

Both alloys used in: Diecasting

Alloy 413.0—Chemical Composition Limits

Si	Fe	Cu	Mn	Mg	Cr	Ni	Zn	Ag	Ti	Sn	Other Elements Each	Other Elements Total
11.0–13.0	2.0	1.0	0.35	0.10	–	0.50	0.50	–	–	0.15	–	0.25

Alloy 413.0—Chemical Composition Limits

Si	Fe	Cu	Mn	Mg	Cr	Ni	Zn	Ag	Ti	Sn	Other Elements Each	Other Elements Total
11.0–13.0	1.3	1.0	0.35	0.10	–	0.50	0.50	–	–	0.15	–	0.25

Typical Mechanical Properties

Casting Process and Temper	Tension: Ultimate Strength (ksi)	Tension: Yield Strength (Set 0.2%) (ksi)	Tension: Elongation (% in 2 in.)	Shearing Strength (ksi)	Compressive Yield Strength (Set 0.2%) (ksi)	Brinell Hardness (500 kg load on 10 mm ball)	Endurance Limit (ksi)
Die 413.0—F	43	21	2.5	25	15	80	19
Die A413.0—F	42	19	3.5	25	–	80	19

Alloys 443.0, A443.0 & C443.0

The 443-type alloys are used widely for die castings, where above average ductility is required. They also are popular for sand and permanent mold casting, where moderate strength, excellent pressure tightness and castability are needed.

Die castings of these alloys are found in cookware, pipe fittings and many marine fittings. Permanent mold castings are used for refrigerator shelf brackets and fittings, tire molds, carburetor bodies, and cooking utensils. Sand castings are used in pipe fittings, food handling equipment, cooking utensils, architectural and ornamental parts, and marine fittings.

Castability

Diecasting
Alloy C443.0 is considered to have good fluidity for diecasting. Pressure tightness is good and resistance to hot cracking is very good.

Sand Casting
In sand casting, 443.0 and A443.0 are classed as having excellent pressure tightness and resistance to hot cracking. Solidification shrinkage tendencies are very low.

Permanent Mold Casting
Excellent resistance to hot cracking and a high degree of pressure tightness with very good resistance to shrinkage are characteristics of Alloy 443.0 for this casting method.

Machinability

Machining characteristics of die and permanent mold castings are fair, but they are poor in sand casting. When silicon particles are coarse, they are abrasive, and tools are worn down rapidly. Such wear can be minimized by the use of tungsten carbide tooling.

These alloys are often modified to reduce the size of silicon particles, which improves machining and ductility. Gumminess occurs in the as-cast condition due, somewhat, to the relatively soft nature of the alloys.

Weldability

The weldability of 443-type alloys is rated excellent for sand or permanent mold castings by all standard methods of welding, but weldability is only fair for die castings.

Finishing

Poor mechanical finishes, especially from polishing, are obtained on sand castings in these alloys, but fair polishing is

Alloys 443.0, A443.0 & C443.0 (formerly 43, 43 with 0.30 max Cu & A43)

Used in: Sand and Permanent Mold Casting

Alloy 443.0—Chemical Composition Limits

Si	Fe	Cu	Mn	Mg	Cr	Ni	Zn	Ag	Ti	Sn	Other Elements Each	Other Elements Total
4.5–6.0	0.8	0.6	0.50	0.05	0.25	–	0.50	–	0.25	–	–	0.35

Used in: Sand Casting

Alloy A443.0—Chemical Composition Limits

Si	Fe	Cu	Mn	Mg	Cr	Ni	Zn	Ag	Ti	Sn	Other Elements Each	Other Elements Total
4.5–6.0	0.8	0.30	0.50	0.05	0.25	–	0.50	–	0.25	–	–	0.35

Used in: Diecasting

Alloy C443.0—Chemical Composition Limits

Si	Fe	Cu	Mn	Mg	Cr	Ni	Zn	Ag	Ti	Sn	Other Elements Each	Other Elements Total
4.5–6.0	2.0	0.6	0.35	0.10	–	0.50	0.50	–	–	0.15	–	0.25

Typical Mechanical Properties

Casting Process and Temper	Tension: Ultimate Strength (ksi)	Tension: Yield Strength (Set 0.2%) (ksi)	Tension: Elongation (% in 2 in.)	Shearing Strength (ksi)	Compressive Yield Strength (Set 0.2%) (ksi)	Brinell Hardness (500 kg load on 10 mm ball)	Endurance Limit (ksi)
Sand A443.0—F	19	8	8.0	14	9	40	8
P.M. 443.0—F	23	9	10.0	16	9	45	8
Die C443.0—F	33	14	9.0	19	13	50	17

obtained on die and permanent mold castings. Sand, die and permanent mold castings take electroplating very well. Anodizing has a fair appearance on permanent mold castings but a poor one on sand and die castings.

Corrosion Resistance

The 443-type alloys are considered to have very good resistance to corrosion in sand and permanent mold castings and good resistance in die castings. Chemical conversion coating protection is good in die castings and very good in sand and permanent mold castings.

Thermal Treatment

Sand & Perm. Mold O Annealing (443.0 & A443.0)
To stress relieve and reduce growth, heat castings at 650F (343C) for 2 hr. Air cool to room temperature.

Die Casting Stress Relieving (C443.0)
Heat at 350–500F (177–260C) for 4–6 hr. Cool in still air.

Die Castings O Annealing (C443.0)
To increase ductility, heat at 500–700F (260–371C) for 4–6 hr. Cool in the furnace or still air.

Alloy A444.0

Alloy A444.0 was developed to provide a high level of impact properties and an unusual ability to absorb energy without rupture for use in highway bridge guard castings.

The alloy is of medium strength with very high elongation and is suited to such applications as outboard motor propellers, treadplates, cast furniture and bumper guards.

Castability

The alloy has excellent resistance to hot cracking and has excellent fluidity and solidification shrinkage characteristics. It is similar to 356-type alloys.

Machinability

The high ductility of this alloy gives it only fair machining characteristics. Sharp tools with high rakes and clearances are recommended.

Weldability

Common welding practices are excellent for joining this alloy. Brazing is not recommended.

Finishing

Fair finishes are obtained by polishing. The appearance of anodized finishes is gray and is rated only as fair.

Corrosion Resistance

The corrosion resistance of this alloy is rated very good in the solution heat treated condition.

Thermal Treatment

T4 Solution Heat Treatment
Hold at 990–1010F (532–543C) for 8–12 hr, then quench in 150–212F (66–100C) water.

Alloy A444.0 (formerly A344)

Used in: Sand and Permanent Mold Casting

Chemical Composition Limits

Si	Fe	Cu	Mn	Mg	Cr	Ni	Zn	Ag	Ti	Sn	Other Elements Each	Other Elements Total
6.5–7.5	0.20	0.10	0.10	0.05	–	–	0.10	–	0.20	–	0.05	0.15

Typical Mechanical Properties

Casting Process and Temper	Tension: Ultimate Strength (ksi)	Tension: Yield Strength (Set 0.2%) (ksi)	Tension: Elongation (% in 2 in.)	Shearing Strength (ksi)	Compressive Yield Strength (Set 0.2%) (ksi)	Brinell Hardness (500 kg load on 10 mm ball)	Endurance Limit (ksi)
Sand—F	21	9	9.0	–	–	–	–
Sand—T4	23	9	12.0	–	–	–	–
P.M.—F	24	11	13.0	–	–	–	–
P.M.—T4	23	10	23.0	16	11	45	8.0

Alloy 511.0

Because 511.0 gives a more uniform appearance after anodizing (its silicon content is about one-third that of Alloy 512.0), this modification of 514.0 is used in ornamental hardware and architectural castings. Improved foundry characteristics also are noted.

Castability

This alloy is rated poor to fair in fluidity, pressure tightness, hot cracking resistance and solidification shrinkage tendency.

Finishing

Alloy 511.0 is noted for its high-quality appearance after anodizing and is commonly used for sand castings that must color-match certain anodized wrought aluminum alloys.

Mechanical finishes are very satisfactory. Electroplating is only fair as a finish.

Machinability

The alloy has excellent machinability.

Weldability

Very good results are obtained with arc, gas and resistance welding methods. Brazing is not recommended.

Corrosion Resistance

Although it has excellent inherent resistance to corrosion, the alloy accepts chemical conversion coatings readily for maximum protection.

Thermal Treatment

O Annealing
To relieve stresses and reduce growth, heat at 650F (343C) for 2 hr. Air cool to room temperature.

Alloy 511.0 (formerly F514.0 & F214)

Used in: Sand Casting

Chemical Composition Limits

Si	Fe	Cu	Mn	Mg	Cr	Ni	Zn	Ag	Ti	Sn	Other Elements Each	Other Elements Total
0.30–0.7	0.50	0.15	0.35	3.5–4.5	–	–	0.15	–	0.25	–	0.05	0.15

Typical Mechanical Properties

Casting Process and Temper	Tension: Ultimate Strength (ksi)	Tension: Yield Strength (Set 0.2%) (ksi)	Tension: Elongation (% in 2 in.)	Shearing Strength (ksi)	Compressive Yield Strength (Set 0.2%) (ksi)	Brinell Hardness (500 kg load on 10 mm ball)	Endurance Limit (ksi)
Sand—F	21	12	3.0	17	13	50	8

Alloy 512.0

Increased silicon content gives Alloy 512.0 improved sand casting characteristics over 514.0, although the mechanical properties are reduced. Its good resistance to tarnishing and corrosion makes the alloy popular for castings such as cooking utensils and pipe fittings for marine applications as well as for general-use castings.

Castability

Alloy 512.0 has poor to fair fluidity and resistance to hot cracking. It is rated fair for pressure tightness and solidification shrinkage.

Finishing

Mechanical finishes are very good for this alloy. Chemical conversion coatings are excellent in appearance. Anodizing for appearance is very good. However, this alloy's silicon content (approximately 1.8%) is great enough to darken the casting to a gray color when it is anodized. Electroplating is considered fair.

Machinability

The machinability of 512.0 is excellent. Although standard tooling can be used successfully, common precautions recommended for all aluminum will improve results.

Weldability

Results from arc welding methods are excellent; resistance and gas methods are less satisfactory but still very good.

Corrosion Resistance

Alloy 512.0 shows excellent resistance to tarnishing or most corrosive atmospheres. Since it anodizes readily and is excellent for chemical coatings, its natural resistance can be improved greatly.

Thermal Treatment

O Annealing
To relieve stresses and reduce growth, heat at 650F (343C) for 2 hr. Air cool to room temperature.

Alloy 512.0 (formerly B514.0 & B214)

Used in: Sand Casting

Chemical Composition Limits

Si	Fe	Cu	Mn	Mg	Cr	Ni	Zn	Ag	Ti	Sn	Other Elements Each	Other Elements Total
1.4–2.2	0.6	0.35	0.8	3.5–4.5	0.25	–	0.35	–	0.25	–	0.05	0.15

Typical Mechanical Properties

Casting Process and Temper	Tension: Ultimate Strength (ksi)	Tension: Yield Strength (Set 0.2%) (ksi)	Tension: Elongation (% in 2 in.)	Shearing Strength (ksi)	Compressive Yield Strength (Set 0.2%) (ksi)	Brinell Hardness (500 kg load on 10 mm ball)	Endurance Limit (ksi)
Sand—F	20	13	2.0	17	14	50	8.5

Alloy 513.0

The addition of zinc to 513.0 makes the alloy popular for permanent mold casting of ornamental hardware and architectural fittings.

Castability

This alloy is rated poor to fair in pressure tightness and in fluidity. Alloy 513.0 has fair resistance to hot cracking, but its solidification shrinkage tendency is rated poor.

Machinability

This alloy has excellent machinability when sharp, polished tools with adequate rakes and clearances are used. Speeds and feeds should be moderate to rapid.

Weldability

Very good results are obtained using arc, resistance and gas welding methods. Brazing is not employed.

Finishing

Anodizing is excellent, both in appearance and protection. Mechanical finishes are excellent. Electroplating is rated poor.

Corrosion Resistance

Resistance to most common corrosive atmospheres and to many strong corrosive environments is very good. Chemical conversion coatings provide excellent protection.

Thermal Treatment

O Annealing
To relieve stresses and reduce growth of castings, heat at 650F (343C) for 2 hr. Air cool to room temperature.

Alloy 513.0 (formerly A514.0 & A214)

Used in: Permanent Mold Casting

Chemical Composition Limits

Si	Fe	Cu	Mn	Mg	Cr	Ni	Zn	Ag	Ti	Sn	Other Elements	
											Each	Total
0.30	0.40	0.10	0.30	3.5–4.5	–	–	1.4–2.2	–	0.20	–	0.05	0.15

Typical Mechanical Properties

Casting Process and Temper	Tension			Shearing Strength (ksi)	Compressive Yield Strength (Set 0.2%) (ksi)	Brinell Hardness (500 kg load on 10 mm ball)
	Ultimate Strength (ksi)	Yield Strength (Set 0.2%) (ksi)	Elongation (% in 2 in.)			
P.M.—F	27	16	7.0	22	17	60

Alloy 514.0

This alloy is used where excellent resistance to corrosion and tarnish are required. Alloy 514.0 has moderate strength and good ductility, and it anodizes readily.

Typical uses for this alloy include fittings for chemical and sewage use, dairy and food handling equipment, tire molds, cooking utensils, hardware, ornamental housings, and architectural applications.

Castability

The casting characteristics are not good, primarily because 514.0 is poor in pressure tightness and solidification shrinkage. Fluidity and resistance to hot cracking are rated poor to fair.

Machinability

The alloy has excellent machining qualities, especially when simple precautions, helpful in aluminum metal removal, are taken.

Weldability

Alloy 514.0 is rated very good for welding with arc, gas and resistance spot and seam methods.

Finishing

Anodizing appearances and protection are excellent on this alloy. The alloy takes an excellent polished finish, but electroplating is rated as poor.

Corrosion Resistance

Excellent resistance to all common and many forms of corrosive atmospheres is exhibited by this versatile alloy. In addition, chemical conversion coatings are well received and offer the maximum protection.

Thermal Treatment

O Annealing

To relieve stresses and reduce growth, heat at 650F (343C) for 2 hr. Air cool to room temperature.

Alloy 514.0 Typical Tensile Properties at Elevated Temperatures*

Test Temp (°F)	Ultimate Strength (ksi)	Yield Strength (Set 0.2%) (ksi)	Elongation (% in 2 in.)
	514.0—F		
75	25	12	9.0
212	24	12	9.0
300	22	12	7.0
400	18	12	9.0
500	13	8	12.0
600	9	4	17.0
700	5	2	35.0

* Lowest strengths during 10,000 hours of heating at test temperatures.

Alloy 514.0 (formerly 214)

Used in: Sand Casting

Chemical Composition Limits

Si	Fe	Cu	Mn	Mg	Cr	Ni	Zn	Ag	Ti	Sn	Other Elements Each	Other Elements Total
0.35	0.50	0.15	0.35	3.5–4.5	–	–	0.15	–	0.25	–	0.05	0.15

Typical Mechanical Properties

Casting Process and Temper	Tension: Ultimate Strength (ksi)	Tension: Yield Strength (Set 0.2%) (ksi)	Tension: Elongation (% in 2 in.)	Shearing Strength (ksi)	Compressive Yield Strength (Set 0.2%) (ksi)	Brinell Hardness (500 kg load on 10 mm ball)	Endurance Limit (ksi)
Sand—F	25	12	9.0	20	12	50	7

Alloy 518.0

Alloy 518.0 has excellent resistance to tarnish and corrosion, and high strength and ductility. However, it does not cast as well as Alloy 413.0 in small, intricate parts, nor does it have the casting qualities of 360.0 and A380.0. The alloy responds well to anodizing and mechanical finishing.

Typical uses for 518.0 include architectural and ornamental items; food and dairy equipment; escalator parts; conveyor components; aircraft and marine hardware and fittings; and similar items requiring high strength, ductility and resistance to corrosive environments.

Castability

Fluidity, resistance to hot cracking and pressure tightness all are poor. Problems may occur because of die soldering and because of the poor feeding characteristics of a simple binary composition. The alloy also exhibits high drossing tendencies.

Machinability

The machining qualities of 518.0 are excellent, probably the best of the diecasting alloys. Speeds and feeds can approach maximum, with sharp, polished, wear-resistant tools, high positive rakes and adequate coolant.

Weldability

Alloy 518.0 has only fair weldability when using the standard methods. However, when welding is necessary, satisfactory results can be obtained with careful procedures and suitable equipment. This alloy is not brazed.

Finishing

Alloy 518.0 is excellent for polished finishes. It takes anodizing very well, leaving a natural aluminum finish without discoloration. Chemical conversion coatings give excellent protection, better than that for Alloy 360.0. Electroplating characteristics are poor.

Corrosion Resistance

The highest corrosive resistant characteristics of the diecasting alloys are shown by 518.0. Furthermore, for severe environments, its excellent receptiveness to chemical conversion coatings gives it added protection, making it very popular for such applications.

Thermal Treatment

Stress Relieving

To reduce internal stresses, heat at 350–500F (177–260C) for 4–6 hr. Cool in still air.

O Annealing

To increase ductility, heat at 500–700F (260–371C) for 4–6 hr. Cool in furnace or still air.

Alloy 518.0 Typical Tensile Properties at Elevated Temperatures*

Test Temp (°F)	Ultimate Strength (ksi)	Yield Strength (Set 0.2%) (ksi)	Elongation (% in 2 in.)
518.0—F			
75	45	27	8.0
212	40	25	8.0
300	32	21	25.0
400	21	15	40.0
500	13	9	45.0
600	8.5	4.5	45.0
700	5	2.5	45.0

* Lowest strengths during 10,000 hours of heating at test temperatures.

Alloy 518.0 (formerly 218)

Used in: Diecasting

Chemical Composition Limits

Si	Fe	Cu	Mn	Mg	Cr	Ni	Zn	Ag	Ti	Sn	Other Elements Each	Other Elements Total
0.35	1.8	0.25	0.35	7.5–8.5	–	0.15	0.15	–	–	0.15	–	0.25

Typical Mechanical Properties

Casting Process and Temper	Tension: Ultimate Strength (ksi)	Tension: Yield Strength (Set 0.2%) (ksi)	Tension: Elongation (% in 2 in.)	Shearing Strength (ksi)	Compressive Yield Strength (Set 0.2%) (ksi)	Brinell Hardness (500 kg load on 10 mm ball)	Endurance Limit (ksi)
Die—F	45	27	8.0	29	23	80	20

Alloy 520.0

Requiring special foundry practice, Alloy 520.0 is used principally for sand casting where excellent machinability, resistance to corrosion, and high strength and elongation are required. In strength and corrosion resistance, 520.0 rates as one of the highest of all sand casting alloys.

Parts that must be shock resistant as well as strong are often made from this alloy. The alloy is not recommended for use at operating temperatures exceeding about 250F (121C), however.

Typical applications for 520.0 are in aircraft fittings, railway passenger car frames, truck and bus frame sections, levers, brackets, and similar parts that will be subjected to severe usage.

Castability

The alloy has poor to fair fluidity and resistance to hot cracking. Pressure tightness and solidification shrinkage tendency are poor.

Alloy 520.0 has a high tendency to microporosity and sand reaction. To get completely sound castings, foundries must use the best foundry practices.

Machinability

The machinability of Alloy 520.0 in the T4 temper is rated excellent.

Weldability

Resistance spot and seam welding are very good methods for welding this alloy. Arc welding produces good results; gas methods are fair. Brazing is not usually used.

Finishing

Excellent finishes are obtained on the alloy by mechanical polishing or anodizing. Chemical conversion coatings result in high protection. Electroplating gives fair results.

Corrosion Resistance

The inherent corrosion resistant qualities of 520.0 are excellent. Additional protection may be provided by chemical conversion coatings, which are received very well.

Thermal Treatment

T4 Solution Heat Treatment
Soak at 810F (432C) for 18 hr; quench in water at 150–212F (66–100C) for a controlled time of 10–20 sec.

Alloy 520.0 Typical Tensile Properties at Elevated Temperatures*

Test Temp (°F)	Ultimate Strength (ksi)	Yield Strength (Set 0.2%) (ksi)	Elongation (% in 2 in.)
520.0—T4			
75	46	25	12.0
212	40	23	12.0
300	36	19	13.0
400	23	12	30.0
500	15	7	50.0
600	11	3.5	60.0
700	6.5	2	—

* After prolonged heating at test temperatures.

Alloy 520.0 (formerly 220)

Used in: Sand Casting

Chemical Composition Limits

Si	Fe	Cu	Mn	Mg	Cr	Ni	Zn	Ag	Ti	Sn	Other Elements Each	Other Elements Total
0.25	0.30	0.25	0.15	9.5–10.6	–	–	0.15	–	0.25	–	0.05	0.15

Typical Mechanical Properties

Casting Process and Temper	Tension			Shearing Strength (ksi)	Compressive Yield Strength (Set 0.2%) (ksi)	Brinell Hardness (500 kg load on 10 mm ball)	Endurance Limit (ksi)
	Ultimate Strength (ksi)	Yield Strength (Set 0.2%) (ksi)	Elongation (% in 2 in.)				
Sand—T4	48	26	16.0	34	27	75	8

Alloy 535.0

Alloy 535.0 is an aluminum-magnesium type alloy possessing a high and stable combination of strength, shock resistance and ductility. It is ideally suited for parts in instruments and computing devices where dimensional stability is of major importance.

Fatigue failures of many castings produced from heat treated aluminum alloys have been traced to residual internal stresses caused by the quenching that follows the solution heat treatment. Heat treating also often causes warping. Elimination of these causes of casting defects as well as the expense of heat treating are attractive features of this alloy.

When heated in the as-cast condition to 450F (232C), some silicon alloys are subject to a volume increase of 0.001 in./in. Although stress relief can be obtained by heat treatments, low physical properties often result from the annealing. Alloy 535.0, when heated to 700–800F (371–427C) for 5 hr and cooled in still air, becomes dimensionally stable and, at the same time, retains its full hardness, physical properties and machinability. Because the alloy does not require stress-inducing solution heat treatments, it is often found that no stress relieving is needed, even in those parts calling for high dimensional stability.

In addition to the high ductility and tensile strength of 535.0, the Charpy impact is from 10 to 12 lb, which makes it suitable for shock-resistant applications. Brackets, C-clamps and machined parts that need strength as well as impellers, optical equipment and similar applications requiring a high polish or anodized finish are typical uses.

In many cases, this alloy has replaced gray iron and malleable iron because its use reduces weight without sacrificing strength.

Castability

The alloy has fair casting characteristics and attains its high physical and mechanical properties immediately upon casting. This fact is important to remember because most high-strength aluminum alloys change their properties as a result of age hardening. These properties remain constant for 535.0 within the entire range of temperatures from –76 to +225F (–60 to +107C). The alloy shows fair fluidity with little serious tendency toward hot tearing.

Machinability

The machinability of this alloy is considered excellent. Despite its high ductility, there is little tool drag. Machining is possible immediately following casting.

Because of its high dimensional stability due to the almost complete absence of precipitation, dimensional tolerances are maintained up to 250F (121C). At higher temperatures, the growth is less than one-third that of Alloy 356.0, for instance. An excellent surface finish is produced, especially when machined with carbide tools at maximum speeds.

Weldability

Alloy 535.0 can be welded by any of the inert gas methods or by shielded arc methods using filler material of 5356, 5183 or 1100 welding wire or rod or using clean scraps or cast pieces of 535.0.

Fluxes generally should be avoided. Consequently, fusion welding processes requiring fluxes are not recommended.

Finishing

Because of this alloy's high resistance to corrosion, castings produced from it will not require any further surface treatment for most applications. However, it can be satin-anodized or chrome-plated for additional corrosion protection.

Alloy 535.0 also can be anodized in a wide variety of colors. Lack of silicon in the alloy gives highly satisfactory coloration.

Alloy 535.0 (formerly Almag 35)

Used in: Sand Casting

Chemical Composition Limits

Si	Fe	Cu	Mn	Mg	Cr	Ni	Zn	Ag	Ti	Sn	Other Elements	
											Each*	Total
0.15	0.15	0.05	0.10–0.25	6.2–7.5	–	–	–	–	0.10–0.25	–	0.05	0.15

Typical Mechanical Properties

Casting Process and Temper	Tension			Shearing Strength (ksi)	Compressive Yield Strength (Set 0.2%) (ksi)	Brinell Hardness (500 kg load on 10 mm ball)	Endurance Limit (ksi)
	Ultimate Strength (ksi)	Yield Strength (Set 0.2%) (ksi)	Elongation (% in 2 in.)				
Sand—F	40	20	13	27.45	35.5	70	10

* Beryllium 0.003–0.007; boron, 0.005 max.

Corrosion Resistance

Because of the almost complete absence of heavy metals, the corrosion resistance of Alloy 535.0 is extremely high. In addition, its relatively high magnesium content gives it further protection against corrosion from mild alkalis or salt spray.

Thermal Treatment

Stress Relieving for Dimensional Stability
Heat 535.0 castings to 700–800F (371–427C) for 5 hr. Cool in still air.

Alloy 535.0 Typical Tensile Properties at Elevated Temperatures

Test Temp (°F)	Tensile Strength (ksi)	Elongation (% in 2 in.)
300	37.8	11.3
350	34.4	13.8
400	32.3	14.0
500	26.7	12.9
600	20.75	13.0
700	15.35	12.0

Alloy 710.0

This self-aging alloy is used regularly for miscellaneous general purpose castings that require subsequent brazing.

Alloy 710.0 has high strength and ductility without heat treatment, although ductility is reduced on room temperature aging.

Castability

This alloy shows fair castability with greater shrinkage tendencies than shown by aluminum-copper alloys. Pressure tightness is good, but resistance to hot cracking is poor.

Machinability

Machinability is excellent, particularly when sharp tools with high rake and clearance angles are used at moderate to fast speeds and feeds.

Weldability

Alloy 710.0 has fair weldability when gas or arc methods are used. It is excellent for brazing. Recommended brazing filler metal is welding and brazing rod 4047.

Finishing

Mechanical finishes are excellent and electroplate and anodized finishes are very good.

Corrosion Resistance

Very good resistance to most common corrosive environments is shown by this alloy. Good additional protection is provided by chemical conversion coatings.

Thermal Treatment

O Annealing
To relieve stresses and reduce growth in castings made of 710.0 alloy, heat at 650F (343C) for 2 hr. Air cool to room temperature.

Alloy 710.0 (formerly A712.0 & A612)

Used in: Sand Casting

Chemical Composition Limits

Si	Fe	Cu	Mn	Mg	Cr	Ni	Zn	Ag	Ti	Sn	Other Elements Each	Other Elements Total
0.15	0.50	0.35– 0.65	0.05	0.6-0.8	–	–	6.0–7.0	–	0.25	–	0.05	0.15

Typical Mechanical Properties

Casting Process and Temper	Tension: Ultimate Strength (ksi)	Tension: Yield Strength (Set 0.2%) (ksi)	Tension: Elongation (% in 2 in.)	Shearing Strength (ksi)	Compressive Yield Strength (Set 0.2%) (ksi)	Brinell Hardness (500 kg load on 10 mm ball)	Endurance Limit (ksi)
Sand—F	35*	25*	5.0*	26	25	75*	8

* Tests made 30 days after casting.

Alloy 711.0

Alloy 711.0 has good strength and ductility without heat treatment. This alloy is widely used for casting torque converter impeller blades and furniture parts requiring subsequent brazing.

Castability

The castability of Alloy 711.0 is considered fair. Pressure tightness and fluidity are rated fair. Resistance to hot cracking and the tendency toward solidification shrinkage are rated poor.

Machinability

This alloy has excellent machinability. Although standard, high-carbon tools can be used satisfactorily, best results and greater economy are obtained by using wear-resistant tools at moderate to fast speeds and feeds.

Weldability

Most common welding methods produce good joints. Alloy 711.0 is recommended for brazing. Welding and brazing rod 4047 is recommended for use in brazing.

Finishing

Most standard finishes for aluminum may be used successfully. Mechanical and anodized finishes are excellent.

Corrosion Resistance

This brazing alloy has very good resistance to most common sources of corrosion. The additional protection obtained from chemical conversion coatings also is very good.

Alloy 711.0 (formerly C712.0 & C612)

Used in: Permanent Mold Casting

Chemical Composition Limits

Si	Fe	Cu	Mn	Mg	Cr	Ni	Zn	Ag	Ti	Sn	Other Elements Each	Other Elements Total
0.30	0.7–1.4	0.35–0.65	0.05	0.25–0.45	–	–	6.0–7.0	–	0.20	–	0.05	0.15

Typical Mechanical Properties

Casting Process and Temper	Tension: Ultimate Strength (ksi)	Tension: Yield Strength (Set 0.2%) (ksi)	Tension: Elongation (% in 2 in.)	Brinell Hardness (500 kg load on 10 mm ball)	Endurance Limit (ksi)
P.M.—F	35*	18*	8.0*	70*	11

* Tests made approximately 30 days after casting.

Alloy 712.0

Alloy 712.0 is employed when a combination of good mechanical properties without heat treatment is needed. It also shows good shock and corrosion resistance and good machinability and dimensional stability.

No distortion is exhibited when Alloy 712.0 is heated. After brazing, the alloy will regain its original strength by natural aging.

The alloy is used widely for marine castings, farm machinery, machine tool parts, and other applications where the part must have good strength or impact resistance.

Castability

Alloy 712.0 has fair to good castability. Although its pressure tightness and resistance to hot cracking are only fair, the alloy's fluidity and solidification shrinkage tendency are rated as good.

Machinability

The machining characteristics of this alloy are excellent. Fast speeds and feeds and wear-resistant tools correctly ground for cutting aluminum are recommended.

Finishing

Polished finishes are good. The anodized finish is very good and the color is medium golden yellow.

Corrosion Resistance

The alloy has good natural resistance to corrosion and good additional protection is received from chemical conversion coatings.

Weldability

Welding characteristics are fair for most common methods. The alloy is recommended for brazing.

Thermal Treatment

O Annealing

To relieve stresses, heat at 650F (343C) for 2 hr. Air cool to room temperature.

Alloy 712.0 (formerly D712.0, D612 & 40E)

Used in: Sand Casting

Chemical Composition Limits

Si	Fe	Cu	Mn	Mg	Cr	Ni	Zn	Ag	Ti	Sn	Other Elements Each	Other Elements Total
0.30	0.50	0.25	0.10	0.50–0.65	0.40–0.60	–	5.0–6.5	–	0.15–0.25	–	0.05	0.20

Typical Mechanical Properties

Casting Process and Temper	Tension: Ultimate Strength (ksi)	Tension: Yield Strength (Set 0.2%) (ksi)	Tension: Elongation (% in 2 in.)	Shearing Strength (ksi)	Compressive Yield Strength (Set 0.2%) (ksi)	Brinell Hardness (500 kg load on 10 mm ball)	Endurance Limit (ksi)
Sand—F	35*	25*	5.0*	26	17	75*	9

* Tests made 30 days after casting.

Alloy 713.0

High-strength Alloy 713.0 is an aluminum-zinc-magnesium alloy that ages at room temperature and is ideal for castings too intricate for heat treatment. Its high yield strength, ductility, impact resistance and other favorable final mechanical properties are achieved by self-aging for 21 days at room temperature.

When faster aging is necessary, the same final properties can be obtained by heating castings for 16 hr at 250F (121C). Because no solution heat treating, quenching or artificial aging strengthening treatments (which often produce internal stresses) are required, larger and more intricate castings can be produced with 713.0 than with most other alloys that are not self-aging. It achieves tensile and yield strengths comparable to non-self-aging alloys.

Alloy 713.0 is dimensionally stable and will not expand or grow during aging as is common with other alloys containing copper and silicon as major alloying elements. After brazing, natural aging will return the alloy's original strength.

Automotive parts, pumps, trailer parts and mining equipment are regularly cast of this alloy.

Castability

Castability of 713.0 is fair to good; it is similar to that of other zinc-magnesium type aluminum alloys. No particular foundry techniques other than normal good practices are necessary.

This alloy has a narrower solidification range than most common foundry alloys. Heavier gates and risers should be used to compensate for its higher shrinkage rate. When gates and risers are adequate, pressure-tight parts can be cast successfully.

Machinability

Alloy 713.0 has excellent machining characteristics.

Weldability

Common welding methods are fair for joining. The alloy is recommended for brazing, and techniques standard for brazing wrought alloys can be employed.

Finishing

Excellent finishes are produced when Alloy 713.0 is polished. The alloy also takes a good anodized finish that is light golden yellow in appearance.

Corrosion Resistance

Corrosion resistance of this alloy is rated good.

Thermal Treatment

T5 Artificial Aging
Final properties approximately equal to those attained after 21 days of natural aging at room temperature can be obtained faster by artificial aging.

The artificial aging procedure consists of heating at 250F (121C) for 16 hr.

Alloy 713.0 (formerly 613, Tenzaloy)

Used in: Sand and Permanent Mold Casting

Chemical Composition Limits

Si	Fe	Cu	Mn	Mg	Cr	Ni	Zn	Ag	Ti	Sn	Other Elements	
											Each	Total
0.25	1.1	0.40–1.0	0.6	0.20–0.50	0.35	0.15	7.0–8.0	–	0.25	–	0.10	0.25

Typical Mechanical Properties

Casting Process and Temper	Tension			Brinell Hardness (500 kg load on 10 mm ball)
	Ultimate Strength (ksi)	Yield Strength (Set 0.2%) (ksi)	Elongation (% in 2 in.)	
Sand A-C	26–29	12–16	6–10	50–60
Sand —F	32	22	3.0	74
Sand —T5	32	22	3.0	74
P.M.—T5	32	–	4.0	–

Alloys 771.0 & 772.0

Alloys 771.0 and 772.0 are high-strength, shock-resistant aluminum sand casting alloys that develop a high combination of physical and mechanical properties in the as-cast and room-temperature-aged conditions. Without heat treatment, mechanical properties for both alloys are superior to many conventional aluminum casting alloys after solution heat treatment.

For castings that may require subsequent straightening, 772.0 is very suitable. The day a casting is made, immediately after a 700F (371C) stress-relieving heat treatment, or following a brazing treatment at 1125F (607C), the low yield strength of about 16,000 psi readily facilitates the straightening of a part.

These alloys are completely dimensionally stable to 0.00001 in./in. after stress relieving. Those parts requiring a high degree of dimensional stability may be stress relieved by heat treatment at 700F (371C) for 4 hr, followed by cooling in still air. No subsequent treatment is required.

Castability

The castability of Alloys 771.0 and 772.0 is considered good. Fluidity and resistance to solidification shrinkage are above average for the 7XX system.

Machinability

The machinability of both alloys is classed as excellent. Both alloys are dimensionally stable and machinable on the day they are cast or stress relieved. Some self-aging alloys require an aging period of up to 21 days before being dimensionally stable or hard enough for machining.

Because no silicon is present, the life of cutting tools is substantially extended, and machining operations can be performed at maximum speeds. Although standard tools may be used, high rake and clearance angles will aid in machining operations.

Brazing

The alloys, in contrast to most others, are not permanently affected by high temperatures (1125F, 607C) of dip or furnace brazing. After brazing, the castings undergo a room temperature aging cycle similar to the regular stress-relieving treatment at 700F (371C) to restore them to their full physical and mechanical properties.

When fully aged after brazing, castings made from these alloys are dimensionally stable and completely stress relieved and have an excellent combination of toughness and high strength to withstand heavy shock, high static, and severe fatigue or endurance loading.

Finishing

Polishing castings of these alloys produces a fine chrome-white finish without the blue tint of many conventional alloys containing copper or silicon. They have unusually low drag resistance and will not smear on a polishing wheel.

The alloys anodize to a satin white finish when the sulfuric acid process is used and dye brilliantly in pastel shades. Because of such qualities, these alloys are used widely for ornamental or decorative purposes.

Corrosion Resistance

Resistance to most forms of corrosion is excellent. In accelerated salt spray tests, the alloys proved to have a resistance equal to that of high-purity 443.0. In addition, 771.0 and 772.0 retain the chrome-white color under corrosive conditions, an advantage for architectural and ornamental castings that are exposed to the atmosphere.

Alloys 771.0 (formerly 71A) & 772.0 (formerly B771.0, precedent 71B)

Both alloys used in: Sand Casting

Alloy 771.0—Chemical Composition Limits

Si	Fe	Cu	Mn	Mg	Cr	Ni	Zn	Ag	Ti	Sn	Other Elements	
											Each	Total
0.15	0.15	0.10	0.10	0.8–1.0	0.06—0.20	–	6.5–7.5	–	0.10—0.20	–	0.05	0.15

Alloy 772.0—Chemical Composition Limits

Si	Fe	Cu	Mn	Mg	Cr	Ni	Zn	Ag	Ti	Sn	Other Elements	
											Each	Total
0.15	0.15	0.10	0.10	0.6–0.8	0.06– 0.20	–	6.0–7.0	–	0.10– 0.20	–	0.05	0.15

Thermal Treatment

The heat treating of these alloys is often varied to attain special properties. Heat treating specifications should conform to the recommendations of the producer of the alloys. Those given below are only general.

T6 Solution Heat Treatment & Precipitation
Solution heat treatment is critical. Careful composition and temperature control are required in order to attain adequate solution and still avoid incipient melting. Soak at 1080–1100F (582–593C) for 6 hr; quench in water at 150–212F (66–100C). Age at 265F (129C) for 3 hr.

Alloys 771.0 & 772.0

Typical Mechanical Properties

Age of Casting	Tension			Brinell Hardness (500 kg load on 10 mm ball)	Tension			Brinell Hardness (500 kg load on 10 mm ball)
	Ultimate Strength (ksi)	Yield Strength (Set 0.2%) (ksi)	Elongation (% in 2 in.)		Ultimate Strength (ksi)	Yield Strength (Set 0.2%) (ksi)	Elongation (% in 2 in.)	
	Alloy 771.0 As Cast				Alloy 772.0 As Cast			
1 day	36	20	11	70	34	16	15	84
1 month	43	34	4	90	39	31	8	85
3 months	44	36	3	93	40	32	7	88
6 months	44	36	3	93	40	32	7	88
1 year	44	36	3	93	40	32	7	88
	Alloy 771.0 Stress Relieved, 700F				Alloy 772.0 Stress Relieved, 700F			
1 day	32	16	12	61	30	15	14	60
4 days	37	24	8	75	35	22	9	72
1 week	38	26	7	78	37	25	8	76
2 weeks	39	27	7	79	38	27	7	77
1 month	40	30	5	89	39	28	6	84
2 months	42	33	4	90	40	30	6	87
3 months	43	33	3	96	40	31	6	90
6 months	43	33	3	96	40	31	6	90
1 year	43	33	3	96	40	31	6	90
	Alloy 771.0 —T6				Alloy 772.0 —T6			
1 day	50	40	9	93	45	35	10	89
6 months	50	40	9	93	45	35	10	89
1 year	50	40	9	93	45	35	10	89

Minimum Properties

Age of Casting	Tension			Brinell Hardness (500 kg load on 10 mm ball)	Tension			Brinell Hardness (500 kg load on 10 mm ball)
	Ultimate Strength (ksi)	Yield Strength (Set 0.2%) (ksi)	Elongation (% in 2 in.)		Ultimate Strength (ksi)	Yield Strength (Set 0.2%) (ksi)	Elongation (% in 2 in.)	
	Alloy 771.0 As Cast				Alloy 772.0 As Cast			
1 day	32	16	10	66	25	13	11	55
1 month	39	32	2	88	36	26	5	80
1 year	37	28	2	88	33	2	5	80
	Alloy 771.0 Stress Relieved, 700F				Alloy 772.0 Stress Relieved, 700F			
1 day	28	14	10	58	25	13	11	55
1 month	36	27	2	88	36	26	5	80
1 year	37	28	2	88	37	27	5	80
	Alloy 771.0 —T6				Alloy 772.0 —T6			
1 day	44	37.6	6	85	42	32	6	83

Alloy 850.0

Alloys 850.0, 851.0 and 852.0 are most commonly used to produce bearings. These alloys all have excellent bearing qualities and good compressive yield strengths.

Major applications for Alloy 850.0 include bushings and journal bearings for railroads.

Castability

Alloy 850.0 has fair to poor castability. Fluidity and solidification shrinkage tendency are rated fair, but pressure tightness and resistance to hot cracking are poor.

Machinability

Machining characteristics are rated excellent for 850.0.

Finishing

Mechanical polishing produces very good finishes with this alloy. Other finishing processes are seldom used for bearings and bushings.

Weldability

Arc and gas welding produce only fair results with 850.0.

Corrosion Resistance

The corrosion resistance is good for this alloy.

Thermal Treatment

Sand & Perm. Mold—T5 Precipitation Treatment
Age artificially to increase strength by heating at 430F (221C) for 7–9 hr.

Alloy 850.0 (formerly 750)

Used in: Sand and Permanent Mold Casting

Chemical Composition Limits

Si	Fe	Cu	Mn	Mg	Cr	Ni	Zn	Ag	Ti	Sn	Other Elements Each	Other Elements Total
0.7	0.7	0.7–1.3	0.10	0.10	–	0.7–1.3	–	–	0.20	5.5–7.0	–	0.30

Typical Mechanical Properties

Casting Process and Temper	Tension: Ultimate Strength (ksi)	Tension: Yield Strength (Set 0.2%) (ksi)	Tension: Elongation (% in 2 in.)	Shearing Strength (ksi)	Compressive Yield Strength (Set 0.2%) (ksi)	Brinell Hardness (500 kg load on 10 mm ball)	Endurance Limit (ksi)
Sand—T5	20	11	8.0	14	11	45	–
P.M.—T5	23	11	12.0	15	11	45	9

Alloy 851.0

Alloy 851.0, with less nickel and a great deal more silicon, has better castability and less hot shortness than does 850.0, but it still shows the excellent bearing qualities and good compressive yield strength of 850.0. Fabricating characteristics and corrosion resistance for Alloy 851.0 are the same as for Alloy 850.0. Rolling mill bearings are typical uses for the alloy.

Machinability

Machining characteristics are excellent for this alloy.

Finishing

Mechanical polishing produces very good finishes with 851.0. Other finishing processes are seldom used for bearings and bushings.

Weldability

Arc and gas welding produce only fair results with 851.0.

Corrosion Resistance

The alloy shows good corrosion resistance.

Thermal Treatment

Sand & Perm. Mold—T5 Precipitation Treatment
Age artificially to increase strength by heating at 430F (221C) for 7–9 hr.

Alloy 851.0 (formerly A850.0 & A750)

Used in: Sand and Permanent Mold Casting

Chemical Composition Limits

Si	Fe	Cu	Mn	Mg	Cr	Ni	Zn	Ag	Ti	Sn	Other Elements Each	Other Elements Total
2.0–3.0	0.7	0.7–1.3	0.10	0.10	–	0.30–0.7	–	–	0.20	5.5–7.0	–	0.30

Typical Mechanical Properties

Casting Process and Temper	Tension: Ultimate Strength (ksi)	Tension: Yield Strength (Set 0.2%) (ksi)	Tension: Elongation (% in 2 in.)	Shearing Strength (ksi)	Compressive Yield Strength (Set 0.2%) (ksi)	Brinell Hardness (500 kg load on 10 mm ball)	Endurance Limit (ksi)
Sand—T5	20	11	5.0	14	11	45	–
P.M.—T5	20	11	5.0	14	11	45	9

Alloy 852.0

The castability and fabricating characteristics of Alloy 852.0 are similar to those of 850.0.

Castability

Alloy 852.0 has fair to poor castability. Fluidity and solidification shrinkage tendency are rated fair, but pressure tightness and resistance to hot cracking are poor.

Machinability

Machining characteristics are rated excellent for 852.0.

Finishing

Mechanical polishing produces very good finishes with this alloy. Other finishing processes are seldom used for bearings and bushings.

Weldability

Arc and gas welding produce only fair results with 852.0.

Corrosion Resistance

The corrosion resistance is good for this alloy.

Thermal Treatment

Sand & Perm. Mold Precipitation Treatment
To increase strength, age artificially by heating at 430F (221C) for 7–9 hr.

Thermal Treatment

Sand & Perm. Mold—T5 Heat Treatment
Age at 430F (221C) for 8 hr.

Sand & Perm. Mold—T61 Solution Heat Treatment
Soak at 900F (482C) for 15 hr; quench in hot water (212F, 100C). Age at 430F (221C) for 8 hr.

Sand & Perm. Mold—T61A Solution Heat Treatment
Soak at 900F (482C) for 15 hr; fan quench. Age at 430F (221C) for 8 hr.

Sand & Perm. Mold—T7 Solution Heat Treatment
Soak at 900F (482C) for 15 hr; quench in hot water (212F, 100C). Age at 430F (221C) for 8 hr.

Alloy 852.0 (formerly B850.0 & B750)

Used in: Sand and Permanent Mold Casting

Chemical Composition Limits

Si	Fe	Cu	Mn	Mg	Cr	Ni	Zn	Ag	Ti	Sn	Other Elements Each	Other Elements Total
0.40	0.70	1.7–2.3	0.10	0.6–0.9	–	0.9–1.5	–	–	0.20	5.5–7.0	–	0.30

Typical Mechanical Properties

Casting Process and Temper	Tension: Ultimate Strength (ksi)	Tension: Yield Strength (Set 0.2%) (ksi)	Tension: Elongation (% in 2 in.)	Shearing Strength (ksi)	Compressive Yield Strength (Set 0.2%) (ksi)	Brinell Hardness (500 kg load on 10 mm ball)	Endurance Limit (ksi)
Sand—T5	27	22	2.0	18	22	65	10
P.M.—T5	32	23	5.0	21	23	70	11

Alloy 853.0

The experimental alloy 853.0 has been used to a limited extent for heavy-duty bushing applications in diesel engines. By comparison to 850-type aluminum bearing alloys, the good performance in a master bearing-knuckle pin bushing has been attributed to high hardness and the chemical composition of 853.0.

Although there is not sufficient production test data to provide typical or guaranteed minimum property values, the results of laboratory tests show approximate levels of properties for test bars of 853.0 cast in sand and permanent molds.

The T5 temper has not been used because the low strength is unsatisfactory for heavy-duty diesel engine bushings. The T7 temper has shown good performance, but the higher hardness of the T61 temper may help avoid failures under conditions of occasional overloads or failure of lubrication.

Thermal Treatment

Sand & Perm. Mold—T5 Heat Treatment
Age at 430F (221C) for 8 hr.

Sand & Perm. Mold—T61 Solution Heat Treatment
Soak at 900F (482C) for 15 hr; quench in hot water (212F, 100C). Age at 430F (221C) for 8 hr.

Sand & Perm. Mold—T61A Solution Heat Treatment
Soak at 900F (482C) for 15 hr; fan quench. Age at 430F (221C) for 8 hr.

Sand & Perm. Mold—T7 Solution Heat Treatment
Soak at 900F (482C) for 15 hr; quench in hot water (212F, 100C). Age at 430F (221C) for 8 hr.

Alloy 853.0 (formerly XC750)

Used in: Sand and Permanent Mold Casting

Chemical Composition Limits

Si	Fe	Cu	Mn	Mg	Cr	Ni	Zn	Ag	Ti	Sn	Other Elements Each	Other Elements Total
5.5–6.5	0.7	3.0–4.0	0.5	–	–	–	–	–	0.20	5.5–7.0	–	0.30

Typical Mechanical Properties

Casting Process and Temper	Tension			Brinell Hardness
	Ultimate Strength (ksi)	Yield Strength (ksi)	Elongation (%)	
Sand—T5	26	15	2.0	63
P.M.—T5	25	13	2.5	64
Sand—T61	38	36	0.5	100
P.M.—T61	40	33	2.0	95
Sand—T61A	36	30	1.5	90
P.M.—T61A	36	24	5.0	80
Sand—T7	34	32	1.0	90
P.M.—T7	38	30	2.0	85

CHAPTER 4 APPENDIX

The following is reprinted with permission from the "Pink Sheets" published by the Aluminum Association.

REGISTRATION RECORD OF ALUMINUM ASSOCIATION ALLOY DESIGNATIONS AND CHEMICAL COMPOSITION LIMITS FOR ALUMINUM ALLOYS IN THE FORM OF CASTINGS AND INGOT

REVISED: January 1989 (Amended April 5, 1991)

SUPERSEDES: September 1987

Foreword

Listed herein are designations and chemical composition limits for aluminum alloys in the form of castings or ingot registered with the Aluminum Association. This list is maintained by the Technical Committee of the Aluminum Association. Additions may be made as required, and alloys will be deleted when no longer in commercial use. In considering requests for additions, the following criteria apply:

1. The aluminum or aluminum alloy shall be offered for sale currently and shall have been sold within the 12 months immediately preceding the date of registration request, in both cases in commercial quantities.
2. The complete chemical composition limits must be disclosed and judged to be reasonable and technically correct by the Technical Committee on Product Standards of the Aluminum Association.
3. The composition must be significantly different from that of any aluminum or aluminum alloy for which a numerical designation for castings and ingot has already been assigned.
4. Except in the case of rotor and aluminizing alloy ingot, composition limits for ingot will not be accepted for registration in the absence of limits for castings of the same designation.
5. Ingot designated XXX.1 has chemical composition limits identical to those assigned to the casting (XXX.0) except grain refining elements and except for the following provisions:

Maximum Iron Percentage:

for Sand and Permanent Mold Castings	for Ingot
Up thru 0.15	0.03 less than castings
Over 0.15 thru 0.25	0.05 less than castings
Over 0.25 thru 0.6	0.10 less than castings
Over 0.6 thru 1.0	0.2 less than castings
Over 1.0	0.3 less than castings

for Die Castings	for Ingot
Up thru 1.3	0.3 Less than castings
Over 1.3	1.1 maximum

Minimum Magnesium Percentage:

for All Castings	for Ingot
Less than 0.50	0.05 more than castings[a]
0.50 and greater	0.1 more than castings[a]

Maximum Zinc Percentage:

for Die Castings	for Ingot
Over 025 thru 0.6	0.10 less than castings
Over 0.6	0.1 less than castings

[a] Applicable only to the extent that the resulting magnesium range will not be less than 0.15%.

Ingot designated XXX.2 has chemical composition limits which differ from, but fall within, those prescribed for XXX.1 ingot.

6. The product (Sand, Permanent Mold or Die Casting) must be shown at the time of registration. More than one product may be indicated only when no conflict occurs in determining the appropriate ingot (XXX.1) composition. Other products may pertain even when not listed.
7. Composition limit changes:

   7.1 Composition limits changes, including the net of changes made sequentially, shall be restricted to the extent allowed by the alloy modification rules of ANSI H35.1. Also, no change shall be permitted if it affects the validity of any registered alloy or modification.

   7.2 The limits of nonexperimental compositions may be changed only by formal ballot of the Technical Committee on Product Standards.

   7.3 Composition limit change proposals for experimental ("X") aluminum or aluminum alloys are subject to review for 30 days by the Technical Committee on Product Standards of the Aluminum Association. However, the registrant has the final authority to change the composition, as desired, provided the change complies with the rule of 7.1 above. Changes proposed by those other than the registrant are acceptable only with the approval of the registrant.

   7.4 A composition shall not be designated as experimental (with prefix "X") for more than five years. During the period that an alloy is designated as experimental, the registrant shall confirm to the Technical Committee on Product Standards of the Aluminum Association at intervals not exceeding two years the reason(s) experimental designation is still required.

In addition to the above criteria, registration of an alloy by the Aluminum Association Technical Committee requires that the proposed chemical composition limits must be judged to be technically correct by formal ballot of the Technical Committee on Product Standards. Aluminum Association Technical Committee registration shall not be applied to an alloy designated as experimental (with prefix "X").

The numerical designations of these compositions were assigned in conformance with the system adopted by the Aluminum Association in 1954 and approved by the American Standards Association (now American National Standards Institute Incorporated) as an American Standard in 1957 (see Aluminum Association Standards and Data, or American National Standard Alloy and Temper Designation Systems for Aluminum ANSI H35.1—1988.)

Some of the registered alloys may be the subject of a U.S. patent or patent application, and their listing herein is not to be construed in any way as the granting of a license under such patent rights.

CHEMICAL COMPOSITION LIMITS [1,2]

Only composition limits that are identical to these listed herein or are registered with The Aluminum Association should be designated as "AA" alloys.

REGISTERED ALLOYS IN THE FORM OF XXX.0 CASTINGS, XXX.1 INGOT AND XXX.2 INGOT.

AA No.	FORMER DESIGNATION	REGISTERED DATE	PRODUCTS [5]	Si	Fe	Cu	Mn	Mg	Cr	Ni	Zn	Sn	Ti	OTHERS [31] Each	Total [3]	ALUMINUM min[4]
100.1 *	—	6/30/70	Ingot	0.15	0.6-0.8	0.10	[20]	—	[20]	—	0.05	—	[20]	0.03 [20]	0.10	99.00
130.1 *	—	6/30/70	Ingot	[19]	[19]	0.10	[20]	—	[20]	—	0.05	—	[20]	0.03 [20]	0.10	99.30
150.1 *	—	6/30/70	Ingot	[21]	[21]	0.05	[20]	—	[20]	—	0.05	—	[20]	0.03 [20]	0.10	99.50
160.1	—	1/28/76	Ingot	0.10 [21]	0.25 [21]	—	[20]	—	[20]	—	0.05	—	[20]	0.03 [20]	0.10	99.60
170.1 *	—	6/30/70	Ingot	[22]	[22]	—	[20]	—	[20]	—	0.05	—	[20]	0.03 [20]	0.10	99.70
201.0	—	4/17/68	S	0.10	0.15	4.0-5.2	0.20-0.50	0.15-0.55	—	—	—	—	0.15-0.35	0.05 [7]	0.10	Remainder
201.2	—	4/17/68	Ingot	0.10	0.10	4.0-5.2	0.20-0.50	0.20-0.55	—	—	—	—	0.15-0.35	0.05 [7]	0.10	Remainder
A201.0	—	10/9/70	S	0.05	0.10	4.0-5.0	0.20-0.40	0.15-0.35	—	—	—	—	0.15-0.35	0.03 [7]	0.10	Remainder
A201.1	A201.2	10/9/70	Ingot	0.05	0.07	4.0-5.0	0.20-0.40	0.20-0.35	—	—	—	—	0.15-0.35	0.03 [7]	0.10	Remainder
B201.0	—	9/21/84	S	0.05	0.05	4.5-5.0	0.20-0.50	0.25-0.35	—	—	—	—	0.15-0.35	0.05 [30]	0.15	Remainder
203.0	Hiduminium 350	12/2/72	S	0.30	0.50	4.5-5.5	0.20-0.30	0.10	—	1.3-1.7	0.10	—	0.15-0.25 [24]	0.05 [23]	0.20	Remainder
203.2	Hiduminium 350	12/2/72	Ingot	0.20	0.35	4.8-5.2	0.20-0.30	0.10	—	1.3-1.7	0.10	—	0.15-0.25 [24]	0.05 [23]	0.20	Remainder
204.0	A-U5GT	10/1/74	S&P	0.20	0.35	4.2-5.0	0.10	0.15-0.35	—	0.05	0.10	0.05	0.15-0.30	0.05	0.15	Remainder
204.2	A-U5GT	10/1/74	Ingot	0.15	0.10-0.20	4.2-4.9	0.05	0.20-0.35	—	0.03	0.05	0.05	0.15-0.25	0.05	0.15	Remainder
206.0	—	4/23/76	S&P	0.10	0.15	4.2-5.0	0.20-0.50	0.15-0.35	—	0.05	0.10	0.05	0.15-0.30	0.05	0.15	Remainder
206.2	—	4/23/76	Ingot	0.10	0.10	4.2-5.0	0.20-0.50	0.20-0.35	—	0.03	0.05	0.05	0.15-0.25	0.05	0.15	Remainder
A206.0	—	4/23/76	S&P	0.05	0.10	4.2-5.0	0.20-0.50	0.15-0.35	—	0.05	0.10	0.05	0.15-0.30	0.05	0.15	Remainder
A206.2	—	4/23/76	Ingot	0.05	0.07	4.2-5.0	0.20-0.50	0.20-0.35	—	0.03	0.05	0.05	0.15-0.25	0.05	0.15	Remainder
208.0	108	—	S&P	2.5-3.5	1.2	3.5-4.5	0.50	0.10	—	0.35	1.0	—	0.25	—	0.50	Remainder
208.1	108	—	Ingot	2.5-3.5	0.9	3.5-4.5	0.50	0.10	—	0.35	1.0	—	0.25	—	0.50	Remainder
208.2	108	—	Ingot	2.5-3.5	0.8	3.5-4.5	0.30	0.03	—	—	0.20	—	0.20	—	0.30	Remainder
213.0	C113	—	S&P	1.0-3.0	1.2	6.0-8.0	0.6	0.10	—	0.35	2.5	—	0.25	—	0.50	Remainder
213.1	C113	—	Ingot	1.0-3.0	0.9	6.0-8.0	0.6	0.10	—	0.35	2.5	—	0.25	—	0.50	Remainder
222.0	122	—	S&P	2.0	1.5	9.2-10.7	0.50	0.15-0.35	—	0.50	0.8	—	0.25	—	0.35	Remainder
222.1	122	—	Ingot	2.0	1.2	9.2-10.7	0.50	0.20-0.35	—	0.50	0.8	—	0.25	—	0.35	Remainder
224.0	—	4/2/69	S&P	0.06	0.10	4.5-5.5	0.20-0.50	—	—	—	—	—	0.35	0.03 [13]	0.10	Remainder
224.2	—	4/2/69	Ingot	0.02	0.04	4.5-5.5	0.20-0.50	—	—	—	—	—	0.25	0.03 [13]	0.10	Remainder
240.0	A240.0 (A140)	—	S	0.50	0.50	7.0-9.0	0.30-0.7	5.5-6.5	—	0.30-0.7	0.10	—	0.20	0.05	0.15	Remainder
240.1	A240.1 (A140)	—	Ingot	0.50	0.40	7.0-9.0	0.30-0.7	5.6-6.5	—	0.30-0.7	0.10	—	0.20	0.05	0.15	Remainder
242.0	142	—	S&P	0.7	1.0	3.5-4.5	0.35	1.2-1.8	0.25	1.7-2.3	0.35	—	0.25	0.05	0.15	Remainder
242.1	142	—	Ingot	0.7	0.8	3.5-4.5	0.35	1.3-1.8	0.25	1.7-2.3	0.35	—	0.25	0.05	0.15	Remainder
242.2	142	—	Ingot	0.6	0.6	3.5-4.5	0.10	1.3-1.8	—	1.7-2.3	0.10	—	0.20	0.05	0.15	Remainder
A242.0	A142	—	S	0.6	0.8	3.7-4.5	0.10	1.2-1.7	0.15-0.25	1.8-2.3	0.10	—	0.07-0.20	0.05	0.15	Remainder
A242.1	A142	—	Ingot	0.6	0.6	3.7.4.5	0.10	1.3-1.7	0.15-0.25	1.8-2.3	0.10	—	0.07-0.20	0.05	0.15	Remainder
A242.2	A142	—	Ingot	0.35	0.6	3.7-4.5	0.10	1.3-1.7	0.15-0.25	1.8-2.3	0.10	—	0.07-0.20	0.05	0.15	Remainder
243.0	ML	4/21/78	S	0.35	0.40	3.5-4.5	0.15-0.45	1.8-2.3	0.20-0.40	1.9-2.3	0.05	—	0.06-0.20	0.05 [26]	0.15	Remainder
243.1	ML	4/21/78	Ingot	0.35	0.30	3.5-4.5	0.15-0.45	1.9-2.3	0.20-0.40	1.9-2.3	0.05	—	0.06-0.20	0.05 [26]	0.15	Remainder
295.0	195	—	S	0.7-1.5	1.0	4.0-5.0	0.35	0.03	—	—	0.35	—	0.25	0.05	0.15	Remainder
295.1	195	—	Ingot	0.7-1.5	0.8	4.0-5.0	0.35	0.03	—	—	0.35	—	0.25	0.05	0.15	Remainder
295.2	195	—	Ingot	0.7-1.2	0.8	4.0-5.0	0.30	0.03	—	—	0.30	—	0.20	0.05	0.15	Remainder

CHEMICAL COMPOSITION LIMITS [1,2] (cont'd)

AA No.	FORMER DESIGNATION	REGISTERED DATE	PRODUCTS [5]	Si	Fe	Cu	Mn	Mg	Cr	Ni	Zn	Sn	Ti	OTHERS [31] Each	Total [3]	ALUMINUM min[4]
296.0	B295.0 (B195)	—	P	2.0-3.0	1.2	4.0-5.0	0.35	0.05	—	0.35	0.50	—	0.25	—	0.35	Remainder
296.1	B295.1 (B195)	—	Ingot	2.0-3.0	0.9	4.0-5.0	0.35	0.05	—	0.35	0.50	—	0.25	—	0.35	Remainder
296.2	B295.2 (B195)	—	Ingot	2.0-3.0	0.8	4.0-5.0	0.30	0.35	—	—	0.30	—	0.20	0.05	0.15	Remainder
305.0	—	11/5/74	S&P	4.5-5.5	0.6	1.0-1.5	0.50	0.10	0.25	—	0.35	—	0.25	0.05	0.15	Remainder
305.2	—	9/24/73	Ingot	4.5-5.5	0.14-0.25	1.0-1.5	0.05	—	—	—	0.05	—	0.20	0.05	0.15	Remainder
A305.0	—	11/5/74	S&P	4.5-5.5	0.20	1.0-1.5	0.10	0.10	—	—	0.10	—	0.20	0.05	0.15	Remainder
A305.1	—	6/4/74	Ingot	4.5-5.5	0.15	1.0-1.5	0.10	0.10	—	—	0.10	—	0.20	0.05	0.15	Remainder
A305.2	—	9/24/73	Ingot	4.5-5.5	0.13	1.0-1.5	0.05	—	—	—	0.05	—	0.20	0.05	0.15	Remainder
308.0	A108	—	S&P	5.0-60	1.0	4.0-5.0	0.50	0.10	—	—	1.0	—	0.25	—	0.50	Remainder
308.1	A108	—	Ingot	5.0-6.0	0.8	4.0-5.0	0.50	0.10	—	—	1.0	—	0.25	—	0.50	Remainder
308.2	A108	—	Ingot	5.0-6.0	0.8	4.0-5.0	0.30	0.10	—	—	0.50	—	0.20	—	0.50	Remainder
318.0	—	1/29/91	S&P	5.5-6.5	1.0	3.0-4.0	0.50	0.10-0.6	—	0.35	1.0	—	0.25	—	0.50	Remainder
318.1	—	1/29/91	Ingot	5.5-6.5	0.8	3.0-4.0	0.50	0.15-0.6	—	0.35	0.9	—	0.25	—	0.50	Remainder
319.0	319, AllCast	—	S&P	5.5-6.5	1.0	3.0-4.0	0.50	0.10	—	0.35	1.0	—	0.25	—	0.50	Remainder
319.1	319, AllCast	—	Ingot	5.5-6.5	0.8	3.0-4.0	0.50	0.10	—	0.35	1.0	—	0.25	—	0.50	Remainder
319.2	319, AllCast	—	Ingot	5.5-6.5	0.6	3.0-4.0	0.10	0.10	—	0.10	0.10	—	0.20	—	0.20	Remainder
A319.0	—	8/28/70	S&P	5.5-6.5	1.0	3.0-4.0	0.50	0.10	—	0.35	3.0	—	0.25	—	0.50	Remainder
A319.1	—	8/28/70	Ingot	5.5-6.5	0.8	3.0-4.0	0.50	0.10	—	0.35	3.0	—	0.25	—	0.50	Remainder
B319.0	SAE 329	10/30/81	S&P	5.5-6.5	1.2	3.0-4.0	0.8	0.10-0.50	—	0.50	1.0	—	0.25	—	0.50	Remainder
B319.1	—	10/30/81	Ingot	5.5-6.5	0.9	3.0-4.0	0.8	0.15-0.50	—	0.50	1.0	—	0.25	—	0.50	Remainder
320.0	—	4/8/82	S&P	5.0-8.0	1.2	2.0-4.0	0.8	0.05-0.6	—	0.35	3.0	—	0.25	—	0.50	Remainder
320.1	—	4/8/82	Ingot	5.0-8.0	0.9	2.0-4.0	0.8	0.10-0.6	—	0.35	3.0	—	0.25	—	0.50	Remainder
324.0	324	—	P	7.0-8.0	1.2	0.40-0.6	0.50	0.40-0.7	—	0.30	1.0	—	0.20	0.15	0.20	Remainder
324.1	324	—	Ingot	7.0-8.0	0.9	0.40-0.6	0.50	0.45-0.7	—	0.30	1.0	—	0.20	0.15	0.20	Remainder
324.2	324	1/26/72	Ingot	7.0-8.0	0.6	0.40-0.6	0.10	0.45-0.7	—	0.10	0.10	—	0.20	0.05	0.15	Remainder
328.0	Red X-8	—	S	7.5-8.5	1.0	1.0-2.0	0.20-0.6	0.20-0.6	0.35	0.25	1.5	—	0.25	—	0.50	Remainder
328.1	Red X-8	—	Ingot	7.5-8.5	0.8	1.0-2.0	0.20-0.6	0.25-0.6	0.35	0.25	1.5	—	0.25	—	0.50	Remainder
332.0	F332.0 (F132)	—	P	8.5-10.5	1.2	2.0-4.0	0.50	0.50-1.5	—	0.50	1.0	—	0.25	—	0.50	Remainder
332.1	F332.1 (F132)	—	Ingot	8.5-10.5	0.9	2.0-4.0	0.50	0.6-1.5	—	0.50	1.0	—	0.25	—	0.50	Remainder
332.2	F332.2 (F132)	—	Ingot	8.5-10.0	0.6	2.0-4.0	0.10	0.9-1.3	—	0.10	0.10	—	0.20	—	0.30	Remainder
333.0	333	—	P	8.0-10.0	1.0	3.0-4.0	0.50	0.05-0.50	—	0.50	1.0	—	0.25	—	0.50	Remainder
333.1	333		Ingot	8.0-10.0	0.8	3.0-4.0	0.50	0.10-0.50	—	0.50	1.0	—	0.25	—	0.50	Remainder
A333.0	—	8/28/70	P	8.0-10.0	1.0	3.0-4.0	0.50	0.05-0.50	—	0.50	3.0	—	0.25	—	0.50	Remainder
A333.1	—	8/28/70	Ingot	8.0-10.0	0.8	3.0-4.0	0.50	0.10-0.50	—	0.50	3.0	—	0.25	—	0.50	Remainder
336.0	A332.0 (A132)	—	P	11.0-13.0	1.2	0.50-1.5	0.35	0.7-1.3	—	2.0-3.0	0.35	—	0.25	0.05	—	Remainder
336.1	A332.1 (A132)	—	Ingot	11.0-13.0	0.9	0.50-1.5	0.35	0.8-1.3	—	2.0-3.0	0.35	—	0.25	0.05	—	Remainder
336.2	A332.2 (A132)	—	Ingot	11.0-13.0	0.9	0.50-1.5	0.10	0.9-1.3	—	2.0-3.0	0.10	—	0.20	0.05	0.15	Remainder
339.0	Z332.0 (Z132)	—	P	11.0-13.0	1.2	1.5-3.0	0.50	0.50-1.5	—	0.50-1.5	1.0	—	0.25	—	0.50	Remainder
339.1	Z332.1 (Z132)	—	Ingot	11.0-13.0	0.9	1.5-3.0	0.50	0.6-1.5	—	0.50-1.5	1.0	—	0.25	—	0.50	Remainder
343.0	X443Z	10/27/72	D	6.7-7.7	1.2	0.50-0.9	0.50	0.10	0.10	—	1.2-2.0	0.50	—	0.10	0.35	Remainder
343.1	X443Z	10/27/72	Ingot	6.7-7.7	0.9	0.50-0.9	0.50	0.10	0.10	—	1.2-1.9	0.50	—	0.10	0.35	Remainder
354.0	354	—	P	8.6-9.4	0.20	1.6-2.0	0.10	0.40-0.6	—	—	0.10	—	0.20	0.05	0.15	Remainder
354.1	354	—	Ingot	8.6-9.4	0.15	1.6-2.0	0.10	0.45-0.6	—	—	0.10	—	0.20	0.05	0.15	Remainder

CHEMICAL COMPOSITION LIMITS [1,2] (cont'd)

AA No.	FORMER DESIGNATION	REGISTERED DATE	PRODUCTS [5]	Si	Fe	Cu	Mn	Mg	Cr	Ni	Zn	Sn	Ti	OTHERS [31] Each	Total [3]	ALUMINUM min [4]
355.0	355	—	S&P	4.5-5.5	0.6 [10]	1.0-1.5	0.50 [10]	0.40-0.6	0.25	—	0.35	—	0.25	0.05	0.15	Remainder
355.1	355	—	Ingot	4.5-5.5	0.50 [10]	1.0-1.5	0.50 [10]	0.45-0.6	0.25	—	0.35	—	0.25	0.05	0.15	Remainder
355.2	355	—	Ingot	4.5-5.5	0.14-0.25	1.0-1.5	0.05	0.50-0.6	—	—	0.05	—	0.20	0.05	0.15	Remainder
A355.0	—	9/17/81	S&P	4.5-5.5	0.09	1.0-1.5	0.05	0.45-0.6	—	—	0.05	—	0.04-0.20	0.05	0.15	Remainder
A355.2	—	9/17/81	Ingot	4.5-5.5	0.06	1.0-1.5	0.03	0.50-0.6	—	—	0.03	—	0.04-0.20	0.03	0.10	Remainder
C355.0	C355	—	S&P	4.5-5.5	0.20	1.0-1.5	0.10	0.40-0.6	—	—	0.10	—	0.20	0.05	0.15	Remainder
C355.1	—	6/4/74	Ingot	4.5-5.5	0.15	1.0-1.5	0.10	0.45-0.6	—	—	0.10	—	0.20	0.05	0.15	Remainder
C355.2	C355	—	Ingot	4.5-5.5	0.13	1.0-1.5	0.05	0.50-0.6	—	—	0.05	—	0.20	0.05	0.15	Remainder
356.0	356	—	S&P	6.5-7.5	0.6 [10]	0.25	0.35 [10]	0.20-0.45	—	—	0.35	—	0.25	0.05	0.15	Remainder
356.1	356	—	Ingot	6.5-7.5	0.50 [10]	0.25	0.35 [10]	0.25-0.45	—	—	0.35	—	0.25	0.05	0.15	Remainder
356.2	356	—	Ingot	6.5-7.5	0.13-0.25	0.10	0.05	0.30-0.45	—	—	0.05	—	0.20	0.05	0.15	Remainder
A356.0	A356	—	S&P	6.5-7.5	0.20	0.20	0.10	0.25-0.45	—	—	0.10	—	0.20	0.05	0.15	Remainder
A356.1	—	6/4/74	Ingot	6.5-7.5	0.15	0.20	0.10	0.30-0.45	—	—	0.10	—	0.20	0.05	0.15	Remainder
A356.2	A356	—	Ingot	6.5-7.5	0.12	0.10	0.05	0.30-0.45	—	—	0.05	—	0.20	0.05	0.15	Remainder
B356.0	—	9/17/81	S&P	6.5-7.5	0.09	0.05	0.05	0.25-0.45	—	—	0.05	—	0.04-0.20	0.05	0.15	Remainder
B356.2	—	9/17/81	Ingot	6.5-7.5	0.06	0.03	0.03	0.30-0.45	—	—	0.03	—	0.04-0.20	0.03	0.10	Remainder
C356.0	—	5/30/85	S&P	6.5-7.5	0.07	0.05	0.05	0.25-0.45	—	—	0.05	—	0.04-0.20	0.05	0.15	Remainder
C356.2	—	5/30/85	Ingot	6.5-7.5	0.04	0.03	0.03	0.30-0.45	—	—	0.03	—	0.04-0.20	0.03	0.10	Remainder
F356.0	—	10/20/71	S&P	6.5-7.5	0.20	0.20	0.10	0.17-0.25	—	—	0.10	—	0.04-0.20	0.05	0.15	Remainder
F356.2	—	10/20/71	Ingot	6.5-7.5	0.12	0.10	0.05	0.17-0.25	—	—	0.05	—	0.04-0.20	0.05	0.15	Remainder
357.0	357	—	S&P	6.5-7.5	0.15	0.05	0.03	0.45-0.6	—	—	0.05	—	0.20	0.05	0.15	Remainder
357.1	357	—	Ingot	6.5-7.5	0.12	0.05	0.03	0.45-0.6	—	—	0.05	—	0.20	0.05	0.15	Remainder
A357.0	A357	—	S&P	6.5-7.5	0.20	0.20	0.10	0.40-0.7	—	—	0.10	—	0.04-0.20	0.05 [6]	0.15	Remainder
A357.2	A357	—	Ingot	6.5-7.5	0.12	0.10	0.05	0.45-0.7	—	—	0.05	—	0.04-0.20	0.03 [6]	0.10	Remainder
B357.0	—	9/17/81	S&P	6.5-7.5	0.09	0.05	0.05	0.40-0.6	—	—	0.05	—	0.04-0.20	0.05	0.15	Remainder
B357.2	—	9/17/81	Ingot	6.5-7.5	0.06	0.03	0.03	0.45-0.6	—	—	0.03	—	0.04-0.20	0.03	0.10	Remainder
C357.0	—	9/17/81	S&P	6.5-7.5	0.09	0.05	0.05	0.45-0.7	—	—	0.05	—	0.04-0.20	0.05 [6]	0.15	Remainder
C357.2	—	9/17/81	Ingot	6.5-7.5	0.06	0.03	0.03	0.50-0.7	—	—	0.03	—	0.04-0.20	0.03 [6]	0.10	Remainder
D357.0	—	9/21/84	S	6.5-7.5	0.20	—	0.10	0.55-0.6	—	—	—	—	0.10-0.20	0.05 [6]	0.15	Remainder
358.0	B358.0 (Tens-50)	—	S&P	7.6-8.6	0.30	0.20	0.20	0.40-0.6	0.20	—	0.20	—	0.10-0.20	0.05 [15]	0.15	Remainder
358.2	B358.2 (Tens-50)	—	Ingot	7.6-8.6	0.20	0.10	0.10	0.45-0.6	0.05	—	0.10	—	0.12-0.20	0.05 [16]	0.15	Remainder
359.0	359	—	S&P	8.5-9.5	0.20	0.20	0.10	0.50-0.7	—	—	0.10	—	0.20	0.05	0.15	Remainder
359.2	359	—	Ingot	8.5-9.5	0.12	0.10	0.10	0.55-0.7	—	—	0.10	—	0.20	0.05	0.15	Remainder
360.0 [11]	360	—	D	9.0-10.0	2.0	0.6	0.35	0.40-0.6	—	0.50	0.50	0.15	—	—	0.25	Remainder
360.2	360	—	Ingot	9.0-10.0	0.7-1.1	0.10	0.10	0.45-0.6	—	0.10	0.10	0.10	—	—	0.20	Remainder
A360.0 [11]	A360	—	D	9.0-10.0	1.3	0.6	0.35	0.40-0.6	—	0.50	0.50	0.15	—	—	0.25	Remainder
A360.1 [11]	A360	—	Ingot	9.0-10.0	1.0	0.6	0.35	0.45-0.6	—	0.50	0.40	0.15	—	—	0.25	Remainder
A360.2	A360	—	Ingot	9.0-10.0	0.6	0.10	0.05	0.45-0.6	—	—	0.05	—	—	0.05	0.15	Remainder
361.0	—	6/30/78	D	9.5-10.5	1.1	0.50	0.25	0.40-0.6	0.20-0.30	0.20-0.30	0.50	0.10	0.20	0.05	0.15	Remainder
361.1	—	6/30/78	Ingot	9.5-10.5	0.8	0.50	0.25	0.45-0.6	0.20-0.30	0.20-0.30	0.40	0.10	0.20	0.05	0.15	Remainder
363.0	363	1/16/70	S&P	4.5-6.0	1.1	2.5-3.5	[17]	0.15-0.40	[17]	0.25	3.0-4.5	0.25	0.20	[18]	0.30	Remainder
363.1	363	1/16/70	Ingot	4.5-6.0	0.8	2.5-3.5	[17]	0.20-0.40	[17]	0.25	3.0-4.5	0.25	0.20	[18]	0.30	Remainder

CHEMICAL COMPOSITION LIMITS [1,2] (cont'd)

AA No.	FORMER DESIGNATION	REGISTERED DATE	PRODUCTS [5]	Si	Fe	Cu	Mn	Mg	Cr	Ni	Zn	Sn	Ti	OTHERS [31] Each	Total [3]	ALUMINUM min[4]
364.0	364	—	D	7.5-9.5	1.5	0.20	0.10	0.20-0.40	0.25-0.50	0.15	0.15	0.15	—	0.05 [12]	0.15	Remainder
364.2	364	—	Ingot	7.5-9.5	0.7-1.1	0.20	0.10	0.25-0.40	0.25-0.50	0.15	0.15	0.15	—	0.05 [12]	0.15	Remainder
369.0	Special K-9	4/4/78	D	11.0-12.0	1.3	0.50	0.35	0.25-0.45	0.30-0.40	0.05	1.0	0.10	—	0.05	0.15	Remainder
369.1	Special K-9	4/4/78	Ingot	11.0-12.0	1.0	0.50	0.35	0.30-0.45	0.30-0.40	0.05	0.9	0.10	—	0.05	0.15	Remainder
380.0 [11]	380	—	D	7.5-9.5	2.0	3.0-4.0	0.50	0.10	—	0.50	3.0	0.35	—	—	0.50	Remainder
380.2	380	—	Ingot	7.5-9.5	0.7-1.1	3.0-4.0	0.10	0.10	—	0.10	0.10	0.10	—	—	0.20	Remainder
A380.0 [11]	A380	—	D	7.5-9.5	1.3	3.0-4.0	0.50	0.10	—	0.50	3.0	0.35	—	—	0.50	Remainder
A380.1 [11]	A380	—	Ingot	7.5-9.5	1.0	3.0-4.0	0.50	0.10	—	0.50	2.9	0.35	—	—	0.50	Remainder
A380.2	A380		Ingot	7.5-9.5	0.6	3.0-4.0	0.10	0.10	—	0.10	0.10	—	—	0.05	0.15	Remainder
B380.0	A380	—	D	7.5-9.5	1.3	3.0-4.0	0.50	0.10	—	0.50	1.0	0.35	—	—	0.50	Remainder
B380.1	A380	—	Ingot	7.5.9.5	1.0	3.0-4.0	0.50	0.10	—	0.50	0.9	0.35	—	—	0.50	Remainder
C380.0	—	1/29/91	D	7.5-9.5	1.3	3.0-4.0	0.50	0.10-0.30	—	0.50	3.0	0.35	—	—	0.50	Remainder
C380.1	—	1/29/91	Ingot	7.5-9.5	1.0	3.0-4.0	0.50	0.15-0.30	—	0.50	2.9	0.35	—	—	0.50	Remainder
D380.0	—	1/29/91	D	7.5-9.5	1.3	3.0-4.0	0.50	0.10-0.30	—	0.50	1.0	0.35	—	—	0.50	Remainder
D380.1	—	1/29/91	Ingot	7.5-9.5	1.0	3.0-4.0	0.50	0.15-0.30	—	0.50	0.9	0.35	—	—	0.50	Remainder
383.0	—	—	D	9.5-11.5	1.3	2.0-3.0	0.50	0.10	—	0.30	3.0	0.15	—	—	0.50	Remainder
383.1	—	—	Ingot	9.5-11.5	1.0	2.0-3.0	0.50	0.10	—	0.30	2.9	0.15	—	—	0.50	Remainder
383.2	—	—	Ingot	9.5-11.5	0.6-1.0	2.0-3.0	0.10	0.10	—	0.10	0.10	0.10	—	—	0.20	Remainder
A383.0	—	1/29/91	D	9.5-11.5	1.3	2.0-3.0	0.50	0.10-0.30	—	0.30	3.0	0.15	—	—	0.50	Remainder
A383.1	—	1/29/91	Ingot	9.5-11.5	1.0	2.0-3.0	0.50	0.15-0.30	—	0.30	2.9	0.15	—	—	0.50	Remainder
384.0	384	—	D	10.5-12.0	1.3	3.0-4.5	0.50	0.10	—	0.50	3.0	0.35	—	—	0.50	Remainder
384.1	384	—	Ingot	10.5-12.0	1.0	3.0-4.5	0.50	0.10	—	0.50	2.9	0.35	—	—	0.50	Remainder
384.2	384	—	Ingot	10.5-12.0	0.6-1.0	3.0-4.5	0.10	0.10	—	0.10	0.10	0.10	—	—	0.20	Remainder
A384.0	384	—	D	10.5-12.0	1.3	3.0-4.5	0.50	0.10	—	0.50	1.0	0.35	—	—	0.50	Remainder
A384.1	384	—	Ingot	10.5-12.0	1.0	3.0-4.5	0.50	0.10	—	0.50	0.9	0.35	—	—	0.50	Remainder
B384.0	—	1/29/91	D	10.5-12.0	1.3	3.0-4.5	0.50	0.10-0.30	—	0.50	1.0	0.35	—	—	0.50	Remainder
B384.1	—	1/29/91	Ingot	10.5-12.0	1.0	3.0-4.5	0.50	0.15-0.30	—	0.50	0.9	0.35	—	—	0.50	Remainder
C384.0	—	1/29/91	D	10.5-12.0	1.3	3.0-4.5	0.50	0.10-0.30	—	0.50	3.0	0.35	—	—	0.50	Remainder
C384.1	—	1/29/91	Ingot	10.5-12.0	1.0	3.0-4.5	0.50	0.15-0.30	—	0.50	2.9	0.35	—	—	0.50	Remainder
385.0	B384.0 (384)	1/21/70	D	11.0-13.0	2.0	2.0-4.0	0.50	0.30	—	0.50	3.0	0.30	—	—	0.50	Remainder
385.1	B384.1 (384)	1/21/70	Ingot	11.0-13.0	1.1	2.0-4.0	0.50	0.30	—	0.50	2.9	0.30	—	—	0.50	Remainder
390.0	390	—	D	16.0-18.0	1.3	4.0-5.0	0.10	0.45-0.65	—	—	0.10	—	0.20	0.10	0.20	Remainder
390.2	390	—	Ingot	16.0-18.0	0.6-1.0	4.0-5.0	0.10	0.50-0.65	—	—	0.10	—	0.20	0.10	0.20	Remainder
A390.0	A390	—	S&P	16.0-18.0	0.50	4.0-5.0	0.10	0.45-0.65	—	—	0.10	—	0.20	0.10	0.20	Remainder
A390.1	A390	—	Ingot	16.0-18.0	0.40	4.0-5.0	0.10	0.50-0.65	—	—	0.10	—	0.20	0.10	0.20	Remainder
B390.0	—	3/29/79	D	16.0-18.0	1.3	4.0-5.0	0.50	0.45-0.65	—	0.10	1.5	—	0.20	0.10	0.20	Remainder
B390.1	—	3/29/79	Ingot	16.0-18.0	1.0	4.0-5.0	0.50	0.50-0.65	—	0.10	1.4	—	0.20	0.10	0.20	Remainder
392.0	392	—	D	18.0-20.0	1.5	0.40-0.8	0.20-0.6	0.8-1.2	—	0.50	0.50	0.30	0.20	0.15	0.50	Remainder
392.1	392	—	Ingot	18.0-20.0	1.1	0.40-0.8	0.20-0.6	0.9-1.2	—	0.50	0.40	0.30	0.20	0.15	0.50	Remainder
393.0	Vanasil	—	SP&D	21.0-23.0	1.3	0.7-1.1	0.10	0.7-1.3	—	2.0-2.5	0.10	—	0.10-0.20	0.05 [14]	0.15	Remainder
393.1	Vanasil	—	Ingot	21.0-23.0	1.0	0.7-1.1	0.10	0.8-1.3	—	2.0-2.5	0.10	—	0.10-0.20	0.05 [14]	0.15	Remainder
393.2	Vanasil	—	Ingot	21.0-23.0	0.8	0.7-1.1	0.10	0.8-1.3	—	2.0-2.5	0.10	—	0.10-0.20	0.05 [14]	0.15	Remainder

CHEMICAL COMPOSITION LIMITS [1,2] (cont'd)

AA No.	FORMER DESIGNATION	REGISTERED DATE	PRODUCTS [5]	Si	Fe	Cu	Mn	Mg	Cr	Ni	Zn	Sn	Ti	OTHERS [31] Each	Total [3]	ALUMINUM min[4]
408.2 [25]	—	9/24/73	Ingot	8.5-9.5	0.6-1.3	0.10	0.10	—	—	—	0.10	—	—	0.10	0.20	Remainder
409.2 [25]	—	9/24/73	Ingot	9.0-10.0	0.6-1.3	0.10	0.10	—	—	—	0.10	—	—	0.10	0.20	Remainder
411.2 [25]	—	9/24/73	Ingot	10.0-12.0	0.6-1.3	0.20	0.10	—	—	—	0.10	—	—	0.10	0.20	Remainder
413.0 [11]	13	—	D	11.0-13.0	2.0	1.0	0.35	0.10	—	0.50	0.50	0.15	—	—	0.25	Remainder
413.2	13	—	Ingot	11.0-13.0	0.7-1.1	0.10	0.10	0.07	—	0.10	0.10	0.10	—	—	0.20	Remainder
A413.0 [11]	A13	—	D	11.0-13.0	1.3	1.0	0.35	0.10	—	0.50	0.50	0.15	—	—	0.25	Remainder
A413.1 [11]	A13	—	Ingot	11.0-13.0	1.0	1.0	0.35	0.10	—	0.50	0.40	0.15	—	—	0.25	Remainder
A413.2	A13		Ingot	11.0-13.0	0.6	0.10	0.05	0.05	—	0.05	0.05	0.05	—	—	0.10	Remainder
B413.0	—	11/6/84	S&P	11.0-13.0	0.50	0.10	0.35	0.05	—	0.05	0.10	—	0.25	0.05	0.20	Remainder
B413.1	—	11/6/84	Ingot	11.0-13.0	0.40	0.10	0.35	0.05	—	0.05	0.10	—	0.25	0.05	0.20	Remainder
435.2 [27]	—	12/18/81	Ingot	3.3-3.9	0.40	0.05	0.05	0.05	—	—	0.10	—	—	0.05	0.20	Remainder
443.0	43	—	S&P	4.5-6.0	0.8	0.6	0.50	0.06	0.25	—	0.50	—	0.25	—	0.35	Remainder
443.1	43	—	Ingot	4.5-6.0	0.6	0.6	0.50	0.05	0.25	—	0.50	—	0.25	—	0.35	Remainder
443.2	43	—	Ingot	4.5-6.0	0.6	0.10	0.10	0.05	—	—	0.10	—	0.20	0.05	0.15	Remainder
A443.0	43 (0.30 max Cu)	—	S	4.5-6.0	0.8	0.30	0.50	0.05	0.25	—	0.50	—	0.25	—	0.35	Remainder
A443.1	43 (0.30 max Cu)	—	Ingot	4.5-6.0	0.6	0.30	0.50	0.05	0.25	—	0.50	—	0.25	—	0.35	Remainder
B443.0	43 (0.15 max Cu)	—	S&P	4.5-6.0	0.8	0.15	0.35	0.05	—	—	0.35	—	0.25	0.05	0.15	Remainder
B443.1	43 (0.15 max Cu)	—	Ingot	4.5-6.0	0.6	0.15	0.35	0.05	—	—	0.35	—	0.25	0.05	0.15	Remainder
C443.0	A43	—	D	4.5-6.0	2.0	0.6	0.35	0.10	—	0.50	0.50	0.15	—	—	0.25	Remainder
C443.1	A43	—	Ingot	4.5-6.0	1.1	0.6	0.35	0.10	—	0.50	0.40	0.15	—	—	0.25	Remainder
C443.2	A43	—	Ingot	4.5-6.0	0.7-1.1	0.10	0.10	0.05	—	—	0.10	—	—	0.05	0.15	Remainder
444.0	—	11/5/74	S&P	6.5-7.5	0.6	0.25	0.35	0.10	—	—	0.35	—	0.25	0.05	0.15	Remainder
444.2	—	9/24/73	Ingot	6.5-7.5	0.13-0.25	0.10	0.05	0.05	—	—	0.05	—	0.20	0.05	0.15	Remainder
A444.0	A344	—	P	6.5-7.5	0.20	0.10	0.10	0.05	—	—	0.10	—	0.20	0.05	0.15	Remainder
A444.1	—	6/4/74	Ingot	6.5-7.5	0.15	0.10	0.10	0.05	—	—	0.10	—	0.20	0.05	0.15	Remainder
A444.2	A344	—	Ingot	6.5-7.5	0.12	0.05	0.05	0.05	—	—	0.05	—	0.20	0.05	0.15	Remainder
445.2 [25]	B444.2	9/24/73	Ingot	6.5-7.5	0.6-1.3	0.10	0.10	—	—	—	0.10	—	—	0.10	0.20	Remainder
511.0	F514.0 (F214)	—	S	0.30-0.7	0.50	0.15	0.35	3.5-4.5	—	—	0.15	—	0.25	0.05	0.15	Remainder
511.1	F514.1 (F214)	—	Ingot	0.30-0.7	0.40	0.15	0.35	3.6-4.5	—	—	0.15	—	0.25	0.05	0.15	Remainder
511.2	F514.2 (F214)	—	Ingot	0.30-0.7	0.30	0.10	0.10	3.6-4.5	—	—	0.10	—	0.20	0.05	0.15	Remainder
512.0	B514.0 (B214)	—	S	1.4-2.2	0.6	0.35	0.8	3.5-4.5	0.25	—	0.35	—	0.25	0.05	0.15	Remainder
512.2	B514.2 (B214)	—	Ingot	1.4-2.2	0.30	0.10	0.10	3.6-4.5	—	—	0.10	—	0.20	0.05	0.15	Remainder
513.0	A514.0 (A214)	—	P	0.30	0.40	0.10	0.30	3.5-4.5	—	—	1.4-2.2	—	0.20	0.05	0.15	Remainder
513.2	A514.2 (A214)	—	Ingot	0.30	0.30	0.10	0.10	3.6-4.5	—	—	1.4-2.2	—	0.20	0.05	0.15	Remainder
514.0	214	—	S	0.35	0.50	0.15	0.35	3.5-4.5	—	—	0.15	—	0.25	0.05	0.15	Remainder
514.1	214	—	Ingot	0.35	0.40	0.15	0.35	3.6-4.5	—	—	0.15	—	0.25	0.05	0.15	Remainder
514.2	214	—	Ingot	0.30	0.30	0.10	0.10	3.6-4.5	—	—	0.10	—	0.20	0.05	0.15	Remainder
515.0	L514.0 (L214)	1/2/70	D	0.50-1.0	1.3	0.20	0.40-0.6	2.5-4.0	—	—	0.10	—	—	0.05	0.15	Remainder
515.2	L514.2 (L214)	1/2/70	Ingot	0.50-1.0	0.6-1.0	0.10	0.40-0.6	2.7-4.0	—	—	0.05	—	—	0.05	0.15	Remainder
516.0	—	9/30/83	D	0.30-1.5	0.35-1.0	0.30	0.15-0.40	2.5-4.5	—	0.25-0.40	0.20	0.10	0.10-0.20	0.05 [28]	—	Remainder
516.		9/30/8	Ingo	0.30-1.	0.35-0.	0.3	0.15-0.4	2.6-4.		0.25-0.4	0.2	0.1	0.10-0.2	0.0 [28		Remainde

CHEMICAL COMPOSITION LIMITS [1,2] (cont'd)

AA No.	FORMER DESIGNATION	REGISTERED DATE	PRODUCTS [5]	Si	Fe	Cu	Mn	Mg	Cr	Ni	Zn	Sn	Ti	OTHERS [31] Each	OTHERS [31] Total [3]	ALUMINUM min [4]
518.0	218	—	D	0.35	1.8	0.25	0.35	7.5-8.5	—	0.15	0.15	0.15	—	—	0.25	Remainder
518.1	218	—	Ingot	0.35	1.1	0.25	0.35	7.6-8.5	—	0.15	0.15	0.15	—	—	0.25	Remainder
518.2	218	—	Ingot	0.25	0.7	0.10	0.10	7.6-8.5	—	0.05	—	0.05	—	—	0.10	Remainder
520.0	220	—	S	0.25	0.30	0.25	0.15	9.5-10.6	—	—	0.15	—	0.25	0.05	0.15	Remainder
520.2	220	—	Ingot	0.15	0.20	0.20	0.10	9.6-10.6	—	—	0.10	—	0.20	0.05	0.15	Remainder
535.0	Almag 35	—	S	0.15	0.15	0.05	0.10-0.25	6.2-7.5	—	—	—	—	0.10-0.25	0.05 [8]	0.15	Remainder
535.2	Almag 35	—	Ingot	0.10	0.10	0.05	0.10-0.25	6.6-7.5	—	—	—	—	0.10-0.25	0.05 [29]	0.15	Remainder
A535.0	A218	—	S	0.20	0.20	0.10	0.10-0.25	6.5-7.5	—	—	—	—	0.25	0.05	0.15	Remainder
A535.1	A218	—	Ingot	0.20	0.15	0.10	0.10-0.25	6.6-7.5	—	—	—	—	0.25	0.05	0.15	Remainder
B535.0	B218	—	S	0.15	0.15	0.10	0.05	6.5-7.5	—	—	—	—	0.10-0.25	0.05	0.15	Remainder
B535.2	B218	—	Ingot	0.10	0.12	0.05	0.05	6.6-7.5	—	—	—	—	0.10-0.25	0.05	0.15	Remainder
705.0	603, Ternalloy 5	—	S&P	0.20	0.8	0.20	0.40-0.6	1.4-1.8	0.20-0.40	—	2.7-3.3	—	0.25	0.05	0.15	Remainder
705.1	603, Ternalloy 5	—	Ingot	0.20	0.6	0.20	0.40-0.6	1.5-1.8	0.20-0.40	—	2.7-3.3	—	0.25	0.05	0.15	Remainder
707.0	607, Ternalloy 7	—	S&P	0.20	0.8	0.20	0.40-0.6	1.8-2.4	0.20-0.40	—	4.0-4.5	—	0.25	0.05	0.15	Remainder
707.1	607, Ternalloy 7	—	Ingot	0.20	0.6	0.20	0.40-0.6	1.9-2.4	0.20-0.40	—	4.0-4.5	—	0.25	0.05	0.15	Remainder
710.0	A712.0 (A612)	—	S	0.15	0.50	0.35-0.65	0.05	0.6-0.8	—	—	6.0-7.0	—	0.25	0.05	0.15	Remainder
710.1	A712.1 (A612)	—	Ingot	0.15	0.40	0.35-0.65	0.05	0.65-0.8	—	—	6.0-7.0	—	0.25	0.05	0.15	Remainder
711.0	C712.0 (C612)	—	P	0.30	0.7-1.4	0.35-0.65	0.05	0.25-0.45	—	—	6.0-7.0	—	0.20	0.05	0.15	Remainder
711.1	C712.1 (C612)	—	Ingot	0.30	0.7-1.1	0.35-0.65	0.05	0.30-0.45	—	—	6.0-7.0	—	0.20	0.05	0.15	Remainder
712.0	D712.0 (D612, 40E)	—	S	0.30	0.50	0.25	0.10	0.50-0.65	0.40-0.6	—	5.0-6.5	—	0.15-0.25	0.05	0.20	Remainder
712.2	D712.2 (D612, 40E)	—	Ingot	0.15	0.40	0.25	0.10	0.50-0.65	0.40-0.6	—	5.0-6.5	—	0.15-0.25	0.05	0.20	Remainder
713.0	613, Tenzaloy	—	S&P	0.25	1.1	0.40-1.0	0.6	0.20-0.50	0.35	0.15	7.0-8.0	—	0.25	0.10	0.25	Remainder
713.1	613, Tenzaloy	—	Ingot	0.25	0.8	0.40-1.0	0.6	0.25-0.50	0.35	0.15	7.0-8.0	—	0.25	0.10	0.25	Remainder
771.0	Precedent 71A	—	S	0.15	0.15	0.10	0.10	0.8-1.0	0.06-0.20	—	6.5-7.5	—	0.10-0.20	0.05	0.15	Remainder
771.2	Precedent 71A	—	Ingot	0.10	0.10	0.10	0.10	0.85-1.0	0.06-0.20	—	6.5-7.5	—	0.10-0.20	0.05	0.15	Remainder
772.0	B771.0 (Precedent 71B)	—	S	0.15	0.15	0.10	0.10	0.6-0.8	0.06-0.20	—	6.0-7.0	—	0.10-0.20	0.05	0.15	Remainder
772.2	B771.2 (Precedent 71B)	—	Ingot	0.10	0.10	0.10	0.10	0.65-0.8	0.06-0.20	—	6.0-7.0	—	0.10-0.20	0.05	0.15	Remainder
850.0	750	—	S&P	0.7	0.7	0.7-1.3	0.10	0.10	—	0.7-1.3	—	5.5-7.0	0.20	—	0.30	Remainder
850.1	750	—	Ingot	0.7	0.50	0.7-1.3	0.10	0.10	—	0.7-1.3	—	5.5-7.0	0.20	—	0.30	Remainder
851.0	A850.0 (A750)	—	S&P	2.0-3.0	0.7	0.7-1.3	0.10	0.10	—	0.30-0.7	—	5.5-7.0	0.20	—	0.30	Remainder
851.1	A550.1 (A750)	—	Ingot	2.0-3.0	0.50	0.7-1.3	0.10	0.10	—	0.30-0.7	—	5.5-7.0	0.20	—	0.30	Remainder
852.0	B850.0 (B750)	—	S&P	0.40	0.7	1.7-2.3	0.10	0.6-0.9	—	0.9-1.5	—	5.5-7.0	0.20	—	0.30	Remainder
852.1	B850.1 (B750)	—	Ingot	0.40	0.50	1.7-2.3	0.10	0.7-0.9	—	0.9-1.5	—	5.5-7.0	0.20	—	0.30	Remainder
853.0	XC850.0 (XC750)	—	S&P	5.5-6.5	0.7	3.0-4.0	0.50	—	—	—	—	5.5-7.0	0.20	—	0.30	Remainder
853.2	XC850.2 (XC750)	—	Ingot	5.5-6.5	0.50	3.0-4.0	0.10	—	—	—	—	5.5-7.0	0.20	—	0.30	Remainder

See footnotes on following page.

FOOTNOTES

(1) Composition in weight percent maximum unless shown as a range or a minimum.

Standard limits for alloying elements and impurities are expressed to the following places:

Less than 0.001 percent 0.000X
0.001 but less than 0.01 percent 0.00X
0.01 but less than 0.10 percent
Unalloyed aluminum made by a refining process 0.0XX
Alloys and unalloyed aluminum
not made by a refining process 0.0X
0.10 through 0.55 percent 0.XX
(It is customary to express limits of 0.30 percent through 0.55 percent as 0.X0 or 0.X5)
Over 0.55 percent .... 0.X, X.X, etc.(9)

(2) Except for "Aluminum" and "Others," analysis normally is made for elements for which specific limits are shown. For purposes of determining conformance to these limits, an observed value or calculated value obtained from analysis is rounded off to the nearest unit in the last right-hand place of figures used in expressing the specified limit, in accordance with the following:

- When the figure next beyond the last figure or place to be retained is less than 5, the figure in the last place retained should be kept unchanged.
- When the figure next beyond the last figure or place to be retained is greater than 5, the figure in the last place retained should be increased by 1.
- When the figure next beyond the last figure or place to be retained is 5 and
  (1) there are no figures or only zeros, beyond this 5, if the figure in the last place to be retained is odd, it should be increased by 1; if even, it should be kept unchanged;
  (2) if the 5 next beyond the figure in the last place to be retained is followed by any figures other than zero, the figure in the last place retained should be increased by 1, whether odd or even.

(3) The sum of those "Others" metallic elements 0.010 percent or more each, expressed to the second decimal before determining the sum.

(4) The aluminum content for unalloyed aluminum not made by a refining process is the difference between 100.00 percent and the sum of all other metallic elements present in amounts of 0.010 percent or more each, expressed to the second decimal before determining the sum.

(5) D = Diecasting. P = Permanent Mold. S = Sand. Other products may pertain to the composition shown even though not listed.

(6) Beryllium 0.04-0.07.

(7) Silver 0.40-1.0.

(8) Beryllium 0.003-0.007, boron 0.005 max.

(9) Magnesium percent for some alloys is an exception to this rule.

(10) If iron exceeds 0.45, manganese content shall not be less than one-half iron content.

(11) A360.1, A380.1 and A413.1 ingot is used to produce 360.0 and A360.0; 380.0 and A380.0; 413.0 and A413.0 castings, respectively.

(12) Beryllium 0.02-0.04.

(13) Vanadium 0.05-0.15; zirconium 0.10-0.25.

(14) Vanadium 0.08-0.15.

(15) Beryllium 0.10-0.30.

(16) Beryllium 0.15-0.30.

(17) Manganese + chromium 0.8 max.

(18) Lead 0.25 max.

(19) Iron/silicon ratio 2.5 min.

(20) Manganese + chromium + titanium + vanadium 0.025 max.

(21) Iron/silicon ratio 2.0 min.

(22) Iron/silicon ratio 1.5 min.

(23) Antimony 0.20-0.30; cobalt 0. 20-0.30; zirconium 0. 10-0.30

(24) Titanium + zirconium 0.50 max.

(25) 445.2,408.2,409.2 and 411.2 are used to coat steel.

(26) Vanadium 0.06-0.20.

(27) Used with zinc to coat steel.

(28) Lead 0.10 max.

(29) Beryllium 0.003-0.007, boron 0.002 max.

(30) Silver 0.50-1.0.

(31) Includes listed elements for which no specific limit is shown.

\+ AA number registered since previous issue.
† Composition limits revised since previous issue.
§ "X" removed from AA number since previous issue.

* Rated minimum conductivities:

Ingot	Percent IACS
100.1	54
130.1	55
150.1	57
170.1	59

The rating of ingot metal for minimum conductivity characteristic is based on established relations between electrical conductivity and metal composition.

INACTIVE ALLOYS

Designation	Date Reclassified
202.0	1988-04-22
202.2	1988-04-22
238.0	1988-04-22
238.1	1988-04-22
238.2	1988-04-22
249.0	1988-04-22
249.2	1988-04-22
D355.2	1974-11-18
B356.2	1974-11-18

Chapter 5

DESIGNING ALUMINUM CASTINGS

Three basic aspects of product design should be considered and evaluated before designing an aluminum casting.

The first is *function*. The aim is to produce a casting designed to successfully withstand the service requirements for which it is intended.

The second aspect is often called *foundry features*. These features (which have no adverse effect on the function of the part) can reduce production costs by greatly assisting the designer and builder of the pattern or die and also the producer of the casting. Working together during the design phase, these individuals can include in the design the most advantageous foundry features to make the part easier and cheaper to cast, with reduced scrap losses and higher overall quality. Machining also should be considered in the early stages of design.

The third aspect is concerned with *appearance*, the manner in which the casting harmonizes with other components and intended applications.

This chapter deals primarily with the first and second phases: functional design and foundry design features.

The reader is cautioned that minimum or maximum values suggested herein for the casting processes should not be selected arbitrarily. Plans for casting any part by any process should be evaluated in light of the considerable number of variables involved. The values listed have been obtained readily in normal foundry practice and can be achieved consistently through adherence to good design practices supplemented with an understanding of aluminum casting processes, the casting producer's problems and the inherent limitations of the casting method selected.

INTRODUCTION TO CASTING DESIGN

Many castings are destined for failure while they are still in the early stages of design. Not all designs should be accepted if a foundry is interested in making a quality casting. Once the design is finalized, it is difficult to obtain alterations or concessions—the time for collaboration between the designer and the casting producer is before, not after, the design is firm. Such early evaluation of a job, which saves time, money and castings, is now receiving wide recognition as value engineering.

There are limitations to a foundry's ability to make sound castings, to say nothing of fulfillment of application requirements, if the original design is incorrect. Unless the design engineer has considered the potential casting from the viewpoint of foundry requirements, the end results may be uneconomical or unsatisfactory. Therefore, aluminum casting designers must have a good working knowledge of the particular casting process and the related equipment that will be used for producing the part. If possible, the designer should seek guidance and approval of the casting producers.

When correctly designed, aluminum castings have great advantages over similar parts produced by many other manufacturing methods or cast in other metals.

PREDESIGN CONSIDERATIONS

Before starting any casting design, a preliminary study should be made to determine the casting process that will be used in making the production runs of the particular part. For example, it is not a good practice to plan and design the prototype and early production parts as sand castings with the thought of later changing them to permanent mold or die castings because, in many instances, the sand cast design may require considerable redesign to accommodate either of these other processes. There also is the possibility that the alloy chosen for sand casting may not be suitable for either of the other processes.

Once choice of process has been finalized, the next consideration is to design the part to accommodate it. If the proposed casting would be large and have a complex configuration (requiring costly pattern or mold equipment) and have a high foundry production cost, a study should be made to see if the casting could be made as an assembly of two or more smaller, simpler castings joined by welding or with mechanical fasteners. Lower equipment cost, sounder castings and lower piece cost could result if this were possible.

After the general form of the casting has been established, full consideration should be given to the location of the gates, risers and chills to ensure a sound casting. In sand, considerable flexibility is possible in their locations; however, in a permanent mold, gates and risers are confined to the mold parting lines, and in semipermanent mold castings, chills (if they are necessary) usually are placed in the sand cores.

DESIGN FOR MINIMUM CASTING STRESSES

Aluminum castings contract approximately $\frac{3}{32}$ to $\frac{5}{32}$ in./ft while cooling to room temperature. If the design configuration is such that the mold or core effectively restricts this normal contraction during cooling, the casting may crack or develop hot tears while in the mold. At best, warping, distortion and locked-in casting stresses may be present.

Castings made in metal molds; in hard, dry sand molds; in high-pressure green sand molds; or in green sand molds that contain large, dry sand cores all are more susceptible to cracking in the mold than castings made in green sand alone. Green sand is less rigid and can be compressed or crushed, to some degree, by the contracting metal.

Hot tearing or cracking is also aggravated in castings where thin walls intersect thick walls or locally thick metal sections. Here, the thinner metal sections solidify, cool rapidly and contract away from the thicker metal sections, which are restrained from movement by the mold or core. There also is a possibility that portions of this thicker mass of metal may still be in the mushy stage where it has extremely low mechanical strength and is unable to resist the stresses created by the contracting members.

Tearing may even develop in straight bars that are unrestrained at the ends. As the section cools and contracts, enough resistance may develop from friction between mold and metal to rupture the casting. If the bar is uniform in cross section, no particular region of weakness will develop and the bar will not tear because solidification and contraction occur uniformly across the section. However, if another member is joined to the straight bar, a hot spot is formed, and the solidification pattern is conducive to hot tearing.

DESIGN FOR STRENGTH

Unnecessary bulk and weight were eliminated long ago from every part used by the aircraft industry because it is well known that strength does not increase proportionately with mass. The demand for light weight with maximum performance and strength has spread throughout the commercial industries. More exacting requirements from customers and keen competition from many other fabricating processes and materials should serve as a warning to designers and producers of aluminum castings that excessive bulk and weight will increase the cost of their products.

Designers must achieve strength by configuration of the part rather than by mass, and the foundry must control its process to produce castings that have the mechanical properties and quality necessary to meet the strength requirements of the design. Casting processes have reached an advanced state of development, and it is no longer difficult to achieve high quality in a well-designed casting produced in a well-equipped and properly managed foundry.

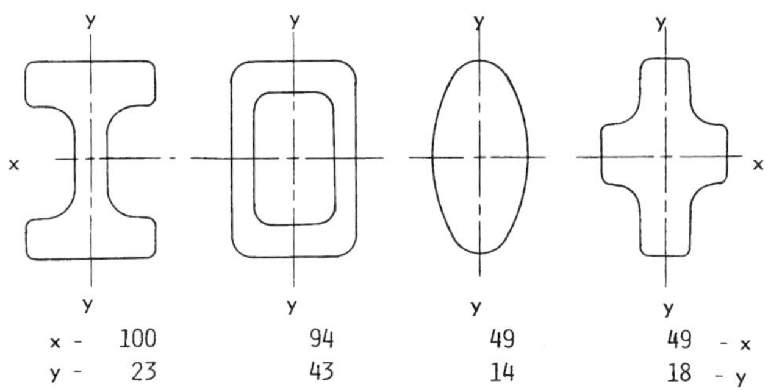

Fig. 5-1. These four commonly used beam cross sections show the importance of metal distribution on the load-carrying ability of a section. All sections have equal area, weight and height, and all (except the oval shape) have equal width. The numbers shown below each shape are dimensionless values comparing the rigidity of that particular section with that of the I-beam about its major axis, which was assigned the value of 100. Note that the rigidity or load-carrying ability of both the oval and the cross shape is essentially half that of the I-beam about the major axis. The thick metal in the central area of these two sections may introduce foundry problems.

CALCULATING SERVICE STRESSES

Many casting designers recommend using stress analysis to make certain the casting will have adequate strength. However, it should be noted that mathematical stress analyses of complex designs are not too reliable. Stress calculations derived from this method usually will predict the performance of an uncomplicated casting in service. If a stress analysis shows the maximum operating stresses to be well within the safe limits applicable to the chosen alloy and heat treatment for a simple and well-designed casting that fails in service, the fault may lie in production process techniques or control.

Unfortunately, structural failures often are corrected by "beefing up" the part, a practice that adds unnecessary weight and cost to the casting. Designers who use analytical techniques to develop their designs often can demonstrate that an excess amount of metal is not necessary to carry a high load. Proper location and sizing of ribs and gussets usually provide the needed strength.

Experimental methods, rather than theoretical mathematical analysis, are preferred for determining stresses in more complex shapes. The brittle lacquer, photoelastic and strain gauge techniques are some of these methods.

Some shape considerations include:

- triangular shapes are more rigid than rectangular ones;
- putting beads around holes to reinforce a section is a common design practice;
- cylindrical shapes are less rigid than cones.

DESIGN FOR DIRECTIONAL SOLIDIFICATION

It is an undisputed fact that progressive or directional solidification is essential to the production of sound castings, and it should be the designer's, patternmaker's and foundryman's responsibility to produce a design that will meet this requirement. The main design considerations that influence directional solidification are as follows.

Thick sections cannot be fed through lighter sections. The members or component parts of castings are joined in various ways (Fig. 5-1). The T-section is selected to represent the series. By inscribing circles, as in Fig. 5-2, it is easy to determine that the region depicted by circle d is a larger mass of metal than those of circles a, b or c. This means that, under normal conditions of cooling, metal at the center of region d is a hot spot, the last to solidify. When it is not convenient, economical or even possible to place a riser at this point, it is necessary to rely on chilling, coring or reducing the section.

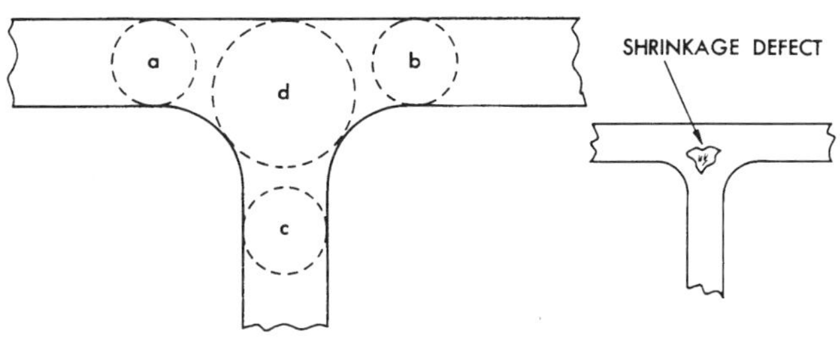

Fig. 5-2. Location of a hot spot in a casting junction can be determined by inscribing circles.

The designer should attempt to limit junctions to as few as possible and to select the least difficult types. The L-joint, for example, is less troublesome and easier to treat than Y or V sections. The L-section is also easier to correct by design alone than any of the others. Sometimes, it is impossible to limit the number of junctions except by casting in segments and joining these together later by welding or bolting.

If possible, sections should taper toward risers. The importance of this principle has clearly been established. Often, a compromise must be made with ideal conditions to meet weight and cost requirements. Within reason, casting soundness obtained by natural solidification is proportional to the degree of tapering allowed. Figure 5-3 illustrates the design of a valve casting. The original design illustrates the usual situation in which no attempt has been made to enhance feeding by natural methods; the revision is correctly designed.

Large, cylindrical castings are difficult to make sound by static methods and are best made by centrifugal casting. To make cylindrical castings of nonferrous metals free of centerline shrinkage by static methods, tapering or some more elaborate expedient is necessary. The cylinder may be tapered and risered in any of the three ways shown in Fig. 5-4.

Isolated hot spots must be avoided. Abnormally thick masses of metal isolated from a feeding source and chills, as well as internal metal areas of a casting substantially surrounded by sand cores, will cool at a slower rate than thinner walls and metal in more open regions. These areas are potentially troublesome for shrinks and tears unless provision is made to adequately feed and chill them (Fig. 5-5).

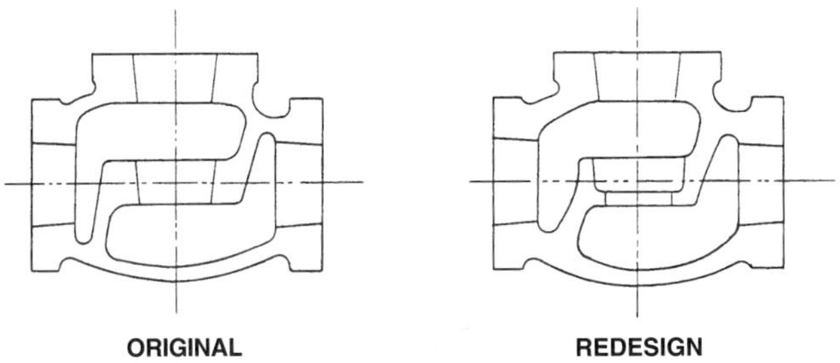

Fig. 5-3. In the original design of this globe valve body, progressive solidification was not possible because the thicker metal sections could not be fed through the thinner adjacent walls. The thick metal in the valve seat area was particularly prone to heavy shrinkage because it was completely surrounded with a sand core and was cut off from any feeding source. The revised design provided progressive solidification and sound metal by reducing the amount of metal in the valve seat area and by tapering or flaring the thinner walls into the thicker flanges, where the risers were located.

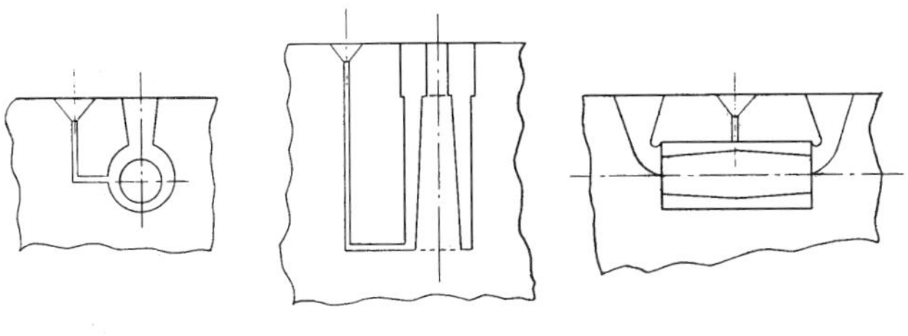

Fig. 5-4. Methods of padding cylindrical castings to ensure directional solidification.

In a casting designed for permanent mold production, the designer should make every effort to specify fairly uniform wall thicknesses. The thickest metal section or wall should be located on the mold parting line where gates and risers are, by necessity, located. While this ideal condition is not always possible, directional solidification can be promoted and hot spots reduced substantially by: 1) varying the metal thickness of the mold; 2) incorporating inserts of copper or other high-conductivity metal at critical areas of the mold; 3) varying the thickness of the mold wash; 4) water-cooling certain mold sections; or 5) directing air jets to the hotter mold areas.

Most of these design problems can be discovered easily using computer simulation designed to model the solidification of castings. Solidification software is available in 2-D and 3-D versions. Evaluations of mold and core materials, chills, and cycles are also features of this type of software.

DESIGN FOR METAL FLOW

Minimum section thicknesses for castings are principally a function of the casting process, metal composition, mold venting, surface area and location in relation to the gating system. In general, sections of sand and permanent mold castings of aluminum alloys should be designed not less than $\frac{1}{8}$ in. thick in small castings or not less than $\frac{3}{16}$ to $\frac{1}{4}$ in. in larger castings. Recommended minimum section thickness for diecasting is 0.05 in.

These considerations are conservative and apply only to average casting practices. Recent innovations in casting methods greatly extend the field for thinner castings, but they are special developments.

Unnecessarily heavy sections increase the problem of feeding. Excessively thin walls increase foundry scrap from misruns and cold shuts. The minimum section thicknesses that can be cast by various processes given in this and other chapters vary considerably with the casting method. In general, for a given casting process, the smaller the total area of the section, the thinner it may be and still be cast successfully. Additionally, the gating and risering system largely determines the minimum sections that can be poured successfully. In general, the minimum section thickness and the gating and risering practices should be such as to permit relatively low pouring temperatures.

It is recognized that intersections of thick and thin metal sections, offsets, projections and complex contours are

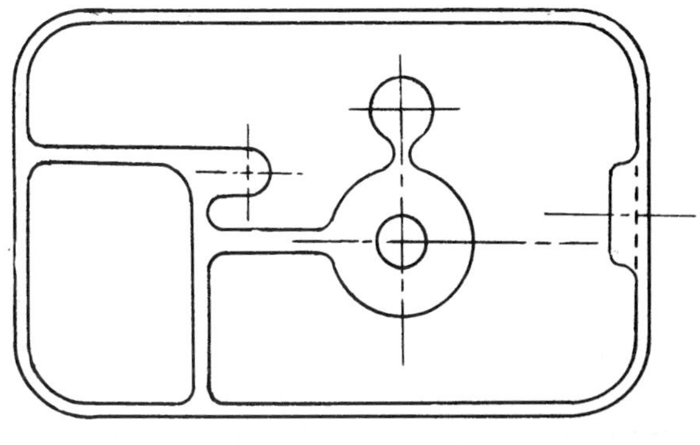

Fig. 5-5. Both a chill and a riser would be required for the large-diameter boss located in the center area of this casting.

responsible for most foundry production problems and scrap, but it also is recognized that these features are inherent characteristics, either singly or collectively, of many casting designs. They can seldom be eliminated, but it is possible to minimize their adverse effect on the production of the casting by carefully modifying the design to accommodate the foundry. This should not necessitate a complete redesign but merely blending or tapering of thin metal sections into thicker sections, coring out excess metal, selecting proper sizes of fillet radii and specifying adequate draft.

All of these features, as well as other equally helpful design recommendations, are shown later in this chapter.

COMPOSITE DESIGN

The tendency of a casting to warp or tear is directly related to its length and complexity. Separating a large and complex casting into parts and welding or mechanically fastening the components together may result in a more sound and more satisfactory part at a lower cost.

It should be noted that composite designs are not restricted to welding or otherwise joining castings to other castings—aluminum forgings and extrusions also can be joined to aluminum castings by either method, and parts made from other metals and fabrication methods may be attached or assembled to the casting with mechanical fasteners. In many cases, assemblies made of different metals improve the service performance of the part by providing a better bearing or wear surface in a specific area of the casting.

SAFETY AND ALLOWABLE DESIGN STRESSES

No aluminum casting nor any part made from any other metal or by any other process should be operated under conditions that will stress the metal close to the published mechanical property values. Most parts are designed to operate at a stress of only a fraction of these values. The ratio of this reduced stress to the ultimate strength of the material is referred to as the safety factor.

The numerical value of the factor chosen for a given part is usually great enough to prevent service failures from severe and unusual service loads, and it also makes allowances for inaccurate stress analysis, poor design features and hidden foundry defects.

In practice, many industries using aluminum castings for critical service applications do not use a safety factor but specify a maximum allowable design stress instead. This value may vary for castings used in different subassemblies of a large piece of equipment or other type of end product. Some factors influencing the allowable design stress for a particular casting are the type and frequency of loading; the effect a failure would have on personal injury, property damage and the operation of the entire unit; the ease and cost of making repairs and replacement; and other considerations that may apply to the particular part.

Also affecting the maximum allowable design stress specification is the type and thoroughness of the in-process and final inspection of the casting. A more stringent quality control and inspection system may justify the use of a higher figure for the maximum allowable design stress.

DESIGN FOR LOW PATTERN AND MOLD COST

After the prints have been studied and the design well-fixed in the designer's mind, there still remains the problem of visualizing the construction of the sand or permanent mold casting equipment. Most design engineers leave this up to the equipment makers, who may or may not consult with the foundry production people who will be responsible for producing sound castings at an acceptable price. This lack of collaboration is not a good practice.

It should be emphasized that a casting design is not good unless it favors low-cost equipment, can be readily cast, meets all service requirements and has an acceptable appearance. It may be too much to expect a designer to be a gifted patternmaker, mold builder and foundryman. However, the designer should work closely with the pattern- or moldmaker and the foundryman to produce the best design. It is not possible to formulate hard and fast rules because each casting design must be considered individually, and it is here that practical experience is necessary.

Figure 5-6 shows how minor design changes substantially reduced the production cost of a cylinder head casting without affecting its appearance or service performance. In the original design, irregular core parting increased pattern cost, slowed core production and made form driers necessary for these cores, which were made from an oil bond core mix.

The high rib extending close to core parting weakened the lower water jacket core to such an extent that many were broken in process. The rib extending into the upper half of the water jacket core slowed production and increased the cost of the core. The draft needed for the upper half of the water jacket core prevented tapering of the thin side walls into the thicker deck. The broad trim surface made flash

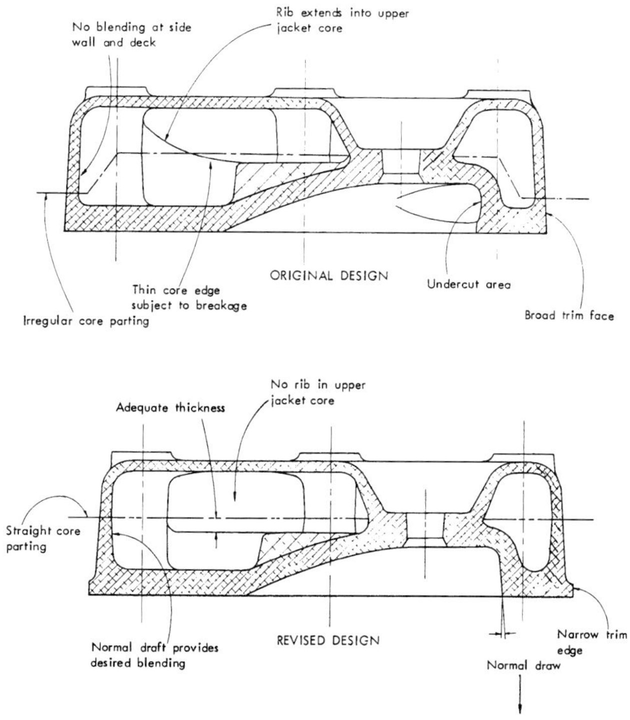

Fig. 5-6. Cross section of the head for an L-head engine. Improved design shows engineering changes that substantially reduce pattern and production costs.

removal difficult. The undercut contour in part of the combustion chamber prevented the normal draw of a one-piece core, again slowing production and increasing cost.

Although many design engineers have no further contact with a job after the design is complete, it often is an advantage when designers can follow their creations through all the final steps. They then can make design changes at early stages, at comparatively low cost, wherever changes would favorably affect the cost or quality of the end products.

MODELMAKING AS A DESIGN AID

Time and money are usually saved when complicated castings are first mocked up as accurately scaled wood or plastic models. Even the best-trained engineers or foundrymen may miss important details of design if all they see is a blueprint. Many costly design changes and many casting repairs are obviated if a model of the part to be cast is made and studied.

Areas of potential shrinkage and/or tearing can be predicted, gates and risers can be placed without trial and error, and money-saving changes in pattern construction can be spotted before working patterns are made. Sometimes it is advantageous to make a model-sized casting that can be cut up and inspected to determine potential problems.

By making sectional models, each part of the casting can be studied. Transparent plastics are excellent for model work; modeling clay can be used to simulate padding and other important features. Models have been used for determining the best designs and procedures for a variety of products.

Simulation of the casting process using computer-aided design (CAD) models and computer-aided engineering (CAE) along with solidification software reduce the need for sample casting during design development.

SUMMARY OF BASIC CASTING DESIGN

Good design is becoming an important requirement for making satisfactory castings. A designer of aluminum castings needs to have a good knowledge of the characteristics of the many available casting alloys and the various heat treatments and their effect on the production and service performance of the casting. Lacking any of this knowledge, the designer should consult with those who do have it, particularly with those who may be asked to produce the casting.

Among the basic rules for satisfactory aluminum casting design are:

1) Limit drastic changes of thickness for minimum casting stresses.
2) Taper sections as liberally as possible toward risers and use chills to help control directional solidification.
3) Do not interpose thin sections between the risers and heavy sections.
4) Avoid extensive horizontal, flat surfaces.
5) Avoid isolated hot spots that are difficult to feed.
6) Determine casting properties and use commensurate safety factors.
7) Simplify a complex casting by dividing it into natural components that can be either welded or mechanically fastened together.
8) Use minimum section thickness necessary for the flow characteristics of the metal, the particular design configuration and the service requirements of the casting.
9) Core out holes, slots and similar surface features to save metal and machining cost, but not to the extent where doing so would increase the cost of the equipment and production.
10) Be sure that specifications are realistic and sensible.
11) Do not compromise on the rules. As stated previously, the more the engineer can help the foundry at the design stage, the better the foundry will be able to make the required castings.

DESIGN CONSIDERATIONS

Locating Points for Machining

Many aluminum castings are machined. The locating points to be used by the machine shop should be marked clearly on the pattern drawing. Locating points should be discussed with the casting producer before machining fixtures are built.

In general, locating points should be as far apart as possible and should be on those outside walls of the casting that have the greatest dimensional accuracy. If possible, these points should be placed on only one and preferably the largest single pattern or mold section. Avoid placing locating points on surfaces formed by separate cores or mold sections because tolerances on dimensions across parting lines are greater than those formed by a single pattern or mold section. The foundry should use these same locating points for checking the dimension of the castings. Obviously, gates, risers and chills should not be placed at these points.

Machining Allowances

Recommended minimum machining allowances for each casting process vary and are given in succeeding sections of this chapter. However, allowances less than the minimum are possible in specific cases where casting and design permit, but the designer and the caster should agree to the reduction.

When possible, the portions of a sand casting to be machined should be cast in the drag, because the drag portion of the casting usually is more sound than the portion cast in the cope. In some cases, machining allowances are greater for cope sections of castings. However, casting producers are aware that if good pouring and gating practices are used, sound cope sections also can be obtained, thereby minimizing the machining allowance and resulting in savings of material in casting and of machining time. Usually, little or no machining is done on die castings.

Sand Cores

The designer should attempt to eliminate cores except where they are required functionally or where their inclusion will reduce the cost of the finished part. Good design also may require coring to provide sections of uniform thickness and to minimize machining.

From the foundry standpoint, the design of cores is a very important aspect of overall casting design. Core materials and coremaking processes are discussed in detail in the chapters devoted to the individual casting processes. It is important to remember that cores always require special tools, extra handling time and more liberal allowances.

It also is important that the minimum dimensions of the core be as large as possible to facilitate its production. Thin or small-diameter cores are extremely fragile and are a cause of both core scrap and casting scrap in the foundry. Where feasible, very small cores should be eliminated, particularly if the hole will be finished by machining.

Parting lines on cores should be avoided, if possible, because they inevitably result in a distinct line on the casting at the juncture of the core parts. Internal cores create a special problem if they are relatively long and thin because they are fragile and difficult to vent, and the extraction of the cores and core wires is difficult. Core prints should be sufficiently large to support the core, thereby avoiding dimensional errors and facilitating the venting of gases to avert core blows.

Chaplets as Core Supports

Chaplets are specially designed small metal pieces that are placed in the mold to assist in supporting cores in their proper positions. This is necessary if core prints of adequate size are not possible. Copper-base alloys should not be used for chaplets. Aluminum is preferred because the chaplet remains in the casting as a structural discontinuity. Consequently, chaplets should not be used, except where essential. They should never be used in castings that must be leak-tight because the space between the chaplet and the aluminum casting will almost certainly allow leakage.

Metal Inserts

In the design of aluminum castings, it frequently is desirable to include inserts of other metals or alloys. This can readily be accomplished. Cast iron and steel inserts are preferred over those of copper-base metals, which are likely to be attacked by the aluminum during the pouring operation unless they are coated with a light mold wash and thoroughly dried. This is particularly true for thin-wall copper tubing, which is occasionally cast-in to serve as a conducting line.

When casting an insert in aluminum, it is important that the insert be free of all finger marks and any organic substance because, if any are present (even to a minute degree), it will cause a blow. It is recommended that inserts be cleaned with a light sand or dust blast just prior to using. Inserts should not be touched with bare fingers and, whenever possible, should be preheated just prior to being placed in the mold.

The designer of any aluminum casting requiring cast-in inserts must provide sufficient metal thickness around the insert to prevent it from cracking the casting. Except in certain patented processes, a metallurgical bond is not obtained between the aluminum and the insert; consequently, it is common practice to knurl, groove or otherwise roughen the surface of the insert to obtain the desired lock or grip. Radiography and ultrasonics are two satisfactory methods of checking for possible blowholes between the insert and the casting. Typical forms of inserts used in sand and in permanent mold castings are shown later in this chapter.

Recording Design Changes

If design modification is necessary, it is important that any changes be noted clearly on the design drawings. In some instances, foundries improvise methods to produce satisfactory aluminum castings by slightly changing the pattern or one or more of the gating, risering and pouring practices.

Such modifications may appear to be of minor importance. However, the same job may be handled again at a later date. If a record of design modification is not made at the time of the original run and details are left to memory for later runs, or if different employees make the castings without being aware of the changes, then considerable time and money may be lost through trial and error. Any changes in design should be "signed off" by the designer, pattern- or moldmaker, and the foundryman.

Calculating Casting Weights

During the planning of each casting, determining the amount of metal that will be required to make each pour is always an important part of the development. There are available publications and computer software that present in detail various methods for determining casting weights. These sources on measuring are valuable aids for the estimator who must use various methods for determining casting weights.

Weighing the Pattern

Although the method is sometimes used, the pattern weighing method has limited use and questionable reliability. Relatively few patterns are suitable for this method because it can be used only for solid patterns and for those with voids and cores with volumes that can be calculated readily. When the weight of the pattern or portion of the pattern forming the actual casting shape has been determined, it is then multiplied by the appropriate factor, depending on the pattern material. These factors are shown in Table 5-1.

There are many reasons for questioning the reliability or practicality of this method. The densities of wood may not always conform closely with the published figures. In instances where the pattern equipment includes a number of coreboxes, this method is impractical.

Displacement Method

The volume of complex cores can be determined by filling the coreboxes with dry sand and then measuring the volume of the fill in a graduated container. The dry sand fill also may be weighed. A variation of this method is to fill a graduated

Table 5-1.
Factors Used When Determining the Weight of Castings

PATTERN MATERIAL	FACTOR
White Pine	5.7
Honduras Mahogany	4.5
Cherry	3.8
Aluminum	1.0
Brass	0.32

beaker or similar container with fine sand and then pour the sand into the cavity until flush. Subtracting the volume of the sand remaining in the beaker from the original amount gives the volume of the cavity. Water, kerosene and other liquids can be used in place of sand with equally satisfactory results.

When using the displacement method, care should be taken to ensure that the sand (or liquid) fills the entire cavity, including undercut areas. It also is important to subtract the volume of any cavity that forms an integral core print.

Calculating Sections of Castings

To accurately calculate the weight of a casting, the design must be divided into simple geometric elements, such as cubes, cylinders, spherical segments, cones, pyramids, etc., and the volumes calculated using the appropriate formulas found in most engineering handbooks.

The sum of the volumes is multiplied by the weight/cu in. of the casting alloy specified. The weight/cu in. for most of the common aluminum alloys ranges from 0.093 lb/cu in. for 220 alloy to 0.107 lb/cu in. for 238 alloy (see Physical Properties Tables in Chapter 4). However, for most purposes, the value of 0.10 lb/cu in. is accurate enough for calculating aluminum casting weights.

GENERAL DESIGN SUGGESTIONS

The primary aims in designing aluminum castings are to plan components to facilitate foundry production and to improve casting soundness. It is important to plan arrangement of sections so that progressive solidification is toward the heavier sections, as described elsewhere in this book. Various design techniques to promote directional solidification are illustrated in this section. In general, these practices are applicable when designing aluminum castings for any casting process, but some are particularly applicable to the production of castings in nonmetallic molds.

The general design principles illustrated in this chapter are intended mainly for the design engineer and not the foundry. This distinction is made because there are two basic aspects that are fundamentally important in the efficient production of aluminum castings.

The first aspect is that the design engineer can aid the foundry and help himself by designing castings so that the optimum balance between performance requirements, foundry economy and appearance of the completed part is achieved. For this reason, many of the design features are presented here with the conviction that their value is twofold—they will help engineers to improve the design, appearance and strength of aluminum castings, and their acceptance by the designer eliminates many problems for the moldmaker and the casting producer, which results in better castings at lower costs.

The second basic aspect in the efficient production of quality aluminum castings includes all the procedures of good aluminum foundry practice. This aspect is mainly the concern of casting producers and includes the technical skills and production knowledge required for moldmaking, proper gating, risering, chilling and other production details that are the subjects of other sections of this book.

The information and suggested design features given in this chapter are meant as general practices for comparative purposes, not as inviolable rules. When there is doubt concerning practical dimensional tolerances for critical areas of casting—areas for which the casting producer may actually be able to supply closer tolerances—the casting designer should consult the casting producer. When the foundry thoroughly understands the designer's objective, sound castings are usually made with better design reproducibility and at lower costs.

Uniform Thickness

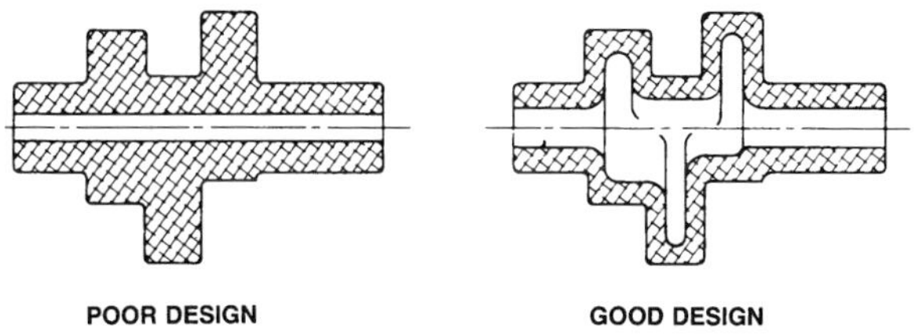

This hydraulic pump component was originally designed with a small-diameter straight core through the center. Excessively heavy sections resulted in shrinkage, which caused a high rate of rejects during machining.

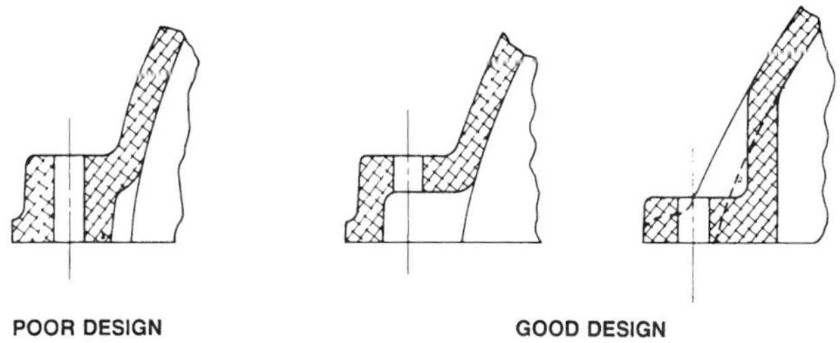

Uniform sections can save weight, material and machining costs and often result in a stronger casting.

Hot Spots and Sharp Internal Corners

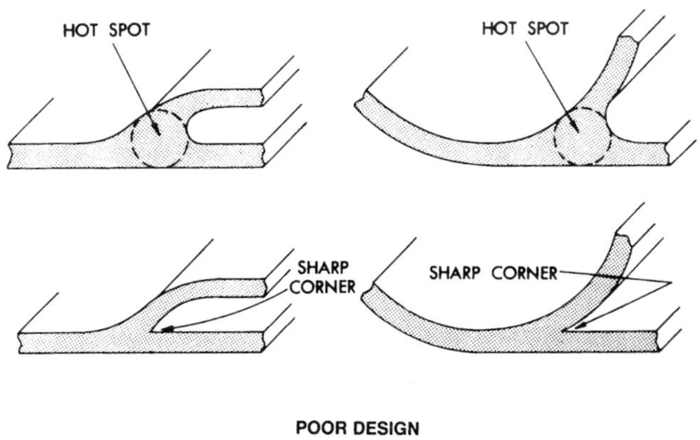

Hot spots and sharp internal corners can cause shrinkage cracks and structural unsoundness. Heavy concentrations of metal and sharp corners can be avoided by proper design of the fillet radii and of the direction of intersecting walls to maintain uniform metal thickness.

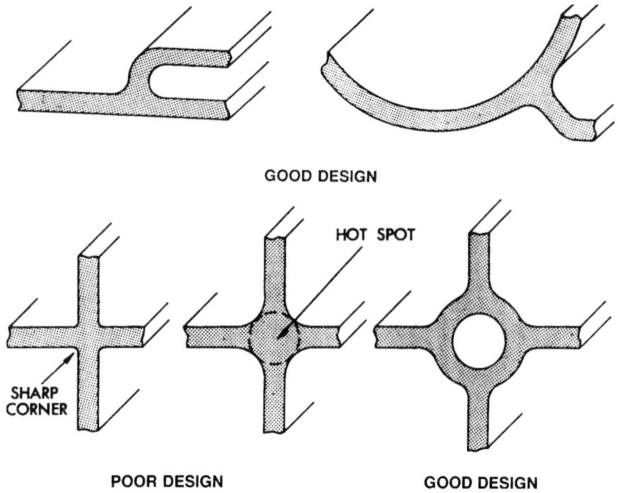

Joining and Changing Sections

Sharp corners should be filleted. Fillets prevent cracks, tears and shrinkage at re-entry angles; make corners more moldable; and reduce stress concentration in the casting when in service. Fillets should be large enough to meet engineering stress requirements and to reduce stress concentration, but not so large that they cause shrinkage.

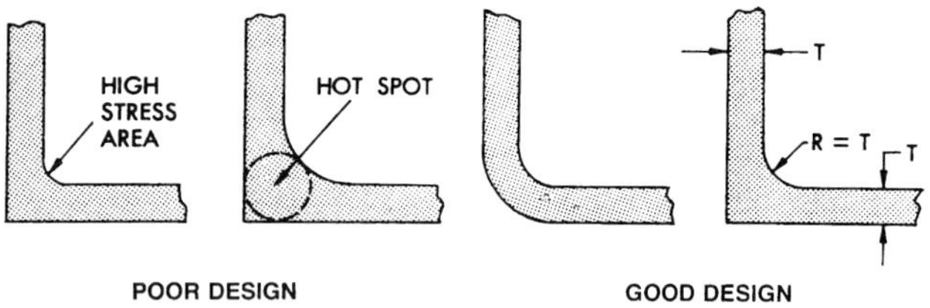

Details for joining equal walls.

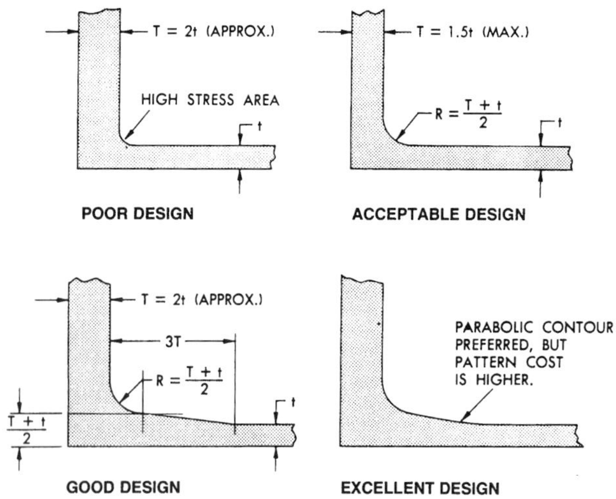

Details for joining unequal walls.

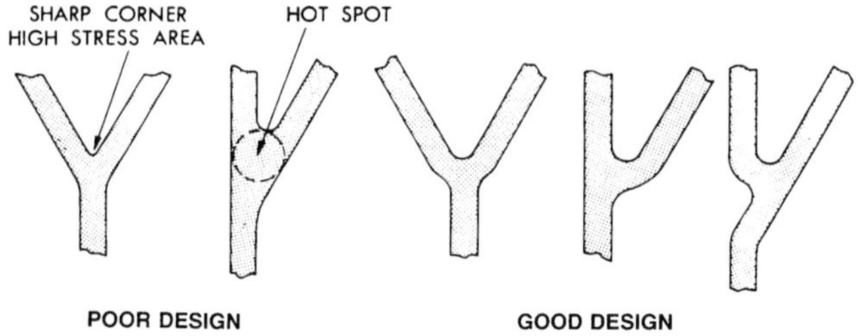

Fillet details for Y or V sections of uniform thickness.

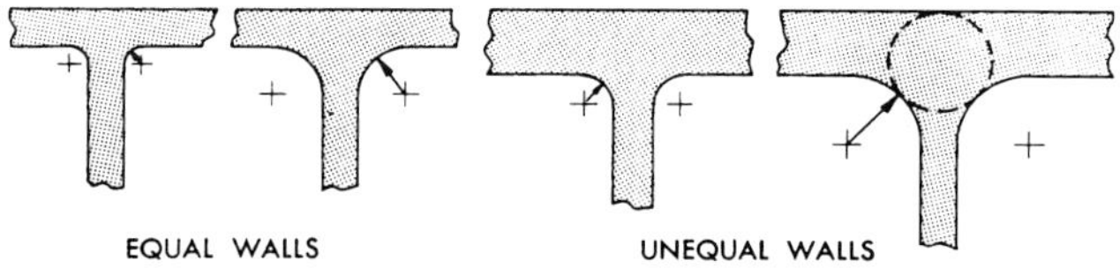

Design of all T-sections shown above are poor because of improper choice of fillet radii.

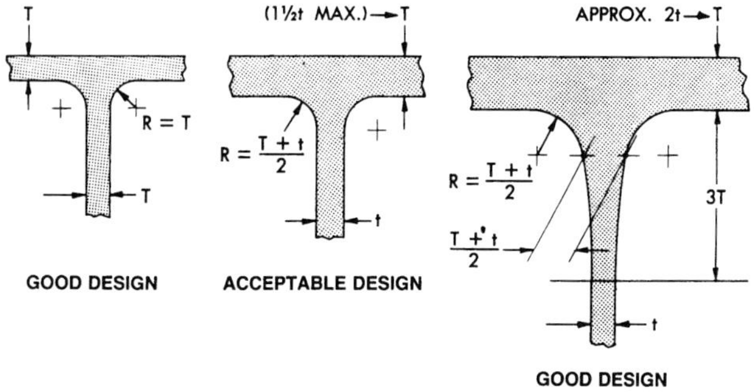

Recommended fillet radii and blending practice for walls intersecting at 90 degrees.

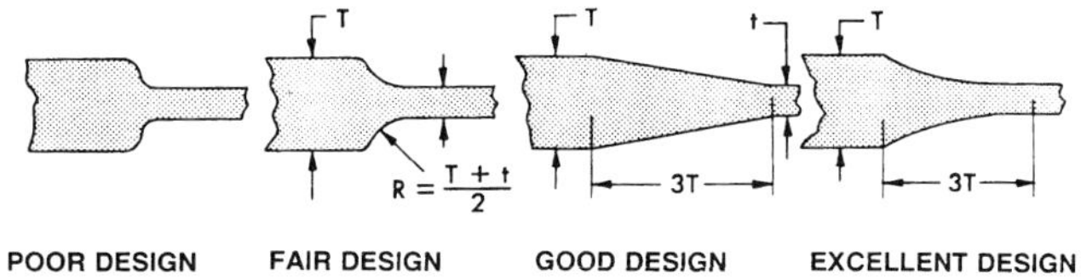

Design details for changing wall thickness.

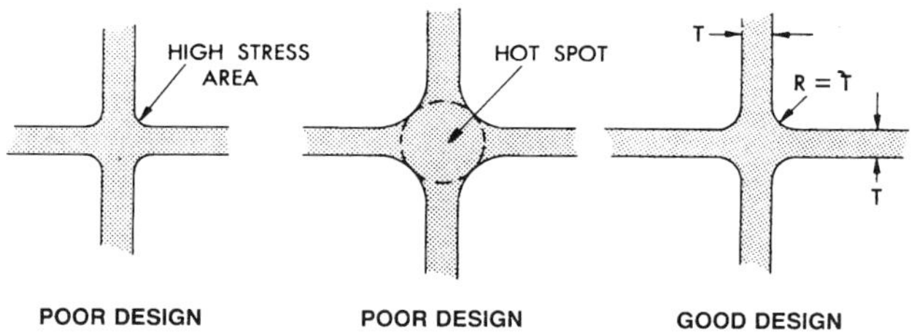

Joining equal intersecting walls.

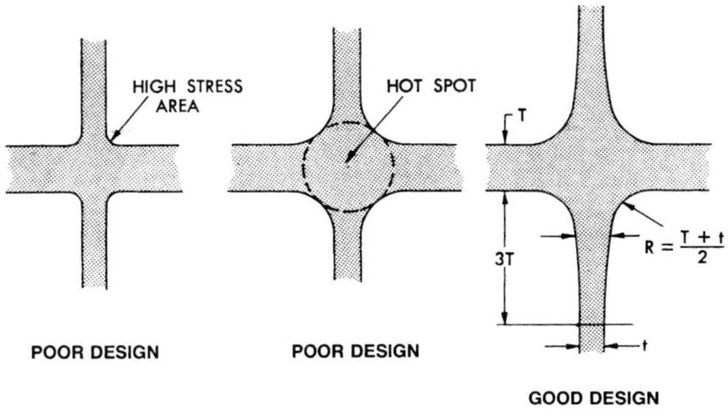

Joining unequal intersecting walls.

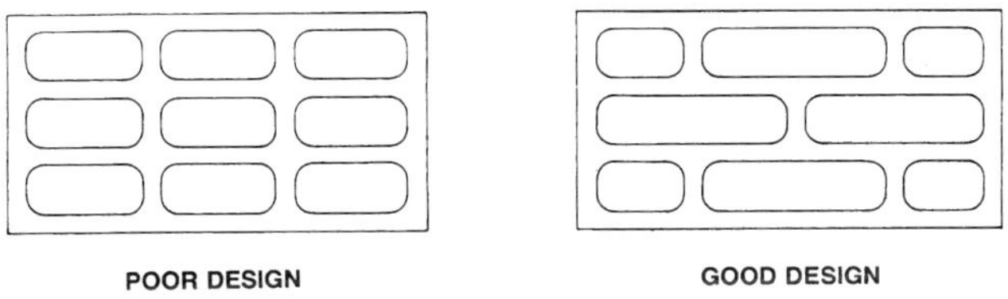

Staggered ribs reduce distortion caused by thermal contraction, reduce metal mass in local areas and minimize the possibility of hot spots.

In a well-designed casting, all sections should be no thicker than necessary to give the casting the desired structural strength. They should be as uniform in thickness as possible to facilitate directional solidification.

Pads and Bosses

Reductions in weight and machining costs often can be realized by including bosses and/or pads in a casting design, but poor location and improper proportions may complicate production by creating hot spots, which could cause shrinkage cracks. Pads and bosses located on walls not normal to the parting line should be "tailed" to eliminate the need for sand cores, loose pieces, split metal cores and multiple mold partings. Auxiliary mold slides should either be eliminated, relocated or redesigned. In a permanent mold, it is advantageous to have ribs on the mold parting line where they can function as feeders.

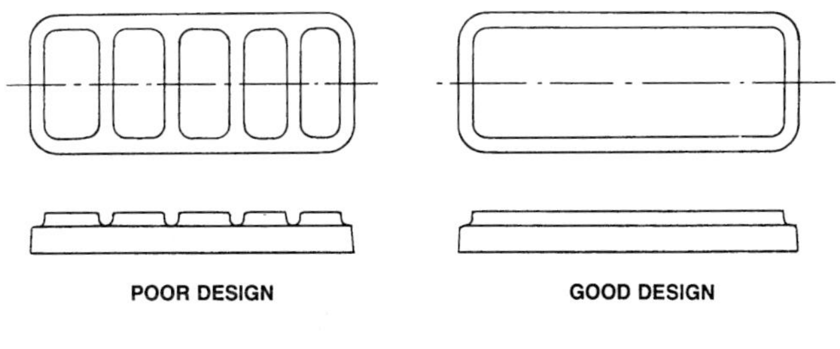

It generally is preferable to keep the height of bosses and pads less than the thickness of the adjacent metal. When there are several bosses in close proximity on one surface of the casting, connecting them may facilitate machining.

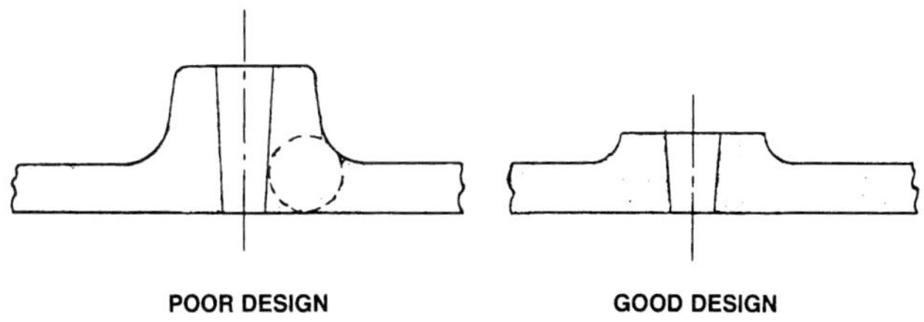

It usually is preferable to keep the height of bosses and pads less than the thickness of the casting section but high enough to permit machining without touching the surfaces of the section.

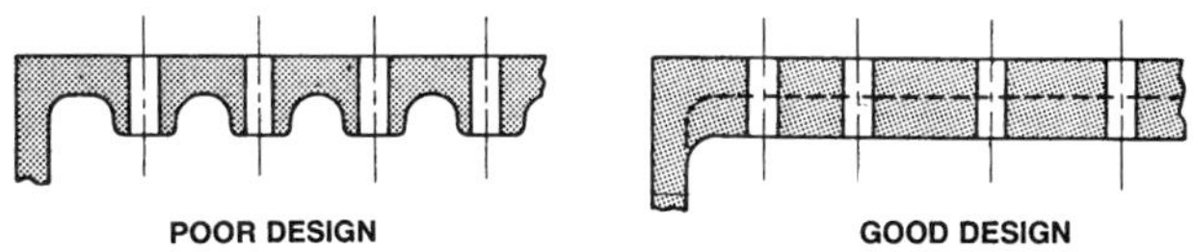

When there are several lugs or bosses on one surface, joining them will facilitate machining.

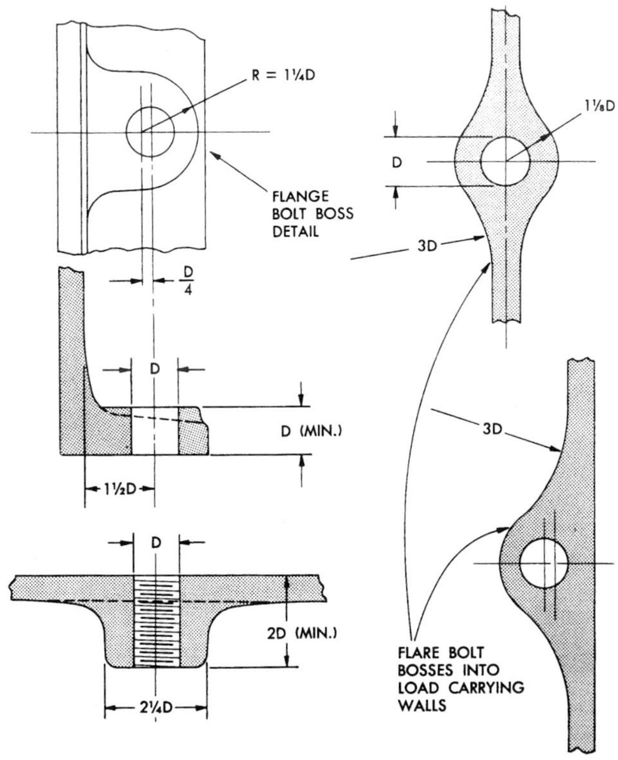

Proportion bolt bosses correctly. The dimensions and proportions given in the sketches above are reliable general guides that may be modified to suit foundry practice and the shape of one section.

The long flared walls are desirable from the designer's aspect because they distribute loads over a wide area, but the casting producer may prefer less blending of wall sections.

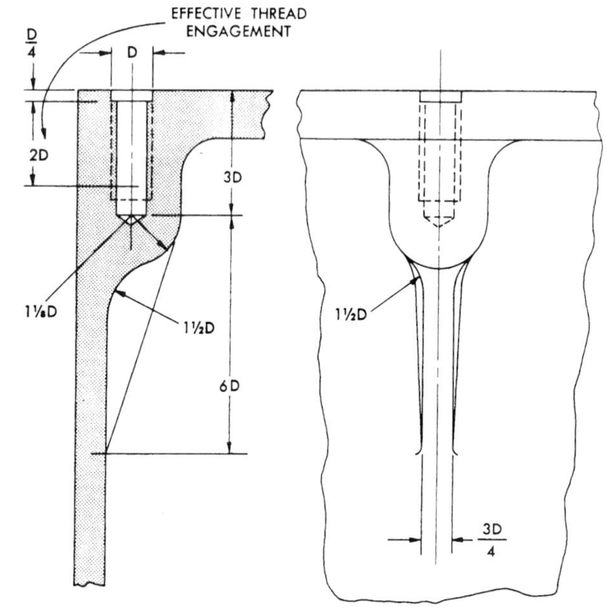

This drawing illustrates the recommended proportions for heavily-loaded threaded bosses tying into side walls.

In general, bosses for fasteners should be designed with enough material to permit liberal tolerances and, if necessary, the use of the next larger size of fastener.

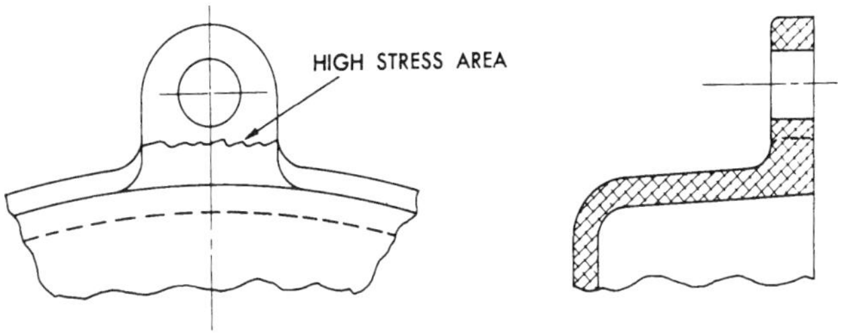

POOR DESIGN

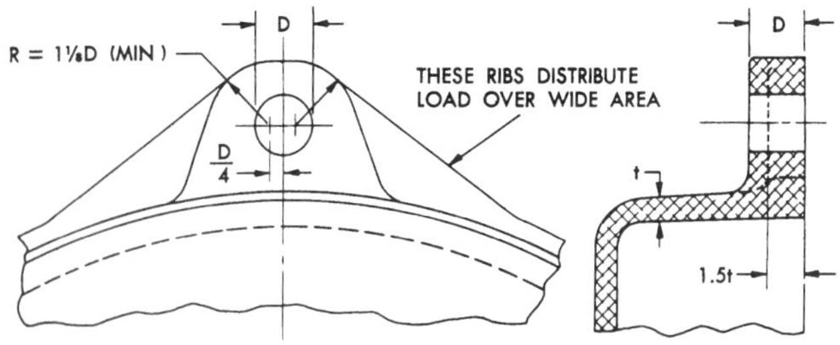

ACCEPTABLE DESIGN
OFFSET BOSS AND HORIZONTAL RIBS

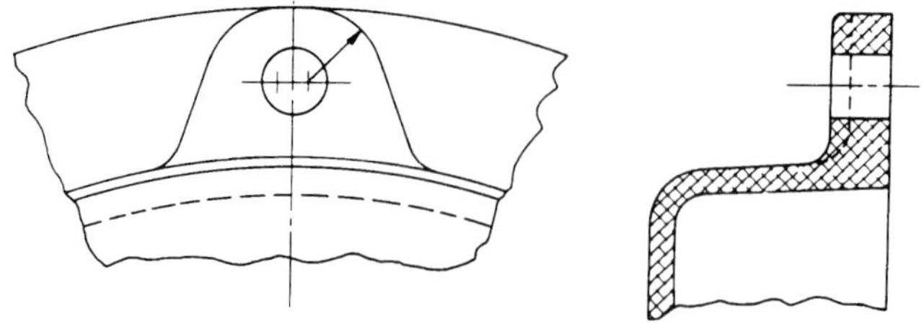

ACCEPTABLE DESIGN
OFFSET BOSS AND CONTINUOUS FLANGE

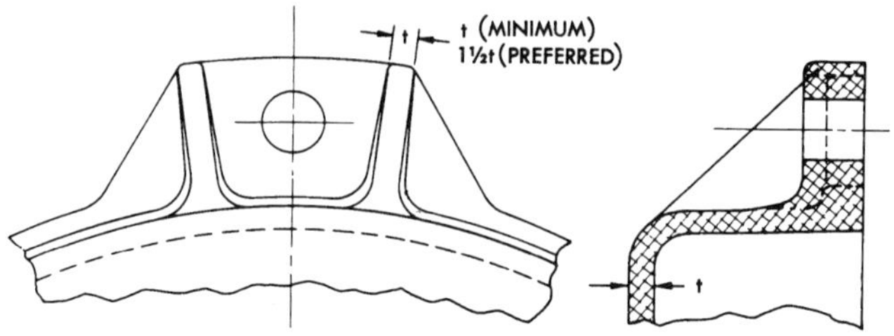

ACCEPTABLE DESIGN
REINFORCED BOSS WITH VERTICAL RIBS

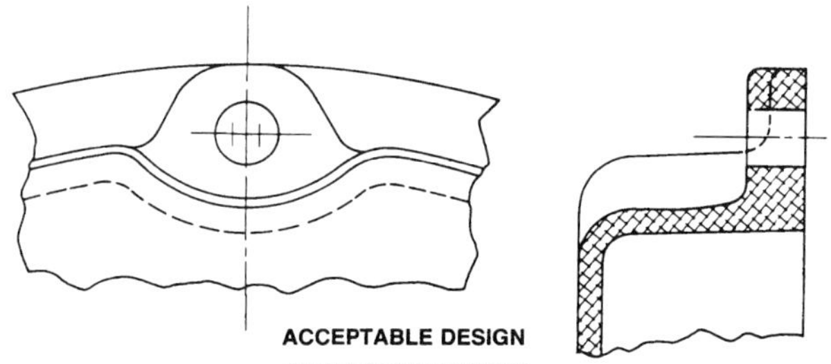

ACCEPTABLE DESIGN
SEMIRECESSED BOSS

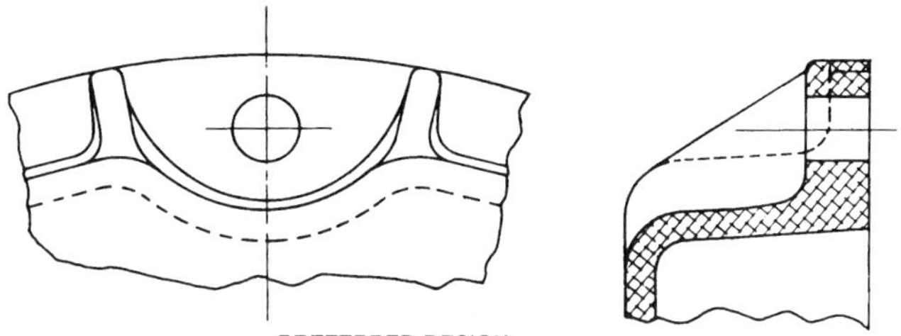

PREFERRED DESIGN
SEMIRECESSED BOSS AND RIBS

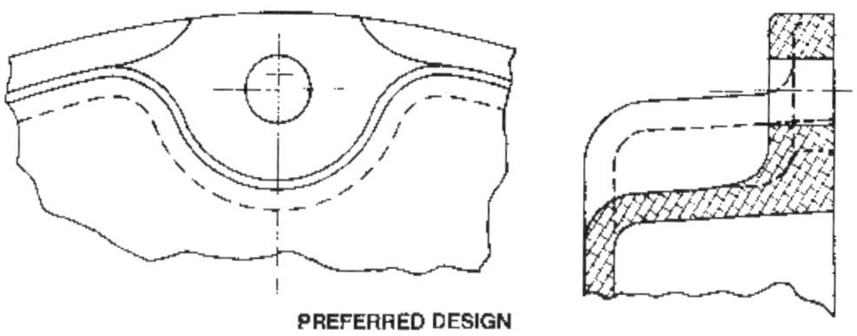

PREFERRED DESIGN
FULLY RECESSED BOSS

These variations of bolt boss design are not arranged in order of importance or desirability. The first, of course, is not recommended. The six remaining sketches are some of the most commonly used forms. The choice for any specific application usually is determined by such factors as die or mold partings, appearance, and clearance problems.

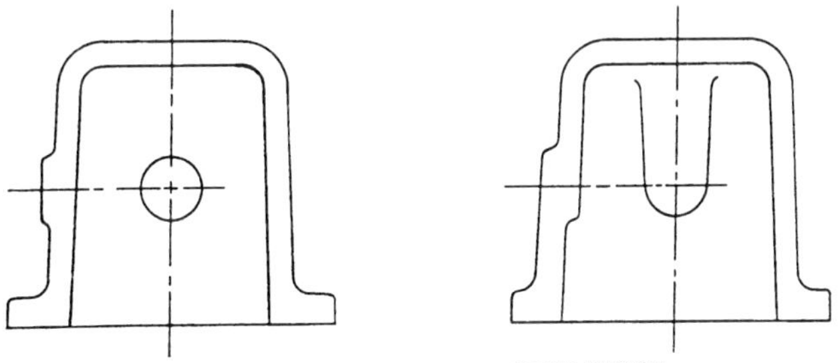

POOR DESIGN
FREE STANDING BOSSES

GOOD DESIGN
"TAILED" BOSSES

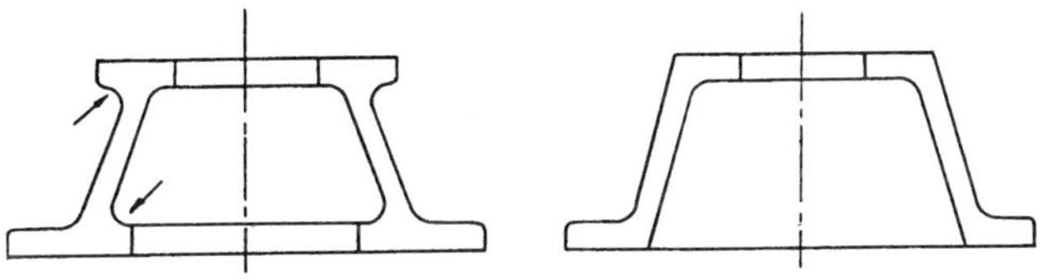

POOR DESIGN
PROJECTING FLANGES AT TOP AND BOTTOM COMPLICATE MOLD DESIGN

GOOD DESIGN

Free-standing internal and external bosses can be cast readily, but to do so requires the use of sand cores, loose pieces or collapsible metal cores.

Using any of these increases production costs and, consequently, the cost of the casting. However, the added expense may be held down by "tailing" the bosses, as shown in the sketch.

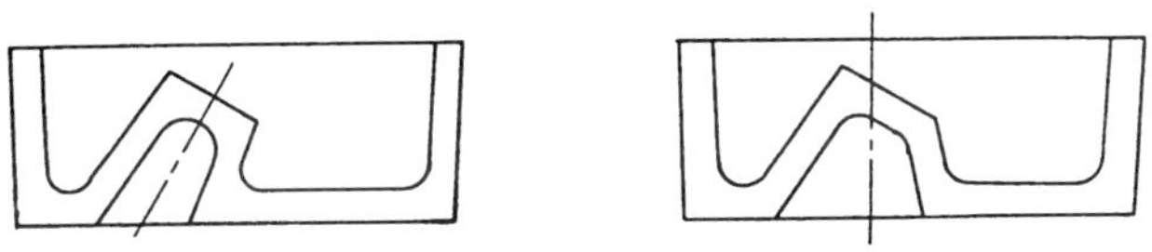

POOR DESIGN
UNDERCUT BOSS IN UPPER MOLD HALF AND CORE PULL DIRECTION NOT NORMAL TO MOLD PARTING PLANE COMPLICATES MOLD DESIGN

GOOD DESIGN

Designing sloping bosses for straight parting of the mold or die will avoid extra expense in mold or die construction.

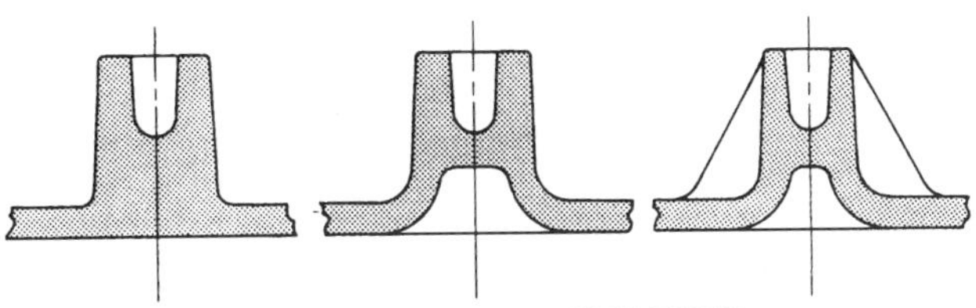

POOR DESIGN

GOOD DESIGN

Ribs and/or fillets improve bosses. Hollow back reduces metal mass. Ribs in sketch on the far right also permit reduction of diameter of the boss without sacrificing strength.

Ribs and Webs

Ribs have two main functions: 1) to improve stiffness and 2) to reduce weight. Correct rib depth and spacing depends on the overall engineering design.

Webs

There is no easy method for computing the proper section thickness of webs. It is important, however, to both the designer and the foundry, that the webs are neither too thick nor too thin for the specific design.

Web thickness is dependent on such variable factors as the web area, the alloy used, the structural requirements, and the number and geometry of joining webs and stiffeners.

Ribs

Ribs are recommended for reinforcing or stiffening most web sections and are almost mandatory for thin webs for the following reasons:

- Ribs can reduce web thickness and save weight.
- Ribs strengthen and stiffen webs, which reduces load deflection.
- Ribs can prevent warping of large, unsupported web areas.
- Ribs can be used to strengthen the casting and prevent hot tearing or cracking by acting as chills.

In the low-pressure permanent mold and the centrifugal casting processes, ribs sometimes can serve as feeders.

It is good design practice to stagger rib junctions in order to avoid the large mass of metal that is present at junctions of multiple sections. This large metal mass at junctions can cause uneven cooling (by preventing directional solidification), which may result in defective castings.

Staggering the ribs will reduce the strength of the section, but the loss both in strength and in the amount of metal can be held to a minimum at the intersections by staggering the rib junctions by a distance no greater than the thickness of the rib.

Ribs that cause undercuts and require cores should be avoided. Cores are necessary for parallel ribs and also are required if two or more ribs are located at the parting line. If only one rib is used, it can be cast right on the parting plane without coring.

It generally is preferred to design ribs that are appreciably deeper than their widths. The recommended width for ribs in aluminum castings is 1½ times the thickness of intersecting walls; minimum rib width is equal to the wall thickness. Ribs in compression offer a greater factor of safety than ribs in tension.

Thin ribs are highly subject to service failures and generally are regarded as improper design. This, however, does not apply for cooling fins used on air-cooled cylinder heads and other applications where fins are made thin for a specific purpose.

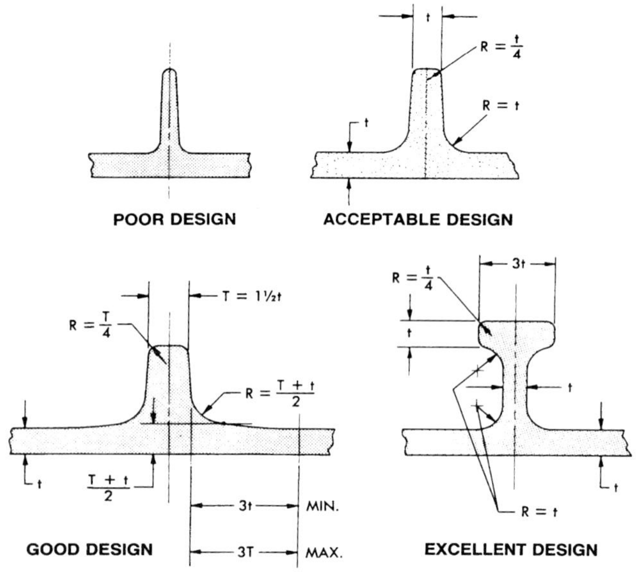

The beaded rib design in the sketch labeled "Excellent" is rated so mainly as a design for load-carrying capability but not necessarily for foundry considerations.

It is recognized that production and engineering recommendations are not always in agreement, and in some cases, the beaded rib may be difficult to cast. However, the design configurations of many castings permit the use of this desirable rib form, and its presence usually will not create additional foundry problems, at least not when the rib is located on the parting plane of a mold or die.

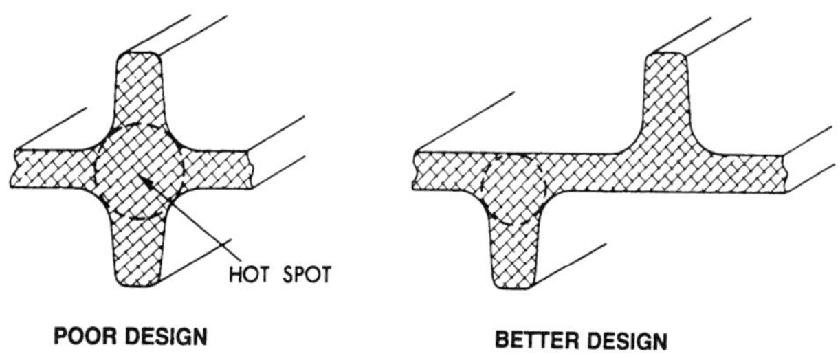

Cross-ribbing, or ribbing on both sides of a casting, is undesirable from the casting producer's standpoint because it generally increases the rate of casting defects and the total cost. Cross-ribbing creates hot spots, makes feeding difficult and may cause shrinkage.

Avoiding complex ribbing simplifies molding procedures and assures more uniform solidification conditions. If the casting wall alone has adequate strength and stiffness, casting stresses are lowered and stress distribution improved when ribs are removed.

Webs can also be stiffened without the use of ribs (which may be objectionable in some designs because of appearance or interference conditions). For these applications, the use of the "wavy" web may be the best solution.

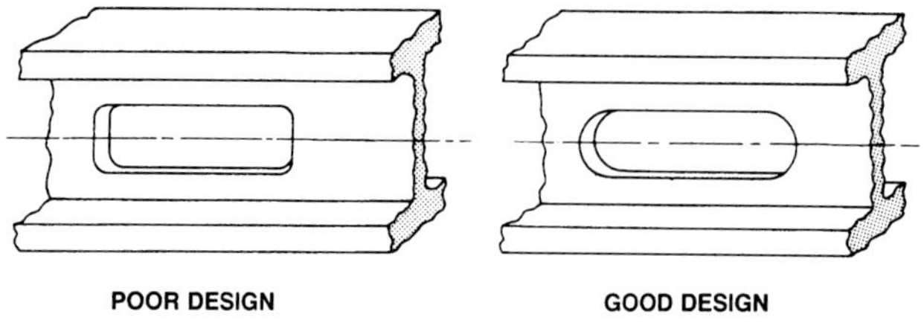

Cored rectangular holes in ribs and webs generally are not desirable. Round or oval holes are preferred, but if other shapes are used, all corners should be well-rounded. It is desirable to bead all cored holes.

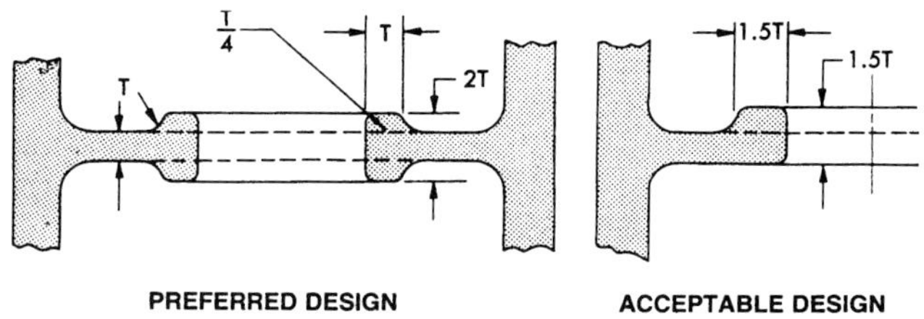

In highly stressed webs, ribs and walls, beading cored holes is important to the service life of the part. However, this desirable design feature may complicate production.

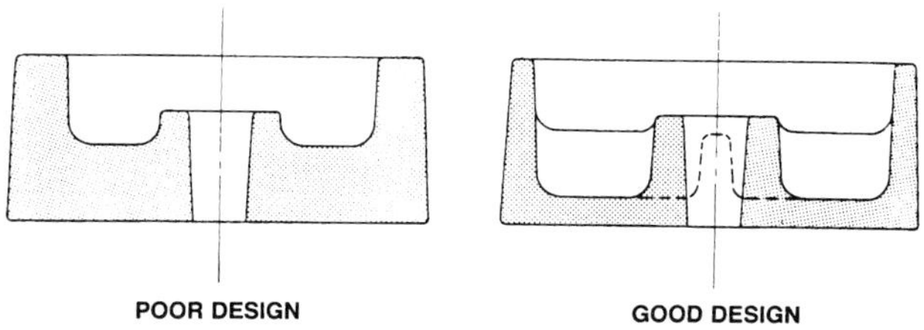

Add ribs for strength where a heavy section has been cored to achieve uniform wall thickness.

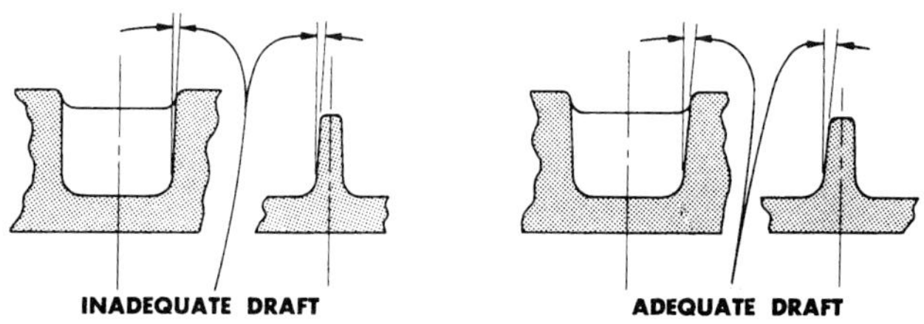

Removal or ejection of deep ribs is made easier by providing adequate draft. See Fig. 5-7 for additional draft data.

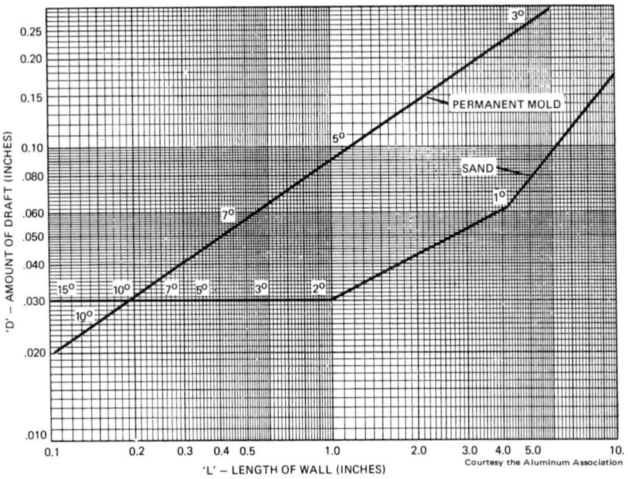

Fig. 5-7. Suggested draft allowances for sand and permanent mold aluminum castings.

Cast-In Metal Inserts

Cast-in metal inserts should be considered by the designer if their use will lower cost and/or improve the performance of the part. Cast-in inserts often can eliminate drilling, tapping, reaming, pressing and assembling steps.

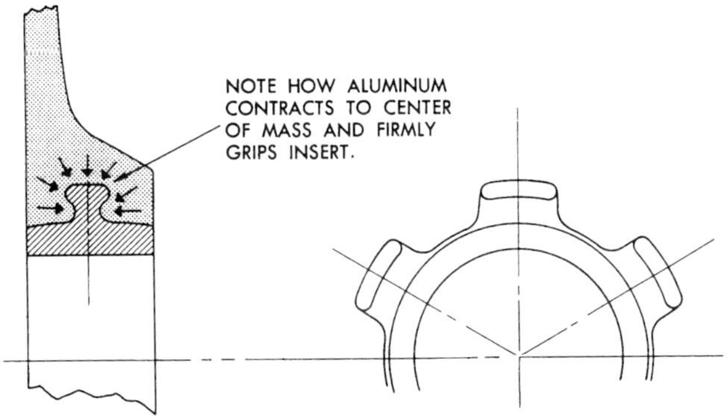

FOR MODERATE LOADS

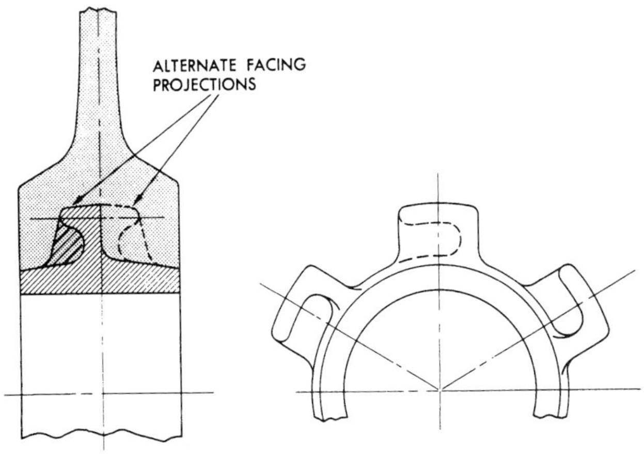

Cast-in metal inserts are designed for aluminum castings that will be subjected to torque or thrust loads in service. The projections on the inserts are shaped to hold the aluminum tightly against expansion at elevated operating temperatures and the area is designed to take torque and/or thrust loads. Castings with these inserts should not be solution heat treated, but they may be artificially aged.

It should be noted that when cast-in inserts are placed horizontally in a sand mold, the core gases are free to pass around the ends of the inserts. If the ears were a continuous band, gases would be trapped on the lower edge, which would prevent close contact of the aluminum to the insert. The rounded edges enable the aluminum to lay close to the insert. If they were sharp internal corners, the aluminum would not fill properly, thereby leaving voids.

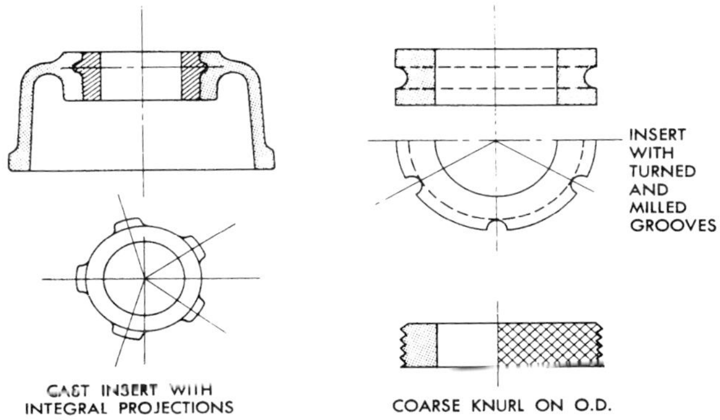

These sketches illustrate some methods used to obtain adequate mechanical bond between aluminum and inserts.

Additional General Design Details

The following suggested design details, like those preceding, should not be accepted as positively applicable in every case. Many points are controversial, for there are sound reasons for variations or exceptions in specific cases.

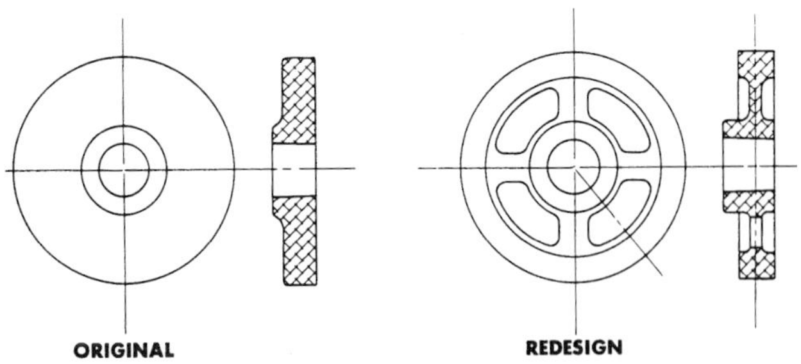

Cored holes and webs can be used to reduce the weight of castings.

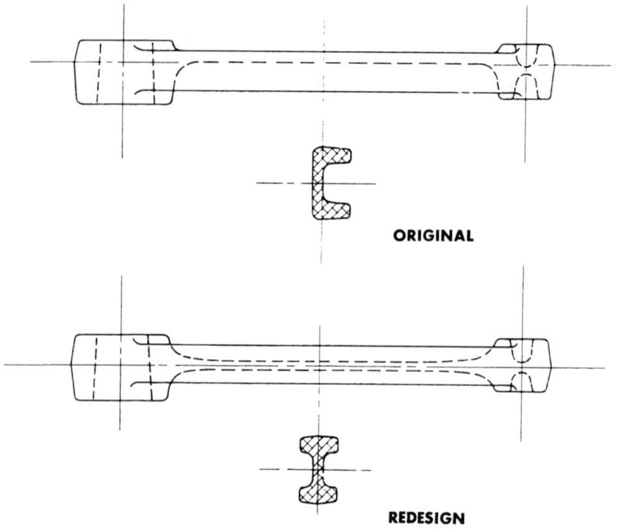

Distortion in long castings can be avoided by distributing metal symmetrically about the longitudinal center line.

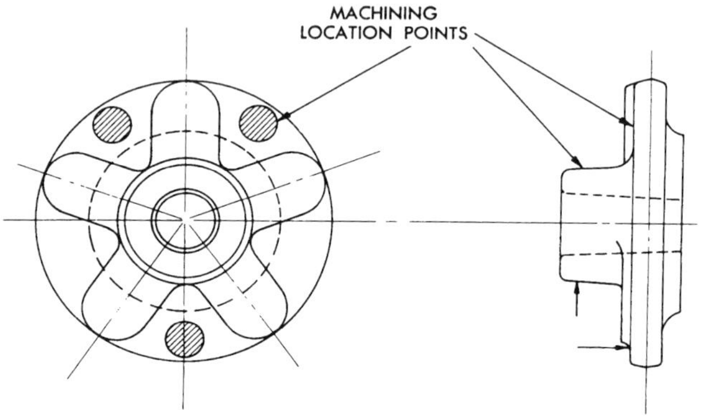

Machine locating target points, as shown in the wheel hub above, should be indicated clearly on the drawing. The same points will be used by the inspector who certifies the casting for machining and also by the machine operators.

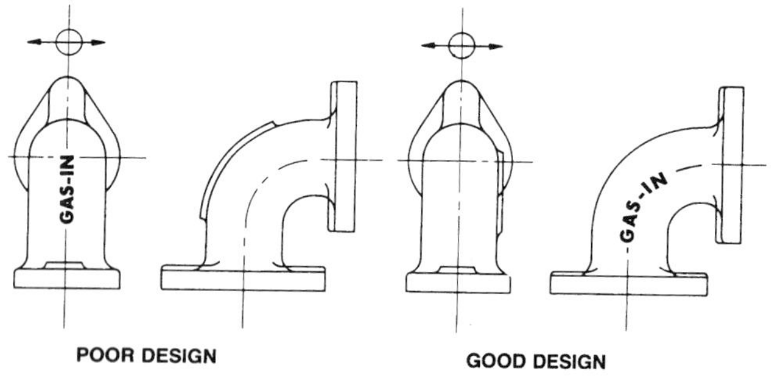

Locate lettering on a casting surface that is normal to the direction of parting. Avoid placing raised letters on surfaces where they will obstruct normal die or mold opening or will require the use of separate sand cores, loose pattern pieces, sliding cores in permanent molds, or other special features that increase tooling and production costs.

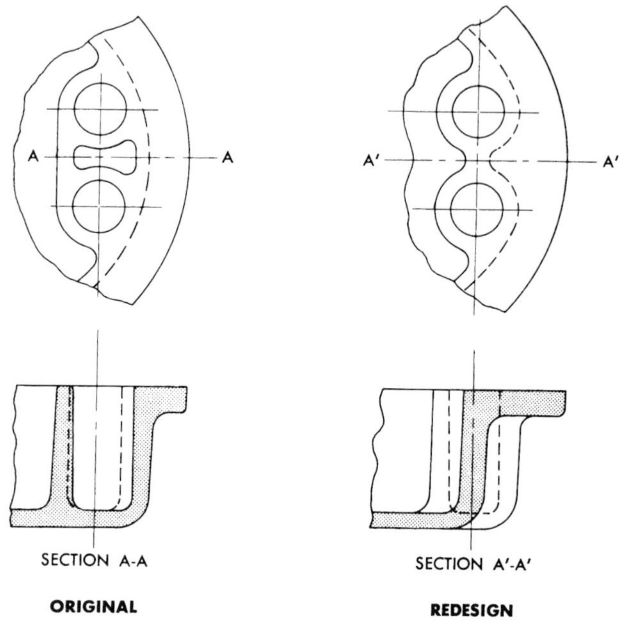

The deep, cored pocket in the original design was eliminated by changing the contour of the casting, which in this case resulted in lower tooling cost, faster production rate and a lighter casting. Total cost of the casting was reduced.

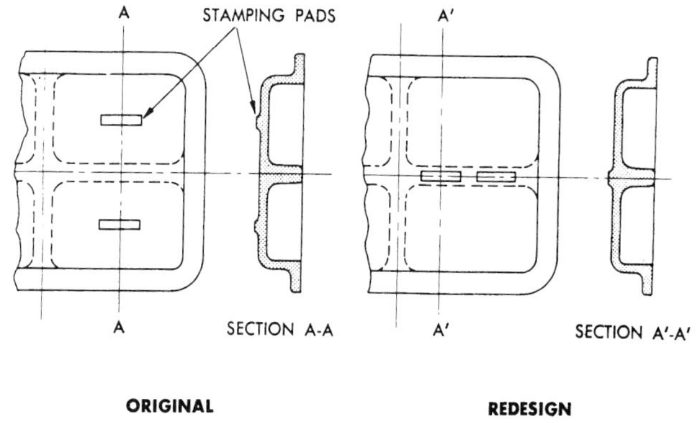

In the original design, stamping pads were located on a thin, unsupported wall. By relocating stamping pads over a deep rib, the possibility of cracking the thin wall was eliminated. This design feature is pertinent, especially to castings that will be pressure tested.

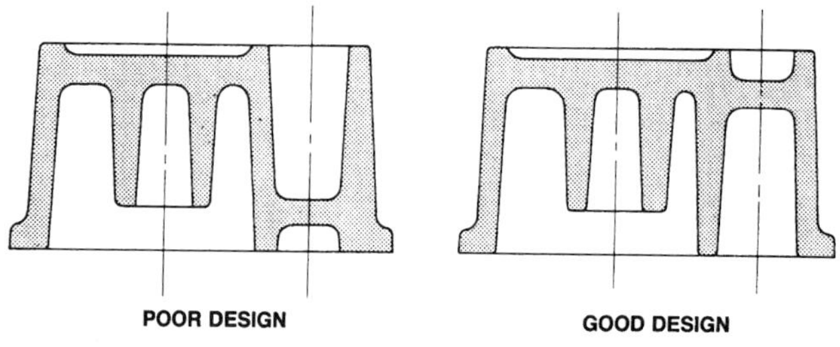

Wherever possible, all deep cavities should be on one side of the casting.

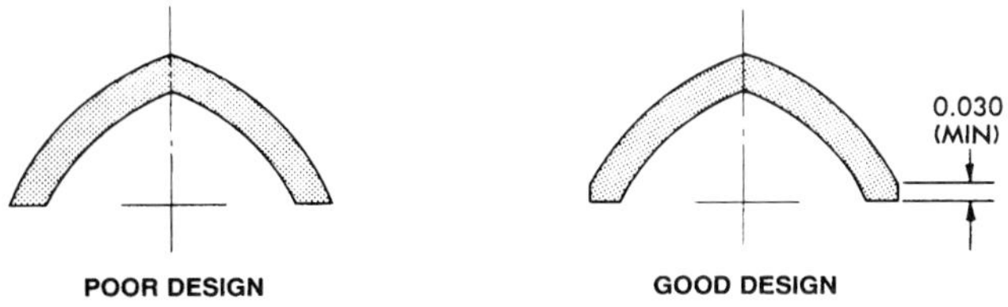

Allow a 0.030 in. minimum flat surface at the parting line for gating and trimming.

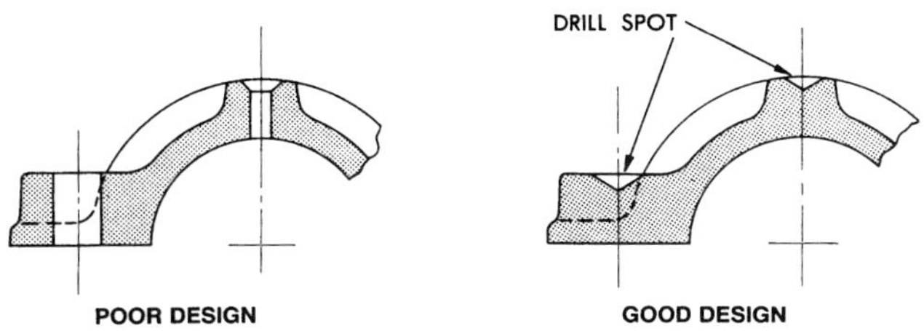

Casting drill spots often is preferable to coring a through-hole when hole center tolerances are not critical. Off-hand drilling from a drill spot is relatively simple.

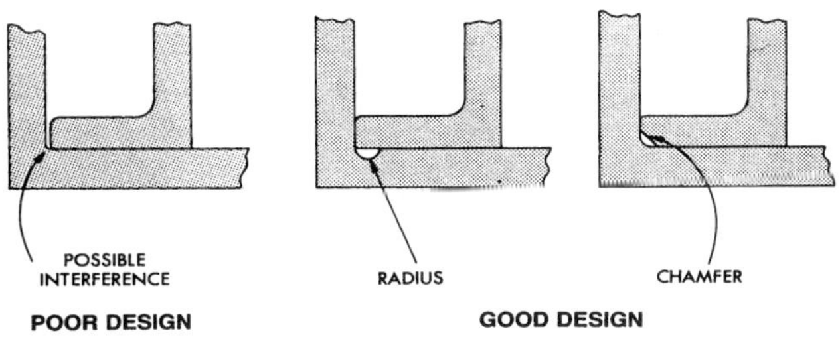

Where corners must fit together on two castings, allow a relief to preclude interference at the corners.

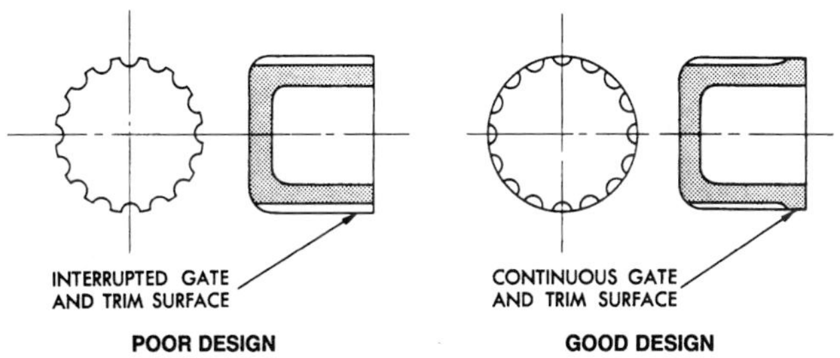

Providing a bead on a scallop simplifies gating and trimming.

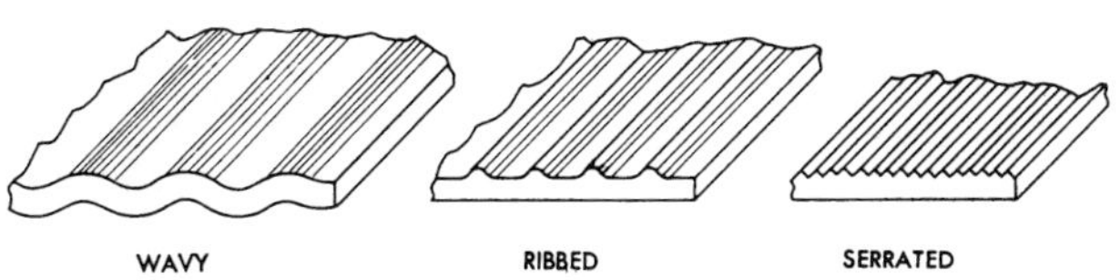

Relatively large areas of plain, thin walls may be difficult to cast and should be designed with waves, ribs or serrations to improve castability.

DESIGNING DIE CASTINGS

Principles for designing aluminum die castings should not be regarded as fixed and, as for all other intricate processes, often must be varied, modified and even violated to meet the individual requirements of specific applications. However, presentation of the many applicable engineering principles and theories lies beyond the scope of this book. Designers who require detailed knowledge of the diecasting process to enable them to design with proper consideration of all relevant factors should consult the various technical texts currently available, as well as metal producers and competent diecasters.

General principles for designing die castings are listed in Table 5-2. Although it must be recognized that there are many exceptions, these principles have proved their value as guides to better casting designs, lower overall costs, extended die life and increased production.

Diecasting Prototypes

Using plaster molded aluminum castings as diecasting prototypes, samples and initial production units often can be of significant value to the customer, the casting designer and the diecaster. This is discussed in Chapter 10.

Design Considerations

General rules, tempered with experience and judgment, generally produce aluminum casting designs that are functional, sound and attainable at the lowest cost. The general rules and suggestions presented in the following pages are not applicable in every case, and there are sound reasons for exceptions in many particular applications. However, the designer will find them worthwhile for checking general design principles.

Some of these rules may seem to be too obvious to mention, but sometimes the most apparent points can be overlooked if they are not rechecked.

Parting Lines

The location of parting lines can result in many problems for the diecaster, can increase die cost and can raise scrap rates. Locating the parting line on an edge of a part is often good practice, but sometimes the gate may break to the inside and damage the section during trimming. A recommended safe approach is for the designer to specify a parting line and then to ask the diecaster to suggest a location, subject to the designer's approval.

Offset parting lines normally add considerably to the cost of producing and maintaining tools (dies) and make it difficult, if not impossible, for the diecaster to operate at top production efficiency. If possible, offset parting lines should be avoided.

Realistic Specified Tolerances

Too often, specified tolerances are much closer than necessary. Realistic casting tolerances should be no closer than can be held on other parts with which the casting is to fit, and certain variables in the process must be considered.

Closest tolerances can be held in single-cavity dies. In multiple-cavity or combination dies, tolerances must be liberalized. Tolerances or linear dimensions should be loosened if casting surfaces are not in a fixed relationship, are not across the same die section or are not on the same die half.

Draft

All surfaces of die castings that are at 90 degrees to the parting line should have draft. Draft of such surfaces can be reduced to the barest minimum as the design demands it, but with the reduction in taper comes increased production problems and casting cost.

NADCA Engineering Product Standards

A fundamental set of standards that clearly defines practical ranges of requirements of die castings for most applications is the "Product Standards" available from the North American Die Casting Association. The standards are realistic and practical as guidelines, and they frequently can be improved upon by the diecaster, often at little or no cost increase, depending on the job.

Die Casting Design Details

Most of the general design suggestions covered earlier in this chapter apply to the design of aluminum die castings. The suggested design points illustrated here should not be regarded as standards for every applicable diecasting design. Many design points are controversial, and rightly so. There are sound reasons for exceptions in specific cases. However, the designer of die castings will find these suggestions of value when used as a partial check of general principles of design.

Cast-In Inserts

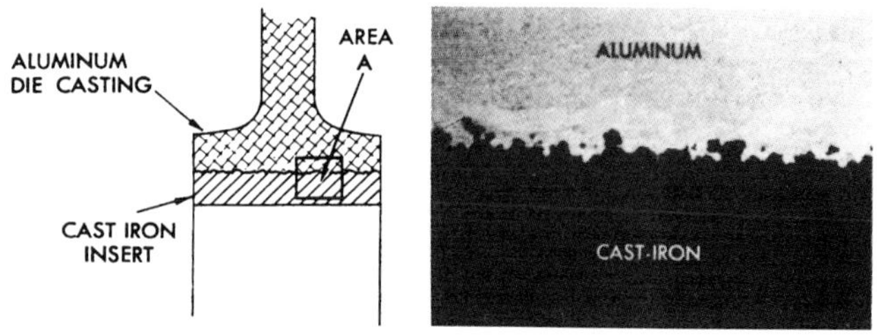

AREA A 10X

In the plain, cylindrical cast iron insert of the type shown, a mechanical bond is obtained by having a rough surface on the insert. With this type of bonding surface (this specific one is called the BMI® bond), the only requirement that must be fulfilled is to obtain a high-strength bond that has excellent thermal conductivity—all voids must be completely filled with cast metal. To ensure complete filling, the metal must be forced into the voids under high pressure. Metal pressures normally obtained in the diecasting process are sufficient.

Table 5-2.
General Design Principles for Diecasting

1) Keep shapes as simple and clean as possible. Avoid projections except when necessary.
2) Sections should be thin, but sufficiently thick to allow proper filling of section. Ribs can aid in providing strength with minimum increase in weight.
3) Uniformity in sections is desirable. If varied, transition should be gradual to avoid concentration of stress.
4) A slight crown is better than large flat surfaces. This is especially true where surface blemishes on finished parts cannot be tolerated.
5) Avoid undercuts that increase either die or operating costs.
6) Unless loose pieces gain a net saving or are unavoidable (because of the nature of the design), they should not be used.
7) Provide fillets at inside corners. Avoid sharp outside corners except at parting lines when dictated by design.
8) Sufficient draft on cores and sidewalls aids removal of the casting from the die with minimum possibility of distortion.
9) If savings in metal and a reduction of overall cost will result, cored holes or recesses should be specified if they do not greatly increase die cost.
10) Cored areas should be designed for easy withdrawal of the core and to avoid complicated die construction and operation.
11) Cored openings should not be excessively deep and narrow because long, thin cores are easily bent or broken. Core replacement can add substantially to cost.
12) Coring of very small holes or of holes in walls that are less than ⅛-in. thick should be avoided unless doing so is less costly than drilling or piercing.
13) Recesses and cored holes should be designed for the minimum core costs that will still satisfy other requirements.
14) Locate ejector pins where pin marks will not mar the appearance and pin pressure will not affect function of the casting. Cooling metal tends to shrink with considerable force against parts of the die. Binding force of this shrinkage must be considered when determining the size and location of ejector pins.
15) When polishing or buffing will be necessary as finishing operations, all areas to receive such treatment should be easily accessible.
16) Design for minimum flash removal to minimize cost. Hold to the least machining possible. Locate flash in a plane where it can be removed most easily.
17) Where machining is necessary, allow enough metal for clean removal but not so much that unnecessarily deep cuts will be required. Provide means for holding castings in machining fixtures.
18) Cast threads in casting when doing so is less expensive than subsequent machining.
19) Integrally cast fastening elements should be specified if they will reduce overall costs by lowering assembly costs.
20) Do not call for closer-than-necessary tolerances.
21) When necessary, permanent inserts should be designed for easy placement and provided with sufficient anchorage to secure them in the casting.
22) When castings influence the final appearance of a product, make them visually attractive and in harmony with adjacent parts.

Perhaps the outstanding feature of the BMI bond is the fact that the bond can withstand high loads, regardless of the direction in which they are applied. This is especially important for cylindrical inserts that are subjected to a combination of axial, radial and torsional loads.

Typical cases are the cast iron inserts that form the wearing surfaces in aluminum engines and bimetallic brake drums. Because of the heat generated in these units, the light metal, with its higher coefficient of expansion, tends to pull away from the cast iron insert. If there is no positive locking or bonding, it is necessary to depend on residual stresses to maintain contact between casting and insert.

At high temperatures, the residual stresses are relieved, and the ability of the unit to withstand loads and to dissipate heat will be impaired. With the BMI bond, this problem is overcome because, even at extremely high temperatures, radial separation will not occur between the two metals.

Threads

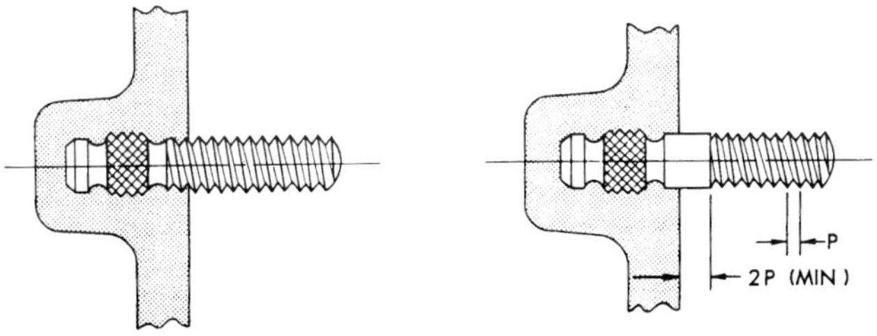

POOR DESIGN GOOD DESIGN

To prevent flash in threads of inserts, terminate threads a distance of at least two full thread widths from the surface of the casting.

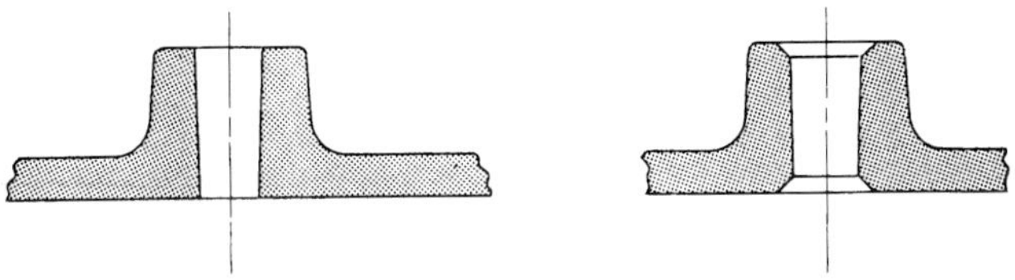

POOR DESIGN GOOD DESIGN

Cored through-holes that will be tapped should be countersunk on both ends.

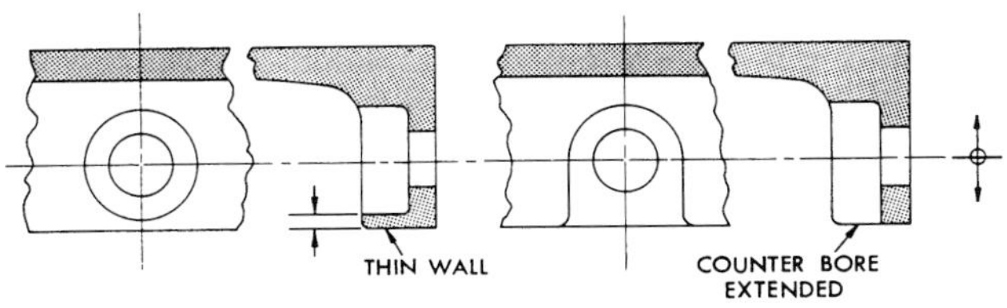

POOR DESIGN GOOD DESIGN

Extending counter bore eliminates thin walls.

DIE AND PERMANENT MOLD CASTING GUIDES

These design recommendations are applicable to the design of both die and permanent mold aluminum castings.

Like most of the other design suggestions shown in this book, these recommendations are not presented as hard and fast standards. However, when appropriately applied, they accomplish the basic objective of any recommended principles, which is improving the design and quality of aluminum castings.

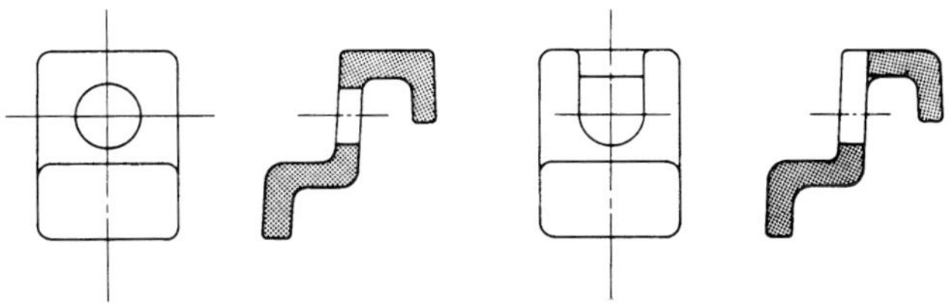

POOR DESIGN GOOD DESIGN

Side cores often can be eliminated by changing the mold design.

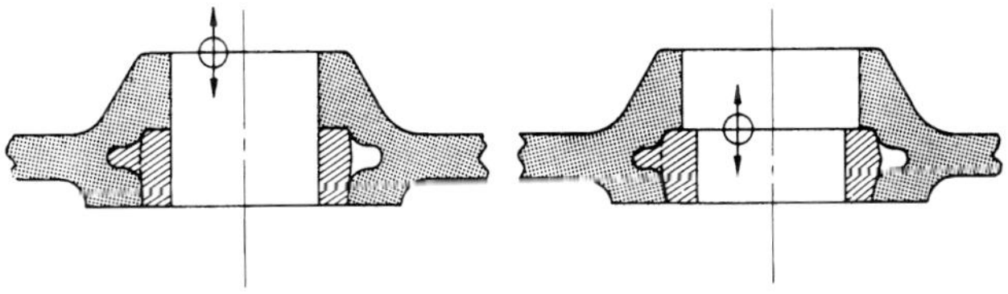

POOR DESIGN GOOD DESIGN

Splitting the core is recommended to hold inserts firmly during casting.

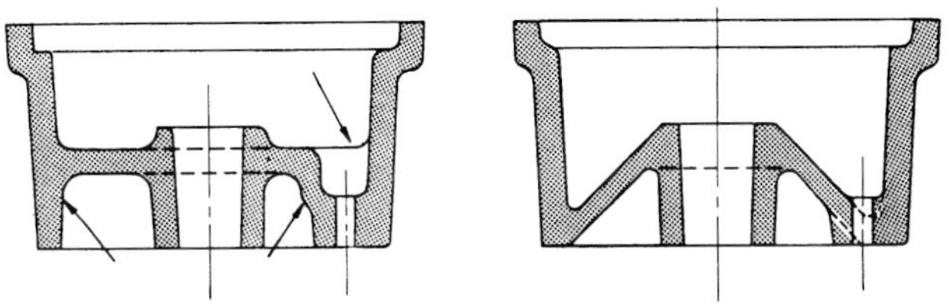

POOR DESIGN GOOD DESIGN

Cavities often can be redesigned to eliminate binding (shown by the arrows) caused by contraction of the casting over cores.

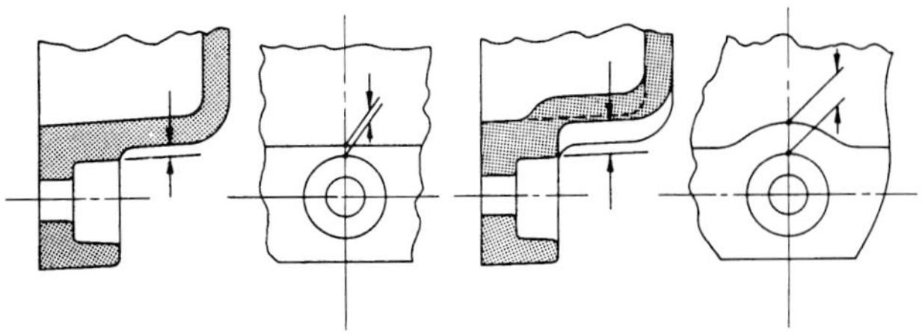

POOR DESIGN GOOD DESIGN

Avoid thin metal sections in the main body of diecasting dies or permanent molds, particularly around core pins. Local thin sections adjacent to substantially thicker die sections are subject to early thermal cracking.

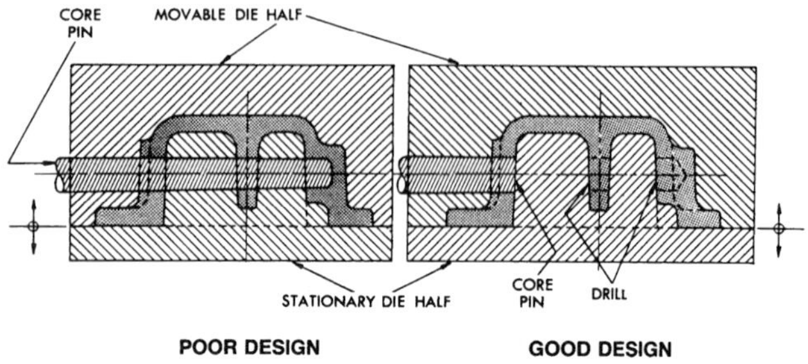

Avoid moving cores because they must fit accurately in both halves of a die or mold. As shown above, drilling holes in inner brackets is better practice.

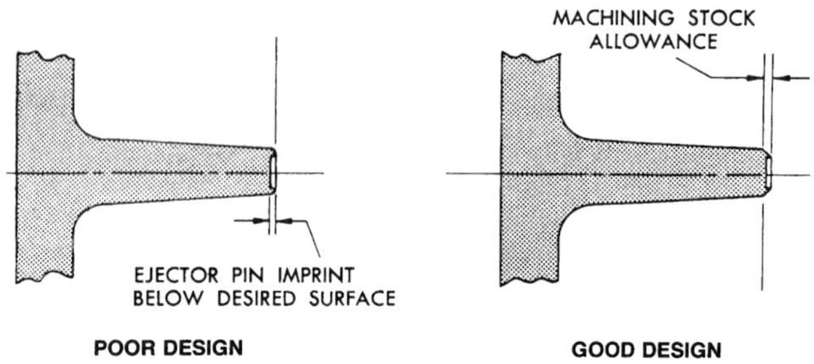

Allow extra machining stock in areas of casting where ejector pins must be located on surfaces that will require machining.

DESIGNING PERMANENT MOLD CASTINGS

The general design rules and suggestions given in this section are applicable to most aluminum permanent mold castings and are adequate when used as guides for the design of relatively simple castings.

When designing complex shapes for permanent mold castings, the casting designer should consult the permanent mold engineer during the design stage to assure that the mold and the process techniques will produce quality castings.

The permanent mold casting process generally is considered suitable for relatively large production runs. If the initial cost of equipment is not included, the cost per permanent mold casting is usually less and production more rapid than with sand casting.

Because of the chilling effect of the metal mold, aluminum permanent mold castings develop highly desirable fine-grained structures with high strength and soundness. Heat treating castings produces even higher mechanical properties. Elongation is 5–100% greater, and other mechanical properties of aluminum permanent mold castings are approximately 20–25% greater than sand castings of the same alloys.

Design Considerations

The designer of aluminum permanent mold castings should not expect to cast key ways, exterior screw or threaded designs, or threaded holes.

Tolerances for the gravity permanent mold process generally are greater than those obtainable in diecasting. As-cast finishes on surfaces formed with sand cores will not be as smooth as surfaces formed by metal molds or cores.

Section Thicknesses

Proper feeding of the molten metal into a permanent mold, a condition that must prevail if consistently good castings are to be produced, is accomplished most easily when wall thicknesses are uniform.

If both light and heavy sections are necessary, provisions must be made in the mold to provide adequate feed and chill of the heavy section. Abrupt section changes should be avoided and an effort made to gradually increase the thickness of the thin section as it approaches the heavy area. It may be desirable to add a feeder pad on a thin section that lies between two thicker areas to properly feed the heavy sections.

If thin ribs or fins are required, it is suggested that the problem be discussed with a permanent mold engineer during the design stage. A wall thickness of 0.060 in. can be obtained for certain castings, and 0.125 in. is practical for castings up to 14 in. in any one dimension.

Beyond the 14 in. linear dimension, minimum wall thickness must be determined by the size and configuration of the individual casting as well as the alloy.

Cored Areas

A cored opening less than ¼ in. in diameter usually is not practical in permanent molds. In some specific cases, however, holes of ⅛ in. diameter or even smaller are possible if formed by stationary core pins.

Normally, it is impossible to core multiple holes close together because the gravity feeding of the permanent mold casting process cannot properly fill the intervening solid sections.

If practical, a design that permits a straight draw of a steel core (with draft) is most economical, both in mold cost and casting cost. If this situation is not possible, an effort should be made to design the opening of the cored area to permit the use of a steel core that is made in sections so it may be collapsed and withdrawn.

The third and generally most costly alternative is to use a core of sand or other destructible material. This method is called *semipermanent mold* casting.

Tolerances

Permanent mold casting tolerances are closer than those required for sand castings but are not as close as those obtainable by the diecasting process.

The tolerances given in Table 5-3 are presented as normal commercial operating tolerances suggested for the most economical production. If closer tolerances are required for certain critical areas, the designer should consult with an experienced permanent mold engineer.

Greater accuracy involving extra close work or care in production should be specified only when and where necessary because additional costs may be involved. Conversely, more liberal values should be indicated, when acceptable, as these tend to keep costs to a minimum. Any variations in these values shall not be binding on the foundry unless accepted in writing.

Fillets

Adequate fillets at all wall intersections and at all angles where bosses or depressions are involved are almost mandatory for the proper flow of metal. Sharp angles weaken the castings and may cause cracks during solidification or during removal from mold.

Table 5-3.
Suggested Dimensional Tolerances for Permanent Mold Castings

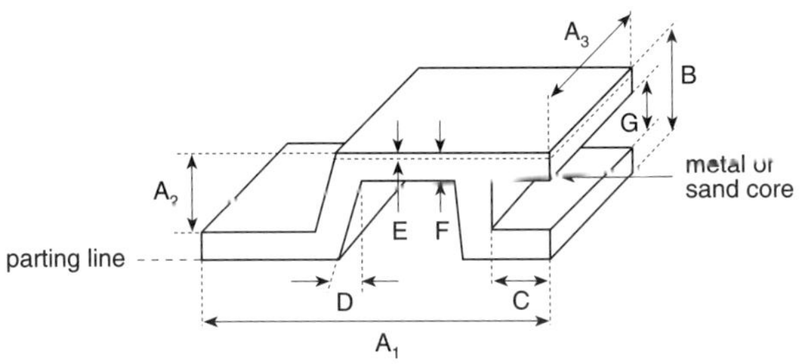

A

Between two points in same part of mold; not affected by parting plane or core.

Specified dimension, in.	Tolerances, in .
up through 1	±0.015
over 1	add ±0.002 in./in.

B

Across parting plane. A-type dimension plus the following chart.

Projected area of casting $A_1 \times A_3$, in.	Additional tolerance for parting plane, in.
up through 10	±0.010
10–49	±0.015
50–99	±0.020
100–249	±0.025
250–500	±0.030

C

Affected by moving parts. A-type dimension plus the following chart.

Projected area of casting affected by moving part $A_3 \times G$, sq in.	Additional tolerance, in. metal core or mold	sand core
up through 10	±0.010	±0.015
10–49	±0.015	±0.025
50–99	±0.015	±0.030
100–249	±0.022	±0.040
250–1000	±0.032	±0.060

D

Dimension: Draft. Normally, the drawing does not show draft. Standard foundry practice is to add draft to the part.

E

Allowance for finish.

maximum dimension in.	nominal allowance, in. metal core or mold	sand core
up through 6	±0.045	±0.060
6–12	±0.060	±0.090
12–18	±0.075	±0.120
18–24	±0.090	±0.180
over 24	consult the foundry	

F

Flatness tolerance, as cast, is the total allowable deviation from a plane. It consists of the total distance between two parallel planes, embracing the entire tolerated surface.

Greatest dimension, in.*	Tolerance, in .
up through 6	within 0.020 in.
each additional in.**	0.003 in./in.

* Section diameter or diagonal
** For castings over 24 in., consult foundry.

Minimum diameter of cored holes is 0.375 in.

Draft Angles

Depending on the casting design, draft angles in the direction of draw on outside surfaces may vary from less than 2° to over 10°. Internal draft angles may vary from less than 2° to 20°. Internal walls may be cast without draft if collapsible metal or destructible cores are used, but doing this generally slows production and increases costs. A casting may be designed so that one surface requires no draft, but the casting designer should consult with a permanent mold engineer.

Undercuts

Avoid undercuts, if possible. They require a collapsible core or a sand core, both of which increase costs. However, if undercuts are essential, they usually can be provided by one of the above-mentioned methods.

Cast-In Inserts

Inserts of aluminum, steel, iron or bronze are sometimes necessary. Commonly inserted are studs, nuts, bushings, pipes, heating elements, iron or steel inserts for threaded areas and compression ring groove inserts for pistons. An interesting use of cast-in inserts is shown in Fig. 5-8, in which aluminum castings of one alloy were cast as inserts into a casting of a different alloy.

When placed in the mold, inserts must be held firmly to prevent movement when the casting is poured. Inserts can be bonded mechanically or metallurgically. Mechanical bonding is usually accomplished by knurling, grooving or keying the inserts for proper anchoring in the aluminum castings.

In one patented metallurgical bonding process, the insert is first heated, then dipped and held in molten aluminum

Fig. 5-8. Three precast aluminum castings are employed as cast-in inserts in this aluminum motor mount for a masonry saw. Inserts are cast of an aluminum-7%, magnesium alloy; motor mount is cast of 308.2 alloy.

(for approx. 5 min) until a coating of an iron-aluminum alloy is formed. The insert is then placed in the mold. To affect the bond with the casting, it is important that the aluminum be poured while the coating on the insert is still in a molten or semi-molten stage.

Machine Locating Points
These points should be determined and their locations approved by the permanent mold engineer before the mold is designed. Otherwise, a gate or riser might be located at a machine locating point, and this would result in costly tooling or mold changes.

Machining Allowances
As a rule, the same surfaces must be machined on a permanent mold casting as on a sand casting, but less machining stock generally needs to be provided on a permanent mold casting. On dimensions up to 10 in., machining allowances range from 1/32 to 1/16 in., but large castings may require as much as 1/8 in. Less than 1/32 in. machining allowance is possible in specific cases where casting and design permit. Machining allowances are in addition to casting and draft tolerances.

Lettering and Part Numbers
Letters and figures should appear raised on a surface rather than depressed. Whenever practical, letters at least 1/8 in. high by a minimum of 0.030 in. deep should be used to obtain sharpness and legibility, and they must appear on a surface where the mold section opens directly away from the casting surface.

As-Cast Finishes
As-cast finishes of permanent mold castings will vary from 125 to 300 rms or more, with the size and configuration of the casting determining any exceptions. A finish of 100 rms or better can be achieved in localized areas of certain castings, but designers should anticipate higher values in normal production.

Permanent Mold Design Standards

These standards, which are suggested as guides for the design of aluminum permanent mold castings, are based on experience gained from several years of permanent mold casting operations and can assist the design engineer in the proper application of the process.

The values, radii and tolerances listed are normal standards for economical production. If critical areas require closer tolerances or values or less draft, thickness or depth, the casting designer should consult with a permanent mold engineer.

Additional design recommendations for the permanent mold process appear in the section "Die and Permanent Mold Casting Guides" earlier in this chapter. The general design suggestions at the beginning of the chapter are also applicable to the permanent mold process.

Draft Requirements for Inside and Outside Walls

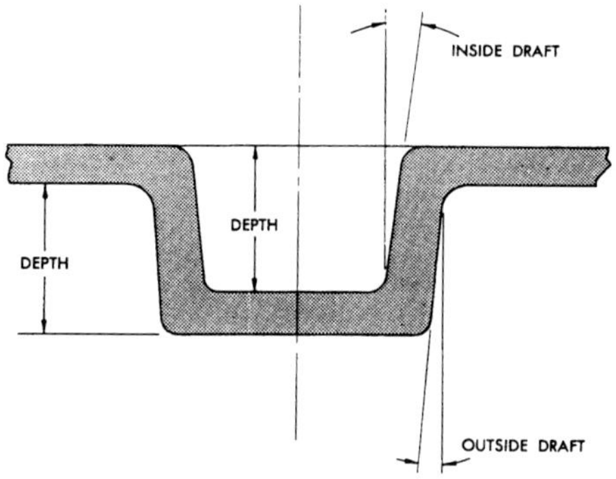

Depth in Inches	Inside Draft	Outside Draft
0.030 to 0.125	20°	10°
0.126 to 0.500	15°	7°
0.501 to 1.00	10°	5°
1.01 to 6.00	5°	3°
6.00 to 12.00	1½°	1½°

Cavities with straight sidewalls or with reverse draft or undercuts can be formed by using collapsible cores.

Recommended Fillet and Outside Corner Radii

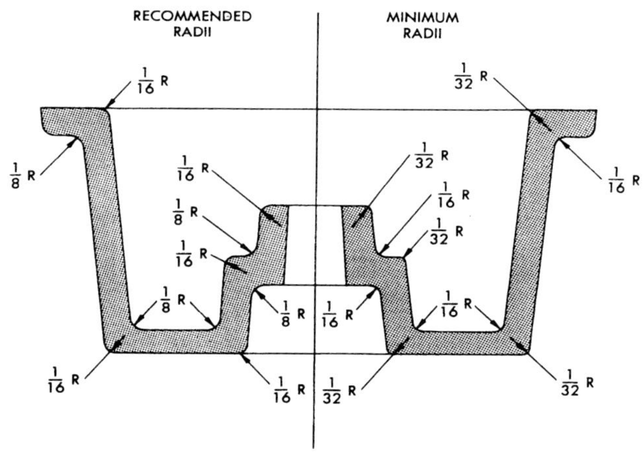

Recommended Radii
Fillet—1/8 in. R
Outside corner—1/16 in. R

Minimum Radii
Fillet—1/16 in. R
Outside corner—1/32 in. R

The minimum casting radii listed above should be specified only when absolutely necessary due to the increased cost involved in maintaining consistent casting quality. Specify generous radii whenever possible.

Cored Hole Diameter vs. Depth vs. Draft (As-Cast)

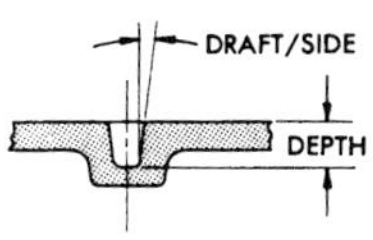

Diameter (inches)	Max Depth (inches)	Draft/Side (degrees)
0.25	0.125	15
0.25 to 0.50	0.25	8
0.50 to 1.00	1.0	5
1 to 2	3.0	3
2 to 4	6.0	2

Depth and draft for holes larger than 4 in. in diameter will be governed by varying conditions. When castings contain large holes, it is recommended that the casting designer obtain the advice of a permanent mold engineer. Use the same draft for blind and through-holes.

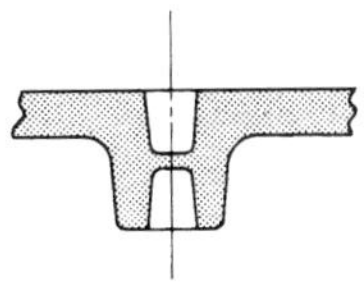

When diameter/depth ratio is greater than recommended, the designer should consider coring the hole from both sides.

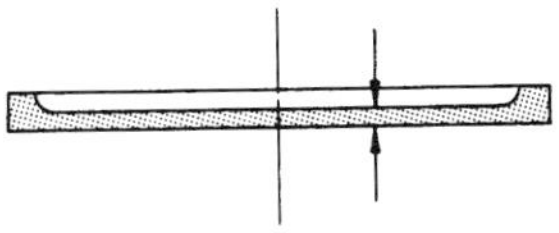

Section Area (sq inches)	Min Wall Thickness (inch)
1.0	0.090
1.0 to 25	0.110
25 to 100	0.125
100 to 200	0.140
200 to 400	0.171

As-cast wall thickness recommendations are shown.

DESIGNING SAND CASTINGS

Generally, aluminum castings are sand cast when they are too large or too complex to be produced economically or with enough structural strength by any other process. Sand casting size and design is limited only by the complexity of the part and the facilities the foundry has available. Aluminum castings that weigh over 2400 lb and others more than 15 ft in length have been sand cast.

The designer of aluminum sand castings should not expect "precision" castings or castings that will not require machining in critical areas. The as-cast surface smoothness of aluminum sand castings will vary from 250–1000 rms, but in the conventional sand casting process, the smoother 250 can be held only locally. It should be noted, however, that aluminum sand castings with as-cast finishes of 90–125 rms (a finish approaching the smoothness of so-called precision castings) are being produced by foundries using waterless green sand molds.

Design Considerations

Cored Holes

The diameter and tolerances of cast-in holes is governed by the overall tolerance, size and length of the aluminum sand casting and by the alloy cast. For the normal foundry process, holes with diameters smaller than ¼ in. generally are not cast but are machined as a postcasting operation.

One special technique for producing small, unlined cored passageways in aluminum sand castings uses what is known as the Dalton® core. Cast-in holes produced by this technique may be as small as ⅛ in. in diameter and far more intricate in shape than holes formed by sand cores.

The Dalton core is a copper tube bent to the desired contour of the cast hole. The tube is sheathed with a braided elastic refractory sleeve of strands of either fibrous glass or stainless steel. The tube does not remain in the casting. It is removed using a chemical agent in which aluminum is insoluble. The refractory sleeve also is insoluble and is removed from the hole manually. A clean sleeve is visual evidence of a clean hole.

Tolerances

The degree of reproducibility and the inherent variations of a process often require a compromise to arrive at practical tolerances. The effects of sand casting inconsistencies may be cumulative on casting dimensions.

General overall tolerances of ±0.030 in. for dimensions up to 6 in. with an additional 0.003 in. for each inch over 6 in. can be held for concentricity, flatness, straightness and mismatch, but this is considered to be the minimum. More liberal tolerances should be specified whenever practical.

Parting Lines

The casting designer should try to keep parting lines straight. Parting lines should be located in a plane coincident with a surface of the casting to permit withdrawal of the pattern without disturbing the mold.

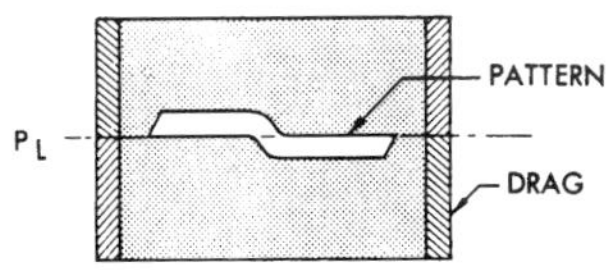

A straight parting line often can be located within the part.

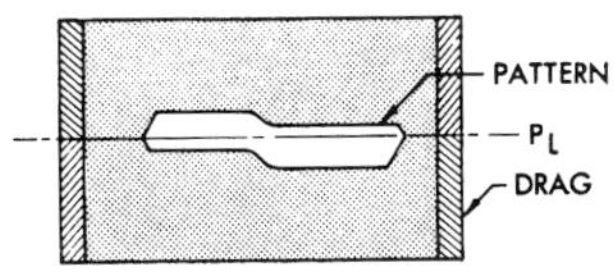

Irregular parting lines result in higher pattern and production costs and often can be avoided. The ingenious designer often can devise a straight parting line for irregular or unsymmetrical parts.

Sand Cores

Designs often require cores to make section thicknesses uniform, to eliminate machining, to reduce weight, and for various other valid reasons. However, the designer should strive to limit or better still, to eliminate coring because cores invariably require special tools, more liberal tolerances and additional handling time. Cores also generally increase the scrap rate and add to cleanup costs. Often they require large holes in the casting to provide for their support and subsequent removal.

Flatness Tolerance, As Cast

Flatness tolerance is the total allowable deviation from a plane. It consists of the total distance between two parallel planes embracing the entire tolerated surface.

Values shown in Table 5-4 represent normal production practice at the most economic level. Greater accuracy should be specified only when necessary because additional cost

Table 5-4.
Suggested Dimensional Tolerances for Sand Castings

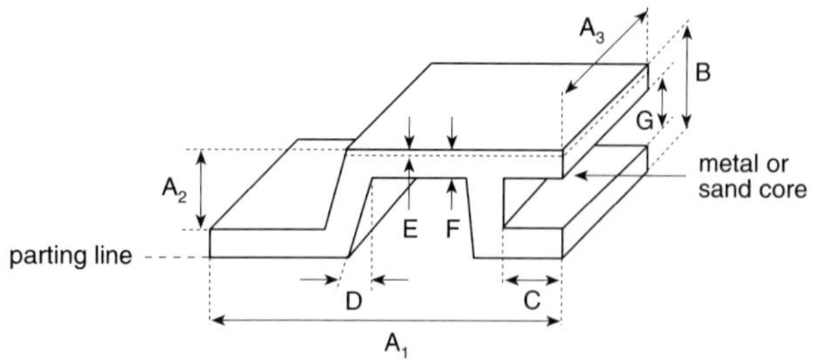

A

Between two points in same part of mold; not affected by parting plane or core.

Specified dimension, in.	Tolerances, in.
up through 6	±0.030
over 6	add ± 0.003 in./in.

B

Across parting plane. A-type dimension plus the following chart.

Projected area of casting $A_1 \times A_3$, in.	Additional tolerance for parting plane, in.
up through 10	±0.020
10–49	±0.035
50–99	±0.045
100–249	±0.060
250–500	±0.090

D

Dimension: Draft. Normally, the drawing does not show draft. Standard foundry practice is to add draft to the part.

C

Affected by core. A-type dimension plus the following chart.

Projected area of casting affected by core, $A_3 \times G$, sq in.	Additional tolerance for core, in.
up through 10	±0.020
10–49	±0.035
50–99	±0.045
100–249	±0.060
250–1000	±0.090
Over 1000	consult foundry

Minimum diameter of cored holes is 0.250 in.

E

Allowance for finish.

Maximum dimension, in.	Nominal allowance, in.
up through 6	±0.060
6–12	±0.90
13–18	±0.120
19–24	±0.150
over 24	consult foundry

may be involved. Conversely, more liberal values should be indicated when acceptable, as these tend to keep costs to a minimum. Any variations in these values shall not be binding to the foundry unless accepted in writing.

Sand Casting Design Details

The general design practices covered earlier in this chapter are applicable to aluminum sand castings. The design features that appear in this section were selected because of their frequent occurrence in many applications. The original designs shown here were reproducible in the sand casting process, but if produced, part costs would have been higher, and in many cases, production would have been difficult, and more rejects (defective castings) would have resulted.

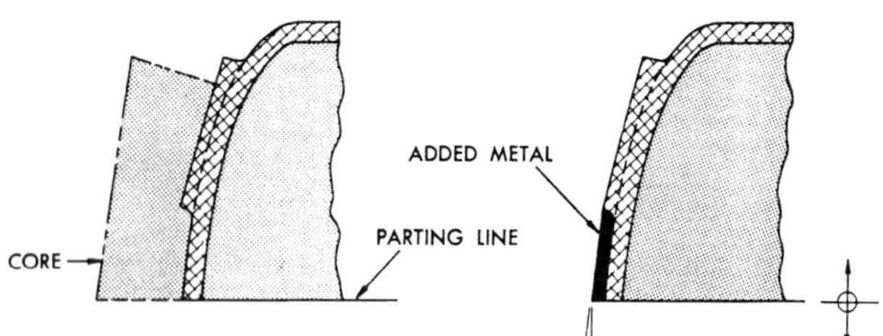

As originally designed, the pad required a separate core. Pad contour redesign eliminated one core and reduced cost.

In aluminum castings, the weight increase due to the addition of a slight metal "fill-in" is insignificant.

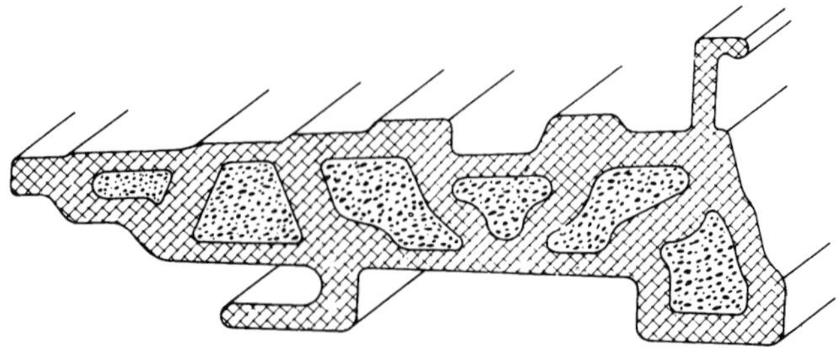

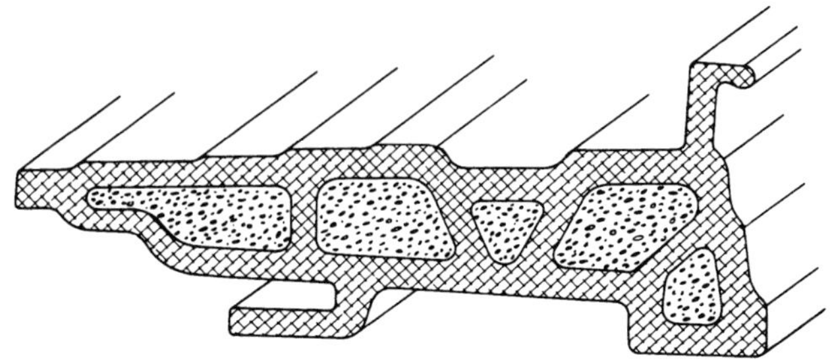

Thick inner sections of castings formed by complex cores will cool more slowly than outer walls and may develop hot spots. The redesign reduces the thickness of the inner sections making them slightly thinner than outer walls, where possible, to improve directional solidification.

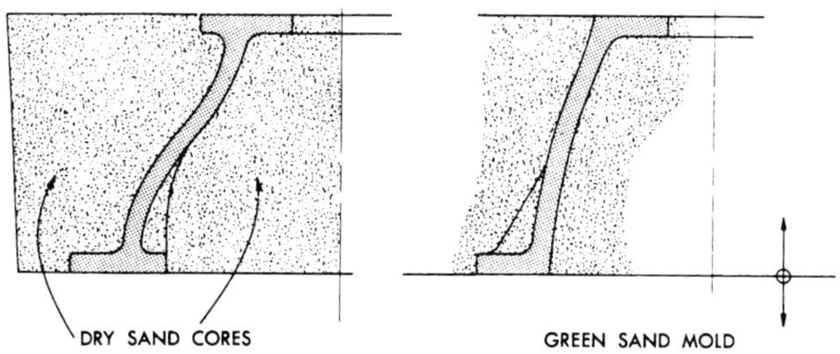

Redesign eliminated undercuts on the outer and inner walls, which required dry sand cores to produce. Production costs for patterns and castings were reduced appreciably without affecting the casting quality or the service reliability.

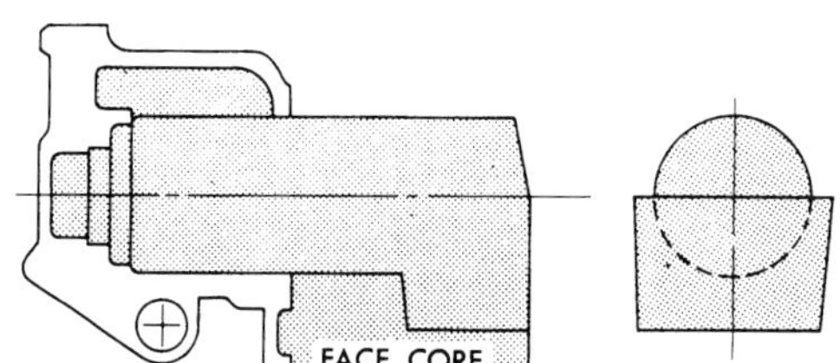

An extra large core print was required to balance a core in the original design, but the print did not guarantee there would be no movement of the core.

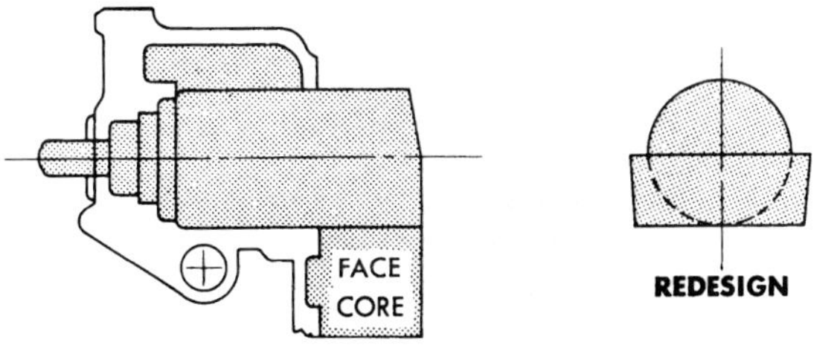

To seat the core in a positive position, an opening through the casting was designed. This eliminated the need for the large core print and provided firm support for holding the core in the mold.

It then was possible to use both a smaller core and a register to locate it. As a general rule, when a large core is used, the balancing problem is best eliminated by having at least a two-seat core position.

DESIGNING SHELL MOLD CASTINGS

When a greater degree of accuracy is required of castings that can best be cast in sand, shell mold methods combined with sand casting techniques will produce castings with closer tolerances and better finishes. Often the advantages of the shell mold process can be fully realized for a particular application only if the foundry engineer, as well as the designer, fully understands the intended use of the part.

Cost of the finished casting is often the basis for choosing the shell process over sand casting. Figure 5-9 compares the cross section of a hub cast in sand with one cast by the shell mold process. Because green sand molding practices require a draft of 10° and ⅛ in. web, considerable finish stock must be machined from the inside diameter. In shell casting, the required draft is only 1 to 2°, and the web is not necessary.

The flange in the sand casting would be cast solid and then a spot-facing and drilling operation would be required on the 17/32-in. holes. In the shell casting, these holes would be cast, thus eliminating the machining operations. Finish stock on the shell casting also is less.

On the basis of these comparisons, the designer and the buyer can determine whether elimination of the machining

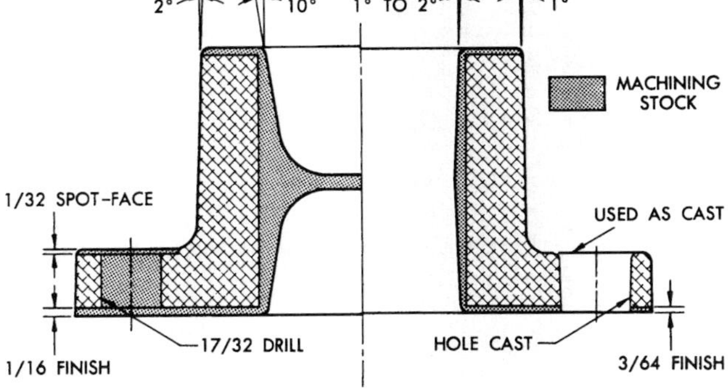

Fig. 5-9. Cross sections comparing normal draft and machining allowances for a sand cast aluminum hub (left) with the same part cast by the shell mold process.

operations and the use of less metal more than offsets the increased cost of the shell casting.

Although there are limitations in shell molding and coring, as there are in every casting process, complex shapes with intricate coring can be cast in shell molds. Cross sections of varying thicknesses, usually difficult to cast in sand molds, may be readily cast in shell molds.

Developments of recent years have made the shell process one of the most important foundry developments of modern times. Some foundries are convinced that, in time, it will be more economical to use the shell process to produce most of the small castings now made by the green sand method.

Design Considerations

The general design suggestions given earlier in this chapter are applicable to the design of aluminum castings for the shell process.

Surface Finishes

Surface finishes of aluminum shell castings can be expected to average between 125 and 300 rms. However, finish will vary somewhat with casting design and alloy used.

Tooling Costs

Tooling costs are higher for shell molds and shell cores than for sand molds and cores. The major reason for the increase in tooling expense is the metal patterns that are necessary to withstand the high investing and curing temperatures used in the shell process. Generally, such costs are in direct proportion to the size and complexity of the casting and to the number of close tolerances specified in the design.

Tolerances

Standard tolerances commonly are ±0.003–0.005 in./in. Unless absolutely necessary, the closer tolerances should not be required unless the extra tooling costs they impose are offset by savings in finish machining costs. Linear dimensions that cross parting lines should be given an additional tolerance of ±0.010 in. Tolerance for mismatch can be held as close as ±0.010 in. Angularity tolerances of 0°15' are usually provided.

Draft

Standard practice of many designers includes a requirement for a draft of 1° maximum on all surfaces normal to the parting line; other designers require only ½° maximum. Under certain conditions, a part may be designed with no draft, but the casting designer should first consult with the foundry engineer.

Section Thickness

Many variables must be considered by the designer before the practical minimum thicknesses of sections of a specific casting can be determined. Under favorable design conditions, it is practical to shell mold cast aluminum parts with sections as thin as 0.10 in. Usually, however, 3/16 in. is regarded as the minimum for wall thickness.

Cored Holes

The general limitations for cored holes in aluminum shell mold castings are:

Minimum diameter	0.375 in.
Maximum depth of blind hole	4 D
Maximum depth of through-hole	10 D

The standard rule for draw of blind holes is twice the diameter of draw depth. However, many exceptions are being taken in actual production.

Minimum Fillet Radius

The area mass where sections join is reduced by the use of small fillet radii. Reducing the mass lightens the part. Fillet radii should be as generous as possible on interior corners of shell molded aluminum castings. Sharp as-cast corners should be avoided. Sharp exterior corners may be readily obtained by adding stock and machining them.

Machining Stock

In shell molding, a machining allowance of 1/16 in. is considered normal. However, because shell molds are usually more accurate than sand molds, the machining allowance for shell castings usually can be 3/64 in. or less in some cases.

Cast-In Threads

Coarse, helical, exterior threads may be readily cast by the shell mold process, but fine exterior threads should be machined. Interior threads cannot be cast.

DESIGNING PLASTER MOLD CASTINGS

Fine surface finish, highly accurate reproduction of details and consistent dimensional accuracy are characteristics of aluminum parts cast in plaster molds. For many years, this process has been used to cast the aluminum tread rings in which automobile truck tires are molded. Precision parts for the electronics industry also are cast in plaster molds.

In the plaster process, it is possible to produce integrally with the casting a pebbled, wood-grained, mottled or just about any other type of decorative surface that is desired. Though comparatively little use is now being made of this feature, it does suggest some interesting possibilities to an architect or stylist.

The as-cast surface finish and appearance and the dimensional accuracy obtainable in plaster mold aluminum castings are almost equal to die castings. For these and other reasons, it is important that the casting designer be aware that the plaster process is often used by many diecasters to provide their customers with a few hundred or more prototype castings to be used for market analyses and for machining and assembly trials.

Chapter 10 of this book gives examples of the use of plaster mold castings in the evaluation and development of die castings. The advantages of using plaster castings as prototypes for die castings include the following. The value of these advantages increases significantly with the increase in size and complexity of the casting.

1) Lead time for the delivery of sample castings is cut drastically. Prototype plaster castings of fairly simple design can be delivered in less than a week. First samples from diecasting dies often take three to six months.
2) Design changes can be resolved during trials conducted with the prototype castings and then incorporated into the final design before or while the diecasting die is built. Common changes, such as the design and location of bosses, flanges and ribs, are made easily in the plaster mold patterns, and the redesigned sample castings can be produced from plaster molds in a few days.
3) The prototype plaster castings can be used to design and test trimming, machining and assembling fixtures and operations. Final die castings can be redesigned to expedite production and to minimize the expense of the postcasting operations.
4) The plaster molded prototype castings can be used for a realistic market evaluation of the cast product. Worthwhile changes resulting from the marketing trials can be designed into the part at only minimal additional cost because the expensive tooling for the die casting has not yet been built.

Sizes and Finish of Plaster Mold Castings

Plaster mold aluminum castings can be produced in almost any size or shape. Typical parts range in weight from less than 1 oz to 2000 lb. A surface finish of 90 rms is generally suitable for plaster mold castings. Finishes of as-cast surfaces range from 32 to 125 rms, but since the finish is largely dependent on the surface quality of the pattern, finishes of 40 to 60 rms can be produced consistently if patterns are of the highest quality. However, for most castings a surface smoothness of 90 rms is suitable, and the designer should specify smoother finishes only when necessary.

Tolerances

Dimensional tolerances often are given as ±0.005–0.010 in./in. However, the following tolerances are being maintained in normal production:

Under 1 in.	±0.005 in.
1 to 2 in.	±0.008 in.
2 to 3 in.	±0.010 in.
3 to 6 in.	±0.012 in.
6 to 9 in.	±0.015 in.
9 to 12 in.	±0.020 in.
12 to 24 in.	±0.025 in.

Add 0.001 in. for every in. over 24.

Designers sometimes prefer to specify tolerances for each different dimensional requirement (Table 5-5).

Table 5-5.
Examples of Tolerances for Dimensional Limits

Dimension	Under 1 in.	Add per in. over 1 in.
Across parting line	±0.010	+0.0015
Between points on same part of mold	±0.005	+0.0015
Between points on core and mold	±0.010	+0.0015

There exists a direct relationship between tolerances and production costs. For example, if the parting line tolerance above is reduced to ±0.010 in. plus only 0.001 in. (instead of +0.0015 in.) for 12-in. dimensions, the production costs would increase by about 14–17%.

Dimensional Limits and Tolerances for Cores

Lengths: Maximum length of cores supported at one end is usually given as five times the diameter of the core.

Diameters: The minimum core diameter is ¼ in. Smaller holes can be cast, but drilling usually is more economical.

Draft: Although zero draft on cores is sometimes permissible, a core draft of ½° is considered minimum, and 2° is preferred.

Section Thickness

The shape and surface area of a section determine the wall or metal thickness of that section, but for most designs, the minimum thickness would be 0.060 ± 0.010 in. Until quite recently, it was not practical to cast walls much thinner than ⅛ in., but with new techniques, wall thicknesses of 0.040–0.060 in. are being cast regularly. Sections as thin as 0.020 in. can be cast but only on localized small areas.

A flatness tolerance of 0.004 in./in. is commonly specified.

DESIGNING CENTRIFUGAL CASTINGS

Aluminum castings can be produced by three methods of centrifugal casting. Centrifugal casting in permanent molds produces a finer grain structure and thinner ribs and webs than can be achieved in ordinary static permanent mold casting. Of particular interest to the designer are the mechanical properties of centrifugal castings, which are superior to those that can be obtained by other casting processes.

Design Considerations

The designer must realize that the use of the centrifugal casting processes is limited by the shape of the part to be cast. However, through the ingenuity of mold designers, many complicated parts have been cast centrifugally. Some shapes cannot be cast centrifugally, and other shapes can be cast by only one of the methods of centrifugal casting. Table 5-6 compares the various methods of the process. The range of sizes shown includes both nonferrous and ferrous metals.

The design suggestions given in this section are representative values for normal production of aluminum castings in permanent molds by any of the three methods of the centrifugal casting process. Specific design and end uses of a casting often will necessitate modifications to normal design procedures, but the experienced casting designer knows that, if design of the part violates basic casting rules, it seldom will be cast successfully or economically.

The general design suggestions given earlier in this chapter are applicable to the design of aluminum castings for the centrifugal casting process.

Minimum Fillet Radius

The minimum fillet radius that may be cast in aluminum alloys without cost penalties is 0.062 in., +0.020 and –0.000 in.

Minimum Outside Corner Radius

Except at the parting line where there should be no radius, minimum radius on outside corners generally is given as 0.062 in., +0.020 and –0.000 in.

Machining Allowances

Finish allowance for all machined surfaces should be at least 0.060 in. Larger castings may require greater allowances for finishing.

Tolerances

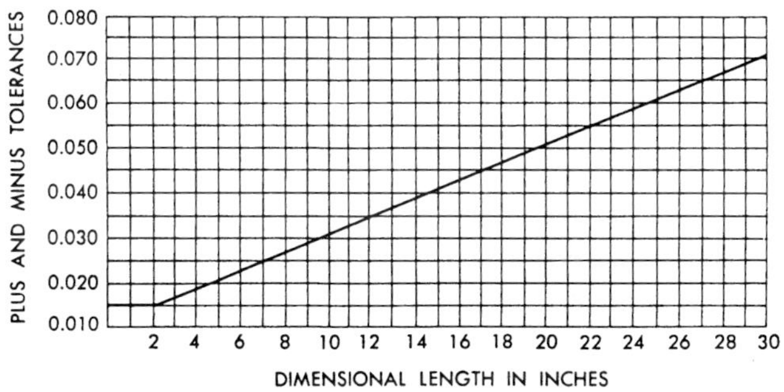

Linear dimensions within the mold that do not cross a parting line may be held to the minimum tolerances given in the above graph. Larger tolerances should be allowed for dimensions when it is feasible.

Table 5-6.
Comparison of Centrifugal Casting Methods

Method	Typical Shapes	Common Size Range
True centrifugal (horizontal)	long, hollow tubes	4 in. D x 4 in. L to 4 ft D x 30 ft L
True centrifugal (vertical)	short, hollow tubes	4 in. D x 4 in. L to 6 ft D x 2 ft L
Semicentrifugal	gears, circular plates	4 in. D x 2 in. L to 6 ft D x 1 ft L
Centrifuge	random shapes—multiple parts on center sprue	less than 1 lb to 150 lb

Linear dimensions that cross a parting line require an additional tolerance of +0.020 in. Mismatch tolerance is usually ±0.010 in., and angularity tolerances commonly given are 0°5' per inch up to 0°30' maximum.

Core Size

Minimum diameter of cores is ⅛ in.

Draft

Draft angles are required on all parts. Minimum angles are as follows:

External surfaces	1 degree
Internal surfaces	2 degrees
Cored holes	1 degree per side

Designing Castings for the Centrifuge Method

The design practices presented in this subsection are guidelines of the engineering department of one foundry that specializes in casting aluminum centrifugally. Although these practices are specifically geared to the centrifuge method, they generally are applicable to all methods of centrifugal casting. The minimum section thicknesses that may be successfully cast by the centrifuge process are as follows:

Square Inches Per Casting	Thickness (Inches)
Up to 4	0.070
5 to 10	0.080
11 to 20	0.090
21 to 30	0.100
31 to 70	0.110
71 to 100	0.120
Over 100	0.156

Heavy sections should not be isolated by sections of these minimum thicknesses. An increase in cross section should not exceed twice the local thickness. Section changes of an abrupt nature should be avoided when in excess of this amount.

The drafts required for metal cores are as follows:

Draft	Minimum (degrees)	Preferred (degrees)
When length equals diameter	½	1½
When length equals 1½ times diameter	1	2
When length is more than 2 times diameter	1½	3

As stated, the draft minimums listed above are for metal cores. Cores made of sand require no draft. The recommended minimum diameter of fixed metal cores is ⅛ in. The recommended minimum diameter of movable metal cores and sand cores is ⅜ in. Portions of the castings that are defined by sand cores should have a tolerance of ±0.030 in. Although this is twice the normal value, it allows for some tolerance on core fit.

The minimum fillet or corner radius is 0.005 in. Usually it is not necessary to taper a web section where it joins another surface, and although a fillet radius of 0.060 in. is preferred here, it should not exceed the web thickness. Corner or edge radii occurring at parting lines will be sharp, and the flash must be removed as in usual foundry operations.

Knurling, cross hatching, lettering, serrations and other fine detail can be cast to a sharpness of 0.005 inch. Curled or grooved inserts, as well as tubular heating elements, can be cast into the part. Pockets and other defects are eliminated because the metal is literally forced around the insert.

Some undercuts and backdrafts can be produced by making the direction of the draw normal to the parting line or by using collapsible metal cores; others require the use of special sand cores. The elimination of backdraft and undercuts simplifies moldmaking and lowers the subsequent cost of the finished part. Whenever possible, it is desirable to design the casting so that it can be produced without using expendable cores.

All high-stress points should be marked clearly on the drawing. Often, the mold can be designed so that the centrifugal casting will develop the highest mechanical properties in stress areas.

Directional solidification will usually occur from the outside surface inward and toward the sprue, if design is correct. Directional solidification is difficult to control if the wall is too thick. In such cases, voids may occur along the plane where the two solidifying sections come together. Sections of weakness are also caused by "banding," a condition that results from excessive vibrations when molds and machines are not properly balanced before rotation begins.

The arrangement of the sections determines the practicability of casting a part centrifugally. The simplest form of gating is the one that uses one gate that also serves as a feed head.

In cases where certain design features may inhibit directional solidification, the condition often can be remedied by using auxiliary gates, pads, blind heads or chills. However, a nonuniform section can require overly elaborate components, which may raise the cleaning cost to a prohibitive figure or make the article unsuitable for centrifugal casting.

Dimensional Tolerances

The tolerances that may be held with aluminum castings made by the centrifuge process using normal production practices are given in Table 5-7.

As-cast tolerances or deviations from a contour check fixture for contoured or rounded surfaces may be held to close limits in the mold:

Up to 2 in.	0.005 in. min	0.010 in. preferred
Over 2 in.	0.010 in. min	0.001 in. for each additional in.; 0.030 in. max

The finished casting tolerances for contoured surfaces depend on the heat treat condition and the straightening required as a result of heat treatment. Table 5-7 gives tolerances for contoured surfaces that can be met for heat treated castings.

Table 5-7.
Tolerances for Centrifuge Aluminum Castings

Type of Tolerance	Tolerance (inches) Minimum	Preferred
Dimensional:		
up to 2 in.	±0.005	±0.010
over 2 in.	±0.010	±0.001 for each additional in., 0.030 max
Mismatch:		
up to 2 in. square	±0.005	±0.010
over 2 in. square	±0.010	±0.001 for each additional in., 0.030 max
Across parting line:		
up to 2 in. square	±0.005	±0.010
over 2 in. square	±0.010	±0.001 for each additional in., 0.030 max
Between points produced by one mold part:		
up to 2 in.	±0.005	±0.010
over 2 in.	±0.010	±0.001 for each additional in., 0.030 max
Between cores and mold parts:		
up to 2 in.	±0.010	±0.015
over 2 in.	±0.015	±0.001 for each additional in., 0.030 max
Angularity:		
up to 1 in.	±5 minutes	±5 minutes for each additional in. up to ±30 minutes
Heat treated:		
—T51 (aged only)	0.010 in.	0.015 in.
—T6 (heat treated and aged)	0.015 in.	0.020 in.

DESIGNING INVESTMENT CASTINGS

The investment casting process generally is considered for the production of small parts in which intricacy of shape or detail and highly accurate dimensional reproducibility are prime considerations. Advances in the process have made it possible to produce aluminum castings as large as 3x3x3 ft.

The designer should remember that the accuracy of this process is high when compared to other casting processes but should not relate casting reproducibility with the accuracy of machining. Investment casting produces parts with good structural values.

Parts in a large variety of shapes and sizes can be investment cast. Investment castings are used in sewing machines, pumps, diesel engines, oil well drilling equipment, food processing machinery, electric motors, hospital instruments and equipment, jet engines, aircraft and missile frames and instruments, and many other commercial and defense applications.

Usually, eight inches is the recommended maximum dimension for an investment cast part from a dimensional tolerance standpoint. The high accuracy advantage of the process is lost as the size of the part increases. There is no minimum size for investment castings.

Great flexibility in design is offered by the investment casting process. Multiple components previously assembled or welded may be produced as an intricately designed investment casting. A real advantage in economy and quality can be had in one-piece cast parts, which often are superior to similar components fabricated of several pieces welded, soldered, riveted, brazed or bolted together. The cast parts generally are stronger, are competitive in cost and have better appearance than fabricated units.

The design principles and suggestions presented earlier in this chapter generally are applicable to the design of aluminum investment castings.

Special Design Advantages of Investment Casting

In the investment casting process, molten metal is poured into hot molds, a condition that permits routine production of castings with extremely thin sections, knife edges and sharp detail. Because hot molds improve metal fluidity, metal can be poured at a comparatively low temperature. Additionally, vacuum, pressure and centrifugal casting techniques used in the investment process facilitate the flow of metal into every cavity of the mold and, further, help reduce porosity in castings.

The monolithic molds used in investment casting do not leave parting lines usually found on parts produced by other casting processes that employ molds of two or more segments. Because there are no parting lines, dimensional accuracy of investment castings is better than can be achieved on most castings produced in segmented molds. Tolerances across mold parting lines cannot be as close as those on a casting produced in a monolithic mold.

Design Considerations

Tooling Costs

Mold requirements will vary with the quantity and sometimes the accuracy required for the specific pattern and part. The use of properly made prototypes facilitates equipment development. When the use of investment casting is anticipated, prototype requirements can be met by machining the pattern of polystyrene or by carving and assembling the pattern of investment wax without making metal molds. This practice assures flexibility during development of design yet produces a part with the same structural properties as will be obtained in the final design.

Expendable polystyrene patterns often are used in the investment process when their use will eliminate tooling, accelerate testing services or cut experimental costs. This temporary method of making patterns is a rapid and efficient means of putting new or revised cast metal parts into service tests. Because no tooling is required, the designer has complete freedom to alter the design. This method also gives manufacturers a comparatively inexpensive method for experimenting with new and more efficient production processes in their attempts to eliminate production steps and costs by redesigning two or more parts into a single investment casting.

Tooling for low- or medium-quantity production runs is often made by machining the part in its true dimension with a shrink factor added to compensate for the shrinkage of the molten alloy. This master becomes a pattern for making a mold cavity in a pattern die.

Designed for use where optimum accuracy and larger quantities are required, the wax or plastic pattern injection mold is made of machined aluminum, steel, brass or magnesium. Machined steel dies are used almost exclusively for plastic patterns. Soft metal dies can be used for wax patterns of simple design, but they are unsatisfactory for complex patterns.

Tolerances

Rules for finished and general dimensional tolerances must take into account factors like size and configuration, which vary so much from part to part. Some castings can be held to much closer tolerances than others. For smaller castings, finished tolerances (on castings blastcleaned and buffed) can reasonably be held to the following values, but such minimums should be specified only when absolutely necessary to meet special requirements.

For nonfunctional areas, the general and finished tolerances should be as liberal as possible. Recommended tolerances are:

FINISHED TOLERANCES

Dimension	Tolerance
Up to 1 in.	±0.005 in.
Over 1 in.	±0.005 in./in.

GENERAL TOLERANCES

Dimension	Tolerance
Less than 2 in.	±1/64 in.
2 in. to 4 in.	±1/32 in.
4 in. to 6 in.	±1/16 in.

An angular tolerance of ±0°30' normally is preferred. A minimum contour band of 0.007 in. is suggested because it is suitable for the economical production of airfoil contours.

Draft

Draft is usually required only when deep draws are necessary to produce the patterns. When deep draws are involved, a maximum draft of ½° per surface is commonly used.

Surface Finishes

Surface finishes attainable on investment cast parts will fall within the following ranges: as-cast finishes, 60–100 rms; finished (blastcleaned and buffed), 40–100 rms. Surface finishes as smooth as 32 rms are achieved by special finishing techniques.

Blind and Through-Holes

In aluminum investment castings, blind holes can be as small as ⅛ in. in diameter, while the minimum diameter of through-holes is 1/32 in. Preformed ceramic cores or ceramic mold investment are the most common methods used for forming holes.

When determining the practicality of casting holes in a casting, the designer must consider such factors as hole diameter, position, the ratio of the hole diameter to its length, the mass around the hole and the parallelism of the walls.

Draft is not usually required for holes of small diameters in aluminum investment castings unless the draw is very long. A draft of 0°15' is recommended for cores for holes larger than 1¼ in. in diameter.

Cast Threads and Serrations

Although cast threads and serrations are details seldom economically justified in aluminum castings, there are instances where special thread forms, serrations or unusual designs are readily obtainable.

The investment mold may consist of a single slip with gypsum. The frozen mercury pattern can produce castings with fine threads or other detail by using the shell investment process. Such detail is not obtainable with conventional wax or plastic patterns and lock investment molds. However, this technique is limited mostly to larger castings and to those special cases unsuited to the wax process.

Section Thickness and Surface Areas

The minimum section thickness for aluminum investment castings generally is considered to be 0.050 in. For areas less than 1.5 in.^{2}, the thickness can be reduced slightly, but it should never be less than 0.035 in.

Uniformly curved or flat surfaces without minor irregularities are difficult to cast. These may be avoided by coring out those sections that are not needed or by breaking the smooth areas with stiffening ribs. If the thickness of the section changes, it is best to join the two thicknesses by tapering them or by using fillets.

Sometimes a heavy section can be redesigned into a suitably ribbed, thin one. Adequate feeding of separated heavy sections is difficult if light sections are used as connectors, and the designer should attempt to keep sections as uniform in thickness as possible.

Using chaplets, chills or other techniques common to the sand foundry is not practical for controlling solidification of investment castings. Minimum fillet radius generally is set at 0.020 in. with tolerances of at least +0.020 in. and –0.010 in. For structural applications, the minimum fillet radius should be 0.060 in.

Machining Allowance

A finishing allowance of at least 0.030 in. should be allowed for surfaces that will be machined. Finish allowances should be shown on the drawing not merely mentioned in the list of drawing notes.

Undercuts

Undercuts may be cast with little difficulty but at considerable increase in tooling costs. As in most other casting processes, it is best to avoid undercuts in investment casting whenever possible.

DESIGNING EXPENDABLE PATTERN CASTINGS

Use of foam patterns in unbonded sand for casting production is known as "lost foam," evaporative pattern casting or expendable pattern casting (EPC). EPC offers design opportunities not available in conventional sand casting. In addition to aluminum- and copper-based alloys, a variety of other metals can be cast, including steel, gray iron and ductile iron.

Relatively complex parts can be produced, incorporating cast-to-size features not feasible in conventional sand casting production. The process also allows designers to consolidate what was formerly made in several parts into a single casting. In most instances, however, simply converting an existing conventional casting design to the EPC process does not result in maximum benefit to the casting user. Castings should be designed to take advantage of opportunities offered to produce a near-net shape part with the EPC process.

Size Range

Most of the recent development[2] of the EPC process has been directed toward the production of castings from injection-molded foam patterns. In aluminum, castings ranging in size from a few ounces to over 40 lb have been cast. The smallest size, at present, is limited more by the capabilities of the foam molding process rather than the inherent limitations in the casting process. The thinnest cross sections routinely available are limited to a size approximately three polystyrene beads thick (Fig. 5-10).

Aluminum alloys permit higher pattern densities, and sections as thin as 0.180 in. (4.5 mm) can be produced. Special molding techniques can sometimes be used to produce sections down to 0.120 in. (3 mm).

Flask size limits the maximum size casting a particular foundry can produce. The flask must be large enough to permit enough sand to support the pattern during pouring. In addition, the weight of the flask and sand must not exceed the load limits of the vibration device used to compact the sand. Most of the current production units are intended to produce castings with a maximum diagonal measurement of about 32 in. (750 mm), but larger castings are possible.

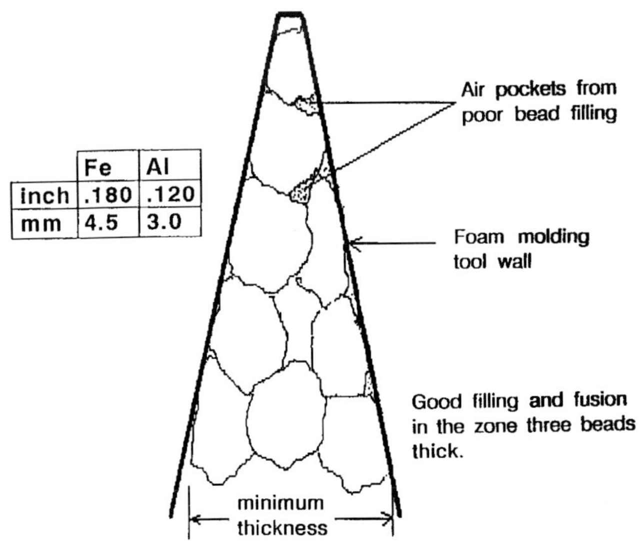

Fig. 5-10. Minimum section thickness for EPC castings.[1]

Appearance

An important characteristic of EPC castings is their surface finish. The bead texture on the surface of the foam patterns is reproduced in the casting process, resulting in a pleasing, uniform surface. The bead texture actually helps improve paint adhesion for aluminum castings where anodizing has been eliminated. The excellent surface finish and good contour of the casting surface result because molten metal touches only the coating that covers the pattern and not the sand.

EPC castings generally have cleaner surfaces and internal passages than those produced by conventional casting methods (Fig. 5-11). Measured surface finish values for aluminum castings as low as 120 micro-inch are possible when special precautions are used to optimize the surface finish.

EPC castings generally are free of extraneous metal such as parting line flash, fins around core prints or the "plus metal" from mold wall movement. Much less grinding is required in the final finishing of castings, resulting in fewer grinding marks. Locating points have higher reliability since grinding normally is limited to the gate contact points.

Patterns

The number and size of cores can be reduced substantially or totally eliminated in EPC casting. The pattern need not be drawn, and any undercuts or easily accessible cavities may simply be filled with unbonded or cold-setting sand. In cases where cores cannot be avoided because of the complicated construction of the casting, the cores are glued to the pattern.

Because EPC is generally a coreless casting process, design engineers have unprecedented freedom. However, the low-density foam is not resistant to deformation, so rigidity of the pattern is a concern. Ribs, gussets and supports are incorporated to give patterns dimensional stability. Wax fillets are not recommended because they create gas during pouring.

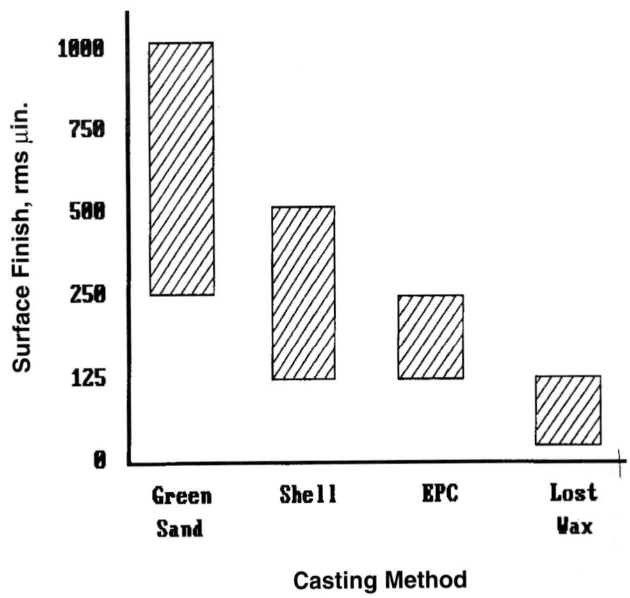

Fig. 5-11. Surface finish comparison for different casting methods.[1]

Internal passages must be designed to permit entry of the wash coating and enough air movement for the wash carrier to escape. Adequate passage dimensions must be allowed to permit the backup sand to enter. Blind openings must not be horizontal or go uphill so that sand can fill the pocket. Most patterns are poured in the vertical position, and multiple gates are common.

Experience has shown that the thinnest pattern sections must be more than two beads wide. While sections as thin as 0.0120 in. (3 mm) have been achieved, it is suggested that sections be at least 0.180 in. (4.5 mm) thick (see Fig. 5-10).

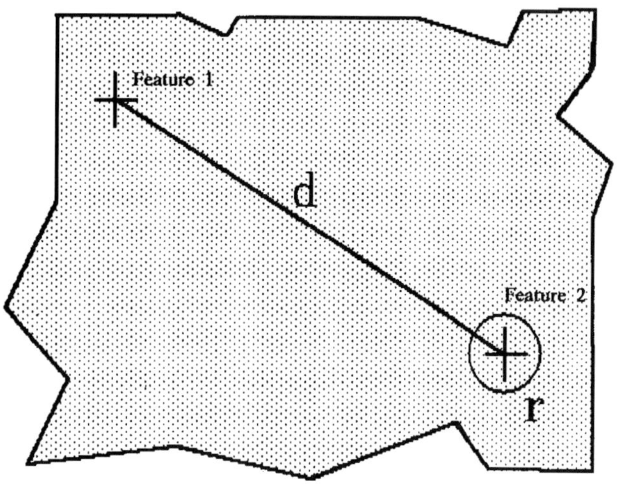

Fig. 5-12. Linear dimensions between features located on the same piece of foam. r = 0.003 + 0.002 (d – 1).[1]

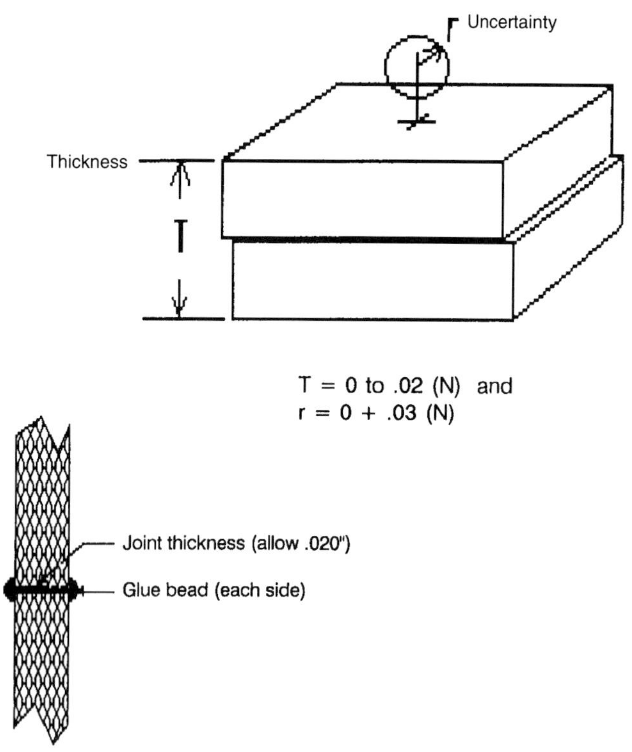

Fig. 5-13. Tolerance for glue joints.[1]

Patterns for small cast parts often are glued in clusters, similar to what is done in investment casting. The lightweight clusters can be handled with a basic automated storage/retrieval system. Linking this retrieval system to a computerized inventory system using bar coding permits full integration of the patternmaking process into just-in-time retrieval and a fully automated pouring sequence.

Dimensional Accuracy

EPC production offers enhanced dimensional accuracy when compared to some conventional casting processes. Expressed in general terms, the castings are more dimensionally accurate than small green sand castings, but not as accurate as precision investment castings. Dimensional accuracy similar to shell casting processes is ordinarily achieved. Standard shrink allowances are used in pattern construction.

Design tolerance rules that generally are achievable can be summarized as follows:

- Linear dimensions between features located on the same piece of foam normally are controlled to ±0.3% for the first inch (25 mm) and ±0.2% for each subsequent inch (Fig. 5-12).
- Linear dimensions across glue joints can be expected to have less accuracy. An additional allowance, typically of 0.020 in. (0.5 mm) variation in thickness, should be made for each joint (Fig. 5-13).
- Positional accuracy across glue joints is affected by potential shifting (or mismatching) of the mating foam pattern pieces. An additional allowance of ±0.030 in. (0.8 mm) should be made for each glue joint.
- Glue seams typically have an extruded bead of glue around the joint. The glue bead can normally be expected to be +0.040 in. (1.0 mm) and –0.0 (Fig. 5-13).
- Cast-in cylindrical holes can be made with no draft (using cylinder pulls in the tool). General tolerances as good as standard drilling operations can be achieved (Table 5-8). In many cases, the holes may be threaded directly using "floating" chucks for the taps. Through-holes as small as 0.192 in. (4.9 mm) in diameter have been cast successfully in bosses 2 in. (50 mm) long and up to 0.60 in. (15 mm) in diameter.

  Blind holes can be made with the EPC process if the holes are oriented appropriately when the patterns are cast. Best results can be obtained if the designer discusses the desired cast-in holes for a particular project with the casting supplier.

Table 5-8.
Typical Oversize Drilling Diameter
(Source: The Metal Cutting Institute)

Drill Diameter (inch)	Average max	Mean	Average min
1/16	0.002	0.0015	0.001
1/8	0.0045	0.003	0.001
1/4	0.0065	0.004	0.0025
1/2	0.008	0.005	0.003
3/4	0.008	0.005	0.003
1	0.009	0.007	0.004

- Typical draft allowances of only ½ degree are adequate for the molding process. Pattern draft can be eliminated on selected surfaces, allowing use of these surfaces as cast.
- Typical stock allowances used for surfaces of aluminum castings that must be machined are 0.060 inch. With special construction methods, as little as 0.045 in. (1.15 mm) is possible.

Design Checklist

The checklist shown in Table 5-9 will remind designers of important questions about the EPC process to consider when designing castings.[3]

ACKNOWLEDGMENT

AFS would like to thank Ray Monroe, Steel Founders' Society of America, for his help with the EPC design section.

REFERENCES

1. J.A. Troxler; "The Prospects for Lost Foam in 1995," *AFS Transactions*, pp 371–378 (1990).
2. A. Wittmoser; "Full Mold Casting," *AFS Transactions*, pp 292–303 (1964).
3. R.W. Monroe, *Expendable Pattern Casting*, AFS, Des Plaines, IL (1992).

Table 5-9.
Casting Design Checklist for Expendable Pattern Casting

1) Is cast component material adequate for service requirements?
2) Is cast component design robust enough for service requirements?
3) Does cast component have machining that can be eliminated or complex core requirements that suggest the use of plastic foam patterns?
4) Could a larger assembly of parts be cast as one component by using plastic foam patterns?
5) Does cast component have sufficient production volume to support the tooling costs of plastic foam pattern production?
6) Do the production volume and quality demands suggest cast-to-shape or fully machined tooling?
7) Should the cast component be divided into more than one foam section and assembled?
8) What should be the tool design for each foam section?
9) What are the critical dimensions and details?
10) What are the restrictions on thin sections and sharp edges?
11) Where should fill guns be placed?
12) Where should ejectors be placed?
13) What about venting?
14) What is the thermal performance of the tool?
15) Where are pulls or retractable sections required?
16) Is the tool stiff enough to resist foam pressure (50 psi) during molding?
17) Is there proper sealing around steam chest, pulls, fill guns, etc.?
18) Are cooling sprays adequate and uniform?
19) Is steam supply adequate and uniform?
20) Can foam section be stored after molding without distortion?
21) What combination of aging, molding and tool dimension will achieve desired casting dimensions?
22) Are glue joints flat and in one place?
23) Are glue joints accurate?
24) Are glue joints minimized?
25) Is assembly tooling accurate?
26) Can gluing be done without shift or distortion?
27) Can troublesome glue joints be inspected?
28) Can assembled pattern cluster be coated?
29) Can coating be applied to all interior cavities and the excess drained?
30) Can the coated cluster be dried, including internal cavities?
31) Is coated cluster drying rack supported to avoid distortion?
32) Is the cluster resistant to distortion during compaction?
33) Can internal cavities fill with compacted sand (no blind uphill cavities)?
34) Does the gating system allow uniform reproducible filling of the cluster?
35) Does the gating system allow uniform reproducible filling of the cluster?
36) Does casting design avoid abrupt section transitions?
37) Does casting design and gating assure the reproducible elimination of the plastic residue?
38) Are isolated heavy sections avoided in casting design?
39) Do gates and risers achieve needed casting soundness?
40) Does gating avoid metal penetration or erosion?
41) Do gating and casting placement avoid sand fluidization?

Chapter 6

DIECASTING

THE PROCESS

Diecasting is a repetitive operation wherein identical parts are cast at maximum production rates by forcing molten metal under considerable pressure into metal dies, which are precision made in two (or more) parts called cavity halves.

Die halves are mounted onto the platens of the diecasting machine. These die sections are held together securely by the diecasting machine clamping mechanism in order to withstand high pressure. Molten metal is injected into the die where it chills rapidly. When the metal has solidified, dies are opened and the hot casting is ejected. The die is then closed again, and the casting cycle is repeated.

Gates are broken off and flash is trimmed from the ejected castings, usually in a press located close to the diecasting machine. Other sequential finishing operations are performed as desired for the individual parts.

Quick chilling in the metal die generally results in castings that have high surface strength but that have somewhat less soundness internally. Thin sections are usually stronger in proportion to section thickness than thick ones. Die castings sometimes tend to be especially porous along the neutral axis of heavy sections.

ECONOMICS OF DIECASTING

The diecasting of aluminum is a proven process and is one of the most rapidly developing metalcasting processes in use today. Domestic aluminum diecasting production has steadily increased over the last several decades, and, with continued automotive weight reduction, production is predicted to continue to grow substantially. The wide range of products that can be diecast, the close tolerances that can be held readily, the accuracy of detail reproduction, and the economy and speed of manufacturing are significant factors in the popularity of diecasting.

The diecasting industry is now capable of producing larger, more complex, sounder, stronger and more dimensionally accurate parts than it ever has before. The result of these improvements is that the economic advantages of the diecasting process are available to an ever-increasing range of applications. In the United States, more than 60% of the aluminum castings produced are die castings. The tremendous growth in the number of aluminum die castings is attributable primarily to the introduction of large diecast structural parts, which the industry currently produces to guaranteed levels of quality.

Diecasting makes rapid production possible, but the dies used must be strong. Consequently, the number of castings required to justify use of the diecasting method can be fairly high. There is no minimum figure that automatically dictates profitable use of the diecasting process, but production runs above 10,000 pieces generally are considered for this method. Where the design is simple and die costs are correspondingly low, or where a die casting can reduce machining and finishing costs or better satisfy requirements, small runs can be economical, especially when unit dies are used.

The versatility of the diecasting process is partially revealed by the following two examples. The 48-in. escalator stair tread shown in Fig. 6-1 is an example of a large aluminum die casting produced to close tolerances. Specifications required nominal thickness of 0.098 in. and flatness held to 0.048 in. total (±0.030 in. in plant control) over a 40-in. length. The casting required a 2000-ton machine with an intensified pressure of 6825 psi. The complete die weighed over 30 tons and required internal heating and cooling during operation. Shot weight was 47 lb, and the die was filled in 0.2 sec.

The die casting shown in Fig. 6-2 was produced almost fully assembled, with all drilling and tapping completed. Only final assembly and painting were required.

While this book makes no attempt to discuss particular costs of die castings, the reader can assume that diecasting is not just a bench and desk operation. It requires expensive machinery; experienced technicians, engineers and metallurgists; and complete facilities to successfully produce castings that are consistently high in quality.

With major emphasis on value engineering and cost reduction, diecasting has an important role to play in nearly every

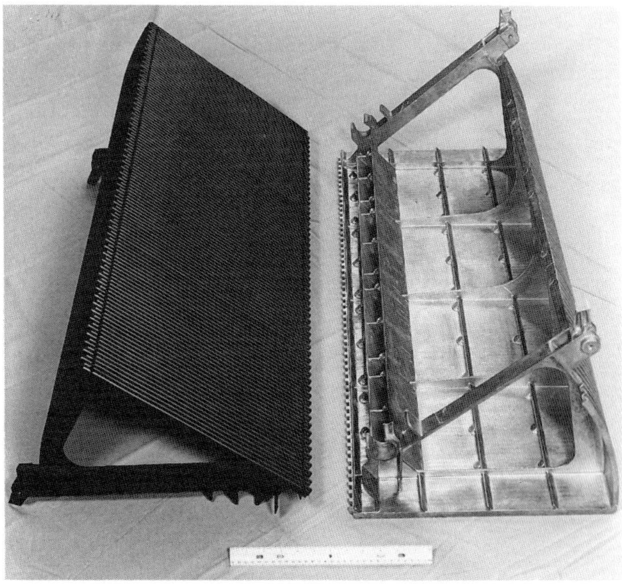

Fig. 6-1. Made of A380.1 alloy, this die casting, a 48-in. escalator stair tread, is cast in one piece. The projected area of the casting is 850 in.^{2}, and the entire surface area at the parting line is 5780 in.^{2}

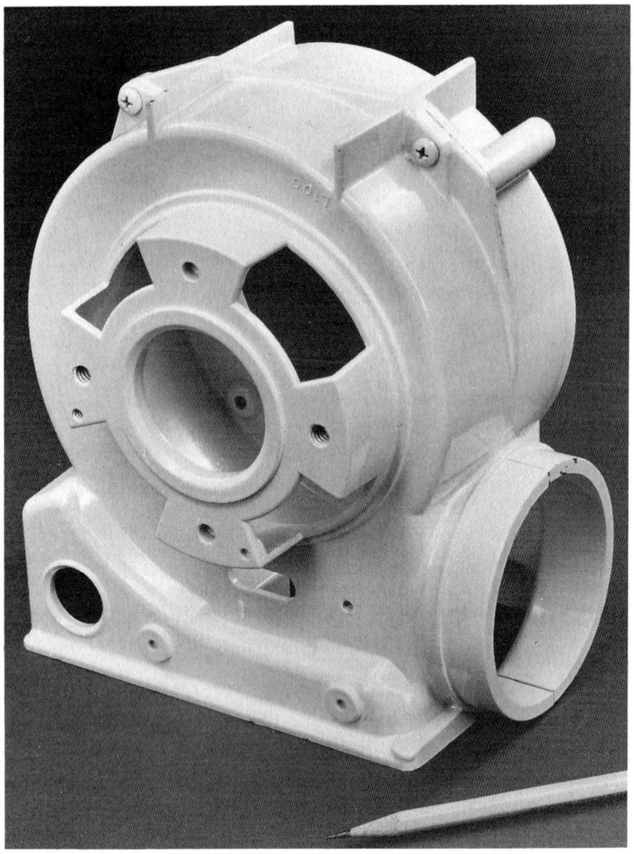

Fig. 6-2. This diecast motor housing saved 55% over the cost of a sand casting. [Photo courtesy of the Newton-New Haven Co.]

industry, provided that process capabilities and economics are understood. Where the potential for the use of die castings has been fully developed, it has usually followed the establishment of a close liaison between the user and a competent diecaster.

The automotive industry has been the greatest single force in the rapid growth in the use of aluminum die castings. Auto manufacturers use aluminum die castings for a variety of functional and decorative components in transmissions, braking systems, steering units, door hinges, wheel covers, trim and engine blocks. Of primary concern to designers are the weight, strength and finishes of the metal components and the cost of producing identical parts in great quantities. Because of its favorable strength-to-weight ratio, excellent finishing qualities and unlimited availability, total poundage of die castings per vehicle is increasing annually.

Although the automotive industry uses more than half the total tonnage of domestically produced aluminum die castings, many other industries also comprise major markets. Product styling, performance advantages, and economy of manufacturing and assembly inherent in the diecasting process because of its versatility have led many diverse industries to rely on diecast components. Industry groups (excluding automotive) in which annual end use of aluminum die castings can be measured in millions of pounds include:

- agriculture, mining and construction equipment;
- transportation (excluding automotive);
- industrial and commercial machinery and tools;
- electronic devices and communications equipment;
- office equipment and business machines;
- plumbing and heating and builder's hardware;
- photographic, optical and recording devices and instruments;
- timing and time-operated devices (clocks);
- home appliances;
- sporting goods, personal goods, toys and jewelry;
- national defense.

Another factor in the remarkable growth of the diecasting process has been the high rate of production of the precision-built equipment used to produce castings. The diecasting process initially was used to produce relatively small castings of thin sections; however, as more experience was gained, larger castings were designed. This, in turn, led to the development of larger diecasting machines.

Machines that have locking forces of over 3500 tons are available (Fig. 6-3), and special equipment with a locking force of 4500 tons has been used to produce exceptionally large aluminum die castings. Dies weighing 30–50 tons and more now produce automatic pinsetter frames, automobile

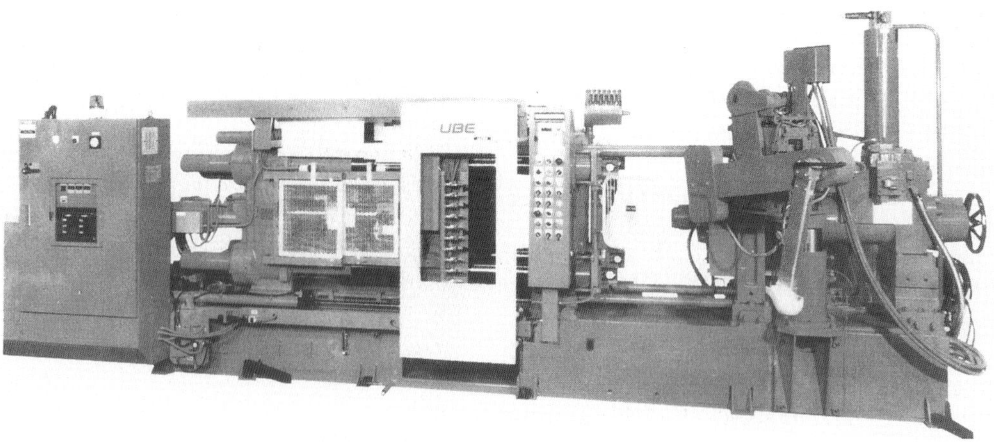

Fig. 6-3. Depending on the manufacturer, diecasting machines can come with locking forces of as little as 25 tons to as large as 4500 tons. [Photo courtesy of UBE Industries (America), Inc.]

transmission cases, escalator stair treads (see Fig. 6-1) and marine engine blocks.

In some instances, diecasting production can be automated to reduce costs. Customers naturally respond favorably to improved products at lower costs. The result is a continually increasing demand for aluminum die castings.

Perhaps the greatest improvement in the diecasting industry has been an improved understanding of the process by the diecasters themselves. Many in the industry now consider specialized technical training and scientific knowledge to be absolute requirements for profitable operation.

Because of the rapid advances made in the diecasting process over recent years, considerable information about it has been and continues to be published. Material in this section, though not exhaustive, has been correlated both for comparative use and to inform the reader of some of the most significant current practices in the important field of aluminum casting.

FACTORS TO CONSIDER ABOUT DIECASTING

Advantages

Choosing a casting method should be based partially upon the benefits that one process offers over other practical processes. Diecasting can offer several advantages over other casting processes.

Diecast parts are near-net shape units that can be cast at high production rates. Dies used in the process, while initially relatively expensive, have relatively long wear life. When large quantities are required, diecast parts generally cost less per casting than those produced by other processes, even when die costs are taken into account. Parts may be designed to reduce or eliminate costly handling, cleanup and assembly operations. Inserts to facilitate assembly, which may be made of most other metals or of some nonmetallic materials, may be cast into a part.

Dimensional tolerances on each diecast part are held more closely than is generally possible with other major casting processes. Recesses, holes and slots can be precisely cored to close tolerances and on accurately-held centers. The high degree of uniformity between castings facilitates automatic assembly or machining operations and helps reduce the amount of machining required.

The surface quality of die castings is excellent. Depending on condition of dies, cast surfaces are usually smoother than those from most other castings. Surface finishes of diecast aluminum parts can vary from 25–125 rms, but 63 rms can generally be obtained quite easily.

The diecasting process offers more freedom to the designer. Castings may be made in a broad variety of shapes and sizes, both simple and intricate. The minimum sizes of diecast parts are usually smaller than those possible using other casting methods. Maximum size is increasing with the development of new machines.

A wide range of aluminum alloys can be diecast. Parts can be cast to resist corrosion; to be machined easily at low cost; to be plated; or to readily receive standard mechanical, organic and electrochemical finishes.

Some of the other advantages of diecasting include:

- Thin sections can be cast. Although 0.05 in. is usually considered minimum thickness, 0.025 in. thick aluminum sections have been diecast.
- Metal loss in diecasting is usually low because sprues, runners, rejects and chips can be reused with little waste.
- Mechanical properties are generally high, and fatigue strength is higher than for most other castings, making die castings suitable for structural applications.
- Little draft is required.
- Both external and internal threads can be diecast, although internal thread casting is limited.

Disadvantages

The rapid growth in the use of diecast parts in all types of industries does not mean that the process is without disadvantages. Unusually large parts that can be cast in sand and permanent molds cannot be diecast. Undercuts in diecast parts are generally limited except when using special machines that use moving die parts and loose pieces that permit diecasting of extremely complex shapes.

One disadvantage of the diecasting process is that the air or gases in the die cavity often become trapped in the casting to form voids. These holes lower the mechanical properties of the casting, and if near the surface, they may cause blistering during subsequent thermal treatment. However, proper gating and venting of the die and correct operating techniques will minimize porosity and produce sound castings. In some instances, a vacuum can be used to evacuate air from the die cavity before each shot.

Other disadvantages of diecasting are:

- costly tool modification;
- tendency to decrease in soundness as sections become heavier—however, vacuum methods can improve soundness;
- coring is restricted to shapes that permit straight core pull;
- a long wait for first production models because building dies is a precise, time-consuming job. However, diecasters have remedied this situation in several respects. One practice is to greatly reduce lead time by providing plaster-molded castings as sample and prototype parts (see Chapter 10).

Control of Metal Temperature

The temperature of the aluminum and the pressure and speed of the shot are the primary factors that affect the structural soundness and surface finish of a die casting. Quality castings can be produced successfully when these factors are controlled and adjusted in the machine.

The importance of maintaining the proper temperature in molten aluminum cannot be overemphasized. Aluminum holding furnaces usually are adjacent to the diecasting machine (Figs. 6-4 and 6-5), and in some instances, they are supplied as an accessory by the builder of the machine. Correct aluminum melting practices and controls, as well as melting and holding furnaces, are discussed in Chapter 2.

Size Limitations

The maximum size of die castings is limited chiefly by economics. For example, large transmission cases and escalator treads are being cast economically on large machines. High volumes cast compensate for higher equipment costs.

Fig. 6-4. A 300-ft long electrically heated molten aluminum launder shows take-off spurs for various low-energy holding furnaces located next to diecasting machines. Launder systems deliver the exact amount of metal needed, on time and at temperature, while minimizing turbulence and oxides. [Photo courtesy of Frank W. Schaefer, Inc.]

Fig. 6-5. This spur from the launder system in Fig. 6-4 brings metal directly to a 2300-lb, low-energy holding furnace. [Photo courtesy of Frank W. Schaefer, Inc.]

Standard diecasting machines are used regularly to produce parts that vary in size from those as small as matchsticks to much larger castings that have surface areas of about 200–300 sq in. Usually, maximum size limitations are given by weight, but many designers now maintain that area and volume also are critical values in determining size. Tooling costs generally increase in proportion to the size and complexity of the part.

Tolerances

Tolerances in die castings can be held unusually close for cast metal parts (0.003–0.005 in./in. for small parts). However, they vary with the size and configuration of the part and the castability of the alloy.

The closest tolerances are obtainable for parts cast in single-cavity dies; more liberal tolerances must be allowed for multiple-cavity and combination dies. In normal practice, dimensions will loosen slightly because of wear on the dies and their locking mechanisms.

COMPARISONS WITH OTHER MANUFACTURING PROCESSES

Among the most important high-quantity production items manufactured today, die castings rank with screw machined, stamped, forged and sand cast products in extent of use and in value. Diecasting is by far the most rapid method of casting because one die can be used for thousands of castings. This is in marked contrast to sand casting, in which a new mold has to be made for each casting or gate of castings. Generally, one person operates one diecasting machine, thereby keeping labor costs low.

Diecasting is appropriate where machining, surface finishing or assembling costs make up a considerable part of the cost of a casting. The process also is appropriate if a large number of parts is involved, if a reduction in machine tool investment is sought, if there is expensive tooling or if floor space is limited.

Aluminum die castings should be chosen based on total cost as compared with products of equivalent quality made either by other production methods or from other less suitable materials. Superior production economy and design are achieved only by using the most suitable fabrication process along with the proper aluminum alloy.

In diecasting, relatively inexpensive prototypes cast in plaster or sand (or produced by some other method) are often made for stress-strain analysis. This eliminates the undue expense often incurred if dies must be changed.

The rapid chilling inherent in the diecasting process produces an unusually fine-grained structure that gives superior mechanical properties without heat treating. This can be considered a benefit because it saves a manufacturing operation and, at the same time, eliminates the tendency of die castings to blister during heat treatment because of interior porosity. Correct die design can relegate porosity to noncritical areas.

In the following paragraphs, die castings are compared with castings made in permanent molds, forgings, screw machine products and plastic moldings.

Sand Castings

Die castings require much less machining than do sand castings. They can be cast with thinner walls; they can have all or nearly all holes cored close to size; and they can be held to much closer dimensions.

Die castings are produced more rapidly in dies that make thousands of castings without replacement, and they do not require new cores for each casting. These castings are easily provided with inserts cast in place; they have smoother surfaces; and they involve lower labor costs.

However, with sand casting, internal shapes presently unobtainable by diecasting can be produced. Sand casting makes a greater maximum size possible. Costs of tooling and setup are generally lower than with diecasting. Small quantities usually can be produced more economically by sand casting.

The advantages of diecasting over sand casting for the appropriate applications can be seen in the aluminum castings shown in Figs. 6-6 and 6-7.

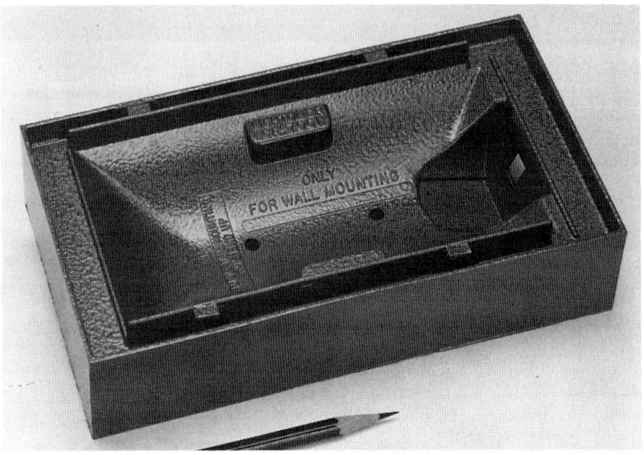

Fig. 6-7. This aluminum receptacle was produced at a 56% savings using diecasting over the sand casting method originally used. The element requires only a simple trimming operation—machining requirements were eliminated completely. [Photo courtesy of the Newton-New Haven Co.]

Permanent Mold Castings

Diecasting makes more rapid production possible, but the dies used must be stronger and hence, more expensive, than the molds used in the permanent mold process. Consequently, the number of castings required to justify use of the diecasting method is higher than for permanent mold casting.

Permanent mold castings are sounder, can be produced at lower tooling costs and can be made with sand cores yielding shapes not now available in diecasting. However, die castings can be made to closer dimensional limits and with thinner sections. When compared to permanent mold castings, die castings can be produced at higher rates with less manual labor, have smoother surfaces, can have smaller cored holes and commonly cost less per casting when the production run is large.

Fig. 6-6. This aluminum distributor housing was diecast at a cost 25% lower than sand casting. One factor in cost reduction was the die design, which allows casting of 14 different size housings on one die using interchangeable die parts. The casting shows excellent surface finish—only flash removal and trimming operations were necessary. [Photo courtesy of the Newton-New Haven Co.]

Forgings

Die castings can be more complex in shape than forgings, may have thinner sections, can be held to closer dimensions and may have coring not feasible in forgings. However, forgings are denser and stronger than die castings, have the properties of wrought alloys, and can be produced in sizes not practical for and alloys unsuitable for diecasting. It should be noted, however, that the fatigue strengths of die castings are approximately equivalent to forgings.

Screw Machine Products

Die castings are often produced more rapidly, involve less waste in scrap and can require fewer operations than screw machine products. They also can be made in shapes difficult to produce from bar or tubular stock.

Advantages of screw machine products include the higher mechanical properties characteristic of wrought metals, lower tooling expenses and the wider choice of alloys (most of which are not suitable for diecasting).

Plastic Moldings

Compared with plastic moldings, die castings are much stronger and more rigid. They usually involve about the same tooling cost and are more stable dimensionally. Die castings are good electrical and thermal conductors and can be finished for appearances not obtainable with plastic moldings.

However, plastic moldings rarely require a subsequent finish and can be held to closer dimensional tolerances. They can be transparent, translucent or opaque in any color and weigh less. Plastic moldings are excellent dielectrics, have low heat conductivity and require little or no draft.

ALUMINUM DIECASTING ALLOYS

The composition, properties, stability and fabricating characteristics of aluminum diecasting alloys are described in detail in Chapter 4. The alloys most commonly diecast are A360.1, 383.1, 384.1, B390.1 and A413.1.

Alloy 385.1 is growing rapidly in importance because its specifications are similar to important specifications in Europe and Japan. Alloys that also are important (although less popular) include A380.1, C443.1, 516.1 and 518.1.

MACHINES FOR DIECASTING ALUMINUM

Aluminum die castings currently are being made from a choice of castable alloys. Production limitations are determined primarily by the machines and methods used in any given case. The two principle techniques of diecasting are known as the hot-chamber process (using a hot-chamber machine) and the cold-chamber process (using a cold-chamber machine).

In the early phases of cold-chamber process development, the hot-chamber (gooseneck) diecasting machine (Fig. 6-8), which operated by the air injection process, was used, but this machine is now used infrequently for casting aluminum. In that early process, the aluminum dissolved the iron, thereby reducing the section thickness of the iron gooseneck and rendering the casting relatively brittle.

The cold-chamber diecasting machine has become the standard for diecasting aluminum (Fig. 6-9). In these machines, the injection plunger and cylinder are not submerged as they are in hot-chamber machines. Instead, molten metal is ladled, either by hand or machine, into the shot cylinder. The hydraulically operated plunger advances forward and forces metal into the die at high pressure.

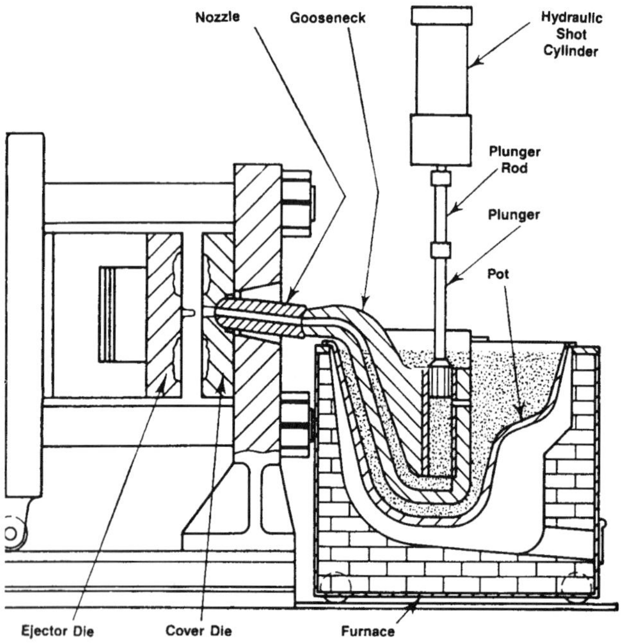

Fig. 6-8. A diagram of a hot-chamber diecasting machine illustrates the plunger mechanism, which is submerged in molten metal. Modern machines are hydraulically operated and equipped with automatic cycling controls and safety devices. [Drawing used with permission of the North American Die Casting Assn.][1]

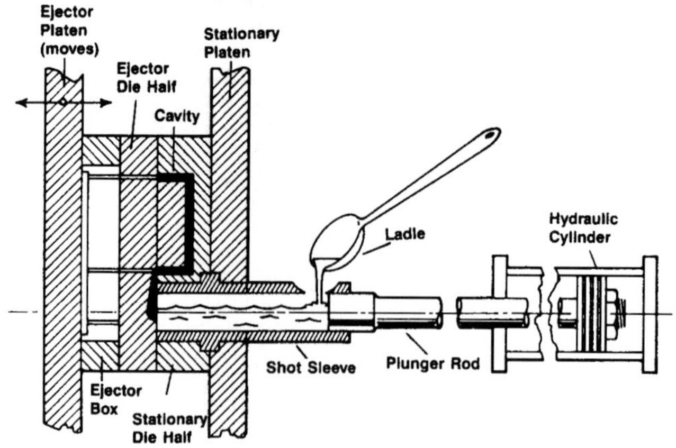

Fig. 6-9. A drawing of a cold-chamber diecasting machine shows the die, cold chamber, and horizontal ram or plunger (in charging position). [Drawing used with permission of the North American Die Casting Assn.][1]

Some manufacturers make dual-process, convertible machines. When equipped with a gooseneck plunger injection unit, these are hot-chamber units for zinc casting, but they can readily be converted for producing aluminum castings by simply removing the hot-chamber injection unit and installing a cold-chamber mechanism. The changeover generally takes 8–16 hours (or more). Such units give the diecasting shop greater flexibility with fewer pieces of capital equipment.

Standard Types and Sizes

The constant increase in the sizes of castings and the demand for better finishes, less porosity, thinner sections and extremely complex shapes of die castings have been instrumental in the development of a wide variety of models, shapes and sizes of diecasting machines to meet the diverse needs of the industry.

Small horizontal diecasting machines that can achieve locking pressures of as little as 25 tons and injection pressures of 1000 psi and under, can produce small, light castings at rates exceeding 600 shots per hour (Fig. 6-10). At the other extreme are large machines that have die locking pressures of 3000+ tons and that inject 100 lb of aluminum per shot (Fig. 6-11). Injection pressures of 40,000 psi or more can be obtained. In a midrange between the smallest and the largest are the vast majority of machines, each of which is economically suitable for producing castings within its own capacity range (Figs. 6-12 and 6-13).

Vertical Diecasting Machines

The design of many aluminum die castings requires the placement of inserts into die cavities before each shot. Vertical models can often eliminate die damage and downtime that can occur with horizontal machines when inserts change

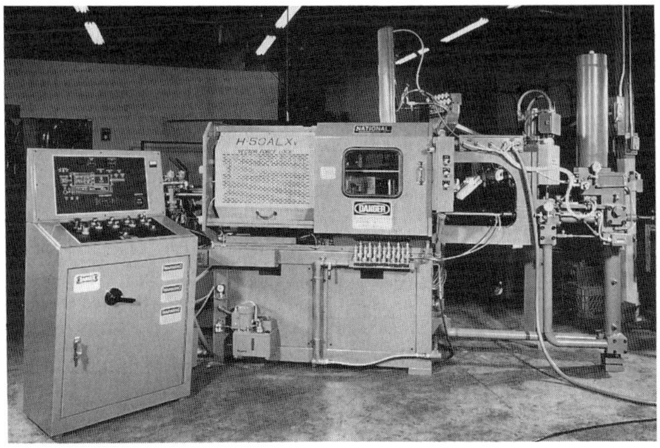

Fig. 6-10. This relatively small diecasting machine can exert 64.8 tons of locking pressure. It averages 1 lb of aluminum per shot but, at times, can achieve up to 3 lb per shot. [Photo courtesy of National Diecasting Machinery.]

Fig. 6-11. This diecasting machine is one of the largest, measuring over 50 ft long and 16 ft high with platens that are 108x108 in. The machine is run by a programmable logic controller (PLC) and can achieve die locking pressures of 3000 tons while injecting as much as 100 lb of aluminum per shot. [Photo courtesy of Prince Machine Corp.]

Fig. 6-12. Shown here is a large, hydraulically controlled machine that exerts a locking force of 2754 tons and can inject 96 lb of aluminum per shot. It can be equipped for fully automated operation. [Photo courtesy of IDRA USA, Inc.]

Fig. 6-13. This medium-sized machine exerts up to 800 tons of locking force and averages 6.6 lb of aluminum per shot with a maximum shot weight of 27 lb. [Photo courtesy of Buhler, Inc.]

position as the die closes. However, diemakers and diecasting designers have greatly advanced their technology to the state where they use automatic core insertion and ejection devices and use auxiliary pressure-locking systems to hold inserts in place in horizontal dies.

Figure 6-14 shows a die inside a vertical machine immediately after the machine opened.

Air-Operated Diecasting Machines

Completely self-contained air-operated machines, primarily in small sizes, are in use in some diecasting plants, but relatively few new ones are being manufactured. Although air-operated machines are extremely fast and usually cost less to install, they often fail to operate properly and, hence, have fallen into disrepute with many diecasters. Some of the other stated disadvantages attributed to the air-operated machines as compared with hydraulic machines are as follows:

Fig. 6-14. A diecast block seen immediately after the die, made of several sections, has opened.

- They require more horsepower to operate, size for size.
- Motion imparted to the operating cylinders is much more violent.
- Air valves, cylinders and other components tend to rust and stick; hence, constant (and costly) maintenance is required.

Hydraulically Operated Machines

Nearly all of the diecasting machines in use today have self-contained hydraulic systems. Most machine builders offer a complete range of machines with standard locking pressures. They also will build machines to special orders. Machines vary widely with respect to such features as locking pressures, rigidity, capacity, electric or electronic controls, locking mechanisms and hydraulic systems with correspondingly varying costs.

The hydraulically operated diecasting machine has been improved to the point where it is far less complicated, much more foolproof and many times safer than in the past. Welding and the use of seamless tubing has eliminated most leaky joints. Larger hydraulic pumps and greater oil volume at high pressure have accelerated the operating cycle.

Hydraulic Fluids

The almost universal use of fire-resistant hydraulic fluids has greatly reduced the number of hydraulic fires experienced each year in diecasting plants. Hydraulic lines on diecasting machines are usually close to potential sources of ignition, such as molten metal and gas flames. Consequently, it is not recommended to use flammable standard hydraulic grades of petroleum oils. Diecasting machines experience considerable vibration and hard usage, which frequently cause hydraulic lines to rupture. A fine spray of flammable hydraulic fluid striking an ignition source can start a major, costly fire.

Older machines can be converted to use fire-resistant hydraulic fluids at modest cost. New machines are designed for fire-resistant hydraulic fluids. Fire-resistant fluids cost more than conventional hydraulic-grade petroleum oils, but eliminating a serious fire hazard makes the increased cost relatively insignificant. Selection of the right type of fire-resistant fluid for each machine is extremely important.

Three of the types of fire-resistant hydraulic fluids used successfully in diecasting machines are:

1) *Water Glycol Solutions* These types of fluids are composed of 35–60% water; glycol; a water-soluble thickener; and several important additives to improve lubrication, to control corrosion and foaming, and to add color for visibility.

   Fire resistance is good only if water content remains in a specified range because the solutions resist fire by the "snuffing" action of water. They lubricate efficiently up to 150F (66C) and at pressures beyond 2000 psi, depending on pump design. Continued use at temperatures high enough to cause rapid evaporation or boiling should not be permitted because the residue that remains after partial evaporation is flammable.

2) *Water-in-Oil (Inverted) Emulsions* These fire-resistant fluids are mixtures of 50% light mineral oil and 50% water. Fire resistance is acceptable when the proper amount of water is present and, as with water-glycol fluids, is lost with water evaporation. Performance is satisfactory at temperatures under 150F (66C).

3) *Straight Synthetic Fluids (Nonaqueous)* Although several variations are available commercially, the most common synthetic type is phosphate ester-based. These fluids are excellent lubricants, and their use is not limited to lower pressures and temperatures by specific types of bearings and hydraulic pumps. Fire resistance is excellent, for it is the nature of the components to resist burning.

DIES FOR ALUMINUM CASTING

A typical cold-chamber die is shown in Fig. 6-15. Dies consist of at least two separate parts so they may be opened to remove the casting. The two parts meet to form the parting line of the casting. The cover portion (commonly called the cover die) is usually fixed to the stationary platen of the diecasting machine. The sprue hole or ram hole, their joining runners, and gates are generally located in the cover half of the die.

The ejector portion of the die (commonly called the ejector die), to which the casting adheres when the die is opened, travels with the movable platen. The ejector die is mounted on a die base or ejector box, generally a steel or iron casting or welded box. This hollow structure separates the die itself from the movable press platen, and it must be strong enough to support the die against the heavy locking pressure applied to it.

Often, ample space must be provided in the ejector box to allow a plate to actuate one or more movable cores that may project through the back of the die. For pulling cores and ejecting the casting, die bases have mechanisms of racks, plates and pinions operated manually or hydraulically.

Ejector pins, used in the ejector half of many dies, actuate after the die opens to push the casting out of the cavity and off any fixed cores. These pins must be sized to eject the casting without breaking, bending or distorting it. Ejector pins, attached to a bumper (ejector) plate, extend through the ejector half of the die. The forward ends of the pins bear on the casting or on ejector pads around the casting.

When ejector pins are in their extended (ejected) position, they are returned to their flush position by hydraulic action or by two or more heavy flush (surface) pins attached to the ejector plate and projecting through to the die parting outside the die cavity wall. These pins, which move forward with the ejector pins and, therefore, are extended when the die is open, meet the cover half of the die as the die is closing and cause the ejector plate (with all its pins) to slide back to the casting position.

Movable cores often are attached to core plates. Other cores (fixed or movable) also are used. When fixed, the core axis must be parallel to the direction of ejection. Ejector pins and cores are discussed in detail in the following sections. Because the casting must adhere to the ejector half as the die is opened and because castings shrink onto cores, if there are cores in both cover and ejector halves of the die, the cores providing the greatest holding force must be in the movable or ejector half.

When the side wall of a casting is undercut, one or more movable slides or side cores must be provided on the die. Though they add to the cost of the die, such slides often make more complex castings possible. They are usually located in the ejector half and are designed so that they can be withdrawn from the casting when the die is opened.

Gates in small diecast parts are made thin so that the casting is easily broken from the runners. Air is expelled from the die through vents at the parting line, which are shallow passages in the die face about 0.005 in. deep.

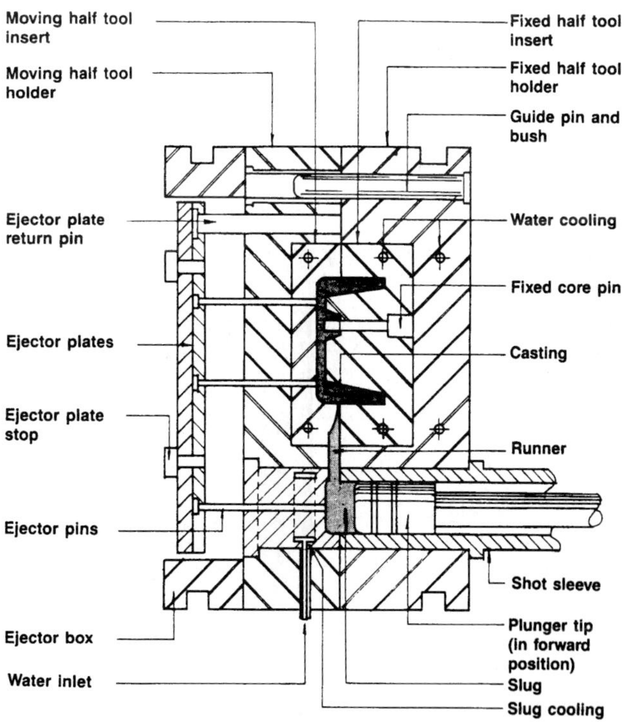

Fig. 6-15. A typical cold-chamber diecasting die. [Drawing used with permission of the North American Die Casting Assn.][1]

True registry between the die halves is assured by guide or dowel pins. In simple two-part dies, to facilitate casting removal, cavities are arranged so that the largest cross dimension or diameter is at the parting line. However, for complex-shaped castings that require the use of slides, the widest portion of the casting need not always be on the normal die parting line. Side walls of the dies and core pins also have a slight draft to make it easier to remove the casting. Because of shrinkage away from the die walls onto core pins or projections, the latter must be designed so that they can be withdrawn easily without damaging the casting.

To maintain the proper operating temperature, excessive heat must be removed. Dies are usually cooled by water passing through channels in the die. The spacing and size of the channels depend on the size of the casting, the amount of heat to be dissipated, the space, and the type of aluminum alloy being cast. Some dies are cooled by conduction to a water-cooled platen. Diecasting machines may be equipped with automatic cooling controls that circulate water when the die is closed but, to prevent overcooling, the cooling process should stop when the die is unlocked.

Types of Dies

Single-Cavity Dies
A single-cavity die is shown in Fig. 6-16a.[1] This type of die makes one casting per shot.

Multiple-Cavity Dies
If a die has more than one cavity, but all are identical, it is called a multiple-cavity die (Fig. 6-16b).[1] These are used whenever possible (especially when production volumes are high) to create more than one casting with each shot.

Combination Dies
A die with two or more cavities of different shapes is called a combination or family die (Fig. 6-16c).[1] These dies frequently are used to produce several different parts of one assembly simultaneously.

Unit Dies
A unit die consists of a die holder into which a number of standard size blocks are fitted (Fig. 6-16d).[1] Each interchangeable die block contains one or more cavities and is connected by a runner to a common opening or sprue hole, permitting simultaneous filling of all cavities. Unit dies allow several parts to be cast at the same time with a high degree of flexibility and quick changeover. These parts can be several parts for an assembly or parts of the same alloy for different customers.

DESIGNING AND BUILDING DIES

Diecasting designers must be familiar with the diecasting process because dies are solid structures that must allow castings to be removed from them with ease. General design rules and specific practices for diecasting are given in detail in Chapter 5 of this book. However, covering all facets of the design and construction of diecasting dies is beyond the scope of this book.

Total cost of a die casting is the sum of the costs of the raw material, labor, initial tooling, die, post-casting machining and finishing, plus overhead. Tooling costs include jigs and fixtures required to trim, machine, polish or plate the part.

The design of the casting affects all the subsequent costs of producing the casting. Designs that minimize complexity and do not impede or slow the casting cycle will result in savings to the diecaster and lower cost for the customer. It should be stressed that the die designer and the die builder, when working together closely and making full use of each other's experiences and techniques, usually give the diecaster the best assurance of receiving a quality die in a reasonable time at a fair cost.

Designing and building a die is generally a time-consuming and costly operation. For comparatively simple dies, about 4–6 extensive layouts usually must be prepared. For large castings, 30 large drawings plus numerous smaller detail drawings may be required for complicated dies. At best, several weeks are required to build, heat treat and try out a die—sometimes it takes longer.

The reputable diecasting die builder, with a background in the diecasting process, is able to provide the best service from the initial quotation for a job through designing, building and servicing after delivery. This person can recognize in advance potential trouble sources, such as inadequate draft angles, poor undercut design and gating irregularities, and thereby help the designer eliminate them. The die builder also is instinctively aware of the refinements in die building essential to the production of quality castings.

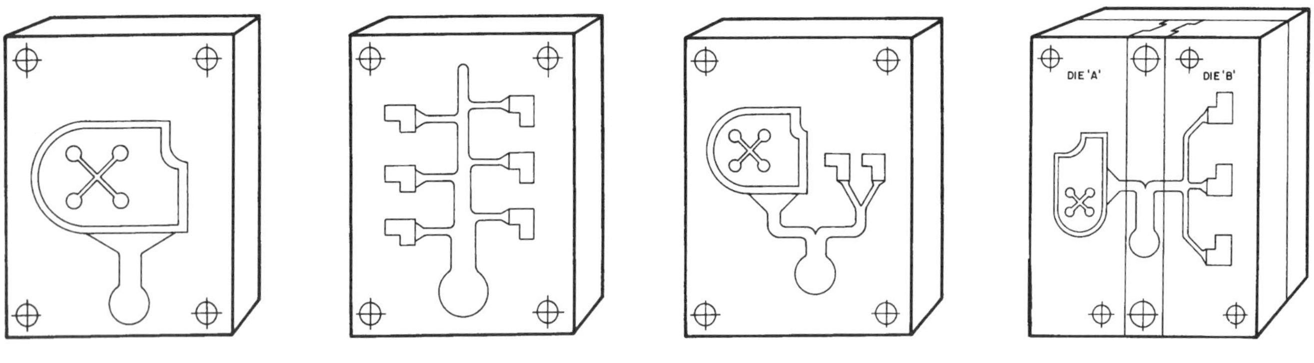

(6-16a) single-cavity die (6-16b) multiple-cavity die (6-16c) combination die (6-16d) unit die

Fig. 6-16. Types of dies available for diecasting.

Some general precautions regarded as basic rules by the die designer, diecaster and die builder are:

- Keep the design of the die as simple as possible.
- Build the die with enough strength to resist the tremendous pressures used in the diecasting process.
- Provide for improvements or additions to the gating and cooling system.
- If the nature of certain details or components in the die will make them subject to breakage or excessive wear, provide ready access to such details.
- Use the simplest method of attaching the die to the machine.
- Be certain the seal around cavities will preclude both rupture and the spewing of molten metal from the die.
- It is always possible that the best die design might be the most difficult and most costly to build. Modify the design when significant reductions in die costs or die building time can be realized with little or no loss of quality in the casting.
- Design the die to minimize such secondary operations as flash trimming. Secondary operations, if overlooked during die design, may cost more than the casting. Parting lines should be properly located so the flash can be trimmed easily with only light shearing action.

Diecasting dies are produced by various processes. Machining is the most widely used method. Two basic types of milling and routing machines can be found in most shops. One is the duplicator, which reproduces from a master part, a master pattern or a template at a ratio of one-to-one. The second type includes two-dimensional and three-dimensional machines, which do work at reduction rates from one-to-one up to whatever ratio is required.

Producing dies with an electrical discharge machine (EDM) is a practical and, under the specific conditions that usually attend the making of dies of 3-D configuration, an economical method. This method has proved extremely valuable for machining difficult shapes, including diecasting die cavity sinking. However, electrical discharge machining produces a highly stressed, hard surface layer on all heat treatable steels, which can adversely affect the performance of dies. Manufacturers of electrical discharge machines can make recommendations for alleviating this condition.

Dies also can be designed and produced using computer-aided design/computer-aided manufacturing (CAD/CAM). CAD systems allow a die designer to create a 3-D model of a die faster than conventional manual methods. The 3-D model can be rotated on the computer screen to view it from different angles, often exposing possible problems early in the design process. Existing models can be modified to create similar dies without having to start from scratch. Once the die has been designed, the data can be used to generate tool path programs for running the computer-numeric controlled (CNC) machines that will make the dies.

The patented Shaw Process can be used for producing diecasting dies. In this process, dies are cast (usually of H13 steel) in ceramic molds rather than machined. This method is proving suitable for difficult cavities and cores that are subject to excessive wear and are expensive to machine. The method is often used when tolerances are relatively liberal (0.010 to 0.015 per linear inch).

Tool Steel for Dies

Dies generally are made of hot-work tool steels developed especially for the economical and efficient diecasting of aluminum alloys in the molten and plastic states. The diecasting operation is unique in many respects; hence, steels used for dies require an unusual combination of special characteristics. Steel properties of particular importance in casting dies include a high degree of resistance both to thermal shock and to softening at elevated temperatures.

Resistance to Heat Checking

Heat checking in a die is the fine cracking resulting from thermal shock caused by alternate heating and cooling in service. High peak temperatures and rapid cooling increase the risk of heat checking. Heat checking starts after a certain number of cycles and grows at an accelerating rate. This fatigue phenomenon follows the laws of fatigue: a starting (initiation) period and a growing (propagation) period.

It has been found that a number of factors are of importance for the occurrence of heat checking.

- *Surface temperature of the die*—When the temperature of the surface layer of the die reaches 1100F (593C), the possibility of heat checking increases. The surface temperature of the die is determined by the preheating temperature and the casting temperature of the metal.
- *Cooling rate*—More rapid cooling of the surface layer of the die causes greater stresses and leads to heat checking at an earlier stage.
- *Homogeneity of the steel microstructure*—Great toughness and high-temperature strength reduce thermal fatigue tendencies. However, fatigue properties and toughness are greatly influenced by slag inclusions and segregation (the purity and homogeneity of the steel). To reduce inclusions and segregation, remelted steel made using electroslag remelt (ESR) or vacuum arc remelt (VAR) techniques should be used to improve the heat checking resistance. These steels are called "premium steels."
- *Heat treatment*—Improved hot hardness reduces heat checking. If the hardening temperature for a die is increased, hot hardness and resistance to tempering increase, and heat checking tendency is thereby reduced. However, the risk of severe cracking (total failure) can increase, so this method should be limited to small dies.

  Similarly, a higher hardness has a positive effect on heat checking, although a hardness exceeding 48 HRC is not recommended for aluminum diecasting. Decarburization and heavy carburization also may cause premature heat checking.

Ultimately, all steels used in hot-work operations will fail by heat checking if they do not fail first from causes such as wear, up-setting or gross cracking. There is no simple solution to the problem of heat-check failures in dies. Repeated thermal stress—the basic reason for heat checking—is difficult to avoid because many of the operating conditions of the diecasting process increase the severity of such stresses. Alloys such as maraging steels and TZM (titanium-zirconium-molybdenum) alloys can help reduce the tendencies of heat checking.

Dies should be hard enough to resist erosion but soft enough and tough enough to resist gross cracking. The wide range of hardnesses that will meet these requirements may be determined through experience. When this range has been established, the hardness of the die alloy can be maintained on the high end of the range to provide maximum wear resistance.

Toughness of Tool Steels

Toughness, an extremely important characteristic, is obtained in hot-work tool steels by keeping the carbon content low, generally 0.30–0.40%. Chromium, molybdenum and nickel are commonly added to increase toughness. The use of a "premium" (remelt) steel and proper heat treatment will further enhance the toughness.

Resistance to Softening

Satisfactory resistance to softening in service is assured by the addition of such elements as tungsten, molybdenum, chromium and vanadium. As the total alloy content increases, there is a corresponding general increase in the optimum hardening and tempering temperature. The greater the resistance to softening by tempering, the greater the ability to retain hardness at working temperatures after tempering.

Good Machinability of Tool Steels

Machinability of die steels generally drops with increasing alloying content. For best results, low annealed hardnesses (217 Bhn or lower) with uniform microstructure are required.

Heat Treatment of Tool Steels

Correct heat treatment is essential for maximum die life. The steel will respond more uniformly and predictably, reducing the risk of distortion and cracking with the proper heat treatment. The following should be considered when specifying heat treatment for tool steels:

- Stress relieve after rough machining to reduce the likelihood of distortion during subsequent machining or heat treating.
- When finish machining, leave sufficient stock to allow for size change.
- Step heat to austenitizing temperature to avoid thermal shock.
- Soak at austenitizing temperature for 30 min/in. of section to equalize tool temperature and produce effective microstructure.
- Quench the die at a minimum of 25°F/min (14°C/min). The cooling rate must be fast to suppress the formation of pearlite. The proper microstructure should be tempered martensite and bainite.
- Tempering (double is mandatory; triple is recommended) should be performed to achieve the proper hardness, to reduce the likelihood of softening and of distortion in service.

Uniformly Sound Structures

Sound microstructures in a die steel depend upon a number of variables that start in the melt shop and include proper control of the composition during the melting process as well as proper temperature control at tap time.

Control of sulfur content also is important to reduce nonmetallic inclusions, which can adversely affect toughness. A maximum sulfur content of 0.005% is specified for premium die steel. Sufficient hot working during processing of the die steel (usually 4:1 reduction minimum) assures elimination of the cast structure and freedom from pipe and bursts.

The use of fast-acting forging presses has resulted in greater work penetration in the centers of large sections and has helped provide freedom from center porosity. The use of electric furnace melting, vacuum degassing during pouring and further refining by remelting (ESR or VAR) results in lower sulfur and gas content, which assures greater uniformity of structure and freedom from impurities.

It is interesting to note that, in the diecasting field, relatively few types of hot-work tool steels have gained wide acceptance. The chromium hot-work tool steels regarded as standard for making aluminum diecasting dies are the air-hardening alloys listed in Table 6-1. Premium H13 is the most popular of the three listed. Cores, slides, ejector pins, inserts and similar die components often are made of commercial H13 steel.

Table 6-1.
Typical Chemistry of Hot-Working Tool Steels for Aluminum Diecasting Dies

AISI-SAE Designation	C	Cr	V	Mo	W
H11	0.38	5.25	0.45	1.35	—
H13	0.39	5.25	1.00	1.40	—
H19	0.39	4.38	2.00	0.45	4.15

Finish and Oxide Coating

Any visible machining or grinding marks left on dies or die tools (components such as cores, slides, pins, etc.) may be the start of a heat check or crack. It is important that such blemishes be eliminated.

Because bright, shiny surfaces tend to encourage galling and sticking, an oxide coating should be applied to dies and die tools before they are put into service. The main function of the oxide coating, however, is to provide a surface that will better retain die lubricant.

An oxide coating may be formed during tempering. If the oxide coating is removed by grinding or polishing, it may be restored by heating the die or die tool for two or more hours at 1000–1200F (538–649C), depending on prior tempering temperatures.

Nitriding Dies and Die Components

There are disagreements on the need for and the value of nitriding. However, it is generally agreed that gas nitriding of dies and die tools made of hot-work tool steels such as H11, H12 and H13 is advantageous where an extra hard wearing surface is needed to resist erosion and soldering. However, although nitriding usually minimizes erosion in early service, some diecasters have found that it may shorten die life.

Spalling of the nitrided surface occurs when the case is too thick. Usually, this spalling results in undercutting sufficient to cause sticking and tearing when the casting is being removed from the die. Nitriding is done after all grinding, polishing and heat treating have been completed. Nitrided H11 and nitriding steel 7140 are sometimes used for sliding die components.

Dies, inserts, slides and core pins of H13 steel are usually subjected to a gas nitriding treatment for 10–12 hours at 950–1000F (510–538C), which will result in a case depth of approximately 0.006–0.008 in. When nitriding, some diecasters and die builders mask the ends of cores and other die components that will be in contact with molten metal to get better resistance of these surfaces to heat checking and spalling.

Determining Die Cavity Dimensions

Shrinkage in castings creates dimensional changes for which the die or pattern designer must compensate. Solid shrinkage at operating temperatures, the so-called "pattern shrinkage," results in a casting with smaller dimensions than the cavity in the die or mold. When designing and constructing dies, an anticipated shrinkage rate is chosen to determine the percentage change of the cooled casting from the original dimensions of the cold die. Cavity dimensions must be greater than the design dimensions of the casting.

Determining the correct allowance (a multiplication factor for increasing die cavity dimensions) to compensate for casting shrinkage is complex and requires experience. Since the amount of solid shrinkage depends not only on the casting itself, but also on its shape, its position, its time in the die, and heat transferred (which will vary with die temperature) while it is in the die, shrinkage can be different at different points on the same casting. For this reason, shrinkage values for certain alloys have little practical value.

When determining the dimensions of a mold cavity, it is the usual practice to multiply the dimension of the finished casting by a factor that enlarges the cavity dimension just enough so that the finished casting will have the desired dimensions after it has been ejected from the die and cooled to room temperature. While this factor can be determined by calculations, it is most common to use values that have been found reliable in commercial practice. An experienced mold or die designer usually can select the correct factor to use for the various cavity dimensions. While approximate values have been fairly well established, they serve only as a starting point. Experienced die designers often find it desirable to change those values for various areas of the die.

For relatively simple shapes that have fairly uniform metal sections and are generally symmetrical, a factor of 1.005–1.007 is a commonly used multiplier. However, if the ratio of width to length is large—about 1:1—then it is common to use 1.005–1.007 for the width dimension and increase the factor for the length dimensions, sometimes to as high as 1.012.

It is necessary to use a factor somewhat beyond the 1.005–1.007 range for relatively thick metal sections, such as bosses, that are free to contract away from the mold without restriction. This also may be true for molds for producing plain, fairly symmetrical castings that have no restraining ribs, bosses, projections, cored holes or other interruptions in surface continuity that would impede normal contraction.

A factor of 1.000, or close to it, is necessary for machining cores to form holes, cavities or pockets in many forms of castings. The reason for this is that the solid cores prevent normal contraction. In effect, they act as a sizing form—the casting dimension is practically the same as the core dimension. When exceptionally close dimensional tolerances must be held, it is customary to size the mold contours in the critical areas so that enough extra metal is left in the mold or on cores for additional machining after pilot castings are run and checked dimensionally.

In long or large castings, all portions may not contract at the same rate (Fig. 6-17). Variation of temperature within the die is a major cause for differences in the rate of contraction. Other factors affecting this are the presence of projections or other interrupted surface continuity in one area of the casting and a relative "clean" area next to it.

It is common practice for designers to use the same machining factor for cutting metal molds for all the aluminum diecasting alloys.

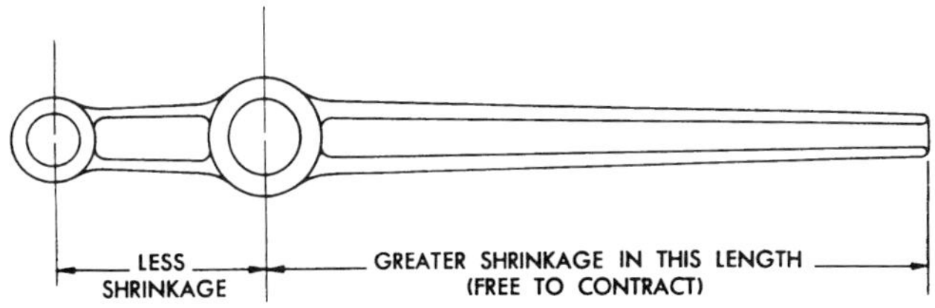

Fig. 6-17. When designing a die cavity, allowances must be made for differences in shrinkage rates in different sections of the die.

RUNNERS AND GATING

Runners commonly are made as a flattened, ellipsoidal section or are made flat on the wide or outer side and rounded on the ends of the section. Their sectional areas should be considerably greater than those of the gate or gates they feed, with a gradual decrease in the area in the direction of flow. Passages should be smooth and should blend into the gate so that eddies (currents in the molten metal running contrary to the main current) will not be created. Curves should be gradual rather than sharp.

The objective of gating is to fill the cavity in an orderly manner without creating undue eddies or trapping air. This is an important consideration in every die casting's design. For many designs, different sizes and arrangements of gates must be tried until satisfactory results are obtained. The gate must not allow gravity flow of metal into the die cavities before the shot is made.

Gates should be designed to minimize nozzle effect and resultant spatter in the cavity. Careful attention should be given to the direction in which the metal enters the cavity. Large gates of uniform thickness without end taper produce good results. In long slender die castings, it is sometimes best to have metal directed across the cavity rather than along it to establish a more orderly filling. For the majority of die castings, a single large gate, if practicable, is recommended. Certain shapes of castings require two or more gates. Many castings can be produced with center gates from which metal fans from a central hole to fill thin sections of a large area.

Dies for producing aluminum castings with thin walls should have gates that are not larger in section than the walls of the casting. When the gates are made thicker than the walls, the wall itself may be broken when the gates are sheared off. In instances where gates must be made thinner than the wall of the die casting, the gates also should be made thinner than the runner, making it possible to break the casting at the gate or thinnest place.

To minimize erosion, turbulence and soldering, the flow of metal through gates should not be directed at cores. Cylindrical die castings with internal cores may have longitudinal parting with ring gating around the lower end of the core feeding vertically from the ring at several points.

Turbulence and air pockets sometimes can be minimized in shell-like die castings by having the metal flow into the cavity in the direction of the casting instead of in an angular or tangential line. This can be accomplished by a semicircular hump over which the metal flows until it is projected down the casting wall.

VENTING

Opinions on the venting of dies vary somewhat. The vents are usually located at the main parting of the die for carrying off gases and entrapped air in the cavity. Ejector pins can sometimes act as vents, as do the clearances for slides and inserts.

Vents are usually about 0.005 inch deep and only wide enough to allow entrapped air to escape. Proper venting helps metal to freeze quickly, forming only a small thin flash for later trimming. Any tendency of vents to "spit" or discharge molten metal is a hazard.

When "straight" venting of dies is not satisfactory or if there are indications that more vents may help to fill out the casting, the volume of the vent is often increased considerably by providing an overflow well just outside the die cavity. This overflow well is a small cavity in the parting face of the die. It is connected to the main cavity by a thick vent. The well also is vented so that the air and some metal can overflow the cavity and fill the well.

Where there are indications that added venting will preclude surface imperfections near the vent, overflow wells are often employed. In some designs, overflow wells can substantially reduce gas porosity and surface imperfections and, consequently, improve casting density. The overflow wells also heat the cool areas of the die.

EJECTOR PINS

Usually, ejector pins are placed at supported ribs or bosses of the casting or on sections that are strong enough to take pressure without deflection. Pins should also be located where the ejector pin marks on the die casting will not be detrimental to the appearance or functioning of the part. When pins bear on overflow wells, flash or runners, no marks will appear on the finished die casting.

In the retracted position, ejector pins usually end flush with up to 0.010 in. protruding from cavity walls. They are attached to a common ejector plate behind the ejector half of the die. The plate may be arranged for manual actuation, but faster cycling and more force are realized when bumper plates are automatically actuated, either hydraulically or mechanically, by the return motion of the movable platen.

In place of ejector pins, a tube or sleeve surrounding a core may be used. However, as the sleeve must fit mating surfaces inside and out, it is highly susceptible to wear. Sleeves are used extensively to eject narrow, circular die castings.

Some dies need no ejector. This is often the case when movable cores are the only means of holding the casting in the ejector die half. After the die is opened, withdrawing the cores releases the casting so that it can be lifted out of the die without sticking. For large castings, hydraulically operated core pullers are often used.

Some die castings are so contoured that it is quite impossible to hold them to the ejector half of the die during die opening. Some castings with thin walls and intricate contours must be held absolutely parallel at the parting line and must not be ejected too quickly or unevenly. Under these conditions, ejector pins are often used as recessed locking and retaining devices. The pins are cut a few diameters short, carefully undercut, and located at runners or overflows. These undercut ejector pins hold the casting to the die face until it is time for the normal ejection operation.

CORES

Wherever possible, diecast parts should be cored to obtain maximum metal economy, provide necessary recesses or holes, and lower casting costs. Also, cores are sometimes necessary to ensure uniform section thickness.

When cores are fixed to a slide or to the main die section, they must be shaped and located to permit ready removal of the casting as it is ejected. Freedom of action also must be considered. In general, the axes of fixed cores must be parallel either to the motion of the slides or to the die motion. Cores may be machined or cast.

The most economical cores are those that are fixed, especially cylindrical ones. If they can be withdrawn, movable cores can be in any position. Some cores and other inserts, sometimes termed loose pieces, are ejected with the die casting and subsequently removed. Such cores, which are useful for forming undercuts, must be positioned in the die cavity each time; this slows down the casting cycle.

When cores have a motion angular to the die section in which they are located, they sometimes require separate acting forces. During the casting cycle, it is necessary that cores be locked against movement. Racks and pinions are commonly used for manual actuation, but there are advantages in having cores operated through hydraulic or air action.

Mechanical operation of cores is often set for automatic action as the die opens. Hydraulically operated cores can be pulled either automatically or by hand-operated controls before or after opening of the die.

Movable cores leave flash at one or both ends of the casting. Where possible, it is good practice to have this flash located where it can be removed when parting flash is taken off. Movable side cores are generally placed in the ejector half of a die and removed before the casting is ejected. If it is necessary to place movable cores in the fixed (cover) portion of the die, interlock is required so the core can be pulled before the die is opened. If a manually operated side core in the fixed half of a die is not pulled before the die is opened, the casting or core will break or the die will be damaged.

DIE RELEASE AGENTS AND LUBRICANTS

A lubricant is any substance used to reduce friction between surfaces in contact. In diecasting, the term "release agent" is used synonymously with lubricant. When correctly applied, release agents that have been formulated from the proper lubricants will form an impenetrable barrier, parting agent or film between molten metal and die metal.

There are two forms of die lubricants: solvent-based and water-based. Because solvent-based die lubricants are used to a minor extent, the following discussion will be based on water-based die lubricants.

The main purposes of die lubricants (release agents) are to ensure easy ejection, to prevent wear, and to eliminate sticking and galling of die parts. A water-based lubricant also can act as a cooling agent on the surface of the die cavity, thus inhibiting soldering of molten aluminum to the die surface. Die lubricants can protect die surfaces from washing (eroding) because of the movement of molten metal under pressure and provide a coating that reduces surface tension, thereby promoting better metal flow and complete filling of the casting.

The parts of the die requiring lubrication are the moving sections (such as sliding cores and the ejector pin mechanism), the injection plunger or piston on cold-chamber machines, and the die cavities themselves. In choosing a die casting release agent, it is best to find one that can be used effectively on all die parts including cores, cavities, slides and the ejector mechanism. However, two or more lubricants can be used on the same die through the use of automatic spray equipment.

Lubricants proven for one metal or alloy are not always suitable for another. For instance, lubrication and release requirements for aluminum, magnesium and brass are much more heavy-duty than those for zinc because of the lower temperatures involved in casting zinc.

Most die lubricants are supplied as concentrates that contain active components emulsified in water. The concentrates are diluted with water by the diecaster prior to application. The choice of materials selected as lubricant components includes petroleum oils, vegetable oils, synthetic oils, waxes, resins, soaps, paintable silicones, graphite, mica, molybdenum disulfide and boron nitride.

The combination of materials selected as the lubricant component of a release agent will affect not only the performance and properties of the release agent, but also the entire diecasting operation. Each lubricant has its advantages and disadvantages. Care must be taken to match a particular die lubricant to the diecasting operating parameters (e.g., alloy type, alloy temperature, die temperature, part geometry) to maximize the productivity of the diecasting operation.

No one lubricant is a cure-all for die problems. Improper venting and gating, insufficient draft for good ejection and incorrect temperatures of the metal or the die can cause problems that cannot be remedied by lubricants.

Application of Die Lubricants

Most die lubricants are sprayed onto the die. There is a large selection of spray equipment and spray nozzle designs available. This includes anything from hand-held spray equipment spraying a single lubricant to complex automatic spray equipment capable of spraying many different lubricants at one time through multiple nozzles.

Automatic spray equipment allows for a controlled, reproducible application time after time. This is important because under- and overapplication have the potential to cause problems. Underapplication can result in increased solder, stuck castings and broken ejector pins. Overapplication can result in buildup on the die and/or casting defects such as surface swirls, stains, cold shuts, porosity and general marring of the surface finish.

Plunger Lubricants

Plunger lubricants are typically oil-based products with viscosities ranging from viscous oils to grease-like pastes. Some contain graphite. Water-based plunger lubricants (emulsions) also are available, although used less often. The choice of plunger lubricant is typically influenced by the application method, the machine mechanical parameters and the particular preference of the diecaster.

Plunger lubricants are applied via brush, spray or drip methods. The latter methods can be automated. Plunger lubricants can be applied in front of the plunger tip in the cold chamber, behind the plunger tip during the tip return, or on the plunger tip when the tip is in the fully returned position. Applying plunger lubricant in front of the plunger tip in the cold chamber is more likely to cause stain and porosity because some of the lubricant will be carried into the die cavity when the metal is injected.

As is the case with die lubricants, care must be taken to control the application of plunger lubricants. Too little lubricant can result in reduced plunger tip life; too much can result in housekeeping problems, stained castings and casting porosity (especially if the plunger lubricant is applied in front of the plunger tip in the cold chamber, as mentioned previously).

The technology of today's die lubricants, plunger lubricants and application equipment is advanced. In order to obtain the greatest benefit of the lubricant technology, one must carefully match the lubricant chemistry and technology to the operating parameters of the diecasting operation.

TRIMMING AND FINISHING

Gates, runners and flash usually are removed from die castings in a trimming press. Special jigs and trim dies often are designed and built to facilitate trimming.

Die castings can be machined with similar tools and in the same manner as other nonferrous products. However, cuts are commonly light and can be made at high speeds. Grinding, polishing, filing and buffing are performed rapidly because as-cast surfaces are usually quite smooth. Carbide tools often are not essential for machining aluminum die castings but often are the most economical.

When machining is required, it usually involves such operations as drilling holes that are too small to be cored economically, tapping holes, and squaring surfaces where the draft required for the casting process cannot be tolerated on the finished product. Sometimes machining is necessary to obtain dimensional accuracy that cannot be provided readily by the die.

In general, die castings can be finished in the same way and with the same basic materials as for other aluminum products. General machining and finishing operations are discussed in detail in Chapter 16.

Fig. 6-18. High-technology diecasting is shown in this computer-based process control system. Computers at both the diecasting machine and process control office communicate over fiber optics. Numerous elements of each machine cycle are loaded, monitored and kept within given parameters by the computer. [Photo courtesy of Paul Krone Diecasting Co.]

AUTOMATING THE DIECASTING PROCESS

Process Control—Shot Monitors and Controllers

The nature of the diecasting process demands that things happen fast. The die cavity fill time is always described in milliseconds. A typical fill time is 100 milliseconds (1/10 sec) in aluminum diecasting. The process is almost instantaneous, but in this short interval, important aspects such as pressure, position, percent fill, solidification, die temperature and velocities (to name a few) are changing. Monitors and controllers are available to perform the lightning-fast calculations necessary for in-line corrections and produce high-technology process control (Fig. 6-18).

Monitoring systems that graphically record all the variables in a casting cycle are frequently referred to as "process control." These systems assist in establishing trends and operating ranges desired for the operation. The records are then used to make adjustments to the process.

Monitors

With the rapid rise in applied electronics, it is now possible to measure changes against time—even down to the ten thousandth of a second, if desired. These measurements have brought about a degree of sophistication not available earlier. Now it is possible to not only measure but to control certain factors in these very short time spans.

Because of the ability to measure functions in such short time periods, the practice now is to exercise control and produce consistent results, where before only failure was evident when attempts were made in particular situations.

Many makes of monitors are available. Most are based on user-friendly computer graphics. Typical monitors plot hydraulic head pressure, rod end pressure, and position and velocity of the plunger. Many can collate and apply statistical procedures to variables and produce summary reports.

Controllers

Recently, more sophisticated machines are available that incorporate closed-loop, feedback controllers. There are also retrofit packages available to bring older machines up to this level of control. Basically, transducer attachments provide signal conversions that feed information to the monitors. Preset programs analyze the sequels against set points and make midcourse corrections or, in some cases, make corrections on following shots by averaging several previous shots.

The necessity of such controllers is due to the inherent changes that take place due to heat, pressure and variations in friction. Long-term studies, for example, have shown that a set slow-shot speed will vary as much as 100% unless some means of compensation is used.

Most monitors and controllers have the following features:

- programmability;
- high-speed response;
- data acquisition;
- real-time control;
- extensive memory;
- profile displays;
- networking capability.

Automatic pyrometric control systems on some highly automated machines protect the dies from overheating yet keep die temperatures constant for each casting cycle by regulating the flow of cooling water. Such systems increase the flow of cooling water during the injection phase and reduce it during the die lubricating phase.

Casting Machine Robots

In an increasing number of diecasting and mechanized permanent mold aluminum casting plants, high production rates are being achieved and maintained by using reliale industrial robots. Their main function is to extract castings from the die or mold in the ejecting phase of the casting cycle. The average robot takes from 3–6 seconds to extract a casting.

Some units also perform die lubrication, go-no-go initial inspection at the time the casting is extracted from the die, quenching, delivery to flash trim, and other tasks. The jobs generally performed are either repetitive or hazardous. The use of robots frees experienced personnel for "human-judgment" tasks.

Robots are available in several types and degrees of sophistication. They range from single-purpose casting extractors to multi-purpose units offering "off-the-shelf" automation for a variety of industrial tasks. Each can be set or programmed to perform repetitive or nonrepetitive jobs within its capabilities. Some are controlled primarily by electromechanical means; others are controlled by computer components as well.

Previously, power sources were strictly hydraulic or pneumatic. Today, the majority of robots are electrically driven using both open loop and servo-control systems. The servosystems can be either AC or DC type.

Diecasting robots typically are equipped with microprocessor control to allow for precise positioning of the robot arm and smooth operation of the mechanical system. Robots can be equipped with a wide array of memory and program storage options that range from rudimentary single-sequence operation to offline, adaptive-control systems.

Robots can be designed to be either free-standing or diecasting machine-mounted. Free-standing units provide more flexibility and easier maintenance than machine-mounted units. Machine-mounted units tend to be less expensive than free-standing units; however, the difference is usually slight.

A precision robot casting extractor is shown in Fig. 6-19. Typical return on investment for such automatic equipment is reported in the 20–30% range, with capital payout in 9–18 months in some cases where volume is high. Two-shift operation is generally considered a basic criterion. Through use of a robot, payback can be improved substantially based on the size of castings, physical shape of the casting or the length of the cycle time.

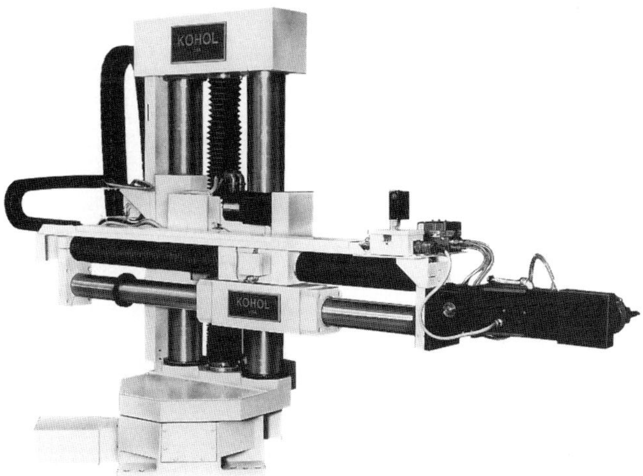

Fig. 6-19. This extractor robot is a free-standing AC electric/servo-controlled unit. The load capacity is 110 lb with a repeatability of ±0.020 in. (0.5 mm). It can be operated easily by a touch-screen CRT display. [Photo courtesy of Kohol Systems, Inc.]

Automated Machines

Diecasting machines are available in which most phases of the machine cycle are fully automated. Hydraulic core actuation, die face and core slide cleaning and lubrication, ladling of molten metal into the shot sleeve, removal of the completed casting from the die, and transfer to quenching and trimming stations are fully automatic. Huge machines can be run by one operator who initiates the casting sequence by pushing a button.

Some machines run continuously, cycle after cycle, with only intermittent monitoring by an operator. Often, these machines are 800 ton and smaller, but larger machines also can be run continuously.

Figure 6-20 shows a diecasting machine designed to function automatically and continuously without an operator attending the unit. This 1200-ton machine can inject up to 35 lb of aluminum per shot and produce one casting every minute.

Mounted on top of the machine is a reciprocator. When a cycle begins, the reciprocator drops down and blows off and sprays lubricant on the die. It raises, and the die closes.

In a fully automated sequence, an automatic ladle pours molten metal into the cold chamber (not shown). In a semiautomated sequence, an operator would pour the metal into the chamber. The hydraulically operated plunger advances, injecting the molten metal into the die.

After a predetermined length of time (long enough for the casting to solidify), the die halves separate, any movable cores are retracted, and the part is ejected forward. The extractor robot then reaches into the die and grasps the casting by the biscuit. As the robot extracts the casting and rotates, it trips heat or visual sensors to begin a new cycle of the diecasting machine.

After removing the casting from the machine, the robot first quenches it (quenching area not shown) then loads the part into a die in the trim press. When the robot arm retracts from the trim die, the robot returns to a "home" position in front of the diecasting machine and waits for the next casting to be ejected.

The trim press closes around the casting, cutting off the gating system and flash. The trimmed part then falls onto a conveyor to be carried away or onto an unloader tray mounted on a cylinder that pulls out and dumps the casting.

DESIGNING BETTER DIECASTING MACHINES

The diecasting process, found to be ideal for completely automatic operation, has been affected greatly by new ideas. The use of vacuum processes, color anodizing of aluminum and other innovations have had some impact on the industry.

Of great importance are the many built-in design features of the present-day diecasting machines, which save die instal-

Fig. 6-20. Fully automated, this diecasting machine (center right, shown with no die installed), extractor robot (center left) and trim press (left, shown without a trim die) can run unattended. [Photo courtesy of Prince Machine Corp.]

lation time and substantially reduce the time, labor and reject rate of the process. In response to the never-ending demand for reduced costs and improved product quality, the builders of diecasting machines, in cooperation with the diecasters themselves, have made remarkable improvements that, over time, have become standards.

Control of Injection Speed and Pressure

In most diecasting machines, injection pressures are completely adjustable over wide ranges to suit the type of casting being produced. It has been discovered that the quality of aluminum castings is affected adversely when the injection pressure drops near the end of the stroke of the plunger.

Machine designers have developed pressure intensifiers that are now part of most diecasting machines. These devices usually are entirely self-contained and automatic. They function as part of the injection cylinder unit without presetting or adjustment, and they increase the ending pressure to two or six or more times the starting pressure. This great increase in pressure at the end of the injection cycle packs metal into the die with tremendous force, increasing the density of the casting and, consequently, improving its quality and mechanical strength.

Many variations on the pressure intensifier unit have been developed. Hence, the diecaster may choose the type that offers the best performance for the specific application.

Injection speeds can be regulated over a wide range from slow to very fast. The plunger speeds of modern diecasting machines are in the range of 600–1200 ft per min (120–240 in./sec). Programmed shot control is available on diecasting machines and works as follows.

Initial plunger speed is relatively slow until metal has completely filled the shot cylinder and runners, allowing air to move ahead of metal with a minimum of turbulence. Then full plunger speed and force are applied as metal flows through the gate into the cavity. On hydraulically operated machines used to cast aluminum, high-speed plunger action is achieved by using accumulator bottles charged with compressed gas (Fig. 6-21). On pneumatically operated machines, the velocity imparted to the injection plunger can be regulated by a valve.

Die Height Adjustment

When a die is too large and projects above the top of the platen, it overbalances the platen and, if used in this position, greatly reduces the locking strength of the machine. This condition occurs frequently in aluminum diecasting where cavities are normally located above the injection point. However, it is possible, on machines with a die height adjusting device, to correct this condition simply by positioning the die near the center of the platen to provide a more central die locating point, which will use the machine's full locking strength.

Lowering the die makes it necessary to adjust the injection position. This is done easily on the machine by one of two methods: raising the entire die plate toggle assembly or lowering the cold-chamber mechanism.

Fig. 6-21. This hydraulic cold-chamber machine can exert 1600 tons of locking force. The vertical accumulator bottles on the left are used for rapid injection of aluminum into the die. [Photo courtesy of Prince Machine Corp.]

Variable Shot Stroke Control

Production of deeper castings in thicker dies requires longer stroke distances (die opening). To make the same machine adaptable and profitable for producing shallow castings, die openings are adjusted by moving an electric limit switch that adjusts the die platen and sets the opening stroke for the exact distance desired. The timing of these movements is controlled electrically by pushbuttons with integral limit-switch safety devices that assure proper sequence in timing the operation.

Hydraulic Ejection and Retraction

Castings are normally ejected automatically by an ejector bumper plate mounted in a fixed position behind the movable platen. However, when extremely high injection pressures are used to make deep-draw castings, extra force is needed to start the initial ejection movement of the casting from the cavity. This extra force is provided by an additional powerful hydraulic ejector mechanism available for most diecasting machines. In cases where side core pins interfere with ejector pins in the cavity, this mechanism can be used to withdraw ejector pins automatically before the die closes.

Strength and Rigidity

Designers and builders of diecasting machines have greatly improved the basic machine construction by using better materials, enlarging sizes of members that absorb pressure, using more bushings, designing wear plates to be replaceable and building in more liberal safety factors.

Heavy, well-ribbed steel plate die platens with long, large bearings offer greater resistance to movement and to wear on tie bars and toggle pins. Many tie bars are now made of heat treatable alloy steel that has a tensile strength of 160,000 psi and ample ductility to resist stress damage.

Other Aspects

Die-locking toggles, considered best by many designers because of their powerful and positive mechanical action, have been made larger and redesigned to better withstand the higher locking pressures of modern machines. Steel castings are used by some machine builders for the toggle links, and large pins of heat treated alloy steel slide in bronze bushings.

The plunger tips of cold-chamber machines usually are made of tough, long-lasting nitrided "Nitralloy" or H13 steels. Beryllium copper also is used for plunger tips. A precision cast plunger tip can have cast-in threads; the user only has to machine the OD to fit the shot sleeve of the diecasting machine.

Core pull cylinders are necessary for many casting operations, and the hydraulic systems of most modern machines are equipped with the valves necessary for automatic timer or pushbutton operation of the core pull cylinders during the casting cycle. Reasonable die safety is ensured by providing electric limit switches on the core pull cylinder—for then the core pins are pulled or pushed at exactly the right moment, and the die is not damaged. Manual insertion or removal of core pins has resulted in irreparable damage to costly dies due to human error.

The electrical systems on modern diecasting machines are vastly improved. All wiring, relays, timers, switches and other components are installed in a single panel box that is sealed against dirt, water, oil and molten metal.

Modern diecasting machines are considerably safer than early machines. The hazards of molten metal being shot while the die is open, the die closing while the operator's hand is between the two die faces, and the holocaust of flaming hydraulic petroleum oil have been eliminated by improved design and engineering practices.

VARIATIONS TO TRADITIONAL DIECASTING

Squeeze Process

There are several alternative squeeze casting processes in use in diecasting. Some are distinct processes; others are variations of existing processes.

In general, during the squeeze process, a latent force is applied to the metal in the die cavity after filling takes place. This force is usually applied in one of three ways by a moving die component (Fig. 6-22).[2]

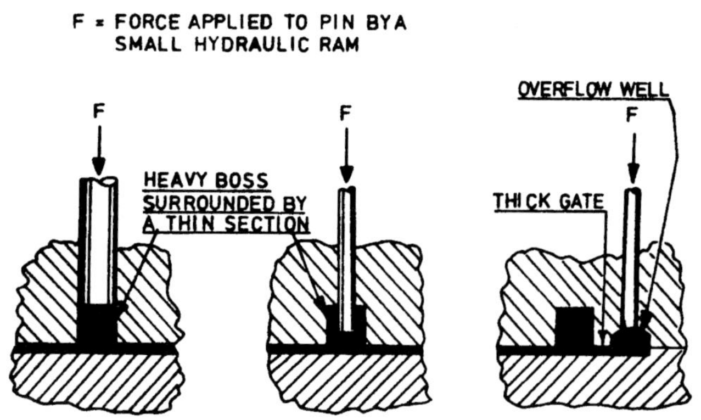

Fig. 6-22. Three methods of applying pressure to the die cavity during solidification in the squeeze process.[3]

In methods in which squeeze casting is a variation of diecasting, the metal is squeezed further as solidification progresses. Standard diecasting machines can be retrofitted with an accumulator, valves and electrical controls to energize and control the hydraulics.

While the metal in the die solidifies, the direct pressure displaces shrinkage voids and feeds them from reservoirs. Large sections that would otherwise be porous can be made sound. Advantages of this process, to name a few, are:

- to eliminate visual porosity opened up by machining;
- to eliminate leakers;
- to eliminate sinks or surface depressions;
- to improve machining tolerances where tools are misguided by porosity;
- to add strength and soundness where required.

In the past, the success of a process such as the squeeze process was limited because diecasters were unable to accurately measure pressures and positions in a time sequence. The direct pressure must be applied free-standing, with only the metal to resist the force of the die component. Because temperature and friction changes affect the timing, as do variations in the shot cycle, it has been difficult to trigger the timing in relation to the time of solidification.

With the electronic advances and practical working monitors, an operator or engineer can now make adjustments to consistently control the process to get the desired results. These tools, when used properly, can meet the diecaster's need for soundness.

Acurad Process

The Acurad process was announced in 1966 as a revolution in diecasting. There are four features, unique to the process, that differ from conventional diecasting:

1) a predetermined thermal analysis of the die, performed to control cooling and force directional solidification;
2) larger, thicker gates;
3) significantly slower injection velocities;
4) a secondary injection plunger.

Almost invariably, parts must be gated from the bottom. When all elements are applied properly, directional solidification takes place from the top of the casting to the biscuit. The larger gate and lower injection velocities create minimal turbulence during filling accompanied by reduced hydraulic pressures. When the main outer plunger begins to stall (due to a freezing effect around the walls of the cold chamber), the secondary inner plunger is activated, applying pressure to the last liquid metal to freeze.

Even though the advantages of lower porosity and higher density appear attractive, few custom diecasters use the Acurad process, at least in part, because of premature cold shuts and the length of time necessary to make the thermal analyses. However, the thermal analysis element of the process has made diecasters aware of the need to improve temperature control in their dies.

Pore-Free Process

Another process that offers improved soundness and mechanical properties in die castings is the Pore-Free process. In brief, the process consists of displacing the air in the mold cavity with an active gas, such as oxygen, which easily combines chemically with the injected molten metal to form a solid compound. The problem of entrapped air is apparently solved by this action.

The advantages of this process are:

- elimination of entrapped air and gases;
- the ability to solution heat treat castings attended with improved mechanical properties;
- reduced injection pressure, hence utilization of smaller machines for the same casting projected area;
- feasibility for welding of castings;
- higher temperature applications for castings.

The major disadvantages of the process are:

- a significant cost to retrofit a machine to accommodate the process;
- redesigned gate entrances produce turbulent flow, which can decrease die life;
- the necessity for a special plunger lubricant and die release agent;
- the necessity for a royalty contract agreement with process developers.

CORRECTING DEFECTS

Defects in a die casting may appear either when a new die is tested or later in production runs. As in most such processes, a production run of die castings normally is not begun on a new or reconditioned die until the die has been tried out. While die tryout methods vary, before heat treating and delivery to the diecaster, the die builder usually tries out the die to check dimensional accuracy of the casting, gating irregularities, venting, ejector pin location and operation, and water lines.

The die can be proved by running a few dozen sample castings on a diecasting machine and correcting irregularities during this pilot run. However, some diecasters prefer to inject a plastic resin or low-melting-point lead into the locked die, where it solidifies into the configuration of the casting.

The latter method precludes the harm that could be done to die cavity surfaces if molten metal were used at this stage.

Usually, many shots must be made before the surfaces of the die cavity have the run-in or cured finish characteristic of a die ready for full, steady production. The finishes and tolerances specified for the castings usually form the basis for determining whether acceptable castings will be obtained by running the die with the oxidized surface formed in the cavity during heat treating.

If castings are not acceptable, additional cleanup machining and polishing usually are required to give the castings the finish and dimensional accuracy specified. If the die cavities are machined and polished after heat treating, it is often regarded as good practice to produce a light oxide film on the cavity surfaces by retempering the die or giving it a steam bluing treatment.

Little anchorage for the release agents (lubricants) is provided on a highly polished die surface unless the surface has acquired a thin, uniform oxide film. This particular oxide layer, built up further during the diecasting process because of the heat and contact of the surfaces by molten metal and release agents, acts as a protective film that inhibits erosion and pitting of die surfaces and minimizes the tendency of castings to stick in the die cavity. Products that can greatly reduce curing time are available.

Before quality aluminum die castings can be produced, it must be determined that the die has been proved and cured, the diecasting machine is operating properly, and the proper molten aluminum alloy is clean and at the right ladling temperature. Poor quality and excessively high scrap losses usually can be expected when the design and construction of the die are faulty and when the melting and diecasting practices are uncontrolled and incompatible with quality production. Preventing defects in castings is usually much more economical than removing defects after the parts are cast.

The three most common causes for rejecting castings are incorrect dimensions, visual defects and leaking during pressure testing. Correcting dimensions usually requires careful check of die construction and components and of the actual operation of the die and the diecasting machine. When the problem is variation in dimensions of castings produced in the same run (i.e., some castings are well within specified tolerances while others are grossly out of tolerance), the causes may be one or more of the following:

- excessive variations in die temperature;
- alloy temperature is too high;
- alloy temperature varies excessively for individual shots;
- machine cycles are not timed properly.

Variations in die temperature may be caused by unsatisfactory layout of runner channels, poor water cooling, bad location of gates or excessive heat in thin die sections. Successive checks can be made with pads of heat-sensitive, cobalt-impregnated paper attached to wrapping paper and placed between die halves. The hottest portions of the die will show after a few seconds of exposure. Three or four tests usually suffice to show a pattern of varying temperature as it occurs in the die. These tests will then guide the rearrangement of cooling channels. Runners often can be adapted to feed metal over portions of the die that cool too rapidly. Dummy runner channels are often helpful in this respect, too.

If excessive or fluctuating alloy temperatures are suspected, a series of short trial runs at various temperature levels with positive temperature control is recommended. The temperature control instruments should be checked and recalibrated if necessary. After the trial runs, temperature control should be set at a temperature that will consistently produce castings within specified dimensional tolerances.

Consistency in the machine cycle is important. Diecasters usually determine the best regular speed of operation and adhere to it, trying to avoid fluctuations in die operating cycles and in time between shots. Consistently good results are obtained by acquiring and maintaining a smooth, uninterrupted operating rhythm wherein each function in each cycle is performed with almost the regularity of clockwork.

When few, if any, castings in the run are within specified tolerance limits, or when shrinkage is evident, causes vary from alloy to alloy. Some of the most common causes are:

- temperature of the molten metal is not correct;
- temperature of the die is incorrect;
- core withdrawal or ejection of casting is badly timed;
- metal flow within the cavity is uneven;
- bosses or cores are located improperly;
- draft angles are too small.

If all the operating factors listed above check out satisfactorily and are unvarying through successive casting cycles but dimensional problems still exist, then it is necessary to stop the run completely to recheck die and part dimensions. In some cases, it may be most practical (and hopefully permissible) to adjust noncritical tolerances or to make minor modifications in the design of the part without lowering its quality. However, experienced designers of both the part and the die, when given ample time to do their jobs well, will build a quality die, prove it in trials, and deliver it to the diecaster ready to use.

Shrink Cracks and Shrinkage

Most problems in the quality of die castings are not "off-tolerance" dimensional; rather, they are visual defects usually resulting from non-fill, shrink or porosity. When deep cuts are to be machined in aluminum castings, subsurface porosity can become a major cause for rejection. The diecaster can check castings for inner porosity by x-raying them or by machining sample castings taken from each shift's production run. The suggestions in Table 6-2 are offered to alleviate the most common causes of visual defects in castings.

When long, thin sections cored in the back or from the cover half remain in the die too long, shrinkage cracks often occur because the core pins restrict the natural rate of contraction. Speeding up core withdrawal and ejection cycles will usually cure the problem. If neither of these is possible, a partial cure can be realized by reducing water flow in the ejector half of the die. This will increase die temperature and thereby retard contraction. Defects in castings also can be caused by:

- not enough metal reaching the die cavity due to faulty plunger operation;
- excessive flash at the parting line, which results in loss of metal or pressure from cavity;

Table 6-2.
Remedies for Common Visual Defects in Die Castings

Defect	Suggested Action
Improper fill (non-fill) of casting	Adjust injection (shot) speed and/or pressure. Enlarge size of bothersome bosses and locating pads, and give them liberal draft angles. Check for and, if possible, eliminate sharp corners. Increase radii and fillets where possible. Long, thin-walled sections should be eliminated.
Shrinks	Revise die cooling arrangements to provide even heating of the die. Use gradual, rather than abrupt, changes in the casting sections.
Porosity	Examine the gating system for possible improvements. Add more venting or overflow wells to the die. Check to see if the die is receiving excessive lubrication. Check to see if ladling operations (especially if performed manually) are causing excessive agitation of the melt. Adjust the speed of the shot. Check the metal temperature. It must be controlled carefully.

- an insufficient amount of molten metal ladled into the shot sleeve;
- local hot spots in die causing bubble formation (a shiny depression or holes in the casting's surface);
- in some designs, the flush coating and backfill method of gating can induce porosity. The operations of the die and the machine should be examined first;
- poor venting, allowing die cavity atmosphere to be entrapped in the metal.

DIECASTING PURE ALUMINUM

One of the few applications in which pure aluminum is used in the cast condition is for rotors for squirrel cage induction motors. Rotors are diecast for motors ranging in size from a fraction of a horsepower to as high as 200–300 horsepower.

In this method, the assembled core of steel laminations, with or without the shaft of the rotor, is clamped in a die, and aluminum is injected under pressure to fill the conductor bar slots and form the end rings and fan vanes. When the aluminum has solidified, it not only provides necessary electrical circuits of the rotor, but also binds the entire assembly together, thus eliminating more expensive assembly methods.

The effects that various elements have on pure aluminum used for diecasting can be grouped into four broad classes:

1) *Effects are negligible* with zinc, gallium and nickel.
2) *Effects are moderate but quantities are comparatively large* with silicon and iron.
3) *Effects are moderate but quantities small* in primary aluminum with copper and magnesium.
4) *Effects are pronounced even in small quantities* with manganese, titanium, vanadium, chromium and zirconium.

Because of the more potent effects of elements in the last two classes, it is important to avoid contamination of the rotor alloys with other aluminum alloys. Moreover, take special care not to use rotor alloy ingot that has been produced partially from scrap metal unless the electrical conductivity of each lot of ingot is checked carefully.

ACKNOWLEDGMENTS

AFS would like to thank the following individuals and companies for their assistance in revising certain parts of this chapter:

- John Puckett, consultant, for the sections on squeeze casting and process control;
- Marc Salenger, Thyssen Specialty Steels, Inc., for the section on tool steels;
- James Graff, Chem-Trend, Inc., for the section on die release agents and lubricants;
- Jim Costello, Chem-Trend, for the section on hydraulic diecasting machines and hydraulic fluids.

REFERENCES

1. "Introduction to Die Casting," a booklet published by the American Die Casting Institute (now NADCA) (1977).
2. *New Technology for Die Casting Zinc,* ILZRO Manual, vol 2—Zinc Die Casting, International Lead Zinc Research Organization, Inc., p 40 (1972).

Chapter 7

PERMANENT MOLD CASTING

THE PROCESS

The permanent mold casting process involves the production of castings by pouring molten metal into permanent metal molds using gravity, low pressure, vacuum, centrifugal pressure or tilt pouring. Basically, the process involves the following steps:

1) A refractory wash or mold coating is brushed or sprayed onto those surfaces of the preheated mold (usually 250–500F, 121–260C) that will be in direct contact with the molten metal alloy. The proper operating temperature for each mold and each alloy being cast usually can be estimated from experience or determined experimentally.
2) Cores, if applicable, are inserted, and the mold is closed either manually or mechanically.
3) The alloy, heated above its melting temperature, is poured into the mold through the gating system.
4) After the casting has been allowed to solidify, cores and other loose mold members are withdrawn, the mold is opened, and the casting is removed.
5) Steps 2–4 are repeated until repair of the refractory walls is required, at which time Step 1 is repeated to the extent necessary.
6) The usual foundry practice is followed for trimming gates and risers from the castings.

Permanent mold casting can, in some cases, be slower than sand casting and is slower than diecasting. The basic difference between sand casting and permanent mold casting is the metal (or permanent) mold used in the latter process. Permanent mold castings usually have better mechanical properties than sand castings because solidification is more rapid. However, permanent mold castings cannot be as complex as sand castings because the gating system is machined into the metal molds.

Metal molds (or dies) (Fig. 7-1) usually are made of high-alloy iron or steel and have a production life of 10,000–120,000 or more castings. However, for some special designs, the permanent molds used are good for only as few as 200–500 castings. The permanent mold process still can be the most economical in such instances.

It is desirable and generally most economical to use permanent steel cores to form cavities in a permanent mold casting. When the casting has re-entrant surfaces or cavities from which one-piece permanent metal cores cannot be withdrawn, destructible cores made of sand, shell, plaster and other materials are used. This variation is called the semipermanent mold method. Sectional steel cores are used in some instances.

Because they are chill castings, permanent mold castings are sounder than sand castings. For pressure-tight and nonporous castings, the permanent mold process is excellent. Permanent mold castings are stronger than die or sand castings and less porous than die castings.

In a variation referred to as "centrifugal permanent mold casting," the permanent mold is rotated so the metal is forced into the cavity by centrifugal force. Metal to feed solidification shrinkage is assisted by this force. Centrifugal casting, thus, generally requires less gating and risers than conventional permanent mold casting. The techniques involved and the results obtained differ so much from the ordinary permanent mold process that it is logical to regard the centrifugal casting method as an entirely different process. Chapter 11 covers the process in more detail.

Permanent mold aluminum castings generally range from less than 1 oz to more than 100 lb (Fig. 7-2). Aluminum castings weighing several hundred pounds have been produced commercially by the permanent mold process.

Fig. 7-1. An open permanent mold shows the gating and venting systems.

Fig. 7-2. Assorted castings show the wide range of sizes and shapes that can be cast using the permanent mold process. [Photo courtesy of Stahl Specialty Co.]

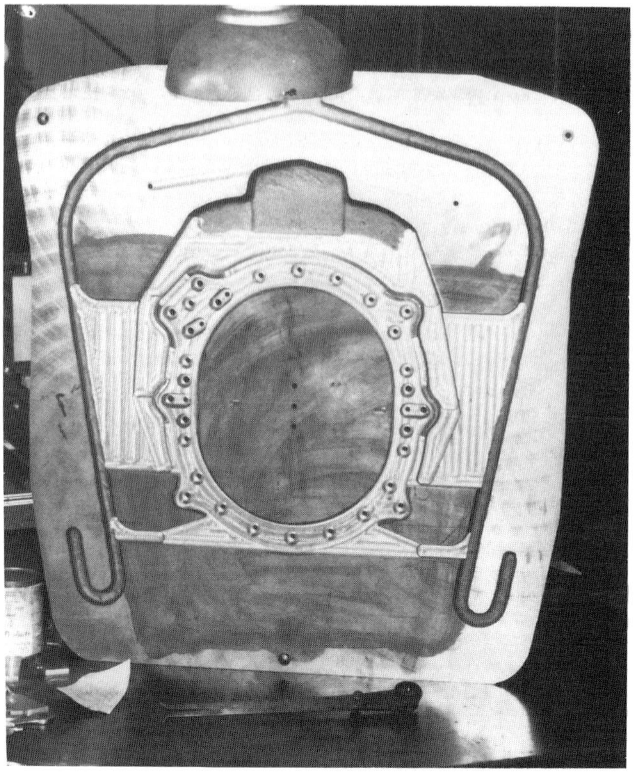

Fig. 7-4. This mold for a nut ring used on furl cells is made using many inserts. [Photo courtesy of LA Aluminum Casting Co.]

Permanent mold castings usually need less finish machining and polishing than sand castings. If the parting lines on the outside of a casting can be held to a minimum, high quality as-cast finishes approaching 100 rms can be achieved and many aluminum castings with these good as-cast finishes are used without subsequent finishing for cooking utensils, hardware items, automotive parts, highway bridge railings and ornamental work. Still, the average surface finish of these aluminum castings varies from 125–750 rms, depending upon the mold, alloy and process.

If possible, positioning of gates and risers should not interfere with locating points for machining. If there is a conflict, machining points should be repositioned, but only after all pertinent aspects are considered. By locating gating on surfaces that will be machined, evidence of gating locations can be eliminated (Fig. 7-3).

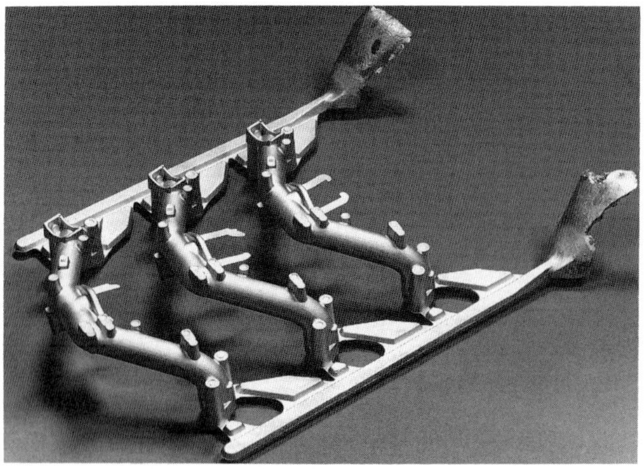

Fig. 7-3. A small engine intake manifold from a three-cavity mold has gating on surfaces to be machined, thereby removing traces of the gating system. [Photo courtesy of Stahl Specialty Co.]

Studs, nuts, bushings, pipes and other inserts may be cast as integral parts of the casting. These must be held in positive position in the mold to prevent movement during casting (Fig. 7-4). To make certain that the aluminum will lock mechanically to the inserts, the latter are often scored, knurled, or undercut to provide a rough surface. Cast-in inserts may be made of aluminum, brass, iron, steel or other suitable material.

Lettering on a casting preferably should be sunk in the mold cavity (raised on the casting) to minimize machining costs on the mold cavity. For trade names and similar identification, letter cavities should be at least 0.050 in. deep by 0.125 in. high and should have approximately 15° draft.

ADVANTAGES AND DISADVANTAGES

Because of the various factors that must be considered when choosing a process, the decision to use permanent molding should be based on thorough engineering and production cost studies for each casting.

Castings produced in permanent molds have finer dendrite arm spacing (DAS) and grain structure and better strength properties than those cast from a similar alloy in sand. Therefore, the casting often can be designed with thinner walls and lower weight. In addition, the cast surfaces are generally smoother than sand castings and closer dimensional tolerances can be maintained.

When properly produced, permanent mold castings can be subject to less shrinkage and gas porosity than sand castings. They do not ordinarily contain entrapped gas (such as in die castings); thus, they are superior to die castings in soundness and pressure tightness. For these reasons, permanent molding is sometimes chosen for better quality.

However, there are limitations to the permanent mold process: if part complexity, surface finish and tolerance requirements are critical, the part might be produced better using investment casting. Extremely complex or irregular shapes may be difficult or even impossible to cast in permanent molds.

There are limitations to the size of castings that can be produced in permanent molds. Many permanent mold castings weigh under 20 lb, although castings of 50–100 lb and more are produced. Designing and operating a permanent mold for large castings may be awkward and difficult unless using the semiautomated tilt-pour process.

The casting design may be so complex that it is impossible to produce it in a permanent mold. A casting's exterior shape might require so many movable mold segments that the operation becomes too complex. Complicated inner cavities or surface undercuts may require a great deal of sand core work. However, extensive use of sand cores is likely to eliminate many of the strength advantages of permanent mold casting. Furthermore, the gating and feeding of complicated shapes may not be practical in permanent molds. If these factors exist, it may be more practical to produce the castings in sand.

Economic Aspects

The economic advantages of permanent-mold casting are most fully exploited when the number of castings produced is large—usually 3000 or more annually. From the economy standpoint, if cast parts are small enough where a large number can be gated on a matchplate, sand molding using high-production coremaking and molding machines may be more economical than permanent molding.

If the quantity of parts required is very large and if strength requirements can be met, a better production process may be diecasting. Diecasting usually results in lower cost than permanent molding, as well as better dimensional uniformity and smoother surface finish. Also, the diecasting process is capable of producing thinner walls, sharper inside corners and more abrupt changes in cross section.

Permanent molds generally have a life of 10,000–120,000 castings or more. For many designs, however, permanent mold casting is the most economical method for production quantities ranging from 400–5000 parts.

If the initial cost of equipment is excluded, cost per permanent mold casting is generally less than that of a sand casting. Initial cost of permanent mold equipment is usually less than that of equivalent diecasting dies, but more than that of metal pattern equipment for sand casting. If the volume is large enough and the size and design of the casting permits it, multiple cavities or impressions may be machined into a mold to increase production and lower unit cost.

Whenever end use requirements dictate and when the number of castings required justifies the cost of the metal mold, permanent mold casting is more economical than sand casting. Reduced machining costs contribute to savings because the castings have better dimensional uniformity, machining allowances are smaller, and holes may be accurately located and dimensioned with metal cores. Thus, the permanent mold process is frequently chosen over sand casting because of the permanent molding's reduced postcasting machining costs.

APPLICATIONS FOR PERMANENT MOLDING

Aluminum permanent mold castings are used widely throughout industry. Pistons, culinder heads and other functional parts of internal combustion engines for the automotive, trucking, diesel, and marine industries are typical applications.

Other major uses for permanent mold castings include: internal and accessory parts for reciprocating and jet-type aviation engines; missiles; forms for concrete; textile machine parts; electric motor housings and parts for portable tools; nozzles, fan cases, and housings for vacuum cleaners; agitators and gear cases for washing machines; support members for outdoor light standards; drums and feed tables for duplicating machines; electric griddles and kitchen pots and pans; rotary lawn mower chassis (Fig. 7-5); supercharger rotor housings; motorized carts; and junction boxes for electric wires.

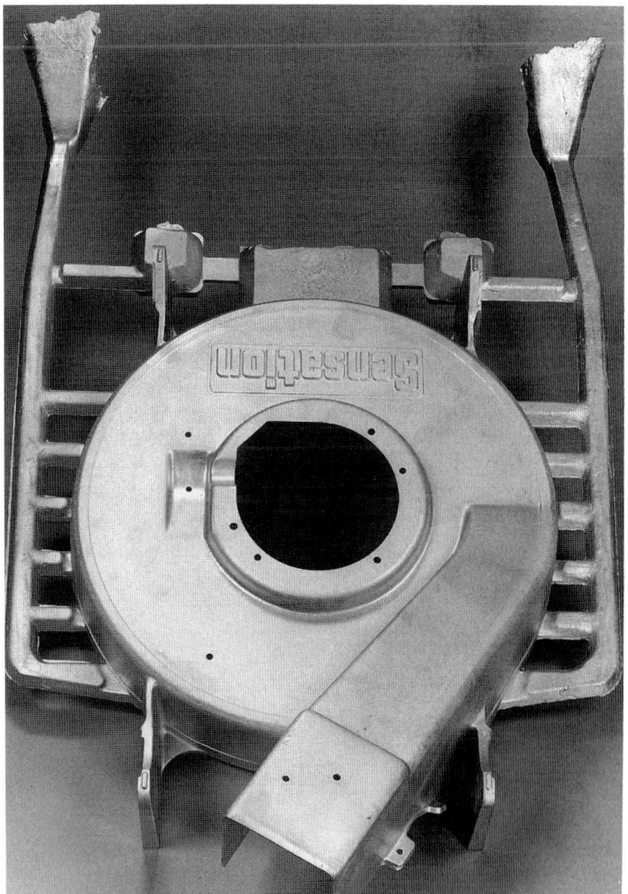

Fig. 7-5. A lawn mower housing is shown with gating still attached. [Photo courtesy of Stahl Specialty Co.]

ALLOYS USED FOR PERMANENT MOLDING

Alloys used for permanent mold casting have better mold-filling characteristics than do those generally used for sand casting. They also are more resistant to hot cracking. Silicon improves both of these characteristics. The alloys most readily cast in permanent mold contain 5% or more of silicon. The properties, composition, foundry and fabricating characteristics, and other information for the most common casting alloys are given in Chapter 4.

Alloy 443 and its variants are used where high ductility and/or corrosion resistance are required but where high strength is not. The copper-silicon Alloy 319 and its variations are commonly used where moderate mechanical properties, developed without heat-treatment, are satisfactory.

Alloys A333 and 333 have unusually good casting characteristics in permanent molds and develop somewhat better as-cast mechanical properties than the 319-type alloys. Those copper-silicon alloys that also contain some magnesium, such as B319, 320 and the 333 alloys, can be heat treated to further improve mechanical properties.

Heat-treatable Alloys 355 and 356 are used where good tensile properties are required. Both have excellent castability. Alloy 356 also has excellent corrosion resistance; 355 has slightly higher tensile strength than 356. Decreasing the impurities, especially iron, in these alloys to produce C355 and A356 greatly improves the mechanical properties over the standard grades. Alloys C355 and A356 are often used for aircraft components and other critical applications.

The casting properties of Alloy 296 are less desirable than the C355 and A356, but a satisfactory combination of high tensile strength and high ductility is obtainable by heat treating. For many applications, 296 has been replaced by 355 or 356-type alloys because of Alloy 296's relatively poor castability.

Straight aluminum-magnesium alloys, such as 514 and 520, cannot always be cast satisfactorily in permanent molds. Addition of about 2% zinc, as in 513, makes it more adaptable to permanent mold casting. Alloy 513 can be used for simple castings where outstanding corrosion resistance and good surface finish are desired.

High-silicon piston alloys 336 and 332 are cast in permanent molds to get the highest mechanical properties and economical production. The no-nickel, higher-copper alloy 332 was used extensively in the manufacture of automotive pistons, but is rapidly being replaced by 336 or the hypereutectic 390-type alloys.

EFFECT OF COOLING RATE

The more rapid chilling of the alloy in permanent molds generally gives finer DAS and grain structure and higher mechanical properties than those that are obtained in sand casting. It is frequently assumed that this effect is sufficient to render grain refinement or silicon modification unnecessary. This is not true, however, because the degree of chilling and the resulting effect on structure and properties vary according to the thickness and to metal flow conditions.

For example, in moderately thin sections (¼ in. or under), rapid chilling gives fine grain structure and modifies the silicon eutectic satisfactorily. For such applications, it often is not necessary to add grain-refining agents or to modify the high-silicon alloys. When the section is greater than ⅜ in., however, the cooling rate alone may not be rapid enough to reap the full benefits of chill casting. In such castings, developing the best grain structure and physical properties may require addition of grain-refining and modifying agents.

In modifying the silicon eutectic, the cooling rate is still faster than in sand casting, even though the permanent mold sections are heavy. While it may not completely modify the eutectic, it does decrease the amount of sodium or strontium required. Thus, while 0.01–0.02% sodium or strontium is generally proper for modification of high-silicon alloys in sand castings, about 0.0015–0.005% sodium or strontium maintained in piston alloys for permanent-mold casting diesel pistons is sufficient to obtain the silicon modification necessary for best mechanical properties and machinability. Because of oxidation, the amount by weight of sodium added to the melt is approximately 10 times greater than that ultimately retained in the casting.

DESIGN FACTORS THAT AFFECT SOUNDNESS

Several factors must be considered when designing the mold if sound permanent mold castings are to be produced efficiently and economically. These factors include:

- the mold vents, which allow the air and gas to escape ahead of the metal;
- the gating system, which carries the metal into the mold cavity at the proper locations and rates;
- the riser (feeding) system, which contributes to temperature gradients and provides the molten reservoir, so that, wherever metal is solidifying in the casting, it will always have access to liquid metal, which will feed back under the influence of atmospheric pressure and capillarity to fill voids that tend to form as a result of solidification shrinkage;
- chills, which are sometimes inserted in the mold to initiate and hasten solidification in desired locations and directions, and thus supplement the effects of feeders and risers in controlling temperature gradients.

Venting

All the air and gas in the mold cavity and gating system must escape rapidly as the metal is flowing in or the mold will not fill in the proper manner, and gases may be trapped inside the cast metal. The natural escape outlets for gas (such as the mold parting, slight clearance spaces around sliding cores, and spaces at the ejector pins and the risers) are not always adequate for all parts of the mold cavity, and more positive means usually must be provided. Venting methods most commonly used include:

- slots, possibly 0.005–0.010-in. deep, cut along the parting face of the mold and leading either to the outside of the mold or to vent holes;
- pin vents formed by drilling a small hole and inserting a square pin (sometimes used in metal cores);
- plug vents formed by drilling holes of various sizes and inserting plugs with slots about 0.010-in. deep

filed in their sides. These are used in various sizes over large, flat areas that require venting. Large recesses remote from other means of venting are usually plug-vented.

Gating

As molten aluminum enters the permanent mold, it loses heat rapidly compared to sand molds and has less time to seek its natural distribution before solidifying. Therefore, it is often necessary to bring the metal into the mold rather close to its final disposition areas. Rapid cooling also necessitates rapid filling. If all the metal came through a single small inlet, it could cause turbulence resulting in dross formation. In addition, a hot spot could form in the casting and cause shrinkage. These difficulties can be avoided by dividing the flow among a number of inlets, or by spreading it out in a web inlet to avoid overheating the mold walls at any local area and to get the metal in quickly without high velocity.

The system for gating vertical castings shown in Fig. 7-6 is recommended on the basis of the results obtained in an AFS-sponsored study. The sprue should be tapered to prevent aspirating air in the molten metal stream. For rectangular sprues commonly used to pour permanent molds, the width of the sprue may be constant throughout its length, and the taper may be formed in its thickness.

This simple tapering method in metal molds entails only a gradual decrease in the cutting depth of the milling cutter or shaper tool. A taper of about 0.008 in./in. of sprue length is sufficient to smooth the metal flow.

The longer sprues usually required for vertical pouring, by their nature, increase the exit velocity of the molten metal. Under these conditions it is practically impossible to prevent a jet effect and excessive turbulence if the sprue discharges the metal directly into the mold cavity, although there is considerably less turbulence if the metal enters the cavity through the bottom rather than the side.

In vertical gating, the riser can be connected, throughout its entire height, to the cavity by a narrow web or slot. This arrangement is shown in Fig. 7-6. The area of the riser and the size of the web opening must be large enough so that the metal level in the riser will not rise appreciably higher than the metal level in the cavity. If the metal level in the riser is substantially higher, the molten metal will cascade turbulently into the mold cavity and entrap air.

It is preferred to have the sprue discharge first into a well and then into a horizontal runner. Flow rate can be reduced approximately 50% if the cross sectional area of the runner is twice that of the bottom of the sprue. Further enlargement of the section of the runner directly under the sprue (the well) will reduce the rate of flow even more and will help wash out the air carried down with the first metal that is poured. For wide, shallow runners this well should extend below the bottom of the runner to help remove trapped air, but this is unnecessary when the runner is deep and narrow.

The entrance end of the gate leading from the runner to the riser should have an area at least equal to that of the runner, and the area at the gate's exit (riser) end should be flared too so that it approaches the area of the riser. Flaring the gate at the riser will further reduce metal flow. Extending the runner beyond the gate provides a reservoir which traps the first metal of the pour, likely to contain dross formed by entrapped air. Venting the runner extension is advisable as it helps ensure prompt filling. For vertical molds, horizontal runners should be under the mold's cavity.

The actual dimensions of the mold components will depend on the weight and dimensions of the casting to be poured. In the experimental work, a runner extension about 5 in. long trapped and prevented backwashing of the dross-contaminated first metal of the pour.

For more information on gating design and sizing, refer to Chapter 14.

Feeding the Casting

The vertical riser in Fig. 7-6 serves as a side riser or feeder. The inlet gate connection between this feeder and the casting should be short and thick enough so as not to freeze off prematurely before feeding is complete. The casting and its gating system shown in Fig. 7-7 are common for a gear blank, a pulley or a wheel. This method of rigging feeds the casting from both sides and utilizes a top riser.

In permanent mold casting, it is important that the risers be large enough—but no larger than the job requires. A riser is adequate when it eliminates solidification shrinkage. In some cases an oversized riser can overheat an area of the

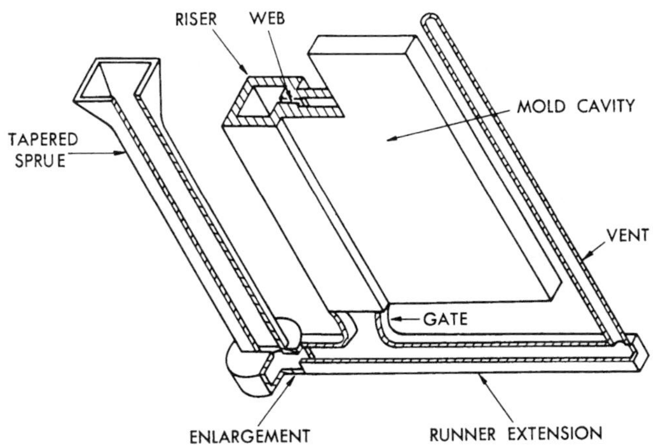

Fig. 7-6. Sketch of recommended vertical gating system.

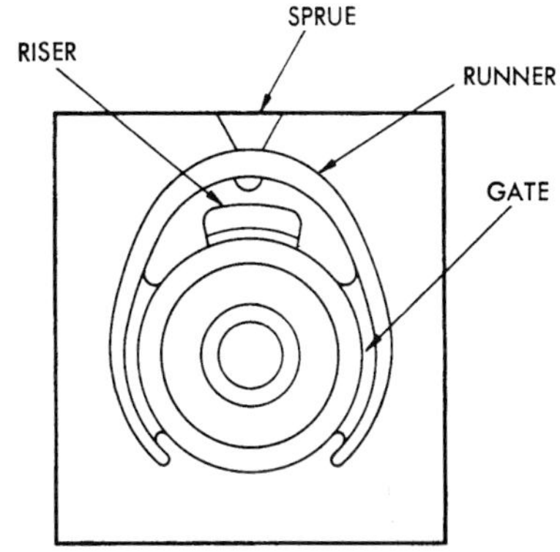

Fig. 7-7. Gating system for a gear blank or wheel-type casting.

mold and thus cause a shrink. Furthermore, a riser in which metal freezes too slowly may delay opening of the mold until excessive contraction stresses have developed in the casting. An oversized riser may indirectly cause hot cracking and is directly responsible for slow production rates.

At the other extreme, the dimensional limitations of the mold may make it physically impossible to make the riser large enough for the feed requirement. In such cases, providing a thicker wash (coating) on the mold surface may increase the feeding efficiency by slowing the chill rate, thereby keeping the metal molten longer.

For simple, thin-walled castings of rather uniform thickness, such as cover plates, cooking utensils or meter boxes, shrinkage is usually not enough of a problem to require feeders. Here, because the walls freeze rapidly, filling the cavity quickly without excessive turbulence is a major consideration. This requires strategic placement of multiple cavity ingates and the inclusion of ample venting.

For more information on risering, refer to Chapter 14.

Chilling

Directional (or progressive) solidification is the freezing of metal in the casting so that molten metal always feeds the portion of the casting just solidifying. Solidification is controlled by providing conditions that will assure correct temperature gradients in all sections of the casting during solidification, so that solidification proceeds from the casting extremities (lowest temperature gradient, sometimes called end effect) toward the riser (highest gradient, called heating effect), without allowing hot spots to form between.

As metal solidifies (freezes), it contracts. A riser provides a reservoir of molten metal to feed solidification shrinkage. Risers, or feeders, then function as the heating effect (hottest end) to increase or prolong high temperature. Desirable thermal gradients can be established by extracting heat from the casting on the low temperature end; that is, by chilling.

In the absence of other variables, the thin sections naturally solidify first and the heaviest sections solidify last. Wherever the shape permits, it is preferable to position the casting in the mold so that solidification starts in thinner sections at the bottom and proceeds upward towards heavier sections at the top which can be fed with risers.

With many shapes of castings, it may not be easy to control solidification. For example, a heavier section sometimes must be located where it cannot be reached by a riser or feeder. It may be isolated by thinner sections that solidify first, leaving the center of the heavy section still molten and a hot spot. Some relief may be obtained by using ribs to introduce more feed metal into the heavy section. However, this method is not always effective, and there still may be too much shrinkage. If so, the remedy is to cool the heavy section so that it will solidify quickly while the connecting sections are still molten and supplying feed metal. A number of methods are available, and the choice depends on whether the chilling is localized or broad, rapid or more moderate.

Installing plug inserts through the mold walls will localize chilling. For moderate chilling, the mild steel insert has higher thermal conductivity than cast iron. Copper plugs have much higher conductivity than steel and are used for quicker chilling. Their effects can be increased by extending them further outside the mold and placing cooling fins on them. In some cases, cooling is aided by circulating water or air through the chill plug.

If cores enter the heavy sections they provide some natural cooling action, which is greater if the cores are steel instead of cast iron. If the cores are loose, their chilling action can be increased further by removing them from the mold between pours and dipping briefly in water to cool them. Aiming a jet of compressed air at the outside of the mold or by periodic application of a wet sponge or mist coolant (water and air) can localize moderate chilling. Care must be taken to assure that all moisture is removed prior to pouring.

Fins on the mold walls increase heat dissipation to the atmosphere cooling moderately over broad areas. This chilling method ban be enhanced by blowing a stream of air across the fins. Jackets in or around the mold walls and circulating air or water through the jackets also will cool broad areas.

Moderate control of solidification can be accomplished by varying the mold coating thickness. Customarily, applying a heavier coating the sprues, gates and feeders helps conserve heat in these areas while thinning out the coating on surfaces provides for more rapid cooling. Variations in coating thickness should not be depended upon extensively for controlling solidification, however, because constantly applying coatings seriously slows casting. Varying the composition and thickness of the mold coatings should be regarded only as supplemental to good mold design.

The casting design itself should take into account the necessity of directional solidification to avoid shrinking and weak spots. Figure 7-8 shows the effect of a sharp inside corner while the metal is solidifying. The arrows indicate heat flow into the corner of the mold and show how the heat concentrates to overheat the mold corner and retard cooling of the aluminum in this area. The dotted lines indicate the type of temperature contours that develop. Because of the contours' unbalance, the skin of the casting remains thin and weak at the inside corner while the rest of the casting is solidifying. A shrink will develop at this point.

This situation could be remedied by chilling the corner, but it is better to round the corner as indicated in Fig. 7-9. Rounding off sharp corners also extends mold life, because overheating of the sharp mold corner, which would eventually cause cracking or eroding of the mold at this corner, is precluded.

PREVENTING HOT CRACKING

The most common causes of hot cracks or hot tears in aluminum permanent mold castings are the internal stresses acting on the casting during solidification and subsequent contraction at temperatures near the solidus. The effects of the mold are the only significant external causes of hot cracking. In nonferrous castings, tearing can usually be prevented by proper mold and casting design.

To minimize cracking by mold restraint, the operator should remove the casting as soon as it has solidified. Incorporating certain features in the design of the mold will in aid the operation. The risers should not be oversized, which increases

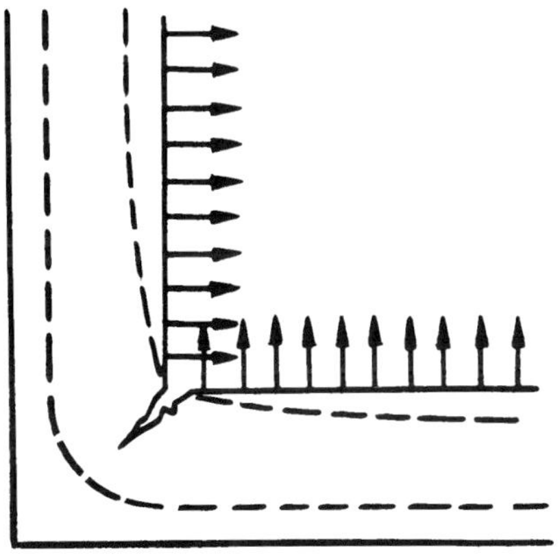

Fig. 7-8. Sharp internal corners on castings cause cracks in the casting and may also result in premature mold failure.

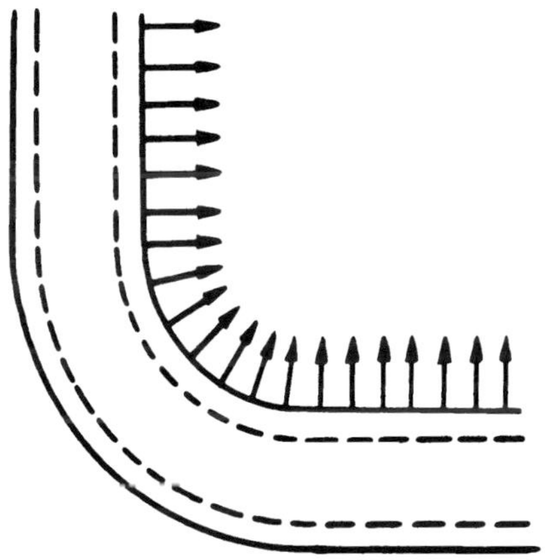

Fig. 7-9. Properly designed fillets at corners help prevent shrinkage cracks and tears in castings and extend mold life.

solidification time and delays removal of the casting, and sharp interior corners should be avoided because contraction stresses are likely to concentrate in such sharp corners.

Cracking also is possible in other areas where cooling lags or solidification is not directional and the area fails to develop the strength required to resist the stress quickly enough. Such difficulties can usually be remedied by chilling these areas to increase the relative strength earlier in the cycle.

The casting design may be such that contraction stresses will develop in cross sections too thin or too small to resist them, even though they may solidify early. It may be necessary to "beef up" these areas or to incorporate some strengthening ribs. If they are near the outside, it may be feasible to cast some attached fins exterior to the casting to provide some additional strength. The fins can be removed later when the casting is trimmed.

DETERMINING MOLD CAVITY DIMENSIONS

When designing and building molds for permanent mold aluminum castings, cavity dimensions must be enlarged to compensate for the shrinkage of the casting as it cools. The factors used for dimensioning cavities in metal permanent molds are generally the same as those given for diecasting dies in Chapter 5 of this book.

Table 7-1.
Recommended Permanent Mold Materials for Aluminum

Casting Size	Number of Pours		
	1000	10,000	100,000
Small castings 1-in. max. dimension	Gray iron*	Gray iron,* H11 die steel	Gray iron* with H14 inserts
Medium & large castings, up to 36-in. max. dimension	Gray iron*	Gray iron*	Gray iron,* H11, H13, H14

* Meehanite types also suitable.

Engineering permanent molds requires careful attention to minute details. The design of sprues, runners, gates and other basic mold elements is influenced by an almost limitless combination of shapes, forms and materials.

MOLD MATERIALS

Mold material for permanent mold casting of aluminum is usually chosen on the basis of the number of castings to be produced in the mold and the cost of the mold material plus machining. As shown in Table 7-1, gray iron is the usual mold material for 1000 pours or less and may even be used for long production runs of castings that do not require smooth as-cast surfaces.

Gray irons are but one type of the iron alloys commonly called "cast iron." It differs from other common cast irons such as pig, white, nodular, chilled and malleable iron. Gray irons usually contain 1.7–4.5% carbon and 1–3% silicon. For permanent molds and cores, graphite size and density of the gray iron should be closely controlled.

Various types of Meehanite metal—comparatively low-silicon cast irons that have been ladle-inoculated to refine the grain and to give fine graphite distribution—are frequently used for molds. These metals offer good machinability, heat resistance and dimensional stability.

Machining the cavity often is a significant cost factor. Gray iron die blocks can usually be machined for about half the cost of cast tool steels.

Mold temperature will vary with mold thickness and with the weight and section thickness of the casting. The magnitude of these effects in aluminum casting is indicated in Fig. 7-10 from a study of temperatures in permanent molds. The molds were composed of rectangular cast iron plates of different thicknesses. The thickness of the castings was also varied by employing interchangeable cavity spacers of different thicknesses.

The mold thickness versus solidification times for three different casting thicknesses and several operating cycles are indicated in Fig. 7-11. At a constant casting rate, the solidification rate increases with increasing mold thickness. (A higher solidification rate is usually desirable because of the finer structure and better mechanical properties in the casting.) Thick molds require a greater number of casting cycles and more time to reach equilibrium than thin molds; however, they cool less rapidly when the operation is interrupted.

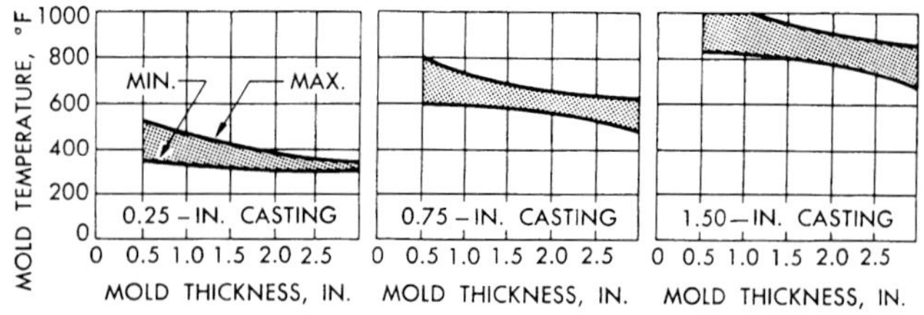

Fig. 7-10. Effect of mold thickness on the maximum and minimum temperatures of molds operated on an 8-min cycle in the permanent mold casting of aluminum.

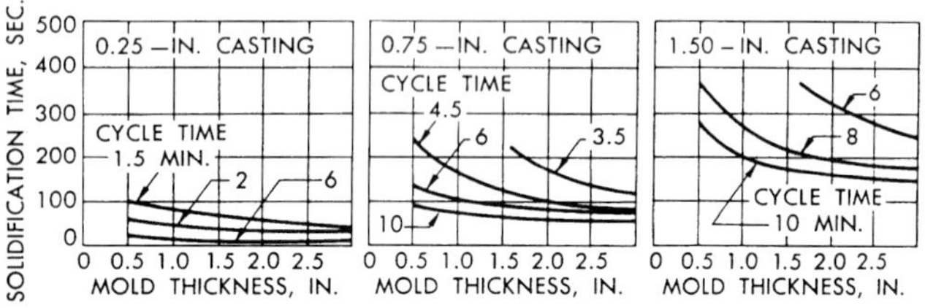

Fig. 7-11. Effect of mold thickness on the solidification time of aluminum permanent mold castings.

The temperature of the mold and the quantity of heat to be removed vary not only with the weight and thickness of the casting, but also with the alloy being cast. Cavity configuration and other factors are more important bases for determining the thickness of mold walls than is the anticipated life of the mold. The range in temperature gradients from the outside to the inside of thin molds is less drastic and, therefore, more preferable, for there will be less heat checking.

Figure 7-12a summarizes typical relationships between mold wall thickness and casting thickness for water-cooled molds for aluminum. The relationship of mold wall thickness to weight of poured metal for castings with a surface area of about 1000 sq in. is shown in Fig. 7-12b.

Mold Inserts

Full or partial mold cavity inserts of the same material as the mold, or of a different material, are sometimes used to lengthen mold life or to simplify machining, handling or replacement. Inserts also may be used for venting, cooling thin walls, and heating certain portions of the mold or the full cavity area.

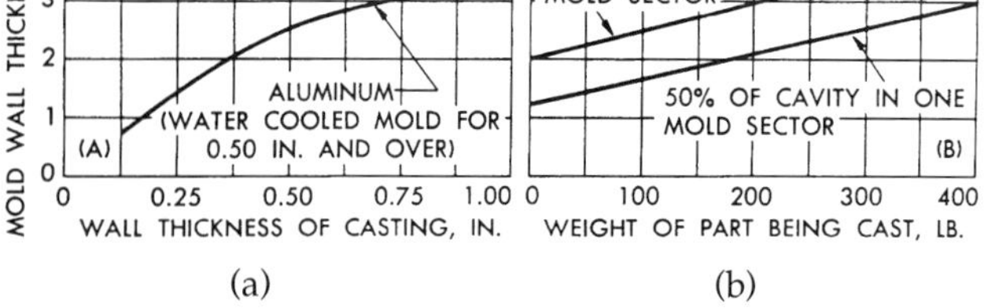

Fig. 7-12. Relations of mold wall thickness (a) to casting wall thickness and (b) to the weight of the aluminum casting.

Inserts that are made of cast-to-shape gray iron (used for casting complex aluminum and magnesium parts with surface area from 50 in.^{2} with tolerances of ±0.030 in. to surface area of 450 in.^{2} with tolerances of ±0.060 in.) will last for 5000–20,000 pours, depending on the complexity of the cast parts.

Tolerances

In producing molds for making aluminum permanent mold castings, cast low-alloy steels and tool steels are the preferred materials for molds that require uncommonly close dimensional tolerances. Normally, tolerances on cast parts are held to ±0.015 in., with ±0.005 in. being the minimum attainable.

When selecting mold materials to meet specified tolerances, one can expect degrees of accuracy as follows (from the lowest to the highest): gray iron, H13 and H14.

CORES IN PERMANENT MOLDS

Cores are used to form the holes and recesses of castings (Fig. 7-13). Permanent metal cores are the most economical. When a casting has cavities from which straight cores cannot be withdrawn, expendable (destructible) cores are used. In some cases, if design and economy permit, collapsible steel cores may be used instead of expendable cores.

Cores should be of H14 or equivalent tool steel where part tolerances are minimum or production run is maximum. Expendable cores of sand or plaster are recommended for cores with minimum draft.

Materials for Cores

Materials listed in this section are recommended on the basis of performance over a wide range of coring requirements. For small cores less than 3 in. in diameter by 10 in. long, recommended materials (in order of preference) are sand, plaster, H11 die steel or equivalent hardened to Rockwell C

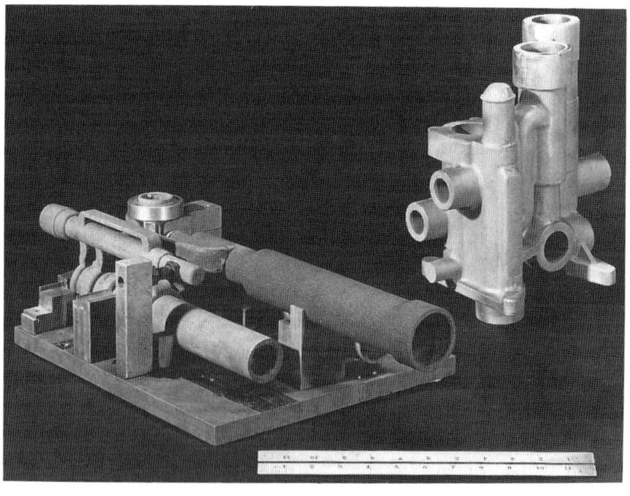

Fig. 7-13. A core assembly fixture (left) is required for casting a heavily cored hydraulic manifold. [Photo courtesy of LA Aluminum Casting Co.]

Table 7-2.
Materials for Large Cores in Permanent Mold Casting of Aluminum (Cores over 3 in. diameter by 10 in. long)

Number of pours		
1000	10,000	100,000
Gray iron*, sand, plaster	Gray iron*, gray iron with H11 inserts,* sand, plaster	Gray iron*, gray iron with H11 inserts, H11 tool steel

* Except for complex shapes that require expendable cores

42–45, and carbon. Graphite cores normally are used for relatively few pours. Materials recommended for large cores are given in Table 7-2 in order of preference.

An expendable core is used when its location or shape makes removing it from the casting impossible or when the expendable core can most economically provide an intricate design (Figs. 7-14 and 7-15). The order of preference in the use of destructible core materials is shell sand, other sand (resin-bonded sand or CO_2 silicate), plaster and graphite.

Mold parts (cores, segments or inserts) can be cast to shape when the castings to be made in the resulting mold are not complicated. These can replace machined gray iron or H11 die steel where recommended in Table 7-1. Cast molds and parts are most frequently used when production is large and mold replacement or shipping schedules require several of the same molds.

The simplest core is of uniform cross section or tapers with the larger end toward the outside of the casting. If the axis of the core is at 90 degrees to a parting line, it may be fixed permanently to a mold segment and the casting ejected off the core after the mold is opened. At any other angle, the core is frequently withdrawn before the mold is opened. It usually slides through the mold wall in a close fitting sleeve long enough to maintain accurate alignment. The core may be moved manually with a pry bar, a rack and pinion or toggle arrangement, or mechanically with an air- or hydraulically-actuated cylinder.

The desirable draft in metal cores is a minimum of 2° although as little as ½ degree has been used. The preferred minimum diameter is ⅜ in., although smaller diameters are possible with very short cores. Long thin cores are likely to require frequent replacement. In general practice, the maximum length of a core supported at one end is ten times the diameter. Some designers prefer a maximum of eight times the diameter on larger cores and not over four times on cores of less than ⅝-in. diameter.

If an interior space is so shaped that it will not permit straight line removal of a core, then a collapsible metal core may be used if the separate pieces can be loosened from the casting's interior and if the hole leading to the interior is large enough to withdraw the collapsed core. The core shown in Fig. 7-16 is used to cast a piston.

Undercuts on the outside of a casting may be formed by sliding cores, if they are uniform in cross section or are properly drafted. If the shape precludes a sliding core, a loose piece may be set in a suitable supporting print in the mold cavity, removed with the casting, and then withdrawn from the casting and returned to the mold for a subsequent pour.

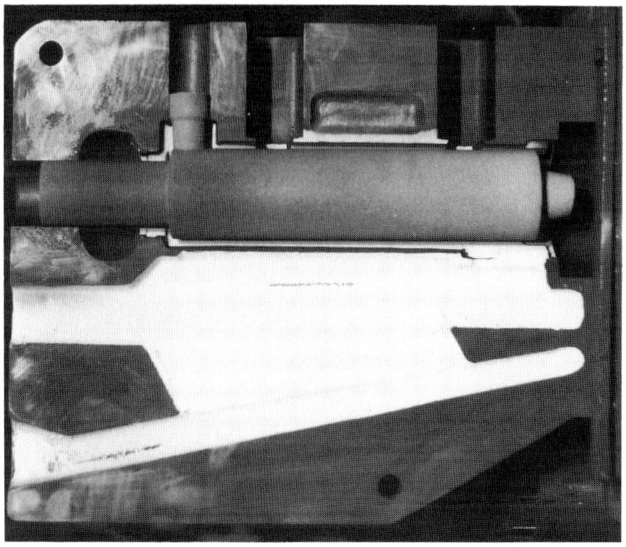

Fig. 7-14. A shell core forms the internal contour of a preheater for diesel engines. [Photo courtesy of LA Aluminum Casting Co.]

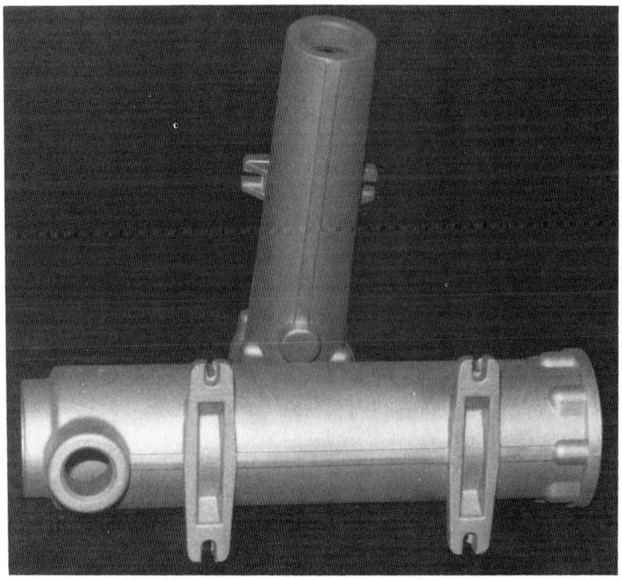

Fig. 7-15. The casting in the foreground was made with the coring arrangement shown in Fig. 7-14. The casting in the background was made using a metal core. [Photo courtesy of LA Aluminum Casting Co.]

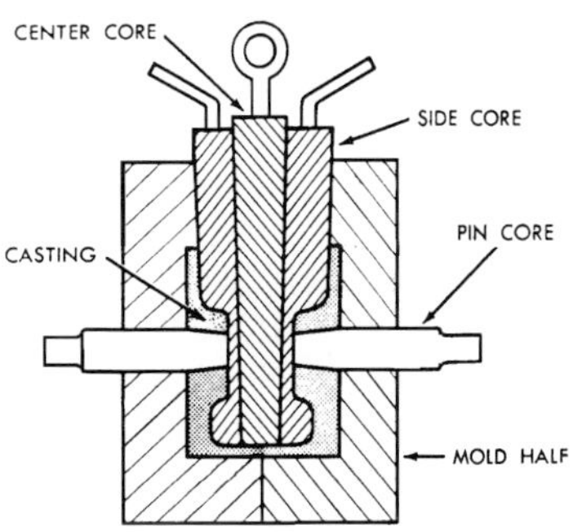

Sequence of Operations:
1. Pull pin cores.
2. Pull center core.
3. Pull one side core.
4. Pull other side core.
5. Open mold.
6. Remove casting.

Fig. 7-16. Cross section of collapsible metal core shown in a permanent mold with casting poured. The core was designed to cast aluminum pistons.

The loose piece must fit closely in its supporting seat to prevent metal from flowing around it and locking it in the casting.

An expendable core of sand or other material is used if an interior cavity's shape does not permit a metal core or if an exterior surface undercut does not allow the use of a loose piece. Sand cores are produced in the same way as those used in sand casting, but the sand should have high permeability. Also, the binder should be a low gas producer, since sand cores may make gas venting difficult. The core should soften more readily for sand removal because there is a shorter time for absorption of heat from the casting. It should be noted that expendable cores will lengthen the casting cycle.

MOLD OPERATIONS

The mechanical operation of the mold should fulfill the following requirements. The specific mechanical arrangement of a given mold is chosen on the basis of the type and complexity of the casting to be made.

- It should be possible to remove cores, open the mold, and remove the casting promptly after the metal has solidified, so that cores will not "seize" and the casting can be freed quickly from the mold.
- When the mold is opened and the casting removed, the mold should not create enough stress on the casting or on the ejection mechanism to distort the casting or abrade the mold coating.
- It should be possible to reassemble and lock the mold rapidly with little manual effort and without unusual skill in aligning all parts.

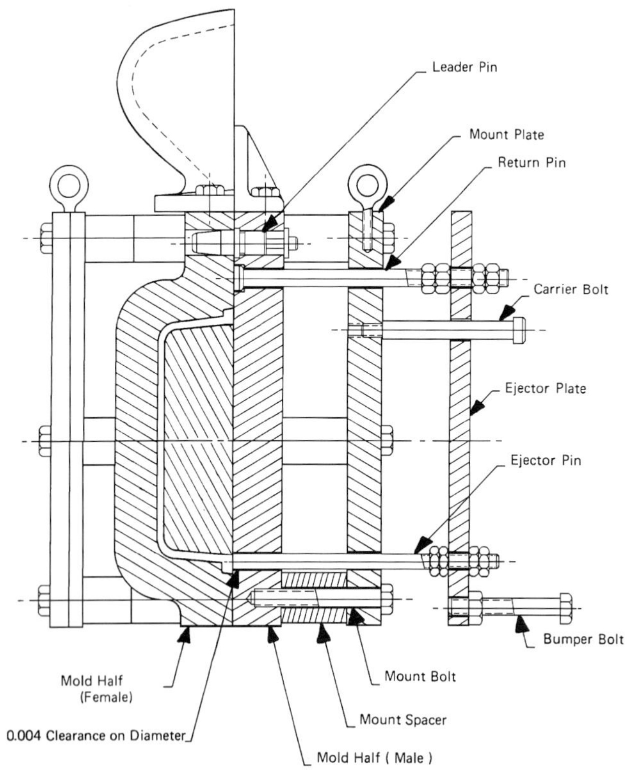

Fig. 7-17. A typical permanent mold assembly.

A simple type of two-part mold may be hinged on one side so the movable half swings away horizontally, or it may be hinged at the bottom so it swings forward and down, balanced by a counterweight. Tapers on the cavities are designed so that the casting remains in the stationary half, from which it is loosened by ejector pins actuated by an ejector plate at the back (Fig. 7-17).

For castings that require deep mold cavities with straight sides and little taper, the movable half of a two-piece mold is supported independently so that it will have straight-line movement guided by keyways on a frame or base plate attached rigidly to the stationary half of the mold (Fig. 7-18). Multiple movable segments should slide straight away horizontally at different angles, supported by a base and guided by keyways.

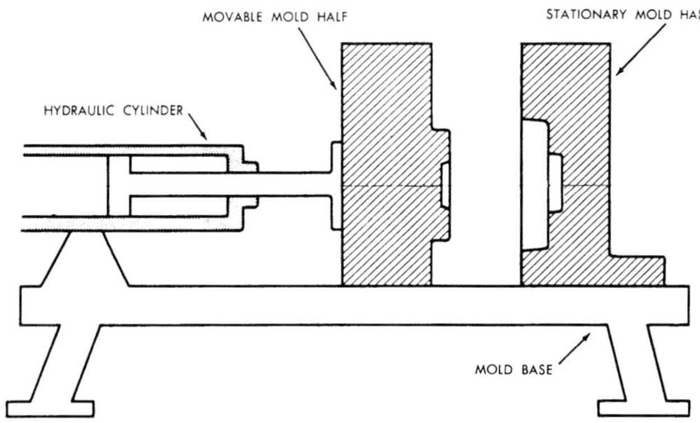

Fig. 7-18. Arrangement for sliding straight-line operation of two-piece deep cavity permanent mold.

In any case, the mold cavities in the various segments must register accurately when the mold is closed. There must be no wobbling of segments in the early part of opening; this could abrade the mold coating and distort or even crack the casting. For this reason, close-fitting dowel pins are installed in the parting faces of the mold segments to ensure perfect alignment when the mold is closed and during the initial part of the opening movement.

Opening and closing can be done manually by levers or toggles with smaller molds or by air or hydraulic cylinders with larger molds. Quick-acting hand clamps are usually applied after the mold is assembled to ensure positive closure. No extraneous material such as dirt or mold coating should be allowed on the flat mold parting faces because this could change the casting dimensions or cause unwanted flash at the parting line by holding the mold partly open.

Although mold and core materials are the same, the mold and metal cores usually last about twice as long if the mold is operated automatically instead of by hand. Normal abuse lessens the life of the tools on hand-operated equipment.

The casting is comparatively weak when it is removed. Therefore, the ejection mechanism should be designed carefully to avoid distortion during removal. The ejector plate should have a positive straight line motion, and the ejector pins should bear only on the more rigid parts of the casting (Fig. 7-17), where they will not cause distortion. There should be enough ejector pins to distribute the load.

Coating the Mold

With sodium silicate as the binding agent and water as the carrier, colloidal graphite, vermiculite, diatomaceous earth, aluminum oxide, fine silocel, clay, and other ingredients are commonly used as mold coatings or washes. The coatings control chill and solidification rates of the casting (affecting its structure and soundness), reduce thermal shock to the mold, and help prevent impingement (soldering of the molten aluminum to the surfaces of the mold).

When the coating of refractory wash is thick, the rate of chill is slowed, and a riser action results. Progressively thinner coatings result in progressively faster chill rates. Wiping off all the refractory wash and applying a graphite coating chills the fastest. Progressive solidification of the casting sections can be controlled by employing proper techniques during wash application.

Many foundries believe that the most important factor for a mold's successful operation is the selection and proper application of mold coatings. When a new mold is to be put into service, a special gas burner should be designed and built to heat it uniformly. The new mold should be clamped as if for casting, heated slowly to near the operating temperature of 700–800F (371–427C), and maintained at this temperature for about 48 hr to relieve stresses. If any warping is found, the mold fit should be readjusted.

The cavity should be given a light sand blast to improve the coating adherence and should be reheated nearly to the operating temperature (at least above 300F, 149C) before coating. Coating applied to a cold surface runs and streaks; painted on a hot surface with a brush, it boils up and becomes rough. The best method is to apply coating lightly as a spray to the hot surface while keeping too much from building up in one place (Fig. 7-19). Under these conditions, the coating dries instantly and smoothly and adheres well.

Several layers can be applied to specific areas that require thick insulated coatings. However, the entire thickness built up this way would be rather soft and easily abraded. It is better to put on a light coating, rub with steel wool to smooth and densify it, and then repeat the spraying and rubbing for the desired thickness. Masking is used to coat specific areas selectively.

Fig. 7-19. A coating is sprayed onto the surface of the mold cavity.

Coating thickness should be 0.002–0.005 in. in mold cavities. It should be heavier in gates, risers and thin sections where slower heat transfer is desired, and it should be lighter at isolated heavier sections of the casting where heat dissipates quickly.

At corners of projections in the mold cavity, which form recesses in the casting, it may be desirable to leave the coating off. However, the absence of protection for the mold must be kept in mind. Metal cores are not usually covered with insulating coating but are lubricated by dipping or spraying with a mixture of graphite and water. Prints for sand cores or loose pieces are not coated. Vents should be checked after the coating is applied to make sure they are open.

A satisfactorily applied mold coating will usually last through 24 hr of continuous use with possible occasional minor repair, depending upon casting design, draft and other factors. When the coating wears too thin, misruns may occur; recoating is then necessary. If the base coating is adherent, it may be roughened slightly with steel wool, and additional coating can be sprayed on. However, this type of renewal eventually makes the coating irregular or causes it to peel and, therefore, cannot be repeated many times. It is necessary to remove the old coating completely with wire brushes or light sandblasting before a complete recoating.

The coating has a different expansion coefficient than the metal mold and is likely to loosen and peel if the mold is allowed to cool down and is then reheated. Thus, if production is delayed and is expected to resume with the same coating, the mold should be kept warm during the delay; otherwise, it may be necessary to recoat the mold.

Removal of the coating by wire brushing or sandblasting wears the mold cavity slightly and will eventually enlarge it. In some cases, if there is excessive wear on the cavity, the mold may become unserviceable early, even though it is structurally sound. For this reason, it is important to take all precautions to minimize the frequency of recoating.

MELTING PRACTICES

Furnace equipment, melting and metal handling are covered in detail in Chapter 2, but some practices particularly applicable to permanent mold casting are emphasized here.

The molten metal supply should be stationed close to the mold for easy, swift transfer (Fig. 7-20). Stationary 250–600 lb capacity holding or melting furnaces with automatic temperature control are common, although some foundries use reverberatory furnaces with capacities up to 10,000 lb. As molten aluminum is used, it is replenished periodically either by adding ingot or by transferring molten metal from a larger melting furnace. There are some limitations in each method.

If the furnace is replenished with ingot, the metal cools down somewhat while the ingot is melting. It is important to note that the furnace capacity should be great enough to prevent excessive temperature changes. Faster production and more even metal temperature in the holding furnace can be achieved by melting the metal in large furnaces and transferring it to the holding furnace at the casting station.

Two-chamber melting and holding furnaces, in which the dry hearth chamber is charged with metal and the other is in the operator's ladling reservoir, are used in some foundries

Fig. 7-20. A reverberatory furnace is stationed near the tilt-pour machine that it supplies. [Photo courtesy of Stahl Specialty Co.]

at the casting station. In this manner, the two-chamber or twin hearth furnace can control metal temperature and keep metal free of dross and gas.

Fine scrap, such as chips and borings, should not be returned to the holding furnace because these carry oxides that may become suspended in the melt. Such material is likely to make the melt gassy because of moisture absorbed by or combined with the oxides. Light foundry scrap should be remelted in a separate furnace by puddling into a molten heel to minimize oxidation. The melt should be fluxed, degassed and thoroughly skimmed, then either used directly or cast into ingot molds, skimming the surface of the poured metal again before it freezes. This ingot can then be handled like new material.

Although not recommended, moderately heavy, clean scrap such as gates and risers can be returned directly to the holding furnace if desired, but it should be preheated first to drive off moisture. However, if this material has been stored long before remelting, it could contain enough oxides with accompanying moisture to cause some gas pickup, and it should be remelted separately as recommended above for chips and borings.

POURING INTO PERMANENT MOLDS

The mold is not usually at an even temperature throughout during the casting cycle. In fact, temperature gradients must exist within the cavity to produce the desired progressive solidification. It is important, however, to keep the average temperature fairly constant by controlling the pouring temperature of the metal and duplicating the timing of each successive cycle's phase.

The best temperature must be determined experimentally for each new mold. The only general guide is to pour the aluminum at the lowest practical temperature to minimize the possibility of cracks and porosity. The most practical temperatures range from 1275–1500F (690–815C). Once it has been determined, the temperature should be standard for that particular mold.

If heating of the holding furnace is controlled manually, the furnace should at least have a thermocouple and temperature indicator to guide the control. It is preferable, however, to control furnace temperature automatically. Expense of the automatic installation is usually repaid many times by improved quality and less scrap.

There are two phases in the casting cycle in which the mold is subject to drastic temperature change:

1) when the molten metal is solidifying and giving up heat to the mold;
2) after the casting has been removed, when the mold is being cooled and reassembled for the next pour.

The first phase is purposely terminated to remove the casting as soon as it sets and to prevent seizure of cores and formation of hot cracks. The second part of the cycle is determined by the time necessary to reassemble the mold. After establishing the correct pouring temperature, results can be duplicated by exact timing of each phase of the cycle. This timing—sometimes called "rhythm"— is almost mandatory to produce quality castings consistently. The entire casting cycle may vary from ¾ to 1½ min for small castings and from 6–7 min for medium- to large-sized castings.

Average temperature of the mold is likely to range from 600–750F (315–399C), indicated by inserting a light gauge thermocouple at the back of the mold or testing the mold surface with a heat-indicating crayon. If a thermocouple is installed, the well should be small and its bottom should be at least ⅜ in. away from the cavity face in order to avoid creating a hot spot.

Automatic timing devices are practical when the cores and molds are actuated mechanically. In some cases, the entire cycle, including pouring, is controlled automatically. Such operation offers labor economy and also can increase production of uniform quality castings.

It is important to handle hot castings carefully because they have low rigidity and are easily distorted. They should be stacked so as not to deform them while they are hot. Also, they should be placed where they will cool uniformly and not too rapidly. Placing them flat on an iron plate or cold floor, for example, may set up stresses due to uneven cooling that can cause cracking. Load them on edge in a tote box or hang them on hooks on a cooling rack, especially prior to heat treating.

Tilt-Pour Permanent Mold Casting

Machines that are designed to use tilt-pouring (Fig. 7-21) eliminate turbulence that is created when metal is poured directly down a vertical sprue. One type of machine (the self-pouring, hydraulically-operated tilt-pour machine) is shown in Fig. 7-22. Advantages of this machine include eliminating turbulence with simple gating, identical pours for successive castings, preventing flash at bottom of mold, and progressive solidification. Figures 7-23, 7-24 and 7-25 show steps of the tilt-pour operation.

In a highly automated system, the operator starts the cycle by pushing a button on the control panel of the power unit, pours molten metal into pouring basins, and then can leave the machine unattended. Succeeding phases of the cycle, including ejection and removal of the casting, are automatic. Standard power and control units can make it possible for one man to operate two or more casting machines on the same or separate cycles.

Some changes are necessary when tilt-pouring aluminum permanent mold castings. These changes will be discussed in the following sections.

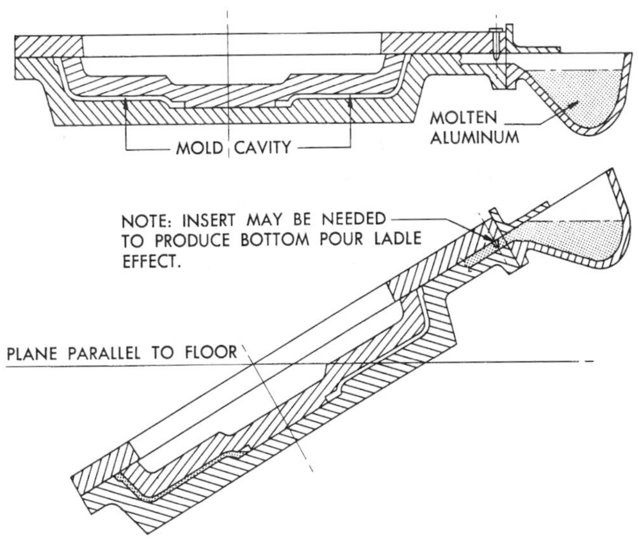

Fig. 7-21. Cross-sections of mold and pouring basin of a tilt-pouring machine. Top: pouring basin is filled when mold is horizontal. Bottom: mold is tilted to fill cavity.

Gating

With tilt pouring, the metal flows gently and without turbulence into the mold cavity if the mold design is as follows:

1) The runner half in the stationary side of the mold should be the same depth as that on the moving side and should furnish material first to the bottom of the mold cavity, then upward toward the riser. Abrupt changes in metal flow direction can cause turbulence; this runner half should fill the mold cavity properly.
2) The runner half on the moving side of the mold helps fill the mold, but it serves primarily as the feeding mechanism. Enlarged feeding areas can be used in the movable mold half runner and gating system without causing turbulence.

Fig. 7-22. A mold for an oil pan is held in the upright position of a tilt-pour permanent mold machine. [Photo courtesy of Stahl Specialty Co.]

Fig. 7-23. A worker fills the pouring basins in a tilt-pour machine with molten aluminum. [Photo courtesy of Stahl Specialty Co.]

Fig. 7-24. The tilt-pour machine rotates to allow molten aluminum to flow into the mold. [Photo courtesy of Stahl Specialty Co.]

Fig. 7-25. A casting is extracted manually from a mold. [Photo courtesy of Stahl Specialty Co.]

3) Ingates should be larger at the lower areas of the mold. Reduced head pressure that reduces turbulence also slows down mold filling capability.
4) In general, the bottom half of the mold runs cooler than the top half, allowing good progressive solidification toward risers and sometimes permitting heavy sections to be placed at the bottom of the mold. This also allows thinner sections at the top of the mold to solidify closely behind the heavier sections. Thus, greater efficiency is achieved through reduced riser sizes (less total solidification time); casting integrity is improved as thinner sections can be removed more quickly from the mold, minimizing hot cracking. Careful positioning of the mold's casting cavities, to improve filling, chilling and feeding is just as important as in other permanent mold casting methods.
5) Tilt pouring's bottom ladle pouring effect leaves surface-floating dross in the pour cup at the end of pour, eliminating the need for goosenecks.
6) Direct gating into the mold cavities of some casting designs is possible, while still producing high-quality castings.

Venting

Venting mold cavities during tilt-pour deserves special consideration. Some foundries use flats on ejector pins, which are mostly self-cleaning, and screw-type vents that are easily cleaned. Some additional foundry practice procedures also are useful. Some foundries have developed detailed techniques to cast permanent mold castings and have use a foundry process sheet to document the controllable variables to maintain better casting quality consistency.

Mold Temperature

The operating cycle is timed primarily to maintain an optimum mold temperature. In an effort to maintain casting quality consistency, some foundries have placed thermocouples in molds, using digital readout meters.

After examining the controllable variables and developing a casting plan, the mold is operated by maintaining maximum and minimum mold temperatures manually or by automatically changing the equipment timer settings. Ambient air temperature changes in the foundry, both through the course of a day and through the change of seasons, can decidedly affect the mold temperature, casting quality and, subsequently, the total cycle time.

Timing

Optimum tilt-pour time must be developed for each mold. Pouring too slow for thin-walled castings can result in misruns, through too much loss of metal superheat. Pouring too fast can cause turbulence and venting problems, indicated by air and gas bubbling into the pour cup through the gating system. For large castings, cam-operated tilt pour systems can vary the speed of pour throughout, provide more head pressure at the start and better venting toward the end of the tilt-pour.

The casting can solidify while the mold is pouring, creating elliptical solidification patterns and, in the case of round castings, reducing hot cracking.

VARIATIONS TO THE PROCESS

Low-Pressure Permanent Mold

In the low-pressure process, a permanent mold is mounted over an air-tight furnace containing molten metal (Fig. 7-26). A refractory-lined tube, known as the stalk, is mated or joined to the nozzle at the bottom of the mold and extends into the molten metal in the furnace. Heat is usually applied externally to the nozzle. With the mold closed, low-pressure air (between 2 and 15 psi) is admitted into the furnace. The air pressure forces the surface of the melt downward and causes molten metal to rise up the stalk and enter the mold cavity.

The casting is fed by pressure on the metal in the sprue opening. This eliminates the need for risers because the die cavity is filled and fed from the bottom. This simplifies the gating system and greatly increases yield. After enough time has elapsed for the casting to solidify, pressure is released and the still-molten metal in the stalk runs back into the furnace. The mold is opened, and the casting is extracted. Because metal enters the mold slowly, air is not entrapped.

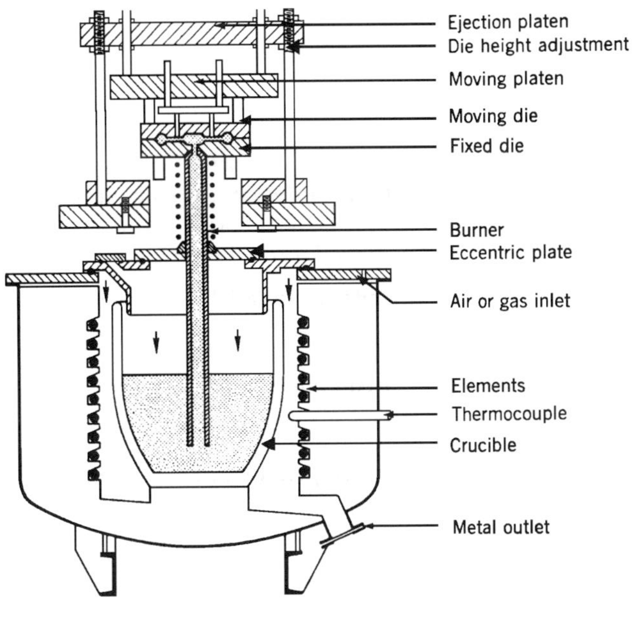

Fig. 7-26. Principal components of a low-pressure casting unit. A hydraulic system is used to open and close the top (ejector) half of the mold for casting removal.

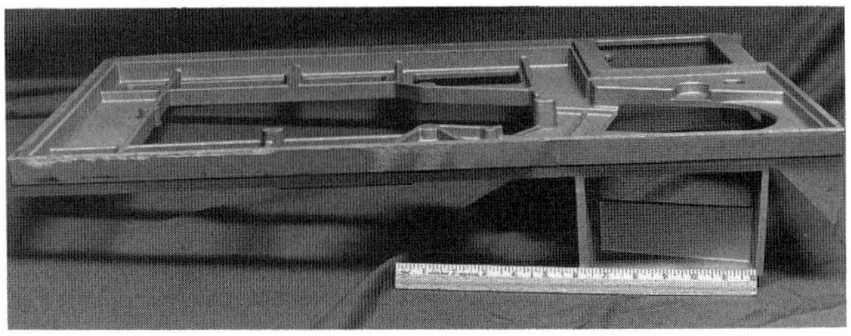

Fig. 7-27. Low pressure provides a way to gate this computer frame casting made of 356 alloy while allowing the outside configuration to be produced with 0.5° draft.

Most low-pressure permanent mold castings require only minimal vent-flash trim. Tensile yield and ultimate strengths are slightly greater for castings of the same alloys used in gravity pour molds. Sand and shell cored internal undercuts can be produced reliably. A low-pressure permanent mold aluminum casting is shown in Fig. 7-27.

Improved control of temperature through die design and integral devices permit thin, uniform coating in many cases, resulting in closer tolerances. Very thin sections also can be successfully cast in suitable low-pressure-process dies with the aid of a 13–15 in. vacuum to ensure complete filling of all extremely narrow cavities.

QUALITY CONTROL

Economy in production is related to quality. For every poor-quality casting that is rejected, an extra one must be cast. The most economical place to reject a casting is at the casting station before any further work is done. Operators should immediately discard those with obvious defects. In larger casting departments, it is desirable to supplement the operator's check with a routine hot inspection by a different inspector who looks for less obvious flaws, such as cracks and shrinks.

If much expensive machining and finishing is specified or if the castings must meet critical service conditions (as in aircraft, missile or ordnance applications), they should be examined more thoroughly. A number of nondestructive methods are available. These include conventional and real-time x-ray for detecting internal flaws and the liquid penetrant methods for detecting porosity and cracks that are open to the surface but not visible.

DEFECTS IN PERMANENT MOLD CASTINGS

The types of defects that are likely to cause rejection of a casting can be classified broadly into groups. Some of the defects occurring in permanent mold castings, their probable cause, and suggested methods for eliminating them are given in Table 7-3.

One group of defects can be termed mold defects—those related to casting or mold design. Once these defects are corrected, they are not likely to recur. It is important, therefore, to search out and remedy such defects promptly in the initial runs of the mold.

If a permanent mold has been machined accurately, the permanent mold foundry will have little difficulty with dimensional errors (with the exception of mold warping) or with defects resulting from faulty mold preparation, which often plague the sand foundry. Instead, the permanent mold foundry must be concerned mainly with defects attributable directly to metallurgical practices and permanent mold technology. These primarily include: dross and associated gas holes, blow holes, pinhole porosity, shrinks of various types, misruns and cold shuts, and cracks. Most of these can be readily observed by the mold operator who can correct or adjust melting and operating practices and eliminate defects at the casting station.

ACKNOWLEDGMENTS

AFS would like to thank Sandi Knepp, Bodine Aluminum, Inc., and Roger Reamer, Progress Casting Group, Inc., for reviewing this chapter.

Table 7-3.
Checklist of Permanent Mold Casting Defects

Defect	Possible Causes	Suggested Action
Dross and associated gas holes	Dirty metal from excess agitation in transferring from melting furnace to holding furnace.	Avoid fast pouring rates. Pour at a steady rate with minimum fall of metal. Fluxing is the best way to remove suspended dross.
	Dross picked up in pouring ladle when dipped in holding furnace.	Skim metal frequently. Push back surface when dipping ladle. Avoid splashing or pushing dross under the surface. Ladle only enough metal to fill the mold.
	Air dragged into sprue with metal while pouring.	Use a tapered sprue, and keep it full while pouring.
	Excessive agitation of metal in the gating system and in entering the mold cavity.	Use recommended gating practices. In vertical castings, increase the width of the web feeding cavity from the side riser.
Trapped gas	Dirt in metal may increase viscosity, so trapped air or gas does not escape as it should.	A) Avoid fast pouring rates. Pour at a steady rate with minimum fall of metal. Fluxing is the best way to remove suspended dross. B) Skim metal frequently. Push back surface when dipping ladle. Avoid splashing or pushing dross under the surface. Ladle only enough metal to fill the mold.
	Deficiency in venting may prevent air escape as metal enters the mold cavity.	Check venting at specific areas where gas holes occur.
Pinhole porosity in heavier sections, indicating gas in metal	A) Melt temperature excessive. B) Use of damp ingot or wet, oily scrap in melt.	Check melting practice. Degas if necessary.
	Excessive agitation in transferring molten metal to holding furnace.	Avoid fast pouring rates. Pour at a steady rate with minimum fall of metal. Fluxing is the best way to remove suspended dross.
	Excessive pouring temperature overheats mold, retards solidification an increases gas porosity.	If pouring temperature must be high to avoid misruns, check causes for misruns.
	Excess mold coating retards solidification.	Clean mold and recoat with thinner layer.
Solidification shrinkage including surface porosity and interior porosity at specific areas	Lack of directional solidification. Proper temperature gradients not established.	Change gating so solidification proceeds directionally toward riser. Check thickness of mold coating, increasing it on feeders and thinning it where more chill is required. If condition persists, in crease feeder size and provide special chilling. Mold temperature control may be needed.
	Heavy section not receiving enough feed because it is isolated between thin sections.	If heavy section cannot be on parting line, try to improve feed by redesigning sections and using more generous fillets or by using ribs. If these are not practical, provide special chilling.
	Pouring temperature is too high.	If pouring temperature must be high to avoid misruns, check causes for misruns.
	Alloy composition.	If shrink continues after correcting problems above, try a grain refiner. Check chemical analysis of the alloy against specifications. Consider using a different alloy. Sometimes it is permissible to use slightly gassy metal. A small amount of evolved gas substitutes a fine and generally distributed (and fairly harmless) gas porosity for the much more serious shrinkage defects.

Defect	Possible Causes	Suggested Action
Misruns	Pouring temperature too low.	Determine and use the best temperature.
	Excessive heat carried to supporting frame through large contact area.	Decrease contact area to frame or interpose light thermal insulation.
	Mold filling rate too slow.	Check pouring rate. If necessary, open gates or provide more gates to fill faster without turbulence. Provide more venting.
	Mold coating improperly applied or trimmed.	If feasible, increase coating in misrun area.
	Insufficient venting causing metal to freeze around air before it all escapes.	Check vents in area to see that they are open andthat their size is adequate.
	Dirt in metal making it "sluggish" so it does not fill mold completely before solidifying.	A) Avoid fast pouring rates to decrease metal agitation. Pour at a steady rate with minimum fall of metal. Fluxing is the best way to remove suspended dross. B) Skim metal frequently. Push back surface when dipping ladle. Avoid splashing or pushing dross under the surface. Ladle only enough metal to fill the mold. C) Use recommended gating practices to decrease agitation of metal. In vertical castings, increase the width of the web feeding cavity from the side riser.
Cracks	Removal of casting too slow because of poor mechanical operation or improper cycling.	See that all opening and ejecting mechanisms operate smoothly, that cycle is properly timed, and that mold is at proper temperature.
	Casing removal delayed because oversized risers or feeders solidify slowly.	Hasten freezing rate in risers and feeders by thinning mold coating or decrease their size and depend more on chilling or cooling.
	Solidification uncontrolled, leaving unfed hot spots while the rest of the casting is contracting.	Use chills at hot spots or change risering to feed these areas and to control directional solidification.
	Sharp internal corners or abrupt section changes in a design may cause hot spots or weak areas, or design may cause early contraction stresses in thinner sections.	Use ample fillets at inside corners. Increase thickness of thin sections as they approach heavier ones. Add ribs to thin sections or add removable exterior fins.
	Ejection pin is bearing on a weak section of the casting or is exerting pressure on an extended section paralleling parting lines.	Pins should bear on sections vertical to parting lines, which are most inclined to seize in the mold cavity.
	Stresses may result from faulty use of tongs or hoists in removing castings.	Tongs or lift hooks should grip a stronger part of the casting.
	Rapid or uneven cooling after removal or faulty stacking while hot.	Hot castings should not be laid flat on metal plates or cold floors and should not be stacked.

Chapter 8

SAND CASTING

INTRODUCTION TO SAND CASTING

This introduction gives information on sand casting in general. The rest of the chapter is broken down into two separate sections: 1) green sand molding and 2) chemically bonded sand molding and coremaking.

Both of these methods are used for binding sand grains together into molds; however, the green sand method uses clay and water to bond the grains, while chemically bonded sand methods use various chemical binder systems to do so. The procedures and components used to bind the sand grains in these methods are different enough to justify covering them separately.

Aluminum sand castings represent about 11% of the total weight of aluminum castings produced in the world. These castings are used in almost all industries, including aerospace (Figs. 8-1 and 8-2) and automotive (Fig. 8-3).

Sand casting is the simplest and most versatile of the processes used to cast aluminum. Sand molds generally are quite easy to make, and if trial runs show that modifications of the mold are required, it is relatively simple to change the pattern. The general foundry methods commonly used to cast other metals in sand molds are adaptable for aluminum alloys. Sand casting can be used for short-run or high-production of quality castings.

Aluminum sand castings weighing from ounces to hundreds of pounds are successfully produced. Parts in excess of 16 ft in length have been sand cast. The sand casting process can be used to produce simple castings (Fig. 8-4) or complex castings requiring intricate cores (Figs. 8-5 and 8-6).

Close casting tolerances can be held in aluminum foundries with know-how and experience with the process. When the process is properly used, the result is a high-quality casting with an excellent finish requiring little (if any) cleaning and finishing.

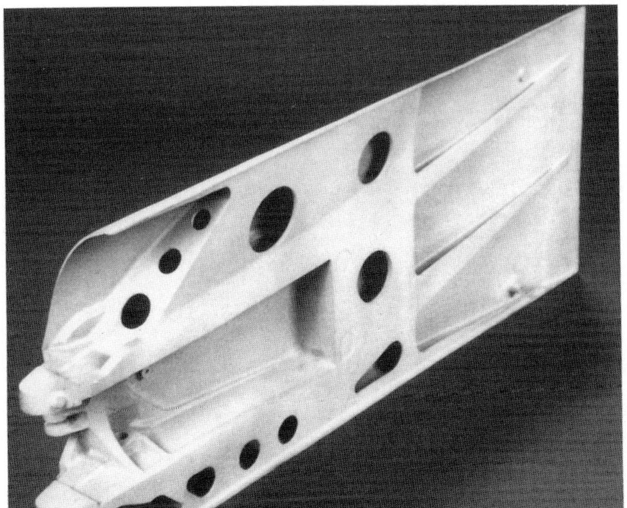

Fig. 8-1. This aluminum sand casting is a speed brake for an F-16 aircraft. [Photo courtesy of Hitchcock Industries, Inc.]

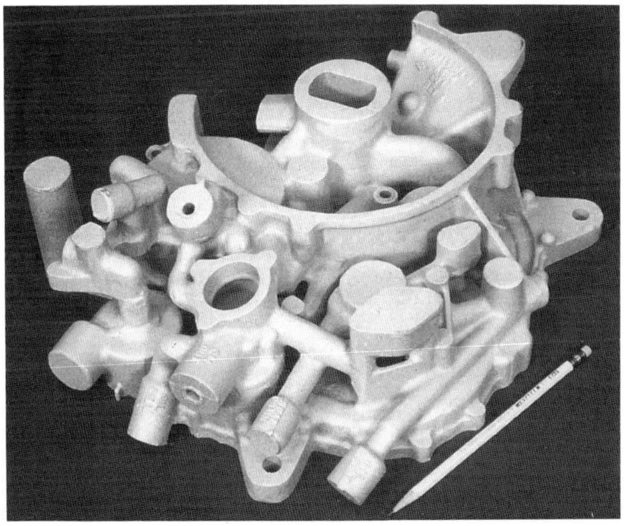

Fig. 8-2. This casting is an auxiliary power unit used in a NASA space shuttle. [Photo courtesy of Hitchcock Industries, Inc.]

Fig. 8-3. Sand castings used in the automotive industry include V-8 engine blocks such as this one.

Fig. 8-4. This green sand casting is a 4-in. internal valve made using a shell core. [Casting courtesy of Knappco and Superior Aluminum Castings, Inc.]

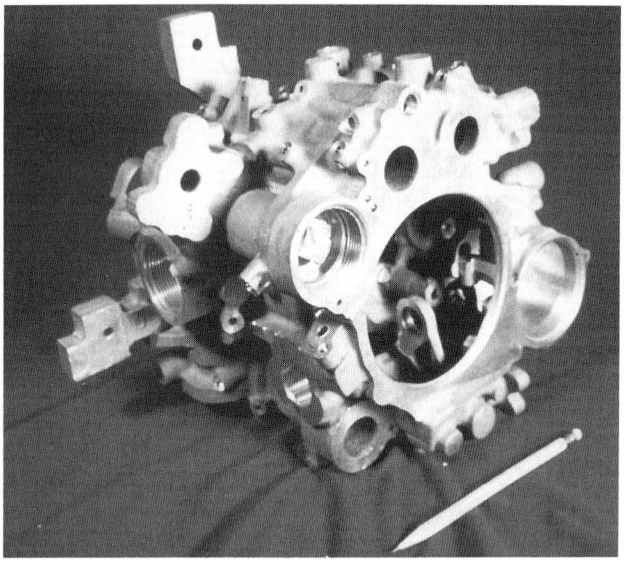

Fig. 8-5. This casting is a fuel control for a commercial jet engine. [Photo courtesy of Hitchcock Industries, Inc.]

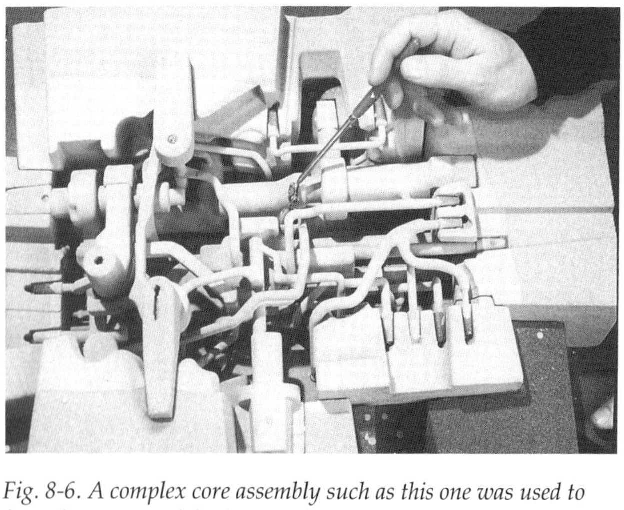

Fig. 8-6. A complex core assembly such as this one was used to form the interior of the fuel control shown in Fig. 8-5. [Photo courtesy of Hitchcock Industries, Inc.]

ALLOYS TYPICALLY USED FOR SAND CASTING

The composition, mechanical properties, thermal treatments, tensile properties and fabricating characteristics of aluminum casting alloys can be found in Chapter 4. Some of the aluminum alloys commonly cast in sand molds include 201, 206, 242, 295, 319, 355, 356, 357, 443, 444, 511, 512, 520, 535, 712, 713, 771, 772, 850, 851, 852 and their variations.

AUTOMATED FOUNDRY SYSTEMS

Mechanized, high-production molding equipment can be used to produce molds at very competitive unit costs. With automated systems, foundry or casting yields often are well above the average for manual operations.

Automatic molding and mold handling systems (Fig. 8-7) offer high production rates, good casting quality, a high degree of casting uniformity, and multiple-shift continuous casting. Pouring, shakeout, core setting and sand renewal systems also may be automated under the appropriate conditions.

As must be done before the purchase and operation of any sophisticated production equipment, capital expenditures for sand foundry automation should be weighed against manual labor costs, potential product volume and versatility, and changeover time required to start a different product run. Other important factors are the costs of misruns and of quality assurance.

Continuous casting with an automated system requires a plentiful supply of molten metal, an adequately sized sand system, and either mechanized pouring or a sufficiently large furnace located close to a manual pouring station. Shakeout and cleaning systems, too, must be capable of handling the volume of castings produced by an automated system.

Fig. 8-7. With this automated matchplate molding machine, molds move from the molding machine (front left) to a turntable. Weights and jackets are set automatically, and the molds are indexed to a pouring station. After initial cooling, molds are elevated and transferred to one of three cooling levels. Finally, the molds are ejected down a mold chute (back right) onto a shakeout conveyor. [Photo courtesy of Hunter Automated Machinery Corp.]

Green Sand Molding

INTRODUCTION

Green sand molding is a process by which molten aluminum is poured directly into a mold cavity formed from a simple mixture of sand, clay and water. There is no correlation between the term green and the mold color. "Green" refers to a water-tempered, uncured sand molding system, resulting in a moldable sand mixture with strength and other physical properties suitable for moldmaking. Green sand molds are not baked or dried.

Green sand molding is used widely for small to medium-sized aluminum castings. This casting process often is favored by foundries because of its economic production of quality castings in both short-run and high-production environments.

Aluminum green sand systems may be composed of many components other than sand, clay and water. These other ingredients are commonly referred to as additives. Systems can vary from simply sand, clay and water mixtures to very complex compositions for special casting applications.

DESCRIPTION OF THE PROCESS

Figure 8-8 illustrates the green sand molding process. The process starts with a pattern, which is a replica of the finished casting, except that it is slightly larger to allow for aluminum shrinkage during solidification and cooling. The pattern (made of wood, plastic or metal) can either be loose or attached to a plate.

A flask is placed around the pattern to contain the prepared, free-flowing molding sand that is placed on the pattern. Pressure is applied to the sand to compact it firmly against the face of the pattern. When it is compacted, the green sand mixture exhibits physical properties that allow it to hold the exact shape of the pattern after the pattern is drawn from the mold.

After cores (if any) are placed in the bottom (drag) half of the mold, the top (cope) section is set in place. The closed mold is then ready for pouring.

As the name of the system implies, the mold will be poured in a green (moist) state. When the aluminum casting has solidified, the sand is removed from the casting during a process called shakeout and is returned to storage bins for reconditioning for later use. An important and economical aspect of the green sand molding process is that the sand can be reused many times with only small additions of new sand, clay and water.

The green sand process is ideally suited for mechanization, which will be discussed later in this section.

SAND FOR GREEN SAND MOLDING

The major component of any sand mold are billions of individual sand grains. These grains will have pronounced effects on aluminum casting quality if they are not understood and controlled within acceptable limits.

Sand is a granular material resulting from the disintegration of crystalline rocks. The shapes and characteristics of sand grains were determined by the mode with which the parent rocks were broken down. Further breakdown, refining, sorting and initial sizing occurred during transportation from one location to another by many natural forces. The mode of transportation, the impact and abrasion encountered, and the distance traveled all served to determine the size, shape and surface conditions of the individual grains within a given deposit.

Where both clay minerals and sand deposits were formed simultaneously, a mixture of sand and clay resulted. This unadulterated mixture of sand and clay is referred to as "naturally bonded" molding sands. Clay must be added to sand deposits that contain little or none, forming what is called "synthetic" sand.

Properties of Sands

There are all types of sands covering the earth. Only a few are acceptable for use in foundries. A good foundry sand must retain its physical and chemical properties at temperatures beyond that of molten aluminum alloys. The sand must allow free passage for escaping gases formed by the evaporation of water and combustion of molding additives.

Sand grains should compact densely enough to create physical properties that will retain a mold cavity's dimensions in green, dry and hot situations. However, ample space should be provided to allow for the natural movement of sand grains when heated, so overcompaction also is a consideration.

Foundry sands must remove heat from the cooling aluminum casting. Grain size, shape and distribution should result in a smooth casting surface finish. The sand also must retain its properties through several molding cycles. Finally, a sand that meets these requirements also must be economical.

Grain Size

Metalcasters define sand as any mineral matter from 1/12 to 1/500 in. in diameter, or 10–250 mesh. It is important to note that the definition only defines grain size. According to the definition, sand can be of any chemical makeup.

The grain size of sand is determined using the screen analysis test. The American Foundrymen's Society grain fineness number (AFS GFN) and screen distribution can be calculated from data derived from this test.

During screen analysis, a representative sample of sand is placed on a standard foundry nest of sieves (Fig. 8-9). A sieve is a woven wire mesh with a prescribed number of openings per linear inch. The sieve numbers and size of openings are given in Table 8-1. As sieve numbers get larger, the openings in the sieve get smaller.

The sieves are stacked with the coarsest (the one with the largest openings) on top. The openings in each screen below

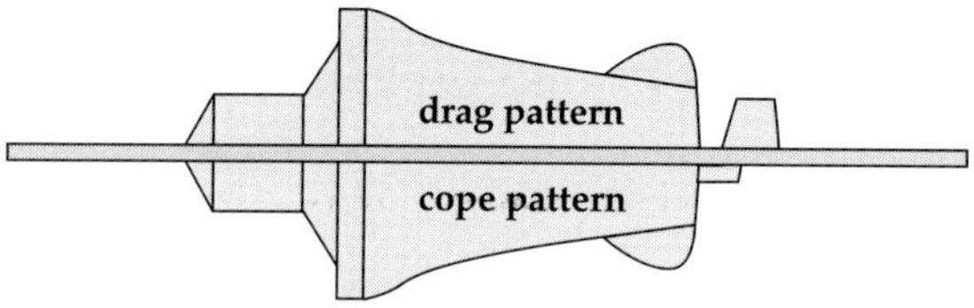

1. In this example, a cope pattern and a drag pattern are mounted onto a plate to form a matchplate.

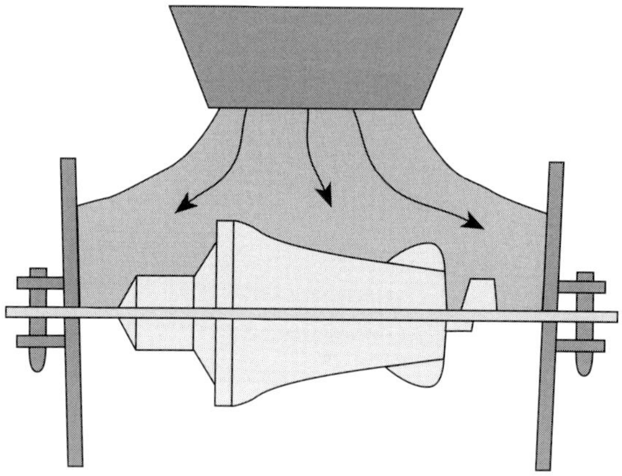

2. A flask is placed around the matchplate, and prepared sand is dumped or blown in on top of it.

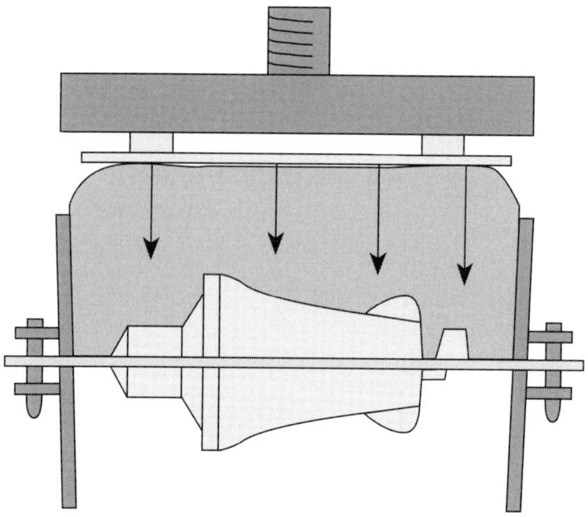

3. A squeeze board is placed over the sand, and the sand is compacted around the pattern.

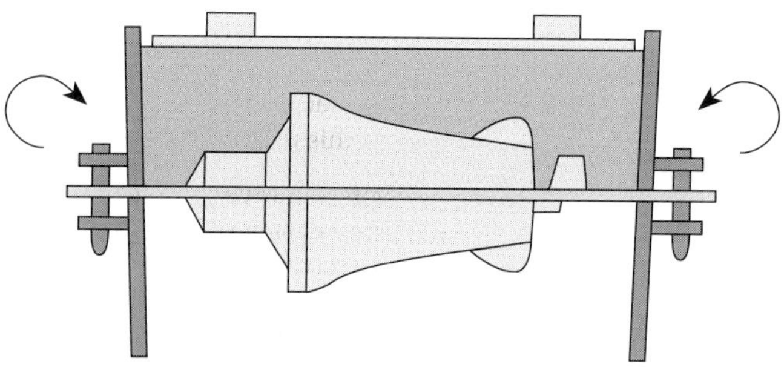

4. After compaction, the flask is turned over.

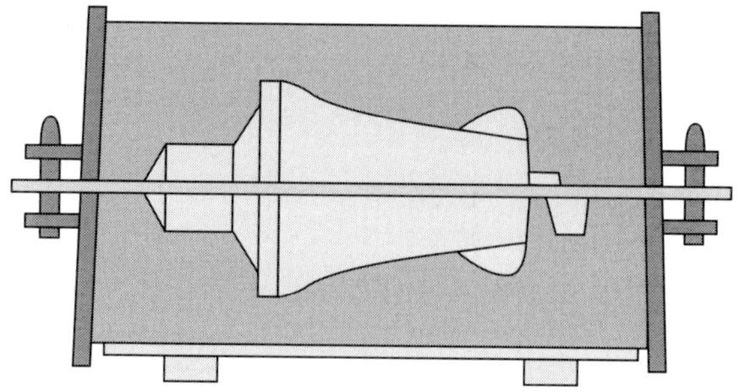

5. After rollover, steps 2 and 3 are repeated for the cope half of the mold.

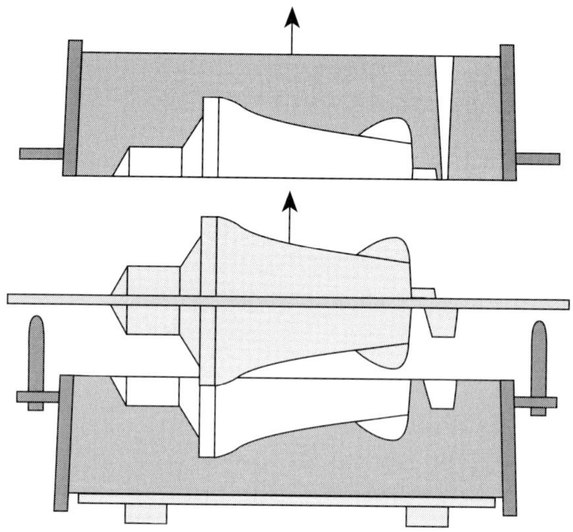

6. A sprue is cut into the cope. The cope then is lifted carefully off the drag, and the matchplate is withdrawn from the mold.

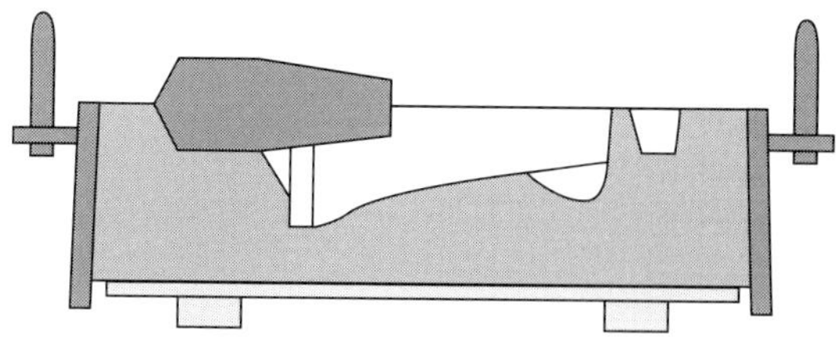

7. Cores (if any) are placed in core prints in the drag half of the mold.

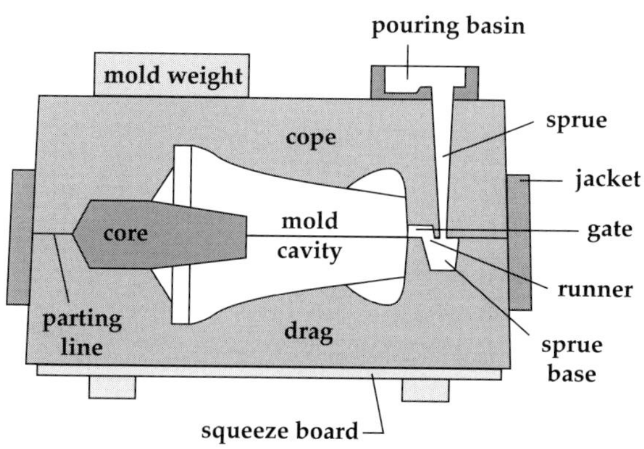

8. The cope is placed on top of the drag, and the flask is replaced with a mold jacket. Weights are placed on the assembled mold, and a pouring basin is added. The mold is ready for pouring.

Fig. 8-8. These steps illustrate how a green sand "matchplate" mold typically is made.

it become smaller, and a pan is located on the bottom. The nest of sieves is agitated for 15 min to allow the sand sample to migrate through the sieves until each grain reaches a sieve it cannot pass through.

Sand on the various screens is weighed to the nearest tenth of a gram. The resulting data are used to calculate an average or mean grain size referred to as the AFS grain fineness number. Table 8-2 illustrates a GFN calculation. As the grain fineness number gets larger, the average size of the sand grains becomes smaller. The average size of sand grains in a base sand can have a pronounced effect on the surface finish of a aluminum casting.

Screen Distribution

Although the average grain size is a valuable tool in determining and describing the general appearance of a sand, taken alone, it can be misleading. Many extremely different sands can have similar average grain sizes (Table 8-3). Most physical characteristics of a compacted sand mold are determined not only by the GFN, but also by the screen distribution of the raw sand.

A sand's distribution is calculated by counting the number of adjacent screens in the screen analysis that retained 10% or more of the sample. In Table 8-3, sand A is considered a 60 grain fineness, five-screen sand, while sand B is a 60 grain

Table 8-1.
Screen Scale Sieves

U.S. Series Equivalent No.	Tyler Screen Scale Sieves (meshes per linear in.)	Openings (mm)	Openings (in.) ratio $\sqrt{2}$ or 1.414*	Possible Variations in Avg. Opening (%)	Diam. wire (decimal in.)	Mesh Openings (microns)
4	4	4.699	0.1870	± 3	0.0650	4760
6	6	3.327	0.1320	± 3	0.0360	3327
8	8	2.362	0.0937	± 3	0.0350	2362
12	10	1.651	0.0661	± 3	0.0320	1651
16	14	1.167	0.0469	± 3	0.0250	1167
20	20	0.833	0.0331	± 5	0.0172	833
30	28	0.589	0.0232	± 5	0.0125	589
40	35	0.414	0.0165	± 5	0.0122	414
50	48	0.295	0.0117	± 5	0.0092	295
70	65	0.208	0.0083	± 5	0.0072	208
100	100	0.147	0.0059	± 6	0.0042	147
140	150	0.104	0.0041	± 6	0.0026	104
200	200	0.074	0.0029	± 7	0.0021	74
270	270	0.053	0.0021	± 7	0.0016	53

* The width of each successive opening is exactly $\sqrt{2}$ or 1.414 times the opening of the previous sieve.

Fig. 8-9. A nest of standard foundry sieves.

Table 8-2.
Example Calculation of AFS GFN

ASTM E11 Sieve Size	Weight Retained	Percent Retained	×	Multiplier	=	Product
6	0.0	(0.0	×	0.03)	=	0.00
12	0.0	(0.0	×	0.05)	=	0.00
20	0.0	(0.0	×	0.10)	=	0.00
30	0.0	(0.0	×	0.20)	=	0.00
40	0.3	(0.4	×	0.30)	=	0.12
50	1.5	(1.7	×	0.40)	=	0.68
70	10.3	(11.9	×	0.50)	=	5.95
100	39.3	(45.6	×	0.70)	=	31.92
140	22.7	(26.3	×	1.00)	=	26.30
200	9.4	(10.9	×	1.40)	=	15.26
270	1.9	(2.2	×	2.00)	=	4.40
PAN	0.8	(0.9	×	3.00)	=	2.70
TOTALS	86.2	99.9				87.33
AFS GFN = 87.33						

Table 8-3.
Sieve Analysis of Two 60 AFS GFN Foundry Sands

Percent Retained on Sieves		
U.S. Sieve No.	**Sand A**	**Sand B**
20	0.0	0.0
30	1.0	0.0
40	24.0	1.0
50	22.0	24.0
70	16.0	41.0
100	17.0	24.0
140	14.0	7.0
200	4.0	2.0
270	1.7	0.0
PAN	0.3	1.0
TOTALS	100.0	100.0
AFS GFN	60.0	60.0

fineness, three-screen sand. Although the sands are similar, they would yield unique molding characteristics because of the different screen distributions.

Generally speaking, the broadest or widest screen distribution will develop higher physical properties when sands are mulled and clay-bonded under the same conditions. Wide sand distributions that form bell-shaped curves when plotted on a graph will supply the best overall working characteristics for aluminum castings.

Surface Area

One important function of a sand department is to coat sand grains with some form of binder. The amount of energy, binder and time required to coat sand grains is relative to the amount of sand grain surface to be coated. Therefore, any change in a sand's surface area can necessitate changes to maintain consistent system sand properties.

An important factor in green sand molding is the relationship between surface area and water. Changes in surface area will result in increased or decreased water demands to reach and maintain certain working properties of the sand. Using aluminum systems sands with extremely high or low water levels can be detrimental to casting quality.

Table 8-4 illustrates the substantial increases in surface area with decreasing sand sizes. Thus, the smaller the sand, the greater the surface area (per weight).

The effects of surface area also can be seen in the sand laboratory. Sand samples used for tests concerned with adhering materials or the reactivity of sand grains should be obtained using standard AFS procedures.

Grain Shape

All foundry sands are grouped into four categories of grain shapes: rounded, subangular, angular and compound (Fig. 8-10). A sand deposit may consist of almost entirely one grain shape or combinations of all.

Grain shapes will affect the physical properties a mold can obtain. The shapes also can affect many process parameters,

Table 8-4.
Surface Area of 1 lb of Sand Grains at Various Grain Diameters

Grain Size (screen)	Surface Area (sq in./lb)
40	4300
50	6100
70	8600
100	12,000
140	17,000
200	24,000

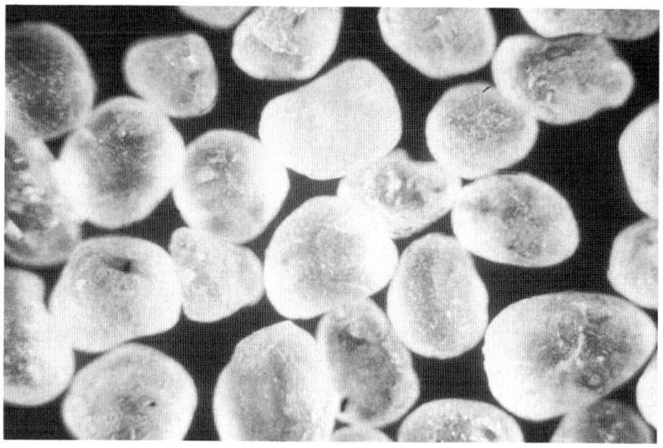

(8-10a) rounded sand grains

(8-10b) angular sand grains

(8-10c) compound sand grains

Fig. 8-10. Three of the general categories of sand grains.

including mull time, binder percentage, water percentage, mulling energy, flowability and response to compaction.

A *rounded* sand grain will require the least amount of clay and water when compared to equaled-sized angular and subangular grains because the rounded grains have less surface area. They also lend themselves to rapid coating with clay during mulling. Because rounded sand grains do not have any sharp corners and brittle edges that can be easily broken away, they tend to be more stable, retaining their original size and shape through many molding cycles.

Angular sands tend to break down during the foundry processes. The majority of this deterioration occurs early, upon its introduction into the system. As angular sands remain in a sand system, they become more rounded and show characteristics closer to those of rounded sands.

Very fine angular grains are most difficult to coat with clay for several reasons. One is that their size allows them to be sheltered in the voids between larger sand grains during mulling. Another is that their irregular shapes result in edge-to-surface contacts that are relatively inefficient when transporting or smearing clay from one sand grain to the next.

Rounded sand grains show a greater response to compaction because the mechanical interlocking nature of angular sand grains resists compaction. This resistance to compaction forces reduces the attainable mold strengths, by physically reducing the number of contacts between grains, resulting in lower strengths (Fig. 8-11).

Molds that are too dense (overcompacted) can result in a reduction of the venting characteristics and an increase in expansion-type defects. Both types of defects are a result of a severe reduction in the voids between sand grains, which allow for the passage of mold gases and the natural movement of heated sand. Many aluminum foundries will use fine angular sand grain mixtures to improve casting surface finish without sacrificing molding permeability.

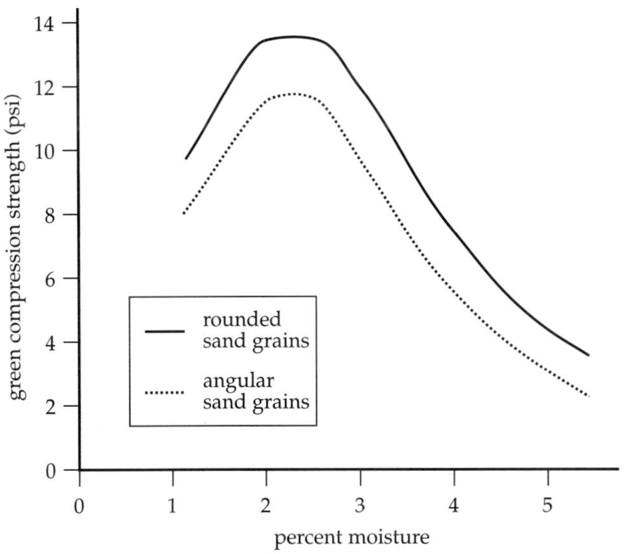

Fig. 8-11. Green compression strength based on grain shape and percent moisture at constant clay percentage.

Subangular grains look much like angular grains with the sharper edges rounded off. They exhibit properties between those of rounded and angular grains.

Compounded grains should be avoided. These are clusters of small sand grains glued loosely together. They are fragile and will break apart when subjected to mechanical and thermal stresses. When compound sand grains do break apart, an increase in fines results.

Permeability of Raw Sands

Permeability is a mold's ability to exhaust or vent the many gases formed in it by the introduction of molten aluminum.

Finer sand (with a higher GFN) results in reduced permeability. This drop is not caused by a reduction in the volume of voids within the mold but by a change from a few large voids to many smaller ones. The decrease in permeability is caused mainly by the friction between gases and sand grains as the gases migrate from the mold cavity to atmosphere.

Poor screen distribution affects permeability by reducing the total void volume between sand grains that make up a mold. Broad sand distribution (more screens) generally lowers the void volume and, therefore, permeability (any fine material that fits between compacted grains will reduce its permeability).

SAND SEGREGATION

The storage and testing of raw foundry sands (those with no binder present) should receive special attention. These sands have a natural tendency to segregate by size.

The surface area of a sand grain is directly related to its grain size. Based on sand weight (not the number of grains), a certain quantity of large sand grains will have less surface area than the same weight of similarly shaped smaller grains. Therefore, any natural segregation that causes nonrepresentative discharge from raw sand bins induces variation on surface area demands in a green sand system, resulting in constantly changing water and clay requirements to reach predetermined molding sand properties.

Segregation also can affect the average grain size and distribution of a base sand at any given time. The change, if sufficient, will make a marked change in the way a sand rams in a mold and behaves at elevated temperatures and can also change casting quality.

Sand segregation has been a problem in foundries since the introduction of sands for molding. Many papers and books discuss storage and sampling methods to reduce the negative effects of this phenomenon.

CLAY CHARACTERISTICS

Clays are the natural "glues" used to hold sand molds together in green, dry and hot conditions. The strength a mold can attain in all three states depends on the properties of the aggregate, the type of clay holding the aggregate together, the amount of clay present and the ratio of moisture to clay.

Clays give green sand mixtures several of their molding properties, including green, dry, hot and wet strengths; deformation; and flowability.

Green Strength

Green strength is needed to retain mold cavity dimensions and ensure undamaged molds prior to pouring. Green strength also is needed to allow for proper pattern draw.

Dry Strength

Even when molds are poured in a green state, a certain amount of dry strength is necessary because, as molten aluminum enters the mold cavity, it dries out the surfaces of the runners, gates and mold walls. The strength of these dry portions must be high enough to resist erosion by the metal stream. An unnecessarily high dry strength, however, may lead to difficulty at shakeout.

Hot Strength

A certain amount of hot strength is needed to maintain the dimensions of the mold cavity during aluminum solidification. If a sand mold has too much hot strength, problems with hot tears and cracked and broken castings may result. A mold that retains too much rigidity during solidification and contraction will cause the still-weak solid aluminum skin to fail, resulting in a cracked or broken casting.

Hot strength helps eliminate erosion and sand-expansion defects. Shakeout characteristics also are influenced by the amount of hot strength possessed by a molding sand.

Deformation

A deformation test measures the change in linear dimensions that a sand mass allows before rupturing under a load. Although this property rarely is measured physically in aluminum foundries, molding sand must have sufficient deformability to ensure adequate pattern stripping. A mold with low deformation would be considered brittle and would be difficult to strip.

Flowability

Flowability is defined as the ability of a sand mixture to fill pattern recesses and to move, under pressure, in any direction against pattern surfaces. If a green sand mixture exhibits poor flowability characteristics, it will cause several problems including: sticking and clogging in hoppers, bins and elevators; balling in mullers; plugging blow chambers; and poor definition of patterns. Casting finishes are better with sands that have good flowability. Molding sands with poor flowability often are referred to as "stiff."

Wet Strength

When a mold face is heated by molten aluminum, moisture in the mold evaporates, and most of it recondenses in the sand about 0.25 in. behind the mold face, forming a weak bond between the clay and sand grains. This overtempered layer of sand is called the "wet layer" or "condensation zone." The low strength of this layer has a direct relationship on the formation of many sand-related casting defects (such as drops, inclusions, buckles and expansion scabs, and rough surfaces).

TYPES OF CLAYS USED IN GREEN SAND

There are two distinct types of clays used in the foundry industry: the kaolinites and the montmorillonites (bentonites). In each clay type are several clay families. Three of these families—fireclay, calcium bentonite and sodium bentonite—are important to the metalcaster.

Fireclay

Kaolinite (fireclay) is the predominant clay in naturally bonded molding sands. Aluminum foundries using synthetic sand mixtures tend not to use fireclay because of the extremely high hot strengths attained by its addition. However, it can be used in small amounts to help give sand a high moisture latitude.

Calcium Bentonite

Calcium bentonites are the primary clay used in aluminum system sands. Small amounts of sodium bentonites may be considered for certain casting geometries.

Calcium bentonite imparts good flowability to a green sand mixture. At equal clay and moisture levels, calcium bentonites exhibit higher green strengths with practical mull times when compared to sodium bentonites.

Calcium bentonite exhibits lower hot and dry strengths than sodium bentonite at equal clay and moisture percentages. Extremely high hot strengths are not necessary with aluminum alloys because temperatures usually reach only 1400F (760C). The lower hot strength helps the sand mass separate from the solidified casting at shakeout, and it reduces the chance for hot tears or cracked or broken castings.

A sand that retains too much strength at shakeout can cause severe control problems from sand system depletion. When using calcium bentonite, sand lumps and chunks break down readily into individual grains that can be transported back to storage bins for reuse. This factor is important because it is expensive to dispose of large, baked chunks of molding sand and to replace sand and additives that are lost this way.

Sodium Bentonite

Sodium bentonite often is called the "swelling clay." In the presence of water, sodium bentonites swell to 10–15 times their original volume. These swelling and balling tendencies result in poor flowability characteristics of sodium bentonite, compared to calcium bentonite. They also can result in poor compaction and can cause sand to stick in sand hoppers.

Sodium bentonite supplies the dry and hot strengths that are sometimes necessary to prevent erosion, cuts, washes and expansion-type defects. However, the bulk of an aluminum green sand mold never reaches a temperature high enough to drive out the moisture retained in the sodium bentonite.

Because a large portion of each sodium bentonite sand mold retains moisture, the sand exhibits poor characteristics at shakeout, causing higher labor costs, more sand to be lost from the system, more cleaning time and extreme wear of equipment.

A small amount of sodium bentonite may be needed for large or special casting geometries to increase hot properties at ingates or flat, expansive surfaces. Sodium bentonite can be used to enhance green deformation in a sand mixture, which promotes easier molding and better lifting of the mold from the pattern. Typically, a blend of calcium and sodium bentonites is used if sodium bentonite is added to a system.

Sodium bentonite exhibits higher strength in the wet layer of the mold. This can help reduce many sand-related defects that occur without such strength. However, the strength in the wet layer is also a function of the amount of moisture present and the total percentage of clay in the layer, whether it be calcium or sodium bentonite.

WATER IN GREEN SAND MIXTURES

With the correct amount of water and proper mulling, a clay/sand mixture can develop adequate strength to maintain the dimensions of the molded cavity. The water in molding sands often is referred to as temper water.

Water is the most misused and the least understood of all the green sand components. It also is the least expensive, which may account for its seemingly unimportant role to the untrained. The misuse of this additive can be attributed in some way to almost all sand-related casting defects. The effects of water are seen in every property of a molding sand, except for grain fineness and distribution.

The binding power of water is caused by electrostatic attractions due to the exchangeable ions and electrical charges found on the surfaces of bentonite platelets and of individual sand grains. Too little water will yield a brittle, unworkable molding sand that will be difficult to mold and extremely sensitive to sand erosion if metal is poured against it. Excessive amounts of moisture will cause the clay/water film to thicken, which weakens the electrostatic forces and decreases obtainable green strengths (Fig. 8-12).

The optimum percentage of water is unique for each molding sand mixture. Moisture contents of aluminum system sands can range from 1.5–5.5% by weight.

Maintaining a system sand at the optimum temper point is not always recommended. Extreme caution must be taken to ensure that the sand does not become too moist. Excessive moisture in a mulled sand mixture will form steam easily during pouring, which causes many casting difficulties. However, in some instances, moisture levels above the optimum level may be required at the muller discharge either because of evaporation from hot sand or because of long transportation distances between mulling and molding stations.

The percentage of water required to properly temper a molding sand depends on: 1) the type and amount of clay within the sand; 2) the grain fineness of the sand (surface area); 3) any specialty additives (cellulose and cereal); and 4) any other fine material, including ash and dead clay. Live clays will require the most water of those mentioned.

It is important to note that the amount of water required to temper a sand to a given compactibility does not remain constant. In fact, it is dynamic and requires constant attention. As the composition of a system sand changes, so too do the moisture requirements to reach predetermined compactibility ranges.

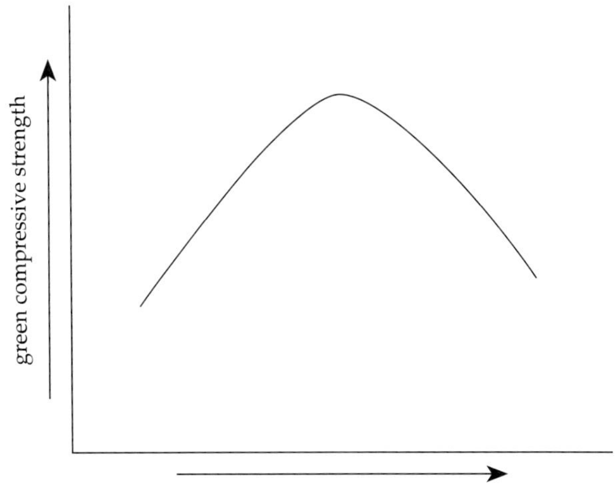

Fig. 8-12. The effect of water on the green compressive strength of molding sand.

OIL-BONDED MOLDING SANDS

In green sand molding, the steam generated by the mold when molten aluminum is poured in it can cause quality problems. Aluminum metalcasters evacuate excessive amounts of moisture (in the form of steam) by taking advantage of the natural permeability of the sand and by using proper venting practices. A coarser sand can be used to enhance the venting characteristics of the sand, but the quality of casting surface finishes is diminished.

To avoid the disadvantages of moisture in sands, water-free molding sands can be used successfully in aluminum foundry practice. These water-free sands use oil and special oil-activated binders instead of water and clay to give sand the proper molding properties.

A major advantage to using oil-bonded sands is that, because oils have a much higher boiling temperature than water, much smaller volumes of gas are evolved. Because less gas is formed, finer molding sands with lower permeability can be used. Sand with grain fineness numbers of 120 to 180 has been used successfully in many aluminum applications. Thus, a much finer casting surface finish can be attained.

Oil-bonded sands can be reused several times with only occasional maintenance and mulling, saving both time and money. However, the oil sand mixtures typically yield low green and hot strengths and tend to smoke during and after pouring, compared to conventional clay-bonded molds. These factors combined with cost limit their widespread use.

NATURAL AND SYNTHETIC GREEN SANDS

Some sands, typically silica, occur in nature mixed with an amount of clay (kaolinite) that makes them suitable for use as molding sands. The sands that contain useful amounts of clay bond are called naturally bonded molding sands.

Other sands contain little to no clay when mined. Selected clays and additives are then added by the foundry in controlled proportion. These are called synthetic sands.

Naturally bonded sands are preferred to synthetic sands for much aluminum foundry work because their comparatively close grains and high clay contents produce smoother cast surfaces. The initial cost for natural sand is less if there is a suitable local source because resulting transportation costs would be low.

However, advocates of synthetic sand often can make a good case for long-range savings because of cheaper reprocessing and reuse of the synthetic type. Because natural sand deposits have become depleted in many areas and because of sand variations within deposits, many foundries have found it practical to shift to synthetic sands.

Naturally Bonded Sands

Properties of natural sands vary widely—for example, from 48–275 in grain fineness, from 11.4–28.2% clay content and from 5.3–185 in green permeability ratings. When casting aluminum, sand properties need to be controlled closely to achieve quality end products. An aluminum foundry selects sand to best suit the work in its shop, and the properties of the preferred sand might vary considerably.

No specific limits for properties of natural sand can be set up as the best for all aluminum foundry work. However, in general, the characteristics of a sand satisfactory for most aluminum castings are likely to be within the following limits:

AFS grain fineness number	130–200
AFS clay content (%)	12–18
recommended moisture (%)	5–7
green permeability	10–25
green compressive strength (psi)	5–10

For lightweight castings that cool quickly without generating much steam, a smoother finish can be obtained by using a finer sand (such as 200–250 grain fineness) with green permeability as low as 5–10 and a slightly higher clay content.

In any case, the permeability and green strength reach optimum values with specific percentages of moisture. It is important, therefore, to determine the most favorable moisture content and to control it quite closely. For the same reason, it is necessary that the moisture be evenly distributed. This requires thorough mixing with simpler types of equipment, although the intensive action of a sand muller is advantageous.

As natural sand is used, the clay gradually loses its binding power resulting in decreased green strength. Because of wear between grains, the fines increase, and permeability decreases. If permeability loss will not be too great, the sand may be revived by moderate addition of clay to increase green strength. The clay additive should be a bentonite, which improves bonding best with the least addition. Blending of the clay addition requires much more intensive mixing than is usually used when merely adjusting the moisture content of natural sand. If the permeability loss and the decrease in strength is too severe, the only simple means of restoration is to mix in new sand.

Properties of naturally bonded molding sands can vary between quarries and between different strata from the same beds. This variation, unacceptable in many high-quality aluminum casting operations, precludes its widespread use.

Synthetic Sands

Silica

There are many types of aggregates that can be used to successfully make a mold. Silica, an oxide of the element silicon (SiO_2 in the mineral form of quartz), is the most common mineral used in aluminum foundry sands. There are many reasons why silica is so widely used in the production of molds. Silica sands:

- are the most abundant sands in nature;
- are easily mined in deposits universally located;
- have satisfactory hardness and abrasion resistance;
- are available in a wide variety of grain size and distribution;
- exhibit no high-temperature reactions with aluminum alloys;
- are excellent refractory materials.

The purity of silica sand deposits can range widely. Typically any silica-based sand that contains 98% or higher SiO_2 is referred to as silica sand, or white sand.

As the purity falls below 98%, due to contamination during formation or transportation, they are sometimes referred to by terms such as bank sand, dune sand, lake sand and river sand. These sands tend to be more yellowish brown in color. The chemical composition of these sands should be understood because the composition can affect the sand's refractory properties and hamper its ability to bond with many of today's core binder systems.

One of the greatest drawbacks to silica-based aggregates in aluminum foundries is its extremely high nonlinear expansion characteristics. When low-temperature silica (alpha quartz) is heated to approximately 1062F (572C), it quickly transforms into a high-temperature form of silica (beta quartz). This transformation causes silica sand grains to expand about 1.3% in size (Fig. 8-13). If the mold cannot accommodate this expansion (i.e., it does not possess a high degree of hot deformation), this abrupt change in volume can cause stress in the mold, causing a variety of defects, including expansion scabs, buckles, rattails, veining, mold spalling and sand inclusions.

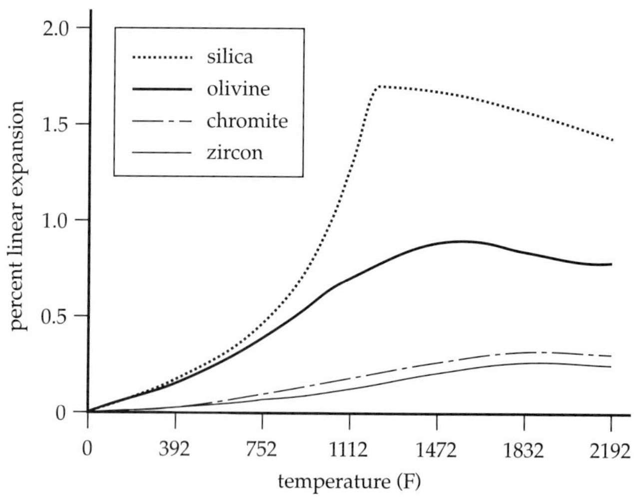

Fig. 8-13. Thermal expansion of synthetic sands.

Silica will revert back to low-temperature form when it cools and will cycle through the expansion every time it is heated. Only when silica sands reach temperatures considerably above that of liquid aluminum will they change into more stable forms of quartz.

Olivine

Olivine is another base sand frequently used in aluminum foundries. Its name comes from its olive-green color. Chemically, it is a solid solution of forsterite (Mg_2SiO_4) and fayalite (Fe_2SiO_4). Deposits of olivine are located in Washington state, North Carolina, Norway and Russia. Many aluminum foundries located in the Northwest tend to use more olivine than silica because major deposits of the mineral are found in the Cascade Mountain range. Also, deposits of silica on the West Coast are limited, and foundries are forced to pay high transportation cost for silica.

Initial research concerning the technical feasibility of using olivine sand for foundry applications was initiated because of environmental problems associated with using silica aggregates. Very small particles of silica (1–10 microns) that become airborne form a health hazard (silicosis). Olivine has a very low free silica content; therefore, the risk of silicosis can be greatly reduced. However, good housekeeping has almost eliminated the problem of silicosis in silica-based foundries.

Because olivine has to be blasted, mined, crushed and screened to size, individual grains tend to be extremely angular and sharp-edged. These sands have characteristics similar to the angular sand grains described earlier. However, older geological olivine deposits require intensive washing processes to dull sharp edges somewhat.

This angular shape has been exploited to the metalcaster's advantage in the production of aluminum castings. Because of the mechanical interlocking nature of its angular grains, olivine sands can be rammed extremely hard without problems that arise from overcompaction. An increase in permeability also is seen when compared to equaled-sized rounded sand grains, allowing for the use of extremely fine olivine sand to improve as-cast surface finishes.

Older geological deposits of olivine exhibit a loss of weight on ignition, due to the removal of water of serpentinization (water associated with the Mg_2SiO_4 mineral). If the steam driven off at ignition is excessive, gas problems may arise. This is a special concern for aluminum casting operations.

Another advantage to olivine when compared to silica is its favorable thermal expansion characteristics. Although olivine does expand upon heating, its rate of expansion is more linear compared to the nonlinear expansion associated with silica-based aggregates. Casting defects, such as rattails, buckles and scabs are minimized with olivine sand.

Olivine sand is basic (alkaline), which is compatible with foundry bentonite clays. This relationship allows olivine foundries to maintain good sand properties at slightly lower clay levels than those used with similar silica-based sands.

Zircon and Chromite

Both zircon and chromite sands can be used in special facing sand operations in an aluminum foundry. Their use as a base or system sand typically is not feasible economically. Both sands have extremely high refractory temperatures and are sometimes called refractory sands.

Each sand exhibits a high thermal diffusivity (the ability to quickly remove heat from a solidifying aluminum casting). They both offer about four times the cooling rate of silica. Chromite is sometimes called a moldable chill. These aggregates can be used to replace metal chills that can cause casting and process problems.

Both zircon and chromite will promote direction solidification towards risers and will improve as-cast surface finishes. The rapid solidification offered by these aggregates also will slightly increase the as-cast mechanical properties on the surface of aluminum castings.

Because both aggregates offer a low thermal expansion, they can be used in certain expansion-prone areas in a casting geometry.

Zircon and chromite are not readily wetted by molten aluminum. Mechanical penetration and rough surface finishes can be reduced by their use. Molten aluminum flows across zircon and chromite much like mercury on a glass table. Zircon flour finds many applications in refractory coating mixes because of this nonwetting characteristic. Their use is limited by their high price and availability.

SPECIAL ADDITIVES FOR GREEN SANDS

Whenever possible, composition of green sands should be kept simple. When making a sand formulation, only ingredients that improve casting quality should be used. More additives in a system result in more aspects to control, more chances for variation, and, thus, more defective castings.

The sand system should be kept simple. If a simple mixture of clay, water and sand will produce quality aluminum castings, it should be used. If certain casting geometries require improved sand characteristics to meet customer specifications, then a special additive may be warranted. The more that is added, the more factors must be controlled.

Cellulose and cereal are two specialty additives typically used in aluminum system sands. If they are understood and used correctly, these additives can be extremely effective in controlling many molding problems and casting defects.

Cellulose

Cellulose is any cellular material derived from a plant or plant by-product. Many forms of cellulose are available to the foundry industry, including crushed corn cobs, ground oat hulls, wood flour, ground nut shells and rice hulls. Cellulose additions in the order of 0.5–2.0% are common in aluminum operations.

The main purpose for adding cellulose is to act as a cushioning material between sand grains. Usually, cellulose additives are similar in grain size to the base sand. When molten aluminum is poured into a mold, it rapidly burns out the cellulose between sand grains at temperatures well below the critical expansion temperature of silica. When the silica reaches its expansion temperature of 1062F (572C), there is ample room left from the burnt cellulose for its natural

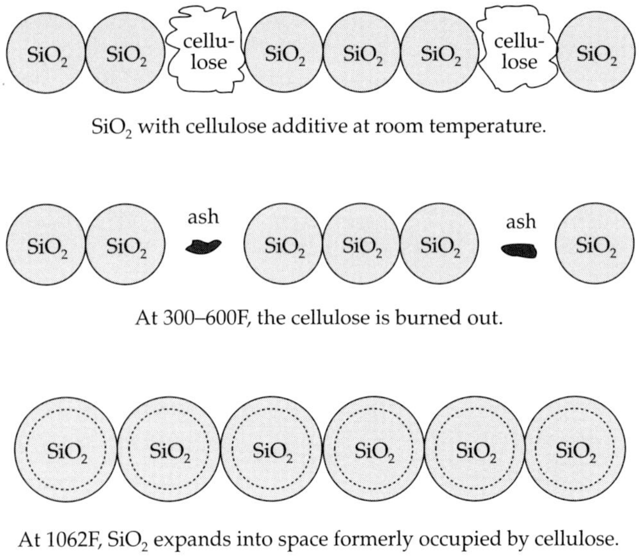

Fig. 8-14. The expansion of SiO_2 with a cellulose additive as temperatures increase.

movement (Fig. 8-14). This lowers the sand's expansion stresses on the mold and reducing expansion-type defects.

Other reasons for adding cellulose include:

- increasing mold collapsibility (better shakeout);
- reducing hot tears and cracks;
- improving sand flowability.

Problems with molds (such as brittle sand, friable mold edges and premature mold breakdown) can occur with excessive quantities of cellulose. Casting defects that can occur include mechanical metal penetration, rough surface, erosion, cuts, washes, sand inclusions and gas-related defects.

Cereal

Cereal and starch additives (normally referred to as flours) are natural organic carbohydrates produced from various grains, typically corn. They are generally added to aluminum system sands to increase the green deformation (ductility) of sand mixtures, making it easier to draw poorly maintained or difficult patterns. They can also be used to eliminate severe sand expansion problems.

Cereals are generally not added to aluminum sand mixtures because problems associated with spoiling, flowability, shakeout and control outweigh their benefits. However, with certain heavy-section aluminum castings and poorly maintained patterns, small additions may be beneficial.

MULLING GREEN SANDS

Mulling is an extension of mixing. It involves the application of compressing and shearing forces beyond the capabilities of simple paddles, blades or plows.

Mulling may be likened to kneading dough (Fig. 8-15). The process entails forcefully compressing a liquid-solid mixture between opposing hard surfaces and then smearing the compacted material apart, reassembling it and recompressing it. Reassembly must combine less-kneaded and most-kneaded sand portions so that (with sufficient time) all parts of a mixture will be subjected to essentially like amounts of energy input or mulling time.

Mullers for the preparation of clay-bonded sands are machines wherein plows continuously blend compressed sand and position the mix for recompression by wheel faces pressing downward or outward against containing bottoms or walls. Most muller designs cause plows and wheels to follow circular paths within cylindrical containers (Fig. 8-16). This permits the positioning of sand mixtures over wide areas of the muller bottom and walls and expedites assembling and disassembling of the compressed and uncompressed portions that are being continuously plowed in and out of wheel paths.

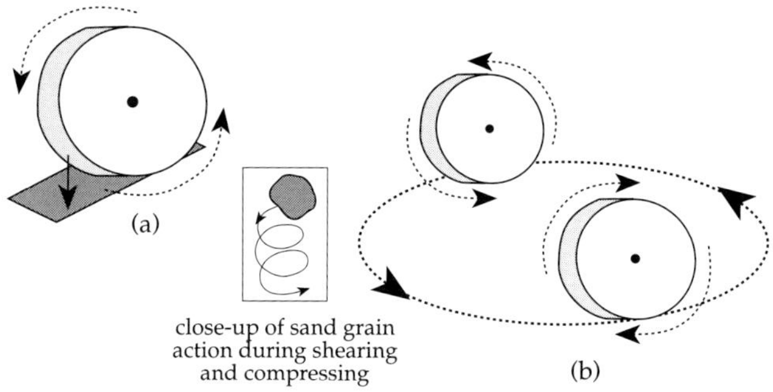

Fig. 8-15. Mulling is compressing plus shearing. (a) During mulling, sand grains are compressed by the muller wheels. (b) The grains also are kneaded during the shearing action that takes place as the wheels rotate on their axes and revolve inside the muller.

Fig. 8-16. The inside of a muller shows the wheels that shear and knead sand during mulling. [Photo courtesy of National Engineering Co.]

Aluminum molding sands primarily are bonded with 100% calcium bentonite. This type of bentonite develops physical properties rapidly compared to other foundry clays. Because aluminum system sands typically operate at low levels of clay and small amounts of new clay and additives, care must be taken not to overmull these sands. This is especially true in olivine sands, which can break down easily because of the very angular nature of the sand grains.

The majority of aluminum sands are mulled with either a horizontal wheel or vertical wheel muller. A number of foundries use a double-sided ring muller, while others prefer a continuous trough muller. No matter which unit is used, wheel clearance from the bottom or wall, wheel weight, plow settings, and general maintenance considerations are important in maintaining quality molding sand.

The action of a muller spreads clay and additives over the surface of the sand grains and, when water is added, produces the clay-water bond that holds sand grains together.

If sands are not properly mulled, a substantial loss in physical properties will occur. Therefore, retention or mulling times should be of adequate length. The amount of time needed to activate and disperse clay will vary depending on the type of mulling machinery used. Minimum mull times for each type of muller are recommended by the manufacturers. These times should be followed closely. Failure to properly mull sand mixtures will result in disaster.

EFFECT OF HEAT ON SYSTEM COMPONENTS

Heat transfer from molten aluminum to the mold begins immediately after liquid metal is introduced. The rate at which heat moves from hot to cold is a function of the thermal gradients between the mold and the aluminum. This characteristic is not extremely critical in controlling sand. Of more importance in sand control is the total amount of heat (in Btu's) introduced and the depth of heat penetration into the molding media, which is a function of pouring temperature and time.

Silica sands are extremely poor conductors of heat. Only a thin perimeter around the casting will receive enough heat energy to thermally destroy clays and special additives. For most aluminum castings, this layer is only ⅛ to ¼ in. thick. It is in this layer, as the temperature of moist clay is raised, that a series of changes takes place. Some of these changes are reversible and some are not.

- First, water is removed in the form of steam. If a tempered clay is raised to 212F (100C), all the free surface water is removed. The amount of free water removed can be estimated by the moisture content determined in the sand lab. This change is reversible. If the addition of heat is stopped, an addition of water and remulling would restore the sand-clay mixture back to normal working properties.
- At slightly higher temperatures (212–575F/100–302C), a further small amount of more rigidly held water is removed from the clays surface. This change also is reversible.
- At a higher temperature (approximately 600F/316C for calcium bentonites and 1200F/649C for sodium bentonites), a chemically combined or crystalline water is removed from the lattice structure of the bentonite, causing a total breakdown of the bentonite structure. This stage is irreversible, and the bonding power of the bentonite is lost. When sections of a sand mold reach these critical temperatures, bentonites within the affected area are rendered useless and are sometimes called dead clay. A certain amount of new bentonite will have to be added after every cycle of a system sand to replace the heat-affected or dead clay.
- Special additives, such as cellulose and cereal, burn out at low temperatures. All are destroyed at 400F (204C).

MAINTAINING THE SYSTEM

As aluminum green sand mixtures cycle through the foundry process or system, they will change in composition. A certain amount of clay and additives will be thermally destroyed in every cycle. Sand grains will be broken down both by thermal and mechanical forces. Extremely fine sand grains will be removed from the system by efficient dust collection or become airborne and blow away. Influxes of calcined core sand will be introduced at shakeout, causing variations in grain fineness numbers and screen distributions, especially in molds where the core sands are coarser than the base sands. Therefore, new materials must be constantly added to the system to maintain consistent sand properties.

Adding new sand helps to maintain desired grain fineness and screen distributions of base sands. Their introduction also helps dilute many impurities that build within a system's sand (such as fine dead clay and ash).

New sand must be added as consistently as possible. The best place to add new sand is at shakeout prior to returning the sand to the sand storage bin. This method yields better blending of the used and new sands prior to mulling for optimum control. Recommended new sand addition for aluminum foundries is 100 lb of new sand per ton of aluminum poured. This number varies, depending on the amount of calcined core sand returning to the system.

For a typical aluminum sand system, the amount of calcium bentonite needed to replace that thermally destroyed is approximately 14 lb per ton of metal poured. This number is a good starting point. Adjustments to this amount should be determined for proper coating of new sand additions and incoming calcined core sands. To further specialize this clay addition for a particular sand system, corrections can be determined using sand test data.

SAND TESTING FOR CLAY-BONDED SYSTEMS

When controlling aluminum system sands, metalcasters typically make decisions on system composition based on a series of standard green sand test procedures and equipment. Their decisions for corrective action are based on the numbers generated. The object of this section is not to make the readers proficient sand testers, but to lead them toward an understanding and interpretation of the data generated from these tests.

The physical act of sand testing and plotting statistical process control (SPC) wall charts does not constitute a valid sand control program. In order to maximize the benefits of testing clay-bonded sand, one must correlate sand testing data with corresponding sand quality problems. Metalcasters can use this information to reduce system variation and maintain uniformity within an acceptable range.

During any one cycle, molding sand undergoes many stresses that will change its composition. For instance, the introduction of core sand at shakeout can change the bonding characteristics of foundry bentonites. Heat deactivation of many additives that make up the sand (including bentonite, cellulose and cereal) changes sand composition. Base sand grains constantly deteriorate from coarse to fine because of mechanical and heat-induced fracturing.

These are only a few examples. There are many other sources of variation that will affect the composition of foundry sand. Each variation must be controlled as tightly as is physically and economically feasible. Controlling and maintaining consistency is the purpose for testing clay-bonded sand.

The physical properties and composition of prepared sand are routinely monitored in most aluminum foundries. A description and evaluation of the tests more commonly used are covered in this section. More details on these and other test procedures can be found in publications such as the AFS *Mold & Core Test Handbook.*

Compactibility

The compactibility test is considered one of the most important control tests. One feature that makes this test attractive is the ease with which the test can be performed. Because the test is based on a volumetric sample, the need for expensive and delicate scales is eliminated. The equipment is durable and easy to use, so the test can be performed on the foundry floor or mulling deck. This gives the metalcaster real-time numbers and allows for corrections prior to molding.

The compactibility test (Fig. 8-17) is performed by filling a standard 2 in. dia, 4.75 in. high specimen tube with prepared sand riddled through a No. 4 screen. Excess sand is struck off. The filled tube is transferred to a standard rammer, which compacts the sand by subjecting it to three blows from a 14-lb weight dropped exactly 2 in. The test measures the percent decrease in height from the original level of loose sand within the filled specimen tube. The number generated from this test is referred to as the percent compactibility.

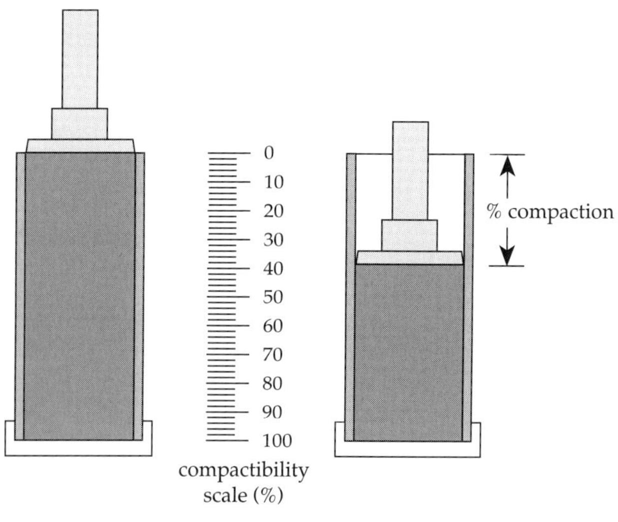

Fig. 8-17. This schematic illustrates the measurement of the percent compaction of a molding sand.

Compactibility data is important for controlling moisture in molding sand. Percent compactibility is affected primarily by the ratio of clay to water within a mulled sand and, to a lesser extent, the amount of mulling. As water increases while clay remains constant, an increase in compactibility will result. Likewise, as water decreases while clay remains constant, a decrease in compactibility can be expected. At constant compactibility, any change in moisture content indicates a change in the moisture requirements of the sand.

If compactibility numbers get too high, the molding sand could be over-tempered or too high in moisture. Over-tempered sand can cause problems such as sticking to hoppers and patterns. Casting quality problems due to gas and expansion defects could also result. Poor casting shakeout also may be anticipated.

If compactibility numbers are too low, the molding sand could be under-tempered or too low in moisture. Quality problems that could arise include difficult molding, friable sand, broken molds, mold erosion and sand inclusions within the castings.

Moisture

There are several ways to determine the moisture content of molding sand, including using infrared heat lamps, drying ovens and speedy moisture tellers (which determine moisture content by creating a controlled reaction between the water present in a molding sand and calcium carbonate). Forced air moisture tellers (Fig. 8-18) are the most widely used testing device for determining moisture content. This equipment dries sand by blowing heated air through a sample contained in a permeable brass pan. The difference between the weight of the tempered and of the dried sample indicates the amount of moisture in the tempered sand.

If moisture content is too low, the activation, (or bonding action) of the clay is greatly reduced. This is expected because the plasticity of clay is dependent on water.

If moisture is too high, the increase in water will interfere with the bonding mechanism of the clays. This causes a severe drop in green strength.

Probably of much more concern to the aluminum metalcaster is the detrimental effects of hydrogen gas pickup from the excess water in the molding sand. High-moisture sands in aluminum shops can cause severe gas porosity just below the mold-metal interface around the entire casting or throughout the entire casting geometry (Fig. 8-19). High moisture also tends to increase the tendency for expansion-type defects.

Methylene Blue Clay

Although the methylene blue (MB) clay test has been under scrutiny since its development in the late 1960s, it has provided a crucial missing ingredient to a foundry sand lab. Prior to development of this test, determining the total amount of available bentonite within a molding sand (the amount of

Fig. 8-18. A forced air moisture teller with a brass sample pan. [Photo courtesy of George Fischer Foundry Systems, Inc.]

clay capable of forming a clay-water bond) was virtually impossible. The MB clay test is now considered one of the key green sand tests because it can provide critical information about the amount of bentonite within a system sand at any given time (Fig. 8-20).

The test determines the amount of ion exchangeable bentonite present in a sand. Bentonite that still has the ability to exchange ions will contribute to green, dry and hot properties of a molding sand.

It is generally accepted that the MB clay test is difficult to perform correctly. Therefore, proper test procedures and care must be exercised to ensure the test is carried out the same way each time.

Green Compression Strength

The green compression strength test for a prepared foundry sand is a measurement of the maximum compressive load (in psi) that can be applied to a standard sand specimen before causing failure. The sand sample is a cylindrical specimen (2 in. dia by $2 \pm \frac{1}{32}$ in. high) that is prepared with a standard specimen tube and sand rammer (Fig. 8-21).

After preparation, the specimen can be broken on either the 401 strength machine or 405 universal strength machine

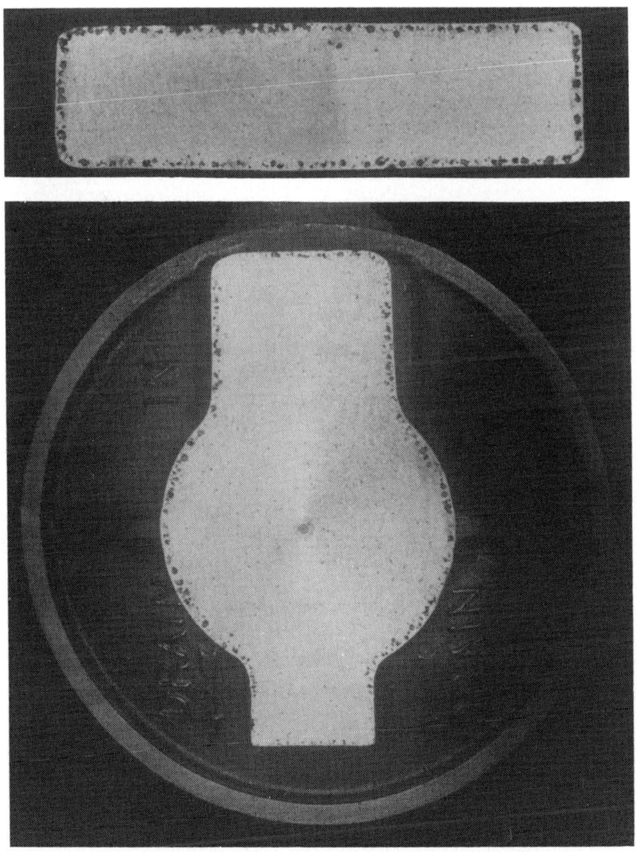

Fig. 8-19. These photos show two cross sections of a "halo defect" in an aluminum casting that was produced in a mold with excessive moisture. [Photo courtesy of Alcoa.]

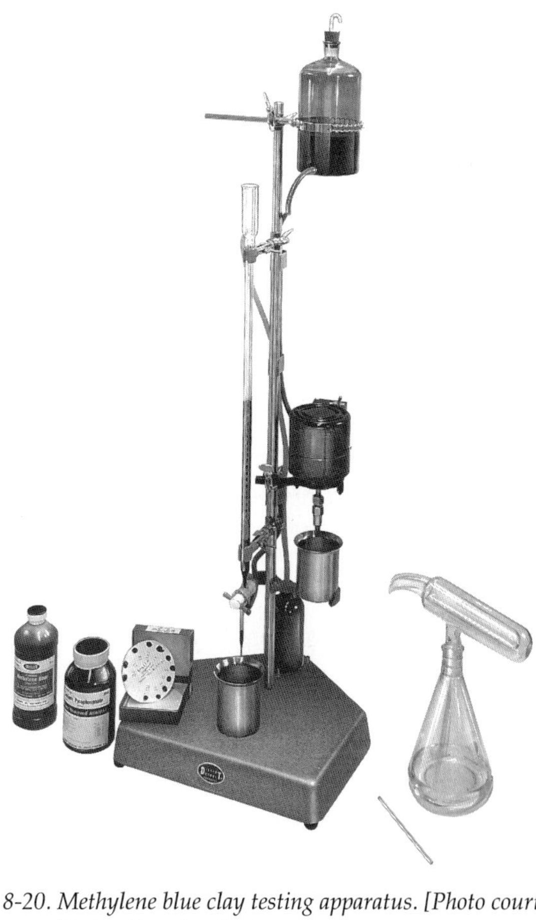

Fig. 8-20. Methylene blue clay testing apparatus. [Photo courtesy of George Fischer Foundry Systems, Inc.]

Fig. 8-21. A standard sand rammer is used to make 2x2 in. sand specimens and to run compactibility tests.

Fig. 8-22. A 405 universal strength machine. [Photo courtesy of George Fischer Foundry Systems, Inc.]

Fig. 8-23. A 2x2 in. specimen is loaded in the splitting strength apparatus or a 405 universal strength machine.

(Fig. 8-22). Many other strength tests can be run with this equipment, including dry compressive strength, green tensile strength, green shear strength and splitting strength.

The splitting test was developed as a replacement for green tensile testing. Because of extremely low green tensile strength on molding sand and poor sensitivity of the tensile test, green tensile data is difficult to reproduce. The splitting test, on the other hand, is simple, reproducible, and less sensitive to human and equipment errors (Fig. 8-23)

Prior to the introduction of the methylene blue test, the green compression strength test was used widely as the primary control tool for measuring the rate of clay additions. Unfortunately, green compression strength is sensitive not only to clay level, but also to many other factors, such as degree of mulling, sand-to-metal ratio and moisture. Today, compression strength data are useful as indicators of how effectively a foundry is activating and mixing sand additives at the muller. A sudden drop in green compression strength might indicate a problem in the preparation of molding sand.

Specimen Weight

The weight of sand used to make a standard 2 in. dia by 2 in. high cylindrical specimen (usually recorded in grams) can alert a metalcaster to possible composition changes of a system sand. Variation in weight can indicate a possible quality problem.

Permeability

Permeability is the physical property of a molding sand that allows gases to pass from the casting cavity to the atmosphere outside the mold. An increase in permeability usually indicates a more open structure between the sand grains that make up a mold.

Although the use of high-pressure molding machines has almost nullified the importance of a foundry sand's natural permeability for venting mold gases, permeability data are still a valuable tool for quickly detecting increases in fine material.

A decreasing permeability can indicate that the structure between sand grains has become clogged with other sand grains, additives or impurities. The grain size, shape and distribution of foundry sands also affect permeability.

An increase in extremely fine inert particles can be detrimental to an aluminum system sand. With an increase in fine material, the sand's water requirements to reach a predetermined compactibility will increase. When the amount of water is increased for moldmaking, steam generation at pouring is increased, and the mold's ability to exhaust steam is lowered, resulting in blows, porosity or even gas-roughened surfaces. An increase in the difficulty of removing sand from castings might occur because of increased retained strength caused physically by the fines themselves and, indirectly, by their increased demand for water (Fig. 8-24).

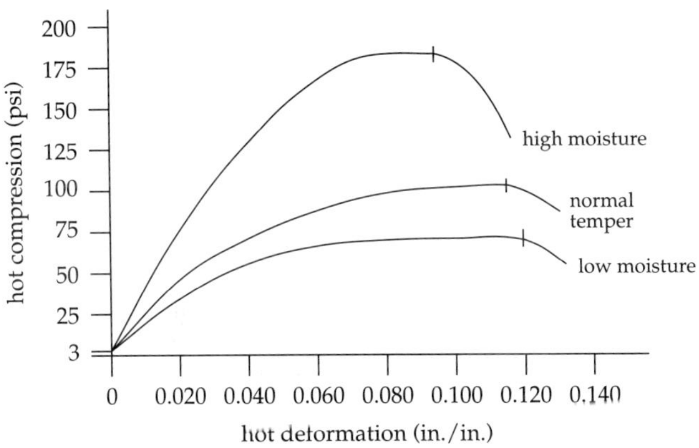

Fig. 8-24. Hot deformation and hot strength as affected by moisture.

Loss on Ignition (LOI)

In a green sand system, the LOI test measures the combustible material in a system. After preparation, a 50 g dried sample is placed into an 1800F furnace for two hours or until the sample reaches a constant weight. The percent loss in weight is then calculated.

Data generated from an LOI test is an indicator of the amount of carbonaceous material in a system sand. In an aluminum green sand, the test can be used to control and monitor cellulose and cereal-type additives. An increase in these additives results in a rise in LOI values. Too much of either additive can lead to gas and penetration-related quality problems. In particular, an increase in cereal can result in poor green sand flowability and shakeout.

For olivine system sand, special equipment is needed to run a loss on ignition test correctly. Olivine aggregates, when heated to 1800F (982C), combine with oxygen and gain an appreciable amount of weight. This weight gain can be greater than the loss, thereby skewing the true combustible content.

AFS Clay and 25 Micron Clay Tests

Both the AFS clay and 25 micron clay tests determine the amount of materials within a system sand that are smaller than 25 microns. The main difference between the two tests is the method each uses to separate these materials from the sand mass.

Each test begins with a 50 g sample of dried system sand that has been thoroughly mixed in 500 ml of a solution of distilled water and tetrasodium pyrophosphate within a rapid sand washer. This step ensures complete dispersion of all clay and other fine materials into the solution. At this point, the two tests differ in their separation of fine particles.

AFS Clay

This test uses the rate at which particles settle as a means of separation and is performed with an automatic clay determination device. The procedure determines the percentage of clay and other particles in a foundry sand that are less than 20 microns or that settle at a rate of less than 1 in. per minute.

Because of the physics of this test, it not only records live clay, but also particles such as dead clay, cellulose, cereal and any fractured sand grains smaller than 20 microns. Therefore, the name is a little misleading because it is sensitive to much more than just live clay.

The AFS clay test generally is performed on clay-bonded system sands prior to a screen analysis test. If this test is not performed first, the screen analysis of the grain size and distribution of the base sand will be skewed. The clay, regardless of moisture content, will not allow sand grains to fully separate from small clusters to individual grains. Data generated from the sieve test after wash are helpful in tracking base sand changes that can occur from core sand influxes, dust collection, transportation, new sand addition and heat-induced fracturing.

The washed fineness of the base sand does not represent the sand that will eventually come in contact with molten aluminum. Thus, an air-dried fineness test should be considered. This test can alert a foundry to possible contamination (such as clay balls, core butts, trash or metallics) in the system. The test is run using a 50 g sample of unwashed, prepared molding sand that is air dried to a compactibility of <20%. After air drying, sieve analysis is performed using the usual procedure with the exception of a 2-min agitation.

Changes in AFS clay content will generally vary with MB clay percentage. However, the difference between both numbers should remain fairly constant. A large spread between these test results could indicate an increase in inert fine material, which will increase water demands of the system and reduce permeability. If too many fines are removed from the system by overefficient dust collection, metal penetration, rough casting surfaces and gas-related defects may result.

25 Micron Clay Test

Data generated from the 25 micron clay test are used in the same manner as AFS clay data. The speed with which this test can be performed is the main difference between the tests.

This procedure uses a 25 micron sieve basket in an ultrasonic scrubber to remove any particle under 25 microns. The residue sand can then be quickly dried in an infrared dryer. The difference between the starting and ending weights is determined and referred as the percent 25 micron clay. Dried residue sand can be transferred to a sieve tester in about 20–30 minutes.

On-Line Testing

The role of on-line mold testing has been questioned for many years. Performance of both the laboratory and on-line tests can seem redundant to the untrained; however, on-line mold testing reveals important information that cannot be attained from laboratory data alone.

Mold strength, mold hardness and mold permeability are three common aspects tested in an aluminum foundry. Mold strength tests are more sensitive to changes when used on the extremely hard molds that are produced on many of today's molding machines. Mold hardness testers (Fig. 8-25) tend to work best on softer molds (under 90 mold hardness). The strength testers have electronic and mechanical versions available.

Another on-line mold test that is growing rapidly in popularity is the mold quality indicator (MQI) (Fig. 8-26). This instrument has been useful in determining the density or degree of compaction of a green sand mold.

Standard laboratory tests are helpful in determining changes in sand composition, such as changes in the contents of clay, moisture and cellulose. On-line mold tests are sensitive to changes in both composition and process. The data generated from laboratory and on-line testing can be used together to better determine the exact cause of changes in a molding sand.

For example, these tests can be used to determine the cause for a sudden drop in mold hardness. The first thing to consider is whether the variation is due to a change in sand composition or to a change in squeeze pressure at the molding machine. With only laboratory data, which tracks only sand composition, it cannot be determined whether the variation is caused by changes in composition, in the process or a combination of both.

Armed with both laboratory data and on-line data, the true cause for the sudden variation can be established and the problem corrected. If, for instance, laboratory data remain constant while mold tests show variation, the cause for changes in mold hardness can be traced to the process (squeeze pressure) and not composition. Without both laboratory and on-line data, these conclusions could not have been drawn.

Fig. 8-25. A "B" scale mold hardness tester.

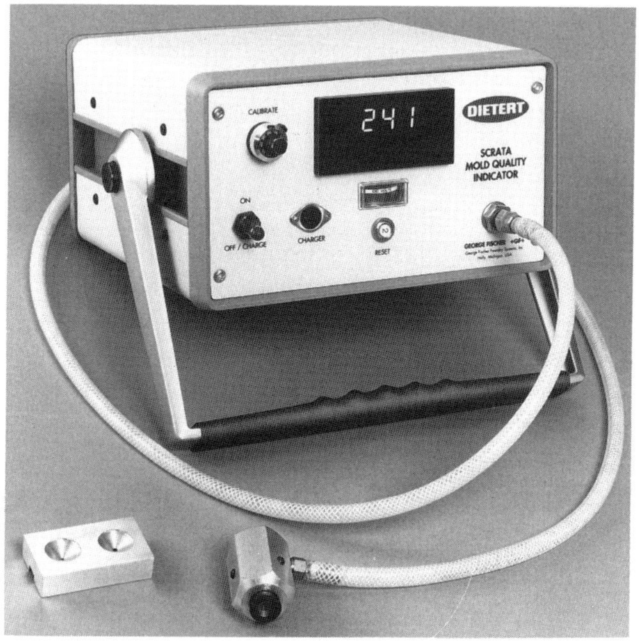

Fig. 8-26. A battery-powered mold quality indicator (MQI). [Photo courtesy of George Fischer Foundry Systems, Inc.]

CALIBRATION OF TESTING EQUIPMENT

No sand testing program is valid without recognizing the importance of calibrating testing equipment. Data derived from standard green sand tests are only as reliable and precise as the equipment used.

The most appropriate time to begin any sand testing program is when a foundry process is running smoothly. Many times, aluminum foundries find themselves chasing problems without a full understanding of what the process limits are when the operation is performing successfully. A good foundry will develop complete historical records, noting characteristics of system sand when the process is most productive. Then, if problems do arise, changes in sand characteristics are easily noted and quickly corrected.

If test equipment is not kept within specifications, then any variation in data could be caused either by sand or by incorrect readings from the sand testing equipment. An intelligent decision cannot be based on inaccurate information. It would be better to eliminate the sand testing program if the equipment is poorly maintained and the testing procedures are not clearly defined and followed.

Chemically Bonded Sand Molding and Coremaking

RESIN BINDER MATERIALS AND PRACTICES

Virtually every resin binder ever made has been used to produce aluminum castings. Today, every type of resin binder system sold is used by someone to make aluminum castings.

The "best" binder system for aluminum was and still is a matter of personal opinion. Which system a foundry selects might be based on the suitability of its tooling and the availability of the equipment; however, more than likely, the choice will be made based on what the boss thinks is best.

BINDER SYSTEM DEVELOPMENT

In the 1940s, sand cores were made with oven-baked, oil-bonded sand. The molding processes during this era consisted of green sand for smaller castings and skin-dried green sand or "dry sand" (oven-baked pitch and clay-bonded molds) for larger castings. The oil sand process was used to construct "core molds" by fastening sections of cores together in ways that would form molds. These core molds produced castings with better dimensional properties than those made in clay-bonded sands.

Following World War II, drastic changes occurred in the core and moldmaking processes used for casting aluminum. Heat-cured "shell resin" sand systems (which produced hollow, shell-like cores) and carbon dioxide gas-hardened silicate systems (which made the more conventional solid cores) were introduced during the 1950s. In the decades that followed, heat-cured hotbox binders, nobake systems (which chemically cured without application of heat), vapor-catalyzed coldbox processes, and vapor-reacted resins were introduced.

The 22 binder processes introduced during the last three decades (Table 8-5) forced the design and development of new and innovative equipment needed for making cores and molds and also for coating and handling the sand. This equipment was designed to take advantage of the specific, improved properties that these new systems offered. Binder improvements and innovative equipment developments continue to improve resin-bonded core and moldmaking.

BINDER IDENTIFICATION & CLASSIFICATION

There are four ways to categorize different binder systems.

1) *Inorganic* or *organic*—Inorganic binder systems are those that do not contain carbon in the binder molecules. Unfortunately, inorganic systems are the least reactive; achieve the least strength; and require the most attention during the coremaking, moldmaking and casting processes. These systems are based on silicate and phosphate/metal oxide technology. All of the other systems are organic.

2) *Catalyst mechanism*—Systems also can be categorized based on how they are chemically catalyzed (Table 8-6). Acid-type binder systems ($pH < 7$) are cured by adding an alkaline material, and basic-type binder systems ($pH > 7$) are cured by adding an acidic material. Still other systems are cured by various chemical mechanisms that are independent of a distinct pH shift.

3) *Physical curing mechanism*—Binders also can be categorized based on the way that the core and mold sand physically hardens (cures) (Table 8-7). Resins cure in several ways: by being in contact with a vaporized catalyst (coldbox); by being exposed to various forms of heat (heat-cured); and by being mixed with a liquid catalyst that reacts with the binder at ambient temperatures (nobake).

   When a binder has fully reacted (polymerized) and reached its final cure stage, it has become a thermally stable, infusible material called a thermoset.

4) *Heat-activated (thermally or thermal-set)*—In addition to activating binders in this classification, heat can be a way to accelerate the cure rates of all binder systems.

   Shell, core oil, hotbox and furan warmbox binder systems are cured by applying heat to the resin-coated sand. Although various forms of catalysts and/or accelerators also may be added to the mixture, heat is the primary curing agent in these systems.

Different suppliers often refer to the same or similar binder systems by different names. Table 8-8 cross references generic chemical binder names with the trade names used by major U.S. resin suppliers. This should help readers who know the trade names but not the chemical names of the binder systems they use.

Table 8-5.
Approximate Time of Commercial Introduction of Core and Mold Processes in North America

Date	Process
pre 1940	core oil
1950	shell: liquid and flake
1952	silicate/CO_2
1953	airset oils
1958	phenolic acid-catalyzed nobake
1960	furan hotbox
1962	phenolic hotbox
1965	oil urethane nobake
1968	phenolic/urethane/amine coldbox
1968	silicate ester-catalyzed nobake
1970	phenolic urethane nobake
1974	alumina phosphate nobake
1977	furan SO_2
1978	polyol urethane nobake
1978	warmbox
1982	FRC SO_2
1983	epoxy SO_2
1984	phenolic ester nobake
1985	phenolic ester coldbox
1992	phosphate/metal oxide

Table 8-6.
Binder Systems Classified by Catalyst Curing Mechanism

Acidic	Basic (Alkaline)	Other
silicate CO_2 plus dehydration	phenolic ester nobake	shell (neutral)—hexa addition
warmbox	phenolic ester coldbox (addition)	silicate nobake—saponification plus dehydration
hotbox	oil urethane nobake plus oxidation	core oil (neutral)—oxidation
SO_2 furan	phenolic urethane nobake and coldbox	phosphate/metal oxide
SO_2 acrylic epoxy and free radical cure	polyol urethane nobake	

Table 8-7.
Categories of Resin Core/Mold Processes

Coldbox	Nobake	Heat-Activated
phenolic/urethane/amine	furan/acid	shell
silicate/CO_2	phenolic/acid	core oil
furan/SO_2	phenolic/ester	phenolic hotbox
acrylic/epoxy/SO_2	oil urethane	furan hotbox
(acrylic) FRC/SO_2	silicate/ester	urea formaldehyde hotbox
phenolic/ester (methyl formate)	phenolic urethane	warmbox
phenolic/CO_2	phosphate/metal oxide	
	polyol urethane	

Table 8-8.
Trade Names and Manufacturers of Chemical Binder Systems

Process	Chemical Name	Trade Name	Manufacturer
coldbox	phenolic/urethane/amine (PUA)	Sigma Cure, Acme Flow	Acme/Borden
		Isocure	Ashland
		CRC	Capital
		UniCure	United Erie
	furan SO_2	Isoset F	Ashland
	acrylic SO_2	Isoset FRC	Ashland
	epoxy/acrylic/SO_2	Isoset	Ashland
	phenolic/methyl formate	Beta Set	Acme/Borden
		NovaSet-CB	Ashland
	phenolic/CO_2	Veloset	Foseco
	sodium silicate/CO_2	CO_2 Set	Acme/Borden
		Accoset, Adcosil	Ashland
		Steinex	Carver
		Silicate	Delta
		Cast Clean	J.B. DeVenne
		Solasil	Foseco
		Insil	Interstate
		Gas Bond	United Erie
		ICP	ICP
		West Cast	West Cast

Table 8-8 (continued).
Trade Names and Manufacturers of Chemical Binder Systems

Process	Chemical Name	Trade Name	Manufacturer
core oil		LinOil	Ashland
		CRC	Capital
		Delta Core Oil	Delta
		Korbond	United Erie
shell		Plasti-Flake	Acme/Borden
		G.P.	Georgia Pacific
hotbox	furan and pehnolic	Acme Hot Box	Acme/Borden
		Chem Rez	Ashland
		Delta Hot Box, San Set	Delta
		Pac Set	ICP
		Chem Cast	Southern Foundry Resins
warmbox		Warm Box	Acme/Borden
		Chem Rez	Ashland
		Warm Box	Delta
nobake	alkyd oil/urethane	CRC	Capital
		Linocure	Ashland
		Alkyd resin	Delta
		UniSet	United Erie
	ester/phenolic	Alpha-Set	Acme/Borden
		Nova-Set	Ashland
	furan and phenolic acid	Super Set	Acme/Borden
		Chem Rez	Ashland
		CRC	Capital
		Airkure	Delta
		Pac-Set	ICP
		Foundrez	Reichold Chem
		Chem Cast	Southern Foundry Resins
		Dry Set	United Erie
	phosphate metal oxide	Ecoset	Ashland
	phenolic urethane	Acme Flow, Sigma Cure	Acme/Borden
		PepSet, Novathane	Ashland
		CRC	Capital
		Delta Set	Delta
		UniSet	United Erie
	polyol/urethane	PepSet	Ashland
		CRC	Capital
		Aluma Cast	Delta
		UniSet	United Erie
	silicate/ester	Chem Bond	Acme/Borden
		Accoset, Chem Rez	Ashland
		Nobake Silicate	Delta
		Cast Clean	J.B. DeVenne
		Veloset	Foseco
		Chem Cast S	Southern Foundry Resins
		Sil Bond	United Erie

HEAT-ACTIVATED SYSTEMS

Shell Cores

The shell process, dating back to World War II Germany, remains one of the most convenient forms of core production. Foundries can use storable, precoated sand that can be supplied by commercial coaters, or they can coat their own sand. A number of foundries use the shell process for molding in addition to coremaking. (For detailed coverage of the process, see Chapter 9.)

Although coldbox processes are used to produce more aluminum castings (by tonnage), the shell core process may be used by a larger number of aluminum foundries. The shell process offers a combination of operating convenience, satisfactory productivity and excellent casting results.

Shell sand can be coated using two different processes:

- In the "hot coating process," a flaked phenolic novolak resin is mixed with sand heated to about 350F (177C). The resin melts and coats the sand. The hot (275F/135C), resin-coated sand is "quenched" with a water solution of hexamethylene tetramine and mulled until the sand mass breaks down. Sand is discharged then screened and aerated to further cool and to particulate it.

  The hot coating process is used to produce the majority of shell process coated sand. This process produces a mix that has less peelback (a core/mold imperfection referring to uncured sand falling away from cured sand at the surface).
- In the "warm coating process," calcium stearate, hexapowder and a water/alcohol solution of novolak resin is added to sand heated to about 200F (93C). When the coated sand mass breaks down after the carrier evaporates, the coated sand is discharged to an aerator to be cooled and particulated.

  Warm-coated sand builds up the sand shell more quickly, potentially reducing the core machine cycle.

Both coating operations produce a free-flowing, storable sand. For aluminum casting operations, the shell sand resin level typically is less than 2.5%.

A test called the "stick point test" is used to determine the melting point of a shell sand's resin coating. The higher the stick-point of a sand, the lower its tendency to suffer from peelback. However, sand with a higher stick-point takes longer to develop shell thickness.

The popularity of the shell process peaked in the early 1970s. Natural gas shortages, combined with various environmental regulations, compelled foundries to consider alternative core/mold processes. Then, refinements in energy-efficient nobake and faster-cycling coldbox processes forced use of the shell process into a state of decline. Shell process production, however, is expected to remain at current levels through the 1990s.

The shell process offers the distinct economic advantage of producing a hollow core that leads to more convenient shakeout. Castings with excellent surface finish are produced because of the superior flowability and density of the sand. The coated sand also may contain additives—such as kaolin-type clay, potassium borofluorate (KBF_4) and Vinsol (a thermoplastic, powdered material obtained from tree stumps)—to prevent surface defects in castings.

Core Oils

The core oil process begins with cereal and other dry additives being mixed with the sand. Water is added to "activate" the cereal (to give it green strength for handling). A kerosene-type oil is then added to the sand near the end of the coating cycle to improve flowability and release.

After coating, the sand mix is blown or rammed into a suitable pattern. The green strength property of the mix allows the wet, pliable core to be removed easily from the pattern, which is placed onto a suitable support for subsequent drying.

Core oil "dries" after the water evaporates from the sand and oxygen from the oven's heated air works to oxidize the vegetable oils and polymers in the core oil. To achieve maximum productivity without burning thin sections of the core, the oven must have a drying atmosphere that is low in humidity and high in convection air volume. As shown in Fig. 8-27, 350F (177C) is the best temperature for core oil baking, but most ovens operate at about 500F (260C).

The core oil process produces cores quickly, even with the poorest tooling. There are some limitations in terms of tight dimensional tolerances because of handling and drying of the pliable core shape.

The core oil process will continue to be used widely to produce aluminum castings simply because foundries already own the necessary tooling and because the system makes cores that can perform as required.

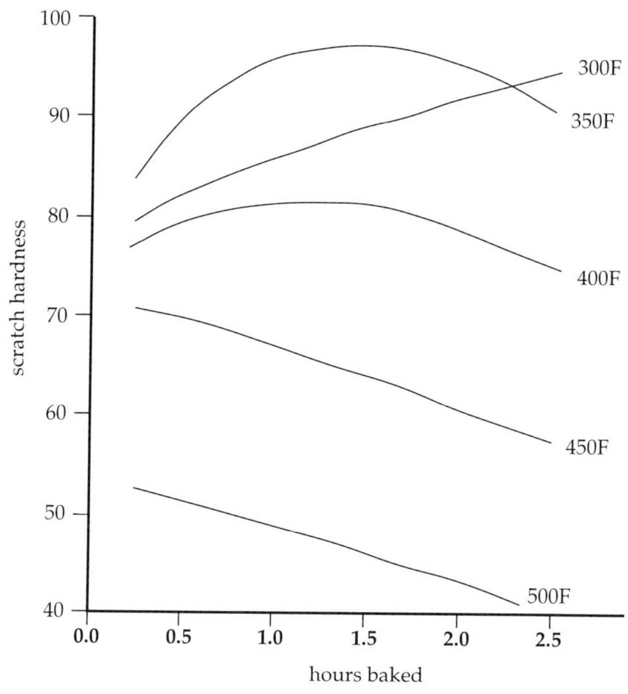

Fig. 8-27. Effect of baking time and temperature on scratch hardness for a typical linseed-type oil sand core with 1.25% binder and 2% water.

Hotbox Systems

There are three distinct hotbox binder systems, based on three types of resins: furan, phenolic and urea formaldehyde. Although the resins can be used individually (along with the proper catalyst), the resins are sometimes blended to optimize properties.

One significant drawback to the hotbox systems is that formaldehyde is given off at several points in the process, making careful venting and fume control necessary. Under certain circumstances, detectable formaldehyde might be measured in the areas around the sand coating operation, around the machine while the core is curing at elevated temperature in the corebox, and around the core after it has been removed from the corebox until the core cools to ambient temperature.

Formaldehyde also can be liberated from the hotbox binder during oven drying of refractory coatings. Various formaldehyde scavengers (formaldehyde neutralizers) lower free formaldehyde resins, and process modifications are minimizing the formaldehyde-related problems.

Urea Formaldehyde

This system is simply made up of a water solution of urea and formaldehyde. It often is modified with up to 30% furan to improve cure speed and strength. This heat-cured system is ideally suited for dielectric oven curing.

Productivity with this system is quite fast, but every step of the coremaking and casting process liberates formaldehyde into the workplace. As formaldehyde emission regulations become more stringent, the already limited use of urea formaldehyde in coremaking for aluminum castings will further decrease.

Furan and Phenolic

Only a few foundries currently use the furan and phenolic hotbox processes for producing aluminum castings.

Furan and phenolic hotbox systems are catalyzed with a water solution of acid salt, based primarily on ammonium nitrate or ammonium chloride. The catalyst, upon being exposed to the elevated pattern temperature, decomposes to form ammonia and a mild acid that cures the binder.

It is important to realize that the hotbox core continues to cure after being removed from the heated pattern. The residual heat in the sand plus a mild exothermic reaction (which causes an increase in temperature) from the resin polymerization continues to make the catalyst decompose and cure the sand until the temperature drops to less than 125F (52C).

Bench life of the coated sand will vary from 1–4 hours, depending on the type of resin and the temperature. Furan binders have a much longer bench life than the phenolics.

Warmbox Systems

The warmbox system is composed of furan resin catalyzed with a copper salt/water/methanol solution of aromatic sulfonic acid. An advantage to using warmbox rather than hotbox is that the warmbox will cure effectively at a processing temperature about 100°F (56°C) lower than hotbox systems.

Warmbox systems with very low free formaldehyde have been developed. These systems also are phenol-free. These two aspects are distinct environmental advantages compared to hotbox systems. As with the hotbox process, few aluminum castings are produced using this process.

COLDBOX SYSTEMS

Coldbox systems are used in most high-production aluminum foundries. The coldbox classification includes silicate/CO_2; phenolic/urethane/amine; acrylic/epoxy/SO_2; phenolic/methyl formate; and phenolic/CO_2.

Phenolic/Urethane/Amine (PUA)

The phenolic urethane coldbox system is comprised of two resin parts. Part 1 is a phenolic resin dissolved in an organic solvent, and part 2 is an isocyanate component dissolved in a similar solvent.

For making cores for aluminum castings, the binder in this system typically is used at a level of about 1% of the sand weight. The part 1:part 2 ratio (by weight) is usually offset to 60:40 to improve shakeout, reduce cost and minimize gas-related defects. A clean silica sand, olivine sand or zircon sand can be used for both coremaking and molding. For producing aluminum castings, versions of the PUA system specifically designed for aluminum should be used.

After the sand is coated with the two-part binder system, it is blown into the tooling cavity through entry ports (blow tubes or blow slots) located in one half of the tooling. The resin coating is then cured by a vaporized amine catalyst, which is introduced into the tooling cavity through "cope vents" and through the sand entry ports.

The vents are specifically placed, and the openings are sized to ensure that the amine catalyst gas develops a positive pressure in the tooling cavity. To this effect, the open area of the exhaust vents is limited to about 80% of that of the gas inlet area. The pressurization causes the amine to travel to all parts of the pattern cavity before it exits through the exhaust vents on the opposite half of the pattern.

After the amine has been introduced into the cavity, a very dry "purge air" is introduced. This volume of air travels through the core and out the exhaust vents, distributing the amine catalyst to all cavities of the corebox and then flushing the amine and its residual odors from the core. The dry purge air often is heated to increase its effectiveness.

After the amine-laden air exits the corebox through the exhaust vents, it travels through the exhaust plenum, into large-diameter piping (piping three times the diameter of the input piping) and, finally, into an exhaust scrubber. The scrubber contains an acidic solution to neutralize the amine and form an acid salt.

The most distinctive feature of the PUA system is the amine catalyst. There are two types of amines that are commonly used as the catalyst—triethylamine (TEA) and dimethylethylamine (DMEA). DMEA is the more effective catalyst, but, because of its stronger odor, higher cost and only marginal improvement in total machine cycle (with properly rigged tooling), most of the larger aluminum operations use TEA.

Recent work has led to economical recycling of the acid salt scrubber solution. Both the TEA and DMEA amines can be recovered and sent back to the original user as a reusable catalyst.

Trimethylamine (TMA) has been used in gassing hood applications, and dimethyl isopropanol amine (with curing properties between TEA and DMEA) is used in Europe. Neither type is currently used in North America.

All polymeric resin binders shrink as they cure, and PUA is no exception. When the core is removed from the corebox, it is about 70–80% cured. The cure is completed as the solvent evaporates from the binder coating on the sand. Figure 8-28 gives a general range for this contraction as measured toward the center of the sand's mass. The amount of shrinkage depends on the density of the core, the type of binder and aggregate, and the age of the core.

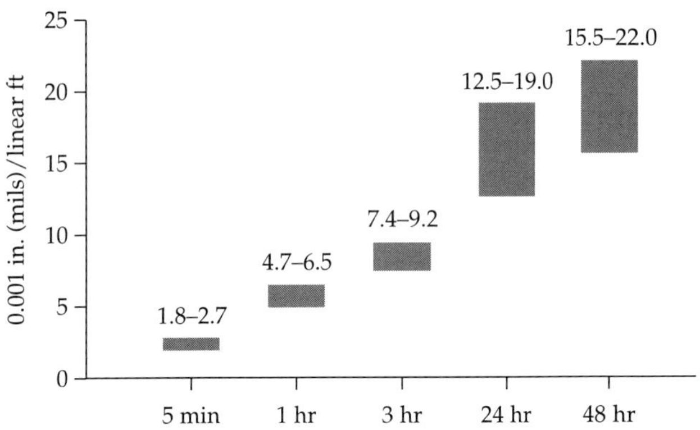

Fig. 8-28. Amount of linear shrinkage in a PUA coldbox core with 1.5% binder level and part 1:part 2 ratio of 55:45.

The keys to a successful PUA coldbox process are 1) using the resin-coated sand as soon after coating as possible and 2) correct rigging of the tooling. Most productivity problems can be traced to improperly rigged tooling, which promotes sticking and poor cycles, or to using coated sand that sat in the hopper too long after coating before it was used for coremaking.

Since its introduction more than 20 years ago, the PUA coldbox system has become the highest volume binder system in the foundry industry. It also is used to produce a greater tonnage of aluminum castings than any other binder system. The tonnage of aluminum castings made with this system will continue to grow during the 1990s because the automotive industry uses PUA molds and cores for manufacturing advanced aluminum power train components.

Furan/SO_2

The furan/SO_2 system originated in Europe around the mid 1970s. The system is comprised of three components: 1) a standard-type furan resin, 2) a peroxide oxidizer, and 3) silane, which is added during the mixing cycle to promote better binder adhesion and give higher strength. The mix, which has outstanding bench life, is either blown into a corebox for gassing or is hand rammed into a pattern where it can be cured effectively in a gassing hood.

In this system, when the SO_2 gas comes in contact with the organic peroxide, it forms an acidic material that cures the furan resin. The reaction is rather slow, and it benefits from the addition of heated purge air and from warming the tooling surface to temperatures above 120F (49C). The reaction is impaired by the presence of moisture in the sand and from sand with a positive acid demand value. As might be expected, the cure rate is negatively affected by cold sand and cold operating conditions.

Even though a sulfur-type catalyst is present in this system, the color staining problem associated with the liquid sulfonic acid-cured furan nobake resin system is not a problem. Outstanding bench life and extraordinary shakeout properties makes this the best coldbox system for producing aluminum castings.

However, the furan/SO_2 system suffers from a difficulty in core production. The furan binders used in this system have a problem known as "excess resin wipe-off," which coats the tooling surface during coremaking. The wipe-off results in sticking and in plugged vents, and frequent shutdowns are required to clean the resin from the vents and the tooling surfaces.

When the tooling is clean, the core is strong and uniformly dense, and most importantly, the furan/SO_2 system shakes out better than any other unmodified binder system (Fig. 8-29). There are a few foundries that produce aluminum castings using this system. They put up with the dirty tooling problems because of the advantages the system offers in high casting quality and easy shakeout.

Acrylic/SO_2

Improvements to the furan/SO_2 process led to the development of the foundry industry's fastest-curing binder system—the acrylic SO_2 system. The rapid cure, outstanding bench life and good shakeout properties of this system qualified it as an ideal coremaking system for aluminum castings. However, the system has relatively poor hot strength, so its use eventually faded in all but a few aluminum foundries where small castings are produced.

Epoxy/Acrylic/SO_2

Resin formulators recognized that acrylic binders offered outstanding cure speed but poor hot strength and that epoxy binders had superior hot strength but needed long gassing cycles because of their slow cure rates. Therefore, the epoxy/acrylic/SO_2 system, using a blend of epoxy and acrylic, was developed to optimize both the handling and casting properties. This system was introduced to the foundry industry in the 1980s.

The epoxy/acrylic/SO_2 system contains acrylic resin for rapid cure speed and epoxy resin for higher handling strength and better hot strength. This SO_2-gassed resin system usually uses a ratio of 65 parts of epoxy to 35 parts of acrylic, with peroxide added to the epoxy component. However, because the shakeout properties of this system can be difficult for

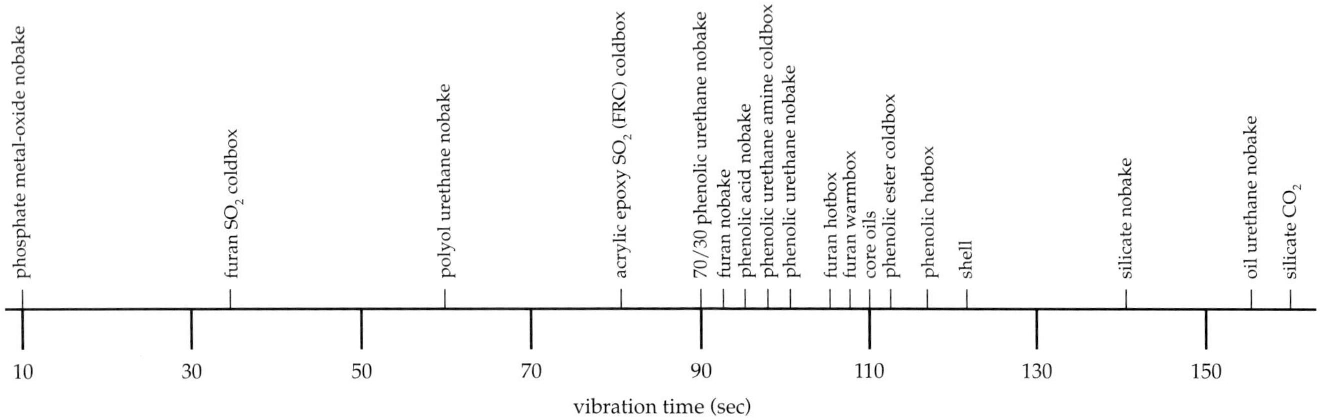

Fig. 8-29. Comparison of shakeout times for chemical binder systems used to cast Alloy 356.

aluminum operations, a 50/50 blend is used to improve the shakeout characteristics.

Phenolic/Methyl Formate

This system was introduced to the foundry industry in 1984. It is based on a high-water-content phenolic resole resin that reacts with a vaporized methyl formate coreactant. The sweet-smelling, flammable methyl formate typically is used in amounts of 30–40 parts per 100 parts of resin (by weight).

Because the methyl formate is a coreactant and not a catalyst, the resin binder *must* be exposed to and react with the proper amount of the vaporized liquid material for a complete chemical reaction to occur. This factor sometimes leads to overgassing to make sure that a complete cure takes place.

This water-based system offers water cleanup and is low in smoke and odor compared to other organic coldbox systems. Because of the low strength of cores produced with this system, the phenolic/methyl formate system is not used in aluminum applications where delicate cores are needed. However, the lower strength is adequate in heavier core sections.

There is a significant increase in the strength of the coated sand when it is heated by molten aluminum, which contributes to shakeout problems. The system has nonetheless been successful when used for cores in aluminum casting.

Phenolic/CO_2

The phenolic/CO_2 system is based on the same type of high-water-content phenolic resole resin used in the phenolic/methyl formate system. The difference is that carbon dioxide gas replaces the methyl formate coreactant.

Because the CO_2 gas does not have to be collected and disposed of, this system offers definite advantages. The same advantages and disadvantages found in the phenolic/methyl formate binder are apparent in this system, with the obvious difference that the flammable methyl formate is replaced with the inoffensive carbon dioxide. Although this new system is relatively weak, it offers the potential for a clean-running aluminum casting binder.

Silicate/CO_2

This is the classic inorganic binder system that is cured with carbon dioxide gas. Because the gas is odor-free and nonirritating, it does not need to be collected for disposal. The inorganic classification of the binder and the harmless gas catalyst help make this system one of the first choices for aluminum casting production.

The chemistry of the sodium silicate was first described in a 1898 British patent. The silicate/CO_2 system was originally used as an oven-heat cured system. Its earliest use as a coldbox system was in a Czechoslovakian foundry at the end of World War II. Sodium silicate/CO_2 was introduced into the U.S. as America's first true coldbox foundry binder system around 1952.

The properties of the finished core depend on the ratio of silicate to soda. The available ratios range from 1.90–3.22 parts SiO_2 to one part Na_2O. The lower ratios (2.0–2.2) result in higher final strength. The higher ratios (> 2.7) give lower final strength but are stronger initially. For optimum properties, most formulators select a ratio of about 2.5:1.

The binder's strength and the ability to store the finished core or mold is based on a specific amount of carbon dioxide gas reaching the binder-coated sand surface (Fig. 8-30). The complex curing mechanism involves first gelation and then dehydration.

It is better to overgas cores if they are going to be used the same day because they have no humidity resistance. It tends to be better to undergas the cores when they are going to be stored. The cores that are undergassed and stored continue to cure, rather than to lose strength in storage. In most cases, it is preferable to undergas a silicate/CO_2 core or mold.

Special attention must be paid to thin core sections that accompany thick sections. These thin sections will probably be overgassed when thick sections are gassed adequately.

The silicate sand's binder level for casting aluminum is typically in the range of 2.0–3.0%, based on the weight of the sand. This high binder level reduces flowability.

Silicate-coated sand is very sensitive to cold temperatures, from the coating and reactivity standpoints (Fig. 8-31).

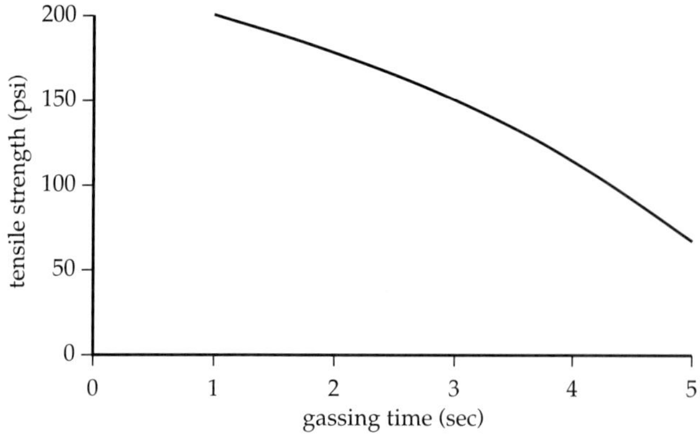

Fig. 8-30. The 48-hour tensile strength of 2.5/1 SiO_2/Na_2O silicate binder at various CO_2 gassing times.

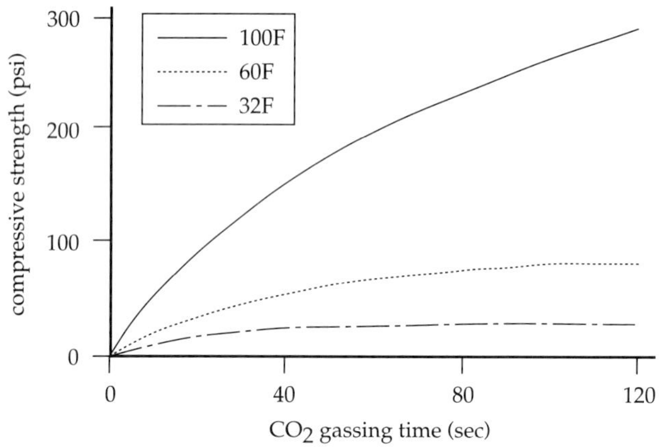

Fig. 8-31. Effects of sand temperature on compressive strength of CO_2 silicate cores with 4% binder.

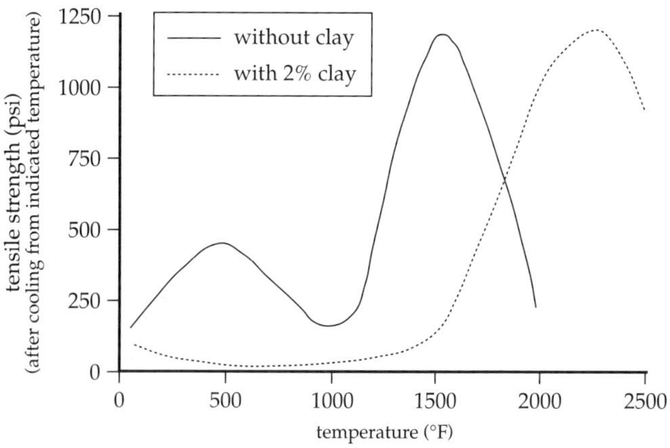

Fig. 8-32. Cold tensile strength after exposure to elevated temperatures and the effects of adding 2% kaolinite-type clay to 4% sodium silicate sand, CO_2 gassed version.

Figure 8-32 illustrates how the strength of the coated sand changes as it is subjected to temperature changes during the metalcasting process.

Figure 8-32 also shows how the addition of various additives can improve shakeout. Additives are rarely added to aluminum core and mold sand, but silicate systems often benefit from this type of addition. In aluminum casting operations, the low metal pouring temperatures and the minimal thermal energy contained in the lightweight metal may not cause the bad shakeout properties normally associated with this system in ferrous casting operations.

Reclamation of the silicate binder often is cited as one of the negative aspects of this system. If the coating is vitrified by exposure to high metal pouring temperatures, the glasslike silicate coating does not lend itself to mechanical or thermal reclamation. It is possible, however, to reclaim it using wet reclamation techniques.

NOBAKE SYSTEMS

The systems in the nobake group that are used for aluminum casting are listed in Table 8-7. All of them are capable of completely curing at ambient temperature. The work time/strip time ratios given in the table indicate how long each system can be worked into the pattern and how long each must remain in the pattern before removal.

The nobake systems use accelerators, various coreactants, and numerous catalyst types and levels to control the rate of the chemical reaction. The rate of cure determines the work time, strip time and core/mold dimensional stability.

The work time (how long after coating the sand with a reactive resin catalyst that the coated sand remains workable, flowable and capable of being made more dense by being compacted) is the most important operator-controllable element in nobake core/mold production.

Any nobake process, as the name implies, is a binder system that will cure at ambient temperature with a predicable and controllable strip rate. This rate is based on the type of resin, the amount and type of catalyst used, and especially the temperature and humidity.

Some binders, such as the acid-cured furans, actually generate water as they cure. This water must evaporate to allow the resin to cure fully. High ambient relative humidity slows the rate of evaporation and extends strip time. Because undercured cores and residual solvents produce excessive amounts of hydrogen gas, care must be taken to optimize the time between stripping and pouring to minimize gas-related defects and to maintain dimensional integrity.

Temperature also is an important controllable factor in nobake systems. Cold sand, cold equipment and low ambient temperatures slow down every part of the core- and moldmaking operations. Conversely, heat can speed up each of the processes. Temperatures under 50F (10C) prevent effective sand coating and necessitate special adjustments to core and mold production. Ideally, process temperature operating parameters for nobake cores and molds should be in the 75–85F (24–29C) range. Lower and upper temperature limits are 50F (10C) and 110F (43C), respectively.

A good rule to keep in mind is the "10° Centigrade or 18° Fahrenheit Rule." This rule states that, for each 18°F (10°C) increase or decrease in temperature, the speed of the chemical reaction takes twice as long to occur (if cooler) or happens twice as fast (if warmer). This means that a reaction with a work time of 10 min and a strip time of 20 min at 72F (22C) would have a 5 min work time and a 10 min strip time at 90F (32C) and a 20 min work time and 40 min strip time at 54F (12C). Temperature is one of the primary things that controls such things as bench life, work time and strip time.

The effects of humidity and temperature control are as important to a nobake binder system as the type and amount of catalyst. In many ways, they are even more important than the effects of the chemicals used in the process.

Silicate/Ester Nobake

Ester-cured silicate nobake binder systems were introduced in 1954. Silicate binder, when used as a nobake system, uses a liquid, ester-type catalyst or hardening agent to about 10 parts of resin (by weight). The liquid catalyst eliminates the effects of under/overgassing, which can be a problem in the silicate/CO_2 system. Thus, nobake silicates have better humidity resistance and core storage capabilities than the gassed version.

As with silicate/CO_2 systems (as was shown in Fig. 8-32), shakeout problems normally associated with this system in ferrous casting production may not be of concern in aluminum casting operations because of lower pouring temperatures. The hot strength and shakeout properties of silicate binders can be altered by the use of various additives mentioned previously in the silicate/CO_2 section.

Relatively poor productivity, humidity resistance problems, burn-in, and difficult shakeout and reclamation are potential problems with this system. However, silicate nobakes are successfully used with all types of sand to successfully make cores and molds in a large number of aluminum foundries.

Phosphate/Metal Oxide Nobake

This inorganic binder system was introduced in 1992. It consists of an acidic, water-soluble, liquid phosphate binder and a powdered metal oxide hardener.

Like other inorganics, this binder suffers from slow cure, low strength and poor humidity resistance. However, it exhibits excellent veining resistance and, like all inorganics, virtually no smoke or odor. The system offers shakeout that is better than any other nobake system, and it is easily reclaimable using existing mechanical systems.

Alkyd Oil/Urethane Nobake

This system dates back to about 1965, and it is the first of the modern nobake binder systems. The alkyd oil/urethane nobake system is the one preferred for producing very large aluminum castings. Its long work time and excellent stripping characteristics allow it to be used for very complex or large castings or with very poor patterns.

This system is comprised of three parts. Part A is an alkyd oil-type resin used in a range of 0.8–1.2% based on sand. Part B is a liquid amine/metallic catalyst adjusted for speed in the range of 2–10% by weight of part A. Part C is a polymeric

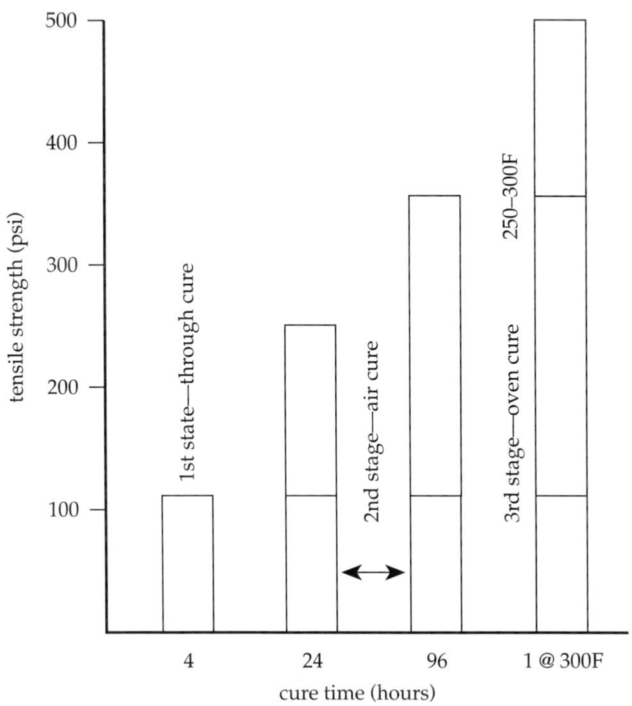

Fig. 8-33. The effect of baking at 300F on oil-urethane system tensile strength development, even after up to 96 hours of room-temperature curing. 1.2% (based on sand) alkyd resin (with 0.6 phr catalyst) + 20% (based on resin) isocyanate → cured resin.

MDI-type isocyanate (the urethane component), which must be used at a ratio of 20–100 parts by weight of part A.

This three-part system can be supplied as a two-part system by blending part A (resin) with part B (catalyst). The unique three-stage curing mechanisms (Fig. 8-33) illustrates how the strength of the system can be controlled in the process. These stages are detailed in the following:

- Phase 1 of the curing process begins when part A reacts with part C to form a urethane-type polymer. (This can be controlled effectively by the part B catalyst.) This stage produces enough strength to strip the sand from the pattern. Because the sand is plastic at this early stage, the stripping characteristics are outstanding.
- The phase 2 cure is a cross-linking of the unsaturated double bonds in the alkyd-type oil. This is the so-called oxidation stage where oxygen reacts with the binder to produce additional strength for up to 72 hours.
- Phase 3 is an optional part of the process where the core or mold is oven-baked to provide maximum physical properties and to minimize gas.

The system is affected negatively by moisture in the sand.

Phenolic Urethane Nobake (PUNB)

This system is standard for making highly accurate, small to medium-sized cores and molds. It is highly productive and offers precise, variable work and strip times. Figure 8-34 compares the typical rates of strength development for the nobake systems and illustrates why phenolic urethane offers such high productivity advantages.

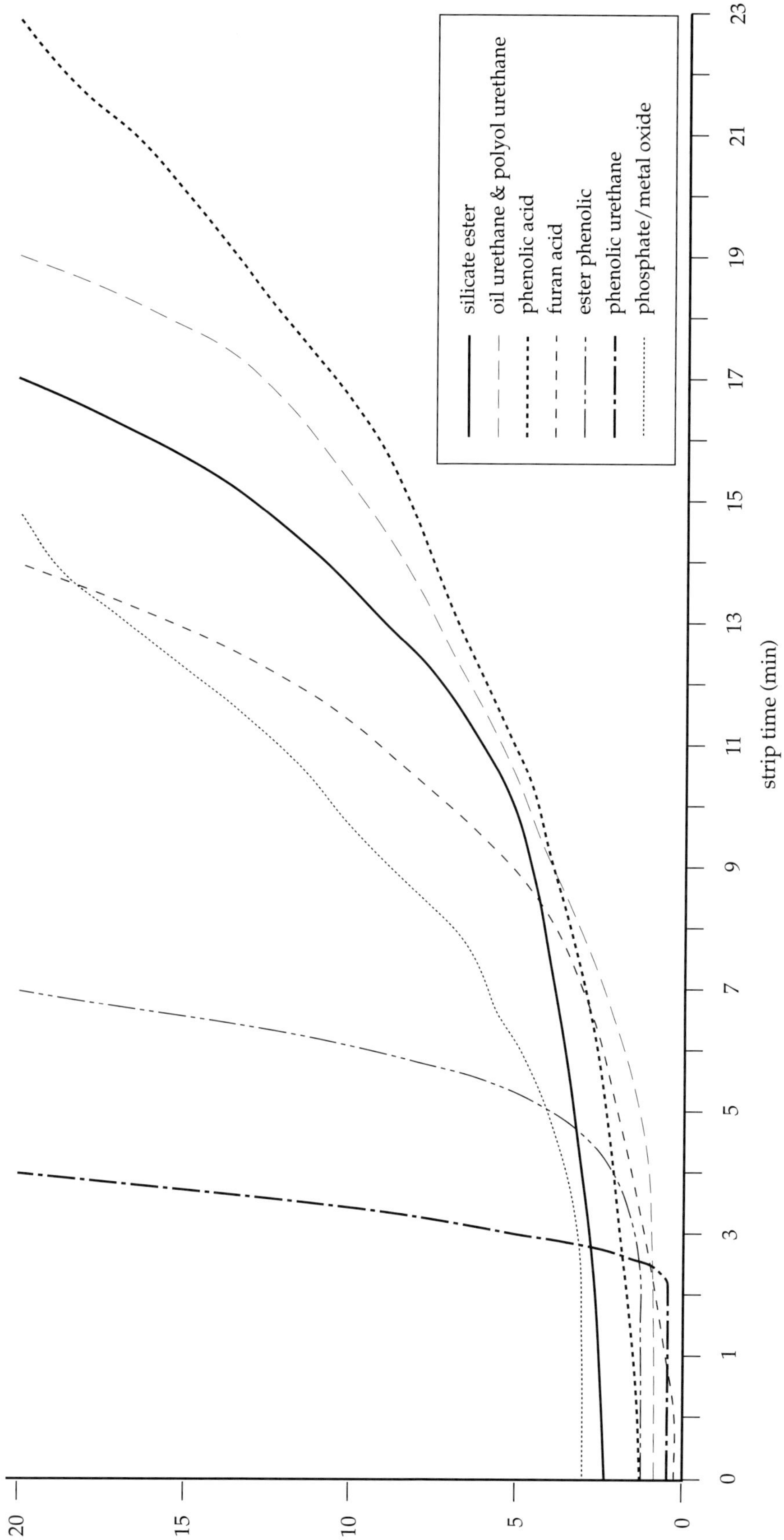

Fig. 8-34. Compressive strength development of nobake binder systems.

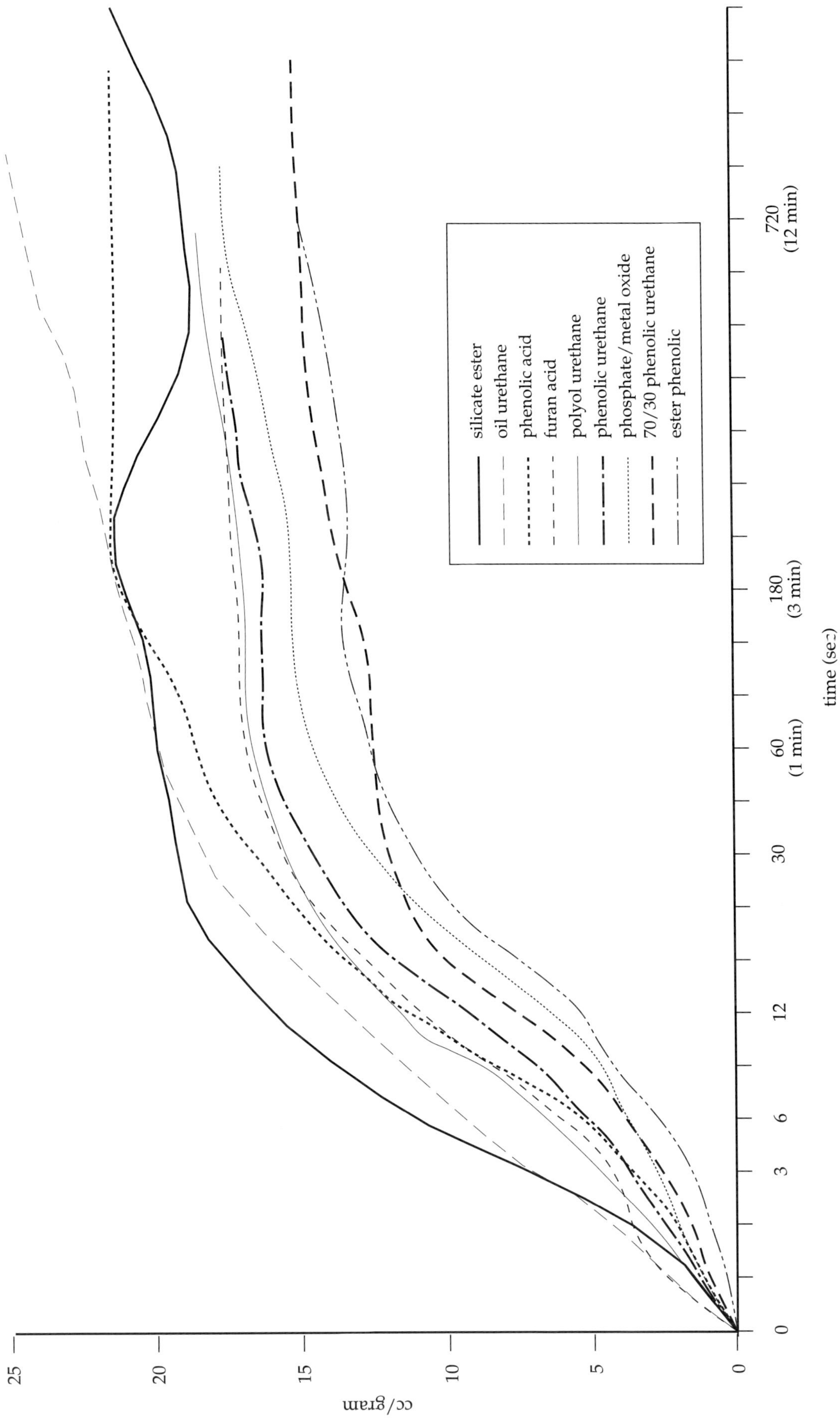

Fig. 8-35. Gas evolution of nobake binder systems at 1800F/982C. (nitrogen atmosphere, no predrying)

The PUNB system was introduced in 1970. The chemical components of the system are similar to the polyol urethane system mentioned previously, with the exception of a liquid catalyst being substituted for the vaporized amine. An advantage of the nobake over the coldbox version is that this system has much better humidity degradation resistance.

This system has benefited from significant innovations in core and moldmaking equipment, along with various improvements to the chemical formulations.

One notable chemical improvement to the PUNB system has resulted in a unique version that uses a part 1:part 2 ratio of 70:30. This enhancement offers a system that requires much less binder to achieve a given strength. This results in less gas, better flowability, less smoke during pouring and shakeout, and greatly improved shakeout characteristics—all desirable properties in aluminum casting applications. Typical binder levels with this system are 0.8–1.0%.

Figure 8-35 compares the low gas evolution of 70/30 phenolic urethane and ester phenolic to other nobake systems.

Polyol Urethane Nobake

This is a three-part system that is similar to the phenolic urethane system. The main difference between these two systems is that, in the polyol urethane nobake system, the resin is formulated to contain a reacted polyol, which gives less hot strength than the phenolic.

The decrease in hot strength helps shakeout in aluminum, which is somewhat of a problem with the urethane nobakes. The polyol urethane system is an excellent alternative to both the alkyd oil/urethane and phenolic urethane systems in aluminum casting operations.

Ester Phenolic Nobake

The ester-cured phenolic nobake binder system, introduced in the U.S. in 1984, is an alkaline-phenolic resin used with an ester coreactant that controls the rates of cure and strip.

The resin is mixed with a variety of ester coreactants at 20–25%, based on the weight of the binder. Speed of the reaction can be adjusted somewhat by increasing the coreactant, but it is better to select an alternative coreactant if big adjustments are needed in strip and cure speeds. Strip times vary from 2–80 min, depending on which coreactant is selected.

This system has a number of processing advantages: no sulfur, low odor, excellent erosion resistance, no lustrous carbon (which eliminates aluminum staining, a frequent problem with the acid-catalyzed systems), low gas evolution, and good expansion defect resistance. There is very little smoke during pouring, and binder cleanup can be accomplished with water. When compared to silicates, this system has better flowability, higher strength and improved handling properties.

The ester-cured phenolic nobake system does not contain furfural alcohol and is not compatible with acid-cured systems. It typically is used at a binder level of 1.5–2.0%, based on the weight of the sand, which is somewhat higher than other organic systems. Acid demand value is not especially important with this system, and it runs well with olivine sand.

During the casting operation, the strength of the ester-cured phenolic core or mold increases dramatically because of a post-cure caused by the heat of the aluminum. This leads to potential shakeout problems. These problems are not as serious in mold shakeout as in core shakeout.

Another drawback to the system is that the resin viscosity increases during storage. This can be a problem in hot weather.

Weak strength has led to handling problems with delicate cores, and shakeout problems have limited the system's use in large-sized core applications. However, the ester phenolic system offers distinct advantages as a molding medium for aluminum casting production because of its relatively clean environmental properties. This system has gained acceptance in Europe, and its use will increase in North American aluminum foundries.

Furan/Acid Nobake

The acid-cured furan nobake binders were introduced in the U.S. market around 1958. Various types of acid catalysts are used to cure the resin (Table 8-9). The type and amount of the acid determines the strip and cure rates. High acid demand value sand, moisture from any source and low temperatures significantly impair the reaction.

The exothermic (heat-producing) chemical reaction between furan and the acid produces water as a by-product. This water slows down the cure rate and makes the core or mold curing process appear to proceed from the side exposed to the air toward the pattern face. This explains the so-called "from the outside in" cure characteristic of furan nobakes.

This system is noted for its satisfactory reactivity with all types of acid catalysts, excellent erosion resistance, good hot strength and outstanding shakeout properties.

The sulfonic acid-cured furan nobake tends to stain aluminum castings a greenish-blue color. This staining does not appear to be a problem in phosphoric acid-cured systems. Because of the staining problem, foundries often are reluctant to use furan nobakes. Regardless, the system often is used for making aluminum castings.

Phenolic/Acid Nobake

The acid-cured phenolic binders were introduced in the early 1960s. Because the phenolic resin is less reactive than the furan, it requires the stronger sulfonic-type acid catalyst for curing. Phenolic nobakes offer a low-cost alternative to furan

Table 8-9.
Acid-Cured Nobake Catalysts

Type of Catalyst	Solvent	Flashpoint	App. wt/gal
BSA	water/alcohol	50/>200	10.5
XSA	—	—	9.7
TSA	—	—	10.1
85% Phosphate	water	>200	13.8
75% Phosphate	—	>200	13.5

nobakes, but the system is slower to cure, and the resin has limited storage stability.

This system is used infrequently for aluminum casting.

CORE AND MOLD TESTING FOR RESIN-BONDED SANDS

When compared to clay-bonded sand, there is relatively little testing done with resin-bonded core and mold sands. The procedures to test for specific core and mold properties are well-established.

The type of test and the upper and lower control limits that are used to measure core and mold quality should vary according to job size, complexity, required handling and storage considerations. A meaningful testing procedure should be used as a value for adjusting core and mold strength to compensate for outside factors that affect performance (such as temperature and relative humidity). Test procedures are given in the *AFS Mold and Core Test Handbook.*

Tensile Strength

Virtually every aluminum foundry core room uses tensile strength as the primary control procedure. The test is quick and easy to run, but its values often are too narrowly and strictly interpreted. In aluminum core- and moldmaking operations, the tensile value for the sand cores and molds generally is far less than that established as the upper control limit for ferrous casting operations.

The tensile strength value of core sand should be just high enough to prevent breakage during handling. Molds should only be strong enough to prevent runouts. It's a mistake to believe that most defects can be solved simply by making everything "rock hard." This approach will likely create other defects and should not be considered a realistic alternative.

Transverse Strength

Transverse resin-bonded sand testing is an excellent alternative to simple tensile testing. There are few times a core or mold is subjected to tensile stress in the casting operation. The transverse test apparatus along with the 1x1x8 in. specimen used in the test is shown in Fig. 8-36.

Tensile and transverse tests using a 1-in. thick specimen provide an overall strength measurement that relates to heavy-section cores. However, a weakness of these tests is that they do not reflect the properties of the core surface. A core surface can become weak or degraded due to humidity (especially if cores are stored in humid conditions) while the interior of the core maintains a high overall strength. Because the core surface is more critical to the casting surface that is the interior of the core, it is important to be able to detect this condition. Strength readings obtained on 1-in. thick cross section specimens can be misleading if the interior has high strength but the surface is weak.

An alternative transverse test uses a thin disk specimen (8 mm thick x 50 mm dia). The thin disk specimen more accurately reflects the properties of the core surface. It also relates better to the strength properties of today's complex core configurations with thin sections.

Loss on Ignition (LOI)

The loss on ignition test is run to determine the percentage of combustible materials present in the raw sand, in the core or mold, or in reclaimed sand. It typically is run by measuring the amount of weight loss that occurs when the sand is exposed to temperatures of 1800F (982C) for 1 hour.

Although it is a useful test for checking incoming sand specifications and the binder content of resin-coated sand, the most useful application for the LOI test is in monitoring the amount of combustible residual materials that are present on reclaimed sand.

Gas Evolution

The rate and amount of gas that is evolved from a resin-coated sand during the aluminum casting operation is one of the most important properties of the core and mold. The gas evolution test measures those properties.

The test is run by taking 0.25 g of a coated sand core/mold section and inserting it into a furnace preheated to 1850F (1010C). As gas is generated from the decomposing coating, its quantity is measured at specific time intervals.

It should be noted that the standard procedure for determining gas evolution has a predrying step that can lead to misleading results. In this procedure, the sample is dried for about an hour at 220–235F (104–113C) before the sample is inserted in the heated furnace. Consequently, any moisture or solvent based material that exists on the sand will be driven off during this "preparation" step.

Water and solvents are significant contributors to gas-related defects because they are likely to decompose into monatomic hydrogen and other monatomic gases that are easily dissolved into molten aluminum. Therefore, it is important to quantify the amount of these present to determine the gas-producing potential of a binder system.

Fig. 8-36. Transverse strength testing apparatus.

The standard evolution test has an alternative procedure that does not predry the specimen. This is the procedure that should be run for resin-bonded core and mold sands.

Screen Distribution—AFS Grain Fineness Number

The test for AFS grain fineness number (GFN) measures the relative amount of the coarse, medium and fine sand grains present in the sand. GFN is an excellent test for monitoring reclaimed sand because it can quantify how much gas-producing "fines" is present in the screen distribution.

Because it is nearly impossible to obtain a representative sample of an incoming shipment of sand to run the screen distribution test, it is impractical to run the test in this situation. A sample obtained by the supplier during loading and sent with the bulk load can be tested.

Acid Demand Value (ADV)

This is a test used to determine the acid-soluble elements that are present in the sand. It also can be an indicator of impurities in the sand.

The ADV test is especially useful in predicting how a sand will work with the catalyst materials of the various binder systems. High ADV sands minimize the bench life of phenolic urethane coldbox systems. High ADV sands also require neutralization of the alkaline materials that are present by the acid catalyst in furan and phenolic nobakes.

ADV is a good test to run on a sample of incoming sand. The sample should be obtained as soon as the material begins to unload. This could give an indication of some incompatible material that might have been shipped in the container during the previous load.

Permeability

This test measures the ability of the coated sand to allow gas to pass through it. Permeability is measured with the device shown in Fig. 8-37. As shown in Fig. 8-38, a well-compacted, 51 GFN sand will have a permeability of about 150. That of finer sand will be 100 or less.

Cores and molds for aluminum casting should have as high a permeability as practical, consistent with the requirements for surface finish and minimal gas-related defects.

Sand Density

Sand density is one of the most important properties that can be measured to indicate core and mold quality. Because aluminum wets the surface of the core and mold more readily than ferrous metals, it becomes a critical consideration for making high-quality aluminum castings. In general, sand density of greater than 1.55 cc/g in a 52 GFN silica sand will give a good casting surface. The higher the density, the more resistant the surface is to penetration by molten aluminum.

A device called a mold quality indicator (MQI) (see Fig. 8-26) determines sand density by measuring the amount of pressure air develops as it passes into a core or mold surface. The fairly typical curve in Fig. 8-39 shows how the MQI value correlates with sand density. The device is equally effective in green sand and resin-bonded sand. However, when the sand's GFN or sand type changes, a new curve must be established to determine the sand density based on the MQI value.

Fig. 8-37. A permeability tester.

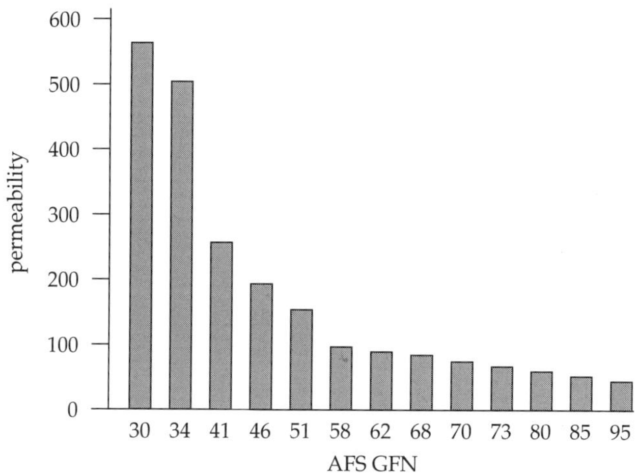

Fig. 8-38. Permeability of round-grain silica sand compared to its AFS grain fineness number.

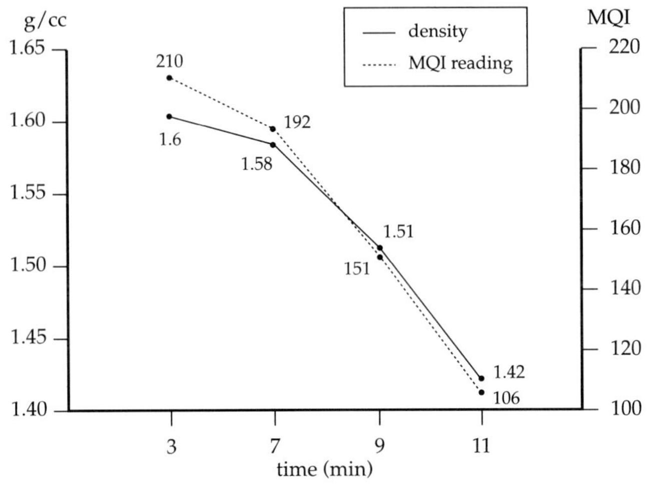

Fig. 8-39. Correlation between density and mold quality indicator (MQI) readings of a 51 GFN PUNB lake sand from the time sand is coated until it is rammed.

SAND FOR PRODUCING MOLDS AND CORES

The silica sand used for aluminum core production typically is a clean, round-grain silica with a three or four screen distribution. The AFS grain fineness number typically ranges from 50–110.

The coarser sands, with the lower GFN, are generally used in high-production casting operations (such as automotive) to minimize binder consumption and gas-related defects. Higher GFN silica sand is used in specialty casting operations to improve surface finish. Sands that contain iron oxide-type contaminants (such as lake and bank sands) should be avoided, but they have been used successfully for producing aluminum castings.

Angular-Shaped Silica Sand

Angular-shaped silica sand or some specialty form of sand (olivine, chromite, carbon-sand or zircon-types) often is used to minimize specific defects or to affect metallurgical properties. An expansion additive (clay or sugar blends) is sometimes added to the round-grain sand to minimize expansion-related defects.

Reclaimed angular sand has very different characteristics from new sand because the grains are rounded during the reclamation process.

Zircon Sand

Zircon sand with 70–110 GFN from Australia and 100–110 GFN from Florida is used for its superior chill properties and its minimal expansion characteristics.

The density of molten aluminum is close to that of zircon sand (about 196 lb/cu ft). This almost eliminates metallostatic buoyancy. The rapid chilling characteristics of zircon (four times that of silica sand) results in excellent mechanical properties on the surface of the metal. This rapid chill also results in outstanding aluminum casting surface properties and improved grain structure within the casting.

Olivine Sand

Olivine sand has lower and more uniform thermal expansion characteristics than silica, and its "uniform" chill characteristics tend to minimize shrinkage defects in aluminum castings. Olivine sand is produced in the Northwestern United States and is the predominant sand used for aluminum casting in that area.

ADDITIVES

Additives are sometimes added to sand to improve surface finish and shakeout and to control certain defects. Kaolin-type clay, fine zircon, carbonates, complex carbohydrates and sugar/clay blends are included for better surface finish. Potassium fluoroborate (KBF_4) is added to minimize gas defects. Lithium carbonate and calcium carbonate also may be added for improved humidity resistance. All of these additives improve shakeout.

Iron oxides should never be added to an aluminum molding sand because of the possibility of forming a thermite reaction ($Fe_2O_3 \rightarrow Al_2O_3 + 2Fe$).

RECLAMATION AND REUSE OF SPENT SAND

Sand reclamation is the removal of the chemical binder coating from the surface of the spent core and mold sand. Reclamation can be performed using mechanical, thermal or water-type processes.

Any serious reclamation project should begin with a study of the sand, binder(s), sand/metal ratio and reclamation process alternatives. After a tentative decision has been made about the method of reclamation (mechanical or thermal), individual vendors should be consulted. Only after a representative sample of sand has gone through the *entire* reclamation process and been recoated with resin at the appropriate binder and catalyst levels a number of times should a final decision be made.

Sand reclamation cannot start at the waste sand pile. It must be an integral part of the foundry process with the design based on consultation with all parties involved—the operating management and personnel, equipment suppliers, the engineering firm, environmental consultants, and relevant permitting agencies.

Mechanical Reclamation and Binder Compatibility

Attempts to describe what happens when various foundry binder systems are mixed together in a common sand reclaimer date back to the 1970s. When core and mold sands that were coated with different binder systems are mixed in the same mechanical reclaimer, unexpected results are likely to occur when the reclaimed sand is recoated with a chemical binder.

The chemistry of a mechanically reclaimed sand surface is no longer that of the original sand. The surface chemistry of the reclaimed sand grains after mechanical reclamation is much like the binder system that the reclaimer tried to remove, and the original physical and chemical nature of the sand has probably been altered dramatically.

Table 8-10.
Compatibility of Chemical Binder Systems Used on High-Purity Silica Sand (No Additives) after Mechanical Reclamation
(0 = incompatible, 10 = compatible)

Binder system used for recoating	Original binder system used to coat sand											
	Shell	**Core Oil**	**Furan & Phenolic Hotbox**[D]	**Warmbox**	**Furan & Phenolic Nobake**[E]	**Oil Urethane Nobake**	**Phenolic Urethane Nobakes & Coldbox**[D]	**Epoxy/ Acrylic SO_2**	**Phenolic Resole Nobake & Coldbox**[F]	**Phenolic Resole/ CO_2**	**Silicate CO_2 & Nobake**[C]	**Inorganic Phosphate Nobake**
Shell	6–10	10	8–10	7–10	5–8	10	10	10	2–8	8	0–5	7
Core Oil	10	10	10	10	10	10	10	10	10	10	10	10
Furan & Phenolic Hotbox	10	10	10	10	10	9	8	10	0	8	5	7
Warmbox	10	10	10	10	10	9	7	10	0	7	1	2
Furan & Phenolic Nobake	5–10	10	10	10	10	9	6	10	0–4	6	5	7
Oil Urethane Nobake	10	10	2–5[A]	1–5	2–5	10	10	10	0–10	8	6	8
Phenolic Urethane Nobakes & Coldbox	7–10	10	2	1–5	1–5	10	10	9	0	8	5	7
Epoxy/Acrylic SO_2	10	10	10	10	10	10	9	10	0–4	6	1–4	3
Phenolic Resole Nobake & Coldbox	10	10	8–10	8–10	10	10	10	10	7–10[B]	8	2	2
Phenolic Resole/CO_2	10	10	4	2	1	10	10	10	0	6	2	2
Silicate CO_2 & Nobake	10	10	2	0	0	10	10	10	0	8	0–8[C]	7
Inorganic Phosphate Nobake	10	10	10	10	10	10	10	10	0	6	7	10

A—Needs extra catalyst B—Needs to have a high percantage of new sand added and has third part additive to help reclamation C—May be difficult to particulate. Recycling accumulates soda and changes SiO_2/Na_2O ratio
D—Might build up nitrogen E—Might build up sulfur F—Will build up potassium and shorten work time

Table 8-10 can be used as a general guide to predict binder system compatibility when recoating mechanically reclaimed sand with the same or different type of binder system. Note that the table is very general. A single chart cannot take into account all variables of mechanical reclamation systems.

Most aluminum foundries that use organically bonded cores and molds can probably meet their requirements with mechanical attrition-types of reclamation. However, simply particulating the sand (returning it to the original grain size and distribution) is not a satisfactory form of reclamation because residual binder will quickly lead to gas problems. Mechanical reclamation benefits from the "thermal embrittlement" of the coating surface caused by exposure to the temperatures of molten metal. Because aluminum has relatively little thermal energy to embrittle the coating, optimum sand:metal ratios and intensive attrition are necessary.

Most sand systems will tolerate a minimum amount of sand coated with a completely incompatible binder system. Some systems do not suffer serious loss of properties in a single recoating cycle but suffer drastically after multiple passes. Also, depending on the physical nature of the base sand, some mechanically reclaimed sands will get stronger, and some will not have the resin-bonded strength of the original sand.

Thermal Reclamation

There are fewer problems with chemical interactions when thermal reclamation is used. Complete thermal reclamation requires about 1,000,000 Btu/ton to thoroughly remove the binder from the silica sand.

Organic binders contribute energy to the calcining process because they burn. Based on the resin content and binder type, 500,000–900,000 Btu/ton will be obtained from the organic resin-bonded sand systems.

Inorganic binders do not burn. They, unlike organic systems, contribute nothing of fuel value for burning the coating off the sand. In addition, any type of inorganic material present on the sand's coating tends to build up on the reclaimed sand, often resulting in curing problems and a significant loss in strength.

Wet Reclamation

Wet reclamation is generally not used because process waste water disposal is extremely expensive. It requires a few hundred gallons of water to wet reclaim a ton of sand that has been coated with a water-soluble binder like silicate.

Reuse of Spent Sand

In the past, resin-coated core and mold sand, after being used for making an aluminum casting has been: 1) metered into clay-bonded molding sand; 2) treated via some reclamation process and recoated with resin; or 3) relegated to a dump site.

A fourth alternative is now available. Studies by many foundry-affiliated organizations have shown that nearly all resin-bonded core/mold sands are nonhazardous by U.S. government standards. Therefore, they can be used as a raw material in diverse applications such as sanitary landfill covers, road beds and road fills. These sands also can be used as one of the ingredients to make bricks and concrete blocks or as a filler for cement, concrete or asphalt.

As waste sand disposal costs continue to escalate, reuse will become a wiser and better economic alternative.

REFRACTORY COATINGS

Refractory coatings are sometimes applied to resin-bonded sand cores and molds for aluminum casting production. When a coating is considered necessary, it usually is applied by dipping certain sections or by brushing the refractory coating onto a small area.

All of the binder systems described in this section benefit from the application of an air-drying (chlorinated solvent-type) or an alcohol light-off (methyl or ethyl alcohol-containing) refractory coating (Fig. 8-40). However, as environmental regulations become more stringent, it is likely that all refractory coatings will eventually be water-based.

When water-based coatings are used, they should not be applied to the heat-activated systems until the molds or cores have cooled to just above room temperature. In coldbox systems, the coatings should be applied as soon as the core is removed from the tooling, and they should be formulated to contain little (if any) penetrating agent so that there is little penetration into the surface of the core or mold. Nobake systems should not be coated until the cure has been completed. It is best to warm the nobakes before applying the

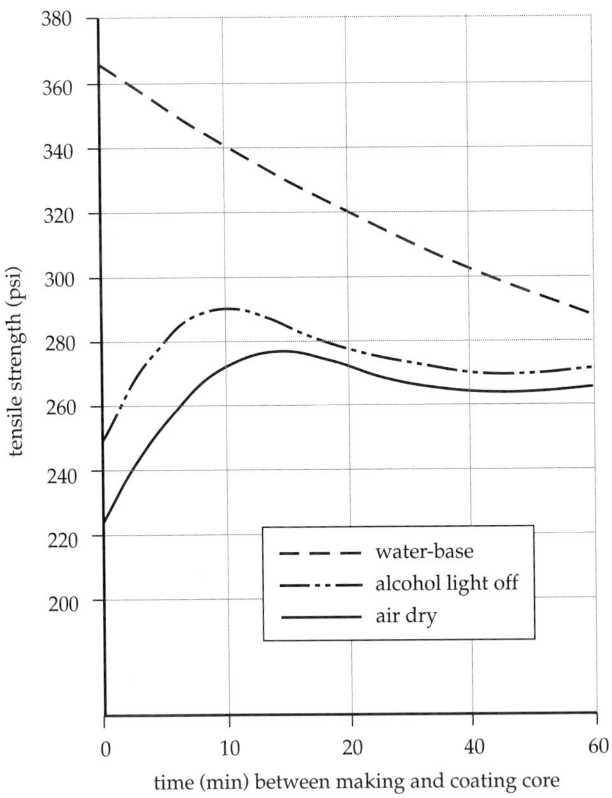

Fig. 8-40. Effects of refractory coating carrier on tensile strengths of organic binder systems.

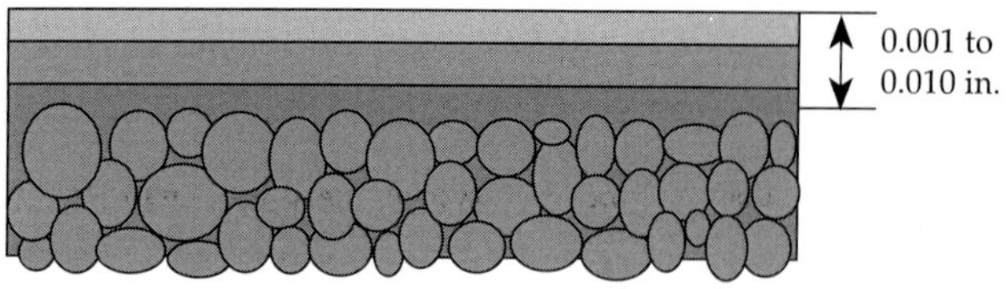

Fig. 8-41. Multiple layers of coatings should be of contrasting colors. A coating layer must be completely dry before the next layer is applied. Total coating thickness should be 1–10 mils above the uppermost sand grains.

coating. Under no circumstances should a refractory coating be applied to cold cores or molds.

In all cases, the applied coating should be 1–10 mils (0.001–0.0010 in.) thick over the surface. The refractory coating must be formulated at the proper Baumé, with just enough wetting agent for it to penetrate no further than 3–4 sand grains into the resin-bonded sand surface. This level of wash penetration anchors the coating and prevents spalling (Fig. 8-41).

VENTING

It is impossible to overvent. Molds should be especially well-vented with vent holes drilled or punched through the cope half of the mold. Cores should have holes drilled into the print areas, wherever practical. Venting of aluminum castings is covered more thoroughly in Chapter 14, Gating and Risering.

EFFECT OF RELATIVE HUMIDITY ON BINDERS

All binder systems lose strength when they are exposed to high relative humidity. How much strength they lose is based on the type of binder and the relative humidity (RH). The higher the RH and the higher the temperature, the greater the effect of RH on the core or mold.

The loss in strength caused by exposure of resin-bonded sand cores and molds to high relative humidity is critical in aluminum casting operations because of the generally low initial core/mold strength used to facilitate shakeout. The most effective countermeasure for humidity degradation is using extra binder. However, aluminum foundries may find that doing so is impractical because of the shakeout problems it creates.

When the upper control strength limit is relatively low and an already weak core loses significant strength for any reason, a serious handling problem results. Figure 8-42 compares the loss in strength for several nobake systems exposed to 100% and 37% relative humidity levels for 24 hours.

Humidity degradation becomes a greater problem as the temperature increases in climates where the relative humidity is high. This factor is true because the air can hold more moisture as the temperature increases and, therefore, can release more moisture as condensation when the air temperature drops.

Core and mold degradation becomes more noticeable in the spring and fall. This happens because, in these seasons, the air is heavily laden with moisture, and the air often cools, losing its ability to retain as much moisture. The excess moisture condenses onto surfaces with high surface areas (like sand molds and cores). When the resin film is thin, as is the case with low resin levels, the moisture may soften and penetrate through the binder film. When the moisture reaches the area between the resin and sand grain surfaces, it can cause adhesive failure. This phenomenon begins on the core/mold surface and progresses inward, partially explaining why thin cores seem so sensitive to humidity degradation.

Humidity degradation reverses somewhat if the RH decreases or if the temperature increases significantly. When this occurs, the cores and molds will regain some of their original strength. When cores are used in green sand molding, they often are sent through a small drying oven just prior to being set in the mold to eliminate the possibility of water from the green sand condensing on the cooler core surface.

The effect of moisture degradation becomes more severe when air pressure is increased, as it is in the blow magazine of a core machine. When using a core blower, "bone dry" blow air must be used to produce strong, storable cores. An

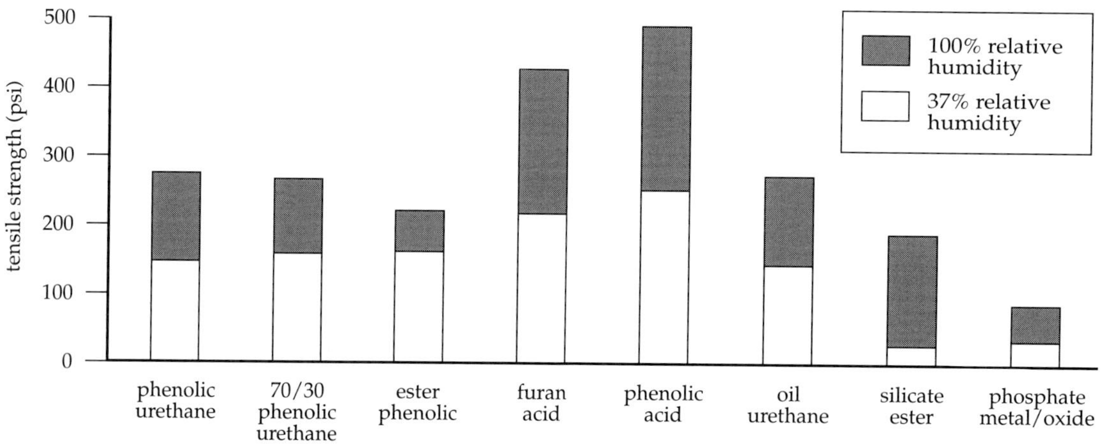

Fig. 8-42. Effect of humidity on tensile strengths. Strengths after 24 hours at 100% relative humidity are compared to those after 24 hours at 37% relative humidity.

atmospheric dew point of –65F (–54C) can be obtained with a regenerative desiccant-type of air dryer. Refrigerant dryers produce a dew point of –25F (–32C), which is considered satisfactory. Deliquescent air dryers (which lower the dew point ten degrees below ambient temperature) do not remove enough moisture from the blow air.[1]

SHAKEOUT

Shakeout is always a problem in aluminum casting production because the molten aluminum has relatively little thermal energy with which to decompose the binder (see Fig. 8-29 for a comparison of shakeout characteristics in aluminum binder systems).

To help shakeout, optimize casting properties and satisfy environmental requirements, the resin-bonded sand and the casting process should satisfy the following requirements:

- Hot strength of the binder should be high enough to prevent expansion defects but low enough to promote adequate shakeout.
- Aluminum "versions" of binder systems should be used when they are available.
- The binder system should generate a minimum amount of gas and should be able to make an acceptable core with minimal binder catalyst.
- Shakeout additives may be added.
- It is helpful to shakeout before the sand has cooled.

CONCLUSIONS

It would be reasonable to assume that the principal reasons for selecting any core- or moldmaking system would be the economics of production and the quality of the castings that could be produced with the process. These certainly are factors in the selection process, but environmental (solid waste and air pollution) and personnel exposure considerations are becoming increasingly important criteria for selecting a binder process.

ACKNOWLEDGMENTS

AFS would like to thank Scott Strobl, AFS/Cast Metals Institute, for writing the section on green sand casting. AFS would also like to thank Paul Carey at Ashland Chemical, Inc. for writing the section on chemically bonded sand systems.

Mr. Carey would like to acknowledge the help of the following individuals: Mark Adamovitz, Jim Archibald, Paula Carey, Tom Dando and Tom Wade, Ashland Chemical; Tom Penko, Foseco, Inc.; Raja Iyer and Dave Jablonski, Acme/Borden Corp.; Bill McKillip, QO Chemicals, Inc.; Ezra Kotzin, AFS; Larry Stahl, GM Powertrain Div.; Tom Piwonka, University of Alabama–Tuscaloosa; Dave Randolph, United Erie; and Dieter Leidel, Tanoak Enterprises.

REFERENCES

1. P. Carey; "The Effects of Humidity on Modern Core Processes," *Modern Casting*, p 52 (Sep 1981).

BIBLIOGRAPHY

P.M. Bralower; "The Use and Misuse of Green Sand Additives, Part 2," *Modern Casting*, pp 46–48 (Oct 1988).

H.W. Dietert; "Processing Molding Sand," *AFS Transactions*, paper 54-98 (Hoyt Lecture) (1954).

H.W. Dietert, A.L. Graham, J.S. Schumacher; "How Mulling Time Affects Sand Properties," *Foundry*, pp 42–47 (July 1971).

H.W. Dietert, A.L. Graham, J.S. Schumacher; "A Sand Control Program that Saves Castings," *AFS Transactions*, pp 329–342 (1974).

R.W. Heine, C.R. Loper, Jr., P.C. Rosenthal; *Principles of Metal Casting*, 2nd ed, McGraw-Hill, New York (1967).

F. Hofmann, H.W. Dietert; "Compactibility Testing, A New Approach in Sand Research," H.W. Dietert Co., Detroit, Mich. (circa 1970).

D. Hoyt; "Back to the Basics of Silica Sand," *Modern Casting*, pp 24–27, (Sep 1987).

W.G. Lawrence; "Factors Influencing Molding Sand Properties," *Foundry*, pp 61–66 (Oct 1961).

R.S. Lee, AFS Committe 80A-2; "Sand System Functions: Requirements and Specifications," *AFS Transactions* (1980).

W.B. Parkes; *Clay-Bonded Foundry Sand*, Applied Science Publishers, Ltd., London (1971).

L.J. Pedicini; "Control of Your Molding Environment Begins with Sand Testing," *Modern Casting*, pp 36–38 (Sep 1988).

L.J. Pedicini, M.B. Mizzi-Krysiak; "Sand Testing and Sand System Design," *AFS Transactions*, pp 373–387 (1989).

C.A. Sanders; *Foundry Sand Practice*, 6th ed, American Colloid Co., Skokie, Ill. (1973).

"Crystal Chemistry of the Clay Minerals," State University of New York, College of Ceramics at Alfred Univ., Ceramic Research Dept., monthly progress report No. 269, vol 23, No. 11 (Nov 1958).

Green Sand Additives: Properties, Applications, and Effects of Misuse, Including Indicative Test Results; AFS, Des Plaines, Ill. (1987).

Principles of Sand Control, Cast Metals Institute, Des Plaines, Ill. (1969).

Processing and Controlling Clay-Bonded Foundry Sands, text course 386, AFS, Des Plaines, Ill. (1971).

Sand Control, H.W. Dietert Co., Detroit, Mich. (1950).

"Simplified Control of Bentonite Bonded Foundry Sands," AIMCOR, Deerfield, Ill. (1987).

J.M. Middleton, Steel Castings Research and Trade Assn.; "Alternatives to Silica for Mould Production," pp 5-1 to 5-19.

R.W. Heine, E.H. King, J.S. Schumacher; "How to Determine Moisture Requirements of Molding Sands," pp 118–122.

Chapter 9

THE SHELL PROCESS

INTRODUCTION

The shell process implements a highly mechanized sand binding system for producing permeable molds and cores from dry sands bonded with phenolic resin (Fig. 9-1 and 9-2). When the shell process is used in aluminum casting, 95% of the time it is used to produce cores; shell molds are used infrequently. With this process, castings with an excellent surface finish—often superior to that of green sand castings—can be produced, and good dimensional control is obtainable. The shells vary from ⅛–½ in. in thickness.

When the process was introduced to the foundry, only a dry-mix method was known. Mulling or mixing equipment was used for dry-mixing the sand with a thermosetting powdered phenolic (phenol formaldehyde) resin. With the introduction of resin-coated sand, which made the production of blown cores possible, changes in the mixing equipment were made. Most of the larger users of the shell process now prepare resin-coated sand. In addition, a number of sand-coating companies now compound and supply resin-coated sands to foundries that lack sand preparation equipment.

Although shapes with intricate cores can be cast in shell molds, there are limitations on shell design, just as there are limitations in every process. As with any casting process, foundry engineers should be consulted before quotations for shell mold castings are requested.

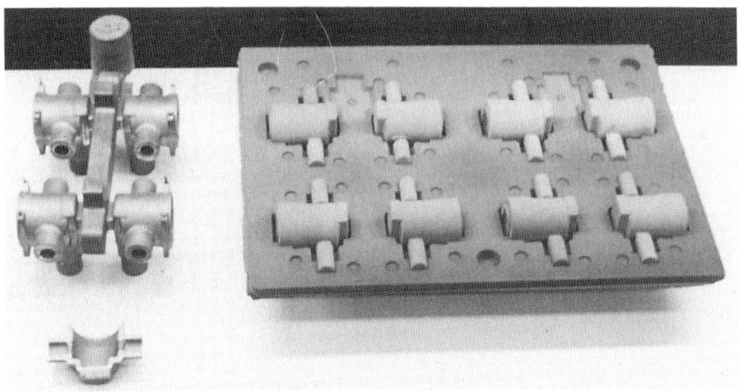

Fig. 9-1. The drag of a shell mold is shown with shell cores in place. Castings produced in a similar mold are shown on the left. [Photo courtesy of Acme/Borden Corp.]

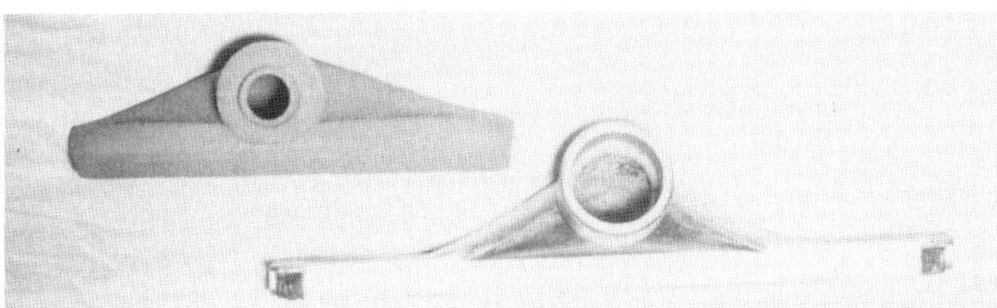

Fig. 9-2. A casting is shown (right) with an example of the shell core that was used to form the interior. [Photo courtesy of Acme/Borden Corp.]

Advantages

Two major advantages of shell molds and cores over conventional green sand molds and cores are their adaptability to mechanization and the reduction in machining costs possible because of greatly improved dimensional control.

Automatic cycling allows rapid production. Molders do not need to be highly skilled to produce excellent cores and molds, and core machines do not have to be cleaned out at the end of shift operations. The process also produces a good surface finish. Aluminum casting surface finishes range between 125 and 300 rms. Casting cleanup is simplified.

Castings with great detail can be produced. The required fillet radius is small, and fairly close tolerances can be held. Pattern reproducibility is consistently good. Thin sections can be cast with ease, and smaller draft angles are required.

The lightweight, self-venting shell cores often can be made an integral part of the mold. No core wash is needed, and conventional core dryers are eliminated. Less core pasting is required.

Shell molds and cores, because they are unaffected by atmospheric conditions, can be stored indefinitely, and the coated sand itself has an infinite shelf life. Precoated sand is available readily on a bag or bulk basis. Sand shows good breakdown ability after casting. Considerably less sand is required per pound of casting weight, and casting yield per pound of melt may be higher.

Disadvantages

Disadvantages of the shell process include expensive raw materials, high corebox costs and time lost for corebox changes. Also, cores removed at high temperature can result in core distortion, and heat retention properties can result in soft metal areas in the casting.

Comparison with Other Processes

A shell mold is more expensive to produce than is a green sand mold of similar design and size. This is primarily because of the higher cost of the binder used in the shell mold process and because the binder only functions for one coremaking (molding) cycle. After casting, sand must be processed through a reclaimer and recoated before it can be reused.

However, higher costs often are offset by the unique ability of this process to produce hollow cores (Fig. 9-3). Most shell cores weigh from ¼–⅓ of the weight of solid cores, making the process economically competitive with most wet sand core systems. When cores cannot be hollowed out because of small section size, wet sand processes become increasingly attractive and competitive.

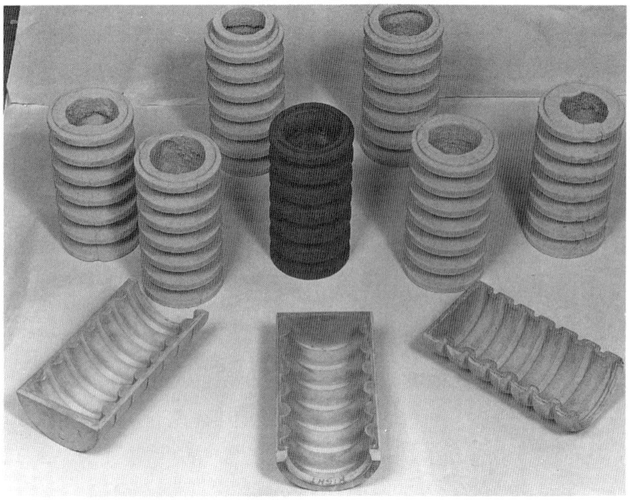

Fig. 9-3. Shell cores made with different sands are shown. Cross sections of castings are shown in the foreground.

Shell mold production costs can be held to a minimum by the practical application of the following rules:

- Choose as coarse a sand as possible because coarse sands require less binder. Usually, many grades of sand within a specific range will produce a finish that meets specifications.
- Use no more binder than necessary.
- Thick molds use more sand but don't make better castings; therefore, the molds should be made as thin as is practical. Shells tend to vary from ⅛–½ in. in thickness.
- Because unproductive areas on molds cost as much as productive areas, the patterns should be designed so that unproductive areas are held to a minimum.

Disadvantages of the shell process compared to other casting processes include the following:

- Compared with green sand molds, pattern cost for shell molding is greater.
- Shell mold castings are more expensive per piece compared to die or permanent mold castings. However, pattern costs are lower.
- Finishes and tolerances are not as good as they are for die castings, but they are generally better than those of green sand castings and are equal to or better than those of permanent mold castings.
- Short runs of aluminum castings are not always economically feasible for shell molding. This method is most suitable for medium to large production runs.

COREBOXES AND PATTERNS

Surface smoothness and tolerances of a casting can be no better than the surface of the shell mold or core. Therefore, to see the benefits of the shell process, coreboxes and patterns must be accurately and carefully built.

Coreboxes and patterns for shell molding must be made of metal because they will be heated to 450–600F (232–316C) and cooled in each molding and coremaking cycle. The type of metal selected will depend on the length of service anticipated. For long production runs, patterns are usually machined of cast iron or steel.

Usually, it is not necessary to build new coreboxes to produce shell cores. Practically any metal box that can be blown and emptied through one end (ordinarily the core print) may be used (Fig. 9-4). Coreboxes may be made of aluminum for short or intermediate runs. Iron or brass is used for high-volume production. Coreboxes should be solid or have a wall thickness of at least ½ in. to retain and transmit heat. They should be equipped with hardened steel guide pins and bushings to ensure positive alignment.

Aluminum patterns are used for short runs, but they tend to warp and grow when exposed to high operating temperatures. When close tolerances are not required, and shell molds are more desirable than sand ones for better consistency of detail and surface finish, aluminum patterns are chosen because they can be produced more economically.

Whenever possible, the cope and drag sections of a mold are machined on the same face of the pattern plate (Fig. 9-5). Each shell will have a cope and drag section, and two shells assembled will form two multiple-cavity molds. In one method, the cope and drag sections are machined on one plate with a V-shaped groove dividing them. The shell mold that is produced is broken easily along the groove, and the two shell sections are booked to make the complete mold.

Machined surfaces of the finished pattern must be smooth. Deep tool marks, scratches, pinholes, undercuts, screw holes and other surface irregularities will impede removal of the shell mold. Repairs cannot be made with wax, patternmaker's putty, solders or other similar materials because their melting points are much lower than the operating temperatures (450–600F, 232–316C) used in the shell process.

Fig. 9-4. A vertically parted shell corebox.

Fig. 9-5. A pattern that has cope and drag on the same plate.

The vertical sections of the patterns should have draft to facilitate ejection of the mold from the pattern. Many designers require a draft of 1° maximum be placed on all surfaces normal to the parting line. A draft of ½° (0.009 in./in.) is usually sufficient. All draft allowances are in addition to the required tolerances or stock allowances for finishing.

Faster, more efficient and more economical production can be achieved if the outer edges of the pattern have a bordering metal frame approximately ¼ in. deep. This framing device should be built as part of the pattern. The inner edges of the frame should be tapered approximately 10°. When the pattern is placed on the dump box, the inner edges of the frame should align with the inner wall surfaces of the dump box.

The purpose of framing a pattern is to provide a cavity in which the shell can build up during the dwell period. If a frame is not used, the shell will build up on the pattern and become embedded in the dump box to a depth equal to its thickness. On removal of the pattern, the edges of the shell tend to hang up on the inside of the dump box ,causing falloff and peel.

Before determining the positions of individual patterns on a base plate, a decision must be made as to whether the mold will be poured in the horizontal or the vertical position. To avoid out-of-round castings, certain round patterns and cylindrical sections should be arranged so the mold parting line parallels the diameter of the sections wherever possible.

Buttons or locators for aligning mold halves must be built into the pattern so that the cope and drag sections will fit perfectly during mold assembly. When the mold has cores, it may be possible to use the core prints to align mold sections.

Shrinkage Allowance

Standard shrink rules apply for most shell mold patterns for small castings and for thin sections in which the molten metal will solidify quickly. For the larger castings, because shrinkage may be less than normally expected, the standard shrink rule should be revised.

Usually the shell mold will shrink on cooling to about the same degree that the pattern expanded during heating. Since mold shrinkage compensates for pattern expansion, it is not necessary to make allowance for expansion when making the pattern. The cooled shell mold will have approximately the same dimensions as the cold metal pattern.

Ejection

Because the shell fits quite snugly on the pattern, some means must be provided for rapid removal. This is easily accomplished by building a knockout or ejector pin mechanism into the pattern. The pins can be either spring-actuated or part of a separate, mechanically operated ejection system. In either case, they are part of the pattern and should be carefully considered when the pattern is designed. Some guidelines for the design of ejector pin mechanisms are as follows:

- minimum diameter of ejector pins should be ⅜ in.;
- ample pins should be provided, particularly around and in the pattern area;
- pins should be as close to vertical sections as possible;
- if straight (headless) pins are used, they should protrude above the pattern surface at least 0.015 in.;
- on ejection, all pins must actuate at the same time;
- pins must be long enough to raise the shell above the pattern.

The best ejection systems for high-speed shell molding machines consist of straight (headless) pins fastened to a mechanically operated bottom ejector plate. Ejector pins retracted by springs are satisfactory for some applications and, because of their lower cost, are quite popular. Spring-actuated pins must have heads on one end and be mounted with the head on the pattern side and the spring on the back side.

There are disadvantages to the spring-actuated arrangement. Springs often lose their temper. Therefore, it is recommended that 17%Cr, 7%Ni stainless steel be used for ejector springs. Although this material is more expensive than the steel commonly used, it will hold its temper to 750F (399C), which greatly extends the service life of the springs. Sand under the heads must be cleaned out after each cycle, or it will cause nonuniform ejection. Pins with one bearing area will cock, wear and also cause nonuniform ejection.

It is important in any ejection system that all pins start the ejection movement against the shell at the same time. Erratic movement of one or more pins can cause: broken or distorted molds or cores; damaged molds or cores in the pattern area; wear on pattern; and defective molds or cores in the pin area.

SHELL PROCESS EQUIPMENT

Although good shell cores and molds are being made on automatic, single-station machines (Fig. 9-6), because of the

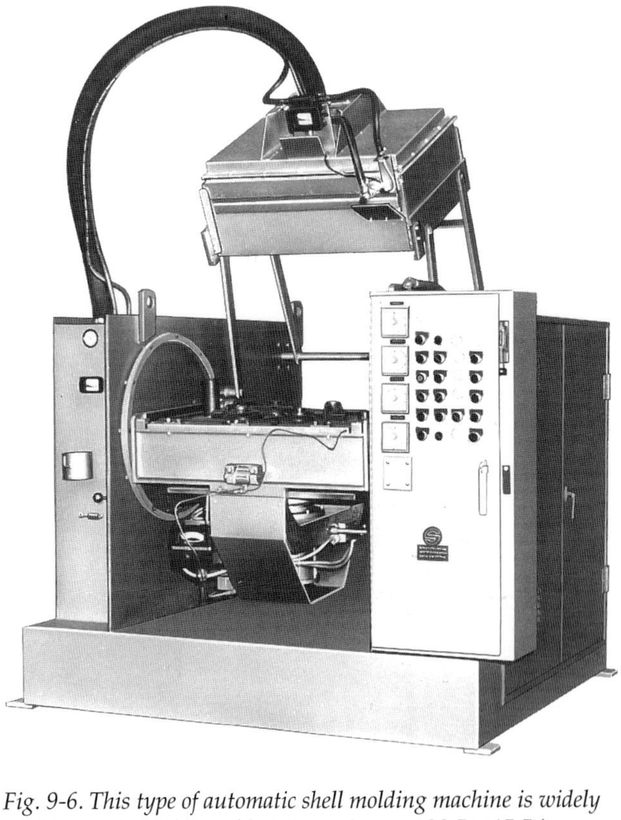

Fig. 9-6. This type of automatic shell molding machine is widely used throughout the world. Pattern size: two 20.5 × 15.5 in. plates or one 20.5 × 31.5 in. plate.

Fig. 9-7. A single-station, fully automatic shell molding machine is capable of producing accurate molds in short cycle times. [Photo courtesy of Dependable Foundry Equipment Co.]

simplicity of the process and its easy adaptability to mechanization, manufacturers have developed several types of high-speed, semiautomatic and fully automatic molding machines with innumerable handling accessories (Fig. 9-7). Production rates on automatic machines vary from 30–200 full molds per hour, depending on the size of the pattern plate and the number of stations used.

Completely automatic shell core blowers (Fig. 9-8) are available in many sizes and styles. Most core machines operate at pressures of 20–80 psi. Some machines function without an

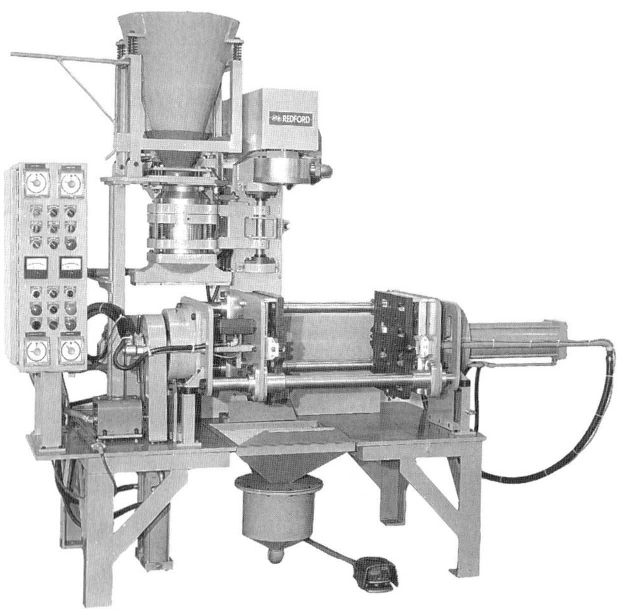

Fig. 9-8. This automatic shell core machine has a blow capacity of 40 lb. [Photo courtesy of Dependable Foundry Equipment Co.]

Fig. 9-9. A fully automatic shell core blower uses the bottom blowing method. Unit features gas-fired heaters; a built-in ejector mechanism; and automatic timing of investing, vibrating, draining and curing operations.

operator—the only labor required is an occasional spraying of coreboxes and sorting of cores (Fig. 9-9). Automatic machines are adaptable to about 95% of shell molding applications, and because of their adjustable cycles, peak production can be obtained with semi-skilled operators. Units are generally air-operated with electrical controls.

Semiautomatic machines also are used. Manual actions on some units are limited to closing the door and operating two levers to engage lock pins and to move the ram.

Multiple-purpose units produce shell cores as well as shell molds. Cores of excellent quality can be produced on manual, semiautomatic and automatic machines, and the individual steps in the molding cycle do not vary. Production with the automatic models, however, is more economical than manual production.

Shell-sand coating plants can be used when production quantities are large enough to justify the costs of equipment, labor, maintenance, emission control and waste disposal. A mixer is the main component of the plant. Auxiliary equipment includes mechanical aids for loading, heating, aerating and discharging the product. Coating plants are equipped with resin measuring and injection devices. No mixing equipment is necessary if precoated, resin-coated sand is used.

MATERIALS

Casting surfaces are direct reproductions of the mold surfaces. Sand is the major constituent of the shell and, therefore, is a major factor in the production of good shell cores and molds. Binders provide the necessary strength while the heat and water resistance and fillers speed solidification and improve finish. For a detailed discussion of sand and binders, see Chapter 8.

Sand

The sand used in the shell process should be of relatively small particle size and be capable of producing a uniform surface. However, costs must be closely monitored; as the particle size decreases, the percentage of resin binder necessary to produce a shell of a given strength will increase. It is, therefore, most economical to choose the coarsest sand that will produce the desired surface finish.

Sands most generally used fall in the AFS fineness range of 60–160. They must be relatively free of clay, moisture and other foreign matter that will reduce the core or mold's strength. Sands with a high content of foreign matter can be used, but the amount of resin required to produce the desired strength characteristics will have to be increased.

Grain shape has a marked influence on strength characteristics. For a given amount of resin binder, higher strengths can be obtained using round grains instead of angular or subangular grains.

Grain distribution is important because it controls core and mold permeability, strength and density. Grain distribution also affects the finish obtained on a casting. Blending two or more sands to obtain proper distribution is often desirable. Besides silica sand, zircon, olivine and chromite have been used successfully for shell cores and molds.

In the dry-mix method, dry sand must be used to obtain maximum strength characteristics with a minimum amount of resin binder. Moisture will cause lumpy mixes and will weaken shell molds. High moisture content or added water causes the finely powdered resin to ball up into much larger particles, thus reducing its bonding efficiency.

The precautions usually observed when handling small particle sand should be taken. Hoods and exhausts should be used to provide good ventilation, especially near the oven and dump box. Dust masks are an added precaution. The hexamethylenetetramine (hexa) catalyst used in the coating can cause dermatitis in some people.

Binders

Many types of resinous bonding materials are available, but the phenol formaldehydes—or phenolics as they are called in the plastics industry—have been found to be the most satisfactory shell process binders. They alone have the strength, the heat and water resistance, and the versatility of cure and flow properties required in the shell process. Phenol formaldehydes, synthetic resins produced by chemically combining phenol and formaldehyde, are thermosetting—heat will cure or harden them.

Phenolic resins can be modified in many ways to produce resins with special characteristics for specific purposes. When used in the shell process, these resins go through three major physical changes. First, the resins melt to a liquid. Second, the liquid changes to a rubbery solid. Third, the rubbery solid becomes progressively harder until it reaches its final form, which is hard, strong and practically insoluble in all solvents.

Phenolic resins also are thermosetting—infusible or incapable of reverting to a liquid stage upon further heating or reheating. At approximately 500F (260C), the material will start to decompose slowly. The rate of decomposition is accelerated by increasing temperature and exposure to air.

Choice of resins is important because resins affect the quality of the shells and castings produced. Resins will vary in the speed with which they build up and cure a shell. Those that are too fluid in the molten state or that have a low melting point may cause the shell to peel from the pattern during rotation of the pattern or corebox. Good resin rigidity helps eject the core or mold without distortion.

Fillers

The primary purpose of fillers is to provide increased heat conductivity, thus increasing the speed of metal solidification. Fillers are sometimes added to sand-resin mixes to improve the surface finish of the casting; however, such additions decrease the strength of the shell. Wherever a minimum strength requirement must be met, more resin binder must be used.

PRODUCING SHELL CORES

The shell coremaking process, an important advance in the foundry industry, was the first process capable of curing the core in its corebox. This process offers dimensional precision, low cost, smooth surface, good collapsibility, substantial burnout and high permeability.

Shell cores may be solid or hollow. They are the only dry sand cores that can be blown and cured on a single machine. Because the process uses a dry material, it is excellent for use in a variety of intricate cores or molds (Fig. 9-10). Shell cores are produced in much the same way as conventional cores except that: 1) the corebox is heated; 2) phenolic resin is used instead of conventional core binders; 3) resin-coated sand is used; 4) shell cores are usually hollow; 5) cores are cured and ready for setting when removed from the corebox; and 6) cores are produced without using drier plates or ovens.

Resin-coated sand is dropped or blown (Figs. 9-11 and 9-12) into a vertically parted, heated, metal corebox where the resin softens, becomes sticky and binds the sand grains. When this reaction has proceeded long enough to penetrate from ⅛–½ in. into the sand (a few seconds), the corebox is inverted, and the unheated sand drains out of the box. This sand is then available for recycling.

Fig. 9-10. Shell molds and cores can be produced in a variety of shapes and sizes. [Photo courtesy of Acme/Borden Corp.]

Fig. 9-11. The supply feed hopper of this core machine is in a position to deposit sand into the open sand magazine. [Photo courtesy of Beardsley & Piper.]

Fig. 9-13. A large shell core is removed from a core machine. [Photo courtesy of Beardsley & Piper.]

Fig. 9-12. A core blowing machine (without guards) in the blow position. [Photo courtesy of Beardsley & Piper.]

Further heating in the box for a minute or two completes the hardening of the two-stage resin. The box opens automatically, and ejection pins release the completely cured core. The result is a hollow core with an extremely smooth surface that is ready to set in the mold.

Shell cores are used in green sand, permanent and shell molds. Excellent permeability is obtained even with fine sands. Intricate details of interior cavities in the casting are readily reproduced. The core, which is fully cured and rigid when removed from the corebox, duplicates the exact contours of the box.

In the case of large cores (Fig. 9-13), substantial material savings are realized through the use of the shell process. A hollow, 3-in. diameter shell core, for example, weighs less than one-sixth of a conventional solid core of the same size.

Surfaces of coreboxes should be smooth and free from undercuts. Common venting methods are used. In some split boxes, a few V-shaped scratch vents approximately 0.010 in. deep along the parting line are sufficient. If it is necessary to provide venting in the box itself, fine screened or slotted vents (about 50 mesh) can be used.

Process Operating Temperatures

Corebox operating temperatures of 475–500F (246–260C) are optimal; the lower temperatures are best suited to the production of complex cores with narrow sections. Higher temperatures are used for larger cores with simple contours.

Coreboxes are usually heated by gas with rows of adjustable burners arranged so the pattern of heating can be adjusted to the needs of the corebox— more heat where the core is heavy, less in lighter sections. Electrical heating elements also may be used, but gas is more versatile and provides quicker response.

Because of the high temperatures, curing cycles range from 15–25 seconds, depending upon the size of cores. A resin content of 1–2.5% is generally sufficient for cores used in aluminum castings.

Ejection

Ejector pins, as used in shell molding, are not always necessary in semiautomatic blowers. With properly constructed coreboxes, cores are readily withdrawn by hand.

To prevent distortion, coreboxes should be opened in a straight line perpendicular to the parting line, and cores should not be removed until fully cured (Figs. 9-14 and 9-15). If the core has natural draft, hinged boxes are practical.

Fig. 9-14. A shell core is ready to be removed from the corebox. [Photo courtesy of Acme/Borden Corp.]

Fig. 9-15. The shell corebox shown in Fig. 9-14 has been opened and the core removed. [Photo courtesy of Acme/Borden Corp.]

Core Assembly

Intricate cores that cannot be blown or drawn in one piece can be made in two or more parts and assembled with resin adhesives in much the same way as shell molds are assembled. Mold assembly will be covered in more detail later in this chapter.

PRODUCING SHELL MOLDS

The correct amount of coated sand is placed in the dump box or hopper of the molding machine. After the pattern is heated and lubricated, the sand is deposited on the hot pattern and allowed to remain there for a predetermined length of time (longer times result in thicker shells). The resin fluxes during the dwell (investment) period.

The pattern is then inverted, and excess sand falls back into the hopper for use in the next cycle. The pattern, with shell intact, is placed in the curing oven. After curing, the finished shell is removed from the pattern.

On automatic shell molding machines, the coated sand mix is fed automatically, and the feeding action usually is continuous. Investment is made by a straight drop that facilitates deep draws and high projections. Coated sand feeding is handled by manual controls on manually operated molding machines.

Process Operating Temperatures

When producing shell molds, it is important that ideal operating temperatures for the equipment and materials in use be established and maintained. Proper operating temperatures vary with the type of equipment and will be influenced by weight and shape of the pattern; pattern material temperature, room temperature, and sand temperature; dwell time; pattern lubricant; and length of time required for ejection. Individual steps and variable factors should be adjusted to reach, or at least approach, a balanced production cycle.

The shell molder's goal is to achieve the fastest production rate possible while still maintaining high quality in the molds produced. To achieve the highest production rate, it is recommended that pattern temperatures be at least 450F (232C) and the oven temperature at least 800F (427C).

Thermostatic controls will keep the oven temperature constant, but the temperature of the pattern during the cycle will fluctuate over a fairly wide range. The pattern temperature cannot be kept equal to the oven temperature because the pattern loses heat whenever it is out of the oven for loading, lubricating and ejection. A substantial heat loss also occurs when the cold sand-resin mix is dumped on the pattern in the investment phase of the production cycle.

Keeping a relatively even pattern temperature under these conditions is possible only by establishing and maintaining a rhythmic and balanced operating cycle. For example, with the oven temperature at 800F (427C), a loading and unloading time totaling about 30 sec, and a curing period of one minute, the maximum constant pattern temperature that can be maintained is approximately 475–500F (246–260C).

When the production cycle is erratic and is allowed to continue in an unbalanced state, great variations in pattern temperatures can cause one or more of the following defects to occur, along with costly interruptions in production: 1) molds will be distorted and warped; 2) in the same shell mold, there will be areas of undercure and areas of overcure; 3) molds will stick to the pattern; and 4) the pattern temperature may drop so low that it will be necessary to stop the operating cycle to reheat it.

Holding a fairly constant pattern temperature is a major requirement for maintaining equilibrium in the production cycle. If the speed of the production cycle is increased, the pattern loses more heat during out-of-oven periods than it can absorb during the curing period, and the production cycle cannot be kept in a balanced state unless more heat is forced into the pattern. Supplemental heating methods commonly used include adding electric heating elements to the pattern; preheating the pattern in an open gas flame during stripping, cleaning and lubrication; and forcing extra heat into the bottom of the pattern while it is in the curing oven.

Heating the Pattern

Heating the pattern is important in the shell process. Pattern temperatures should be checked by surface pyrometers or other temperature-measuring devices. About 450F (232C) is best for most applications, with variations of no more than 30°F (17°C) from the coldest to the hottest part.

Sources of heat may be gas or oil ovens, gas or electric radiant heaters, or improved radiant heaters with parabolic reflectors. Calrod-type heaters attached to the base of the pattern may be used to supplement the heat supply and maintain better control.

Lubricating the Pattern

After it is heated, the pattern is sprayed with a lubricant to ensure a clean release of the shell from the pattern. Many lubricants have been tried, but silicone has been found to work the best. Brushing or dipping can be used, but spraying a diluted silicone emulsion is the most economical method. A fine, atomized spray is the most satisfactory; otherwise, large drops may roll down, creating uneven distribution. For best production results, the pattern should be lubricated before each molding.

When breaking in a new pattern, lubricant should be used to wet the pattern thoroughly. A recommended procedure for the initial lubrication is to: 1) heat the pattern to a predetermined operating temperature; 2) apply a liberal amount of lubricant and rub it in well with a brush or soft pad; 3) blow off excess lubricant with air and reheat the pattern; and 4) apply a light coating of lubricant and make the first mold.

It may help to clean patterns occasionally with cleaning chemicals or mold abrasives. Blasting with glass beads, ground corn husks, or rice is effective. After a pattern has been cleaned, it should be heated to the operating temperature and lubricated as if it were a new pattern.

Investing

Dwell or investment time is the period the sand-resin mixture remains in contact with the surface of the hot pattern in the investment or dump box. With the pattern firmly clamped to the investment box, the sand-resin mix is dropped onto the hot pattern, where the heat of the pattern causes the powdered resin to flux. This binds the grains of sand together to form the mold.

The thickness of the shell is determined by the length of the dwell and the temperature of the pattern. As an example, at a pattern temperature of 450F (232C) and a dwell time of approximately 15 sec, a shell with a wall thickness of about $^{3}/_{16}$ in. will be produced. When the operating temperature is low and shells to be produced are small, dwell time can be reduced to form thinner shells.

Not only should sufficient sand-resin mix be in the box during investment, but care should be taken to keep the mix at a constant level to assure uniform reproducibility. This is particularly important in deep-draw patterns where it often is advisable to use a vibrator to obtain a more densely packed mix. Louvers built into certain production shell molding machines also help to maintain even distribution of the mix on the pattern by retaining the mix until the investment box has approached its 180-degree inversion, then releasing it. Sliding action of the mix as it falls onto the pattern may be prevented by moving the point of axis in order to get an off-center throwing effect.

Blown Molds

Often it is convenient to "blow" molds on a coremaking machine. With high-production units, especially those with automatic core take-away belts, molds can be produced at least as quickly on these as on molding machines. Blown molds have their backs formed by pattern equipment; hence, the shell thickness is always held to close limits. Molds made in this way are particularly useful for stack-molding.

Curing and Ejection

The shell is not removed from the pattern immediately after investment because the resin binder is not yet thoroughly cured. The shell, while still on the pattern, is cured in a curing oven, where the resin binder is transformed chemically from a fusible (capable of being liquefied or melted) to an infusible state (Fig. 9-16).

The length of cure depends on pattern and oven temperatures, oven heating methods, thickness of the shell, and the resin used. Usually, however, with pattern temperatures from 350–500F (177–260C) and oven temperatures from 600–1000F (316–538C), curing times will range from 20 sec to 2½ min, with 20–40 seconds being most common.

After curing, the shell and the pattern are taken from the oven (Fig. 9-17), and the shell is freed from the pattern by operating the ejector pin mechanism. Unless cleaning is required, the pattern is ready for the next cycle.

Assembling the Mold

Although shell molding lends itself to mechanization, Joining the two shell halves sometimes causes a bottleneck. Adhesives and clips are commonly used to provide intimate contact of cope and drag sections. For large molds where the metallostatic pressure tends to distort the mold, adhesives are more effective than clips.

Most shell molding foundries glue the cope and drag shells together instead of using mechanical fasteners. A drawback

Fig. 9-16. A shell mold is ready to be removed from a shell molding machine.

Fig. 9-17. A shell mold, still attached to the pattern, is removed from the curing oven.

Fig. 9-18. Cores are inserted in the mold, glue is applied, and the cope half is lowered onto the drag in preparation for inserting the mold into the bonding machine.

to the use of mechanical fasteners is the amount of manual labor required. For smaller shells and in horizontal pouring, the need for a mold backup is usually eliminated when adhesives are used. Due to the uniform manner of joining plus the fact that human error is largely eliminated, greater consistency in tolerances across the parting line is possible when shells are bonded.

After the adhesive is applied to the drag half (which is mounted on a special plate in the bonding machine), the mating hot cope is then placed over it, and pressure is applied (Figs. 9-18 and 9-19). If cores are used, they are inserted into the mold before it is assembled.

One type of machine uses silicone rubber diaphragms on either side of the assembled shells. Another machine exerts pressure by using pins with flexible spring tips located so that pressure can be applied where needed. In changing shells, a template is used to obtain proper pin location, thereby decreasing setup time.

A satisfactory adhesive for bonding shell mold halves is a phenolic resin similar to that used as the binder in the mold itself. Paste, powder or a dispersion in core oil may be used. One advantage of the paste is that a type that cures at room temperature may be used on cold shells, whereas the powdered resins are thermosetting. A disadvantage of pastes is that they generally require hand application. A powder or dispersion is applied to a hot shell half by means of a tray container with holes that permit the resin to fall into small beads when vibrated.

Where there is sufficient metallostatic pressure to require a backup material, metal shot, gravel or molding sand can be used (Fig. 9-20). When packed, shot and gravel are more permeable than sand; they also pack and handle more easily. Backup materials should be vibrated during loading to ensure good density.

Fig. 9-19. The pneumatically operated gluing press closes and presses the mold halves together while the glue dries.

Fig. 9-20. Molds that are ready to be poured are covered with backing sand to catch molten metal in case of a spill.

Pouring the Metal

Shell molding melting practices call for the same methods as those used in other methods of casting. Pouring temperatures, however, can be reduced by as much as 150°F (83°C) because of the higher insulating properties, smoother surface, and negligible moisture content of shell molds. Pouring temperatures usually range from 100–300°F (56–167°C) above the liquidus temperature of the alloy.

If fins form at the parting line when the mold is poured, one or more of the following may be the cause: 1) improperly assembled mold; 2) inadequate mold backing; 3) improperly bonded mold halves; 4) mold surfaces at parting line too rough to mate tightly, a condition which results from a dirty or damaged pattern; or 5) improperly fitting cores which prevent mold closure.

Shakeout

When casting thin sections of aluminum with internal cores, shakeout difficulties may be reduced by using the minimum permissible resin content in the shell mix. The heat liberated during the solidification of the casting is sufficient to burn out the bond of the shell adjacent to the casting, thus making shakeout easier.

Sand may be recovered from burned-out molds, broken molds, unused molds, and unused sand-resin mixes. However, it should be determined whether the cost of recovery exceeds the cost of new sand. Burning at a high temperature will remove the resin binder. Recovered sand must be treated as new, clean sand, and the same amount of resin is required as when making a new dry mixture.

TESTING AND PROCESS CONTROL

The shell process has evolved into an accepted production method for making both shell molds and shell cores. To achieve optimum production efficiency and high-quality products, it is important to control the basic materials and the working mix. Process control should include testing in the laboratory and production testing at the mixing station.

The development of test equipment especially designed for the shell process has kept pace with the evolution of the process itself. In fact, the prolific growth has resulted in some confusion for the individual selecting equipment for shell casting. Foundry test personnel should evaluate their particular needs and the effectiveness of the quality control program before selecting test equipment.

Shell process tests may be divided into groups based on the material tested and the time required to complete the tests.

Production Control Tests

Tests on the sand mix that can be completed within a few minutes are of more value during production than are laboratory tests. Information from these tests is available quickly enough to adjust the mix before the material advances too far in the actual production process. The tests include: cold shell tensile strength; hot shell tensile strength; and stick point or melt point.

Laboratory Control Tests

Sand Mix

Tests on the prepared mix that require more than five minutes to complete are called laboratory control tests. When these are used to supplement the production control tests, a better understanding of the sand mix and its influence on the final castings can be made. Correlation between these tests and any casting defects provides the key for improving the sand mix and reducing casting losses and cost.

These types of tests include the following:

- shell tensile strength, or shell transverse strength and deflection;
- shell permeability;
- burnoff or total loss on ignition.

Basic Materials

Other tests can be made to check the basic materials before they are used in the mix. These tests, which are valuable in evaluating new materials as well as for controlling the quality of incoming materials, include the following:

- screen analysis;
- specific surface and coefficient of angularity;
- sintering point.

The physical characteristics of shell molds do not indicate the quality to expect in the casting. However, testing for the physical characteristics of the prepared sand provides a basis for determining daily uniformity. Testing and controlling the physical characteristics enables the moldmaker to control the process effectively, resulting in a minimal percentage of scrap.

CASTING DEFECTS

Certain common defects may occur in shells and shell castings. Table 9-1 lists common defects and their likely causes.

To explain surface roughness defects in shell molds, it is important that the fundamental difference between shell molds and green sand molds be understood. Basically, a green sand mold is composed of sand, clay and water. When metal is poured into it, the water provides rapid cooling. A solid metal envelope forms and helps hold the molten metal in place. Meanwhile, the water dries out, and the clay shrinks. Silica sand particles pick up heat and expand into the space vacated by the shrinking clay. As a result, there is considerable movement of materials without any overall change.

A shell is composed of sand and resin binder. Due to the absence of water, the shell becomes a good insulator. When metal is poured into it, heat is transferred at a slower rate than in a green sand mold. Consequently, the metal does not solidify as rapidly, particularly on the surface. It is assumed that the resin binder in the face of the mold or core and in direct contact with molten metal is decomposed almost instantly. A carbon bond is formed. Although weak, this bond still is capable of holding the sand particles together.

As heat is carried through the shell, sand particles become heated and start to expand. There is nothing present in the shell capable of absorbing this expansion. The result is an overall dimensional change. (A shell mold placed in an oven at 600F (316C) for 3 min will expand 0.005 in./in. In the presence of molten gray iron, shell molds have expanded 0.012 in./in.)

The surface roughness on some shell mold castings is caused by mold movement during solidification. A good finish on a casting made in a shell mold is obtained only when the metal solidifies in close contact with the shell surface. This contact with the mold surface can be obtained by making the metal solidify before mold expansion begins and/ or by feeding the cavity during the mold expansion period.

In some small and medium-sized castings, lower pouring temperatures can be used to obtain metal solidification before mold expansion begins. Generally, the greater volume of metal needed to fill a large cavity will cause the expansion of the mold to be completed before the cavity is completely filled. The mass will stay fluid during expansion, with risers feeding the cavity as more metal is required. Proper gates and risers are necessary, regardless of the size of the casting.

ACKNOWLEDGMENT

AFS would like to thank Wayne Rossbacher, Acme/Borden Corp. for his help with this chapter.

Table 9-1.
Shell Process Defects Checklist
(Table courtesy of Acme/Borden Corp.)

Foundry Operating Problems		
Defect	**Possible Causes**	**Suggested Action**
Peelback	Hot or cold spots in corebox or on pattern	Rearrange heaters. For cold spots, add beryllium copper inserts.
	Corebox itself is too hot or too cold.	Adjust the corebox temperature.
	Low melting point sand.	Change resin or refer to section on low melting point sand problems.
	For cores, air pressure is too high or too low.	Periodically check air pressure settings. Use pulsating blow. Change angle of blow.
	Moisture in the sand.	Check moisture traps on air lines going to the blower.
	Cold sand.	Do not store sand in an extremely cold area. Review inventory handling policies.
	Dirty patterns or coreboxes.	Clean patterns and boxes to improve heat transfer.
Veining and thermal shock	Pouring temperatures too high.	Monitor pouring temperatures.
	Stress areas in cores or molds.	Check ejection system. Check uniformity of cooling.
	Cores or molds are overcured.	Reduce resin content.
	Cores or molds are excessively brittle.	Use vinsol in sand to increase thermoplasticity. Reduce hexa content.
	Cores are too strong.	Reduce resin content.
	Base sand does not allow for enough expansion.	Switch to a more angular clay-bearing sand, or add clay to the mix.
	Uneven mold bonding.	Check to see if locators are lining up properly.
	Cores or molds are too cold.	Check storage conditions.
	Core or mold weights are too low.	Increase weights by lengthening investment cycle. The sand itself may be too cold.
	Cores or molds have thin-walled spots.	Check for peelback or lamination.
Uneven buildup	Poor heat distribution.	Check arrangement of heaters.
	Cores drain poorly.	May require vibration.
	Dirty patterns or coreboxes.	Check patterns and boxes regularly.
	Poor blow pattern.	Check location of vents and blow holes.
	Plugged vents.	Check boxes periodically. Change type of vents.
Lack of buildup	Cool box or pattern.	Raise box or pattern temperature.
	Melting point of sand is too high.	Use a lower melting point resin or refer to section on high melting point sand problems.
	Cold sand.	Increase the investment cycle, and review sand storage conditions.
	Too short an investment cycle.	Increase the investment cycle.
Excessive buildup	Pattern or corebox too hot.	Lower the corebox or pattern temperature.
	Melting point of the sand is too low.	Use a higher melting point resin or refer to section on high melting point sand problems.
	Cores not draining properly.	Vibrate during drain.
	Too long an investment cycle.	Decrease the investment cycle.
Poor flowability or blowability	Plugged vents.	Check vents regularly. Change the type of vents.
	Not enough vents.	Add vents.
	Pattern or corebox too hot.	Lower pattern or corebox temperature.
	Low coated sand melting point.	Use a higher melting point resin or refer to section on low melting point sand problems.
	Moisture in air line.	Check moisture traps on a periodic basis.
	Insufficient amount of release agent in sand.	Add more release agent.

Table 9-1 (cont'd).
Shell Process Defects Checklist

Foundry Operating Problems		
Defect	**Possible Causes**	**Suggested Action**
Metal penetration	Core not blown dense.	Increase blow pressure.
	Surface or core or mold is too brittle, indicating overcuring.	Decrease the cure cycle, or reduce corebox or pattern temperature.
	Sand too coarse or too open.	Switch to a finer base sand. Use wash. Add fines (iron oxide, clay).
Casting porosity	Undercured mold or core surface.	Lengthen cure time.
	Excess gas generated at metal/sand interface.	Decrease resin and/or hexa level.
	Permeability of base sand is too low.	Use a coarser base sand.
	Inadequate venting of cores or molds.	Add vents where needed.
	Need for a gas scavenger.	Add iron oxide to the mix.
Sticking	Core or mold is overcured.	Decrease the cure cycle.
	Release agent buildup in the corebox or on the pattern.	Clean the corebox or pattern. Spray release agent less often.
	Scored corebox or pattern.	Repair damaged coreboxes or patterns. Minimize damage.
	Insufficient release agent in sand.	Add more release agent to sand.
	Free resin in the coated sand.	Refer to free resin in sand problems.
	Dusty resin-coated sand.	Sand has been overmulled or has been subjected to abrasion.
Weak cores or molds	Low resin content.	Check resin weigh up and LOI of sand. Increase resin level of sand.
	Low hexa content.	Check hexa weigh up. Increase hexa level.
	Cores or molds over- or undercured.	Check cure cycle and corebox or pattern temperatures.
Odor	Lack of deodorizer in the coated sand.	Use deodorized sand.
	Vinsol present.	Use non-vinsol plasticized sand, if required.

Resin-Coated Sand Problems		
Defect	**Possible Causes**	**Suggested Action**
Low melting point sand	Low starting sand temperature during coating operation.	Check sand temperature controller. Increase heater temperature.
	Time before hexa addition.	Increase initial mulling time.
	High moisture level in coated sand.	Increase starting sand temperature. Reduce the amount of water in the hexa-quench solution.
	Base resin melting point too low.	Switch to a higher melting point resin.
High melting point sand	High starting sand temperature during coating operation.	Check sand temperature controller. Decrease heater temperature.
	Insufficient quench water. This results in the coated sand being discharged hot at the end of the cycle.	Increase quench water.
	Base resin melting point too high.	Switch to a lower melting point resin.
Free resin in the coated sand	Low starting sand temperature during coating.	Check sand temperature controller. Increase heater temperature.
	Cold muller.	Preheat muller before coating.
	Short mulling time before quench addition.	Increase mulling time.
Low hot tensile strength	Low resin content.	Check resin weigh up. Increase resin level.
	Low hexa content.	Check hexa weigh up. Increase hexa level.
	Raw sand has excessive fines and/or clay.	Check with raw sand supplier.

Chapter 10

PLASTER MOLD CASTING

INTRODUCTION

After permanent mold, die and sand casting, the plaster mold process is probably the most widely used in the United States for aluminum foundry products. The development of permeable plasters has been largely responsible for the growth of this process in recent years.

Plaster mold castings cost more per unit than sand castings, so the process usually is not considered competitive with the sand or permanent mold casting processes. Rather, plaster mold casting is regarded as a process for producing castings that are not obtainable with these other methods.

Casting aluminum in plaster molds has wide industrial usage and many practical applications. Plaster molding is a refinement of sand casting in which plaster replaces sand to give the finished casting a smoother surface and better dimensional accuracy. Castings can be made with sections as thin as 0.020 in. and a surface finish of 80–125 microinches rms.

Because metal shrinkage can be controlled accurately in plaster molds, warping and distortion of castings is negligible. One significant features of the plaster molding process is that parts can consistently be produced to close tolerances, which greatly reduces and often even eliminates costly machining and other finishing operations. Good metal properties also are characteristic of plaster mold castings.

Properties and characteristics of some aluminum alloys frequently cast in plaster molds are shown in Table 10-1. Castings made in plaster molds are noted for their uniform hardness and machinability. The slow solidification of the castings, for instance, tends to prevent internal stresses. Because plaster that has been entirely dehydrated has little chilling action, many foundries add silica sand to the mixes to accelerate the solidification rate.

The plaster mold casting process is easily adaptable to production quantities ranging from several large pieces to several thousand small pieces per day. However, production runs for plaster casting usually do not exceed 1500. Weights of castings commonly produced vary from less than an ounce to approximately 2000 lb.

Aluminum castings with extremely thin sections and feather edges can be produced. Small holes can be cast to size ready for tapping. Parts with complex coring as well as undercuts and intricate passages can be produced readily (Fig. 10-1). Reproduction of fine surface detail is excellent—usually better than can be attained from shell molds.

Plaster molding methods are all basically the same, but they use different plaster mixtures, mixing techniques and curing practices. The two general methods discussed in this chapter in some detail use metalcasting plasters (the basic process) and permeable plasters.

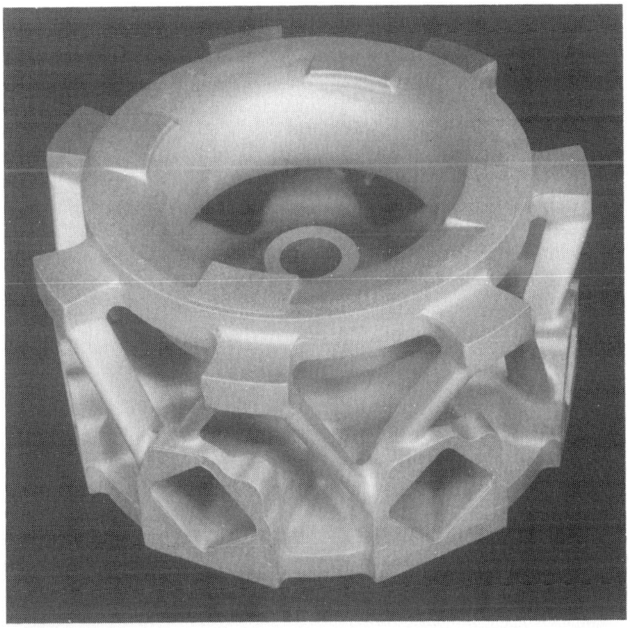

Fig. 10-1. This plaster mold casting, a first-stage compressor housing for a turbine engine, is made with an intricate core assembly.

Table 10-1.
Properties and Characteristics of Plaster Molding Alloys

Alloy No.	Typical Tensile Strength (ksi)	Typical Yield Strength (ksi)	Elongation (% in 2 in.)	Castability	Pressure Tightness	Machinability	Corrosion Resistance
319.0–T6	36	24	2.0	VG	G	E	G
355.0–T6	35	25	3.0	E	E	E	VG
356.0–T6	33	24	3.5	E	E	E	E
A356.0–T51	26	18	3.0	E	E	G	VG
357.0–T6	45	35	3.0	E	E	E	E
712.0	35	25	5.0	G	G	VG	E

Ratings: E = Excellent, VG = very good, G = good

PROTOTYPE AND SIMULATED DIE CASTINGS

Plaster molded aluminum castings often are an economical answer to experimental casting and prototype production. Diecasters and their customers frequently use plaster mold castings as prototype and simulated die castings (Fig. 10-2).

Plaster molded aluminum castings have dimensional accuracy and surface finish appearance nearly identical to those of die castings even though their mechanical properties (especially fatigue properties) are not comparable to those of die castings. Hence, for small runs, expensive dies and equipment can be avoided. Sales and design ideas can be tested with plaster mold castings, which cost less to tool and to change and which have short lead times.

An example of an inexpensive evaluation and development program conducted with plaster mold castings before expensive dies were machined is illustrated in Fig. 10-3.

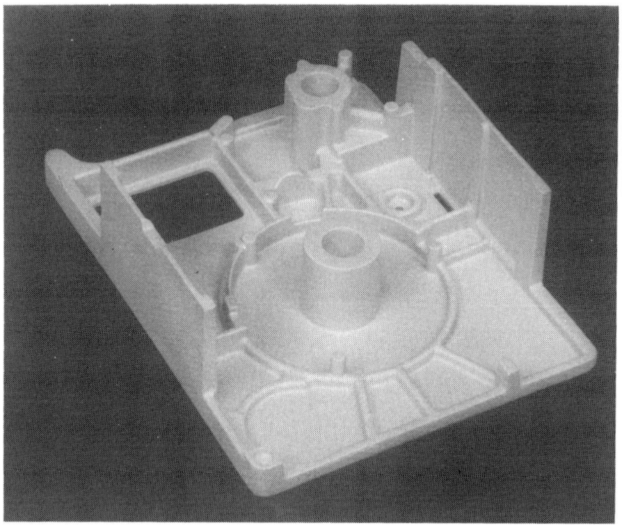

Fig. 10-2. This casting was used as a prototype for a diecast computer disk drive.

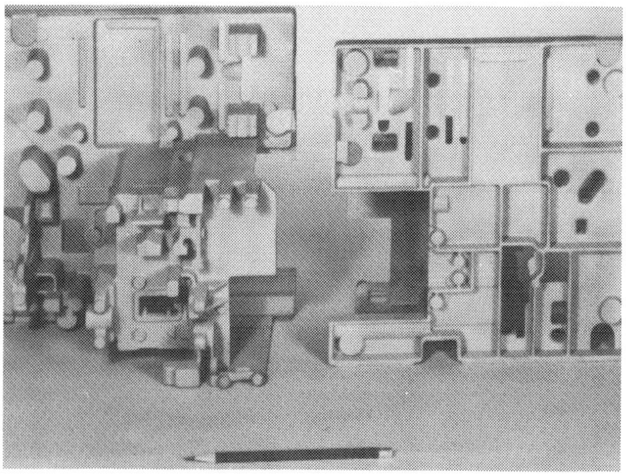

Fig. 10-3. These business machine bases are prototype plaster mold castings made of 356.0–T51 aluminum. Each base weighs 5 lb, and overall dimensions are 10½ by 10½ by 5¾ inches.

Samples of this business machine base casting were produced using plaster molds in just eight weeks. This compares to the 20–24 weeks that passed before delivery of the diecasting dies. Even though the part was designed as and was eventually produced as a die casting, plaster molds were used to produce the first 1000 castings.

DESIGNING PLASTER MOLD CASTINGS

In some respects, the use of plaster molding offers greater design latitude than does sand casting. Draft angles, for example, may be smaller because the plaster mold is denser and more homogeneous. These characteristics of the plaster mold also permit the designer to use sharper edges and smaller fillet radii than are possible for sand castings.

General design principles and details for plaster molding are discussed at length in Chapter 5.

THE PROCESS

The finest engraved detail of the pattern (Fig. 10-4) is reproduced in the plaster. A completed plaster mold that is ready to be dried and assembled is shown in Fig. 10-5.

Plaster molds are made from a mixture of water and metal-casting plaster, talc or other refractory materials. The slurry is poured from a bucket or piped (Fig. 10-6) from the mixer into a flask (usually made of metal) containing a pattern.

Fig. 10-4. These patterns were used to make molds similar to the ones shown in Fig. 10-5.

Fig. 10-5. The two halves of a plaster mold are shown after the pattern was removed. Note the excellent reproduction of detail.

Molds are dried in a continuously circulating air oven at 500F (260C) until there is no moisture present (Fig. 10-7). This drying step (also called burnout) ensures that no gas or steam will form when molten metal is poured into the mold.

When cores are required, they are placed in the drag prior to the assembly of the cope. Cores are baked at the same temperature as the molds and are made of similar plaster mixtures. Core boxes should be highly polished and should be constructed with extreme care. They may be made of aluminum, white metal, brass, or plastics such as phenolic or epoxy resins. After the cores have been placed and the mold halves assembled, molten metal is poured into the mold (Fig. 10-8), and the castings are allowed to solidify.

After the metal has solidified, the plaster is broken away (Fig. 10-9). Cores or adhering plaster are removed with a high-pressure water stream (Fig. 10-10).

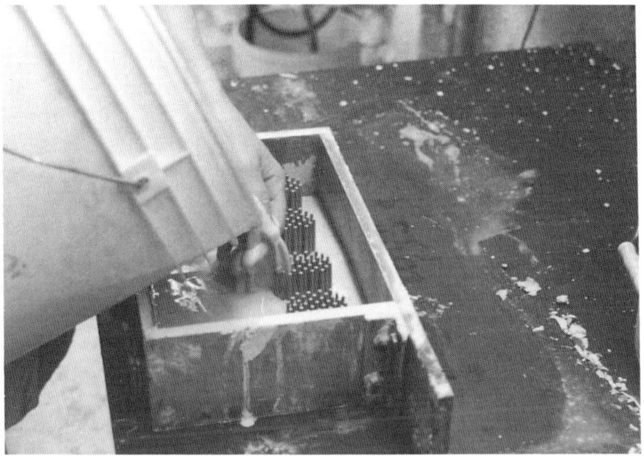

(10-6a)

(10-6b)

Fig. 10-6. Plaster slurry is (a) poured and (b) piped over patterns in metal flasks to form molds.

Fig. 10-7. Plaster molds and cores are loaded into batch ovens and baked to remove moisture.

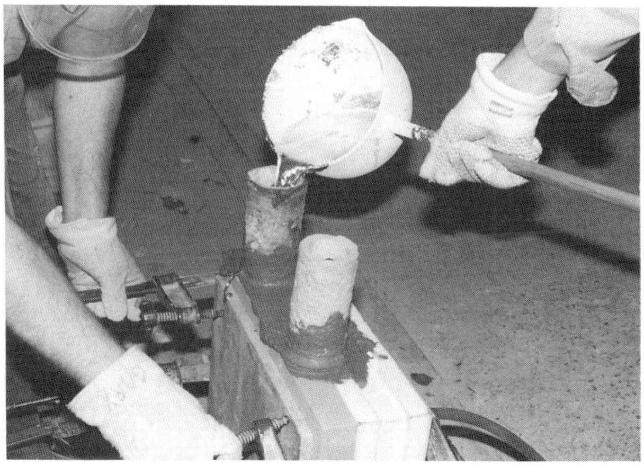

Fig. 10-8. Aluminum is poured into an assembled plaster mold.

Fig. 10-9. Plaster is removed mechanically from a solidified casting.

Fig. 10-10. Any plaster remaining on the casting after knockout will be removed using a high-pressure water jet.

Variations in patternmaking include using plaster cores in sand and plaster molds, plaster cores in semipermanent molds, sand additives instead of talc, and waxes that melt out to form complex cavities with no natural partings.

PATTERN MATERIAL

Both rigid and flexible patterns are used for plaster molding. Patterns are generally cope and drag type. Rigid patterns are usually sealed wood, metal and tooling plaster (usually sealed and lacquered).

Flexible patterns made of polysulfide (thiokol rubber), urethane elastomers and silicone rubbers are used widely to make intricate molds for higher volumes (Fig. 10-11). These patterns have complete recovery after many flexings, are waterproof, and release easily from the mold after it has set.

Fig. 10-11. This cope and drag pattern is made of rubber.

METALCASTING PLASTER PROCESS

Metalcasting plasters, commonly called plaster of Paris, are purchased as special mixtures of calcined gypsum, 20–40% refractory to prevent mold cracking, and various other compounds, such as terra alba or magnesium oxide to hasten setting time or other compounds to retard setting time. Lime, sodium silicates or portland cement are sometimes added to control plaster expansion and to add strength during baking and handling.

Calcined gypsum as received in metalcasting plaster mixes is $CaSO_4 \cdot \frac{1}{2} H_2$. During the initial set, it reacts chemically with the water of the slurry to form $CaSO_4 \cdot 2H_2O$. When dried at temperatures lower than 320F (160C), the gypsum reverts to the half hydrate. At temperatures above 320F (160C), the last water is removed, leaving $CaSO_4$ (anhydrous calcium sulfate).

Molds of conventional metalcasting plasters produce aluminum castings that have intricate and thin sections, good dimensional accuracy and excellent surface finish. However, they have a basic inadequacy in that gypsum as a bonding agent in molds for casting metal is practically impermeable to gases. Molds made of these materials must, because of their low permeability, be completely dehydrated before they are used. The long drying necessary to dehydrate them usually results in some warping and shrinkage of the mold. Consequently, molds made of permeable metalcasting plasters are preferred for many aluminum casting applications.

Plaster Mixing

Plaster and water must be weighed carefully to obtain the correct proportions for the proper consistency. (Consistency is the ratio of water to plaster, expressed as pounds of water per 100 pounds of plaster.) Plaster should be added to the water and then mixed. Procedures and consistency vary according to foundries and requirements. However, 100 lb of metalcasting plaster mixed with 130–160 lb of water are representative consistencies. As consistency increases:

- green strength decreases;
- burnout strength decreases;
- density decreases;
- permeability of burned out mold increases;
- setting expansion decreases;
- burnout shrink decreases;
- more water must be removed in the oven;
- pattern removal becomes easier.

It is advantageous to use the highest consistency that will give adequate strength. Once the proper use consistency has been determined, proper mixing can be accomplished by:

Fig. 10-12. An inexpensive mixer mounted on a shop-built holding fixture is a satisfactory slurry mixer. Mixing cycle is timed to give proper consistency.

1) weighing the plaster accurately; 2) measuring the water carefully (the water must be clear, potable and at a temperature of 90–100F, 32–38C); 3) timing the soaking period; 4) timing the mixing cycle; and 5) using a mixer of proper size in a correctly shaped bucket.

To eliminate air bubbles and subsequent holes in the mold, mechanical mixing is preferred (Fig. 10-12). Mechanical mixing also makes a more uniform slurry. Mixers may be propeller type, turbine type and simple disk type. A 2- or 3-in. three-blade standard pitch propeller or rubber disk, rotating at about 1760 rpm, is recommended.

Rotation should force slurry downward. The center of the propeller or disk should be located about midway between the center and side of the container and should clear the bottom of the mixing bucket by 1 to 1½ in. The mixer shaft should enter the bucket at an angle of 15° to 20° from vertical. Depth of slurry should be 1½ to 2 times the diameter of the bucket. Mixing for 2–5 min is generally long enough for the plaster to cream. If thin mixes are used, molds should not be poured until the mixture has thickened slightly. Mechanical vibration during pouring helps prevent air holes.

Continuous mixers (Fig. 10-13) meter the plaster and water as they are mixed. They pump the plaster to the pouring point. Outputs up to 20 gal/min are available.

Pouring the Plaster Mix

It is important to pour the mix as soon as it is ready and not to delay until it becomes stiff and thick. Detail may be lost if the mix is poured too late.

Fig. 10-13. An automatic mixer can provide continuous metering, mixing and pumping of plaster. [Photo courtesy of Hoge·Warren·Zimmerman Co.]

When pouring the plaster, it is usually best to pour it in such a way that the plaster flows slowly over the face of the pattern. This prevents air from becoming entrapped in small depressions on the pattern and helps ensure the reproduction of fine detail. When pattern design permits, the pour should be made on a noncritical surface. Pouring directly onto the pattern or in one spot can lead to difficulty later on when the mold is burned out since there is a variation in density where the plaster is poured that can cause surface imperfections in the casting.

All plaster molds for the same flask should be made, as nearly as possible, in the same way—e.g., at the same consistency, the same mixing, the same pouring. After the mold segments are made, they should be assembled into the flask as soon as possible. Late assembly into the flask may result in soft molds with greater shrinkage than can be tolerated. If mold segments or green plaster molds in the flasks are allowed to stand, shrinkage sets in. The longer the molds set before going into the oven, the more pronounced such shrinkage will be. It is important to get them into the oven as quickly as possible.

Gating

Gates may be smaller in plaster molds than in sand molds because of the low thermal conductivity of plaster. To establish optimum gating practice, sprues and gates may be cut by hand in the plaster after it sets but before it is dried. Backpressure of mold gas that develops as the metal fills the molds can be relieved by venting high portions of the mold cavity or by using a vacuum during pouring.

When the most appropriate gating practice has been established for a given job, the entire gating system should be made a part of the master pattern. Some foundries cut the gates off the first good casting, polish them, then use them along with the pattern to form gates throughout the run.

Parting Compounds and Separating

Wooden and other porous patterns should be sealed to prevent water from penetrating the pattern. To facilitate its removal, the pattern can be coated with a parting compound. A thin film of potter's soap, lard oil or paraffin oil, or a mixture of 75 parts kerosene, 25 parts light oil, and 1 part wetting agent are all satisfactory.

Another widely used compound is made by putting 6 oz of stearic acid flakes into 3 pints of hot kerosene and adding 1/10 oz of powdered mica. Heavy greases are not recommended because they cause soft mold surfaces and loss of detail.

At room temperature, the plaster develops an initial set, and the pattern can be removed. Compressed air often is used to help release the pattern from the solid plaster mold (Fig. 10-14). After the plaster has set completely, a small wire is pushed through the top mold section. Air is then blown through to separate the mold halves.

Drying or Burnout

After the mold has set, it can usually be handled as an integral self-supporting body. It should be supported uniformly on its edge or placed face-up on a flat, perforated metal plate,

Fig. 10-14. After plaster has completely set, air pressure (admitted through a small hole) is applied to part the cope and drag sections. Parting compounds are used to help facilitate removal.

rigid metal grid, or other level support, then placed in the drying oven. Smooth plates should be covered with a thin layer of talc or other powder to permit free movement between mold and base plate and prevent cracking or warping during drying of the molds.

Time and temperature determine the satisfactory drying and calcination of a plaster mold. Molds are usually dried by placing them in a drying oven and then increasing the temperature to 450–500F (232–260C) for 4–24 hours. They are sometimes cured for casting by heating at a relatively high temperature for a short period. In either case, all water must be expelled from the low-permeability plasters for successful casting.

Several factors determine the length of time necessary to cure a mold. These include temperature and capacity of the heat source, weight and thickness of the mold, exposed surface area, and level and efficiency of circulation of air in the oven. Because of this, a thermocouple should be embedded in the center of the thickest section of the mold (Fig. 10-15). When the reading for the thermocouple exceeds 220F (104C), the mold is dry (i.e., the free moisture has been removed).

As long as water remains in the mold, it is impossible for the inside mold temperature to exceed 220F (104C). Hydrated gypsum in the plaster loses water of crystallization at 262F (128C) to become the semi-hydrate, and it again loses water of crystallization at 335F (168C) to become the anhydrous form. Therefore, the temperature of the mold surface where metal contact will occur is usually heated to about 350F (177C) to a depth of at least ½ in.

If the water of crystallization is not removed, the plaster may dehydrate and release water during pouring because of the high metal temperature, thereby causing casting defects. To prevent cracking from thermal shock, the mold's surface temperature should be reduced slowly to at least 400F (204C) before removing it from the oven.

The temperature of the oven should be controlled to prevent the surface temperature of the mold from exceeding 1000F (538C); otherwise chemical decomposition, loss of burnout strength and excessive shrinkage are likely to result.

In low-temperature drying, metalcasting plaster molds may be dried in core ovens with a maximum temperature of 500F (260C). A center thermocouple should indicate at least 220F (104C). Some foundries bake the molds at a temperature of 400–500F (204–260C) for 14–16 hr, and others use ½ to 1 hr at 1000F (538C) for molds with a maximum thickness of 3 in.

Because the humidity must be kept relatively low throughout the drying cycle, ovens should include a provision for discharging wet air. Any calcined plaster that is allowed to stand at room temperature will reabsorb moisture rapidly. Therefore, care should be used to prevent rehydration of calcined molds prior to casting.

It is best to pour metal into the molds when they are just hot to the touch (150F, 66C). If a delay is necessary, molds should be protected from air by storing them in a container or warm oven. Plaster molds can only be used once; the plaster generally is not recycled for use in making more plaster molds.

Mold Shrinkage

Consistency of the mix and the calcination temperature determine the amount of shrinkage of the plaster mold. At low temperatures, baking shrinkage averages 0.1–0.5%; at high temperatures it will be 1.25–1.50% (Fig. 10-16).

Measuring the distance between gauge marks on the mold before and after calcination, rather than measuring the pattern and casting, is the most satisfactory method for determining mold shrinkage.

Fig. 10-15. A thermocouple embedded in a thick section of a mold during baking measures the electrical conductivity of the mold. The meter indicates the mold's moisture content based on this conductivity reading.

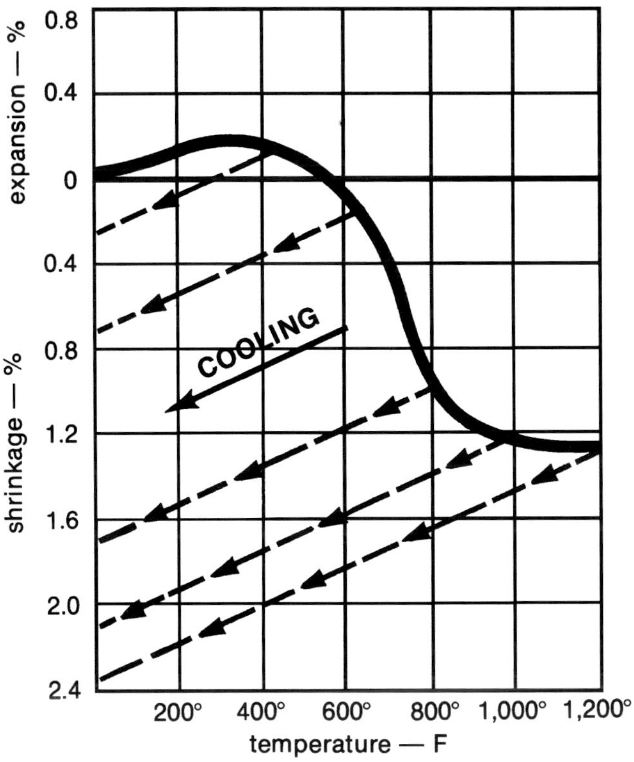

Fig. 10-16. Burnout temperature effect on burnout shrink. [Drawing courtesy of United States Gypsum Co.]

There are other factors that affect shrinking during baking of a plaster mold. Rapid setting of the plaster will result in greater final shrink, while a slow set minimizes shrinkage. The rate of set is determined, to a certain extent, by mixing speed and times—faster mixing speeds and lower mixing time result in quicker setting. Alkalinity reduces shrinkage, while acidity increases it. Small quantities of lime often are added to reduce shrinkage. Hot water minimizes shrinkage, while cold water increases it.

VARIATIONS TO THE METALCASTING PLASTER METHOD

Pressure Cast Process

The pressure cast process is used extensively for producing cast aluminum foundry tooling, such as matchplates, cope and drag pattern plates, coreboxes, and pattern castings. Metalcasting plaster is the molding material, and because it is poured in slurry form, it is able to reproduce precise detail. The process also is used to produce a variety of castings for the plastics industry. Molds for vacuum forming, rotational molding, blow molding and expandable polystyrene molding are made in this process because of the fine detail and smooth surface finish.

Patterns should be well-constructed and made of metal, epoxy or good pattern wood, such as mahogany. Joints should be well-glued because the pattern will be immersed in liquid plaster. Savings can be realized by providing multiple patterns where there are many on the matchplate or mold.

A refractory-lined metal cylinder is attached to the pouring cup. An insulating disk is placed under the cylinder to block the downsprue into the mold. Molten aluminum (Alloy 319 or 356) is poured into the cylinder at approximately 1400F (760C). The metal is then fluxed to remove impurities and dissolved gases. While the metal is being cleaned, it is also cooling to pouring temperature.

When the metal is clean and has reached pouring temperature, the cylinder is capped, and air pressure (3–4 psi) is applied to the cylinder. The pressure ruptures the disk, forcing molten aluminum into the mold. Pressure is maintained on the mold for about 15 min while the metal solidifies.

Other variations of this pouring technique are used. One is where a vacuum is applied to each half of the mold while the metal is poured by gravity. This method usually produces better quality castings with less porosity because better gating methods may be used.

PERMEABLE PLASTER METHODS

The greatest advantage of any plaster mold casting process is that it produces aluminum and other nonferrous castings with intricate and thin sections, fine detail, and excellent surface finish. The permeable metalcasting plasters offer the additional advantages of permeable molds and higher receptivity to chills than ordinary metalcasting plasters.

Antioch Process

A little-used plaster casting technique known as the Antioch Process develops a high degree of permeability in plaster molds. Permeable molds make it easier to produce castings with fine detail and allow moisture and other gases to escape through the mold.

In this process, a slurry is made of water and a dry mixture of 50% sand; 40% gypsum cement; 8% talc; and small amounts of sodium silicate, portland cement and terra alba. Approximately 100 parts of water are mixed with 50 parts of dry ingredients. The slurry is piped by hose into metal coreboxes or cope and drag flasks fitted to special matchplates.

The slurry sets to a solid in about 5 min. Chills can be added during initial set. Patterns are drawn, and the mold is assembled in the green condition. The high set strength, which averages about 70 psi, makes it possible to hold even large metal chills while it being withdrawn from the pattern.

The green plaster mold is dehydrated for 6–12 hr in an autoclave, in an atmosphere of saturated steam, at approximately 15 lb gauge pressure. After the dehydration process, the mold is left to stand at room temperature for approximately 14 hr, during which time the free water in the mold mass is slowly combined with (rehydrates) the half-hydrate, turning the it to the granular form. The mold then is dried to the desirable degree. Regular core ovens that operate at temperatures of 450–470F (232–243C) are used. If the molds are allowed to reach oven temperature, all free water and most of the chemically combined water is removed.

Autoclaving develops a special permeable structure in the mold but greatly reduces dry strength, a desirable condition because it permits the mold to crush easily as the casting cools and shrinks.

Foam Plaster Processes

Plasters mixed by foamed, or aerated, processes produce molds almost as permeable as those obtained in the Antioch process but without the autoclaving step. The permeable plasters are especially useful for making gravity-cast cope and drag molds, cores for semipermanent molds, insulating riser sleeves, and pads and cores for dry sand molds.

In this system, the desired mold structure is obtained by using a mixing procedure that gives uniform density and permeability as well as a strong, smooth face. The mixer used must beat air into the mix and produce cells approximately 0.01 in. in diameter (Fig. 10-17). When the mold is dried, the cells provide channels through which steam or other mold gases may escape or be entrapped during casting. A properly made mold can contain as much as 50% air holes finely distributed throughout the mold, with the exception of a thin layer at the pattern surface.

Large cells must not be formed in mixing because large cells just under the mold surface may allow the metal to break through and cause imperfections in the casting. Large bubbles can be eliminated with different types of mixers, but indications are that the higher the power input per unit of mix, the finer the permeable structure and the smoother the mold surface. For small mixes, a unit similar to a malted milk mixer works satisfactorily.

There are several variables that directly affect the characteristics of the mix, the speed of set and the permeability of the mold. Among these are the size of the bucket, the diameter and style of the mixing disk, the speed of the mixer, and the volume of slurry (Table 10-2). In general, the diameter of the disk should be about ⅔ the diameter of the bottom of the bucket. The mixer speed should be fast enough to create a vortex of air down to within an inch or two of the bottom of the bucket when the mixer shaft is in the vertical position.

Table 10-2.
An Example of Permeability Control

% Vol. Increase	Wet Weight lb/ft^3	AFS dry units perm.[1]
0	90.8	0.5
50	60.6	5.0
61	57.0	8.5
68	53.8	11.5
76	49.4	16.5
100	46.0	30.0

1) At 150 consistency, the permeability of conventional matchplate plaster is 0.75–1.5.

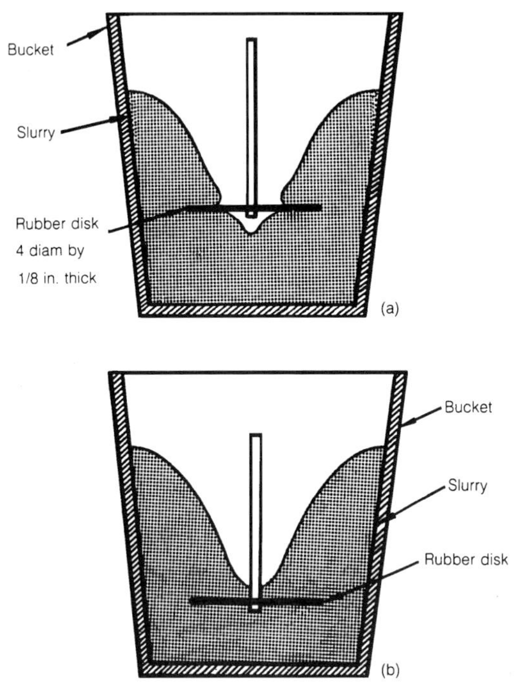

Fig. 10-17. A mixing disk is positioned for a) throwing air into a foamed-plaster slurry, and b) final mixing of the slurry.

Molds may be poured by gravity. Risers and gates are designed in accordance with good foundry practice. Since calcination is unnecessary, high temperature burnout is not needed, and consequently, molds are more true-to-size because mold shrinkage is minimal.

Two common permeable metalcasting plaster processes are the in situ foam and the separately generated foam systems. In the first process, a formulated metalcasting plaster, when combined with the recommended proportions of water and mixed correctly by the proper equipment, produces an extremely smooth, permeable molding material. The second variation requires that the slurry be mixed with a separately generated foam in the correct amount.

Pattern and Equipment

Although mold shrinkage may be disregarded because of the low drying temperatures, the usual allowance for metal shrinkage must be made on the pattern. Because of ease of extraction and freedom from separator problems, flexible patterns are popular. Patterns may also be made from properly sealed plaster, metal, rubber, plastics or wood waterproofed with lacquer.

Foam plasters are rigid, so care must be taken to have smooth, drafted patters with extraction perpendicular to the parting lines.

Gating

In large-scale production, all gates, risers and runners are made a part of the pattern equipment. Because of the plaster's insulating properties, gates and runners may be made smaller than normal for sand molds. On experimental molds or short runs on simple castings, sprues and gates may be carved into the wet mold. Carved surfaces should be slicked down or troweled smooth.

Parting Compounds

To facilitate pattern removal, a parting compound must be applied to rigid patterns. Commercial parting compounds commonly found in matchplate shops are satisfactory if applied sparingly. These include stearine and water-ground mica in liquid parting compounds.

To separate the permeable plaster parts from themselves, a commercial water emulsion wax preparation, such as those hardware stores carry, is satisfactory. All parting compounds should be applied sparingly with a fine brush or spray. Excessive use will cause rough and soft surfaces on the mold, which will produce rough surfaces on the castings.

Setting and Baking Foamed Plaster Molds

The setting and baking steps for this process are similar to those described previously for the metalcasting plaster process, except that baking may be done at lower temperatures and for shorter times. Drying temperatures vary with the type of oven but usually range from 250–500F (121–260C).

Air circulation and velocity, humidity, size of mold, desired degree of permeability, and oven temperature all affect the length of time required to bake the mold. Because evaporation of the water on the mold surfaces is increased when air is moved over the surface, core ovens of the recirculating type will dry molds more rapidly.

Relative mold dryness can be determined by: 1) an electrical conductivity test for free water; 2) loss in mold weight; 3) the temperature of the center of the thickest part of the mold; or 4) time and temperature judgment through previous experience. Of these methods, the third is the most precise and the easiest to conduct. The most practical, after enough experience has been attained to make it reliable, is the fourth.

Molds that contain free water will conduct an electrical current. A self-contained testing unit is recommended to test for electrical conductivity. A two-cell pencil battery is enclosed in the unit with a meter reading to 2,000,000 ohms. Rigid leads long enough to reach the deepest portion of the mold should be used.

In using the weight-loss method, the wet weight of the mold as stripped from the pattern is determined. The mold is then dried in an oven until a loss of 37–42% of the wet weight has occurred. This weight loss percentage is for a mix of 80 parts water and 100 parts plaster, or a mix containing about 44% water. This method is difficult to use in matchplate work where, in the drying operation, the mold is surrounded by a flask.

Temperature measurement of the center of the heaviest section is taken by inserting a 22 gauge chromel-alumel thermocouple with leads long enough to extend outside the oven. The interior of the mold will indicate a rise above 212F (100C) when it is dry. When 220F (104C) is reached, the oven should be turned off.

PLASTER CORES

Plaster cores are especially suitable for casting intricate parts such as impellers for superchargers on internal combustion engines. The plaster provides good dimensional control and imparts smooth surface finish to the impeller vanes, both of

Fig. 10-18. Cored aluminum castings produced by the plaster mold process exhibit fine surface detail.

which are essential to the finished part. The castings shown in Fig. 10-18 were made in plaster molds.

In many instances, the plaster cores can be set individually in the mold. Frequently, however, dimensional requirements or shape and contour of the vanes require a one-piece core to be used for all vanes in the impeller. It then actually becomes necessary to "core the core." This is done by placing replicas of the impeller vanes in a corebox and pouring the plaster mix around the assembly of vanes. After the plaster hardens and cures, the vanes, which are made of a low-melting white metal alloy, are melted out of the plaster core. The core is then placed in the mold, and the impeller is cast, usually centrifugally. This is a type of investment molding process.

When cores are dried at high temperatures, they should not be left in the oven too long after the free water is driven off; otherwise, excessive cracking, shrinking or breakage may occur. Small, unvented cores are best when calcined; if cores contain moisture, steam formed during the metalcasting operation may not escape through the core with rapidly enough. This is especially true when permeability of the core is low.

ACKNOWLEDGMENT

AFS would like to thank John Armstrong, Armstrong Mold Corp., for providing the photographs for Figs. 10-6 through 10-12, 10-14, 10-15 and 10-18.

Chapter 11

CENTRIFUGAL CASTING

INTRODUCTION

In general, centrifugal casting involves rotating or revolving a mold, filled with molten metal, about an axis. The centrifugal pressure forces metal against the interior mold wall or cavity, where it solidifies, producing a sound, dense casting.

Cylindrical parts or circular plates are usually the most preferable shapes for centrifugal casting. Tubular aluminum castings produced in permanent molds by true centrifugal casting have higher structural strengths and more distinct cast impressions than castings produced by the static permanent mold process. Other types of castings made centrifugally do not always develop properties superior to good static castings. Specific castings not adaptable to the centrifugal process probably can be produced with better quality and economy by other casting methods.

Many parts previously made from forgings have been converted to centrifugal castings. In centrifuge and semicentrifugal casting, mechanical properties can be higher than those of other casting methods but are generally lower than those of forgings.

The yield, or ratio of the weight of castings to weight of metal poured, is usually higher in centrifugal casting than in gravity casting processes. Production rates vary considerably, depending upon the size and shape of the castings.

METHODS FOR CENTRIFUGAL CASTING

Castings of various sizes and shapes can be produced by one of three methods of centrifugal casting: true centrifugal, semicentrifugal and centrifuge (Fig. 11-1). In all three, a measured quantity of metal is poured into the central sprue of a mold that is spinning or that spins immediately after pouring.

True Centrifugal Casting

In true centrifugal casting, the mold is spun about its own axis during the solidification. Molten metal is thrown to the mold wall and held there until it solidifies. A casting may have a hole (bore) through the center, the size of which is determined by the amount of metal poured. No central cores are used, and the internal bore is cylindrical.

Because measured amounts of metal are poured and no gates, risers or sprues are needed, the metal-to-casting yield approaches 100%. An important and favorable consideration is that solidification is directional from the outside toward the center of rotation.

Tubular products are readily cast by this process, and the axis of rotation may be horizontal or vertical. Sleeves and liners, bushings and pipe are some of the parts commonly produced by the true centrifugal casting method.

Semicentrifugal Casting

In semicentrifugal casting, central cores are used to form irregular contours inside the central cavity and may be made with or without a gate. Wheel- or dish-shaped parts are commonly cast by this process (Fig. 11-2). Because the mold is spun on the vertical axis, centrifugal force throws the molten metal to the periphery and thus aids in feeding the casting. Risers are usually required.

When the same wheel design is cast using the semicentrifugal versus the conventional permanent mold processes, semicentrifugal casting realizes only 85% of the production

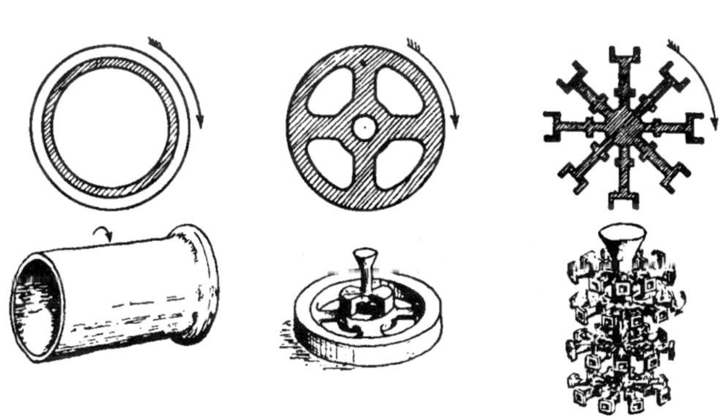

Fig. 11-1. The three methods for centrifugal casting: (l–r) true centrifugal, semicentrifugal and centrifuge.

Fig. 11-2. A semicentrifugally cast passenger car wheel is poured directly down the center riser.

rates of permanent mold casting. However, the better yield and reduced costs of trimming semicentrifugal castings make the total costs for both casting processes about the same.

Unlike true centrifugal casting, solidification is not unidirectional because all casting surfaces contact either mold or core surfaces. For best results, solidification must progress from the maximum outside diameter of the casting toward the center. This is sometimes controlled with auxiliary gates, chills or pads. Directional solidification toward the axis of rotation may also be accomplished by using sand or plaster cores in metal molds or by cooling the exterior surfaces of metal molds. As in true centrifugal casting, the quality of the metal at the outside of semicentrifugal castings is somewhat superior to that of the inside.

Centrifuge Casting

Centrifuge casting is the most versatile of the three processes. Figures 11-3, 11-4 and 11-5 illustrate this method, which is often used for nonsymmetrical shapes that cannot be rotated around their own axis. In one centrifuge method, molds are arranged around a central downgate or sprue on the axis of rotation, or hub, and radial gates feed each mold from the sprue like the spokes of a wheel. Rotation is on a vertical axis centered in the sprue.

Fig. 11-3. This casting (left) shows fractional horsepower motor rotors (with all but one rotor removed) made using the centrifuge method on a vertical centrifugal casting machine. At right is the type of rotor lamination stack and shaft that was inserted into the mold to make the casting.

Fig. 11-4. Steel laminations like the one shown in Fig. 11-3 are set up in the mold of a vertical casting machine to make fractional horsepower motor rotors using the centrifuge method.

The primary value of centrifugal force is filling the mold. Metallurgically, there is little or no difference between a semicentrifugal and a centrifuge aluminum casting. This process will be covered in more detail later in this chapter.

CENTRIFUGAL CASTING MACHINES

In centrifugal casting machines, the axis of rotation can be vertical, horizontal or at any angle. There are advantages in both vertical and horizontal machines (few machines use other angles). Horizontal machines are usually used to make true centrifugal castings. Vertical machines can be used for all three processes.

If a machine is used to make many different sizes and types of castings, as in a jobbing foundry, a wide range of speeds is essential. The machine should be sturdy and of rigid construction to prevent excessive vibration. A continuously variable, easily adjustable speed control also is recommended. A simple method for loading and unloading, safety features, and a brake for rapid deceleration should be provided. Figure 11-6 shows a machine that can be used to cast parts using centrifugal or centrifuge methods. In Fig. 11-7, a similar machine is shown with the castings it produced.

Vertical Axis Casting Machines

In centrifugal casting, vertical machines are limited to producing castings with lengths no greater than twice their diameters. If the length were more than twice the diameter in a vertical mold, the speed needed to force the metal up the mold walls (counteracting the earth's gravity) would be prohibitive, especially if the central walls need to be held to only a slight taper. A vertical centrifugal casting machine is shown in Fig. 11-8.

Fig. 11-5. A group of fractional horsepower motor rotors is removed from the mold of a vertical casting machine.

Fig. 11-6. This vertical machine can be used for centrifugal casting of single motor rotors or centrifuge casting of several fractional motor rotors. The machine can also be set up to centrifugally cast auto wheels, brake drums and brake disks.

Fig. 11-7. This vertical machine is set up to make centrifuge cast fractional horsepower motor rotors like those sitting on the corner of the machine housing.

Fig. 11-8. This machine is set up for vertical centrifugal casting of electric motor rotors.

The actual shape of the inside of a centrifugal casting spun on a vertical axis is parabolic (Fig. 11-9). The rotational speed used determines the angle of taper. If the speed is high enough, the resulting taper (usually ⅛ in. per foot) normally will not be apparent to the eye unless the length of the casting is great.

Horizontal Axis Casting Machines

Horizontal machines are used to produce tubelike shapes with a cylindrical inside diameter and where the length of the part is greater than two times the diameter. In horizontal molds, the inside walls are cast parallel, and speeds are lower than for vertical casting. One drawback is that the castings have lower properties because the metal cannot be poured into horizontal molds as quickly as into vertical molds.

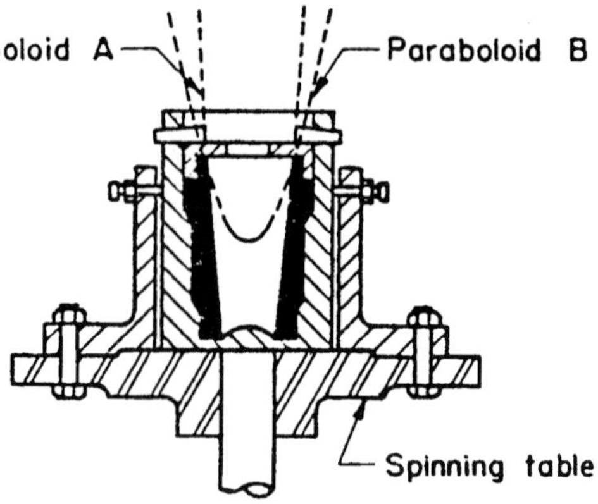

Fig. 11-9. A schematic of vertical centrifugal casting showing the effect of rotational speed on the shape of the inner surface. Paraboloid A shows the shape of the cavity formed by a relatively high rotational speed, and B shows the approximate shape of the cavity that would be formed at a lower speed.

DESIGNING CENTRIFUGAL CASTINGS

Most of the general principles and details in the design of parts produced by other casting processes are equally important in the design of centrifugal castings. General principles for designing aluminum castings are presented in the first sections of Chapter 5.

A separate subsection of Chapter 5 contains specific suggestions and engineering data for the centrifugal processes. When using the centrifuge process, special techniques and sound practical production experience are needed to design the gating properly and to space the individual molds correctly on the periphery around the central downgate.

ALUMINUM ALLOYS & FOUNDRY PRACTICES

Most aluminum alloys cast by the die, permanent mold or sand casting processes can be cast centrifugally, but those with shorter solidification ranges are preferred. As with other aluminum casting methods, the metal should be poured smoothly without interruptions to avoid cold shuts (incomplete filling of the mold).

For best results, molten metal should be poured at temperatures that are approximately 100°F (56°C) below those used in static processes. Although metal should be poured quickly, special pouring and gating techniques may be required for sound quality castings. Sand or permanent mold inspection methods are applicable to centrifugal castings.

THE CENTRIFUGE PROCESS

The centrifuge method of centrifugal casting is especially suitable for aluminum because it can be used to make pieces of many shapes and sizes (Figs. 11-10 and 11-11). Generally, small castings or intricate castings with low yield-to-pour ratios if cast statically are made by the centrifuge process at the rate of up to 10,000 pieces per hour. However, castings weighing more than 100 lb also have been produced by this process.

In another variation of centrifuge casting, a single mold rotates at the end of an arm driven by a vertical shaft. The metal is usually poured directly into the mold sprue while the mold is at rest, and then the mold is rotated immediately.

The metal may be poured while the mold is rotating when the sprue is located at the center of the shaft. In the latter setup, the gate is designed to connect the sprue to the mold at the end of the rotating arm.

Centrifuge casting is unique in the apparently unlimited combinations of intricate shapes that can be cast successfully using one or a combination of molding techniques. The following detailed description of the single-arm innovation is one example of this versatility.

Fig. 11-10. Castings made by the centrifuge or centrifugal methods can vary in size. All of these castings are motor rotors.

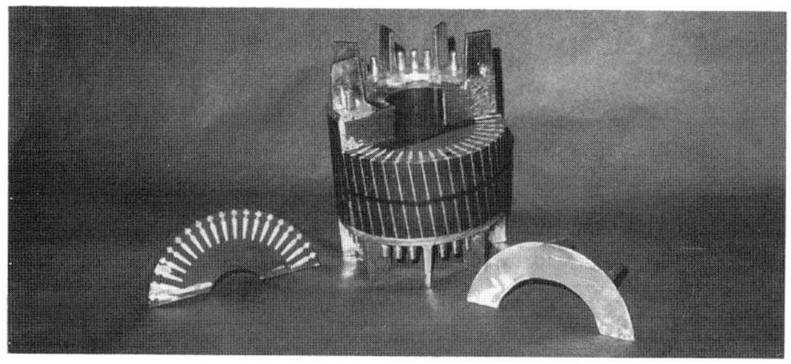

Fig. 11-11. Cutaways of a motor rotor show the intricate shapes possible with centrifugal methods.

With a permanent mold, the pressure of the centrifugal force improves casting soundness. Melting practices, mold and pouring temperatures, rotation time, alloys used, and other factors are closely controlled. As-cast finishes tend to be of excellent quality, and mechanical and physical properties are unusually high, particularly when high-purity, heat treatable aluminum alloys are used.

Small- and medium-sized castings are produced on a centrifuge table mold while a centrifuge pit mold is used for larger parts. There is practically no limit to size and weight of castings that can be produced on the pit type machine.

Because the metal for most of these castings is poured directly into the mold just prior to rotating (centrifuging), the sprue and gating may be quite simple. This minimizes costs for trimming and cleaning the rough castings. Gates usually enter the heaviest sections of the casting, preferably along a plane of minimum angle or incline. Risers, in most cases, are unnecessary. Molds of component parts may be assembled and hinged separately, thereby permitting many surfaces to be cast without draft.

Molds and Cores

Steel, cast iron, green or dry sand, plaster, or graphite molds and cores are used in centrifuge casting. The permanent molds (or dies, as they are often called) are usually machined from solid stock. Walls of molds must be of proper thickness for rapid chilling of the molten metal. Generally, mold wall thickness will average 2½ times the casting thickness.

In baked sand molds for centrifuge casting, patterns are designed with the parts to be cast arranged around a central pouring sprue. A core sand mix is rammed up on the pattern uniformly. Then the green mold is sprayed with silica wash and baked, producing a strong, open dry sand mold.

Green sand molds require tight flasks, preferably round. The flask can be fastened to the revolving table of the machine by various holding fixtures including a device similar to a lathe chuck. Cope and drag halves are clamped together and lowered onto the table, and the clamps are tightened to hold the mold. In another method, two pins are fastened to the table over which the flask is lowered. Then the cope and drag are clamped together in the usual manner.

In a third method, the green sand mold is pushed on a roller conveyor into a snap (hinged) flask, which is fastened to the table. Rollers on the machine facilitate movement of the green sand mold, while hinging allows the flask body to open in half. The cover, holding a runner cup, distributes pressure equally from the clamping arrangement to the mold. The clamping arrangement, a part of the flask body, keeps the cope and drag in firm contact.

Flasks are unnecessary with dry sand molds, which are placed in jackets similar to those used with snap flasks. After components have been set on the table in the machine, clamps hold the mold cover and mold firmly to the table. Dry sand molds can be rolled into the machine on a roller conveyor. When using this type of flask body, dry sand and green sand molds can be used interchangeably in the same machine.

Some advantages are found in graphite as a mold material. Although its heat conductivity is approximately triple that of cast iron or steel, it is only about one-third as heavy. It has good machinability and has proven satisfactory for small to medium production runs when handled carefully.

Rotating Speeds

Spinning speeds vary greatly. An average rotational speed is about 1000 ft per minute peripheral velocity. Excessive speed may crack the periphery of a casting (a form of hot tearing). Cracks also may be caused by too strong or too hard a mold material, abrupt change in section, fins or other factors.

Usually no benefit is gained by increasing the speed over that required for the poured metal to enter the mold cavity quickly. In horizontal casting particularly, rotational speed should be determined by the capability of the mold to accelerate or "pick up" the metal to the spinning speed. A mold with a rough surface permits a slower speed than one with a smooth surface. General practice in horizontal centrifugal casting is to use speeds of 60–75G (i.e., as a result of the centrifugal force, the "weight" of 1 lb of metal is 60–75 lb) for sand-lined molds. Permanent or graphite molds are spun somewhat slower at about 40–60G.

When pouring sand molds (especially with green sand molds), low speeds should be used initially. When the mold is partially poured, the speed can then be increased to that required. Pouring during rotation ensures complete filling of the mold cavity and favorable temperature gradients, with the highest toward the center of rotation.

A rapid, simple arrangement for loading and unloading is essential for economy. Although spinning time varies from 2–10 min, with the average being about 3½ min, efficient braking and handling can shorten the total casting cycle to about 6 min. With smaller molds, less time will be needed in handling and spinning. Spinning times for permanent molds are considerably less than the values given above.

Spray Coatings

A coating (mold wash) is generally required on the walls of metal molds for the centrifuge process. Graphite and sodium silicate in water will produce satisfactory results. Another acceptable mold wash is made by mixing 800 cc of hot water with 9 g of soap flakes, then adding 2 cc hydrochloric acid, 30 cc sodium silicate and 18 cc more water. For each 200 cc of the solution, 300 g zirconium oxide should be added. Solutions should be sprayed on the hot, rotating mold.

A widely used silica wash is made by mixing a bentonite paste consisting of 2 quarts of bentonite in 14.4 quarts of water at 180F (82C). To 4.8 quarts of this paste is added 19.2 quarts of water and 2 quarts of gluten. The Baumé is brought up to between 35 and 40F (2 and 4C) by adding silica flour. The mixture should be kept agitated in the tanks.

Green sand molds are sprayed to aid in the cleaning operations and to ensure a dry surface.

DEFECTS IN CENTRIFUGAL CASTINGS

Defects in centrifugal casting are similar to many that occur in static casting. Most fall into four general types: cold shuts or incomplete filling of the mold; porosity; excessive dross or pits on the inside diameter (raining); and hard spots.

When metal solidifies before the section is filled, *cold shuts* or *incomplete mold filling* usually result. Faster rates of pour, higher temperature, hotter mold, a thicker coat of mold wash (about 1/32 in.), or increasing the spinning speed may alleviate this condition.

Porosity may be caused by absorbed gas, by excessive turbulence during the pour or by incorrect cooling. Once started, pouring should never be interrupted.

Raining may occur when a horizontal machine is rotated too slowly or when metal is poured at an excessively fast rate, causing dross or pits on the inside diameter. If the molten metal is not picked up by or held against the mold wall because the mold surface is too smooth or because the spinning speed is too slow, it will be carried to the top of the mold and then dropped off like rain. Raining can be reduced or eliminated by lowering the pouring temperature, increasing the speed of rotation, or by slowing the pouring rate either at the start or after half of the metal has been poured. Because aluminum forms dross quickly upon oxidation, it should not be permitted to rain.

Hard spots are usually caused by dross or other nonmetallics in the castings. These conditions can be avoided by preventing oxidation and/or inclusion of foreign materials during melting and when handling molten metal. In some instances, hard spots also may be caused by excessively hard intermetallic compounds. The latter usually can be prevented by making certain the alloy composition is within prescribed limits and by avoiding excessively slow solidification.

ACKNOWLEDGMENT

AFS would like to thank The Centrifugal Casting Machine Co., Tulsa, Oklahoma, for providing the photographs used in this chapter.

BIBLIOGRAPHY

N. Janco, *Centrifugal Casting*, American Foundrymen's Society, Inc., Des Plaines, IL (1988).

Chapter 12

INVESTMENT CASTING

BACKGROUND

Although investment casting has been used for thousands of years by artists, it was adopted as an engineering manufacturing technique only in the late 1930s. The technique, formerly known as the lost wax process, was developed gradually in the twentieth century in the dental and jewelry fields.

The investment casting process was used to produce cast-to-shape surgical prosthesis devices in a cobalt-chromium alloy. Although this alloy could not be machined or forged, its excellent high-temperature mechanical and hot corrosion-resistant properties made it desirable for use in aircraft piston engine turbocharger buckets. Subsequently, this alloy found wide application in early gas turbine engines. Because this material had to be cast to shape led to the use of investment casting methods and the subsequent establishment of a new engineering discipline.

The technological growth of this process has been steady and, at times, spectacular. Its main areas are:

- pattern materials and pattern production;
- ceramic binders, refractories and cores;
- melting and casting of metal;
- knockout, cleaning and inspection of castings.

THE PROCESS

A number of casting methods are used to produce precision castings. To avoid confusion, the term investment casting has been adopted to describe the production of engineering castings using a consumable pattern. This term largely replaces such terms as lost wax, cire perdue and precision casting.

In simple terms, the process consists of producing a replica, in wax or plastic, of the part to be produced (Figs. 12-1 and 12-2). This replica (or pattern) is assembled to a gating and runner system of similar material. This assembly is coated (or invested) with a hard-setting refractory slurry.

The pattern and runner system are melted out of the mold to leave an exact cavity. The mold is then heated or fired to burn off the residual pattern material and to develop the high temperature strength of the refractory shell. When the firing cycle is complete, molten metal is cast into a hot mold.

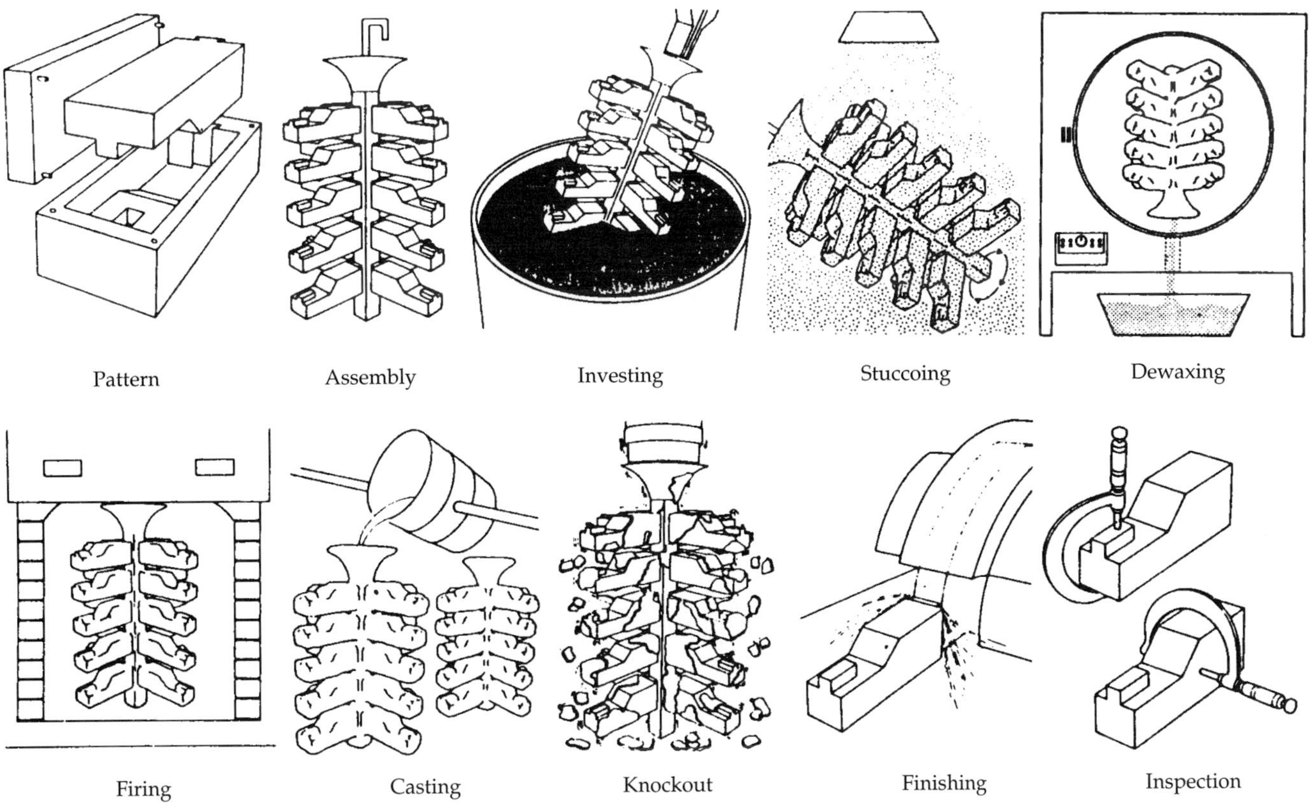

Fig. 12-1. A schematic of the typical ceramic shell process.

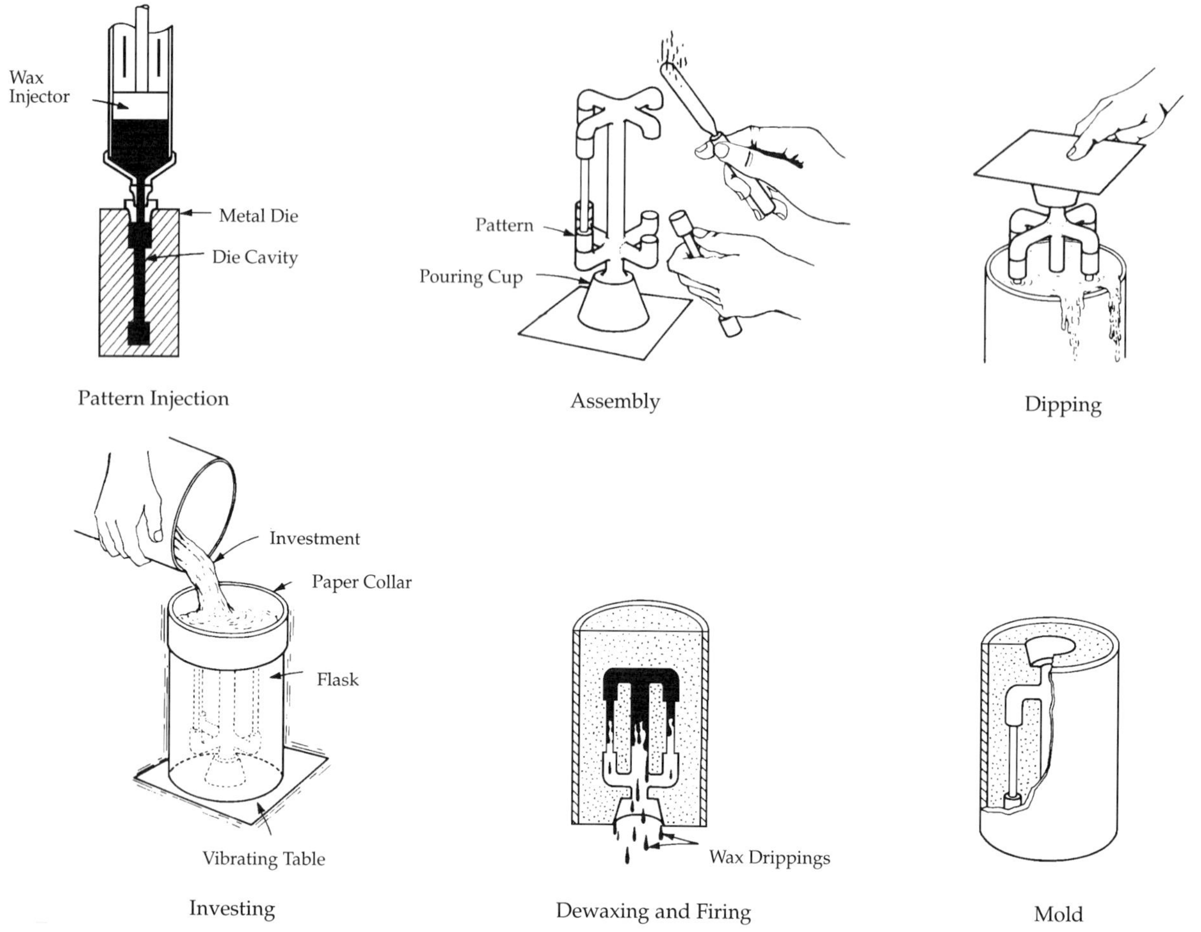

Fig. 12-2. The solid mold investment casting process.

The solid or block mold system (flask method) illustrated in Fig. 12-2 was, for many years, the most successful method of manufacturing. As techniques and technology improved, the ceramic shell method was introduced, permitting an increase in the size of castings. The ceramic shell process was outlined in Fig. 12-1.

Investment castings are produced in a wide range of alloys. Steels, copper-base alloys, aluminum alloys, magnesium alloys, and specialty alloys for corrosion and high-temperature operations are all cast regularly. The castings produced can weigh from a few grams up to 2000 lb.

PATTERNMAKING

Pattern Materials

Consumable pattern materials that are used for investment casting include plastics and wax blends. Alternative materials have been tried but with limited success. Urea-base materials have been used, as have various mixtures of inorganic salts, alone or in combination with a carrier material such as paraffin.

Pattern materials should be easy to join and assemble, nontoxic, and readily available. Irrespective of the pattern material used, the basic properties required are:

- low ash content;
- good fluidity to reproduce detail;
- low contraction and expansion characteristics;
- stability and reproducibility;
- compatibility with process materials;
- adequate strength and hardness.

Pattern Waxes

Criteria for the acceptability of pattern materials have been established. Most of these are based on test procedures such as those listed by AFS or the Investment Casting Institute. Probably the most widely used acceptance test is that for ash content. The pattern material must not leave a large ash residue because this would be included in the casting. Other useful tests include those for determining softening point, thermal expansion, viscosity and percent filler.

There are many types and formulations of pattern waxes. The basic categories are natural waxes, mineral waxes and petroleum waxes.

The most important natural waxes are beeswax, carnauba, ceresin and cadelilla waxes. Natural waxes are normally added to other waxes and resins to obtain desired pattern wax properties.

Montan wax, extracted from brown coal (U.S. and Eastern Europe), is the most important mineral wax. It is used unmodified but has some strength and hardness disadvantages.

Paraffin waxes have been used, but the so-called microcrystalline waxes are the most important class of petroleum waxes. These may be modified to give specific properties.

Filled Waxes

Many types of waxes have a certain percentage of filler in them. The five commonly used fillers are thermoplastic polystyrene, thermosetting polystyrene, bisphenol, isophthalic acid and water. The typical particle size of the four solid fillers is less than 150 microns. Most fillers are added by weight percentage and range from 5–40% of the base wax.

Organic fillers are added to the pattern wax as a shrinkage modifier. Fillers also affect the thermal conductivity and softening point of waxes (Tables 12-1 and 12-2).

Table 12-1.
Melting/Softening Points of Various Wax Fillers

Thermoplastic	290F
Thermosetting	>365F
Bisphenol	310F
Isophthalic Acid	570F*

*Sublimes

Table 12-2.
Normalized Thermal Conductivities of Various Wax Fillers

Wax	1.0
Bisphenol	0.3
Polystyrene	0.2
Isophthalic Acid	4.0
Water	18.3

Pattern Manufacture

Most patterns used in investment casting are produced by injecting the pattern material into a die or mold (Figs. 12-3 and 12-4). Occasionally, patterns are made by fabrication or by machining. Pattern molds are usually machined from metals such as aluminum or steel, but the use of cast soft metal molds, epoxy molds and rubber molds is still fairly common for samples and low-volume production.

The lightness and rigidity of magnesium-aluminum alloys make them very attractive for mold applications. In areas such as slideways for cores, harder wearing inserts are used to minimize wear and maintain accuracy. These inserts may be of brass, bronze or steel.

Automated wax injection presses are being used for investment casting. The automated press maintains pattern-to-pattern cycle consistency and reduces handling damage. A typical automated wax injection press is shown in Fig. 12-5. Molds for plastic injection of patterns normally are machined from steel, which can withstand the high clamping and injection stresses associated with this type of material.

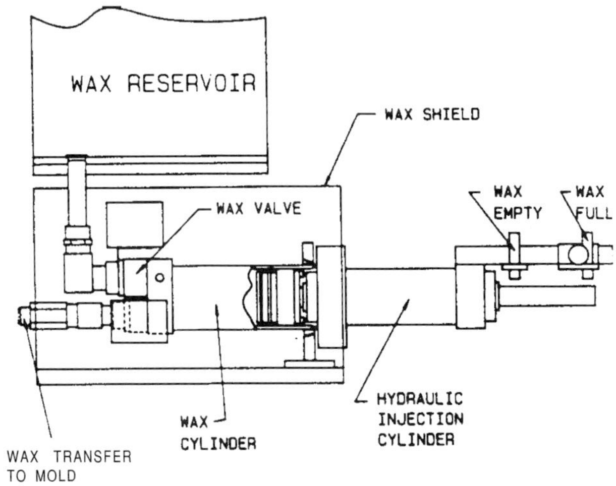

Fig. 12-3. A wax injection unit.

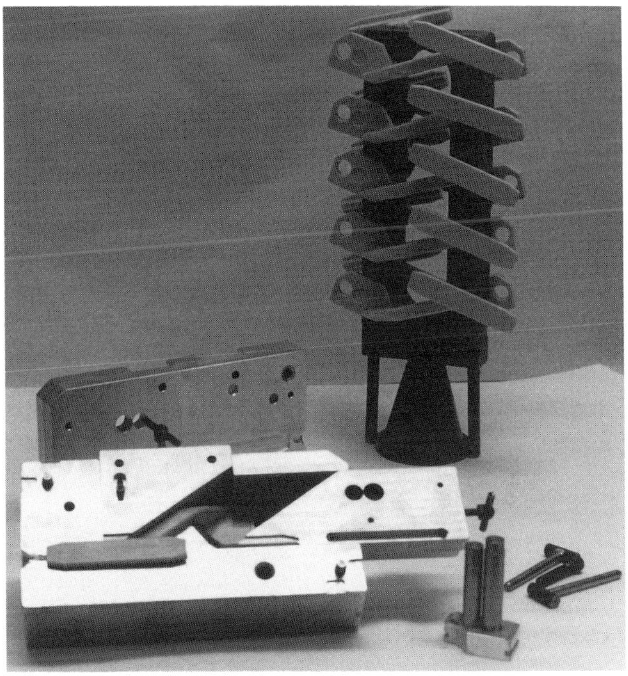

Fig. 12-4. Pattern material has been injected into this mold. A pattern tree assembled from patterns made in this mold is shown in the background.

Dimensions

Dimensioning the mold cavity is critical to the accuracy of the final casting. A number of dimensional changes occur during the investment casting process. When wax is injected into the mold, it contracts upon solidification. When the ceramic shell or solid molds are heated for dewaxing and firing, expansion of the cavity occurs. Finally, the metal cast into those expanded cavities contracts upon solidification.

Dimensional stability is affected by injection machine characteristics (for example, injection pressures), by whether pattern chills are used, by the type of pattern material, and by pattern geometry.

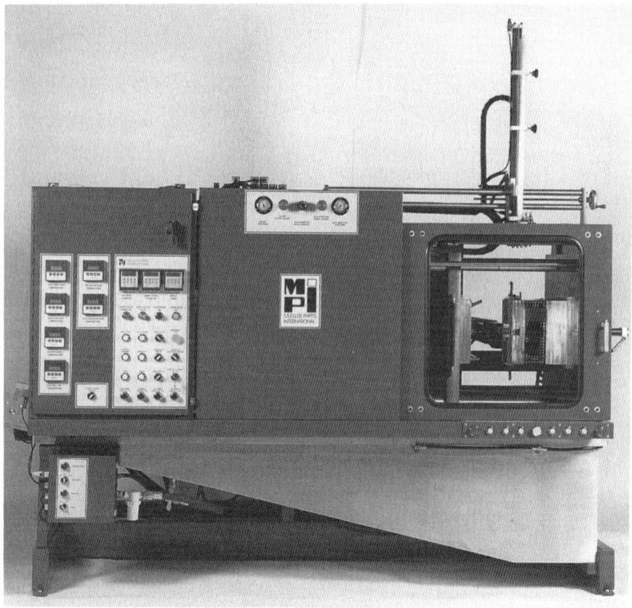

Fig. 12-5. An automated wax injection press with pattern removal conveyor. [Photo courtesy of Mueller Phipps International.]

Forming Interior Shapes

Three types of cores (removable metal, soluble and ceramic) are used to form interior shapes in investment castings (Table 12-3). Methods for using soluble and ceramic cores are described in further detail in the following sections.

Soluble Cores

To make complex interior shapes, soluble cores may be used instead of a complicated removable metal core assembly in the pattern mold. Soluble core waxes are comprised of three main components: a binder that is generally polyethylene glycol (carbowax), a powdered filler (usually silica, mica or sodium chloride) for improving strength and reducing shrinkage, and an effervescing agent (carbonates) to aid in the solubility of the polyethylene glycol.

The soluble core, produced by injecting the core material into suitable molds, is placed in the mold of the pattern to be produced, and pattern wax is injected around the core. The pattern is removed from the die and then is immersed in dilute hydrochloric acid, which leaches out the soluble core, leaving a cavity of the desired shape inside the pattern.

Ceramic Cores

If the exterior shell cannot extend into and form the interior areas, preformed ceramic cores can be used. Manufacturing ceramic cores requires additional tooling, or they can be purchased from a core manufacturer.

The cores are placed in the main pattern tool, as with the soluble type, and the pattern wax is injected around the core. Unlike water-soluble cores, ceramic cores are not removed until after the metal has been poured and the casting formed. Because ceramic cores are difficult to remove from the casting, they are not widely used in casting aluminum.

Pattern Assembly

A typical pattern assembly is shown in Fig. 12-6. The basic components are a pouring cup, runner system, ingate and pattern. Normally, patterns are attached to the runners by

Table 12-3.
Types of Cores Used for Forming Different Types of Holes

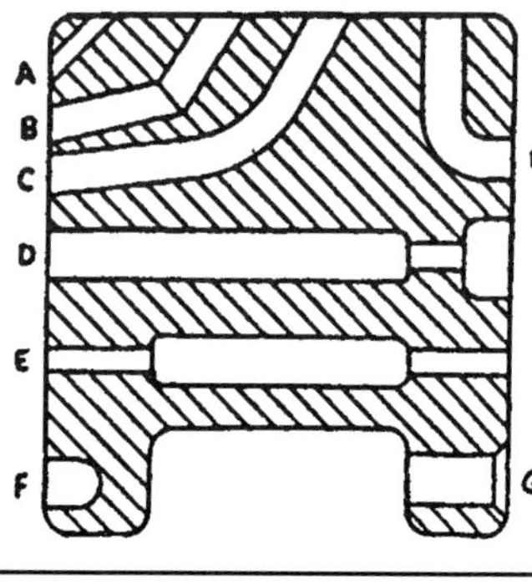

Hole	Metal	Soluble	Ceramic
A	X		
B	X		X
C		X	X
D	X		X
E		X	X
F	X		
G	X		
H		X	X

Fig. 12-6. A worker assembles wax patterns into a tree.

fusion, or cementing, with a tacky wax. For plastic patterns, a solvent cement is used.

When laying out the pattern assembly, a number of factors should be considered. Patterns must be spaced appropriately to assure adequate shell formation. Patterns should not be so close that the slurry bridges the gap between patterns. Bridging leaves a void and a thin shell that can easily crack on casting, allowing metal to run out. Because castings have to be cut off the runner, the cutoff wheel or saw must be able to access the runner without damaging the castings. Casting geometry and gating requirements often dictate the pattern density in an assembly, as do the individual component requirements.

Pattern assemblies should be inspected for the following:

- A good bond should exist between the gate and runner, and the joint line must be well-sealed to avoid formation of a ceramic fin, which could easily become a source of inclusions in the casting.
- The runner system must be adequate to feed castings upon solidification. It also must be structurally capable of carrying the pattern and shell material without distortion or failure during the shell buildup operation.
- Excess wax should not be present on the patterns (such as drops of wax from the joining operation), since these require that extra metal be removed from the casting.
- Patterns must not have wax flash at mold parting lines.

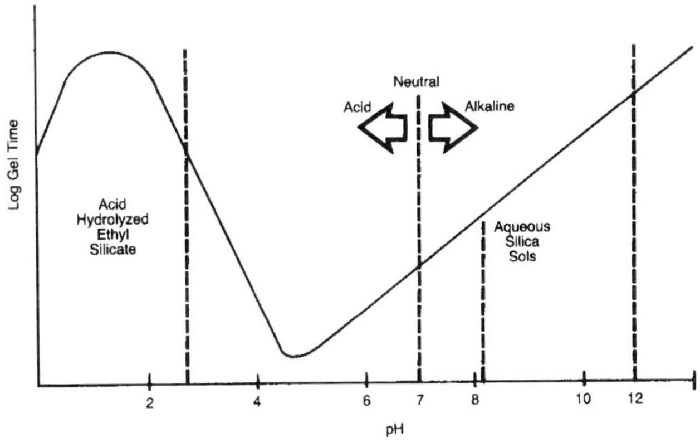

Fig. 12-7. Effect of pH on gel time of silica binders.

MATERIALS AND MOLD MANUFACTURING

This section can be subdivided into refractory binders and refractory materials. The general material in this section is relevant to both solid and ceramic shell mold production. Binders commonly used are colloidal silica and ethyl silicate.

Colloidal Silica Sols

Colloidal silica sols can be defined simply as minute particles of silica suspended in water. These silica particles are so small that they cannot settle out of suspension. The sols most usually employed have a particle size of 14 millimicrons; some "small particle" sols have silica particle diameters of about 7 millimicrons. Typically, colloidal silica has a silica content of 30%.

Colloidal silica binder systems normally are set by drying in air. The water evaporates, resulting in coalescence of the silica particles to form the green bond or gel. The stability of silica sols is affected by the pH of the system. Controlling the pH of colloidal silica is important for a stable product. The pH should be maintained between 9.5 and 10.7, depending on the manufacturer. pH values below 9 or over 11 can cause an unstable product where gelling will take place (Fig. 12-7).

Ethyl Silicate

Ethyl silicate is used by investment casters in the shell and solid moldmaking processes. This binder is used much more predominantly in Europe than in the U.S.

Ethyl silicate bound shells can be built faster than colloidal silica bound shells. This is the major advantage of this binder system. A major disadvantage is the emission of volatile organic compounds during shell drying.

Ethyl silicate is made from the reaction of ethyl alcohol with silicon metal in the presence of a catalyst. This product is then reacted with specific amounts of water in the presence of an organic solvent (usually ethyl alcohol) with an acid or base catalyst to produce a binder. This chemical reaction is called hydrolysis.

Gelling is accomplished by adjusting the pH range to 5–7. This can be done through various methods, such as exposure to ammonia gas or mixture with an ammonium hydroxide solution or amines.

Refractories

A wide selection of refractory materials can be used to make solid and ceramic shell molds. The most popular materials are natural silica, fused silica, alumina silicates, zircon and alumina.

The use of any of these materials will depend upon their availability, cost and compatibility. Availability and cost are perhaps self-evident criteria. Compatibility, however, is harder to determine. For example, the thermal expansion characteristics of refractories (Fig. 12-8) are very important to compatibility. Natural silica has a dramatic thermal expansion curve during both heating and cooling. This expansion can cause surface cracking of the mold. If two refractories are used in conjunction to produce a mold, a large differential expansion between layers of dissimilar materials can cause mold cracks.

As a general guide, refractories should be:

- refractory enough to accommodate the process;
- compatible with the binder system in use;
- compatible with a wide range of alloys;
- chemically and physically consistent;
- available with consistent gradings;
- readily removable from the casting;
- readily available and inexpensive.

By understanding the systems involved and careful selection, it is possible to meet most of these requirements.

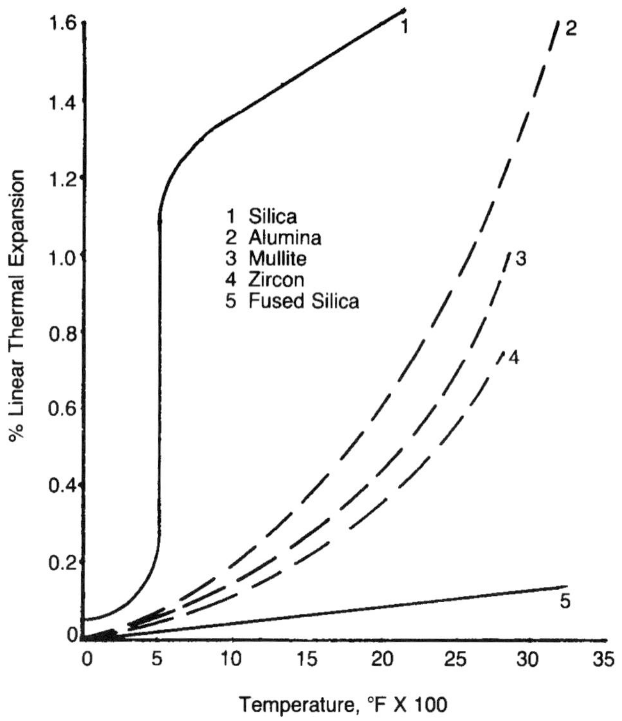

Fig. 12-8. Thermal expansion of investment casting refractories.

Refractories used in slurries range in size from –325 to –120 mesh. These refractories are called flours and, with the silica sol content, make up the percent solids in the slurry mix. Generally, flours of –325 and –200 mesh are used in the primary coat slurries. These fine flours provide an excellent cast surface finish.

The slurries formulated for backup coats are made of coarser-grade flours. Generally, –120 mesh flours are used.

Refractories used for stuccos range in size from –120 to –50 mesh. Finer stuccos are used on the first two primary slurry coats, and coarser stuccos are used on the backup coats.

SOLID MOLD PRODUCTION

Binders

Hydrolyzed ethyl silicate is generally used for solid mold production, and can be obtained in a prehydrolyzed condition. However, many investment casters hydrolyze the material as required.

To speed up the hydrolysis reaction, an acid such as hydrochloric (muriatic) acid may be added as a catalyst. This is usually added in the form of a 1% solution in amounts ranging from 0.03–0.1% of the total volume of liquid.

Table 12-4 provides typical formulations for hydrolysis of ethyl silicates to give specific silica contents in the binder.

The ethyl silicate should be added to the alcohol and water, and mixed thoroughly. There will be a rise in temperature as hydrolysis occurs. Once the binder has been hydrolyzed, it can be stored in a fairly stable condition for hours or even days, depending upon the degree of hydrolysis and the absence or presence of excess water.

Table 12-4.
Ethyl Silicate Hydrolysis Formulations

Silica Content	10	15	20
Vol. ES 40	23.5	35.0	47.0
Vol. Ethyl alcohol (95%)	76.5	62.8	48.2
Vol. H_2O + HCl	—	2.2	4.8
Vol. ES 28	37.8	56.6	75.5
Vol. Ethyl alcohol (95%)	59.0	36.0	12.9
Vol. H_2O + HCl	3.2	7.4	11.6

The control of gel time by varying the pH is widely practiced in the production of solid molds. Hydrolyzed binder is mixed with a suitable refractory, and the gel time is adjusted by adding basic (alkaline) agents such as magnesia (MgO).

Additions of water to the mix assist in fine control of the setting time. In this case, the addition of a few ounces of water to several gallons of binder is all that is required to adjust the gel time. Table 12-5 shows a typical solid mold slurry formulation.

Mixing time is usually quite short, about 2–3 minutes, and the mix consistency is thick. This material must be invested with the vibrators running, so as to fluidize the semi-solid investment. If this is not done, pattern damage will occur.

The molds should be vibrated for about 20–30 minutes to ensure that the investment packs evenly about the pattern. This will allow excess binder to work its way to the top of the mold. If vibration is continued for too long, fine segregation of the refractory will occur, upsetting the composition of the mold material.

After 30–40 minutes, the molds will gel and excess binder material can be cut off the top of the mold. Removal of the excess material is facilitated by disposable extensions on the containing flask, such as rubber sleeves or reinforced gum paper. After the molds are allowed to dry for 12 to 24 hr, they are ready for the burnout cycle.

Gypsum-Bonded Mold Materials

Gypsum-bonded materials consist of calcium sulfate-hemihydrate along with cristobalite, quartz, alumino silicates and various other additives to control properties such as viscosity, working time, thermal expansion and permeability. The

Table 12-5.
Typical Solid Mold Slurry Formulation

Hydrolyzed ethyl silicate		2.75 gallons
Other		160 lb
Silica sand	25%	
Fireclay	50%	
Silica flour	24%	
Magnesia (MgO)	0.4%	

(12-9a)

(12-9b)

Fig. 12-9. Molds are built up by dipping the pattern trees into a refractory slurry, applying a refractory stucco and allowing the coating to dry. This sequence, which is repeated until the desired shell thickness is acquired, can be performed (a) manually or (b) by robot.

procedure for using these materials generally involves mixing the powder with water in a predetermined ratio. The mixture normally is vacuumed to eliminate excess air that may have been entrapped during mixing.

The flask containing the wax pattern is then filled with the investment mixture and again vacuumed to remove air that may be trapped within the pattern. Wax is removed either by low-pressure steam or by low-temperature electric or gas-fired ovens. Wax residue is eliminated by firing to 1250–1300F (677–704C). Molds are then fired to the appropriate temperature for casting.

Applications for this process are in casting aluminum alloys for thin-walled aerospace and commercial castings. The process is used extensively for casting jewelry and various items of brass and other copper-base alloys.

CERAMIC SHELL PRODUCTION

The complete coating cycle consists of dipping the pattern assembly into a refractory slurry, draining off excess slurry, applying a refractory stucco, and allowing the coating to gel or dry (Fig. 12-9). This sequence is repeated until a shell of adequate thickness is achieved— usually about ⅜ in. thick.

The refractory stucco has two functions: 1) to minimize drying stresses in the slurry coating, thus preventing cracking; and 2) to provide a mechanical key for the next dip coat. The stress-relief function of the stucco is most important in the early stages of shell buildup.

When mixing slurries, always add the refractory flour to the binder. This ensures good dispersion of the solids. A wetting agent is recommended for the first coat to ensure good coverage of patterns. If the wetting agent encourages entrapment of air bubbles in the slurry, use an antifoam agent. Table 12-6 shows some typical slurry formulations.

Stucco Materials

Stucco is applied to the wet surface of the slurry to provide a key for the next coat of slurry and to minimize dry stresses. Ideally, the stucco should be similar in physical properties to the slurry refractory. For example, fused silica stucco should be used with fused silica slurries, and alumina silicates with alumina-silicate slurries.

The stucco is applied either by the fluidized bed method or by raining. In the case of raining, as the name implies, the stucco particles are allowed to fall in a raindrop pattern, via a diffuser, onto the wet slurry coating. This technique has some advantages with respect to the shell quality, but it is not as convenient as the fluidized bed method.

Table 12-6.
Typical Slurry Formulations for Nonferrous Metals

First Coat	
Colloidal binder (30% SiO_2)	25 wt %
Fused silica flour, –325 mesh	75 wt %
Wetting agent	10 mils/gallon of binder
Viscosity at 75F (24C)	40 sec No. 5 Zahn cup
Second Coat	
Colloidal binder (30% SiO_2)	30 wt %
Fused silica flour, –200 mesh	70 wt %
Viscosity at 70F (21C)	32 sec No. 5 Zahn cup
Backup Coat	
Colloidal binder (30% SiO_2)	37.5 wt %
Fused silica flour, –150 mesh	62.5 wt %
Viscosity at 75F (24C)	15 sec No. 4 Zahn cup

A fluidized bed consists of a vertical drum container with a porous brick bottom. A high volume of air at low pressure passes through the brick and the bed of stucco material. The effect of this air flow is to impart fluid characteristics to the stucco bed allowing the pattern assembly to be immersed into the stucco material as if it were a liquid. The stucco particles then stick to the wet slurry surface. Fine stucco is used for first and second coats, while coarser stuccos are used for backup coats.

Drying of Ceramic Shell Molds

The drying of shell molds is important. The basic mechanism takes place in two stages. First, a constant-rate period involves removal of surface binder liquid. Then, there is transfer of the binder liquid from the inside of the shell to the surface.

Control of temperature and humidity is crucial if reproducible results are to be obtained. The most important factor to control is the humidity range (±5% is good). In the case of ethyl silicate, the drying rate must be retarded sufficiently to prevent drying cracks. Hence, a relative humidity over 50% is desirable. In the case of a water-base binder, a relative humidity below 50% is desirable.

Constant temperature control is required because relative humidity will vary with temperature. In addition, the pattern will expand or contract with change in temperature, and this size variation can cause shell cracking. Temperature control within ±2°F (1°C) is desirable.

The constant-rate drying period is markedly affected by surrounding conditions. Assuming that air temperature and humidity are constant, the velocity of air flowing over the shell surface is a controlling factor. Industry tests have shown that, if air velocity drops below 200 ft/min, then other local factors, such as convection and radiation, will be more important. Ideally, air velocity should be in the 600–1200 ft/min range for best drying conditions. The first coats should be dried at much lower velocity levels to avoid drying cracks.

Setting or Gelling of Ceramic Shell Molds

When ethyl silicate binders are used for shell manufacture, rapid setting or gelling can be accomplished by adjusting pH to the unstable range. Ceramic shells can be built up rapidly by chemically setting each layer instead of drying. This normally is accomplished by placing the dipped and stuccoed assembly into an ammonia gas atmosphere.

Shells produced in this manner must be dried for 24–48 hr after shell buildup is complete. There is evidence, however, that green strength (strength of the shell up to the firing stage) is lower than air-dried shell. This method is very attractive for mass production. The shell can be completed and then dried elsewhere. If robot dipping is used, the handling problems are minimal.

Control of Ceramic Slurries

The following are a number of simple tests that help control ceramic slurries:

1) *pH.* The pH of a slurry can be measured using a calibrated pH meter.
2) *Viscosity.* The use of viscosimeters, such as the Zahn cup or the Ford cup, allows reliable checks to be carried out rapidly.
3) *Temperature.* Slurry temperature should be kept constant because viscosity varies with temperature.
4) *Modulus of Rupture.* By producing a test piece using a wax-covered form, a modulus of rupture test can be used to determine green strength and cold-fired strength of the shell. This test employs a three-point loading system, as shown in Fig. 12-10.
5) *Thermal Expansion.* Thermal expansion tests using standard dilatometer techniques will establish shell characteristics and may be crucial where precise dimensional control is required. Since most data on refractory systems are freely available, the thermal expansion test is largely academic.
6) *Sub-sieve Particle Size Testing for Refractory Flour.* A number of tests, both simple and sophisticated, can be used for particle size.

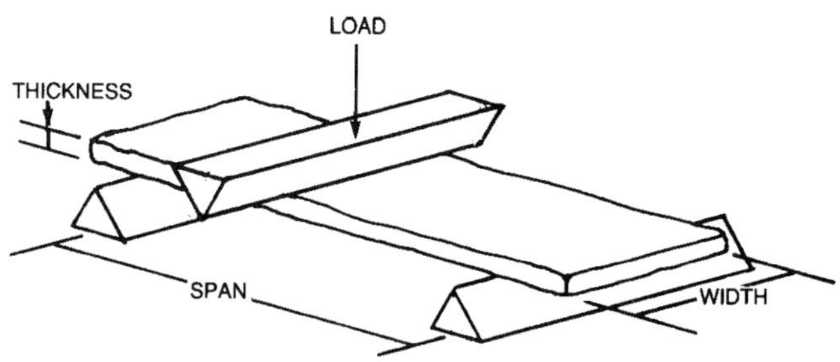

$$\text{Modulus of Rupture} = \frac{3}{2} \times \frac{\text{Load} \times \text{Span}}{\text{Width} \times (\text{Thickness})^2}$$

Fig. 12-10. A three-point loading system used to test for modulus of rupture.

7) *% Binder SiO_2.* Maintaining a controlled percentage of binder solids in the slurry is important. Insufficient concentration of SiO_2 in the binder creates a weak bond in the unfired mold. An excess concentration causes premature gelling of the slurry. The specific gravity of the binder, separated from the slurry, is the typical method used (Fig. 12-11).

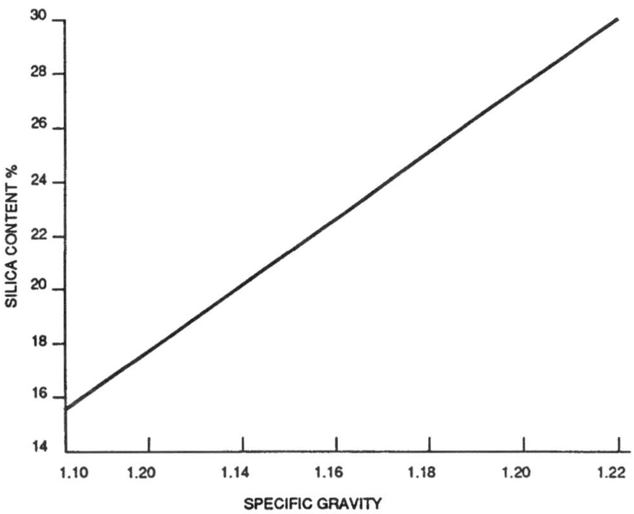

Fig. 12-11. Binder silica content versus specific gravity.

Fig. 12-12. A dewaxing autoclave with a steam accumulator mounted on top and internal and external carts for loading and unloading. Steam pressure of 130 psi can be reached in less than five seconds. [Photo courtesy of WSF Industries, Inc.]

DEWAXING SOLID MOLDS

Solid molds normally are dewaxed during the mold firing cycle. The strength of the solid mold is adequate to withstand the expansion stresses of plastic or wax patterns.

As the mold heats up during the firing sequence, the patterns melt and run out. The pattern material then burns off in the firing furnace. Care must be taken to ensure the presence of an oxidizing atmosphere during burnout so no carbon is left on the mold surface. Additional air should be available to minimize excess smoke discharge from the furnace stack.

DEWAXING CERAMIC SHELL MOLDS

A number of dewaxing techniques are available. The most common are autoclave dewaxing and flash firing.

Regardless of the dewaxing technique used, the system must provide a great quantity of heat to the shell so that the surface of the pattern will melt before the bulk of the pattern material begins to expand. In other words, there must be room for the bulk of the pattern material to expand without imposing to the point of bursting the ceramic shell.

Autoclave Dewaxing

The use of a steam autoclave for dewaxing (Figs. 12-12 and 12-13) has been very successful. The autoclave must reach a pressure of at least 60 psi within eight seconds, otherwise the heat input is inadequate, and shell cracking occurs.

Fig. 12-13. A worker removes shells from an autoclave after dewaxing.

The great advantage of using a steam autoclave is that the entire surface of the shell is uniformly heated. An economic advantage is the fact that the wax melted out can be recovered and reused.

Flash Firing

The basic flash firing process consists of introducing the shell into a high-temperature furnace (1400–1600F, 760–871C). The shell must first be drilled with ⅛ to ¼-in. holes, which helps to minimize shell stress from wax expansion during the initial stages of dewaxing. The intense heat source of the furnace causes rapid melting of the pattern material.

Flash dewaxing depends on radiant heat. Those shells or parts of shells that are overshadowed are not subjected to the same heat as unshaded areas. Consequently, local shell cracking or weakness can occur because of stresses imposed by expanding pattern material.

When flash firing is used, pattern waxes usually are not reclaimed. Some furnaces are designed to allow wax to run through the base of the furnace into a receptacle fitted below the furnace. This wax is often contaminated and partially oxidized. A distinct advantage of flash firing is that, because high-temperature strength of the shell begins to develop at this stage, the shell is somewhat stronger than autoclaved shells after dewaxing.

MOLD FIRING

Once molds are dewaxed they are subjected to a firing operation. They are heated in an oxidizing atmosphere to a temperature of 1600–2000F (871–1093C), depending upon specific foundry requirements. The residual pattern material is burned off and the high-temperature bond of the ceramic system is developed. Once firing is complete, the molds are brought down in temperature for pouring.

The firing cycle for ceramic shell molds is much shorter than that for solid molds because the latter are relatively bulky. Solid molds burn out much more slowly, and residual wax often migrates into the body of the refractory, where it carbonizes. Because carbon is relatively difficult to remove, burnout time must be extended to as much as 24 hr in an oxidizing atmosphere to clear carbon from the mold cavity face. If this is not done, metal-mold reactions can occur easily. In comparison, ceramic shells can be fired completely in as little as an hour.

METAL MELTING AND CASTING

Melting

Coreless induction and fuel-fired furnaces are the two main types of furnaces used in melting for the investment casting industry. These and other furnaces are covered in detail in Chapter 2. The advantages of induction furnaces include cleaner, quieter and faster melting. Its capital and maintenance costs are its disadvantages. Fuel-fired melting is common in nonferrous investment casting. This method is less costly than electric melting.

The vacuum induction melting furnace was developed to meet the requirements of the aerospace industry. Within its size range, it is considered to be unique to investment casting. Well-designed vacuum furnaces will operate at pressures as low as 10^{-5} Torr (0.01 microns).

Any type of aluminum alloy can be poured using investment casting. Most investment cast aluminum components are used for the aerospace industry. The alloys most commonly poured are 355, 356, 357 and 206. Alloys 356 and 357 are used to produce castings for commercial applications.

Pouring

Molten metal can be transferred from the furnace to the hot ceramic mold either by directly pouring from the furnace or by using a transfer ladle. Many high-quality aluminum investment castings are produced using a vacuum assist apparatus on the shell molds.

The vacuum assist device helps keep oxide formation and gas absorption during pouring to a minimum. This vacuum assist device usually consists of a split metal box that is hinged and gasketed. Sight glasses are sealed into the top and sides of the box.

Molten aluminum is transferred into a ladle from the melting furnace. The ladle is placed into the box on a rotating handle that is controlled from outside the vacuum box. The prefired shell is placed on a stabilizing fixture in the vacuum box. Then, the box is closed, a vacuum is applied, and molten aluminum is poured into the mold.

Solid molds can be cast on sand beds or, more usually, on vacuum casting tables (Fig. 12-14). In this method, the mold is placed over a port in the top of the table. This port is connected to a vacuum system and the mold is purged of air ahead of the incoming metal. The application of vacuum is believed to be essential for reproducing fine details on small investment castings.

Centrifugal casting methods have been developed to a high degree of excellence by a number of specialist foundries. The use of centrifugal force ensures excellent fill-out of small castings. Specially designed systems permit several hundred castings to be cast in one mold.

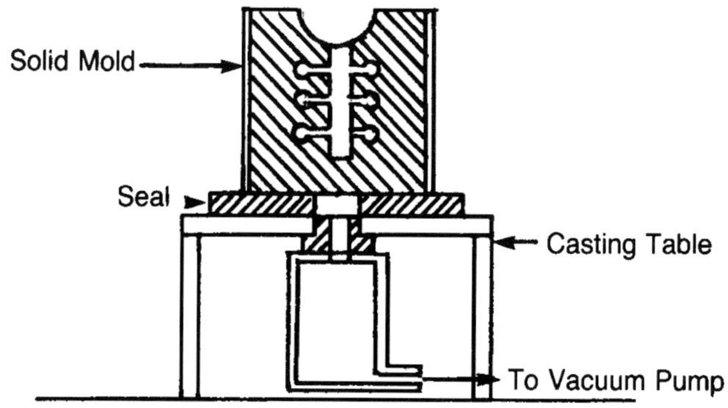

Fig. 12-14. Vacuum casting table for solid molds.

POSTCASTING OPERATIONS

Knockout

After molds have been cast and cooled, vibrating hammers are used to remove the bulk of the mold materials. If the alloy cast is brittle or if the castings can be damaged easily by mechanical vibration, high-pressure water jets may be used instead of the vibrating hammer. Nonferrous investment casters usually use this water blasting method. Residual refractory may be removed by chemical or mechanical cleaning.

Mechanical Cleaning and Finishing

Mechanical cleaning and finishing operations are essentially the same as for sand castings: abrasive blasting, abrasive wheels, friction saws and abrasive belts are all used for completing the investment casting process.

Inspection

Standard inspection procedures include visual inspection, radiography and hardness testing. Dimensional inspection is very applicable to investment castings. Coordinate measuring machines (CMMs), gauges, calipers and micrometers are all standard inspection equipment. The production of airfoil sections requires tailor-made inspection methods including fluidic gauges, profile templates, eddy current bridges, electrical induction measurements for internal passages, etc.

MECHANIZATION OF INVESTMENT CASTING

Mechanization of the investment casting process is gradually gaining momentum. In some cases, the driving force has been the need for equipment to handle larger molds, thus minimizing operator fatigue while providing more reproducible production conditions. Many mass-produced investment castings involve orders as large as hundreds of thousands of pieces.

A number of foundries in the U.S. use robot mechanical aids for readily programmed sequences such as shell buildup (see Fig. 12-9b). A few large foundries have built complex multistation mechanized plants for processing as many as 10 clusters simultaneously at each stage of a multistation unit.

Fig. 12-15. An automated assembly machine.

Such equipment is capable of processing hundreds of thousands of castings a week.

Other areas of automation include assembly (Fig. 12-15), drying, pouring and gate grinding.

The U.S. has been in the forefront of technological developments for the investment casting industry. Research programs in elastic wave core removal and microwave dewaxing allow the U.S. industry to stay ahead of the rest of the world.

ACKNOWLEDGMENT

AFS would like to thank Daniel Twarog, AFS Director of Research, for revising this chapter.

Chapter 13

EXPENDABLE PATTERN CASTING

HISTORY

Expendable pattern casting (EPC) involves the replacement of a low-density foam pattern by liquid metal. The casting that emerges is an exact duplicate of the foam pattern. A U.S. Patent was granted to H.F. Shroyer in 1958 for cavityless casting using bonded sand. In 1964, a patent was issued for a similar method using unbonded sand.

While Shroyer was concerned primarily with using his process to produce art castings, A. Wittmosser from a West German company purchased the patent with an eye toward its industrial applications. After the first EPC commercial casting was produced in West Germany in 1962, a group of Americans made an agreement with the German owners of the Shroyer patent specifically to produce iron and steel automotive stamping dies.

After the 1964 patent expired in 1981, interest intensified in applying the process to smaller, mass-produced parts. Extensive development and experimentation in the design of pattern tooling and coatings has led to use of the process in Europe and North America in a variety of applications.

Proponents of the EPC process argue that it has attracted considerable interest based on cost savings (both capital equipment and operating); greater design freedom; less dimensional variability; and cleaner, quieter working conditions.

The original application of EPC involved the use of foam board that was cut and milled to the same tolerances as conventional patterns, primarily to produce large, one-off stamping dies in bonded sand. This process is still used today to produce intricate, one-off or short-run large castings.

At about the same time, use of expandable polystyrene (EPS) began. First commercially produced in the U.S. in 1954, EPS is manufactured in bead form from hydrocarbons—about 94%C, 4%H_2, various trace elements and a blowing agent (usually pentane)—that expand and fuse when heated in the tool, creating the foam pattern. After the patterns are removed from the molding machine, they are aged to obtain the desired pattern shrinkage. Patterns are then attached to a sprue using a single gating system (cluster). The cluster is coated with a special permeable wash, dried and placed in the flask (Fig. 13-1).

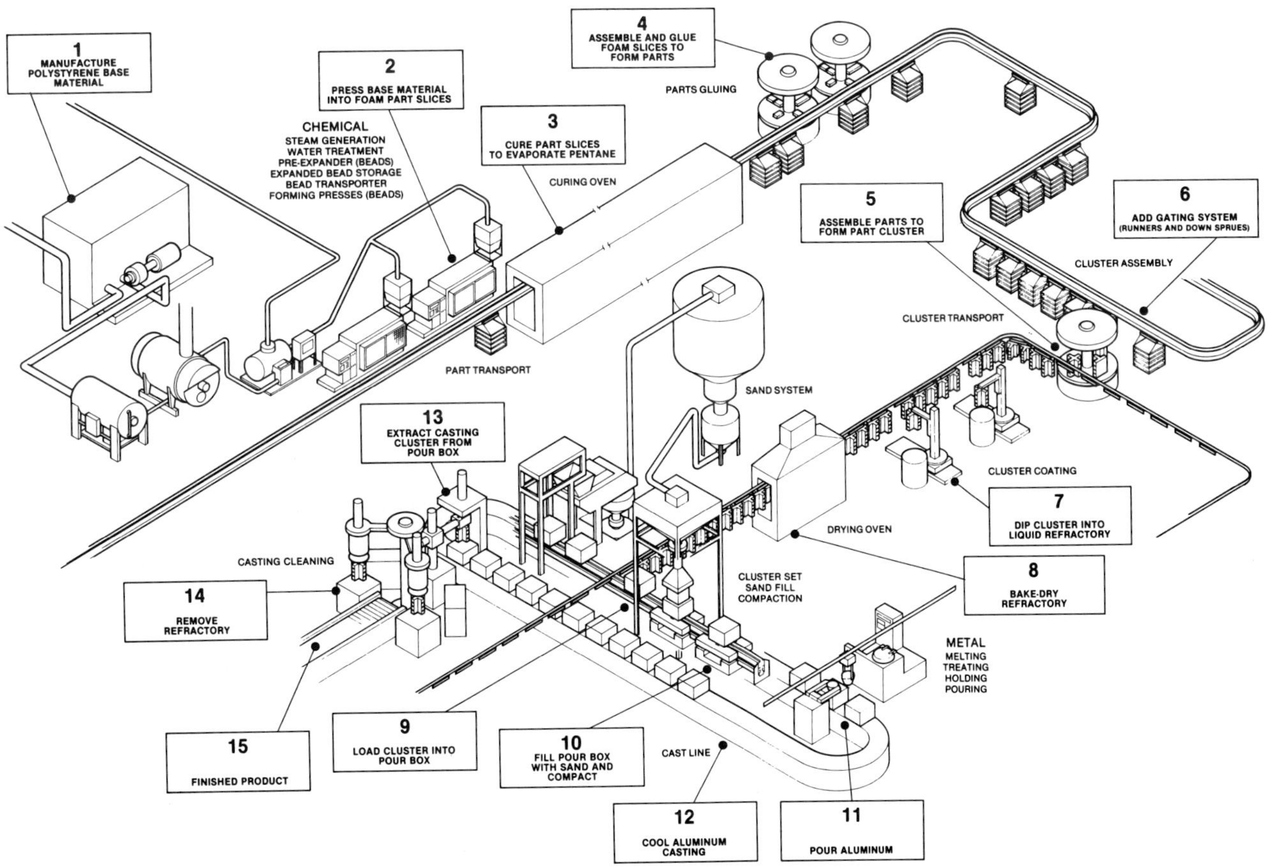

Fig. 13-1. A schematic of an EPC process system. [Drawing courtesy of Jervis B. Webb Co.]

One of the most significant breakthroughs in adapting the EPC process to the mass production of castings was the use of unbonded sand. The process patented in 1964 involves placing the patterns in a simple, one-piece cylindrical flask partially filled with clean sand. Additional unbonded sand is added and compacted using vibration.

EXPANDABLE POLYSTYRENE PATTERNS

Raw and expanded EPS beads come in several sizes (Tables 13-1 and 13-2). The desired surface texture and surface feature definition usually require that foam patterns be made from small beads because these beads allow molding of thin sections. Type "T" beads are the most common for pattern production.

Preexpansion

Preexpansion of the beads is the first critical step, assuring beads of low density. Small, batch-type vacuum preexpanders with an output of 15 kg/hr commonly are used.

A partial vacuum increases the pressure differential between the interior and exterior, creating beads that have considerably lower densities than are possible at atmospheric conditions. The beads are then cooled while still under vacuum (to prevent collapse) and screened to remove lumps.

The prepuff (the preexpanded beads) have densities ranging from 1.5 lb/ft^3 for aluminum casting patterns to 1.2 lb/ft^3 for iron castings. Gas formation during metal pouring is a result of both pour temperature and pattern density. The greater the pattern density, the greater the gas that will be evolved at a constant pour temperature. Higher temperatures result in increased breakdown of the molecules and more gas.

Table 13-1.
Sizes of Raw EPS Beads

Bead Type	U.S. Sieve Size	AFS gfn	Raw Size (in.)
A	10–20	8	0.079–0.033
B	16–30	13	0.047–0.023
C	25–45	24	0.028–0.014
T	36–60	38	0.020–0.010

Table 13-2.
Preexpanded EPS Bead Sizes

Bead Type	1.0 pcf (in.)	1.25 pcf (in.)	1.5 pcf (in.)
A	0.26-0.11	0.25-0.10	0.23-0.09
B	0.16-0.08	0.15-0.07	0.14-0.07
C	0.09-0.05	0.09-0.04	0.08-0.04
T	0.07-0.03	0.06-0.03	0.06-0.03

Timing is critical in handling the expanded beads because they require about 2 hr to stabilize, but foam quality declines after only 24 hr. The beads are stored in nylon mesh bags.

EPS Molding Cycle

The EPS molding process is somewhat analogous to core blowing. In both cases, the granular material is blown into a preheated tool cavity. In EPC, the two sections of the tool are backed with detachable steam chambers (Fig. 13-2) and contoured to ensure an even wall thickness behind the cavity. The steam chest is a chamber that allows steam to flow into the mold cavity through vents on the face to permit either steam or air to enter the mold. The tool parting line is precisely machined and generally sealed.

Unlike conventional core blowers, EPS molding machines use a blowfill gun and an air venturi or nozzle connected to an adapter tube, which provides a path for the beads from the venturi through the steam chest and into the mold cavity. The preexpanded, aged beads are emptied into a hopper on the molding machine. An air-operated piston within the adapter tube retracts to open the tube for the tool filling operation and remains closed for other phases of the cycle.

Obviously, both the air already in the tool and the transport air must be vented, and several varieties of fill guns and venting methods have been developed. In conventional vented molding machines, the air escapes through vents and into the steam chest as the mold fills with beads. The vent area should be about 2% of the total cavity surface area. For added venting, the tools have been preset to remain open a few thousandths of an inch at the parting line.

When the tool cavity has been filled, the air forcing the beads into the cavity is forced back up the fill tube to the hopper. This vacuum reversal in the fill tube (called "blowback") activates a sensor that retracts the cylinder into the mold wall, closing the fill passage.

Because the surface of an EPC casting replicates the foam pattern, the quality of the tooling is critical, both in terms of surface finish and dimensional stability. As the EPS beads absorb heat from the steam, the pentane in the EPS beads expands, forcing the beads to compress against one another. Those beads in contact with the tool surface flatten out and

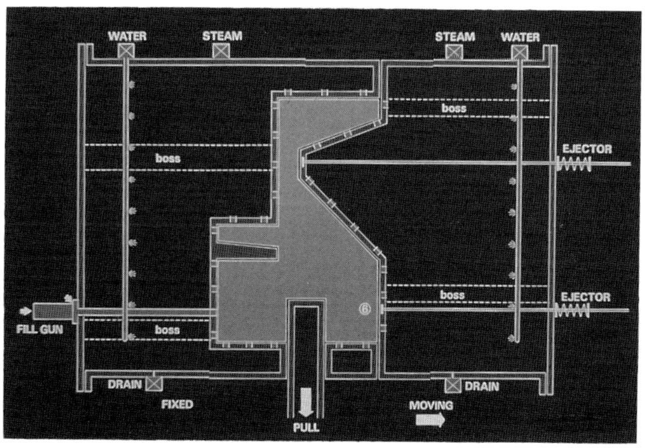

Fig. 13-2. A schematic of a tool design with a steam chest.

fuse, forming the desired shape. Temperature control is critical to ensure uniform fusion without distortion.

Ventless tools use two separate steam systems—one for the tool face and one for the steam chest. Proponents argue that the separate system permits improved steam temperature, pressure and moisture control. They also point out that conventionally vented tools can create watermark defects on the surface of the pattern that end up on the casting.

The cooling cycle stabilizes the pattern to keep it from expanding upon ejection from the tool. Water inlets and drains in the steam chest provide the cooling. The relative cooling of different sections will be a function of thickness; thinner sections cool faster because of faster heat transfer. A vacuum may be applied to reduce the amount of water required for cooling and the amount of pattern shrinkage.

The molding cycle is completed when the patterns are removed from the tool cavity. A variety of methods are used: air, mechanical or vacuum ejection. Cycle times vary, but typical molding cycles are 1.5 min for patterns of 0.5 in. or less in thickness.

At the time of removal, the pattern has no "memory" and extreme care must be used not to leave an indentation on the pattern surface. Automated pattern demolding can help eliminate this problem. In manual removal, because of the continuous possibility for pattern damage, thorough pattern inspection must be performed routinely.

Pattern Storage and Handling

When patterns are first ejected, they are approximately 0.50 in. undersized. Expansion of the pattern occurs during the first four hours, typically 0.2–0.3%, followed by pattern shrinkage of 0.7–0.8% over the next 30 days. Approximately 75% of the total shrinkage occurs in the first seven days, with significant shrinkage ending in the first 18-20 days (Fig. 13-3).

Full 30-day storage is usually not practical because of the large number of patterns produced. Most foundries target a shrinkage "window" and size their patterns accordingly. This reduces storage area requirements but requires more sophisticated control—including pattern dating—of all pattern and casting production procedures.

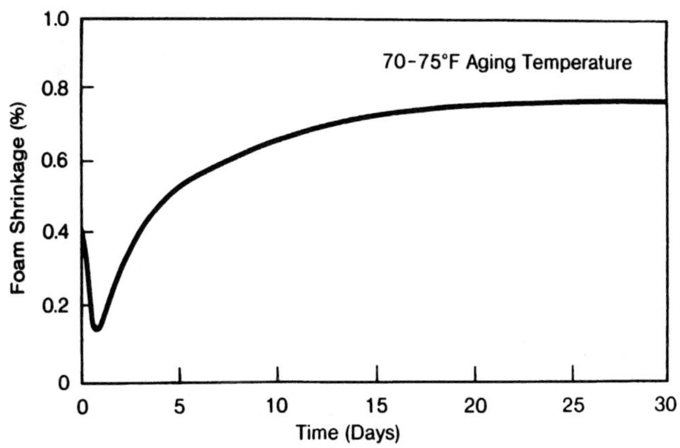

Fig. 13-3. EPS foam shrinkage by days.

EPS shrinks at a predictable rate of between 0.005 and 0.008 in./in., depending on the cross-sectional wall thickness, age of the beads prior to expansion, molding technique and density of the molded pattern. Studies have shown that shrinkage also varies with pattern geometry.

In addition to the normal pattern shrinkage allowance for the particular metal cast, EPC patterns are designed with an additional 0.1 in./ft (0.008 in./in.) of shrinkage. Shrinkage is due to two factors: gas equilibrium and molecular relaxation. New patterns typically have 2% residual pentane, which decreases to 1% after 2–3 days and approximately 0.5% after 30 days.

Relaxation is a phenomenon of molecular stress relief. The stress is incurred during the rapid heating and cooling cycles of preexpansion and molding, which is "frozen" during cooling. Over time at ambient temperatures, the "frozen" state stress relaxes, and the patterns shrink.

Shrinkage also is related to the degree of pattern fusion. Tests of three steam pressures—low, normal and high—showed that high pressure produced the maximum fusion possible, and the lowest shrinkage. It is theorized that high pressure acts in a manner similar to annealing-relieving stress during molding (Fig. 13-4).

A second factor found to contribute to lower pattern shrinkage is the duration of prepuff aging. Extending the aging cycle from four to 24 hours produced a slight but measurable increase. The increased aging permits greater pentane loss and therefore less stress during molding (Fig. 13-5).

If a vacuum is applied during the cooling cycle of the pattern molding process, the residual pentane and stress relaxation is reduced. This reduces the amount of shrinkage, as seen in Fig. 13-6. Density of the pattern also exerts a minor effect on shrinkage (Fig. 13-7).

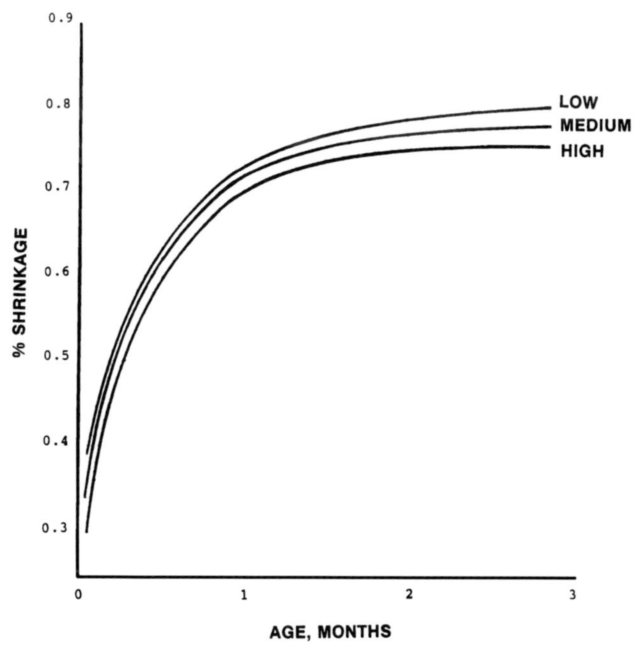

Fig. 13-4. Effect of molding pressure (conventional molding, 4-hr prepuff age, 1.6 pcf density).[1]

Pattern Assembly

Those foundries producing complex patterns with difficult-to-mold internal passages may opt to produce the patterns in sections with locators and recesses, then glue them together. Liquid and hot melt adhesives are frequently used, but foam and air-set compounds also are available. Glue printers and robotic assemblers are used frequently to apply the liquid glues. Excessive amounts of glue at the joint may actually form an external bead, which will be replicated in the casting. Defects also are produced when the molten metal actually stops or passes around the glue joint.

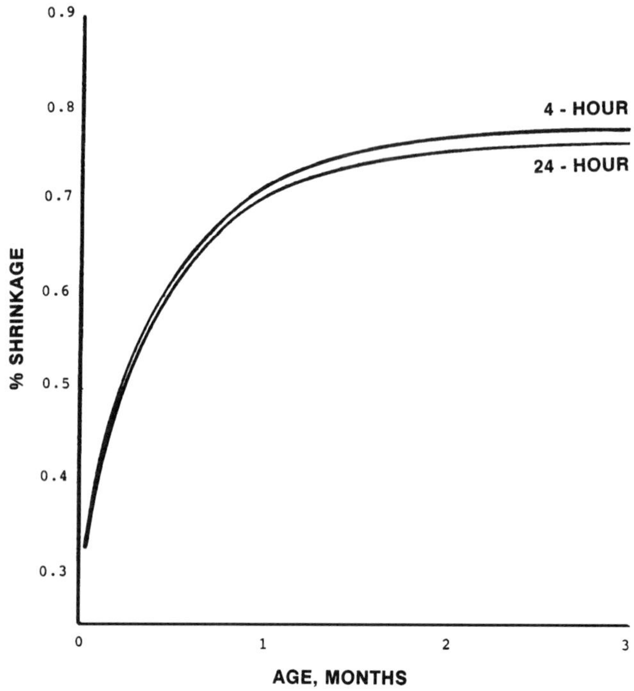

Fig. 13-5. Effect of prepuff age (conventional molding, 1.6 pcf density, standard steam pressure).[1]

Adhesive assembly adds as much as 50% to the total cost of the assembled EPS pattern. In addition, assembly also adds the possibility of misalignment and failure along the glue line, which also may be imparted to the casting. For these reasons, pattern producers have invested in tooling with multiple pulls, collapsible cores and composite actions, eliminating assembly and most or all gluing. It also is advised that if patterns are assembled in sections that the separate pieces be produced in the same tool to minimize instability over the shrinkage period.

Tooling

Tools (molds) for EPS patterns can be cast, although they are generally made of machined wrought aluminum. Cast aluminum tools are not recommended for high production because of the possible presence of porosity, oxides and shrink. Plastic prototype tools also have been used, where the tool life expectancy is 300-500 patterns.

Tool wall thicknesses range from 7–10 mm. Wear on the tooling is considerably less than in conventional systems, because the machines operate at lower pressure and are not subject to the abrasive action of sand. Both the pressures and the abrasive wear on the tooling are far lower than in sand coremaking.

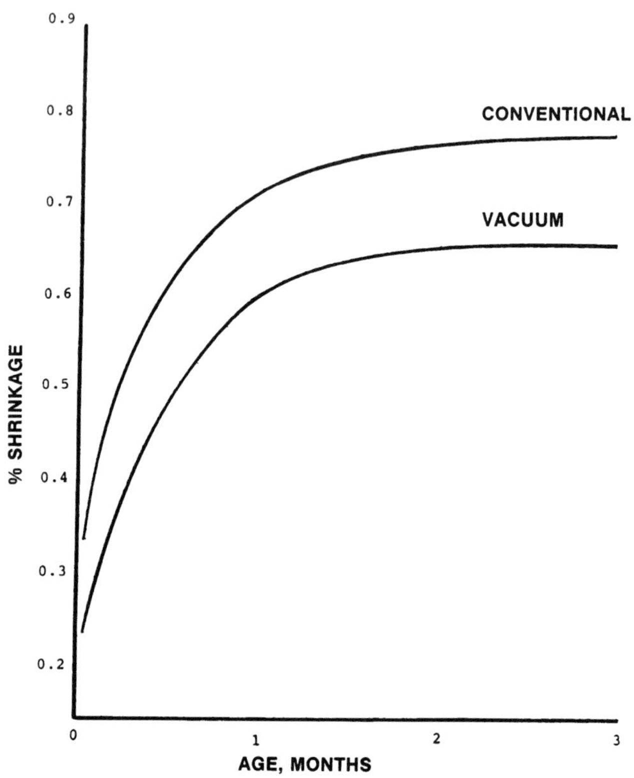

Fig. 13-6. Effect of conventional versus vacuum molding methods (4-hr prepuff age, 1.6 pcf, standard steam pressure).[1]

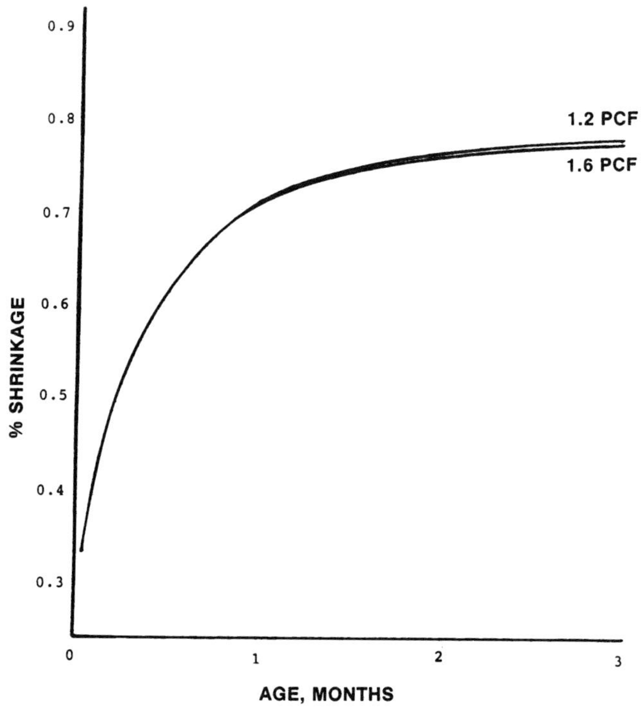

Fig. 13-7. Effect of density on pattern shrinkage (conventional molding, 4-hr prepuff age, standard steam pressure).[1]

Pulls and Pullbacks

Pullbacks must be designed to withstand the heating/cooling cycles that are unique to the EPC process and occur continuously during the molding of the pattern. This may result in considerable apparent misfit of the tool and pulls at ambient temperatures. The tool should be dry-cycled to full operating temperature before bead injection to eliminate the possibility of excessive flash formation, jamming and tool damage.

Tool design must take into consideration the alignment of the pullbacks. A solid pull will not heat or cool in the same manner as the wall of the tool, creating an initial cold surface condition. This will produce poor fusion and surface finish of the pattern. As the number of cycles increase, blistering or shrinkage of the pattern surface away from the hot core will occur. One frequent result is the foam pattern will actually adhere to the hot core and be deformed as the core retracts.

Special attention must be given to curved pulls to ensure that they operate smoothly and repetitively. Pullbacks that meet should be machined accurately to mate perfectly in the extended position to eliminate flash on the foam pattern.

COATINGS

Patterns for aluminum castings are usually coated for two reasons: 1) to eliminate metal penetration and burned-on sand ; and 2) to improve as-cast surfaces. Additionally, coatings reduce the incidence of lap or fold defects in aluminum.

EPC coatings consist of:

- Carriers, usually water, which function to convey the refractory onto the pattern either via dipping, spraying or flooding;
- Refractory materials, which determine the permeability of the coatings, their heat transfer or their insulating properties, providing barriers between the molten metal and backing sand;
- Suspending agents, which ensure even deposition of the refractory coating on the patterns. They provide the correct dilution necessary to adequately coat the pattern;
- Binders, which provide cohesion, imparting dimensional stability and improving pattern resistance while the flask is filled with sand;
- Rheological agents, which impart the required thixotropic properties, in turn determining the amount of coating deposited and the desired dispersion on the pattern;
- Preservatives (and bactericides), which protect the coating from deterioration during aging, and prevent the degradation of any organic ingredients.

Coating Tests

Tests are available for coating density, percent of solids, viscosity and dry coating weight. Before performing any tests, however, the coating should be thoroughly mixed to ensure that no stratification of ingredients has taken place. The mechanical mixer must be capable of creating a vortex in a drum or tank of coating. In fact, continuous mixing of the coating at low speed, except while dipping, is recommended.

A Baumé (specific gravity) reading of the coating slurry cannot be taken accurately with a Baumé gauge because it will indicate both density and viscosity, but not the contribution of each.

Density or specific gravity (the weight in grams divided by the volume in milliliters) is an indication of the consistency of the coating, and can be determined using a weight cup or volumetric container.

Percent of solids (by weight) can be determined by drying a sample of coating to a consistent weight then comparing it to the weight of the original sample. The percentage of percent solids divided by the weight of the original sample is used to maintain the consistency of the coating.

Accurate viscosity measurements, taken with a viscometer, are necessary to insure repeatable coating results. Combined with the percent of solids reading, correct viscosity allows the determination of the coating weight.

However, dry coating weight is the single most important test in controlling and adjusting the slurry coating. The method of determining dry coating weight is essentially experimental: preweighed clusters are dried and reweighed. The weight of the dry coating is then recorded and correlated with the casting results.

Repeated experiments should indicate a range of coating weights, varying with cluster size and surface area, that will produce good castings. This range can be expected to be from 10–20% of the total dry coating weight. For example, a cluster with a dry total coating weight of 500 g should have a range of 50–100 g per cluster, yielding reproducible results of from 475–525 g of dry coating per cluster.

Changes in handling methods and procedures, cluster size, temperature, tank size, mixing equipment and method of application all can influence coating rheology. Knowing the desired coating operating range makes adjusting the rheological properties of the coating considerably easier.

SAND

Sand used in the EPC process can be either washed and dried silica sands, lake sands or other good sand. When pouring aluminum, sand compaction is not as critical as when pouring brass, iron or steel due to the lower-density metal and lower pouring temperature. Round sand grains give a higher permeability compared to subangular sand with similar AFS numbers, but their greater fluidity improves the probability that internal pattern passages will be filled consistently.

One of many advantages to the EPC process is that sand is reusable with minimal reclamation. Aluminum vaporizes polystyrene but does not have a high enough temperature to maintain the vapor, which rapidly recondenses in the sand. If only aluminum is cast, the sand must be reclaimed and/or new sand added based on continuous LOI monitoring.

FLASKS

One of the main advantages of the EPC process is that mated cope and drag flasks are eliminated. Indeed, the flask can be any container rigid enough to withstand vibration and the weight of the sand. In practice, flasks are usually designed for optimum handling, pouring, shakeout and vibrating. In some automated systems, the flask consists of two chambers: a sand chamber and a blow chamber, separated by a blow plate used in a sand conditioning process that takes place before shakeout.

GATES AND RISERS

Gating system development is a critical step in achieving successful casting production. However, no uniform set of gating rules has yet been developed for EPC.

Design of the EPS gating system is considered extremely important. The gating is "upstream" from the casting, and any design shortcomings directly impact on the casting. As in conventional casting, the gating system must provide an adequate path for the metal to both fill the casting and feed its critical sections. Sufficient stiffness must be a part of the design of the gating system in EPC to avoid deflection. The gating system must be robust enough for handling the cluster during coating, drying and compaction.

Gating systems normally are classified according to where the metal is allowed to enter the mold cavity. Patterns can be top, side or bottom gated (Fig. 13-8). Top gates have the fastest filling times and actually require the fastest pouring to avoid collapse. Using side gates results in the slowest filling, while bottom gates are somewhere in between.

Aluminum is commonly top gated because decomposition of the foam is slow enough that the casting cavity will not collapse. It is difficult to control the flow of metal in top or side gating to avoid entrainment of some plastic residue.

Another consideration in gating design is sufficient access for trimming. The gating and part design must optimize the number of parts that can be accommodated by the standardized, cylindrical flask.

POURING

EPC pouring practices differ significantly from conventional foundry methods. While the standard concerns—avoiding turbulence, aspiration of air into the metal stream and erosion of the mold during pouring—are still important, two other concerns must be addressed: supporting the casting cavity wall and progressive elimination of the foam pattern.

EPS begins to break down at approximately 150F (66C), decreasing in volume to about 1/40 of its original size. Decomposition of the molten plastic mass occurs at 400F (204C).

Foundries have found that an extremely fast rate of pouring (keeping the pouring basin full) to achieve a substantial metal head pressure helps contain the gas formed by EPS decomposition by forcing it down into the foam, along the gaseous interface cavity between the incoming metal and solid foam, and eventually out through the flask vents. When pouring aluminum, this has the advantage of reducing sooty combustion, but with metals with higher pouring temperatures, the sand adjacent to the mold "cavity" can be fluidized, resulting in defects.

As the pour temperature increases, so does the rate of foam decomposition. This may require some adjustment in the runners and sprues in order to maintain adequate metal flow and prevent "starving" the pattern. Lighter density metals, such as aluminum, require greater sprue height or possibly a tapered or hollow sprue to compensate for the lower weight.

When pouring molds by hand, fast pouring rates are achieved by dumping a ladle containing exactly enough metal for the fill. The furnace must, of course, be in close proximity to the mold, and an enlarged pouring basin or funnel is required to provide an easy target and little overflow.

An alternative to the lower yields resulting from the larger pouring basin is a superimposed pouring basin, either an expendable core or a metal pour basin that is completely drained of metal in each casting cycle, removed and repositioned on another mold before pouring.

Regardless of the metal poured, the only opening in the mold is the sprue core. No open risers, vents or pop-offs are used because any air that would enter would create combustion and a totally unsatisfactory casting.

Gases emitted by the pattern combustion can be exhausted through the flask vents directly to a vacuum system for treatment, or a standard exhaust hood can be put in place over the pouring station.

Molds are also poured under vacuum, eliminating the combustion problem. The top of the mold is covered with a sheet of paper or polyethylene, and a thin layer of sand is spread over the sheet to prevent burning while pouring.

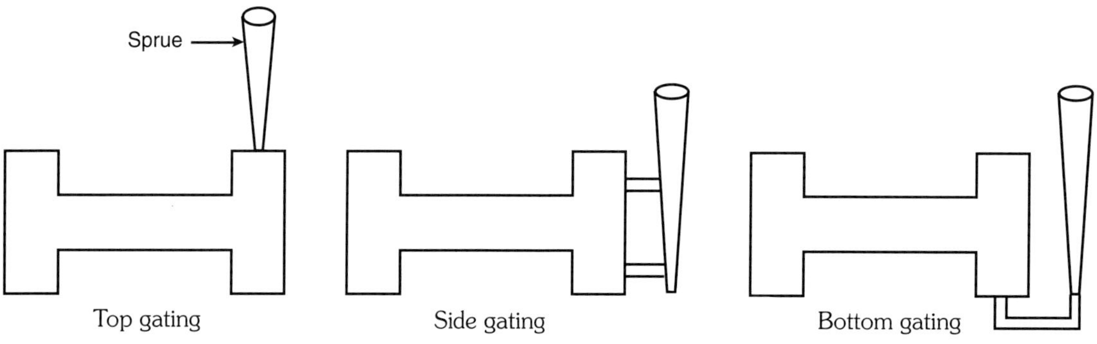

Fig. 13-8. Illustration of top, side and bottom gating.

Mechanized Ladles

A number of different automatic pouring devices are in place at foundries using EPC. Automatic ladling may be purely mechanical or robotic. In either case, the exact amount of metal required for the pour is delivered. To accomplish this, slag, dross and sludge must be prevented from entering the ladle. With conventionally shaped ladles, the skimming action involves lateral movement of the ladle bottom, which brushes aside any slag or dross prior to dipping. The ladle is then rocked over at an angle and quietly immersed into the skimmed metal to a controlled depth.

Pneumatically pressurized pouring vessels are used for direct pouring from the furnace into the pouring basin. This minimizes temperature loss, avoiding the use of an intermediate vessel. Slag and dross traps or filters also are used.

If the entire holding furnace is pressurized, the metal is forced out a teapot spout or vertical stalk into a horizontal launder leading to the mold pouring basin. Control response depends on the volume of air above the metal—the larger the volume, the slower the response.

CASTING QUALITY

Metallurgical quality is used to describe the solidification microstructure and small defects that determine the properties of the casting. The change in cooling because of foam decomposition and the possible entrapment of plastic residue both have raised the question of casting quality in EPC.

The alloys most commonly cast using EPC are the aluminum-silicon alloys 319, 356 and 390. Several evaluations have been made on the influence of EPC on casting microstructure. There also have been studies on EPC defects such as bubbles, folds or laps, black inclusions, and porosity.

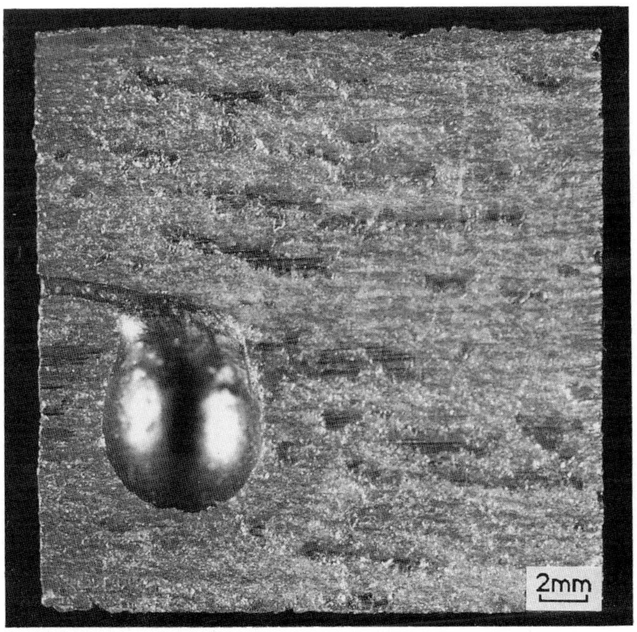

Fig. 13-9. Photograph of a gas bubble connected to the surface through a small passage. This bubble may have originated from liquid degradation products trapped in the metal.[2]

Porosity

Porosity is a major factor in all forms of aluminum casting production. Hydrogen dissolves readily in the molten alloy but is rejected during solidification to form porosity. Control of hydrogen levels in the molten metal supply is important if porosity is to be avoided. Porosity is often found in isolated heavy sections of the casting, associated with areas prone to ordinary shrinkage.

Porosity also has been found at the location of glue joints. The density of the glue is significantly greater than the foam and, consequently, produces more gas. Reducing the amount of adhesive deposited lowers the amount of gas. One adhesive manufacturer reports that 2200–2700 cc of gas is evolved per gram of its adhesive, but it is continuing to study the effect of this on casting quality. Defects also can be produced when the molten metal actually stops at or passes around the glue joint.

Traces of plastic residue from foam or glue can form gas bubbles that also may be trapped in the molten metal. At the metal-mold interface, the products of liquid degradation may accumulate before they are gradually decomposed, forming trapped gases. Gas bubbles also have been associated with folds or laps (Fig. 13-9).

Folds

Folds appear like cold laps, where two metal fronts touch but are not fused (Fig. 13-10). These folds contain some organic residue and oxide films. The presence of folds in critical casting areas can cause leakage in pressure-tight components or inferior mechanical properties.

Folds are not easily detected on the surface. The fracture appearance of a fold is shown in Figs. 13-11 and 13-12.

Folds or laps have been associated with bubbles in aluminum EPC castings. Fold occurrence is much higher in high-silicon alloys, such as 390, compared to Alloys 356 or 319. The folds in 390 also are associated with poor surface texture,

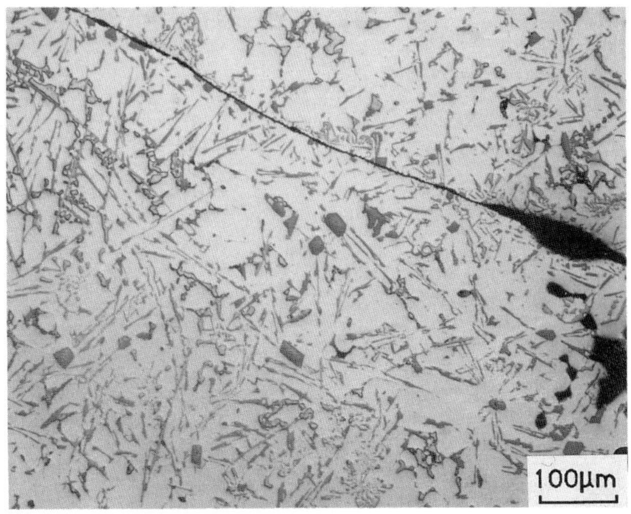

Fig. 13-10. Microstructure of a fold defect in a pattern showing segregation of alloying elements near the two metal fronts.[2]

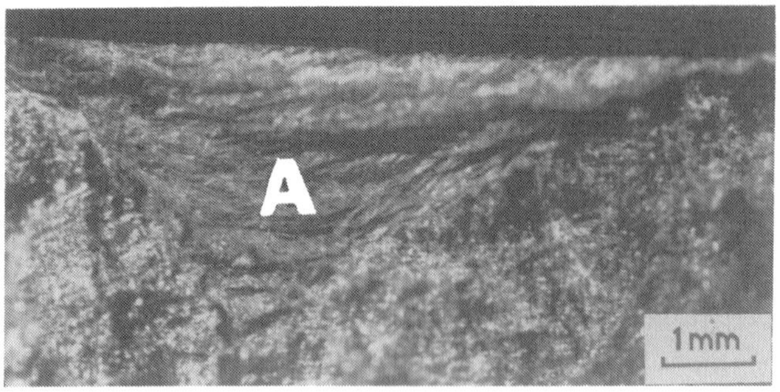

Fig. 13-11. Defects on the fracture surface of a fold (A).[3]

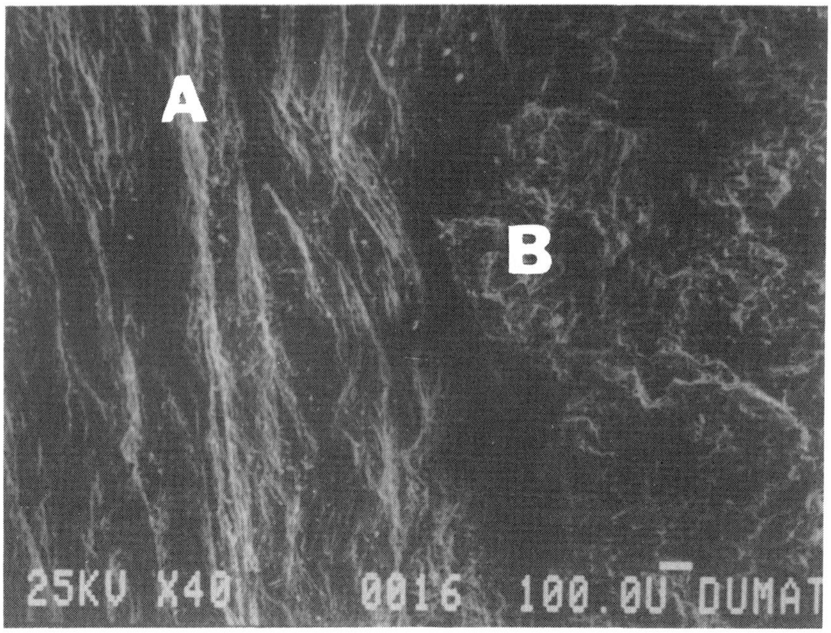

Fig. 13-12. Scanning electron micrograph of a fold area (A). The fracture surface in the matrix (B) is shown for comparison.[3]

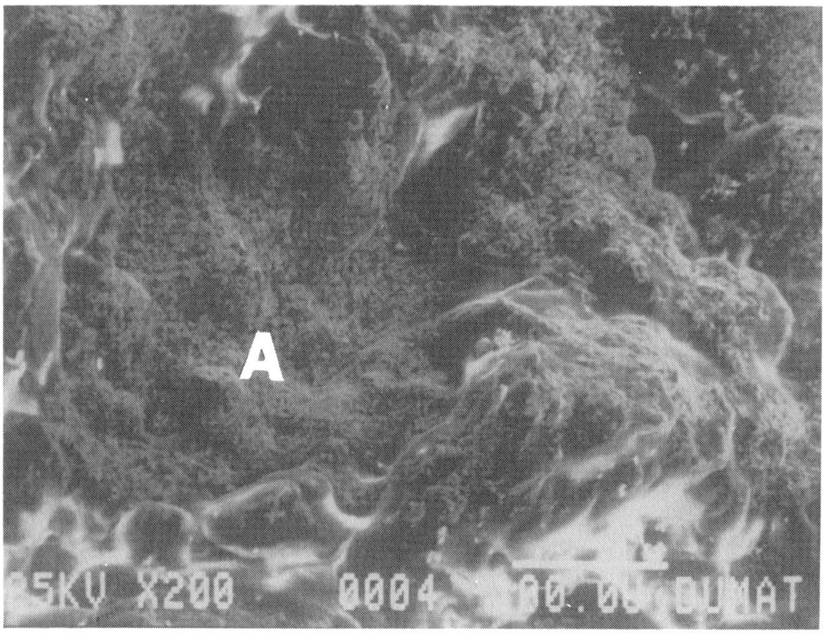

Fig. 13-13. Scanning electron micrograph of a black inclusion (A) observed on a fracture surface.[3]

which appears to result from direct and extended contact of the molten metal with the plastic residue.

While no satisfactory way of explaining folds has been found, it is possible to prevent or reduce their occurrence. Using a low-silicon aluminum alloy greatly reduces the problem. Gating and casting design that avoids abrupt changes in section thickness and the convergence of two or more streams is helpful. Selection of coating also has been identified as an important factor. Finally, bead type and bead fusion can have an influence on fold formation.

Inclusions

Black inclusions are frequently observed on fracture surfaces, generally in sections filled last. These inclusions vary in size from a few millimeters to a centimeter.

In one series of tests, chemical analysis revealed that the inclusions contained powdery depositions of calcium and potassium, which also were present in the coating material. The researchers theorized that the inclusions on fracture surfaces may originate from the coating material. Coating can penetrate two or three EPS beads deep in patterns with poor surface finishes from foam molding. Erosion of this coating can result in inclusions. Plastic residue is likely the cause of the black coloring. An example is given in Fig. 13-13.

ACKNOWLEDGMENT

AFS would like to thank Ray Monroe, Steel Founders' Society of America, for his help in writing this chapter. Illustrations in the chapter that are not referenced are from Mr. Monroe's book *Expendable Pattern Casting*, © 1992, AFS.

REFERENCES

1. E.H. Niemann, G.B. Kelly; "Evaporative Patten Casting: Factors Affecting Expandable Polystyrene (EPS) Pattern Shrinkage," *AFS Transactions*, pp 367–370 (1990).
2. L. Wang, S. Shivkumar, D. Apelian; "Effects of Polymer Degradation on the Quality of Lost Foam Castings," *AFS Transactions*, pp 923–933 (1990).
3. S. Shivkumar, L. Wang, B. Steenhoff; "Metallurgical Quality of Aluminum Castings Produced by the Lost Foam Process," *AFS Transactions*, pp 825–836 (1989).

Chapter 14

GATING & RISERING

A key element in producing quality aluminum castings is the proper design and sizing of the gating and risering systems. A foundry can produce the best quality molds, cores and molten aluminum and still end up with a poor-quality casting by using poorly designed gating and risering systems. This is proven many times over in the foundry by the amount of work done on the redesign and sizing of either the gating system or risers until quality castings are finally produced.

The objective of this chapter is to eliminate the "hit or miss" tactics commonly practiced and to improve chances of getting it right the first time.

VELOCITY, FLOW RATE AND CHOKE

There are several terms that are used frequently when discussing gating and risering systems.

Velocity is the term used to measure how fast the molten aluminum is flowing past a given point in the gating system. In gating systems, this normally is measured in in./sec.

Flow rate, on the other hand, is the quantity or volume of molten aluminum flowing past a given point in the gating system. The flow rate normally is measured in lb/sec, which can be divided by the density to give in.^{3}/sec.

The *choke* is the smallest total cross-sectional area in the gating system. The key word to remember is TOTAL. It is important that the choke be physically located in the gating system. While not normally advisable, the choke could be the pourer, ladle lip or the lip or orifice of an automatic pouring device. The choke is measured in in.^{2}

FLUID FLOW

Before gating practices are covered, the reader should understand basic fluid flow principles.

Fluid flow laws apply to all fluid materials, regardless of their composition. Molten aluminum is a liquid, so it follows fluid flow laws, just as water does. It doesn't matter if the aluminum is flowing in a sand, plaster, ceramic or metal mold—it still follows the same rules. The key things to remember are:

1) Fluid flow laws only apply when the entire gating system is completely filled with molten metal.
2) These laws can be used to a foundry's advantage if the laws are understood and the gating system is designed accordingly.

By using the laws derived from Bernoulli's Theorem, which deals with relationships influencing the behavior of liquids as they flow in a channel, problem areas in gating system design can be anticipated and compensation made for them.

Velocity of Molten Metal

As a material (whether solid or liquid) falls, its velocity (or speed) increases. This is proven by the velocity formula for a free falling body or liquid:

$$\text{Velocity} = (2gH)^{1/2}$$

The only part of this formula that we have control over is "H," which stands for the distance (height) that the molten aluminum falls. The letter "g" stands for the affect of gravity on a falling body or liquid, which is equal to 32 ft/sec^2.

What this velocity formula tells us is that, the farther the liquid aluminum falls, the faster it travels. And the greater the velocity, in either vertical or horizontal flow, the greater the chance for turbulence.

When pouring aluminum alloys, the amount of turbulence in the gating system can be minimized by decreasing the velocity of the metal. The velocity formula tells us that the initial velocity in the gating system is determined by the fall of the molten aluminum. This fall begins where the molten aluminum leaves the pouring device (such as the lip of the ladle) and ends where the vertical drop is changed to a horizontal flow. If the metal is poured directly down the sprue, H is measured from the lip of the ladle to the bottom of the sprue. The pourer can determine the initial velocity of the metal (and, thus, the velocity in the whole gating system) by controlling how high the lip of the ladle is kept above the sprue.

Law of Continuity

The "Law of Continuity" states simply that what goes in one end has to come out the other end. This means that, when pouring a casting, molten aluminum that enters the sprue has to come out somewhere in the mold. What this law also says is that the quantity (Q) of molten aluminum flowing will remain constant throughout the gating system, if the system is completely filled. This law is based on the formula:

$$Q = AV$$

where

Q = quantity (volume) of molten aluminum flowing (in.^{3}/sec),

A = the cross-sectional area at a given point in the gating system (in.^{2}), and

V = the velocity at that point (in./sec).

In a full gating system, Q remains constant. Simple math tells us that if A were small, then V would be larger, and conversely, if A were large, then V would be smaller.

This Law of Continuity is a handy formula to remember because, if you know any two of the three items (Q, V or A), you can calculate the third. As mentioned previously, if the velocity is decreased, the turbulence in the gating system is reduced. Q and A can be adjusted to minimize V.

To fill a mold faster, the flow rate (not the velocity) must be increased. This is done by increasing the size of the choke. To determine where the choke is located in the gating system, the total cross-sectional area at various points must be calculated.

If an inverted sprue (one with the small opening at the top) is used, the quantity of molten metal flowing through the small opening needs to be determined. If this quantity is less than in any other part of the gating system, then this opening is the choke. To calculate this, the velocity at this point and the cross-sectional area must be obtained. Then, using the Q = AV formula, the quantity can be determined and compared to other areas in the gating system.

Next, the total cross-sectional area(s) of the runner(s) that are attached to the base of the sprue need to be calculated. Finally, the total cross-sectional areas of all the gate(s) feeding at the same time are determined. Whichever of these three total cross-sectional area(s) is the smallest is where the choke is located in the gating system. It is at this point that the flow rate is established, and based on the Law of Continuity, it is also the point of highest velocity.

Fluidity of Molten Aluminum

The fluidity or fluid life of molten metal is determined by several factors. Fluidity is the ability of the molten aluminum to flow and completely fill the mold cavity. The factors that have the greatest influence on fluid life include:

- *temperature.* The thinner the casting sections, the hotter the molten aluminum has to be to prevent misruns. Ideally, castings should be poured at a temperature just high enough to prevent misruns. To follow this practice, you have to have excellent control of the pouring temperatures.
- *chemistry.* The chemistry of the aluminum alloy used also plays a role in its fluid life. Of the chemical elements added to aluminum alloys, silicon has the greatest affect on fluid life. The higher the silicon level, the greater the fluid life. The chemistry of the alloy used also affects its solidification range. Pure metals, eutectic alloys and narrow freezing range alloys tend to exhibit the best fluid life. Outside of the Al-Si system, wide freezing range alloys have the worst fluid life.
- *entrained gas and inclusions.* Both entrained gasses and inclusions tend to increase the viscosity and, therefore, reduce the fluid life of aluminum alloys. In all cases, aluminum alloys should be free of inclusions, such as dross and sand.

REQUIREMENTS OF GATING SYSTEMS

Molds can be gated vertically or horizontally. The fundamental requirements of an ideal gating system, whether horizontal or vertical, may be stated as follows:

1) The molten aluminum should flow through the gating system with the least amount of turbulence to prevent the aluminum oxide film from breaking and, thereby, allowing the entrapment of air; mold and core gasses; and inclusions such as dross, sand and other foreign materials.
2) The molten aluminum should enter the mold cavity in a manner that will produce temperature gradients in the casting such that directional solidification is promoted towards the source of feed metal.
3) The gating system should be designed so that the pourer or pouring device can fill the sprue quickly and keep it full during the entire pour.
4) The gating system should be designed so that the chances for misruns and cold shuts in castings are eliminated.
5) The gating system must be capable of realization commercially with a minimum yield but still produce a casting acceptable to the customer.
6) The gating system should be designed in such a way that it can be removed readily from the casting.

HORIZONTAL GATING

This section covers the design and application of a gating system for use in horizontally parted molds. These recommendations hold true for any refractory-type molds such as green sand, dry sand, nobake, shell, plaster or ceramic. The same recommendations apply to metal molds that are parted and poured horizontally. Figure 14-1 shows a horizontal gating system.

The way a gating system is designed will determine if it is pressurized or nonpressurized. In a pressurized gating system, the choke is located at the gate(s). In a nonpressurized gating system, the choke is located either at the sprue outlet or in the runner(s) adjacent to the sprue.

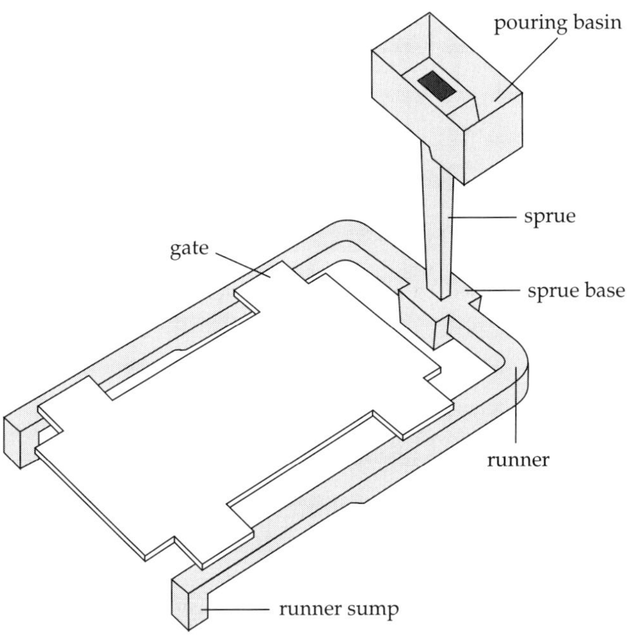

Fig. 14-1. Drawing of a typical horizontal gating system.

Pressurized Gating

Because the choke is the point of highest velocity in a gating system, it is a point at which there can be excessive turbulence. In the pressurized system, with the choke at the gates, the high velocity and turbulence can create dross and/or oxide skin at a point where they can be carried directly into the casting. Aluminum alloys oxidize readily and are, therefore, extremely sensitive to turbulence, so pressurized gating usually should *not* be used for aluminum castings.

Some foundries report producing satisfactory aluminum castings using the pressurized gating system. However, most aluminum gating authorities recommend the use of nonpressurized systems.

Nonpressurized Gating

The term "nonpressurized gating system" is not exactly correct—there has to be pressure in the molten aluminum in the gating system or it will not flow. More technically correct would be a term like "expanding gating system." In this type of system, components of the gating system are larger in total cross-sectional area once the molten aluminum passes the choke. This enlarging of the gating system reduces the velocity of the molten aluminum and, thus, reduces turbulence.

Pouring Basins, Sprues and Sprue Bases

To aid in pouring, an enlargement should be used at the top of the sprue to serve as a *pouring basin.* The pouring basin receives molten aluminum from the pouring ladle or pouring device and directs it into the sprue. The pouring basin also helps the pourer maintain a near-constant metallostatic head, which, in turn, helps to maintain a near-constant initial velocity at the sprue outlet. Figure 14-2 is a cross-sectional view of a typical pouring basin. Using a pouring cup (Fig. 14-3) causes problems because the pourer is tempted to pour directly down the sprue, thereby influencing the initial velocity in the falling stream.

The *sprue* is the vertical passageway that takes the molten aluminum from the pouring basin to the runner(s). The design of the sprue incorporates the physics of a free-falling stream. The velocity formula tells us that a free-falling stream increases in velocity as it falls. The Law of Continuity states that, as the velocity increases, the cross-sectional area decreases, provided a constant flow rate is maintained.

Because the falling stream of metal tends to form a natural taper, it is best to use a tapered sprue with the small end down as opposed to a straight or reversed-tapered sprue (Fig. 14-4). In the straight or reversed-tapered sprue, this natural taper creates air pockets between the sprue wall and the metal stream. In refractory-type molds, which are permeable, the stream of molten aluminum can create an area of low pressure between itself and the sprue wall, which can create undesirable oxides and the entrainment of gases.

In some cases, such as with automatic green sand molding machines, the use of a tapered sprue, small end down, is prohibited. In these cases, a reversed-tapered sprue must be used, even though it is the least desirable sprue to use for

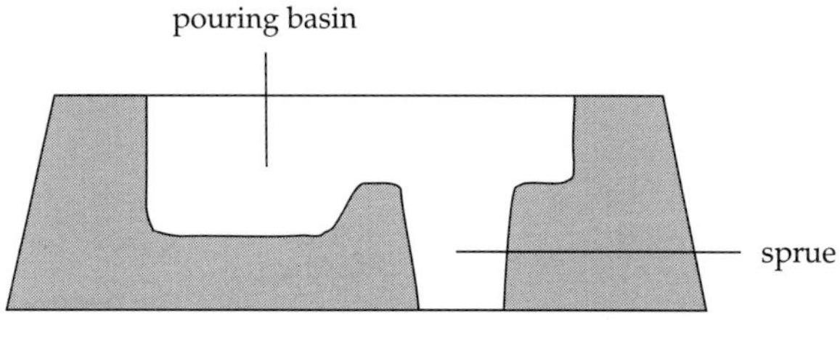

Fig. 14-2. Cross section of a typical pouring basin.

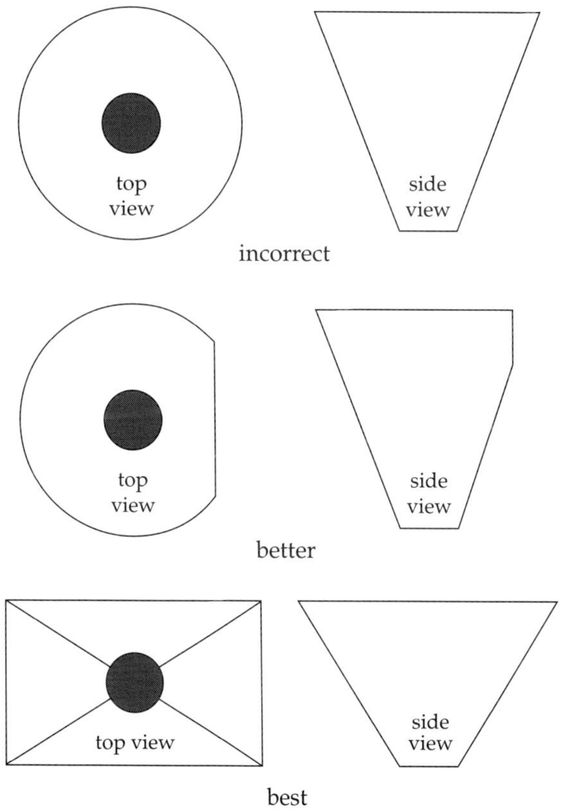

Fig. 14-3. Sketches of typical pouring cups.

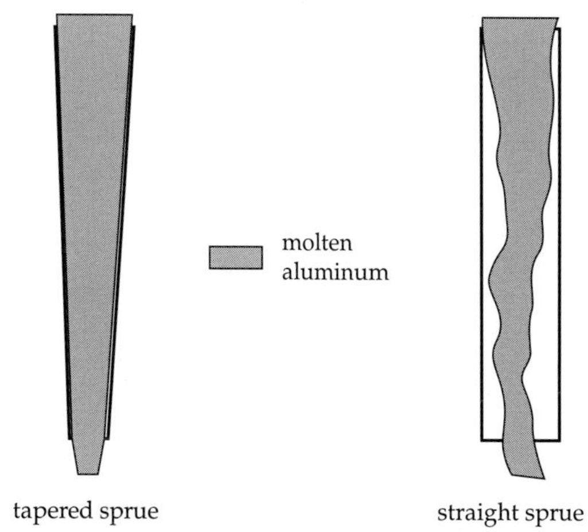

Fig. 14-4. The amount of turbulence in a tapered sprue, small end down, vs. a straight sprue.

aluminum casting. Using a choke core at the bottom of the sprue or reducing the runner(s) cross-sectional area next to the sprue base can partially overcome the problems associated with the reversed-tapered sprue.

The use of a strainer core to act as a choke core is not recommended. A strainer core breaks the falling stream into as many streams as there are holes in the core, which increases the surface area of the stream exposed to air and mold gases. This, in turn, causes oxides to form and get trapped in the flowing stream (Fig. 14-5). Strainer cores do little to strain undesirable nonmetallics from the stream of molten aluminum.

If a tapered sprue, small end down, is being used, the point of highest velocity in the gating system will be at the exit of the sprue. The high velocity occurs because of the fall of the molten aluminum and because the choke is at this point. It is here that steps are taken to decrease the velocity.

A crucial thing to remember is that, at this point, the flow changes from a vertical drop to a horizontal flow. To take care of this situation, a *sprue base* or *well* is attached to the bottom of the sprue (Fig. 14-6). This enlargement helps reduce the velocity of the falling stream by collecting a pool of molten aluminum, which helps to kill the stream's kinetic (moving) energy. This reduces the turbulence caused by the transition from a vertical to a horizontal flow.

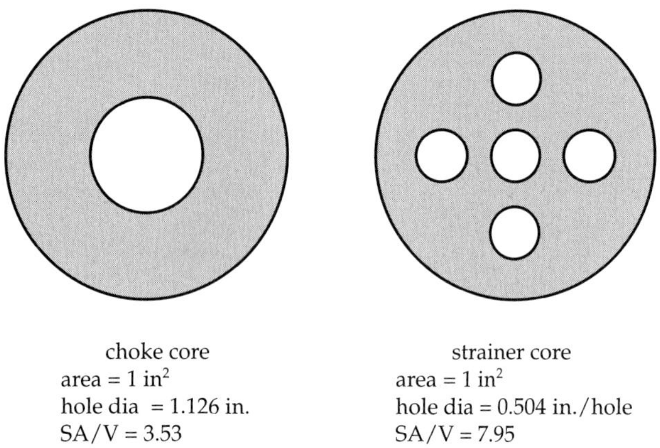

Fig. 14-5. Surface areas of choke cores vs. strainer cores.

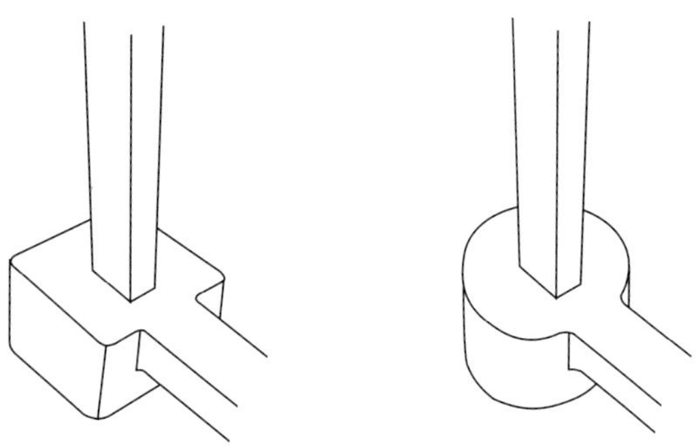

Fig. 14-6. A sprue base can be rectangular or cylindrical. The depth of the sprue base should be at least twice the runner depth.

Note that the bottom of the sprue base is flat, not hemispherical. A hemispherical bottom has been shown to be detrimental to obtaining a smooth transition in the direction of flow. The sprue base can be either round or rectangular in shape, as shown in Fig 14-6. Also note that the runner comes off the upper portion of the sprue base.

Runners

From the sprue base, the molten aluminum next enters the runner(s). The *runner* is located in the drag of the mold and is trapezoidal in cross-section (Fig. 14-7). Because this is a nonpressurized, expanding gating system, the cross-sectional area of the runner at this point is larger than the choke, located at the outlet of the sprue.

Based on past research, a 1:4:4 gating ratio generally is recommended when pouring aluminum alloys. The "1" is the cross-sectional area at the outlet of the sprue or the choke. The first "4" means that the total cross-sectional area of the runner(s) adjacent to the sprue base should be four times that of the sprue outlet or choke. The next "4" represents the total cross-sectional area of the gate or gates. As with the runners, the gate's total cross-sectional area will be four times that of the choke (making it equal to that of the runner).

The gating ratio can be varied to meet a foundry's needs. Much of the decision as to what gating ratio to use depends on quality requirements for the casting. The higher the gating ratio (within reason), the slower the molten aluminum will be flowing and the less turbulence there will be. In one incidence, a 1:8:6 gating ratio was used to produce a high-quality aircraft casting. On the other hand, satisfactory aluminum castings have been poured using a 1:2:2 gating ratio.

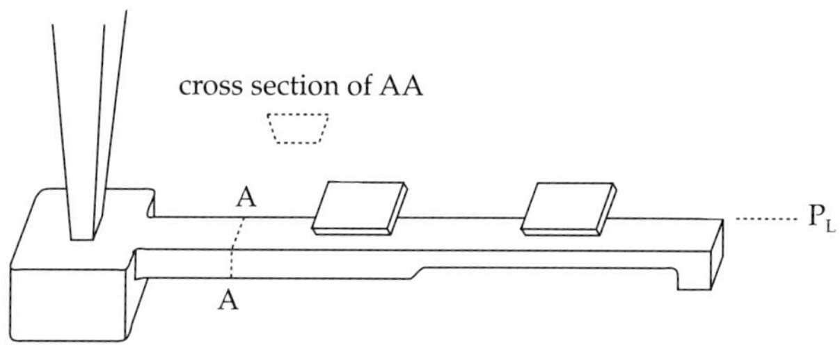

Fig. 14-7. The runner in this gating system is stepped down after passing a gate. Note the location of the parting line.

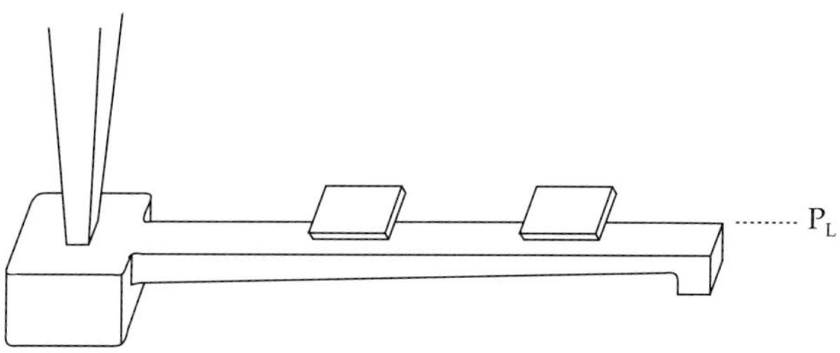

Fig. 14-8. In this gating system, the runner is tapered instead of being stepped down as the one in Fig. 14-7.

The design of the runner(s) must not promote turbulence. When the molten aluminum has to change its direction of flow, it should do so with the least amount of turbulence. To accomplish this, the runner is streamlined. There should be no sharp corners, only radiused corners, to help the molten aluminum change direction. This practice helps keep the oxide envelope on the molten aluminum from being broken.

Two primary requirements of a gating system are to prevent cold shuts and misruns during mold filling and to set up proper thermal gradients in the casting. In castings that require multiple gates, all the gates must be feeding at the same time. This is accomplished by stepping down the runner(s) as each gate is passed. This step down is equal to the cross-sectional area of the gate just passed (Fig 14-7). Some foundries prefer to taper the runner rather than step it down to reduce the chance of creating turbulence. The amount of taper is calculated using the same method as a stepped-down runner (Fig. 14-8).

The next area of concern is the end of the runner. The worst-quality molten aluminum is in the leading edge of the flowing metal stream. This molten aluminum has taken the brunt of the punishment from a fluid flow standpoint, and it also tends to clean the runner(s) of any loose sand and other foreign materials. This partially damaged molten aluminum must not be allowed into the mold cavity. There are two ways in which this can be done. Figure 14-9 shows both a *runner extension* and a *runner sump*.

In the first method, the runner is extended beyond the last gate. This extension must be of the same cross section as the runner under the last gate and should be vented. The length of the runner extension is determined by experimentation. If the damaged metal is too close to the last gate, then the runner extension is lengthened. On the other hand, if too much clean metal passes the last gate after the bad metal, the runner extension can be shortened.

Because most foundries do not allow enough space in the mold to use a properly designed runner extension, a second method may be necessary. In these cases, a runner sump can be used. The runner sump is similar to the sprue basin but does not have to be as large.

Gates

The gates in the gating system direct the molten aluminum from the runner into the mold cavity.

The cross-sectional area of the gates is determined by the gating ratio used. If a 1:4:4 gating ratio is used, then the total cross-sectional area of the gates has to be four times that of the choke. The velocity at the gates would be one-fourth that of the choke. Because of this reduction in velocity, the molten aluminum enters the mold cavity in a calm manner. The gates can be no thicker than the casting section to which they are attached. It usually is best to have the gates slightly thinner than the casting section. This practice leaves a "witness mark," which aids in the removal of the gate connection when grinding.

The gates in the expanding gating system are taken off the cope side of the runner. This allows the runner to fill completely with molten aluminum before any metal passes through a gate. When this method is used, the envelope of aluminum oxide is protected until the last second, when the molten aluminum goes through the gates.

Remember that molten aluminum is not distributed equally through the gates unless the stepped down or tapered runners are completely filled. If the gating system is not completely filled during the entire pour, it will not function correctly, and defective castings can result.

The gates may, but do not have to, completely overlap the runner (Fig. 14-10). The only time complete overlap is necessary is when two gates are used to feed two castings off of each side of the runner at the same location. A precaution here is that the overlaps must not become the choke. Some

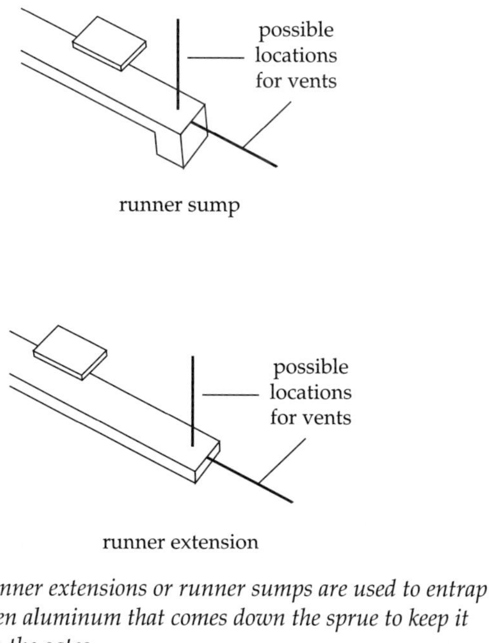

Fig. 14-9. Runner extensions or runner sumps are used to entrap the first molten aluminum that comes down the sprue to keep it from entering the gates.

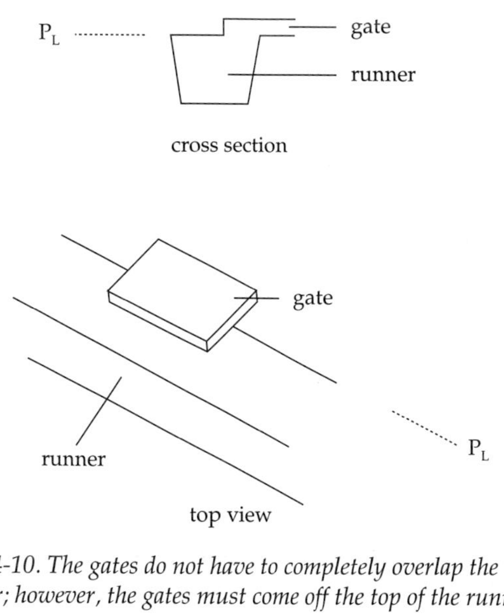

Fig. 14-10. The gates do not have to completely overlap the runner; however, the gates must come off the top of the runner.

illustrations of gating systems show the gates coming off the cope side of the runners with a sweeping radius. This is a desired practice and may be required for very critical castings; however, for most commercial castings, this gate configuration is not required.

The key thing to remember when designing a gating system is that the rules must be obeyed. This is true whether the gating system is an expanding system or a pressurized system. If you cheat, the gating system will not function correctly, and you may end up with defective castings.

VERTICAL GATING

Not all molds are parted horizontally. With permanent molds and some shell molds that are parted vertically, a vertical gating system has to be used. There also are some cases in which horizontally parted molds use a vertical gating system. If the molten aluminum has to fall a great distance into the mold cavity, a vertical gating system is designed to bring the metal up into the mold. For all practical purposes, it is best to have a mold fill from the bottom up. However, when risers are used, this can cause a problem, which will be discussed in detail in the risering section.

Researchers found that the nonpressurized gating system is the best to use when pouring molten aluminum with a vertical gating system. The gating ratio recommended is 1:4:4, the same as for horizontal systems.

The velocity of the falling stream at the exit of the sprue is normally greater in a vertical gating system than a horizontal gating system because of the increased length of the sprue. Thus, it becomes of even greater importance not to pour directly down the sprue.

One advantage of the vertical gating system is that a tapered sprue, small end down, can always be used because the mold is parted vertically. Thus, the choke can always be located at the outlet of the sprue.

Figure 14-11 illustrates a vertical gating system. Note that the design is similar to that of a horizontal gating system. A pouring basin is used along with a tapered sprue and a sprue base. The runners are again attached to the top section of the sprue base. The gates can be oriented horizontally or vertically off the top of the runners, depending on the orientation of the casting in the mold.

When using the type of system shown in Fig. 14-11, the coldest molten metal will be at the top of the casting(s) and the hottest at the bottom. A riser attached to the top of the casting(s) would be undesirable for two reasons: 1) it would be a cold riser, and 2) the effect of directional solidification would be reduced. Using a side riser or "uprunner" with a web gate (Fig. 14-12) or with step gates can help overcome this problem by progressively filling the mold from the side instead of the bottom.

There is another problem that is more pronounced in vertically gated castings than those gated horizontally. As molten aluminum rises in the casting, its metallostatic pressure begins to counteract that of the falling stream in the sprue. This action causes the casting to fill more slowly as the molten aluminum goes higher up into the casting. It is, therefore, important that a sufficient "head" of molten aluminum be maintained in the pouring basin and sprue to overcome the back pressure from inside the casting cavity. If an uprunner is connected to the casting with individual gates instead of a web gate, the gates are made progressively larger in cross section the higher they are located in the mold to maintain the same flow rate through all the gates. Many times, lack of consideration of the laws of fluid flow causes misruns and gas-related defects at the top of castings.

The sizing of the vertical gating system is the same as that of the horizontal gating system. To size this type of gating system, a flow rate is established and a choke sized to provide it. Next, a gating ratio is selected, and the runner and gate cross-sectional areas are calculated. Remember, however, that as the gates are located higher up in the mold, their cross-sectional areas must get progressively larger to attempt to maintain the proper flow rate.

With the exception of tilt-poured permanent molds, the design of the vertical gating system follows fluid flow laws. If any of these laws is violated, the gating system will not produce defect-free castings.

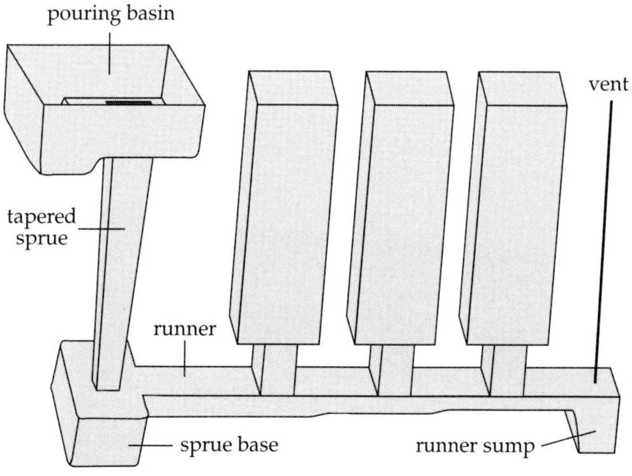

Fig. 14-11. A vertical gating system with multiple cavities and a stepped bottom runner.

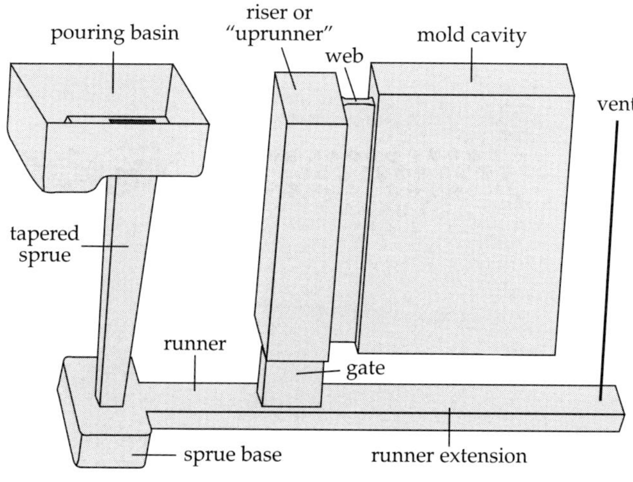

Fig. 14-12. Drawing of a common vertical gating system using an uprunner and a web gate.

Tilt-Pour Permanent Molding

The vertical gating system described above is for static-poured molds—those molds poured while in the vertical position. When permanent molds are tilted during pouring, different fluid flow principles take over.

In the tilt-pour method, molten metal is poured into a pouring cup while the mold is in the horizontal position. Then, as the mold is tilted into the vertical position, the molten metal flows down the sprue or side runners. (For more details on this process, see Chapter 7, Permanent Mold Casting.) The speed at which the mold is tilted determines the velocity of the metal and also the flow rate of the molten metal into the mold cavity. The metallostatic pressure in the gating system and casting remains relatively constant; thus, the gates do not become progressively larger as they are located higher in the mold.

There is no published literature that explains how large a choke is needed in tilt-pouring aluminum castings. Because the flow rate is determined by how fast the mold is tilted, determinations often are made through trial and error. Besides the reduced turbulence during the filling of the mold, another advantage of tilt-pouring permanent molds is that the gating system acts as a natural vent. This occurs because the sprue and gates are never completely filled with molten aluminum until the mold is in a completely vertical position.

FILTERS

An obstacle facing foundries pouring aluminum alloys is that the aluminum oxide formed when melting and pouring molten aluminum is nearly the same or somewhat denser than the molten aluminum. When pouring molten aluminum, one must strive not to create oxides, but when this doesn't work, filters can be used to filter these and other nonmetallics out of the flowing molten aluminum.

Filtering of molten aluminum has been done for many years. Steel wool was placed in or at the base of the sprue. Formica screening and fiberglass cloth also have been used to filter out dross and other nonmetallics. Some aluminum foundries attempt to remove dross by using strainer cores at the base of the sprue. However, the strainer core merely acts as a choke, which helps to fill the sprue quickly but does not filter out anything other than large particles of refractory or dross. In fact, a strainer core can even create oxides.

Research has shown that the use of filters can result in the following benefits:

1) improved mechanical properties, such as tensile strength and elongation, in the casting;
2) improved pressure tightness;
3) improved machinability;
4) improved molten metal fluid life;
5) improved cosmetic appearance of the casting.

Filters can aid even the best designed gating system to deliver clean molten aluminum alloy to the casting. However, the use of filters does not eliminate the need to maintain good melting and handling practices for molten aluminum alloys.

The use of molten metal filters at the outlet of the sprue requires precautions. A filter will remove inclusions from the falling stream; however, as it does so, the cross-sectional area through which the molten aluminum can flow is reduced. At the start of pouring, the filter may not act as the choke, causing the choke to be elsewhere in the gating system (such as at the gates, which would cause the highest velocity to be at the gates). As the filter removes inclusions from the metal, its cross-sectional area for molten aluminum to flow through is reduced, and the filter becomes the choke. The flow rate through a filter may be reduced to the point where it is insufficient to fill the casting properly, and defects, such as misruns and cold shuts, can occur.

Ceramic Filters

In the last few years, ceramic filters have been used more frequently by aluminum foundries. There are several ways that these filters are classified. One way they are classified is by how they are made. There are basically three ways ceramic filters are made: 1) extruded, 2) pressed and 3) foamed (Fig. 14-13). The first two, extruded and pressed, are a series of many small, round or square holes through which the molten aluminum flows. Foamed filters look like sponges in that they have many tortuous channels through which the molten aluminum flows.

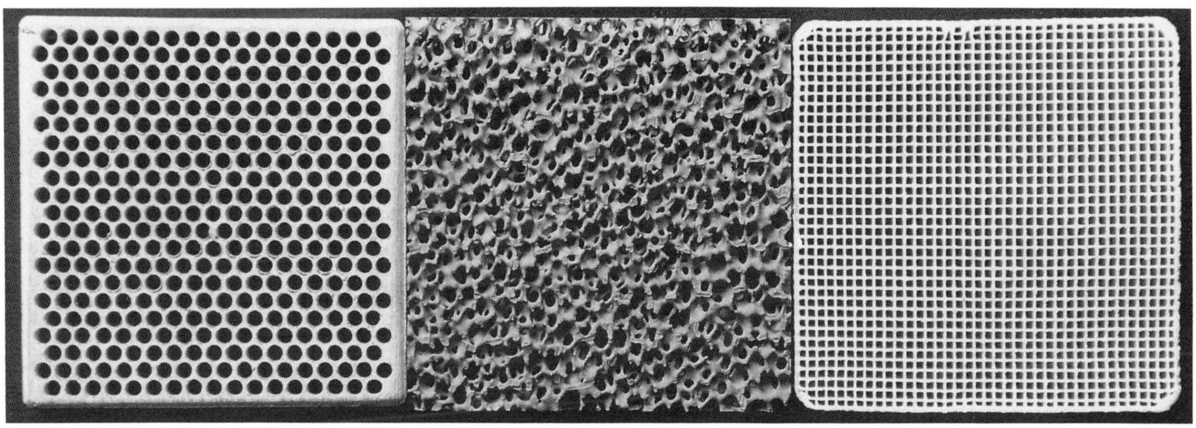

Fig. 14-13. These are the three types of ceramic filters (l–r): pressed, foamed and extruded. Each is 55x55x13 mm (2.17x2.17x0.5 in.). The pressed filter contains 367 holes with hole sizes of 0.090 in. (2.28 mm). The foamed filter has 25 pores/in. (ppi). The extruded monolith filter has 300 cells/sq in. [Photo courtesy of Selee Corp.].

Another way ceramic filters are classified is the method by which they trap inclusions. The three methods are: 1) screening filtration, 2) cake filtration and 3) deep bed filtration.

The first two methods are the basis on which the extruded and pressed ceramic filters work. Screening filtration works by making the holes in the filter so small that they prevent the inclusions from entering the filter (i.e., screening them out). The filter core has many small, round holes, which can be as small as 0.080 in. Cellular filters, on the other hand, have holes that are usually square and can be 50, 100, 150, 200, 300 or 400 cells (holes) per square inch. In either case, the smaller the holes, the more filtration takes place.

Cake filtration mainly occurs when previously captured inclusions do not allow remaining inclusions to enter the filter structure. The presence of inclusions that effectively block the entrance of the filter media is required for cake filtration.

The ceramic foam filter operates primarily by deep bed filtration. The efficiency of this type of filtration is governed by the total three-dimensional surface area of the filter material. Inclusion retention is possible throughout the entire filter thickness. The reticulated structure of the ceramic foam filter allows for a high internal surface area and enhances the deep bed filtration mechanism. Filtration efficiency is improved by reducing the filter pore size, thus increasing the internal surface area.

The sizing of the filters is based on the choke size in the gating system. It is normally recommended that the filter be at least two to four times larger in available cross-sectional area for the molten aluminum to flow through than the choke cross-sectional area. This allows the filter to do its job of filtering out the inclusions but without reducing the available cross-sectional area so much that the filter becomes the choke. Should the filter become the choke, it could be so small that it reduces the flow rate to the point that misruns and/or cold shuts occur.

Figure 14-14 shows two ways in which filters can be placed in the gating system. Besides being placed in the outlet of the sprue, they can be placed in the runners. When the filters are placed in the runners, an enlargement is designed to accommodate the filter. Ideally, the filter should be placed as close to the first gate as possible or at each gate to help minimize the chance of the molten aluminum forming or entrapping more inclusions.

A unique use of ceramic foam filters is in the DYPUR® system (Fig. 14-15). This system uses an insulating sleeve in which a ceramic foam filter is placed. The molten aluminum is poured into the insulating sleeve and then passes through the filter directly into the casting.

This also can be called direct pouring. This method of pouring castings promotes directional solidification, improves casting yield in that the conventional gating system can be eliminated, and inclusions are removed from the molten aluminum at the point where the it enters the casting. In this system, the molten aluminum remaining in the riser sleeve after the casting has been filled acts as a riser. Thus, effectively, the gating system and riser are one.

Casting dimension is critical when using this system. When producing large, chunky castings or castings where the area immediately below the filter is open, the distance between

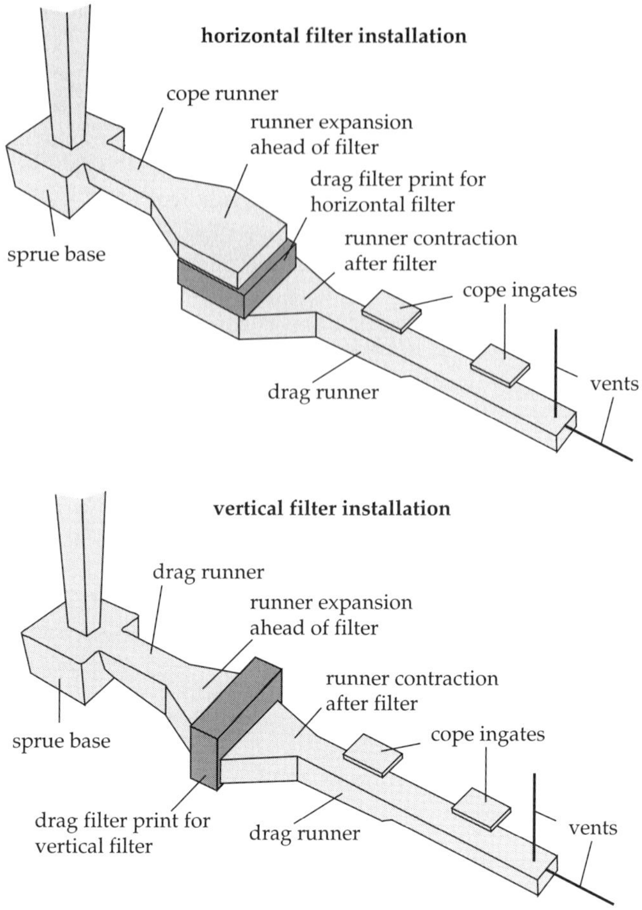

Fig. 14-14. Two ways that filters can be placed in the runner.

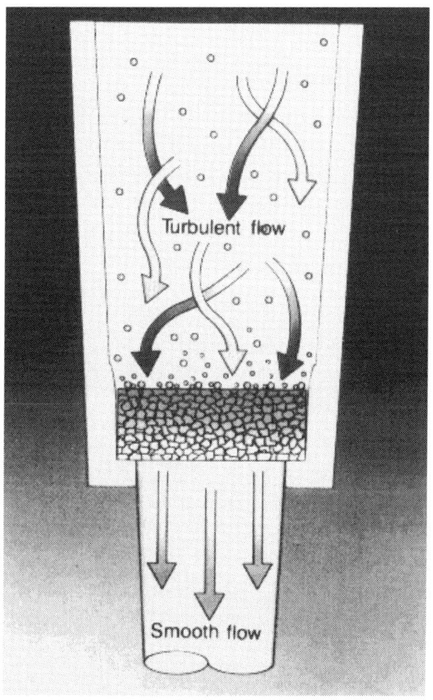

Fig. 14-15. The DYPUR® direct-pouring system. In this method, an insulating sleeve and a ceramic foam filter reduce turbulence and remove inclusions. They also serve as the gating system in many cases, allowing metal to be poured directly in the mold cavity through the top of the casting. [Drawing courtesy of Foseco, Inc.]

the bottom of the filter and the impact area (core or bottom of the mold) should be less than 4 in., for optimum results. This distance can be stretched, in some circumstances, to 6 in. In an open free-fall, gravity will redevelop turbulence in the metal stream. However, when pouring thin-walled castings (0.5 in. or less), the molten metal drops can be extended without increased turbulence during the fall. However, it has been found that even in fairly tight areas, molten metal drops of greater than 12 in. should be avoided.

GATING FOR ALUMINUM COMPOSITES

When the expanding gating system was applied to pouring an aluminum composite alloy with an A356 matrix and 15% by volume silicon carbide reinforcement, poor casting quality resulted. It was shown that conventional gating techniques will not always work for composites. The main problem with the test castings was the appearance of gas bubbles internally and on the surface of the castings. Research into this problem revealed that the placement of a "bubble trap" (open riser) at the point where the molten aluminum entered the runner reduced the occurrences of these types of defects. Further research showed that, if the choke was placed in the runners where the molten aluminum left the bubble trap, the castings were radiographically free of gas bubbles.

Figure 14-16 illustrates the gating system used to produce good aluminum composite castings. In this drawing, the bubble trap is drawn larger than necessary. Researchers found that the cross-sectional area of the bubble trap needs to be only 1–1.5 times the cross-sectional area of the runners. Also, the bubble trap does not have to extend through the cope; however, it must be vented to the atmosphere.

It is very important to place the choke within 0.5 in. after the bubble trap. In addition, it is recommended that the area of the sprue outlet be no larger than 10–12% of the combined cross-sectional areas in the runners beyond the chokes.

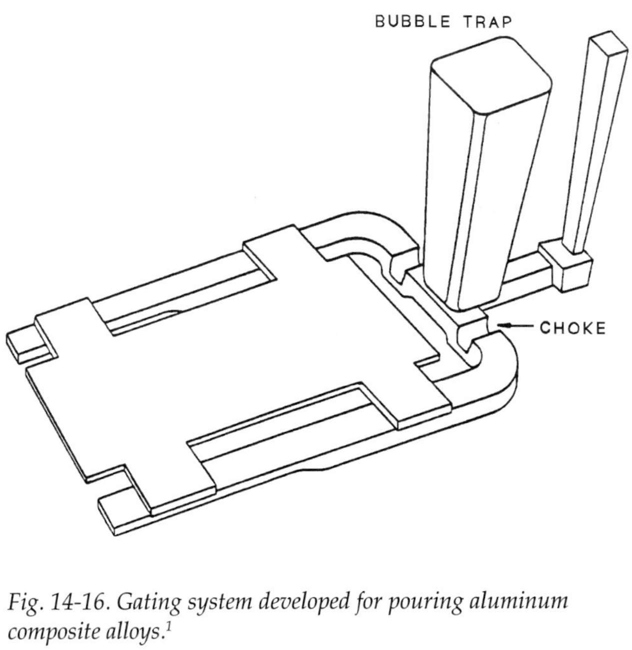

Fig. 14-16. Gating system developed for pouring aluminum composite alloys.[1]

For aluminum composites, the pouring rates are generally best if kept slower than those used for conventional aluminum alloys. Castings weighing 1–5 lb have been successfully poured using a flow rate of 0.5 lb/sec, and castings weighing 5–20 lb at 2.0–2.5 pounds/second. A commonly used flow rate is in the range of 1.3–1.5 lb/sec.

The gating ratios discussed earlier consisted of three numbers. When discussing gating ratios for the aluminum composite alloys, four numbers are used. For instance, a gating ratio may be; 1.55:1:10:10 for a given casting (remember that these numbers represent total cross-sectional areas). The 1.55 represents the sprue(s) outlet cross-sectional area; the 1 is the cross-sectional area of the choke; and the two 10s represent the total cross-sectional areas of the runner(s) and gate(s), respectively.

Figure 14-17 is a sketch of the gating system used to pour a test plate casting that is 0.250 in. thick. Also included in the sketch are the sizes of the various parts of the gating system. The bubble trap does not need to be as large as that shown in the sketch.

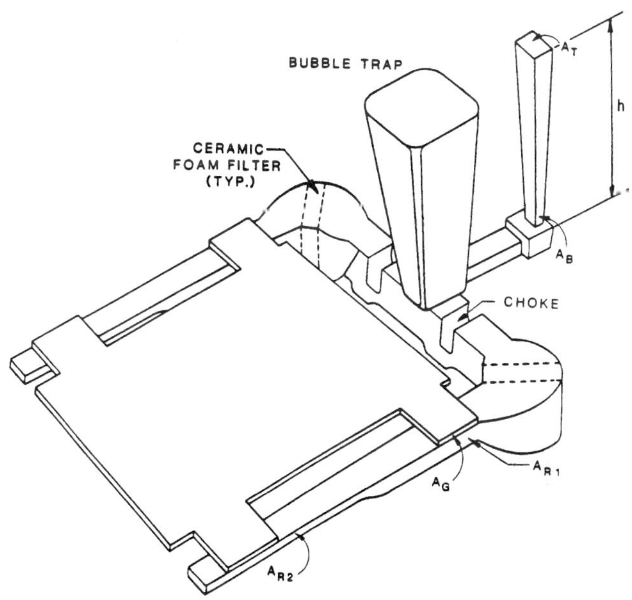

Specifications Used

pouring temp 1335–1350F		A_T = ¾ x 1 in. (0.75 in.^{2})
pouring basin	1.25	A_B = ⅜ x 21/32 in. (0.250 in.^{2})
rectangular tapered sprue		well = 1.5 x 1.5 x 1.5 in. deep
sprue basin (well)		bubble trap = 2 x 2 in.
bubble trap and choke filter	1	choke = 0.100 x 1 in. (0.100 in.^{2}) x 2
U-shaped runner		filter = 3 x 3 in. (10 ppi)
stepped runners w/extensions	10	A_{R1} = 1 x 1 in. (1 in.^{2}) x 2
four ingates		A_{R2} = ½ x 1 in. (0.5 in.^{2})
1.25:1:10:10 gating ratio	10	A_G = ¼ x 2 in. (0.5 in.^{2}) x 4
parts and gates in cope		h = 8 in.
runners in drag		fill rate = 1.0–1.5 lb/sec

Fig. 14-17. Gating system incorporating filters used to pour the best aluminum composite alloy trial casting.[1]

It is recommended, when pouring aluminum composite alloys, that the gates be located at the bottom of the casting, if possible. This will prevent the molten aluminum from dropping into the mold cavity, causing turbulence. The molten aluminum composite alloy must be handled in a nonturbulent manner. If this fundamental fact is respected, then many successful gating systems other than that described here can be engineered, limited only by the capabilities of the foundry and the part to be cast.

SOLIDIFICATION

Once the casting cavity is filled with molten aluminum, the next step in the casting process—solidification—comes under consideration. Solidification is a critical stage in the casting process because the speed at which solidification occurs plays a large part in determining the mechanical properties of the casting. The faster the solidification rate, the finer the solidified structure and the higher the mechanical properties. Conversely, the slower the solidification rate, the coarser the structure and the lower the mechanical properties.

The solidification rate is determined by how fast or slow heat can be removed from the casting. For the casting to solidify (freeze), heat transfer has to take place. The speed at which the heat is transferred is driven by a thermal gradient. Thermal gradients are measured by the difference in temperature between two points and are given in degrees Fahrenheit or centigrade per unit of distance (Fig. 14-18). The steeper (larger) the thermal gradient, the faster heat transfer will take place and the faster the molten metal will solidify. A foundry can control thermal gradients through gating, the use of various mold and core materials, and the use of chills.

Solidification normally begins on the surface of the casting and moves inward toward the center of the casting or toward the source of feed metal (called a riser, which is a reservoir of molten aluminum to feed shrinkage in the casting). Solidification can take place in two directions in the casting. One, progressive solidification, occurs from the sidewalls of the casting toward its center. The other, directional solidification, occurs toward the source of feed metal.

Directional solidification should always be promoted. If progressive solidification moves faster than directional solidification, shrinkage voids will occur in the casting. Figure 14-19 illustrates these two types of solidification and shows how a shrinkage void can occur if directional solidification is not promoted.

There are computer programs available to be used in determining how a casting or casting section will solidify. These programs, available from several sources, can be used on personal computers as well as on large mainframe computers. These valuable tools can aid in the production of quality aluminum castings.

Solidification Ranges

When a casting turns from liquid to solid, it doesn't do it all at once. Solidification occurs during a period of time over a range of temperatures, which is called the freezing or solidification range of an alloy. Each alloy has a different freezing range and different solidification characteristics, which are shown in the cooling curve for the specific alloy. Fig. 14-20

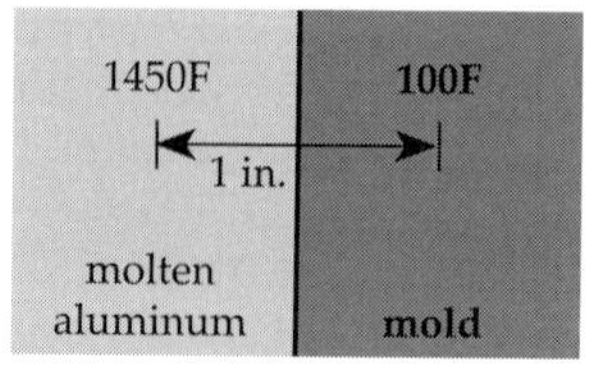

$$\text{thermal gradient} = \frac{1450\text{F} - 100\text{F}}{1 \text{ in.}} = 1350°\text{F/in.}$$

Fig. 14-18. How to measure a thermal gradient in the mold based on distance.

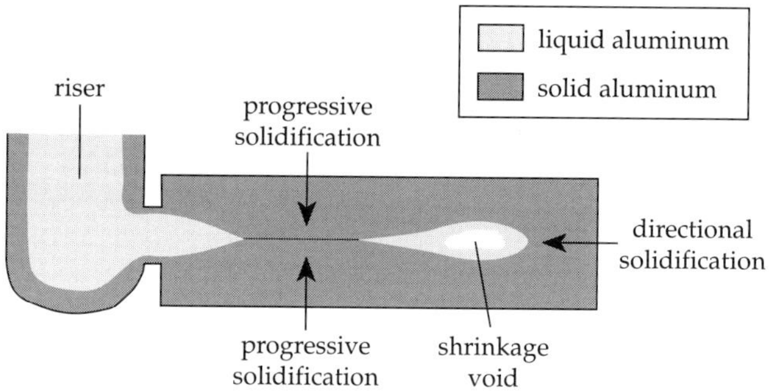

Fig. 14-19. How progressive solidification can cause a shrinkage void when it occurs more rapidly than directional solidification.

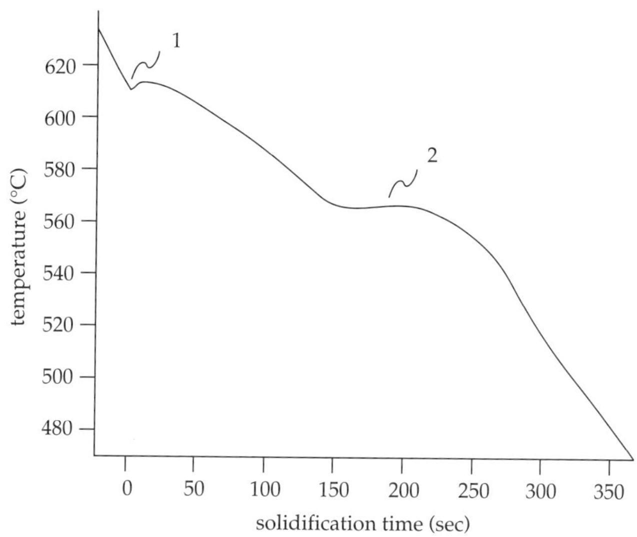

Fig. 14-20. Cooling curve for a 356 alloy specimen. Point 1 shows the start of the solidification process, while point 2 shows the main eutectic reaction. The data plotted here were taken from a thermocouple placed in the center of a 35-mm dia cylinder casting.

shows the cooling curve for Alloy 356. More details on solidification can be found in the section "Isothermal Solidification" in Chapter 4 and in other AFS publications.[2]

Shrinkage

All aluminum alloys shrink when they cool. The amount of shrinkage normally assumed with aluminum alloys is 6%. For example, this means that an aluminum alloy casting with a volume of 600 cu in. in the liquid state would have a volume of 564 cu in. when solidified.

Three types of shrinkage take place in the casting after it has been poured—liquid, solidification and solid.

Liquid shrinkage or contraction takes place when the molten metal cools. Solidification shrinkage takes place from the time the first solid metal appears until the casting is completely solid. The foundry compensates for liquid and solidification shrinkage by using risers.

After the casting has completed solidifying, it cools to ambient temperature, and during this time, solid shrinkage or contraction takes place. Solid shrinkage is equal to the metal's coefficient of thermal expansion multiplied by the temperature difference between the metal's freezing point and ambient temperature. A patternmaker compensates for this shrinkage by making a pattern slightly larger than the finished casting dimensions. Because of this, solid shrinkage is sometimes called "patternmaker's shrinkage" and the ruler used to adjust for it a "shrink rule."

RISERS

Risers are used to feed the liquid contraction and solidification shrinkage that take place in castings. The risers act as reservoirs of molten aluminum from which the casting can draw "feed metal" to compensate for the shrinkage taking place as metal solidifies.

In many cases, the riser is attached to the side of the casting or the section being fed, forming a *side riser*. A *blind riser* is one that does not go entirely through the cope of the mold.

Very seldom does a riser feed an entire casting. What it does feed is a section of the casting that, in turn, may have fed another section. The correct locations for risers can be determined by studying how the casting will solidify and then, if possible, attaching the risers to the sections that will be the last to solidify.

A riser has to be designed and sized so that it has enough liquid metal available to compensate for shrinkage in the casting and so that it completes its solidification after the casting has solidified. For example, if the expected 6% shrinkage will equal 1 lb of aluminum, then the riser has to provide 1 lb of aluminum, which must remain molten until the casting is finished freezing.

A riser is attached to the casting by a riser neck or pad. This area plays a critical role in the riser's ability to do its job because the feed metal passes through it into the casting or section being fed. Therefore, the riser neck has to stay liquid long enough for sufficient feed metal to pass through it.

During solidification, the casting should solidify first, the riser neck second and the riser third (Fig. 14-21).

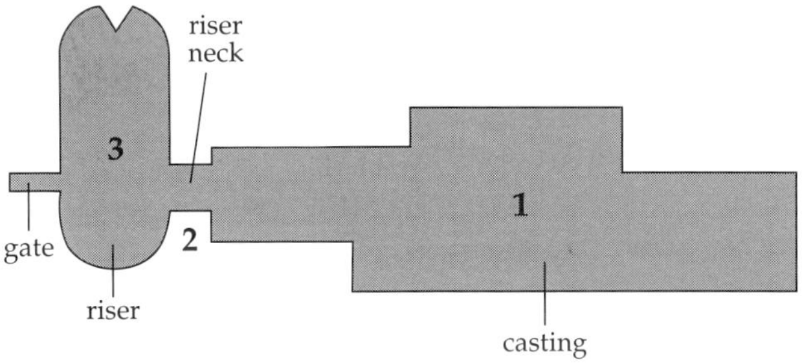

Fig. 14-21. The sequence in which solidification should occur.

Riser Shape

Because the riser has to remain liquid longer than the casting section it is to feed, its geometric shape should retard heat transfer from the riser to the mold. One way of doing this is by adjusting the riser's volume-to-surface-area ratio. For a given volume of molten aluminum, the shape selected should have the least possible amount of surface area. Out of all the geometric shapes, a sphere has the smallest surface area for a given volume. Thus, from a volume-to-surface-area standpoint (Fig. 14-22), a spherical riser would stay liquid longer than a square, rectangular or cylindrical riser (some of the riser shapes used in aluminum foundries). Therefore, in theory, spherical risers should be used.

There is a problem, however, with spherical risers when it comes to horizontally parted molds. When a spherical riser is sufficiently large to have the proper metallostatic head for

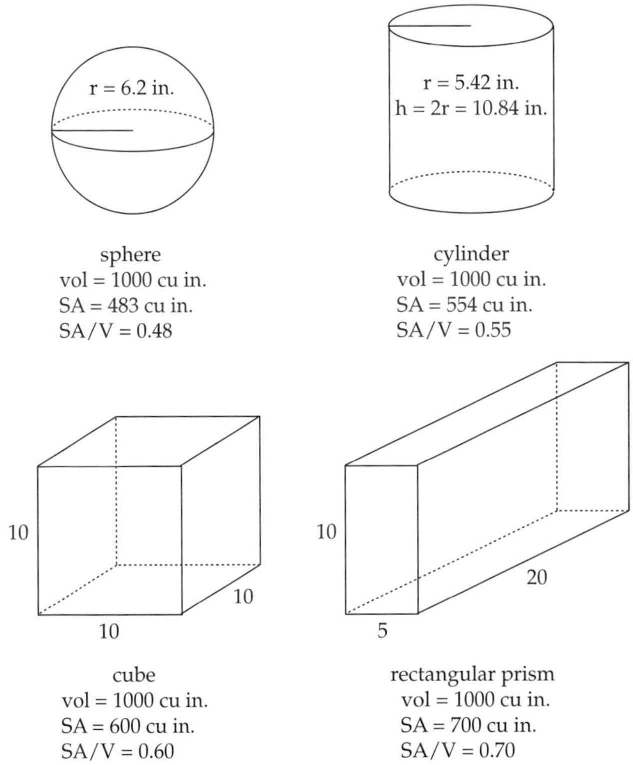

Fig. 14-22. Surface area to volume relationships for risers of constant volume and different geometric shapes.

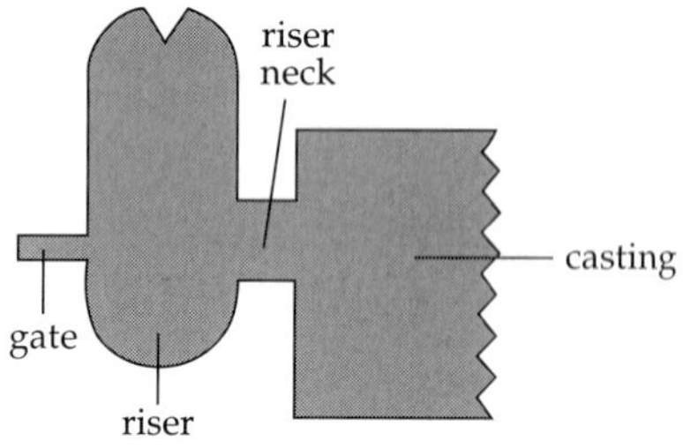

Fig. 14-23. The volume-to-surface-area ratio of a cylindrical riser can be improved by using a hemispherical bottom. For blind risers, a hemispherical top also can be incorporated.

the casting, its centerline may not fall on the parting line of the mold. In this situation, the pattern for the riser cannot be withdrawn from a horizontally parted mold without tearing up the mold. Therefore, cylindrical shaped risers are used in the majority of horizontally parted molds (Fig. 14-23).

There are several things that can be done to a cylindrical riser to improve its volume-to-surface-area ratio.

- It can and should have a hemispherical bottom on it.
- A ratio of 2:1 should be maintained between its height and diameter.
- If a blind riser is used, the top should be hemispherical. The top of the blind riser also should have a wedge in it to help it to collapse as it feeds.

Riser Size

Because of the low density of aluminum, risers that are used to feed aluminum alloy castings should be at least 1–2 in. taller than the section that is being fed in order for the metallostatic head or pressure in the riser to be effective in helping the riser feed. As mentioned previously, the riser's height should be 1.5–2 times its diameter. This is true, provided that this ratio makes the riser taller than the section of the casting it is to feed.

There are several ways to determine the size of a riser besides the "we-think-it-should-be-this-big" method. One of the more popular methods is called the modulus method. Many computer risering programs are based on this method. Basically, the modulus method compares the volume-to-surface-area ratio of the casting or section to be fed to the volume-to-surface-area ratio of the riser. The riser must have a higher modulus (i.e., volume-to-surface-area ratio) than the section being fed. Most literature says that the modulus of the riser should be 20% larger than the modulus of the section being fed. Remember that the riser also has to contain the 6% of molten aluminum needed to feed the shrinkage in the casting section to be fed.

The modulus method also can be used to determine the size of the riser neck. Most authorities agree that the modulus (volume-to-surface-area ratio) of the riser neck should be 10% larger than that of the casting or section to be fed. Another method, based on the type of casting or section being fed, uses the dimensions of the riser to calculate the size of the riser neck.

One method for determining the size of a riser is based on work done by R. Heine at the University of Wisconsin–Madison. This method is based on taking the 6% of feed metal required to feed the casting or section to be fed and building a riser around it. In this method, a thermal wall of molten aluminum is built around the 6% of feed metal to keep it liquid until it can be used to feed the casting section. For instance, to feed a 10 lb section, a thermal wall would be built around a cylinder of molten aluminum weighing 0.6 lb. How this thermal wall is built is described in the available literature.

Riser Aids

Probably the best way to help a riser do its job is to gate through the riser into the casting or section being fed. This ensures that the last molten aluminum poured down the sprue, which is the hottest, will end up in the riser, provided it is a side riser. This practice also aids in setting up directional solidification. This type of riser is often called a hot or live riser.

If using this type of riser is not possible due to the design of the casting or to placement of locating points on a casting, then molten aluminum must pass through the casting into the risers. This situation creates what is called a cold or dead riser. A riser may even be placed on top of the casting or section to be fed, forming a top riser.

When a cold or dead riser is used, some means must be provided to help the riser stay liquid long enough to do its job. With top risers that are open to the atmosphere, the practice of back pouring into the riser is common (Fig. 14-24). This is done by not filling the riser completely, then coming back with a ladle of molten aluminum and pouring it directly into the partially filled riser.

When back pouring into a riser, there initially must be a sufficient amount of molten aluminum in the riser to keep the falling stream from penetrating into the casting proper. If this penetration occurs, dross and gas can find their way into the casting and cause defects. If pouring into a riser causes metal to rise in the sprue or in other risers, this means that metal from the riser is being washed back into the casting cavity.

Molten aluminum loses most of its heat through conduction into the mold material. A way to help a riser stay liquid is to surround it with an insulating material that will reduce the heat lost by conduction. Besides serving as an insulating barrier between the cold riser and molding material, these insulating sleeves also can be used to reduce the size of hot risers. With a cold riser, as much heat as possible must be maintained in the riser so that it can do its job. Figure 14-25 illustrates how insulating sleeves can affect the solidification time of risers.

Aluminum loses very little heat due to radiation because it has poor emissivity (the ability to radiate heat). Many people have burned their hands on aluminum castings that were recently removed from a mold because it is possible to nearly touch a hot aluminum casting before feeling any radiating heat. However, even though aluminum has poor emissivity, when open risers are used, it may be necessary to place an insulating material on top of the molten aluminum to reduce the amount of heat lost due to radiation (Fig. 14-25).

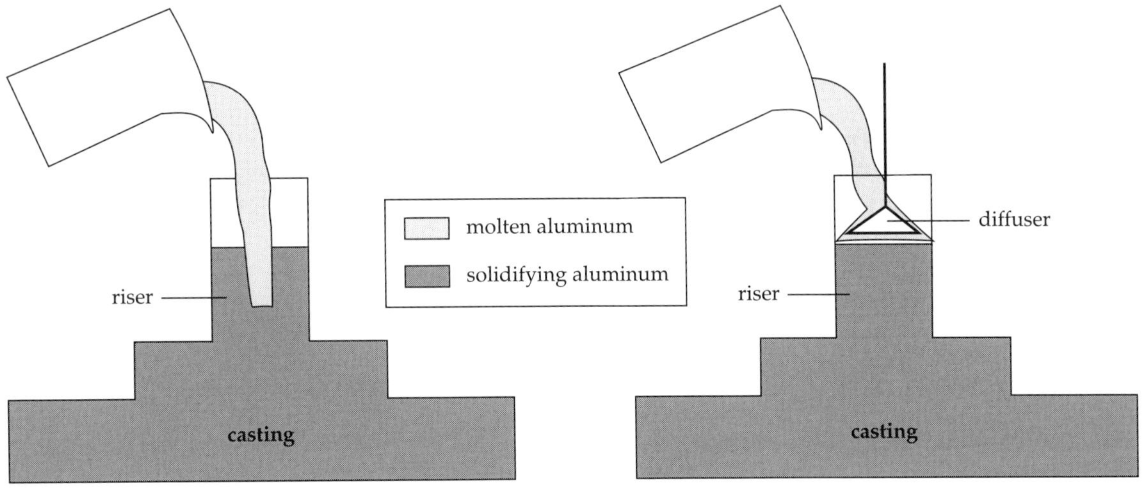

Fig. 14-24. Two methods for back pouring into an open riser. The method on the right is the more desirable way to do this because dross from the top is not forced down the riser toward the casting. As molten metal fills the riser, the diffuser is drawn upward.

This material is sometimes called "hot topping." When using insulating materials (either as sleeves or as hot topping), be certain that they are completely dry.

Feeding Distance

The distance a riser can deliver feed metal into the casting section to which it is attached is called its feeding distance. One reason a riser may seem ineffective when shrinkage occurs is that the feeding distance of the riser has probably been exceeded.

Feeding distances of risers are determined by the alloy being poured and by the thickness of the section being fed (Fig. 14-26). For aluminum castings, a formula to approximate the theoretical feeding distance of a hot riser is:

$$FD = 2T$$

where FD is the feeding distance, and T is the thickness of the section being fed.

In Fig. 14-26, shrinkage has occurred just beyond the 8 in. feeding distance of the hot riser, so the riser will not eliminate the shrinkage problem. In this case, progressive solidification has beaten directional solidification, and the feed metal from the riser is cut off from the shrinkage cavity.

There are several ways this problem can be fixed. One is to change the section thickness. Another is to make the casting section a wedge by making the section thicker next to the riser and tapering it down to the 4 in. section at the end. This would effectively slow down progressive solidification and increase the amount of directional solidification.

Another method would be to increase the "end effect" (Fig. 14-27) by placing a chill on the end of the casting. Note that the end of the bar contains more solid metal than any other part of the bar. This end effect is caused by the ease

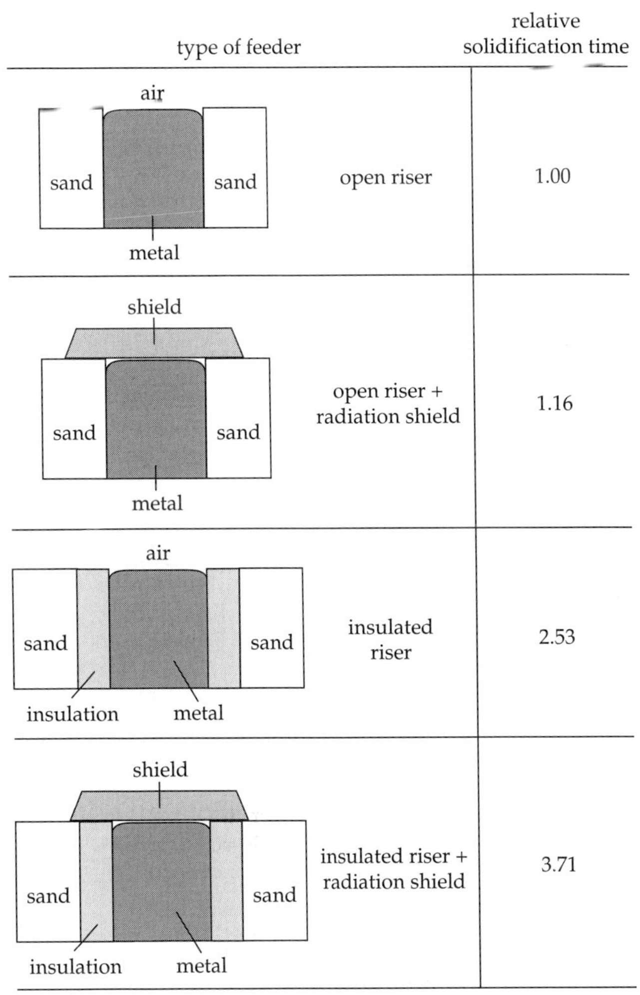

Fig. 14-25. Relative effect of insulating materials and radiation shields on the solidification times of 4x4 in. aluminum alloy cylinders.

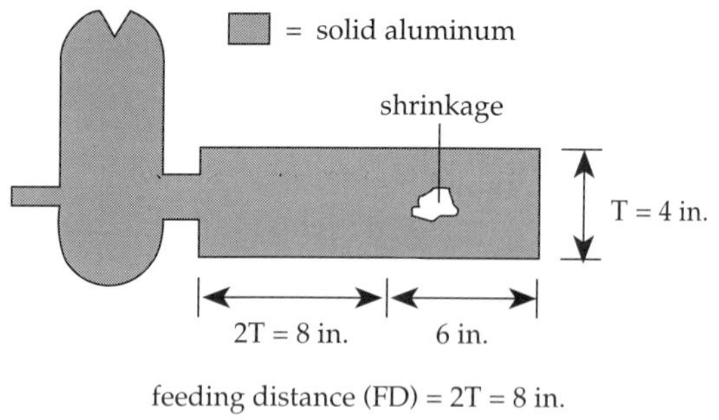

Fig. 14-26. Too large a feeding distance would cause shrinkage in this casting.

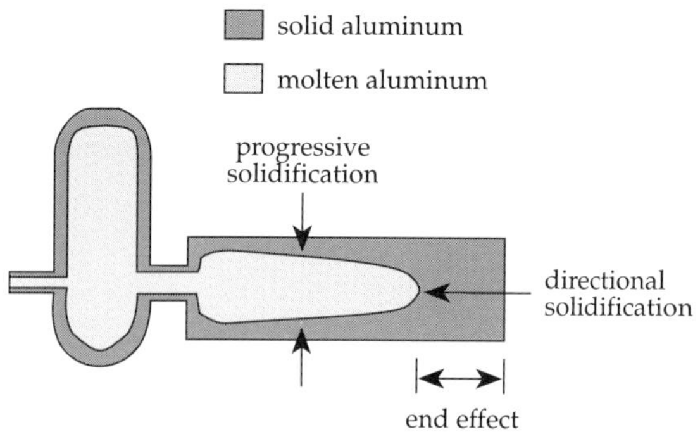

Fig. 14-27. An illustration of the "end effect" that can occur during solidification.

with which the heat can leave the molten aluminum because of the increase in surface area. This end effect can also be called directional solidification because it is headed in the direction of the riser.

Placing a chill against the end of the bar would increase the amount of end effect. By doing this, directional solidification would be increased, and the shrinkage shown in Fig. 14-26 would end up within the feeding distance of the riser.

Still another way to eliminate the shrinkage defect would be to place a riser on the top of the bar halfway from either end (Fig. 14-28a). Thus, there would be two end effects and only half the necessary feeding distance, provided the casting was gated into the side. Or two risers could be placed on the casting, one on each end (Fig. 14-28b). In this case, a chill would be placed, as shown, to promote directional solidification back toward the risers. However, this method would put more metal into the gating and risering system, which may not be the most economical way of eliminating the shrinkage problem.

With very large castings, section thickness can be thermally changed by placing insulating boards in the mold against the sides of the casting sections where progressive solidification needs to be slowed down (Fig. 14-29).

The illustrations in this section are not meant to imply that only one riser can be used to feed a casting. In the case of

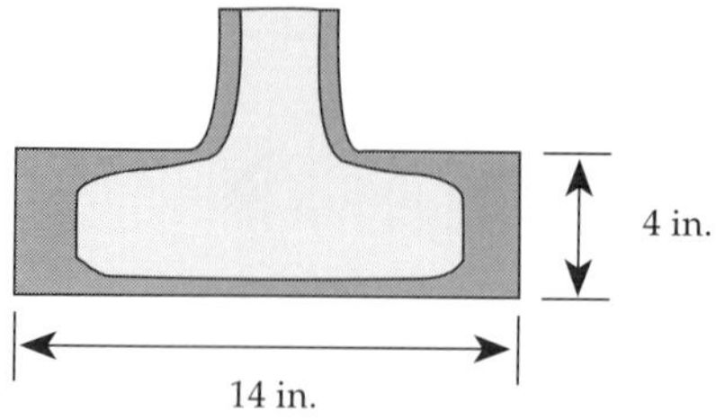

(14-28a)

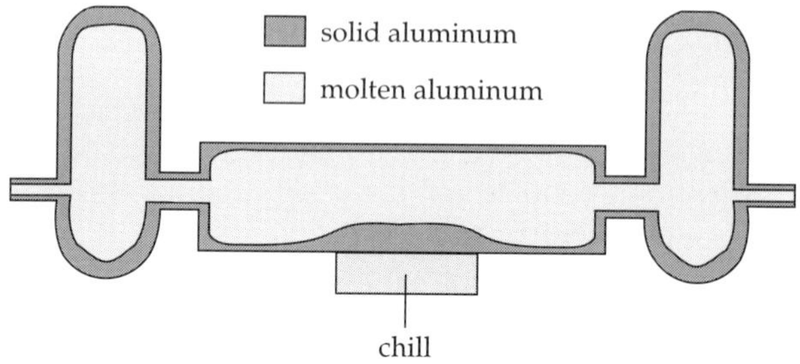

(14-28b)

Fig. 14-28. Two possible ways to eliminate the shrinkage defect that occurred in Fig. 14-26. a) A riser is placed on top of the casting halfway between the ends. b) A riser is placed on each end.

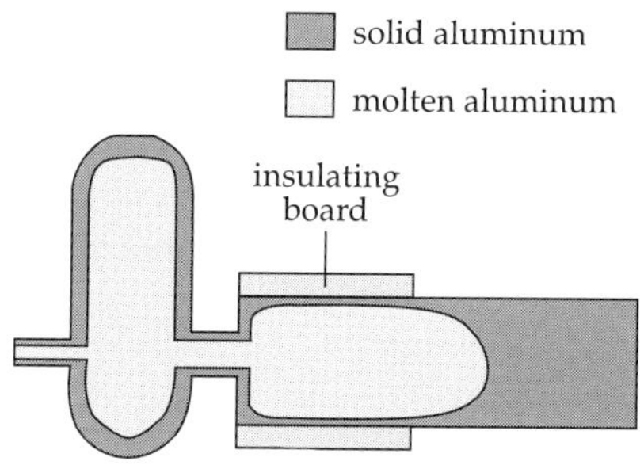

Fig. 14-29. Progressive solidification can be slowed down by placing insulating boards in the mold.

larger and more complicated castings, more than one riser may be required, with each riser assigned a section of the casting to feed. It is, therefore, very important to study carefully how the casting is going to solidify so that risers can be placed where they are needed.

CHILLS

In several instances, the use of chills has been mentioned. As one person said, "Chills are the foundryman's way of doing thermal trickery." By this, it is meant that chills are a means by which a casting can be made to solidify in a desired manner. Just as solidification was slowed down by using insulating materials, it can be sped up by using chills.

By speeding up the solidification of the casting or casting section, the foundry can:

- promote directional solidification;
- help risers feed shrinkage;
- promote steep thermal gradients;
- refine the microstructure;
- reduce dendrite arm spacing;
- help reduce macro gas cavities.

Chills are named by where they are located and the material they are made of. External chills are placed so that they contact the external or internal surface of the casting (see Fig. 14-28b). In most cases, external chills are made of metal such as aluminum, copper, cast iron or steel. Chill aggregates such as zircon, chromite and graphite also can be used as external chills. With sand molds and cores, zircon and chromite aggregate can be used as part of the mold formulation. Graphite blocks can be used in the same way that metal chills are used. Any of these three materials can be used as coatings on cores or mold surfaces.

Internal chills are surrounded by molten aluminum during pouring and become part of the casting. However, the material used for internal chills should have a melting temperature and composition nearly equal to that of the aluminum alloy used in the casting. Thus, for all practical purposes, aluminum alloy internal chills must be used for aluminum alloys.

Chills should have a high volumetric heat capacity. This requirement is determined by using the following formula:

volumetric heat capacity = volume × density × specific heat

At a given volume and material temperature, cast iron proves to be the best metal material to use for external chills, based on this formula. The size of the metal chill can be determined by comparing the volumetric heat capacity of the chill to that of the casting section it will be chilling. Remember, the higher the volumetric heat capacity, the better the chill will work.

The surfaces of both external and internal chills must be clean and free of moisture and oil. The chill's surface should be roughened by methods such as sandblasting or abrasive grinding. A good practice is to knurl or groove the surface of the chill that comes in contact with the molten aluminum.

If necessary, the chill surface in contact with the molten aluminum can be coated. The chill coating should not be so thick that it hinders the ability of the chill to do its job. The chill's temperature should be equal to or slightly higher than that of the mold or core materials. This will help prevent the chill from sweating and causing gas-related defects. Cope chills often cause defects because moisture from the atmosphere condenses on chill surfaces and then "blows" when the molten metal contacts the chill.

External chills should be placed in the drag surface of the mold whenever possible. By placing external chills in the drag, contact with the casting surface is ensured. Chills should not be placed near gates or in the feed paths of risers, or these areas may solidify too soon.

It has been shown that improvements in aluminum alloy castings can be made at a distance of greater than 4T (T = section thickness) from the center of the chill. As discussed earlier, chills can increase the amount of directional solidification and, thereby, help a riser feed shrinkage before progressive solidification closes the feed path.

ACKNOWLEDGMENT

AFS would like to thank Fred Schleg, AFS Cast Metals Institute, for writing this chapter.

REFERENCES

1. D.O. Kennedy, J.C. Church; "Gating Castable Metal Matrix Composites," *AFS Transactions;* paper 91-87 (1981).
2. L. Backerud et al; *Solidification Characteristics of Aluminum Alloys, Vol. 2: Foundry Alloys;* AFS, Des Plaines, Ill. (1990).

BIBLIOGRAPHY

L.W. Eastwood; "Tentative Design of Horizontal Gating Systems for Light Alloys," *AFS Symposium on Principles of Gating* (1951).

K. Grube, L.W. Eastwood; "A Study of the Principles of Gating," *AFS Transactions,* vol 58 (1950).

H. Huang, A.V. Lodhia, J.T. Berry; "Modeling the Effectiveness of Chills During Solidification," *AFS Transactions;* vol 98, paper 90-58, (1990).

R.W. Heine; "Riser Design For Mold Dilation," *AFS Transactions;* paper 65-15 (1965).

K. Nilsson; *Use of Chills in Sand Casting of Aluminum Alloys;* Translation No. T 1646, BCIRA (Available from the AFS Technical Library).

P. Sandford; "Advances in the Production of High Yield Aluminum Castings Using Direct-Pour Technology," *AFS Transactions;* paper 93-63.

J.R. Schmahl, N.J. Davidson; "The Application Of Ceramic Foam Filter Technology In The Aluminum Foundry" (Available from the AFS Technical Library).

S. Strobl; "Good Gating Leads to Good Castings," *Modern Casting* (March 1992).

S. Strobl; "Control Inclusions with Good Gating Design," *Modern Casting* (Apr 1992).

S. Strobl; "Good Sprue Design Eliminates Gas, Oxide Entrainment," *Modern Casting* (May 1992).

S. Strobl; "Gate, Runner Location Critical to Quality Castings," *Modern Casting* (June 1992).

R.E. Swift, J.H. Jackson, L.W. Eastwood; "A Study of the Principles of Gating," *AFS Transactions,* vol 57 (1949).

Aluminum Casting Applications, Selee Corporation.

Basic Principles of Gating; AFS programmed learning textbook (June 1967).

Basic Principles of Gating and Risering, Cast Metals Institute Course Textbook (©1973).

Basic Principles of Risering; AFS programmed learning textbook (June 1968).

"DYPUR® Direct Pouring System For Aluminum and Copper Base Alloys," Technical Information Manual, Foseco, Inc., Foundry Products Division.

"The Function of Risers and Chills in Casting Aluminum," *Modern Casting* (Nov 1983).

Gating and Feeding for Light Metal Castings, AFS Research Report (1962).

Chapter 15

HEAT TREATMENT & STRENGTHENING OF ALUMINUM CASTINGS

Editor's Note: For clarity, the following conventions will be used throughout this chapter: aluminum (the alloy) and silicon (the nearly pure constituent phase) will be spelled out to differentiate them from the elements Al and Si. The matrix phase will be called α-aluminum.

HEAT TREATMENT EFFECTS ON PROPERTIES

Although heat treatment improves properties, many aluminum castings meet requirements as-cast and are not heat treated. Die castings, in particular, are not usually heat treated because they tend to blister. However, in the as-cast condition, high residual stresses may be present.

Depending on the alloy and on the casting process used, as-cast aluminum is susceptible to dimensional growth and low strength, toughness and corrosion resistance. To solve these problems and to enhance strength and ductility, castings are thermally processed by one or more heating and cooling cycles. In aluminum, this entails three basic operations: solution, quench and age. Combinations of these heat treatments, which include stabilization and stress relief, annealing and age hardening are called tempers.

This chapter will discuss aluminum heat treatments. Metallurgical aspects and strengthening will be explained, with emphasis on precipitation hardening. These points will be highlighted in a practical example. Additional information on vacancies, dislocations and deformation, which may be of interest to some, is located in Appendix A of this chapter.

Benefits of Heat Treatment

Aluminum castings are heat treated for:

- homogenization;
- stress relief;
- improved dimensional stability and machinability;
- optimized strength, ductility, toughness and corrosion resistance.

Microstructure of castings in the as-cast state is not homogeneous. This is because alloying and impurity elements tend to segregate into networks of eutectic constituents. Concentration gradients of elements traverse dendritic cells due to nonequilibrium solidification (coring) (Fig. 15-1), the extent of which depends on alloy type and solidification rate. Homogenization is desirable to distribute elements evenly throughout the matrix so properties in the casting will be uniform.

Residual stresses are created during cooling from elevated casting and solution temperatures. Heating the casting to an intermediate temperature can relieve residual stresses, at least to the extent of yield strength at that temperature. Control of residual stresses will be covered more later in this chapter.

Changes in the structure, such as the evolution of transition precipitates or the precipitation of fine silicon particles, can cause castings to grow over time. To maintain tight dimensional tolerances during and after machining, castings should be heat treated to form stable precipitate phases and to deplete the matrix of excess solutes.

The greatest use of heat treatment is to enhance mechanical and corrosion properties. Mechanical properties are improved by spheroidizing constituent phase particles and by precipitation hardening. Corrosion resistance is improved by homogenizing the distribution of alloying elements in solution and by dissolving easily oxidized phases, such as Mg_2Si. Rarely, however, are all of the properties in a casting optimized. Most often, a heat treatment is a compromise, maximizing the important properties at the expense of others. With this strategy, age hardening principles are exploited to tailor heat treatments to each application.

Age Hardening Treatments

Solution Treatment

Three events occur during solution treatment: 1) dissolution of elements that will later cause age hardening, 2) spheroidization of undissolved constituents, and 3) homogenization.

In the as-cast condition, a large proportion of the precipitate-forming elements are contained in constituent phases. Hence, the primary purpose of solution treatment is to dissolve constituent phases and enrich the α-aluminum solid solution in solute.

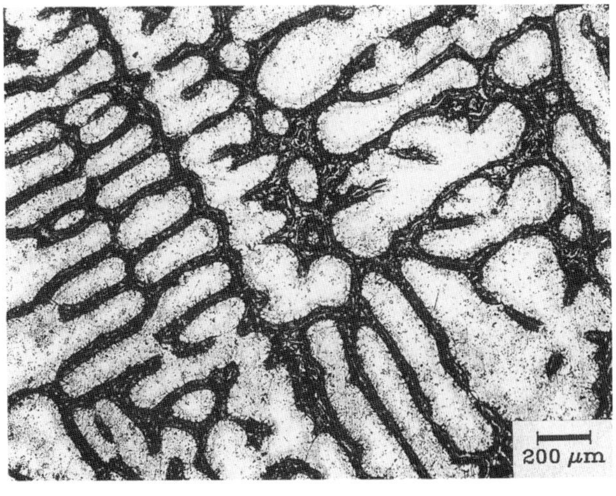

Fig. 15-1. Evidence of coring in a binary Al-4.5Cu alloy in the as-cast condition. Composition gradients are indicated by changes in the response to etching. [Photo courtesy of A.K. Sachdev, General Motors Research Labs.]

The second effect of solution treatment is to spheroidize constituents that are present in quantities that cannot be fully dissolved. In 3XX (Al-Si-Mg and Al-Si-Cu-Mg) alloys, excess silicon is always present. As-cast eutectic silicon forms either flakes, fibers or lamellae (the latter two result from eutectic modification treatments). During solution treatment, silicon particles globularize and begin to coarsen by Ostwald ripening.[1] These events are illustrated in Fig. 15-2. As compared to unmodified silicon, fibrous eutectic-modified silicon rapidly spheroidizes. Regardless of prior shape, silicon particles coarsen as solution treatment continues.

Unmodified | Sr-Modified

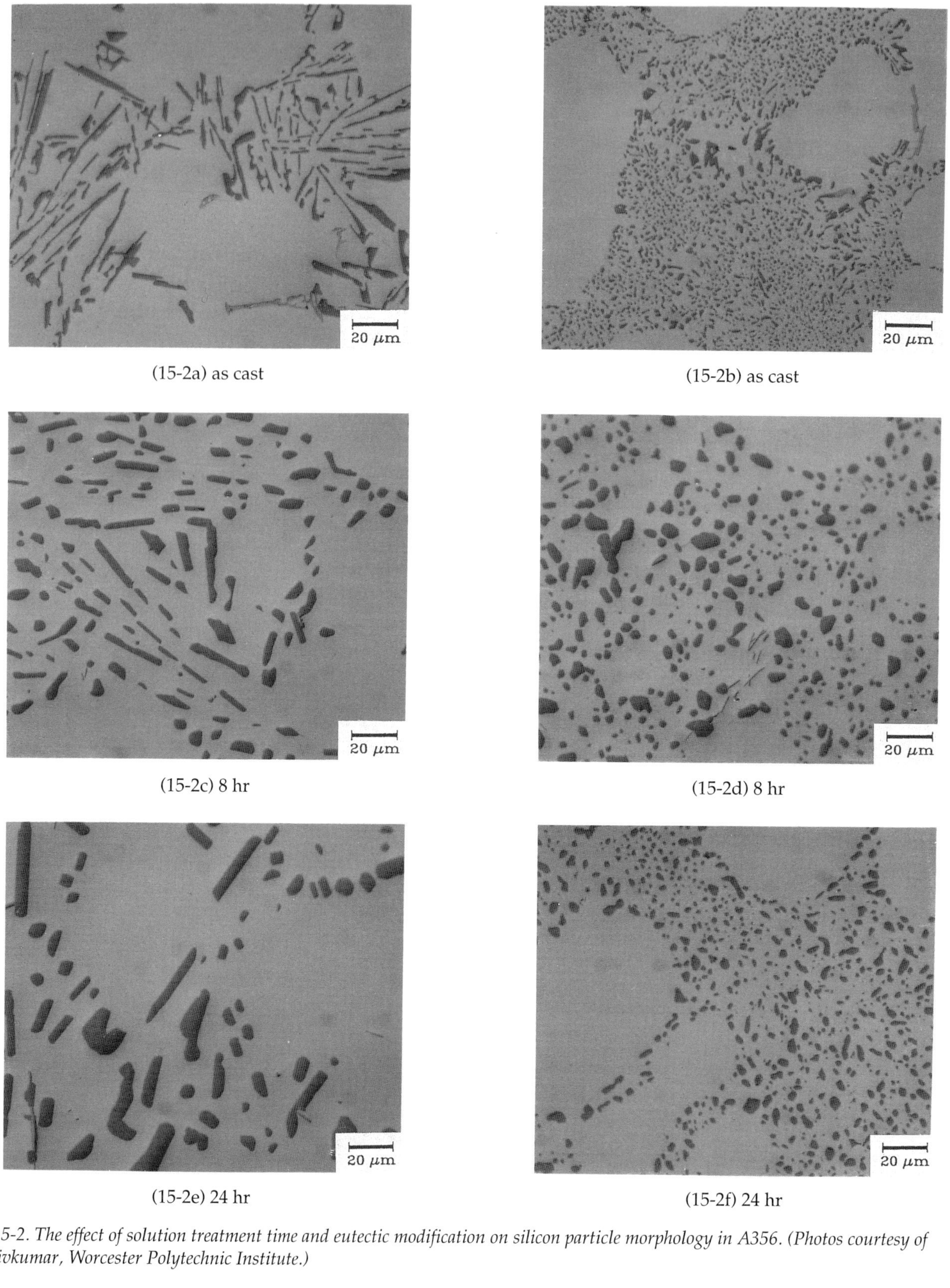

(15-2a) as cast | (15-2b) as cast

(15-2c) 8 hr | (15-2d) 8 hr

(15-2e) 24 hr | (15-2f) 24 hr

Fig. 15-2. The effect of solution treatment time and eutectic modification on silicon particle morphology in A356. (Photos courtesy of S. Shivkumar, Worcester Polytechnic Institute.)

The third effect of solution treatment is to homogenize the microstructure. In addition to eliminating the effects of coring, the elements in constituent phases are redistributed when they dissolve. However, some constituent phases, particularly iron-rich intermetallics, contain insoluble elements and are changed little by solution treatment.

Prior to age hardening, it is desirable to have the greatest possible concentration of precipitate-forming elements in solution. Hence, solution temperatures must be high to facilitate diffusion of alloying element atoms and to increase their solubility.

A potential for localized melting exists because of the presence of nonequilibrium eutectic colonies that form during solidification. These eutectic colonies melt if the liquidus temperature that corresponds to their local composition is exceeded, a process called incipient melting.[2] Expansion is followed by contraction as the region melts and (after approaching equilibrium with the surrounding metal) resolidifies. This creates voids that approximate the breadth of the melted zone.

Although incipient melting of small colonies of the Al-$CuAl_2$-$Cu_2Si_6Mg_8Al_5$ ternary eutectic in Al-Si-Cu-Mg alloys is not dangerous,[2] melting of the eutectic silicon phase can be widespread enough to be damaging. Because incipient melting irreparably embrittles aluminum castings, this condition must be prevented by controlling alloy composition and solution treatment temperature.

Solution treatments may include a two-step heating procedure to avoid incipient melting. Castings are first heated to just below the eutectic temperature to dissolve nonequilibrium eutectic colonies. Then they are safely heated to the final solution temperature to dissolve additional hardening elements and to achieve greater homogenization.

The appropriate solution time is determined by the diffusion rate of alloying elements and the coarseness of the microstructure. Diffusion rate depends on atomic species and solution temperature, but both are usually fixed. However, microstructural coarseness is controlled by solidification rate, which varies throughout the casting and changes with casting process.

For example, permanent mold castings solidify more rapidly than sand castings and contain smaller dendritic cells separated by finer, more closely spaced constituent phase particles. Finer particles dissolve sooner, and smaller dendritic cells and decreased separation between particles reduce diffusion distances. Consequently, permanent mold casting facilitates heat treatment by reducing solution time.

Quench

This operation finishes the solution treatment. When the casting is cooled from the solution temperature, the solubility for hardening elements decreases. Solutes are lost from the enriched α-aluminum solid solution by precipitation and by diffusion to grain boundaries, processes that occur most rapidly between 500 and 750F (260 and 400C).[2] Hence, rapid cooling or quenching is necessary to retain the high concentrations that were in equilibrium at the solution temperature, particularly near grain boundaries. Although rapid quenching increases response to age hardening, it also creates residual stresses and distortion, factors that must be weighed for each casting design.

Age Hardening

The excess solutes that are supersaturated in α-aluminum will eventually precipitate, but this occurs slowly at room temperature. After days and weeks at room temperature, some alloys, such as those belonging to the Al-Zn-Mg (7XX) system, harden appreciably—a process called natural aging. However, aging may be accelerated by heating castings in the as-quenched condition to intermediate temperatures—a process called artificial aging.

After initial stages of aging, the structure is said to be underaged. Increased time at temperature or aging at a greater temperature further evolves the precipitate structure, and hardness increases to a maximum, the peak hardness condition. Further aging decreases hardness, and the structure becomes overaged. Progression of the aging process also affects ductility. Ductility, while high in the as-quenched condition, reduces during hardening. During overaging, loss of hardening mechanisms permits more extensive deformation to occur before fracture (i.e., ductility increases). Annealing, which can be described as an extreme overaging, maximizes ductility.

Change in the precipitate structure halts when the castings are cooled back to room temperature. The structure will remain stable so long as the service temperature does not exceed the artificial aging temperature. Therefore, naturally aged castings are relatively unstable. Conversely, overaged castings have extended thermal stability because their microstructures were obtained at higher aging temperatures.

Tempers

Combinations of heat treatments (or tempers) include solution and age; stress relief and stabilization; and annealing. The quench method is not specified. The first digit of a temper designation represents the heat treatment sequence of time and temperature. One or two additional digits may be added to indicate a variation. The following sections will describe the basic tempers and common variants.

Solution and Age Tempers[2]

These tempers form a maximum hardness curve (Fig. 15-3).

- *T4*—Solution treat and age naturally to a substantially stable condition. Natural aging may continue slowly, particularly at elevated service temperatures, so structural stability may not be satisfactory.
- *T6*—Solution treat and age artificially. In castings, T6 commonly describes optimum strength and ductility. Aging is typically at 293–320F (145–160C) for 3–5 hr.
- *T61*—Solution treat and age artificially for maximum hardness and strength. This variant of T6 yields additional strength and stability but at reduced ductility. Aging is typically at 302–338F (150–170C) for 6–10 hr.
- *T7*—Solution treat and artificially overage or stabilize. This temper improves ductility, thermal stability and resistance to stress corrosion cracking. Aging is typically at 374–437F (190–225C) for 4–6 hr.
- *T71*—Solution treat and artificially overage to a substantially stable condition. This temper further increases thermal stability and resistance to stress corrosion cracking and reduces strength. Aging is typically at 437–500F (225–260C) for 4–6 hr.

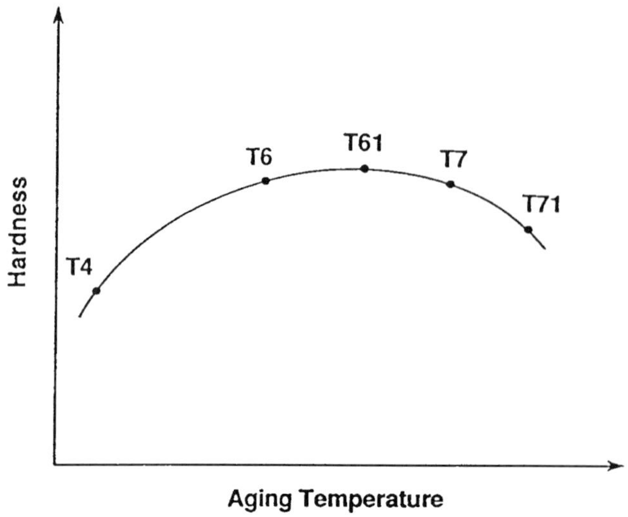

Fig. 15-3. Maximum hardness curve for age-hardened aluminum. Hardness changes with increased aging, brought upon by the increasing temperature of the tempers depicted.

The T4 temper can be thought of as the starting point for other solution and age tempers. Castings that are to be aged artificially can natural age at room temperature after solution treatment. In Al-Si-Mg alloys, room temperature incubation before artificial aging can reduce both strength and ductility. The T4 temper also is used to indicate solution treatment for so-called non-heat treatable alloys, which do not increase hardness by aging.

Solution Treatment for Non-Heat Treatable Alloys

Some non-heat treatable alloys, such as Al-Mg alloy 520 and Al-Si alloy 444, are solution treated although age hardening does not occur. In the case of Alloy 520, solution treatment dissolves Mg, which forms eutectic constituents during solidification. Alloy 520–T4 is strengthened by solid solution hardening, and once the as-cast eutectic networks are removed, it has good ductility. Solution treatment of Alloy 444 spheroidizes the eutectic silicon network, potentially doubling ductility. For these special cases:[2]

- *T4*—Solution treat to dissolve or spheroidize eutectic constituents.

Solution treatment can provide a supersaturated solid solution, so the microstructures of non-heat treatable alloys may be unstable. For example, the 520–T4 alloy will age at temperatures only slightly above room temperature. Precipitation tends to occur at grain boundaries, which reduces strength, ductility and resistance to stress corrosion cracking. Alloy 444 is susceptible to growth.

Stress Relief or Stabilization Treatment

Frequently, the as-cast condition provides acceptable mechanical properties but is accompanied by microstructural instability or undesirable residual stresses. Perhaps the possibility of in-service growth is the only constraint against using a casting in the as-cast state. In each case, the T5 temper is appropriate.[2]

- *T5*—Cool from casting temperature and artificially age or stabilize (without prior solution treatment). This is a common temper for permanent mold castings, such as pistons. Aging is typically at 401–500F (205–260C) for 7–10 hr.

To improve the mechanical properties provided by the T5 temper, castings may be water quenched immediately after casting. This so-called "simplified supersaturation treatment"[3] is very effective and convenient after sand and permanent mold casting. Of the solute that normally precipitates during cooling through the 500–752F (260–400C) temperature range, a portion is retained in solution by quenching and made available for age hardening during the T5 temper.

Variants of the T5 treatment provide additional stability, T51 by increasing aging temperature and T551 by increasing aging time.

Annealing

Castings that have low strength requirements but that need high dimensional stability are annealed. Annealing also substantially reduces residual stress, a need in die castings. Annealing is a severe stabilization treatment and an elevated temperature variant of the T5 temper. Softening occurs because annealing depletes the matrix of solutes, and the precipitates formed are too large to provide hardening.

Summarized below are several annealing practices that may be given to castings with temper designation O:

- *Non-heat treatable Al-Mg castings.* For stress relief, heat to 653–797F (345–425C) for 2–5 hr and air cool.
- *Sand and permanent mold castings.* For stress relief and thermal stability, heat to 500–698F (260–370C) for 2 hr and air cool.
- *Die castings.* For stress relief and maximum ductility, heat castings to 500–698F (260–370C) for 4–6 hr and furnace or air cool.

HEAT TREATMENT PRACTICES

Furnaces

Furnaces for heat treating aluminum must have good temperature control and, for solution treatment, be capable of rapid manipulation and removal of their charges. Because of the good oxidation resistance of aluminum, any fuel may be used. Fluidized bed furnaces and liquid salt baths are useful for high heating rates and precise temperature control. In drop-bottom furnaces, doors form the furnace floor—castings are removed rapidly when the doors open and the castings drop directly into the quench bath.

Castings should be loaded into heat treatment furnaces with care. Jigs or fixtures may be necessary to support castings and to prevent sagging and distortion during solution treatment. Since heat treatment temperatures for aluminum are relatively low (especially during artificial aging), air must circulate freely between castings. Gas-fired furnaces often use excess air in the burners to increase circulation. Fans are often necessary in resistance and radiant tube furnaces.

Temperatures should be controlled to within ~10°F (~5°C). Age hardening response is diminished by insufficient solution temperature, which impedes dissolution of eutectic constituents and reduces as-quenched concentrations of solutes and vacancies. Excessive temperature causes the risk of incipient melting and scrap castings. Consequently, temperature should be precisely determined on critical work with thermocouples strategically placed throughout the load.

The dew point in the furnace atmosphere should be controlled during solution treatment to avoid a defect called high temperature oxidation (HTO). This condition is caused by a reaction between Al and water vapor, which liberates hydrogen. The hydrogen penetrates the surface of the casting and forms blisters and subsurface voids. HTO may be avoided by using a dry atmosphere or by conditioning the furnace atmosphere with boron trifluoride gas or volatile, fluoride-containing salts.[2]

Quench Medium

For aluminum, water near its boiling point is the most frequently used quench medium, or quenchant. Faster cooling may be obtained using water at or below 149F (65C), but the risk of warping and cracking is greater. Agitation of the quench bath increases heat transfer and discourages formation of steam pockets. Castings should be positioned in ways 1) to avoid inverted concave surfaces, which trap air and steam and locally retard cooling; and 2) to provide uniform cooling to lanky sections, which minimizes warpage.

An important factor in this process is the quench interval, the time between opening of the furnace doors and immersion of castings into the quench bath. Although sensitivity to the quench interval differs between alloys, good practice should limit the interval to 10 or 20 sec. In critical castings, quench intervals of less than 10 sec, obtainable only with drop-bottom furnaces, may be required to achieve specified properties. The more severe the quench, the sooner rapid cooling is established in the critical 500–752F (260–400C) range, during which time valuable solute is lost, particularly in the middle of thick sections in the castings.

Control of Residual Stress

Because the surface of a casting cools and contracts before its interior, cooling initially forms tensile stresses at the surface and compressive stresses in the interior. After the interior cools and contracts, the surface stresses become compressive. Similar stresses arise when thin sections cool faster than thick sections or when the quenchant is not uniformly distributed over the casting.

A tangible result of residual stresses is warpage as stresses seek to balance each other. This can occur during machining if stressed metal is removed—the remaining stresses re-equilibrate by causing the casting to distort or to seize tool bits.

Residual stress can be reduced with stress relief or anneal tempers or, after solution treatment, with lower quench severity. Another measure is to reduce the solution temperature by 10°C (18°F) in the last minutes prior to quench. However, slack quenches and lowered solution temperatures must be weighed against the loss of mechanical properties.

Quench severity can be reduced with the use of synthetic heat transfer fluids, water-soluble polymer solutions, or even forced air. Water-soluble polymer solutions have proven, in recent years, to offer the best compromise between residual stress control and property loss.

Cryogenic Treatments

Another method (which is rarely used) to control residual stress is uphill quenching from cryogenic temperatures. Cryogenic treatments have two uses: to reduce residual stresses and to delay the onset of age hardening. Unlike with ferrous alloys, aluminum's metallurgical structure is not changed by these treatments.

A technique called uphill quenching, if done correctly, can reverse the distribution of stresses caused by quenching after solution treatment (i.e., surface under compression, interior under tension as described previously). The casting is cooled to liquid nitrogen temperature, then rapidly heated to room temperature or above with quenchants such as synthetic heat transfer fluid or steam. No loss of properties associated with solute loss occurs. However, control difficulties and the successful application of water-soluble polymer solutions have limited the use of uphill quenching.

A simpler cryogenic treatment is refrigerating solutioned-and-quenched castings to 0F (–20C) or below to postpone artificial aging. This applies to alloys in the T4 temper or alloys that should be aged artificially within 24 hr after quenching. In the case of the T4 temper, this provides time for straightening operations when the casting is most ductile, and for other tempers, a weekend delay between quenching and artificial aging.

METALLURGICAL ASPECTS

Cast aluminum is composed of many different phases in a largely Al matrix known as α-aluminum.[1] Aluminum contains alloying elements to improve its strength and corrosion properties, to counteract detrimental effects of certain impurities, and to improve castability. Portions of these elements combine with each other and with Al and impurities to form constituents that are distinctly different from α-aluminum.

Many of the alloying elements in aluminum alloys are added for a single purpose. For example, Ti is used to refine the grain structure. However, important alloying elements often have detrimental side effects that must be counteracted by the addition of other elements. In another example, A357, which contains Mg, also contains Be to enhance age hardening and to minimize Mg-rich dross films that form on the surface of the liquid and leave oxide folds in the casting.[2]

Aluminum alloys are divided into two classes: heat treatable and non-heat treatable. Heat treatable alloys, the 2XX (Al-Cu), 3XX (Al-Si-Mg and Al-Si- Cu-Mg) and 7XX (Al-Zn-Mg) systems, respond to the full range of age-hardening treatments. Most of this chapter deals with heat treatment of these alloys.

Non-heat treatable alloys do not age harden. This class consists of the 1XX (>99% pure Al), 4XX (Al-Si), 5XX (Al-Mg) and 8XX (Al-Sn) alloy systems. Binary 4XX alloys derive hardness from silicon constituent particles contained in eutectic networks, but large quantities of this phase lead to low ductility. The 5XX alloys are hardened quite effectively by

Mg in solid solution and can be as strong as age-hardened alloys. Since non-heat treatable alloys are frequently annealed, they are also (perhaps more properly) called non-hardenable.

Properties of cast aluminum, as those of other materials, are largely determined by microstructure—the identity, size, shape, distribution and proportion of phases that are present. Although the as-cast structure may be modified by heat treatment, cast aluminum is not mechanically worked to redistribute constituents as is wrought aluminum.

The initial microstructure of aluminum castings is determined in part by the solidification rate during casting and by the subsequent cooling rate. Hence, properties tend to vary throughout castings because different sections (e.g., thin and thick) experience different solidification and cooling rates. The as-cast microstructure can be improved, however, by such methods as grain refinement, eutectic modification and rapid solidification. Heat treatment, on the other hand, can spheroidize certain constituent phases, dissolve others and control most aspects of the precipitate structure.

Constituent Phases in Aluminum Alloys

Phases are physically distinct and separable components in a material. The α-aluminum phase is a solid solution of Al (the solvent) and alloying elements (the solutes). Other phases in cast aluminum contain excess alloying and impurity elements that are not in solution and that comprise the primary constituents, eutectic constituents and precipitates.

Phases other than α-aluminum usually occur as intermetallic compounds. An intermetallic compound is a phase that exists over a narrow range of compositions and is usually referred to by a specific chemical formula, such as $CuAl_2$. However, like solid solutions, some intermetallic compounds have non-stoichiometric compositions and considerable solubility for other elements.

The structures that are visible metallographically are the primary and eutectic phases. For example, A356–T6 contains primary α-aluminum and the eutectic constituents, silicon and $Al_8Mg_3FeSi_6$ (Fig. 15-4). These and other phases in cast aluminum, such as copper aluminide ($CuAl_2$), magnesium silicide (Mg_2Si), and numerous Fe-rich compounds, form during solidification and are generally stable over a wide range of temperatures.

Iron is a major impurity in aluminum alloys, but it is virtually insoluble in α-aluminum. Hence, any Fe that is present is contained in the eutectic constituents that it forms with Si, Cu and Mg. Fe-rich intermetallics often include other insoluble elements such as Mn and Cr, which are also insoluble in α-aluminum. Many of the particles have distinct shapes and morphologies (Fig. 15-5), which may depend on composition and the thermal history of the metal before and during casting. In general, precise identification of Fe-rich intermetallics is complicated because the possible phases often have non-stoichiometric compositions.

Precipitates, on the other hand, form from the α-aluminum solid solution. In α-aluminum, the solubility for most elements decreases when the temperature is reduced. During cooling, if the solubility for a solute falls below its concentration in solution, the solid solution becomes supersaturated. The excess solute leaves the supersaturated solid solution by forming precipitates that are rich in that solute element. Precipitates and their role in strengthening aluminum will be discussed further in a separate section.

Aluminum Crystallography

Atoms in the α-aluminum solid solution are arranged in a pattern, or crystal structure, called face-centered cubic (FCC). The FCC structure is illustrated in Fig. 15-6 in the form of a unit cell, a convention crystallographers use to classify crystal lattices. Atoms are centered in the sides, or faces, of the FCC unit cell. Many individual α-aluminum crystals, or grains, collectively form the matrix of aluminum alloys.

Different phases in aluminum tend to have different crystal structures. Alternatively, constituents may have the same

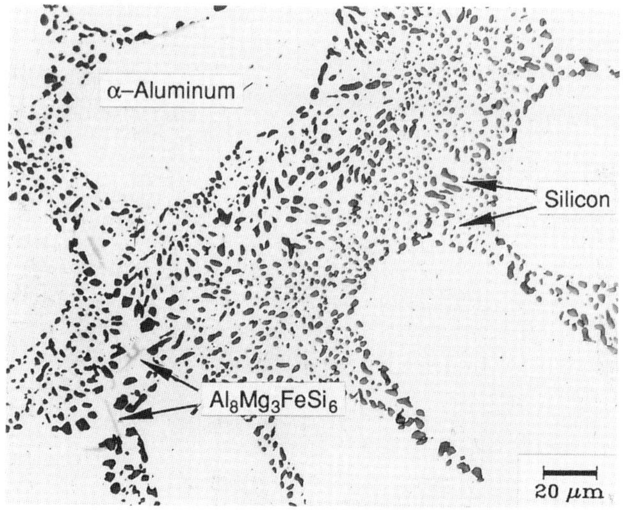

Fig. 15-4. Constituent phases in A356 aluminum that are visible metallographically.

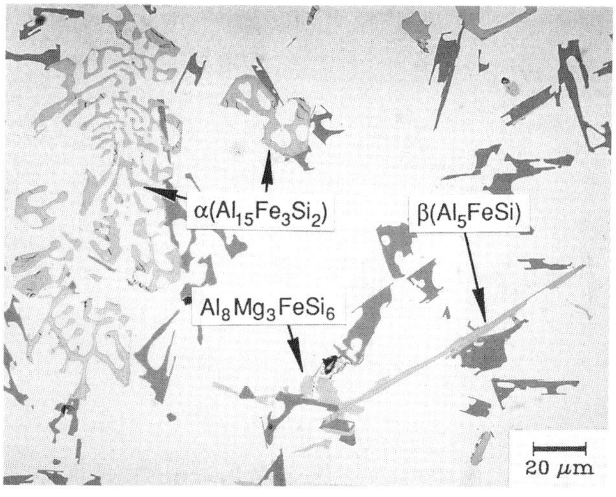

Fig. 15-5. Three Fe-rich intermetallic phases in a cast Al-Si-Cu-Mg alloy.

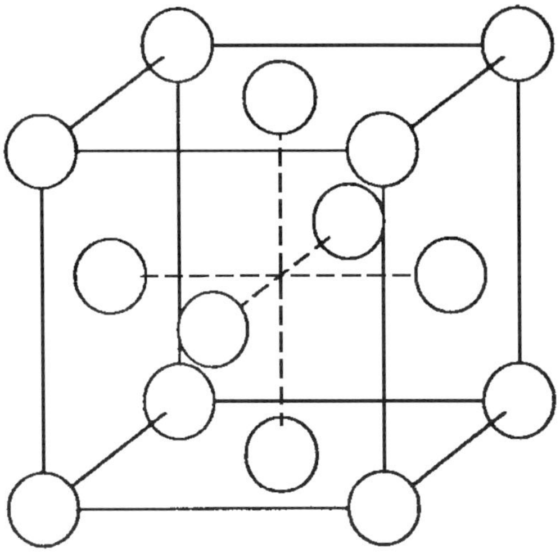

Fig. 15-6. A unit cell of the face-centered cubic (FCC) crystal lattice.

crystal structure as α-aluminum but distinctly different chemical compositions or sizes of their unit cells (measured by lattice parameters).

Precipitate phases that do not initially interrupt the continuity of the α-aluminum crystal lattice are called coherent. However, coherent precipitates typically distort the normal position of atoms in α-aluminum (Fig. 15-7). Such coherency strains increase when coherent precipitates coarsen or enlarge. When precipitates transform into their non-coherent terminal phase, coherency strains disappear.

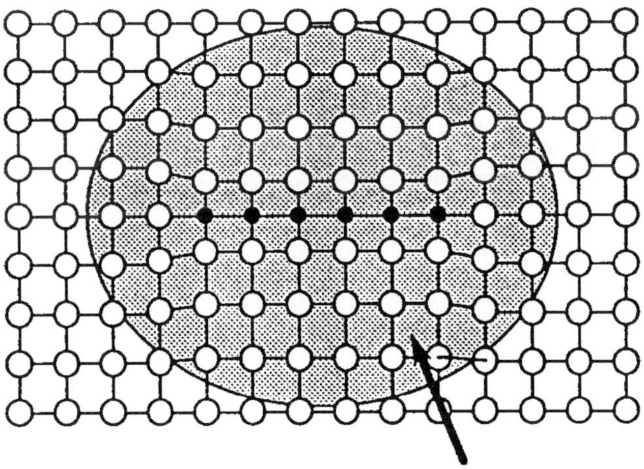

Fig. 15-7. An illustration of a GP zone, a precipitate (black circles) that is coherent with α-aluminum (open circles). The mismatch between the α-aluminum matrix and the precipitate lattice causes coherency strains, a shift of atoms from their normal positions in the lattice. [After Ref. 6.]

STRENGTHENING ALUMINUM

Imperfections in the crystal lattice tend to reduce ductility and increase strength. Ductility is the capacity for a material to sustain permanent or plastic deformation without fracture, and strength is the maximum stress a material can carry without failure. Details on deformation and how it occurs with mobile crystal defects (called dislocations) are located in the appendix of this chapter. Other crystal defects—solute atoms and second-phase particles—will be discussed next.

Solute atoms usually do not fit exactly into the parent lattice. The distortion of the α-aluminum lattice by larger or smaller solute atoms impedes plastic deformation and results in solid solution hardening.

Alloying elements can also harden aluminum by forming second-phase particles. The triaxial state of stress that exists between constituent particles that are macroscopically large and densely spaced restricts plastic deformation. Consequently, networks of eutectic silicon particles increase the strength of the non-heat treatable 4XX alloys. However, the inability to deform plastically causes these alloys to be brittle.

Alternatively, a dense dispersion of fine precipitates formed from alloying elements can physically obstruct the submicroscopic movements of atoms during plastic deformation. The deforming matrix may circumvent precipitates either by deforming them (called shear) or by simply bypassing them (called Orowan looping). Precipitates can also harden aluminum by creating coherency strains (analogous to solid solution hardening), which further resist deformation of the matrix that surrounds them. These strengthening mechanisms are characteristic of precipitation hardening.[4]

Precipitation Hardening

Precipitation hardening, or age hardening, is caused by precipitation from supersaturated solid solution of submicron size particles that are too small to be seen optically.[5] At lower temperatures, precipitates first appear as structured concentrations of solute atoms called Guinier-Preston (GP) zones. These may subsequently transform into a transition precipitate phase. Over time, transition precipitates may also change into another transition phase that has different structure and composition. After one or more transformations, transition precipitates form a stable terminal phase, whereupon further phase changes do not occur. Table 15-1 lists some useful precipitation systems in aluminum alloys.

Grain and phase boundaries are favorable nucleation sites for precipitates.[6] Consequently, solute tends to diffuse from the adjacent α-aluminum to these locations. During slow cooling from elevated temperature, such as a slack quench, this may happen to the extent that regions adjacent to grain and phase boundaries lose supersaturation and the ability to form precipitates. These narrow layers of α-aluminum, called precipitate-free zones (PFZs), remain soft after age hardening, adversely affecting bulk properties. PFZs can be minimized by high solution temperatures followed by rapid cooling, topics that will be discussed further.

Although many other details remain in question, GP zones are known to have crystal structures distinctly different from that of α-aluminum. However, GP zones are small and coherent, initially only a few atomic layers thick and, depending on their shape, larger in length or diameter. They precipitate

Table 15-1.
Precipitation Hardening Systems in Aluminum[1,6]

Alloy System	Aging Sequence[1]	PRECIPITATE INFORMATION Type	Coherency[2]	Structure[3]	Shape	Max. Size[4,5]
Al-Cu	SS→GP→θ″→θ′→θ	GP	C	unknown	disk	6x100 Å
		θ″	C	Tetr	disk	40x1000 Å
		θ′	C→N	Tetr-2	disk	150x6000 Å
	(terminal phase: $CuAl_2$)→	θ	N	BC Tetr		
Al-Cu-Mg	SS→GP→S″→S′→S	GP	C	unknown	unknown	16 Å dia.
		S″	C	unknown–Ord	rod	unknown
		S′	C	Ortho	unknown	100 Å dia.
	(terminal phase: $CuMgAl_2$)→	S	N	FC Ortho		
Al-Mg-Si	SS→GP→β″→β′→β	GP	C	unknown	needle	60x1000 Å
		β″	C	Mono	needle	60x2000 Å
		β′	C	Hex	rod	unknown
	(terminal phase: Mg_2Si)→	β	N	FC C		
Al-Zn-Mg (Zn>Mg)	SS→GP→η′→η	GP	C	unknown	sphere	35 Å dia.
		η′	SC	Mono	sphere	300 Å dia.
	(terminal phase: $MgZn_2$)→	η	N	Hex		
Al-Mg-Zn (Mg>Zn)	SS→GP→T′→T	GP	C	unknown	sphere	1000 Å dia.
		T′	SC	C	unknown	unknown
	(terminal phase: $Mg_3Zn_3Al_2$)→	T	N	BC C		

Notes:
1. SS = supersaturated solid solution, GP = Guinier-Preston zones.
2. Coherency: C = coherent, SC = semi-coherent, N = noncoherent.
3. Crystal Structure: BC = body-centered, FC = face-centered, C = cubic, Hex = hexagonal, Ord = ordered, Ortho = orthorhombic, Mono = monoclinic, Tetr = tetragonal, Tetr-2 = tetragonal with different composition.
4. Compare the precipitate size to the atomic radius of Al, 2.86 Å.
5. The terminal phase does not have a fixed shape or a maximum size.

profusely in α-aluminum because coherent phase boundaries facilitate nucleation. Consequently, GP zones are very closely spaced.

After time elapses, temperature increases—i.e., both the aging progresses and the GP zones coarsen. However, coherency strains, which increase as coherent particles enlarge, inhibit continued growth. Hence, transition precipitates of more favorable composition and structure tend to nucleate at GP zone sites, and the GP zones, in turn, are consumed. The new precipitate phase is usually coherent but of a structure that, for larger particles, reduces coherency strains. If aging is allowed to continue, the precipitates finally coarsen to the extent that coherency cannot be maintained—a condition that favors the formation of the terminal phase.

As the precipitates coarsen, they also become fewer in number and more widely spaced. By a process known as Ostwald ripening, larger precipitates grow, consuming atoms from smaller precipitates that shrink. This leads to many of the characteristics of precipitation hardening that are discussed in the following sections.

Hardening Effect of Precipitates

As mentioned previously, precipitates harden aluminum by several mechanisms. GP zones create coherency strains in α-aluminum that increase internal frictional forces that resist deformation. The GP zones are themselves more difficult to deform or shear than α-aluminum, and this also impedes deformation. Aluminum hardens when GP zones coarsen because coherency strains increase, and larger GP zones are more difficult to shear. Transition precipitates that replace GP zones, for example θ″, are initially coherent and, by the same mechanisms exerted by GP zones, continue to increase the hardness of aluminum as aging progresses.[4]

An alternative deformation mechanism within a material is Orowan looping, in which the precipitates are bypassed and left intact. Because this requires that the deforming metal concentrate between the precipitates, Orowan looping occurs more easily when precipitates are coarse and widely spaced. Hence, Orowan looping differs from precipitate shear in that the required stress decreases as aging progresses,[4] because aging increases precipitate spacing. In age-hardened aluminum, maximum hardness is obtained when the decreasing stresses for Orowan looping equal the increasing stresses for precipitate shearing.

The effect of GP zones and the precipitates that follow is exemplified in the aging curves of an Al-4Cu alloy (Fig. 15-8). Although commercial casting alloys have additional alloying elements that change kinetics, events that occur during precipitation hardening are universal. Hardness tends to increase to a maximum then decrease. Higher aging temperature expedites the transition between each stage of precipitation and can even circumvent the initial GP-zone formation stage. Further, when the aging temperature is lower, the maximum hardness is greater, but the time required to achieve maximum hardness is much longer (in the example in Fig. 15-8, about 50 times longer).

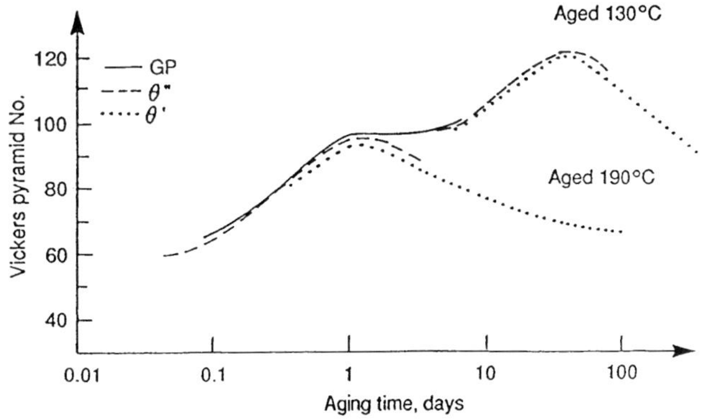

Fig. 15-8. Aging curves for a binary Al-4Cu alloy. For peak hardness, GP zones are replaced by θ″ and θ′. At 374F (190C), GP zones are not detected.[7]

A PRACTICAL EXAMPLE

The concepts discussed previously may be illustrated with the 356 alloy, a heat treatable Al-Si-Mg alloy hardened by β′. The composition of 356 is given in Table 15-2, and the mechanical properties achieved by tempering this alloy are listed Table 15-3.

The 356 alloy solidifies similarly to a binary Al-7Si alloy, the solidification sequence of which (kinetic effects neglected) is illustrated in Fig. 15-9. The first product of solidification is α-aluminum, point A. The primary α-aluminum dendrites form as the casting cools from A to B. During this time the liquid becomes enriched in Si, and its composition moves from A to B″. Next, a eutectic mixture of silicon and α-aluminum solidify from liquid that has the composition given by point B″. The eutectic α-aluminum solidifying at this stage has the composition given by point B′. After solidification and cooling to point C, the equilibrium Si content of α-aluminum (i.e., the solubility of Si in α-aluminum) is indicated by point C′. After further cooling, point D, the solubility for Si is reduced further, point D′.

In addition to Si, Alloy 356 also contains Mg, which enters into both the α-aluminum solid solution and the small eutectic colonies containing β(Mg_2Si).[8] After cooling to room temperature, the α-aluminum matrix retains little supersaturation in Si and Mg, and precipitates are mostly β′ and β(Mg_2Si), which provide moderate strength. The major constituent phase is eutectic silicon in flake, lamellar or fibrous form, depending on the solidification rate and modification treatment.[9]

If 356 castings are to be annealed, they should be heated to 649F (343C) and held for 2 hr. The castings should then be removed from the furnace and still-air cooled to room temperature. In addition to depleting α-aluminum of Si and Mg, this procedure replaces β′ precipitates with β(Mg_2Si), which is coarsened.

Table 15-2.
Chemical Composition of 356[1]

Element	Range wt %
Si	6.5–7.5
Mg	0.2–0.45
Fe	< 0.60
Cu	< 0.25
Mn	< 0.35
Zn	< 0.35
Ti	< 0.25

Table 15-3.
Mechanical Properties of 356 with Various Tempers

Temper	Tensile Strength MPa	Tensile Strength ksi	Yield Strength MPa	Yield Strength ksi	Elongation %
Sand Cast					
T51	172	25	140	20	2.0
T6	228	33	165	24	3.5
T7	234	34	205	30	2.0
T71	193	28	145	21	3.5
Permanent Mold					
T51	185	27	140	20	2.0
T6	262	38	185	27	5.0
T7	221	32	165	24	6.0

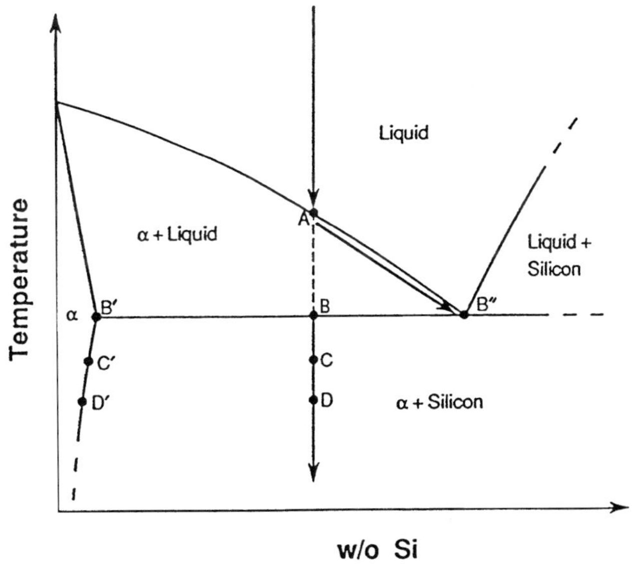

Fig. 15-9. Phase transformations of an Al-7Si alloy cooling from liquid to solid. Kinetic effects are ignored. Primary α-aluminum solidifies between points A and B. The Al-Si eutectic solidifies from Si-enriched liquid at point B″. When the casting cools to point C and then to point D, the equilibrium concentrations of Si in α-aluminum are given by points C′ and D′, respectively.

If 356 castings are to be stabilized (without performing solution treatment), the T51 temper can be used. The castings should be heated to 440F (227C) and held for 7–9 hr. This completes the precipitation reaction and transforms Si and Mg solutes into β'.

After aging, castings should be removed from the furnace and still-air cooled to room temperature. The hardness can be increased, but with some thermal stability lost, by aging at lower temperatures. The hardness also can be increased by water quenching the castings immediately after solidification—the simplified supersaturation treatment, which provides additional Si and Mg for age hardening.

For T6, T7 and T71 tempers, the castings should be solution treated at 1000F (538C) for 12 hr. (The soak time for permanent mold castings, which have a finer structure, may be safely reduced to 4–6 hr since diffusion distances are shorter.) The castings should then be quenched in hot water at 149–212F (65–100C).

Figure 15-4 illustrates the microstructure of A356–T6 that had been treated with Sr to form fibrous eutectic silicon particles during solidification. (The as-quenched and aged conditions cannot be differentiated metallographically.) During the solution treatment, constituent and precipitate $\beta(Mg_2Si)$ particles dissolved, the small amounts of $Al_8Mg_3FeSi_6$ remained unchanged, and the eutectic silicon spheroidized.

Artificial aging differs between the T6, T7 and T71 tempers. For peak hardness (T6), the castings should be aged at 309F (154C) for 3–5 hr; for stabilization (T7), aged at 399F (204C) for 3–5 hr; and for enhanced stabilization (T71), aged at 475F (246C) for 3–5 hr. The resulting structures have not been confirmed but likely vary from β" plus β' (T6) to coarsened β' plus fine $\beta(Mg_2Si)$ (T71).[8]

SUMMARY

Aluminum alloys are divided into two classes—non-heat treatable and heat treatable. Solid solution hardening and dense dispersions of large constituent phase particles strengthen non-heat treatable alloys. In addition to the preceding mechanisms, dense dispersions of fine precipitates strengthen heat treatable alloys.

Heat treatments that are applied to both heat treatable and non-heat treatable alloys are stabilization, which transforms precipitates to a stable phase, and annealing, which precipitates solutes as coarse particles.

Strengthening heat treatments for hardenable alloys consist of three steps: solution, quench and age. Solution treatment dissolves alloying elements, and quenching retains them in solid solution at room temperature. Aging at room or elevated temperature forms precipitates of a reproducible size and composition.

ACKNOWLEDGMENTS

AFS would like to thank Paul N. Crepeau, General Motors, for writing this chapter.

The author would like to thank E.F. Ryntz, J.N. Johnson, A.K. Sachdev, S. Shivkumar, B.L. Tuttle and J.L. Jorstad for reviewing the manuscript.

REFERENCES

1. C.W. Meyers; "Solution Heat Treatment Effects on Ultimate Tensile Strength and Uniform Elongation in A357 Aluminum Alloys," *AFS Transactions,* vol 92, p 511 (1984).
2. J.E. Hatch, ed.; *Aluminum: Properties and Physical Metallurgy,* American Society for Metals, Metals Park, OH (1984).
3. S. Shivkumar, S. Ricci, Jr., D. Apelian; "Influence of Solution Parameters and Simplified Supersaturation Treatments on Tensile Properties of A356 Alloy," *AFS Transactions,* vol 98 (1990).
4. D. Hull, D.J. Bacon; *Introduction to Dislocations,* 3rd Edition, Pergamon Press, New York, 1984.
5. M.E. Fine; "Precipitation Hardening of Aluminum," *Met. Trans. A,* vol 6A, p 625 (1975).
6. D.A. Porter, K.E. Easterling; *Phase Transformations in Metals and Alloys,* Van Nostrand Reinhold Co., Ltd., Wokingham, Berkshire, UK (1983).
7. J.M. Silcock, T.J. Neal, H.K. Hardy; *J. Inst. Metals,* vol 82, p 239 (1953-1954).
8. L.F. Mondolfo; *Aluminum Alloys: Structure and Properties,* Butterworth Inc., Boston, MA (1976).
9. J.E. Gruzleski, B.M. Closset; *The Treatment of Liquid Aluminum-Silicon Alloys,* American Foundrymen's Society, Des Plaines, IL (1990).

CHAPTER 15 APPENDIX A

VACANCIES

In a real crystal, some lattice sites are vacant. Such vacancies behave in much the same manner as solute atoms because the equilibrium concentrations of both increase with temperature. In fact, near the crystal's melting point, vacancies can occupy as much as 0.01–0.1% of the volume.[A1] These vacancies are readily created and lost at free surfaces such as grain and phase boundaries. By providing holes in the lattice into which atoms may jump, vacancies facilitate diffusion of solute atoms and play an important role in precipitation hardening.

As with solutes, α-aluminum can become supersaturated in these unoccupied lattice positions after cooling from elevated temperature. Vacancies that are created at elevated temperatures can be retained in excess concentrations by rapid cooling. The increased diffusion rate caused by vacancies is beneficial to age hardening, so elements are often included in alloys to aid in vacancy retention.[A2] Two examples of this in Al-Cu alloys are Ag in Alloy 201 and Mg in Alloy 206.

High vacancy concentrations decrease the degree of solute supersaturation needed to form GP zones and initiate precipitation hardening. Conversely, reduced vacancy concentrations inhibit precipitation. Hence, the loss of vacancies adjacent to grain boundaries during a slack quench is a second important cause of precipitate-free zones (PFZs).

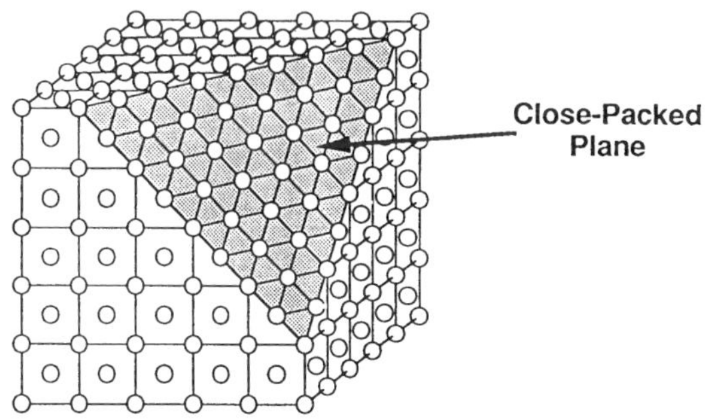

Fig. 15-A1. An illustration of a close-packed (111) plane in the lattice of a face-centered cubic (FCC) metal.

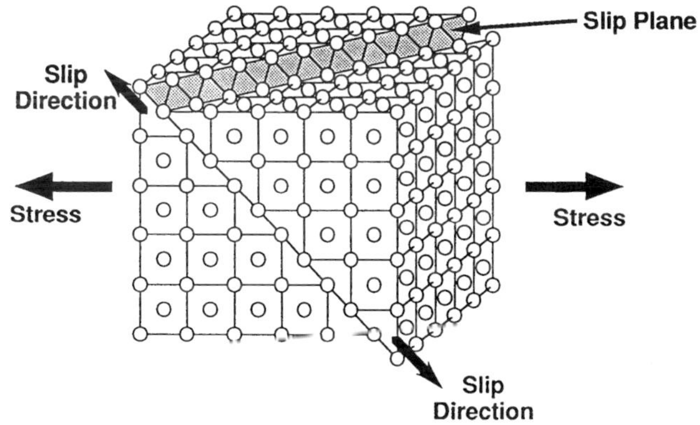

Fig. 15-A2. Slip occurring along a close-packed (111) plane in a face-centered cubic (FCC) metal. [After Ref. A2.]

PHYSICAL METALLURGY OF DEFORMATION

Elastic and Plastic Deformation

All materials deform elastically when they are loaded. Elastic deformation occurs when atoms shift slightly from their normal positions on the crystal lattice yet can return to their original places if the stress is released. Resistance to elastic deformation, the material property called Young's modulus of elasticity, is determined by the strength of atomic bonds.

When the stress is great enough, atoms are forced from their normal lattice positions to adjacent positions. Plastic deformation occurs when atoms move one or more interatomic distances. Plastic deformation is permanent because the atoms cannot return to their original positions. The resistance to the onset of plastic deformation is called yield strength.

Slip Planes

Because of the periodicity of crystal lattices, atoms are arranged in atomic planes. Although an endless number of atomic planes exist, only the planes that contain atoms at their closest spacing participate in plastic deformation. Figure 15-A1 illustrates one of the four close-packed planes, identified by crystallographic orientation as the "(111)" type, in α-aluminum, which is present as crystals of the face-centered cubic (FCC) type.

During plastic deformation of FCC metals, (111) planes translate relative to one another, a phenomenon known as slip (Fig. 15-A2). Hence, (111) planes are the so-called slip planes in FCC metals.

Dislocations

During slip, all of the participating atoms do not move simultaneously. Instead, atoms move in small groups with the aid of a special class of crystal defects that are more mobile in metals than in other materials.

These crystal defects are called dislocations—linear discontinuities in the crystal lattice that are surrounded by complex strain fields. A plane of atoms that ends abruptly forms an edge dislocation (Fig. 15-A3a), and one that contains a twist offset forms a screw dislocation (Fig. 15-A3b). In the general case, dislocations are of mixed character—part edge and part screw. Under stress, dislocations move across close-packed planes leaving rows of atoms offset by exactly one atomic position wherever they pass.

Many dislocations must move for measurable plastic deformation to occur. In fact, metals typically contain between 10^8 and 10^{12} cm of dislocation lines per cubic cm of metal,[A3] more than enough to sustain large deformations. When obstacles restrict free passage of these dislocations, the stress required for their movement increases, and the material is said to be strengthened (i.e., hardened). However, restricted dislocation motion reduces ductility.

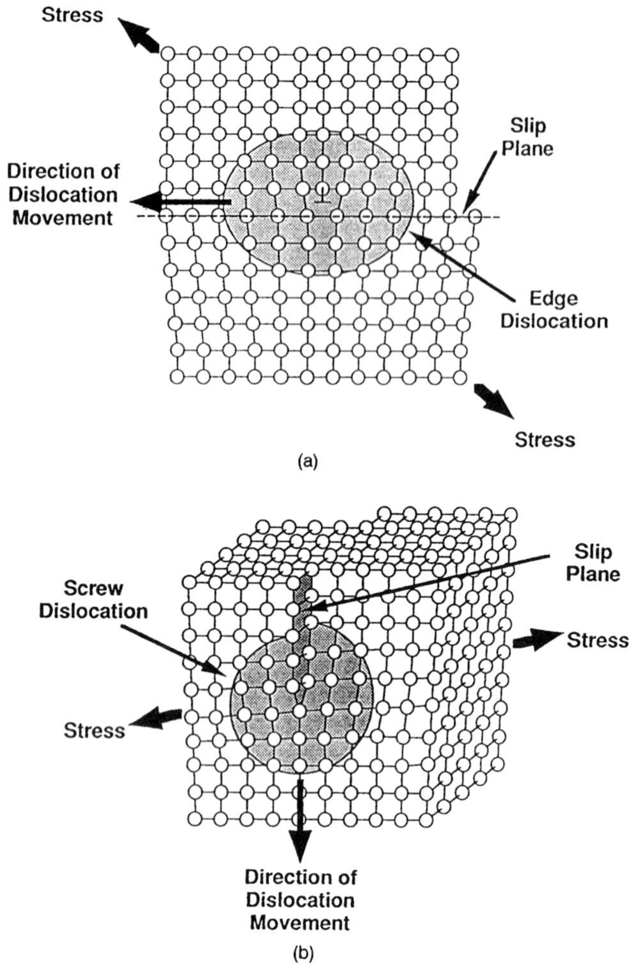

Fig. 15-A3. The basic types of dislocations: (a) an edge dislocation and (b) a screw dislocation.

In addition, the coherency strains caused by coherent precipitates, GP zones in particular, increase frictional forces that oppose dislocation movement.

The shearing resistance of coherent precipitates and GP zones to penetrating dislocations in α-aluminum arises from differences in crystal structure. Hence, the stress for shearing precipitates increases as they coarsen because larger particles are more difficult for dislocations to enter.

Another mechanism dislocations use to bypass precipitates is Orowan looping (Fig. 15-A4). A dislocation that cannot penetrate a precipitate can loop around it instead and leave behind a residual dislocation loop. The stress for Orowan looping decreases as distances between coarsening precipitates increases.

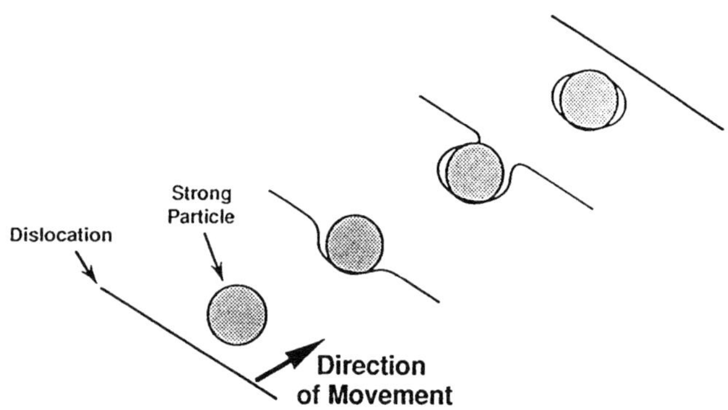

Fig. 15-A4. The mechanism for Orowan looping of a dislocation around an incoherent precipitate.[2]

Role of Dislocations in Strengthening

When two dislocations that are travelling along intersecting slip planes cut across one another, they may become pinned where they meet. Pinning points accumulate rapidly during plastic deformation, effectively shortening each dislocation segment and thereby increasing the stresses needed to move them. Such strengthening by dislocations impeding one another occurs rapidly during room temperature deformation and is called work hardening. For example, after most metals yield, the stress for continued deformation increases because the deforming metal work hardens.

Forces are exerted on dislocations when the body that contains them is stressed. However, certain stress states contain few force components that can cause dislocations to move. Triaxial stresses developed between coarse, densely populated constituent particles are one example in which deformation is inhibited because the forces to move dislocations are limited. This is why the eutectic silicon particles tend to strengthen 4XX aluminum alloys.

Precipitates strengthen aluminum by physically obstructing dislocations. To bypass a precipitate, a dislocation can penetrate it (particle shear) or envelop it (Orowan looping)[A1]

The maximum hardness curve exhibited by aluminum alloys as they progress from underaged to peak hardness to overaged is due to a progression of deformation mechanisms from precipitate shear to Orowan looping. In the as-quenched condition, dislocations are only impeded by the strain fields that surround solute atoms (solid solution strengthening). In the early stages of age hardening, precipitates that form are very small and closely spaced. Although the precipitates impede dislocations, they are readily sheared. At this stage, Orowan looping is precluded because the precipitates are too closely spaced.

As aging progresses, precipitates grow, and the stress required to move dislocations through them increases. Concurrently, precipitate spacing increases, and the stress that will activate Orowan looping decreases. Ultimately, the Orowan looping stress becomes less than the precipitate shear stress, and as aging continues, hardness begins to decrease.

SUMMARY

Plastic or permanent deformation in metals is caused by the movement under stress of many dislocations. The movement of each dislocation causes a slip plane to shear by one atomic distance. Hardening occurs when barriers to dislocation motion are created in the crystalline lattice. Fine precipitates harden aluminum by impeding dislocation motion by the mechanisms discussed above, thereby raising the stress necessary to cause plastic deformation. Precipitates coarsen during aging, initially increasing strength as they become larger and finally decreasing strength as they become more widely spaced.

REFERENCES

A1. D. Hull, D.J. Bacon; *Introduction to Dislocations,* 3rd Edition, Pergamon Press, New York, 1984.

A2. J.E. Hatch, ed.; *Aluminum: Properties and Physical Metallurgy,* American Society for Metals, Metals Park, OH (1984).

A3. R.E. Reed-Hill; *Physical Metallurgy Principles,* Van Nostrand Co., New York (1973).

CHAPTER 15 APPENDIX B

EFFECTS OF AGING ON PROPERTIES OF ALLOYS 355 AND 356

The following 16 graphs show the changes that occurred in properties of Alloys 355 and 356 that were aged at different temperatures for up to 12 hours.

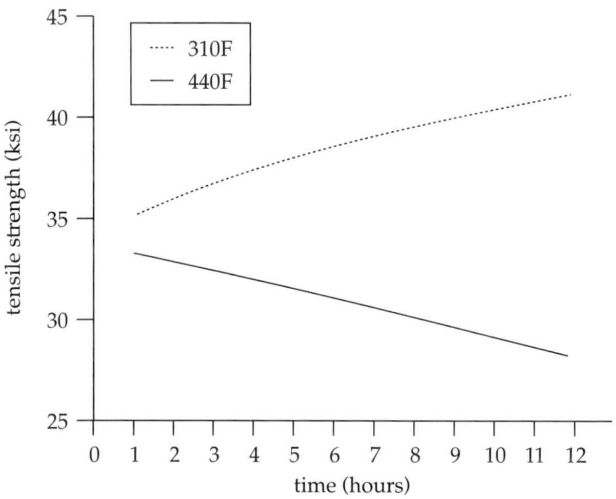

Fig. 15-B1. Tensile strength of permanent mold cast 356–T4.

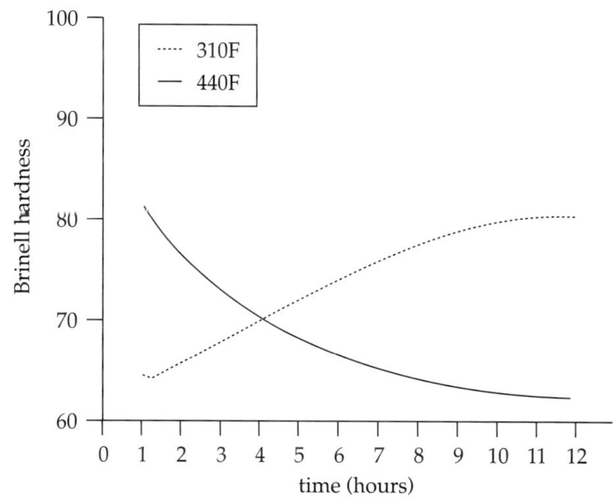

Fig. 15-B2. Brinell hardness of permanent mold cast 356–T4.

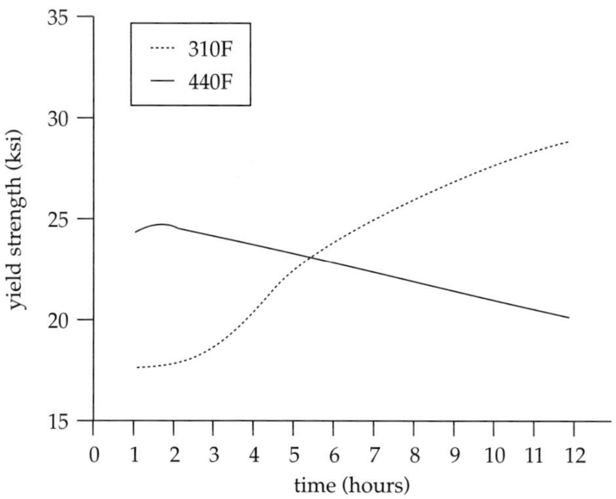

Fig. 15-B3. Yield strength of permanent mold cast 356–T4.

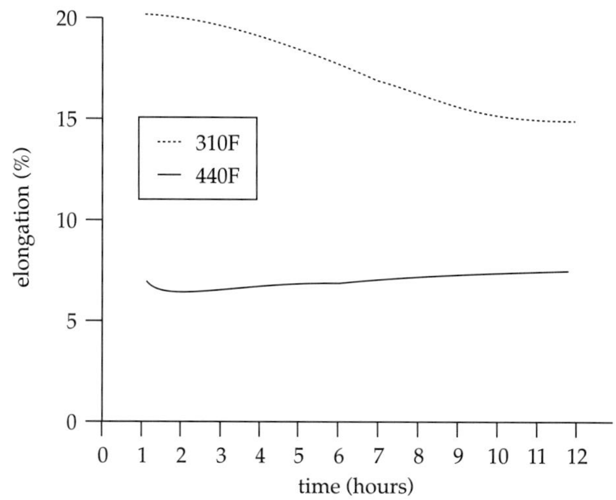

Fig. 15-B4. Elongation of permanent mold cast 356–T4.

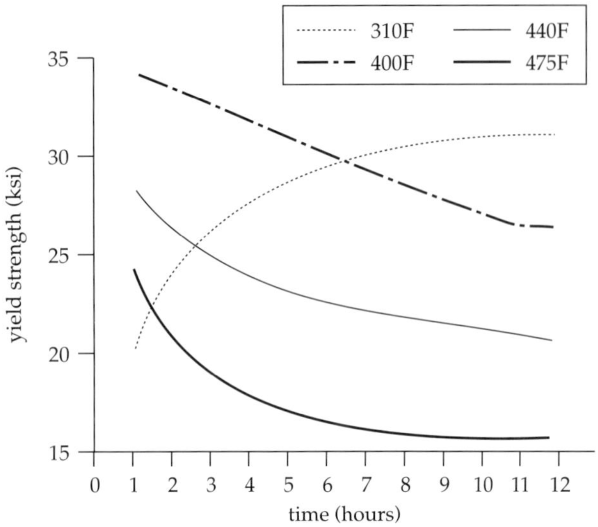

Fig. 15-B5. Yield strength of sand cast 356–T4.

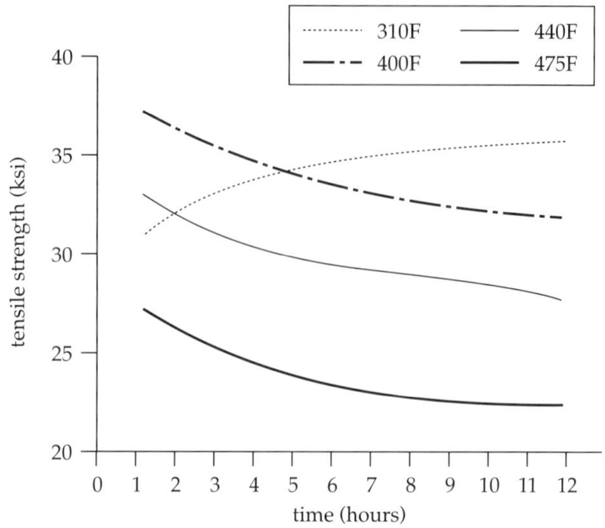

Fig. 15-B6. Tensile strength of sand cast 356–T4.

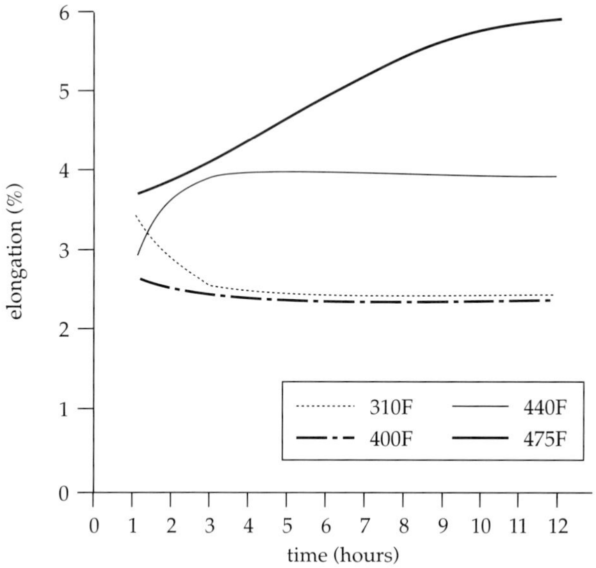

Fig. 15-B7. Elongation of sand cast 356–T4.

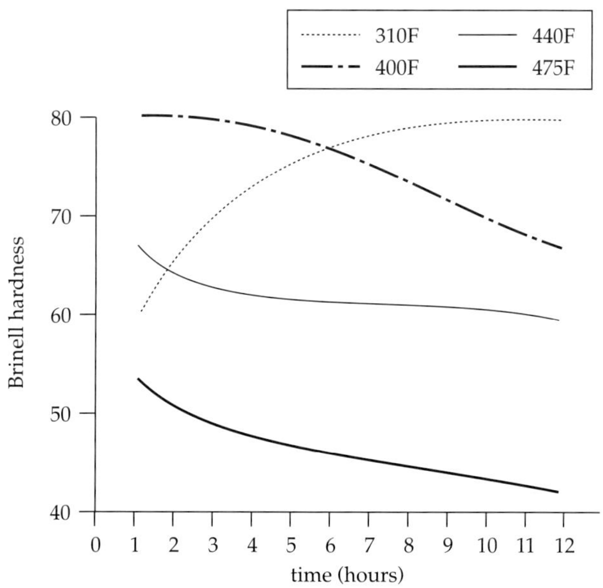

Fig. 15-B8. Brinell hardness of sand cast 356–T4.

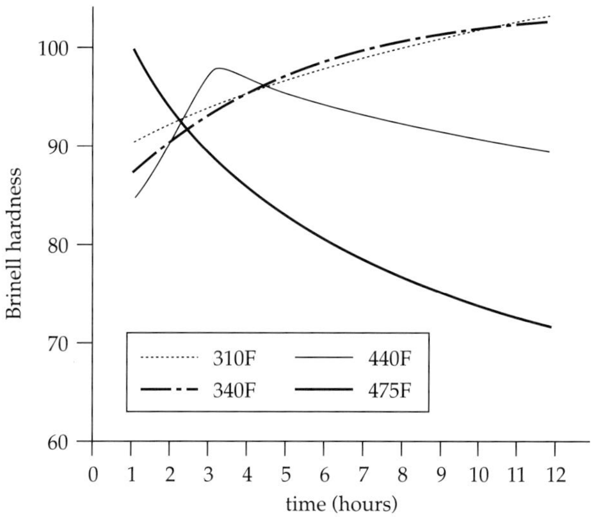

Fig. 15-B9. Brinell hardness of permanent mold cast 355–T4.

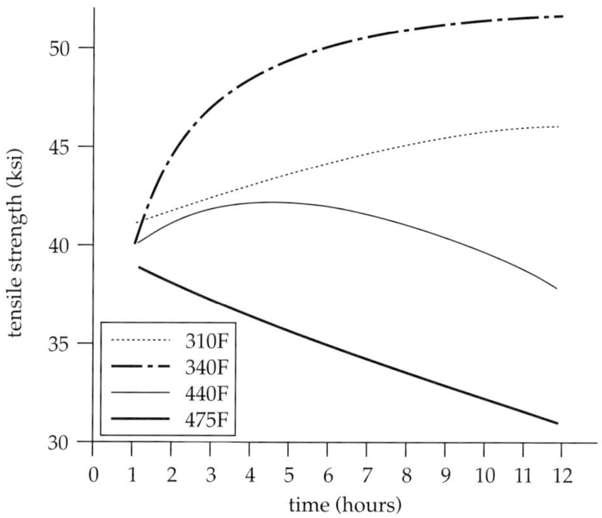

Fig. 15-B10. Tensile strength of permanent mold cast 355–T4.

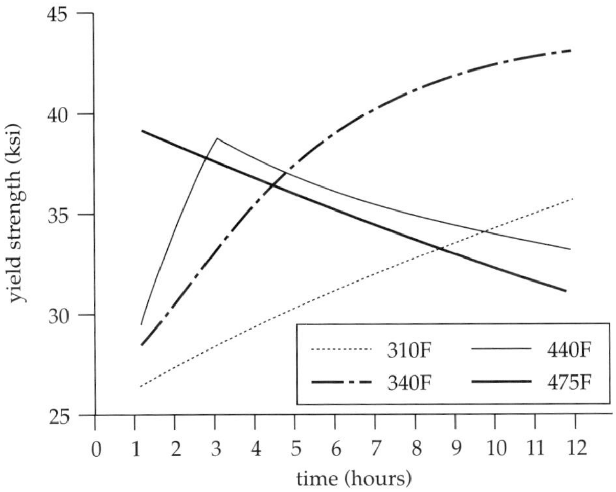

Fig. 15-B11. Yield strength of permanent mold cast 355–T4.

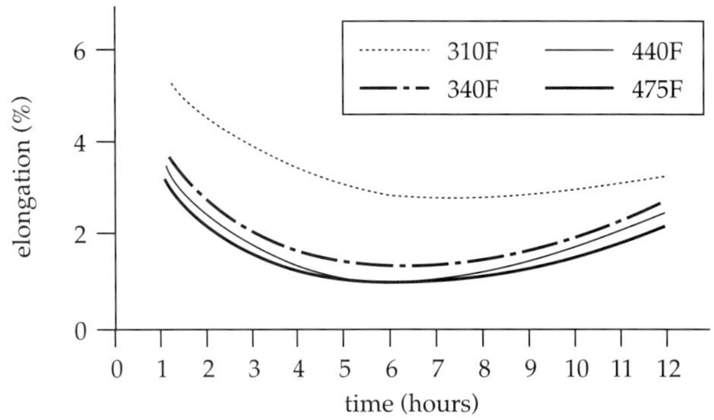

Fig. 15-B12. Elongation of permanent mold cast 355–T4.

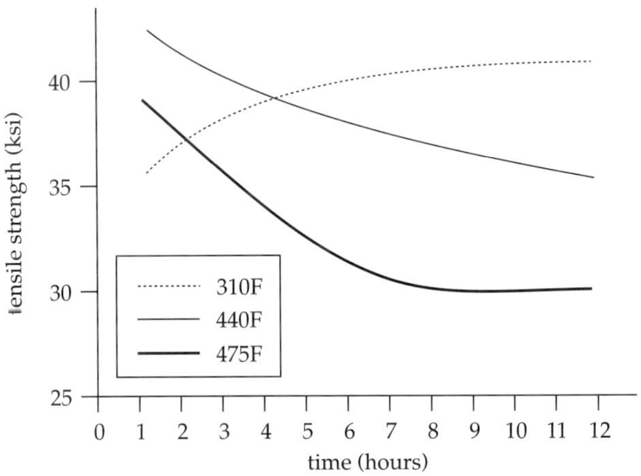

Fig. 15-B13. Tensile strength of sand cast 355–T4.

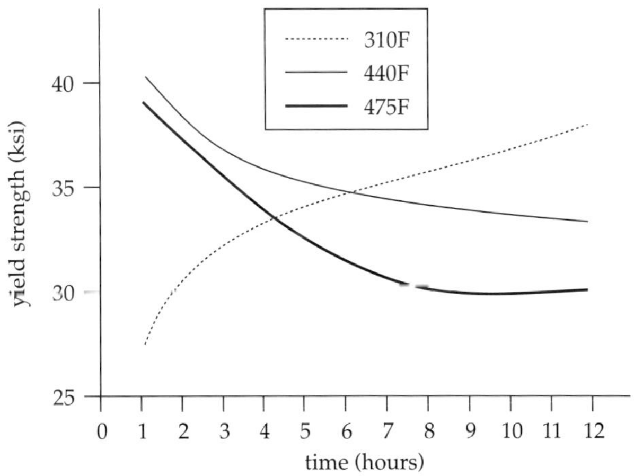

Fig. 15-B14. Yield strength of sand cast 355–T4.

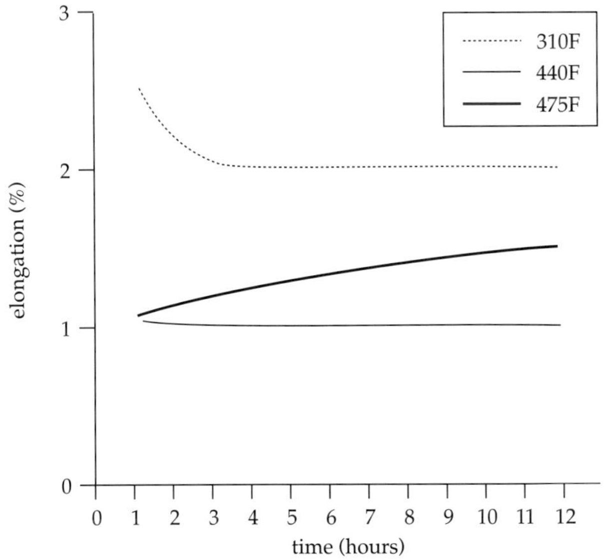

Fig. 15-B15. Elongation of sand cast 355–T4.

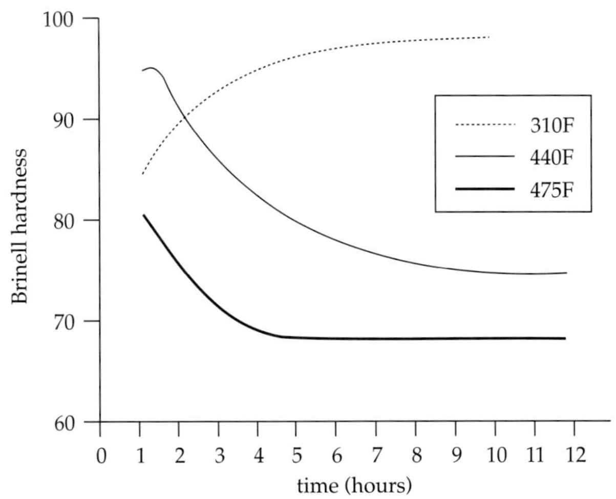

Fig. 15-B16. Brinell hardness of sand cast 355–T4.

Chapter 16

METALLOGRAPHY

INTRODUCTION

Examination of suitably selected and prepared specimens by optical or light metallography can provide important information concerning composition, production and heat treatment of cast aluminum products. Such examinations also can provide information on the effects resulting from mechanical or environmental conditions of service.

With experience, the metallographer can usually identify alloy type, certain kinds of composition deviations, and the presence or absence of voids, inclusions or segregation. From the size of grains, dendrite cells or arms and second-phase microstructural constituents, as well as the degree of soundness, the type of voids, the rate of solidification and possibly the method of casting may be deduced, and some estimate made of the care observed in melting and casting operations. The type of heat treatment applied may be determined, as well as deviations in practice that result in overheating or inadequate solution during solution heat treatments.

The thickness of surface coatings can be measured, and coating type often can be determined, as well as the presence, type and depth of any corrosion. Examination of fractures from test or service may allow classification according to mechanical or environmental conditions chiefly responsible for occurrence, whether the result of fatigue, stress-corrosion cracking, stress exceeding tensile strength or a combination of these possibilities. This information may show that design inadequacies, improper alloy selection, subnormal quality features, poor machining, abuse or accidental damage have contributed to premature or unexpected failure.

Metallographic examinations are usually for process control, quality assurance or investigation of failures from service or test. Information from optical metallography must be supplemented, in many cases, by data from other methods of examination or analysis to provide unequivocal conclusions.

For more complete investigation, examination by other methods to determine chemical composition (bulk and local), soundness, identification of phases or submicroscopic features of the microstructure or of fractures may be required. Among such methods are spectrochemical analysis, the electron microprobe, radiography, transmission and scanning electron microscopy (Fig. 16-1), and x-ray and electron diffraction. Coordinated information from several of these methods may be required for complete diagnosis. This chapter will be restricted to the methods and kinds of information obtainable from optical or light metallography.

SAMPLING

Selection of the specimen(s) and of the orientation of the plane to be examined requires good judgment and planning to ensure that the features revealed are either representative or specific to the inquiry being made. Careful visual inspection should precede cutting or etching. The choice of cutting plane may be governed by the general shape of the part or by the suspected or known form and location of discontinuity.

Complete evaluation of castings may require removal of several sections representing portions differing in thickness or in proximity to chills or risers (which affect solidification rate) or portions showing internal features revealed by exploratory radiographic examination. Penetrant inspection is useful in locating fine cracks. Fracture surfaces should be carefully preserved to guard against abrasion or contamination.

Sample Removal

Specimens for macroscopic or microscopic examination of castings are removed by sawing or by using an abrasive cutoff wheel. Hacksaw or power saw blades should be sharp and lubricated with stick lubricant to minimize the depth of distortion from the cut surface. Fragile or small specimens may require use of a jeweler's hand saw.

MACROSCOPIC EXAMINATION

Macroscopic examinations are made without magnification, although in general usage, this term is applied to magnifications of up to about 10 diameters (10X). Inspection of a fractured surface without any preparation or etching and with little or no magnification can provide evidence of the presence of voids; nonmetallic inclusions; or large, faceted intermetallic phase particles, indications of grain size, and features indicative of the type of fracture.

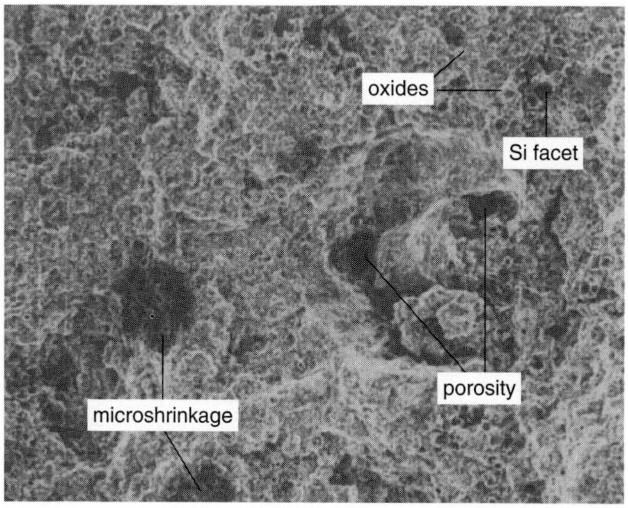

Fig. 16-1. SEM photo showing the fracture surface of permanent mold cast A356.0. Note the gross intergranular shrinkage (100X).

Fatigue fractures or those involving stress-corrosion cracking may reveal progressive features allowing their origins to be established and, in some cases, an estimation of the rate of progression. The initiation site of fatigue cracks frequently can be traced to a specific pre-existing discontinuity such as a nonmetallic inclusion or void or to stress intensifying surface features such as cold shuts, machining marks or notches, inadequate radii, or nicks or gouges from accidental damage. Such preliminary macroscopic examination is essential for selecting appropriate sections and sample removal for further microscopic examination.

Macroscopic examination of properly prepared and etched sections is useful in revealing grain size and shape, as well as the presence of porosity, inclusions and segregation.

Mechanical Preparation

Depending on the size of samples needed for macroscopic examination, further mechanical preparation of the sawed surface, which should be reasonably planar, may be performed by manual operations or by machining. For small sections, the sawed surface roughness and disturbed layer may be removed by careful filing or by grinding the sample on a belt or disc sander.

Further mechanical preparation of the surface may not be required, depending on the detail of observation needed, but it generally is preferable to continue through successive stages of grinding with finer abrasive papers. These may be used manually by hand rubbing or mounted on horizontal wheels. In either case, lubricant/coolant should be used to minimize embedding of abrasive particles in the surface. A clean surface from a 400-grit paper will provide good detail after macroetching.

Machined surfaces provided by lathe, milling or shaper cuts with sharp tools are satisfactory for macroetching and other examination on a macro scale. Care must be taken to avoid dull tools or other conditions that will cause dragging, smearing or distortion, and bridging-over of voids.

If the machining is done properly, with heat treated Al-Cu alloys or Al-Mg alloys, the grain structure should be visible before etching, sometimes clearly, usually only faintly. Cracks or porosity may be apparent before etching, but preliminary light etching may be necessary to remove any bridging or sealing by smeared or dragged metal. Preliminary light caustic etching is recommended when penetrant examination is used to reveal distribution and size of voids.

Etchants for macroscopic examination are generally strong caustic or acid solutions, and the temperature and time of etching must be controlled carefully to avoid overetching (Table 16-1). A water solution of 10% NaOH used at temperatures of 140–158F (60–70C) is a good general-purpose macroetching reagent. When etching in this caustic solution

Table 16-1.
Etchants for Use in Macroscopic Examination of Aluminum Casting Alloys

Etchant [1]	Composition (parts are by volume)	Procedure for Use	Applications
1. caustic etch	10 g NaOH to each 90 ml water	Immerse specimen 5–15 min in solution heated to 140–158F (60–70C),[1] rinse in water, dip in 50% HNO_3 solution to desmut, rinse in water, dry.	General purpose, 1XX, 2XX alloys
2. Tucker's reagent	45 parts HCl (conc), 15 parts HNO_3 (conc), 15 parts HF (48%), 25 parts water	Mix fresh before using. Immerse or swab specimen for 10–15 sec, rinse in warm water, dry, and examine for desired effect. Repeat as much as necessary until desired effect is obtained.	1XX, 5XX alloys
3. __________	1 part HF (48%), 9 parts water	Requires fairly smooth surface. Immerse until desired effect is obtained, hot-water rinse, dry.	3XX, 4XX alloys
4. Poulton's reagent	12 parts HCl (conc), 6 parts HNO_3 (conc), 1 part HF (48%), 1 part water	May be premixed and stored[2] for long periods. Etch by brief immersion or by swabbing. Rinse in cool water, and do not allow either the etchant or the specimen to heat up during etching.	3XX, 4XX alloys for grain contrast
5. __________	50 parts HCl (conc), 15 parts HNO_3 (conc), 3 parts HF (48%), 5 parts $FeCl_3$ soln (conc)	Mix fresh before use. Cool solution to 50–59F (10–15C) with jacket of cold water. Immerse a few seconds, rinse in cold water; repeat until desired effect is obtained.	1XX, 7XX alloys for grain contrast
6. __________	75 parts HCl (conc), 25 parts HNO_3 (conc), 5 parts HF (48%), 25 parts water	Mix fresh before use. Etch by brief immersion or by swabbing.	3XX, 4XX alloys

1) These caustic and strong acid solutions should be prepared and handled only when protective clothing and equipment are worn (safety glasses, rubber gloves, and laboratory coats or aprons). Solution preparation and etching should be carried out under an effective fume hood. In mixing strong acid solutions, the acid should be added to the water slowly with stirring. HF should be handled with extreme care because sores resulting from its attack on skin and under fingernails do not heal readily.
2) Solution should be stored in a vented container, preferably under a fume hood, to prevent buildup of gas pressure. The container either should be made of polyethylene or be lined with wax.

is by immersion, the time may range from 3–10 minutes, and etching by swabbing may require somewhat longer time.

With alloys that contain copper, a dark coating or residue forms on the surface during etching in caustic. Because this coating masks the structure, it must be removed. This is done, after washing in water, by immersion in or flooding with a 50% or stronger solution of HNO_3, followed by washing in water and drying. Grain contrast, particularly for Al-Cu type alloys, may be increased by reimmersing in the caustic solution for a few seconds followed by water washing and drying.

Little or no grain contrast is developed in alloys containing more than about 3% Si by etching in caustic. In fact, grain structure is difficult to reveal by macroetching these higher Si content alloys, but hydrofluoric acid and mixed acid etches are useful, particularly for castings that have been solution heat treated. Poulton's reagent is suitable for this purpose. Grain contrast can be developed by etching in a 10–30% solution of $CuCl_2$ in water with addition of a few drops of HF. Etching is by immersion with continuous swabbing to remove the black coating and promote even attack. Preliminary degreasing in NaOH solution may be necessary to avoid uneven etching in the acid or $CuCl_2$ solutions.

Grain Structure

Grain structures of cast aluminum alloys, as well as size and arrangement of microstructural features within the grains, are highly dependent on the solidification rates and the transient thermal gradients developed during solidification. Under some conditions of high thermal gradients, 1XX (99% aluminum) alloys and relatively dilute or solid solution alloys such as 5XX (aluminum-magnesium) types form columnar grains (Fig. 16-2), but most cast products have nearly equiaxed, essentially isotropic grain structures (Fig. 16-3).

The average grain diameter in sand castings may range from about 0.25–5 mm (0.01–0.2 in.). This corresponds to ASTM macro grain size numbers of M-14 to M-5.5, respectively (ANSI/ASTM E112). For permanent and chilled composite mold or "premium" castings, the range may be from about 0.13–1.3 mm (0.005–0.05 in.). In addition to the factors of solidification rate and temperature gradient, determined by section thickness and mold materials, grain size is substantially affected by melting and casting or pouring temperatures and by the effectiveness of nucleant grain refining additions and procedures employed in using them.

Grain size in die castings is much smaller and of less concern than in sand or permanent mold castings. Grain refinement treatments are not generally needed or used in producing die castings.

Grain refinement is beneficial in reducing hot cracking, may assist feeding, and may improve strength and ductility. Macroetched sections are used to check the effectiveness of grain refinement from the quality control viewpoint and to correlate with other metallographic structure features of cast products for both quality control and failure analysis.

MICROSCOPIC EXAMINATION

The size of specimens employed for microscopic examination will vary with the size, shape and feature of the casting to be examined as well as the objectives of the investigation. Unnecessarily large specimens increase difficulty of attaining a suitable polish, although in infrequent situations in which destructive sectioning is not permissible, it is even possible to prepare a small area on the surface of a cast part without removing a specimen.

Cut specimens approximately 13x13x7 mm (½ x ½ x ¼ in.) or somewhat larger can be prepared easily and rapidly. Smaller specimens, those of irregular shape and those used for examination in cross section of a cast, machined, coated or corroded surface should be mounted in a plastic mounting material.

Mounting

Bakelite or diallyl phthalate resins, opaque or Lucite (a transparent material), generally are used for hot compression molding in tools that produce a cylindrical mount 25–37 mm (1–1.5 in.) in diameter. The molding is done in a metallurgical mounting press at pressures up to about 33 MPa (4.8 ksi) and

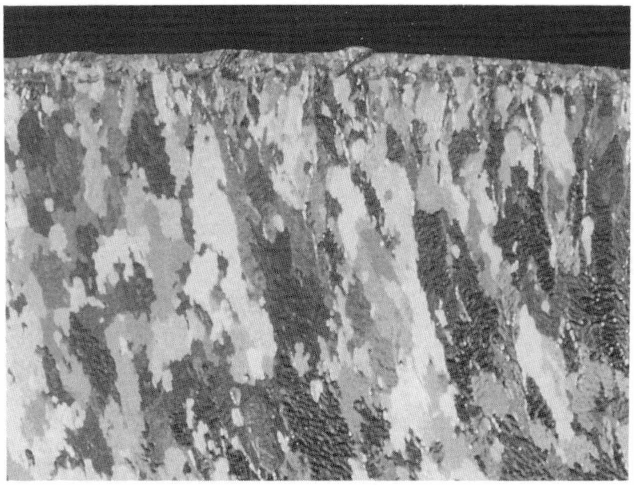

Fig. 16-2. Columnar grain structure (10X).

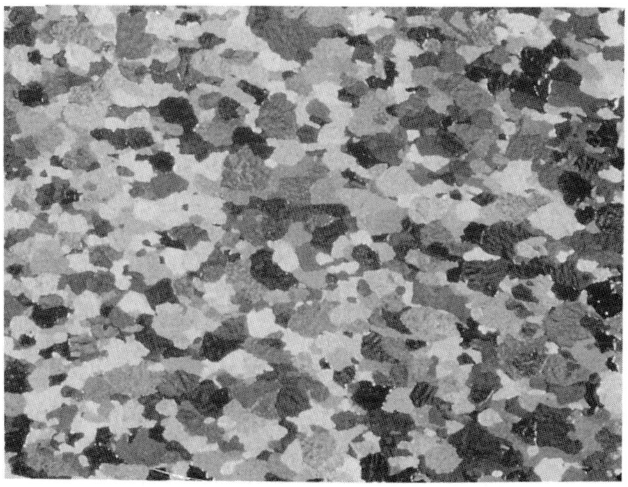

Fig. 16-3. Equiaxed grain structure (10X).

at a temperature of 266–284F (130–140C). Sections that are too large for compression molding can be mounted in a cold molding material, such as epoxy.

Mechanical Preparation

Grinding and polishing remove deformed or disturbed metal from the specimen surface, to reach the desired plane of sectioning, and to provide a clean, smooth and planar surface. Mounted or unmounted specimens can be prepared for polishing by filing, followed by hand rubbing on a series of abrasive papers of successively finer abrasive size, rotating the direction of the specimen 90 degrees between each stage. Alternatively, the grinding can be done on belt or rotary wet grinders.

Between the successive stages, the specimen should be cleaned to prevent carryover of coarse abrasive particles to the next finer stage, and all scratches from the preceding paper should be removed before proceeding to the next paper. Various abrasives may be used, with a typical series being waterproof silicon carbide papers with 240, 320 or 400 grit followed by 600 grit, then metallographic papers to 0 or finer, 2/0 or 3/0.

The transition or rough polishing stage generally employs a cloth lap on a rotating wheel. A 6 μm diamond paste on a nylon or short-nap cloth disc using kerosene or propylene glycol as a lubricant is generally used. This is followed by cleaning and another step using a 1 μm diamond paste. Again, a lubricant of kerosene or propylene is used.

Final polishing can be done using a water suspension of 0.05 μm alumina on a short nap cloth disc. Alternatively, a heavy grade of magnesium oxide powder paste can be used. In this case, paste and heavy pressure are followed by reducing pressure and speed and adding deionized or distilled water to dilute the paste. Rinsing in tap water, wiping with wet cotton, then drying in a hot air blast make up the final step before examination and subsequent etching.

Specimens are examined microscopically in the as-polished condition to be sure that misleading artifacts, such as embedded abrasive particles or pits from failure to remove all scratches from earlier abrasive stages, are not present. Voids, inclusions or segregation should be apparent in the as-polished condition.

To avoid incorrect interpretation, the intermediate polishing operation must be performed with sufficient pressure and time to uncover voids, which, if not opened, may appear as dross or oxide films. A simple test to determine the nature of such linear or irregular features, which have frequently been misinterpreted, consists of pressing the surrounding metal with a needle while viewing the operation under a microscope. If a layer over a void collapses under the needle pressure, the discontinuity is not a dross film.

Etching

Etching solutions for microscopic examination are highly diluted and much less aggressive chemically than those used for macroetching. Specimens are etched by swabbing with a cotton ball saturated with the etching solution or by immersion to remove the last traces of metal flowed by polishing and to reveal more clearly the details of the microstructure.

The hydrofluoric acid etchant has the most general application, and when any other solution is used, its use can follow application of the HF etch without requiring repolishing. Keller's etch, always used by immersion, provides considerable additional information with alloys of the copper-containing 2XX group and also is used for those of the Al-Zn 7XX group. The times shown in Table 16-2 are general guidelines and may be varied to obtain the most informative microstructure appearance.

Table 16-2.
Etchants for Use in Microscopic Examination of Aluminum Casting Alloys

Etchant	Composition (parts are by volume)	Procedure for Use	Applications
1. Hydrofluoric acid etch	1 part HF (48%) 200 parts water	Immerse 30–45 sec or swab 15 sec, 1–2 min.	General purpose. Outlines light-tone constituents. Outlines grain boundaries.
2. Keller's etch	5 parts HNO_3 (conc) 3 parts HCl (conc) 2 parts HF (48%) 190 parts water	Immerse 10–30 sec, wash in warm water, blow dry. Do not remove etching products from surface.	With copper-containing alloys reveals dendritic segregation (coring) before solution heat treatment, grain contrast and absence of coring after solution heat treatment, reduced grain contrast and presence of precipitate after precipitation heat treatment. Outlines grain boundaries and constituents. Increases contrast or colors some constituents.
3. Sulfuric acid etch	20 parts H_2SO_4 (conc) 80 parts water	Immerse 30 sec at 158F (70C) solution temperature, rinse in cold water.	Darkens and dissolves Fe-containing constituents, Al_3Fe, Al_8Fe_2Si and Al_5FeSi.
4. Caustic etch	1 g NaOH 100 ml water	Swab 5–10 sec.	Constituents, Al_2Cu or Al_8Mg_5 unattacked, outlined.

General Characterization

Most of the commercial aluminum casting alloys are hypoeutectic, although among those based primarily on the Al-Si system, some are essentially of 100% eutectic composition and others are hypereutectic. These terms derive from the equilibrium constitution relationships as functions of composition and temperature.

Consider, for example, the Al-Si system, which is one of the simplest of the aluminum alloy systems because no intermetallic phases are formed in the binary alloy, and regardless of the proportions of the two elements, no more than two phases may be present. One is Al, which may have up to 1.65% silicon in solid solution, and the other is Si, which is essentially pure (99.93% Si), having a maximum solubility for aluminum of only 0.07%.

The eutectic or lowest melting temperature mixture of the two phases occurs at approximately 87.4% aluminum, 12.6% silicon (Fig. 16-4). Binary alloys with less than 12.6% silicon are hypoeutectic (Fig. 16-5), and those with a higher silicon content are hypereutectic (Fig. 16-6). Among the commercial alloys, those considered essentially of eutectic composition include 339.0 (Fig. 16-7), 413.0, A413.0, 383.0, 384.0 and B384.0 (10–13% Si). Examples of hypoeutectic commercial alloys are 319.0 (Fig. 16-8), 333.0, 336.0, 355.0, 356.0 and A356.0. Hypereutectic compositions, those with over 13% silicon, are 390.0, A390.0, B390.0 (Fig. 16-9), 392.0 and 393.0.

Representative of the hypoeutectic type of the Al-Si system is the microstructure of an Al-6%Si alloy, such as 319.0 (Fig. 16-10) with a dendritic pattern in which the elongated clear areas are the primary-phase, Al solid solution. These are surrounded by a network consisting of Al solid solution plus particles of Si, the interdendritic eutectic and other impurity phases. The term primary phase refers to the fact that this is the first to solidify as the metal is cooled from the molten, liquid solution, state.

In the range of silicon contents of the hypoeutectic alloys, the volume and area fraction of primary Al solid solution decreases and that of the interdendritic eutectic increases until with about 12% Si the alloy is essentially of eutectic composition characterized by a microstructure of fine Si particles uniformly distributed in Al solid solution matrix. This binary composition solidifies (and melts) at an invariant or fixed temperature, 1071F (577C), and has no primary phase.

In alloys with over 12.6% silicon content, Si solidifies as the primary phase surrounded by eutectic, with microstructures representative of the commercial hypereutectic alloys corresponding to that for 18% silicon.

Regardless of the alloy system, hypoeutectic alloys have a dendritic structure of primary Al solid solution. Eutectics are of the divorced type consisting of discrete second-phase particles in Al solid solution, which is continuous with the primary phase. Second-phase particles may be binary, ternary or polynary intermetallic compounds or of elemental composition. Of the latter, Si is most prominent, while elemental Sn occurs in globular form in certain alloys used for bearing applications (Fig. 16-11).

The primary Al solid solution in the cast condition is of nonuniform composition, increasing in solute concentration from the center to the periphery of the dendrite cells or arms, a condition termed coring. Coring, although always present,

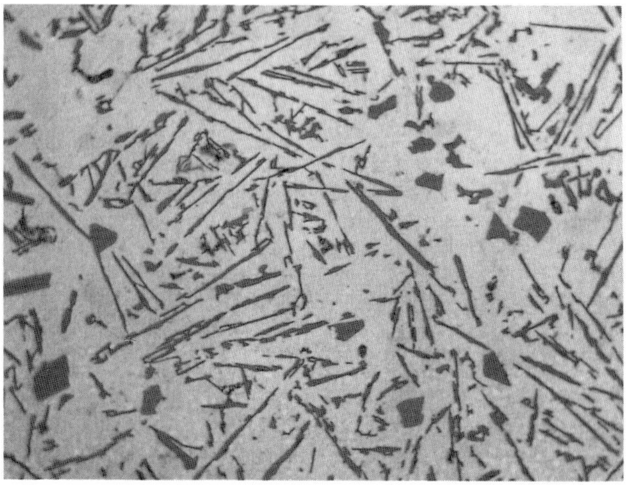

Fig. 16-4. This photomicrograph shows the eutectic structure of a binary Al-Si alloy with 12.6% Si. Unetched, 100X.

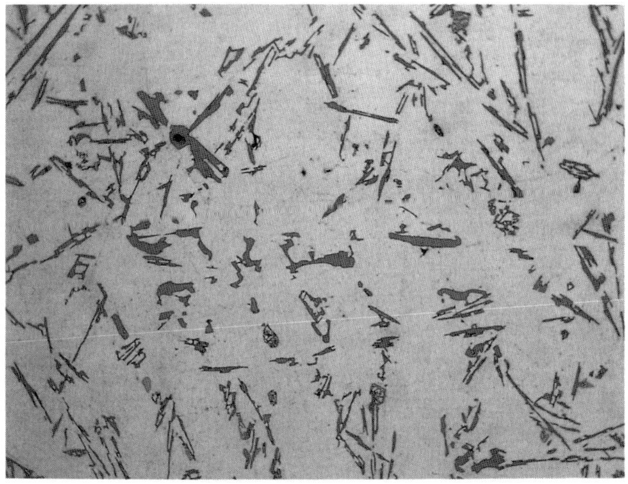

Fig. 16-5. This is the structure of a binary hypoeutectic Al-Si alloy containing 9% Si. Unetched, 100X.

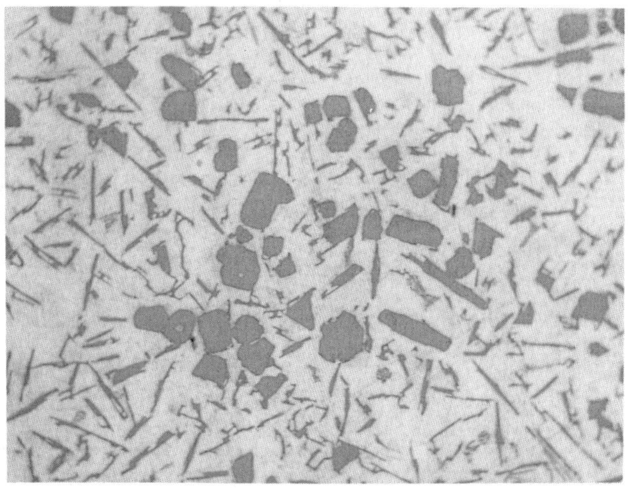

Fig. 16-6. The hypereutectic structure of a binary Al-Si alloy containing 15% Si. 0.5% HF etch, 100X.

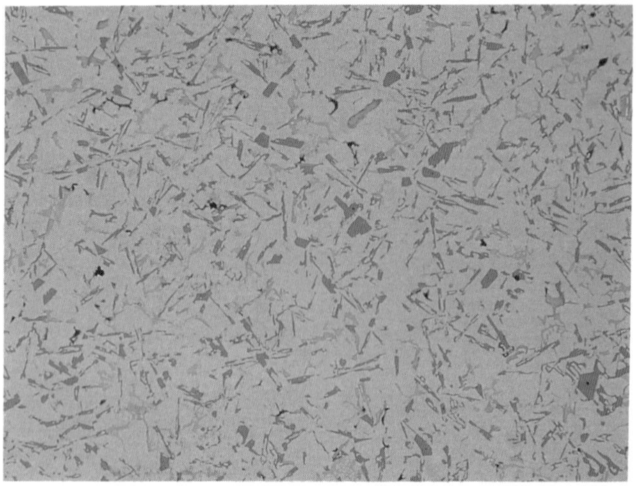

Fig. 16-7. This photomicrograph of permanent mold cast Alloy 339.0 shows a eutectic structure in a commercial alloy (100X).

may not be apparent metallographically because little or no etching effects occur as a result of differences in Mg or Si concentrations in solid solution. However, for alloys containing over about 1% copper and with the zinc-containing alloys, etching with Keller's reagent produces variations in coloration from the center to the periphery of the primary dendrites disclosing the cored structure (Fig. 16-12).

Some of the solute that was in solid solution immediately after solidification may be precipitated during post-solidification cooling. This solid state precipitation normally is greatest in the portion of the primary dendrites having the highest solute concentration, that is, the peripheral zones, and in the aluminum of the eutectic. Although this precipitate is primarily of sub-microscopic size, its presence alters the etching effects, and evidence of coring can be revealed by etching even though precipitation has occurred (Fig. 16-13).

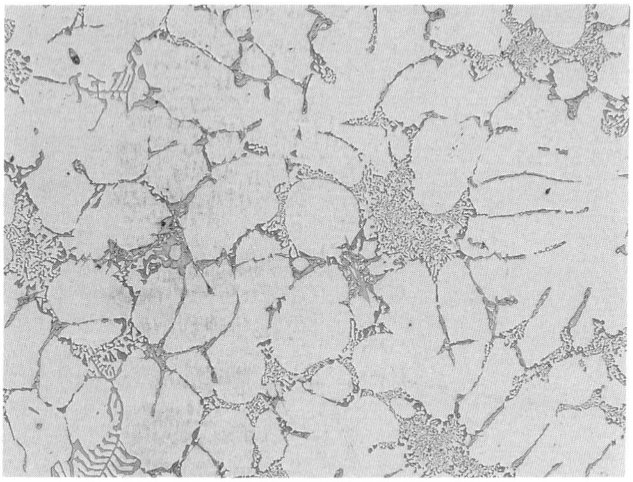

Fig. 16-8. Sand cast Alloy 319.0 is an example of a hypoeutectic structure in a commercial alloy (100X).

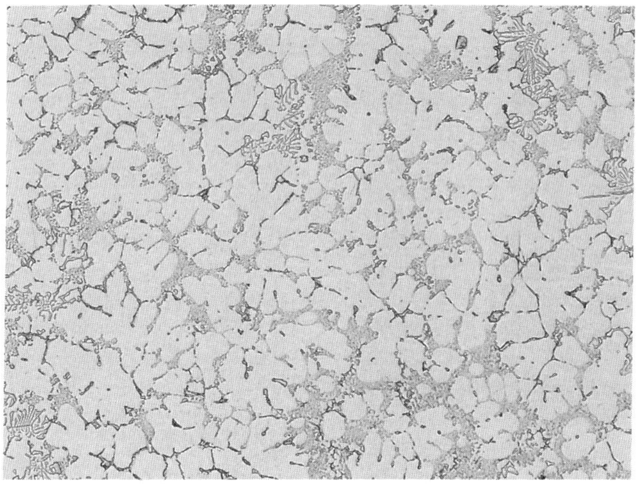

Fig. 16-10. General microstructure of permanent mold cast 319.0 alloy (100X).

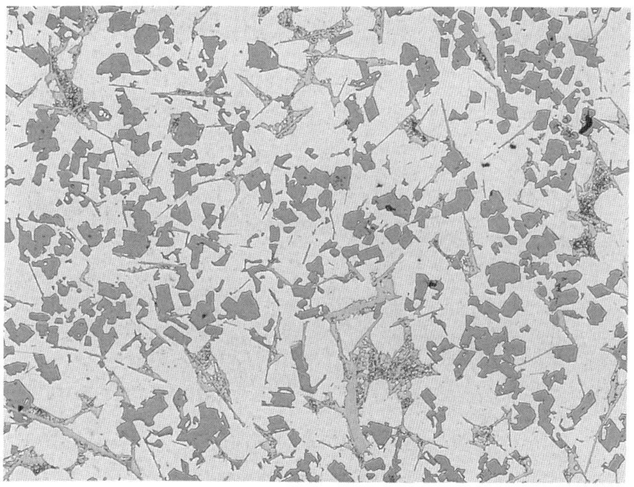

Fig. 16-9. Permanent mold cast B390.0 alloy shows the structure of a hypereutectic Al-Si commercial alloy (100X).

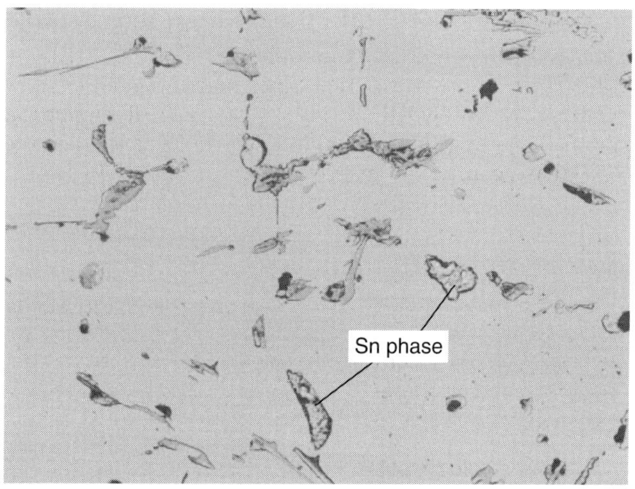

Fig. 16-11. Sand casting of Alloy 851.0 (500X).

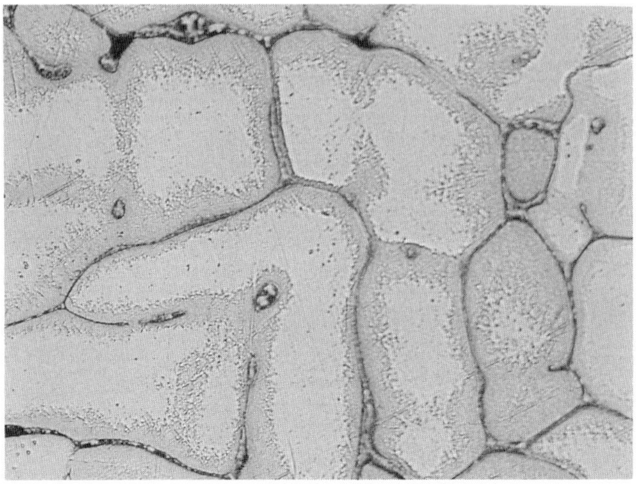

Fig. 16-12. Alloy 295.0–F sand casting showing cored structure. Keller's etch, 250X.

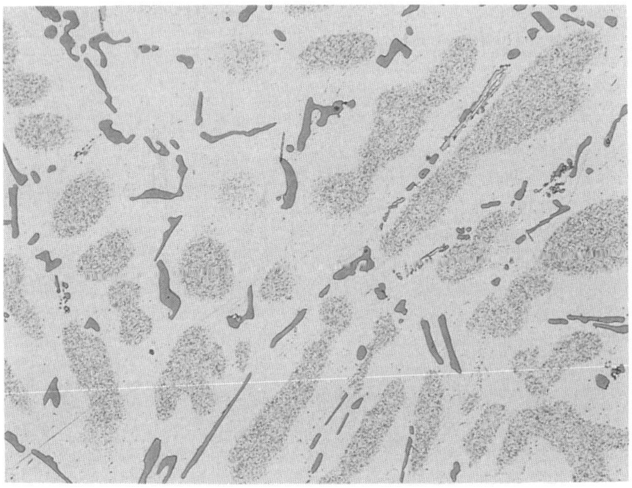

Fig. 16-13. A356 permanent mold casting showing precipitate coring. Keller's etch, 250X.

Second-Phase Constituents

Microstructural features other than the Al solid solution depend primarily on the alloy composition and, for heat treatable alloys, on the state of heat treatment. All commercial alloys (including those listed in Table 16-3) have some second-phase constituent particles. Even those in which the addition elements are largely in solid solution contain phases formed by impurity elements, the most prominent of which is iron.

Because of the multicomponent compositions of many alloys and the metastable conditions of solidification, the microstructures are usually more complex and contain a greater number of phases than is indicated by simple equilibrium diagrams (Figs. 16-14, 16-15 and 16-16).

The morphology (shape) of the second-phase constituent particles is, in some cases, distinctive, but they are affected by solidification rate and heat treatment. Identification by

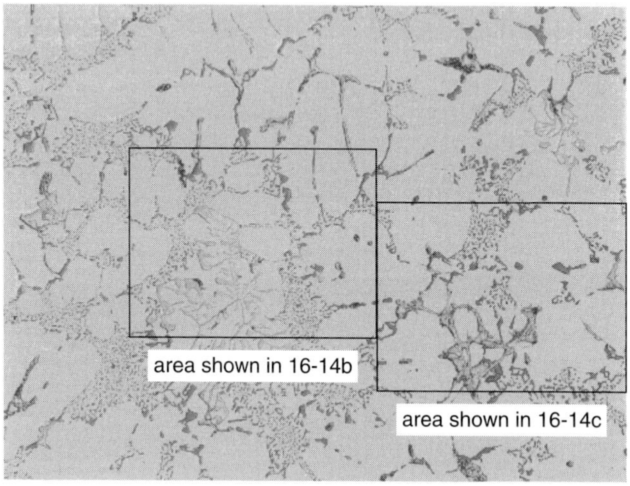

(16-14a) 100X

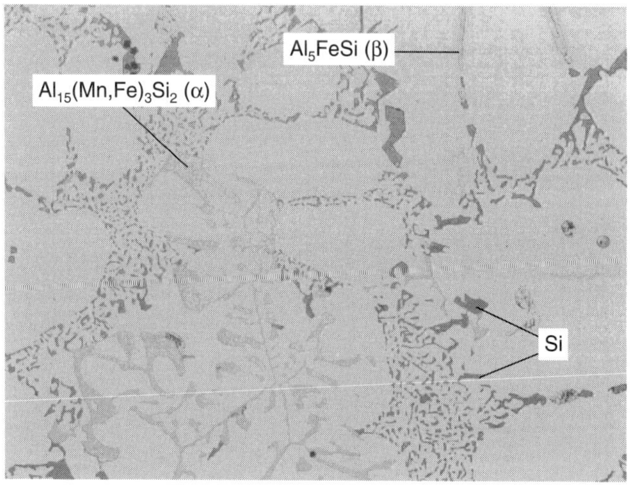

(16-14b) 250X

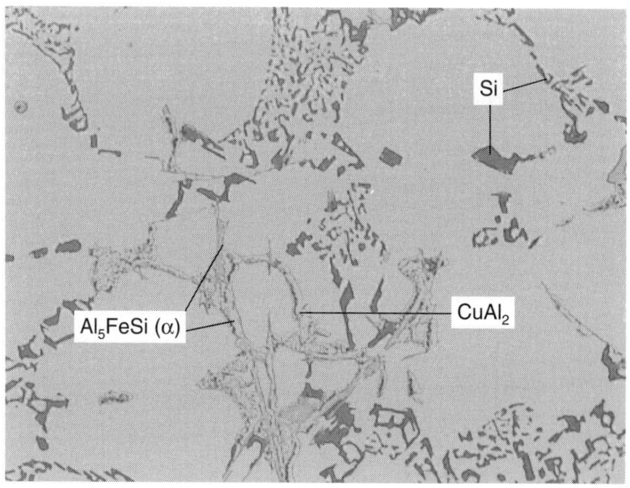

(16-14c) 250X

Fig. 16-14. Sand cast 319.0 alloy. Note the Fe-bearing phases.

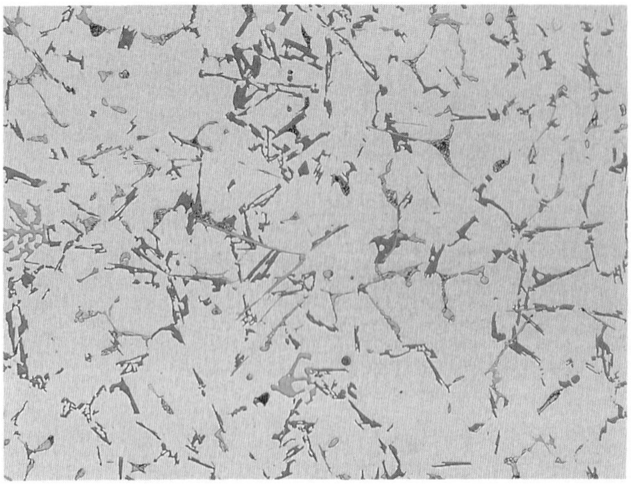

(16-15a) 100X

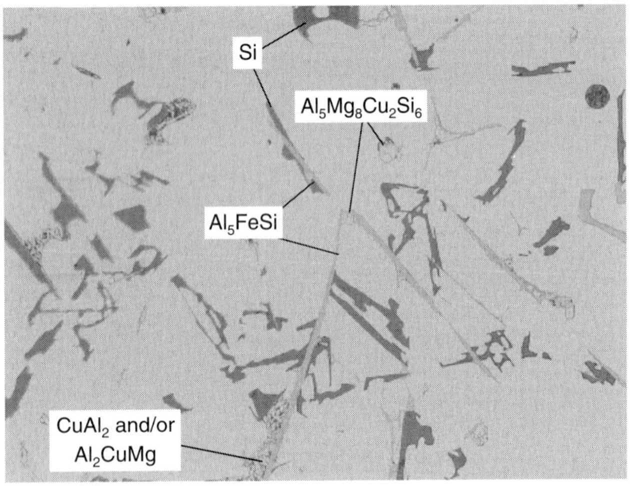

(16-15b) 250X

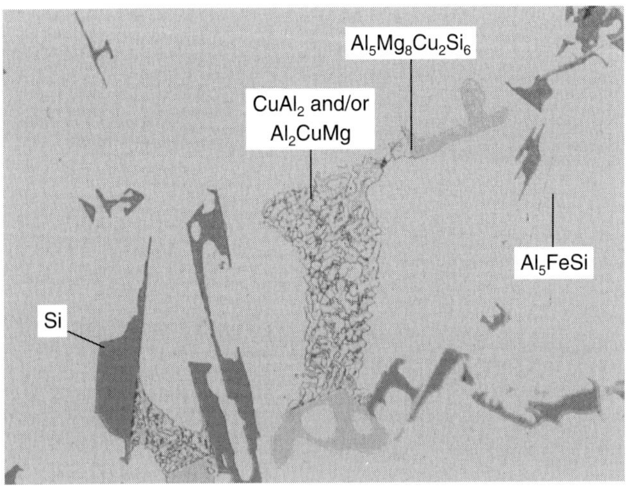

(16-15c) 500X

Fig. 16-15. General microstructure of unmodified, sand cast 355.0.

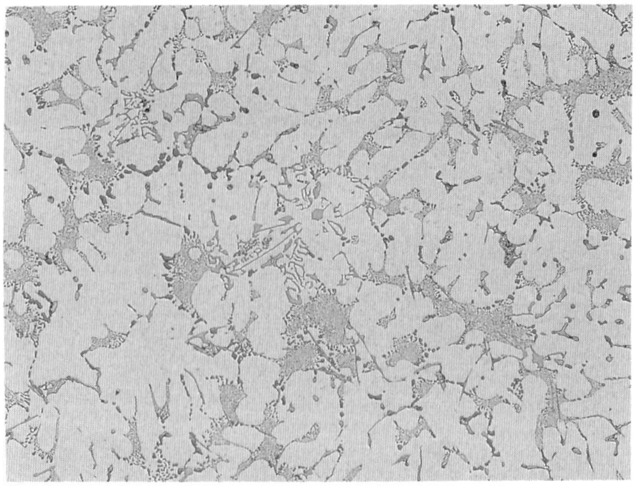

(16-16a) 100X

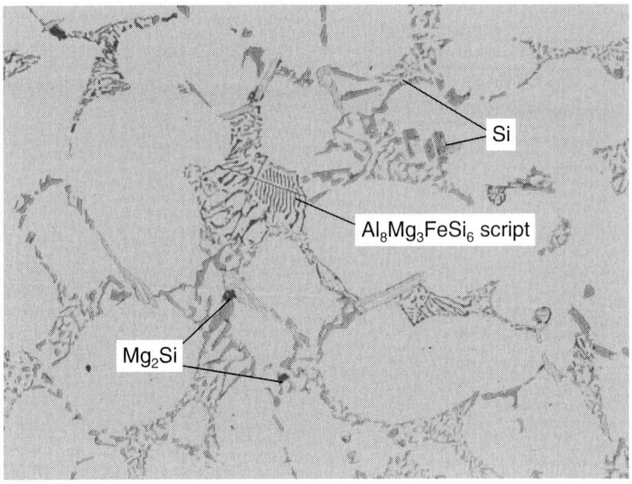

(16-16b) 250X

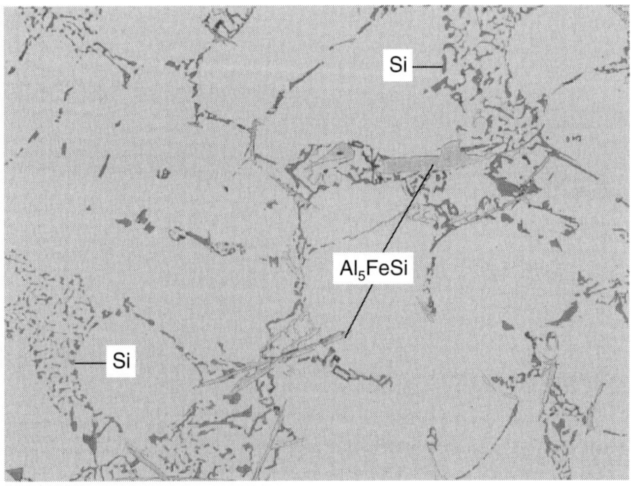

(16-16c) 250X

Fig. 16-16. The general microstructure of sand cast Alloy A356.0.

Table 16-3.
Major Constituents of Casting Alloys

Commercial Designation	Major Constituents
213.0	(Al-Cu), α(Al-Cu-Fe), α(Al-Fe-Si)
295.0	(Al-Cu), α(Al-Fe-Si)
242.0	α(Al-Ni), (Mg-Si), (Al-Cu-Ni), (Al-Cu)
222.0	(Al-Cu), α(Al-Cu-Fe)
443.0	(Si), α(Al-Fe-Si)
355.0	(Si), P(Al-Fe-Si), (Al-Cu)
356.0	(Si), P(Al-Fe-Si), (Mg-Si)
319.0	(Si), (Al-Cu), α(Al-Fe-Si)
514.0	(Mg-Si), α(Al-Fe)
520.0	α(Al-Mg), (Mg-Si), α(Al-Fe)
710.0	(Mg-Si), α(Al-Fe-Si)
850.0	(Sn), α(Al-Ni)
712.0	(Mg-Si), α(Al-Fe-Si)

characteristic shape, color or etching effects is facilitated by a slowly solidified structure and can be more difficult in the case of rapidly solidified material. Information for guidance in phase identification is presented in Table 16-4.

Alloy Type Identification

Alloy type frequently can be determined from the identities and the quantities of second-phase constituents present and the etching characteristics of the aluminum matrix. With increasing content of alloying elements in excess of the solid solubility, increasing volume (and area) fractions of distinctive constituents, such as Si, Al_2Cu or Al_8Mg_5, occur in certain alloy types.

By comparison with specimens of known composition cast under comparable conditions, alloy type identification is possible. The presence of abnormally high or unusual amounts of impurity elements, such as iron, also is usually apparent to the experienced metallographer. These observations are not highly quantitative, and should be considered as estimates requiring quantitative analysis by chemical or spectrochemical methods for verification.

Solidification Rate Effects

The rate at which the metal solidified is reflected in several features of the microstructure. With increasing solidification rate, the size of the dendrite cells (cross sections of the dendrite arms or branches) and their spacing decreases and the size of second-phase particles is reduced. With the increasing solidification rates progressively from sand to permanent mold to die casting, both the size and spacing of the primary dendrite cells decrease, and the Si particles become finer. The same effects are observed with the Al-5%Si-1.30%Cu-0.5%Mg alloy (355.0).

The grain size also tends to decrease with increasing solidification rate, but grain size alone is not a reliable criterion of solidification rate since it is strongly affected by the use of nucleant grain refining additions as well as by melting and casting temperatures.

Correlations have been established between measurements of average secondary dendrite arm spacing and solidification rate indicating a relationship of exponential form:

$$\lambda = KR^{-n}$$

where:

λ = average secondary dendrite arm spacing, μm

R = solidification time, s

K = a constant (which may change with temperature gradient), and

n = exponent (0.3 to 0.4).

The relationship is generally considered to be independent of alloy composition and, therefore, is useful for estimating solidification rate from measured average spacing. Such measurements allow a judgment concerning the type of casting process when this information is not supplied.

The solidification rate may affect the shape as well as the size and spacing of constituent particles, and it can change certain phases that occur in the microstructure. An example of the latter effect would be the iron-bearing phase in 443.0 alloy. The effect of increasing cooling rate is illustrated by the sequence of processes from sand casting (Fig. 16-17a) to permanent mold casting (Fig. 16-17b) to diecasting (Fig. 16-17c). The iron-bearing phase changes from the needlelike Al_5FeSi in the sand cast microstructure to Al_{15} (Mn, $Fe)_3Si_2$ in characteristic Chinese script form in the permanent mold alloy.

An additional factor that enters into the morphology of the $Al_{15}(Mn, Fe)_3Si_2$ phase is the cooling rate. In alloys where the manganese content is high enough, formation of the needlelike Al_5FeSi is suppressed. The effect of increasing cooling rate is to change the $Al_{15}(Mn, Fe)_3Si_2$ from blocky at very low cooling rates to Chinese script at high cooling rates.

The Al_5FeSi phase is usually considered undesirable because it reduces ductility, the fractures exhibit facets from decohesion of the matrix from the plate-like particles, and the particles (in some cases) obstruct feeding during later stages of solidification resulting in shrinkage cavities. Additions of manganese, cobalt or aluminum frequently are made to convert the iron-bearing phase to the more desirable form.

SIMPLE, AS-CAST MICROSTRUCTURES

The simplest alloys, from the standpoint of their optical microstructures, are those that consist essentially of solid solution in the as-cast condition with only small amounts of second-phase constituent particles formed by impurity elements.

The only commercial alloys that fit this description are Alloys 514.0, 513.0 and 710.0, although other 5XX alloys with magnesium contents of 7% or less, such as 535.0 (Fig. 16-18), and other 7XX alloys, such as 713.0 (Fig. 16-19), have only modest amounts of additions—generally less than 1% of elements such as silicon, manganese or chromium—that introduce additional second-phase constituents.

The 5XX alloys with up to 7% magnesium are not heat treated. They depend upon the solid solution for their strength, and they do not exhibit significant natural aging effects. The Al-Zn 7XX alloys solidify with essentially all of

Table 16-4.
Metallographic Identification of Phases in Aluminum Alloys[1]

Basic and alternative phase designations	Elements that enter in solution	External shape	Appearance before etching[2]	Birefringence[3]	Etchants that aid identification[4]
Si	—	Cubic habit; primary particles form isometric polygons; eutectic may form script, blades or very fine lamellae.	Light bluish-gray.	None.	Generally best identified without etching. Etchant 1 (swab) outlines particles and appears to lighten color.
Mg_2Si	—	Cubic habit; eutectic forms script that easily coalesces on heating.	Natural color is darker bluish-gray than Si, but usually tarnishes to bright blue, black or varicolored.	None (when not roughened or tarnished).	Easily identified without etching. Caustic etchant 4 will not attack and may enhance blue color. Acid etchants will attack and dissolve readily.
MgZn or η(Mg-Zn)	Isomorphous series with CuMgAl.	Usually well-rounded or irregular, except in lamellar eutectic or precipitated from solid solution.	White, watery; does not polish.	Slight change from light to dark gray.	Etchant 2 gives a dark gray to black color.
Al_2Cu or θ(Al-Cu) and Al_2CuMg	—	Usually well-rounded or irregular, except when precipitated from solid solution.	Pale pinkish color.	Strong, orange to greenish-blue. Some orientations show little change.	Remains light and clear in etchants 1 and 2.
Al_8Mg_5 or β(Al-Mg)	—	Usually well-rounded or irregular.	White; lighter than aluminum but may tarnish to yellow or tan; not in relief.	None (when not tarnished).	Caustic etchant 4 will not attack or color. Acid etchants generally pit and dissolve it with varying rapidity.
$Al_{15}(Mn,Fe)_3Si_2$ and Al_8Fe_2Si or α(Al-Fe-Si)	Besides the apparent interchangeability of Fe, Cr and Mn, this phase can probably also contain Cu.[5]	Usually well-defined script, especially when silicon is not low. May also form polyhedrons or irregular shapes.	Light metallic gray.	None.	This phase and its variants can give a variety of etching responses to a given etch, depending on its composition and that of the matrix.
Al_5FeSi or β(Al-Fe-Si)	—	Bladelike or acicular.	Light metallic gray.	Moderate; light to dark gray.	Etchant 1 (immerse) will attack and darken to varying degrees, depending on Fe-Si ratio. Etchant 3 will attack and dissolve it out. In both cases, $A1_8Fe_2Si$ is outlined but not appreciably darkened.

1) Phases other than those listed in this table are less common or appear in such small amounts or as such fine particulate that identification can be made only indirectly.
2) Applies to appearance after mechanical polishing.
3) An exceptionally good flat polish with no tarnishing is required because any element of the surface that is not parallel to the plane of the surface (that is, normal to the optical axis) will cause an apparent birefringence that is not due to crystal structure. The sensitivity of this technique will also vary with the quality of the optical system. A rotating stage is necessary.
4) Etchant numbers referred to in this column correspond to etchants that are identified by number in Table 16-2.
5) Formation of Al_8Fe_2Si in high Si alloys is promoted by presence of Mn, Cr or Cu. In absence of these elements, $Al_{10}FeSi$ form is probable. Distinction metallographically is predominantly by shape.

the solute elements (magnesium, zinc and copper) in solid solution and develop additional strength over that existing immediately after casting by natural aging at room temperature over a few weeks.

HEAT TREATMENT

Heat treatments applied to cast material produce microstructural changes that, in many cases, allow the type of treatment to be determined and whether certain conditions of the treatment were normal or outside established limits of temperature or time. The changes from solution heat treatment generally are distinctive.

Those from precipitation heat treatments applied either to the as-cast material (T5-type treatments) or after solution heat treating (T6- or T7-type treatments) are submicrostructural but detectable in some cases by changes in etching

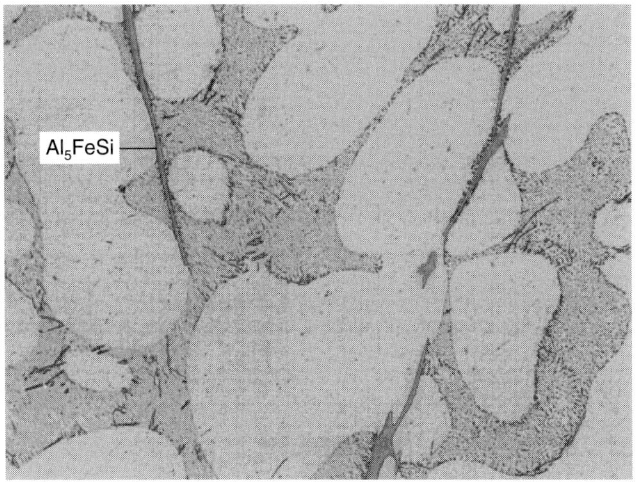

(16-17a) sand cast

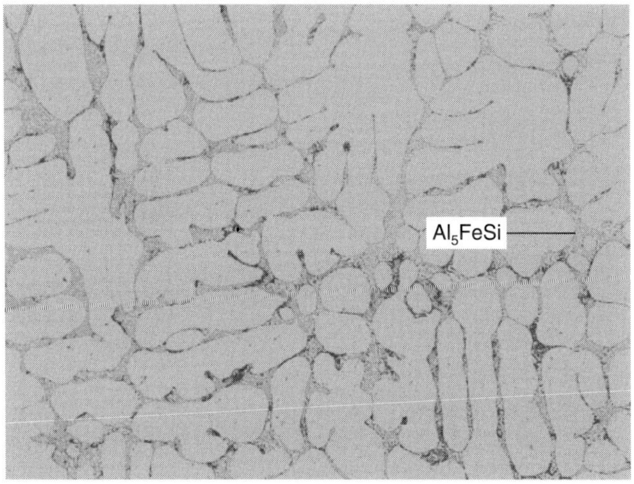

(16-17b) permanent mold cast

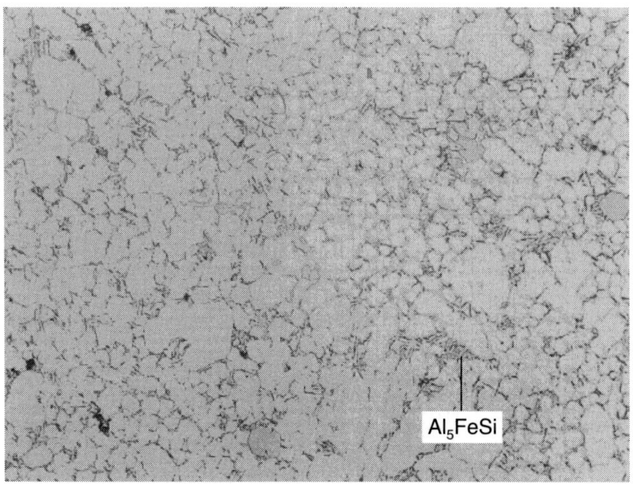

(16-17c) diecast

Fig. 16-17. Comparison of the structures of Alloy 443.0 cast using different processes (250X).

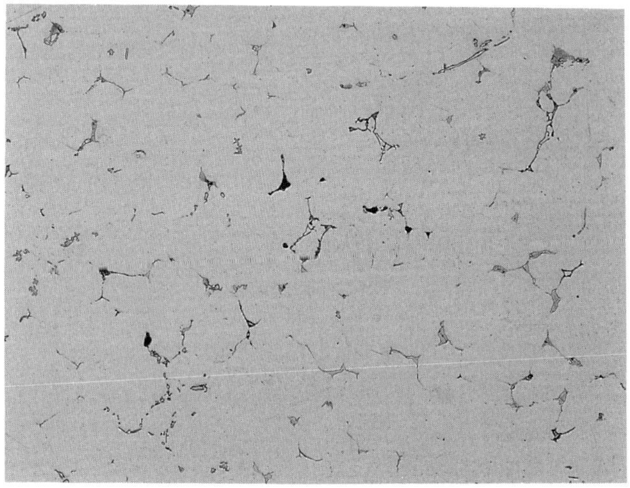

(16-18a) 100X

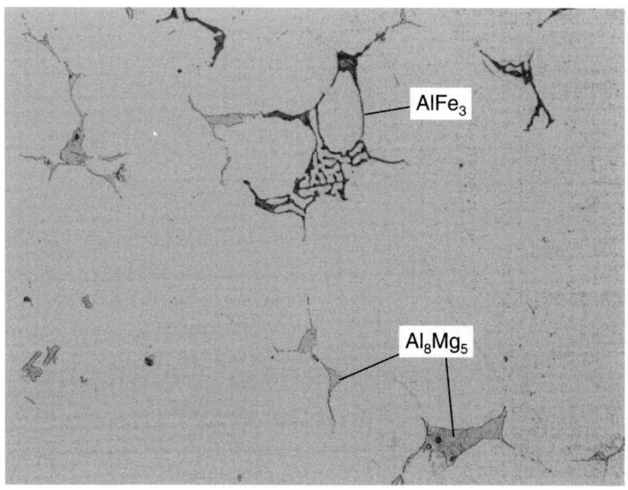

(16-18b) 250X

Fig. 16-18. Sand cast Alloy 535.0.

response or precipitation effects at grain boundaries. These changes all are attributable to the solution-precipitation reactions of the alloying elements that have marked temperature-dependence of solid solubility and are responsible for the mechanical property changes associated with heat treatment.

Solution Heat Treatment

Solution heat treatment and quenching are intended to place into solid solution and retain in that state as much as possible of the soluble alloying elements. This is limited, of course, by the equilibrium solid solubility of the principal solute element or elements of the alloy at the solution heat treating temperature. Some alloys have little or no excess of such elements, while others have amounts that greatly exceed the maximum solid solubility. Changes are specific to the type and composition of the alloy, although the principles are applicable universally.

The simplest cases and those that exhibit the most apparent microstructural changes from solution heat treatment are alloys that are based on binary systems and have little or no excess of solute over the equilibrium solubility. A good example of this is the Al-10% Mg alloy 520.0. Solution heat treatment and quenching to produce the T4 temper takes almost all of the Al_8Mg_5 constituent formed during solidification into solid solution (Fig. 16-20a).

In the F (as-cast) condition, a considerable area fraction of the structure is the constituent Al_8Mg_5, which forms a relatively continuous interdendritic and intergranular network. Because of the relative continuity and brittle character of this constituent (Fig. 16-20b), alloy 520.0–F is brittle, and castings before heat treatment can be cracked or even shattered by dropping them.

Solution heat treatment dissolves most of the Al_8Mg_5 constituent and destroys the network, making the Al solid

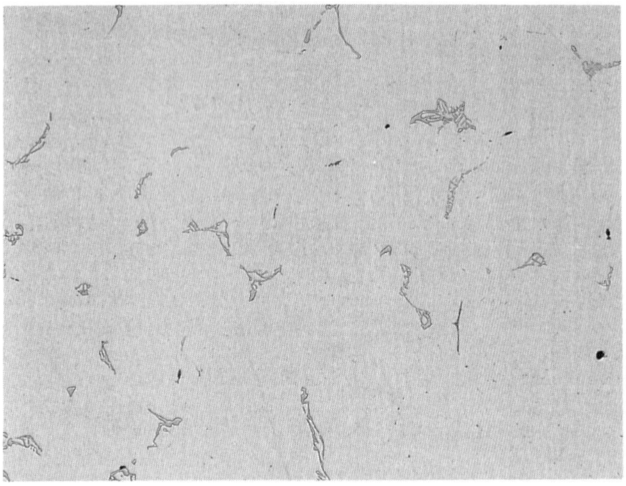

(16-19a) 100X

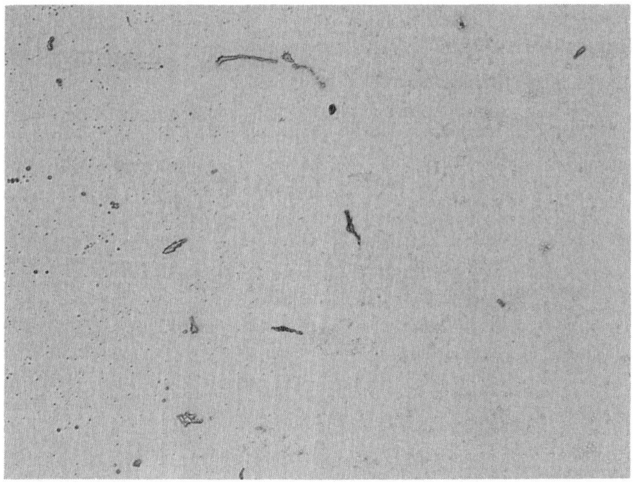

(16-20a) T4 temper

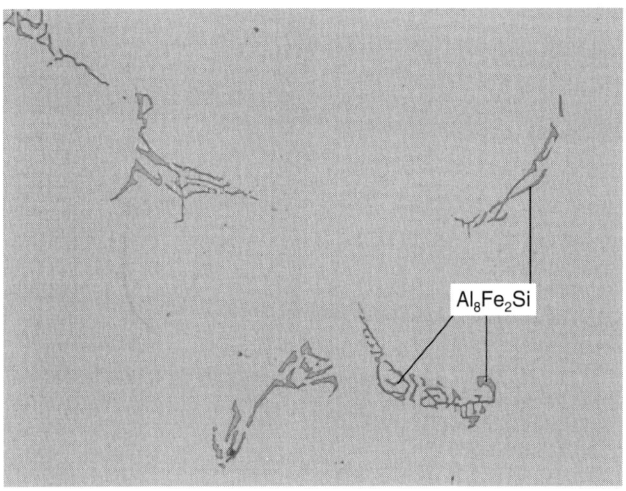

(16-19b) 250X

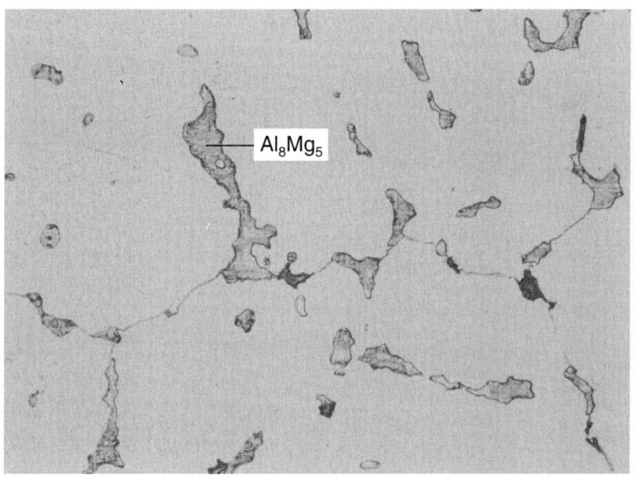

(16-20b) F temper

Fig. 16-19. Sand cast Alloy 713.0.

Fig. 16-20. Sand cast Alloy 520.0 (250X).

solution phase continuous. With this structure, the material has high strength, ductility and fracture toughness. The impurity elements (Fe and Si) form a phase (Al_3Fe) that has very low solid solubility and that remains relatively unchanged by the solution heat treatment.

Another alloy of the class in which most of the solute is soluble at the normal solution temperature is Alloy 295.0 with Al-4.5% Cu-1% Si. The Al_2Cu formed during solidification nearly disappears during solution heat treatment, with the copper being taken into solid solution. Dendritic coring revealed by Keller's etch in the F (as-cast) condition (Fig. 16-21a) is eliminated, and marked grain contrast is not apparent after etching the solution heat treated material in this reagent (Fig. 16-21b).

The contrast between grains is partly a result of angularity differences in the extremely fine scale surface texture developed by etching, the crystalline grains having different orientations with respect to the plane of examination. In addition, with copper-containing alloys, these effects are accentuated by coloration from films of etching reaction product formed by mixed acid reagents such as Keller's.

The microstructural constituent formed by impurities (in this case, silicon and iron) is a ternary intermetallic phase, $Al_5(Mn, Fe)_3Si_2$. The silicon content of the alloy is partitioned between the solid solution and this phase. The amount of the phase and its characteristic script form are unchanged by normal solution heat treatment.

Faulty solution heat treatment may result from inadequate process control or by accidents. Inadequate solution may result from a temperature lower than the normal range or insufficient time. For example, part of the Al_2Cu may remain undissolved (Fig. 16-22). Overheating, the result of exposure to a temperature exceeding not only the normal range for solution heat treatment but also the lowest melting temperature of the alloy, is evidenced by the formation of eutectic as rosettes or at grain boundaries (Fig. 16-23).

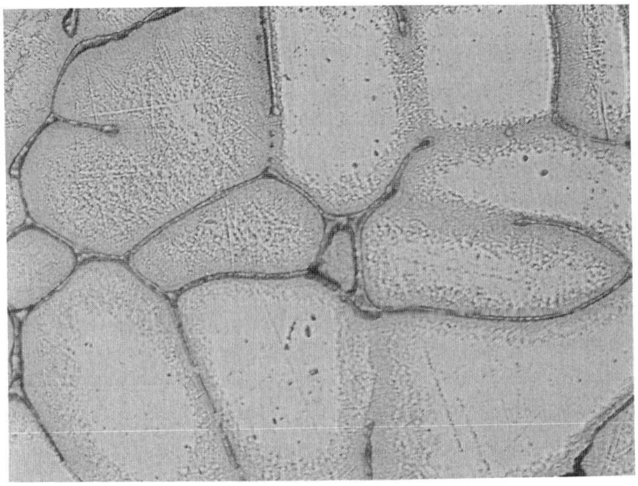

(16-21a) F temper

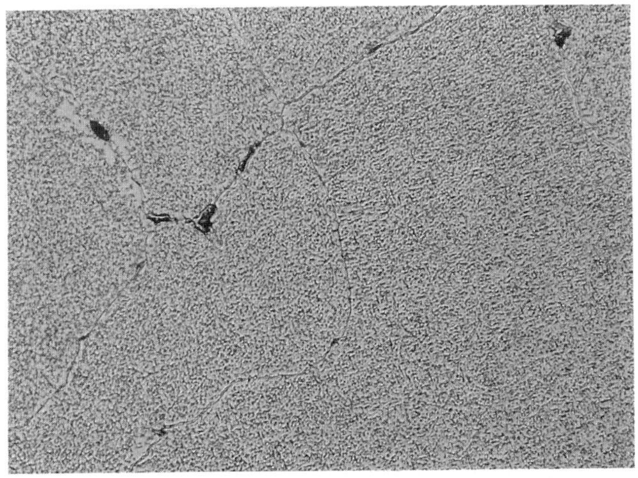

(16-21b) T6 temper

Fig. 16-21. Alloy 295.0 sand casting. Keller's etch, 250X.

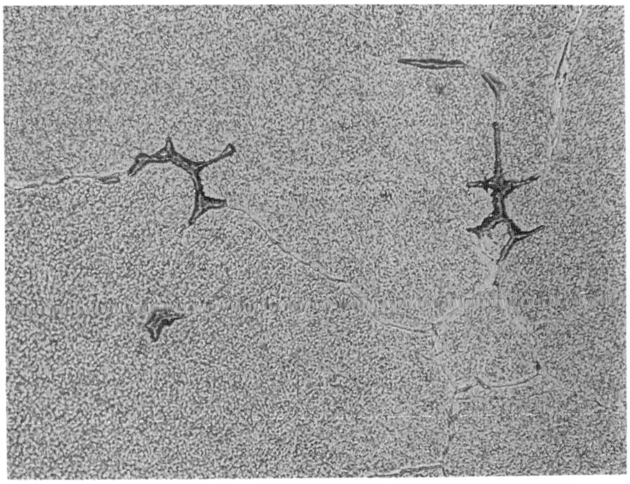

Fig. 16-22. Undissolved Al_2Cu is present in the microstructure of a 295.0 alloy sand casting that was inadequately solution heat treated. Keller's etch, 250X.

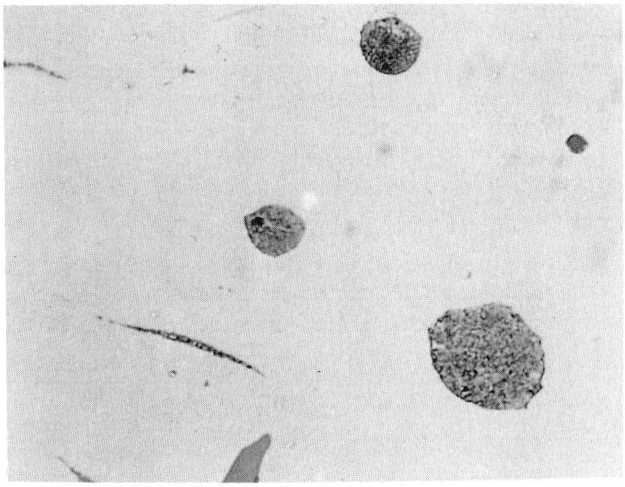

Fig. 16-23. Overheating of a 295.0 alloy sand casting during solution heat treatment caused Al_2Cu rosettes and Al_2Cu at grain boundaries (250X).

In addition, this abnormally high temperature during heat treatment may cause coalescence or agglomeration of the normally insoluble, script form $Al_{15}(Mn, Fe)_3Si_2$ into larger, more massive particles. Microstructures showing these features are the metallographer's clue to overheating, which reduces ductility and, in severe cases, strength. Evidence of overheating is cause for rejection under most specifications for heat treated castings.

The most widely used casting alloys have normal amounts of the partially soluble alloying elements silicon and/or copper that considerably exceed the solid state solubility. This is true of the Al-Si (4XX), Al-Si-Mg and Al-Si-Cu-Mg (3XX) alloys, as well as some of the 2XX types and all the Si containing diecasting alloys, which are not normally solution heat treated. For alloys that are solution heat treated—such as 222.0, 319.0, 333.0, 354.0, 355.0, C355.0, 356.0, A356.0, 357.0, A357.0 and 359.0—the solution heat treatment may not dissolve all of the Si, Al_2Cu or Mg_2Si constituent particles but may substantially alter their size and shape. Figures 16-24 through 16-26 illustrate this for alloys 242.0, 355.0 and 356.0.

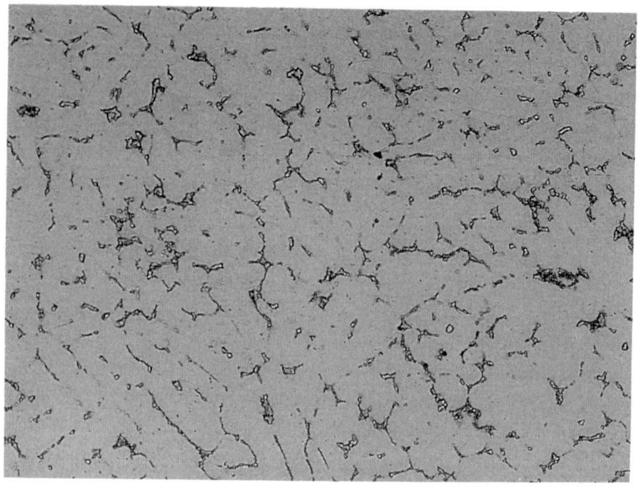

Fig. 16-24. Permanent mold cast Alloy 242.0–T7 (250X).

The F temper (as-cast) microstructure, refined by modification, consists of primary Al solid solution dendrites surrounded by Al-Si eutectic in which the Si particles are very small. Relative volumes (and areas in a section) of the solid solution and eutectic are roughly equal at this silicon content. The weight percent of Mg_2Si—the principal source of strength increase through subsequent precipitation heat treatment—that can be formed in this composition is about 0.5% (Mg_2Si = 1.6xMg). Most of the magnesium and enough silicon to precipitate Mg_2Si are in solid solution in the F temper, and little or no Mg_2Si is apparent in the microstructure.

Further evidence of the magnesium and silicon in solid solution is provided by the increase in strength and hardness that can be obtained by precipitation heat treating the F temper material to a T51 temper. A small amount of insoluble Al_5FeSi, characteristically acicular or bladelike particles, is formed in A356.0 (Fig. 16-27a). Larger amounts may be present in Alloy 356.0 because of the higher allowable iron content (Fig. 16-27b).

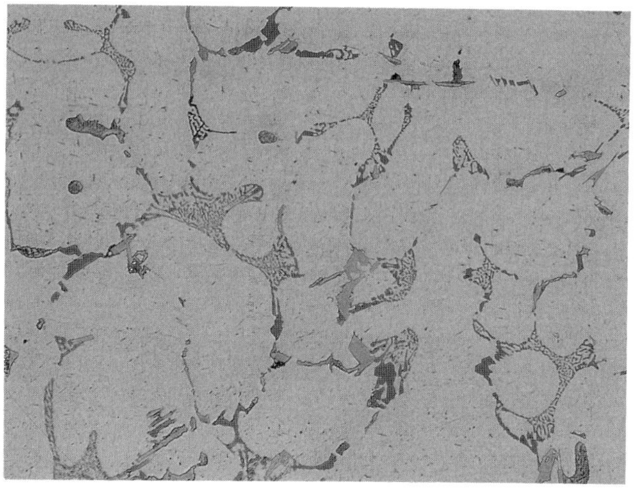

Fig. 16-25. Alloy 355.0–T6. There is no spheroidization because of short solution heat treatment time.

Solution heat treatment increases the solute concentration of the solid solution and eliminates coring, which is normally not apparent microstructurally in this alloy. The increased solid solution concentration and homogenization make possible additional strength increase upon subsequent precipitation heat treatment to T6 or T61 temper. The only evident microstructural changes from the solution heat treatment are increased size and more rounded shapes of the Si particles of the eutectic (Fig. 16-28), which is called spheroidization.

Fine Si particles are dissolved selectively, and larger Si particles coalesce and grow. Inadequate solution is evidenced by an intermediate size distribution, less shape change and more sharply cornered and faceted Si particles. Overheating results in exaggerated Si particle growth and intergranular eutectic melting (Fig. 16-29).

Similar effects are observed with the Al-5% Si-1.3% Cu-0.5% Mg alloy C355.0. Because of the higher magnesium content and the presence of copper, the F temper (as-cast) structure of this alloy may contain small amounts of Mg_2Si and Al_2Cu, and coring will be apparent after etching with Keller's reagent. Higher iron content, as permitted in 355.0, increases the size and amount of Al_5FeSi constituent.

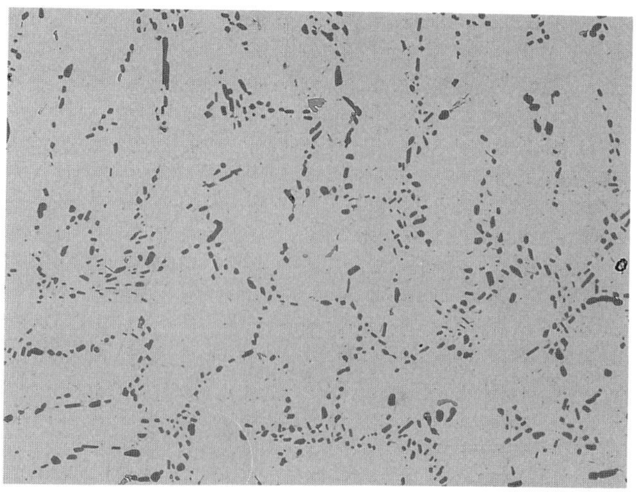

Fig. 16-26. An A356.0–T6 permanent mold casting (250X).

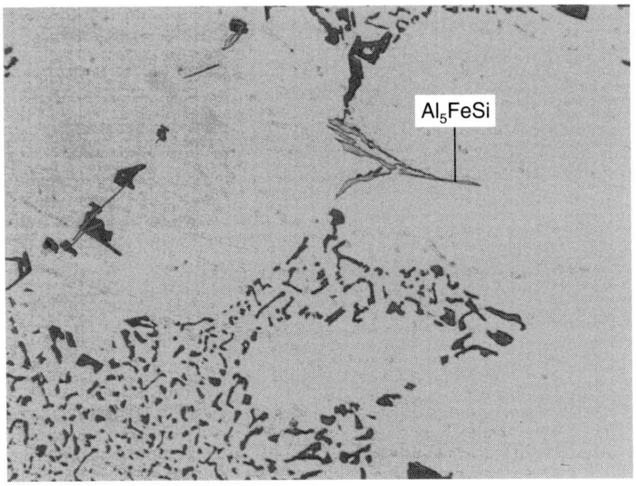

(16-27a) A356.0 alloy

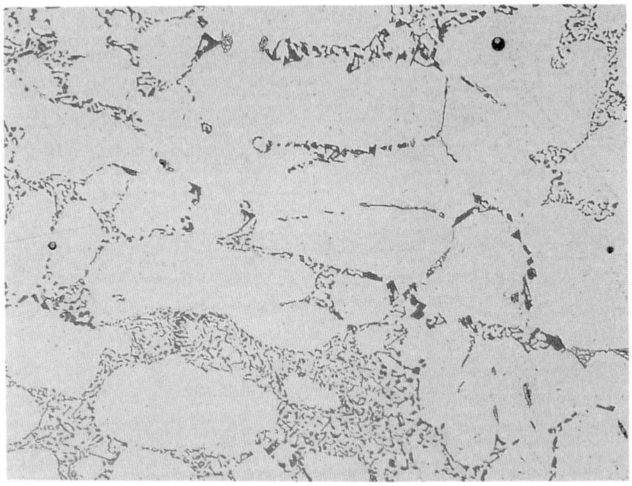

(16-28a) F temper

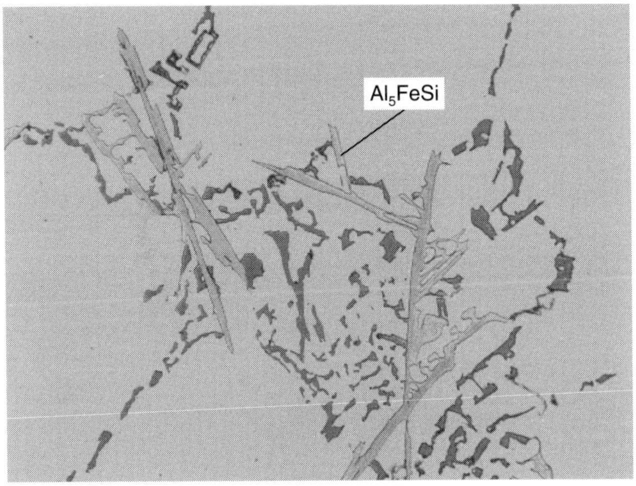

(16-27b) 356.0 alloy

Fig. 16-27. Fe phases a) in an Alloy A356.0 permanent mold casting and b) in sand cast Alloy 356.0 (500X).

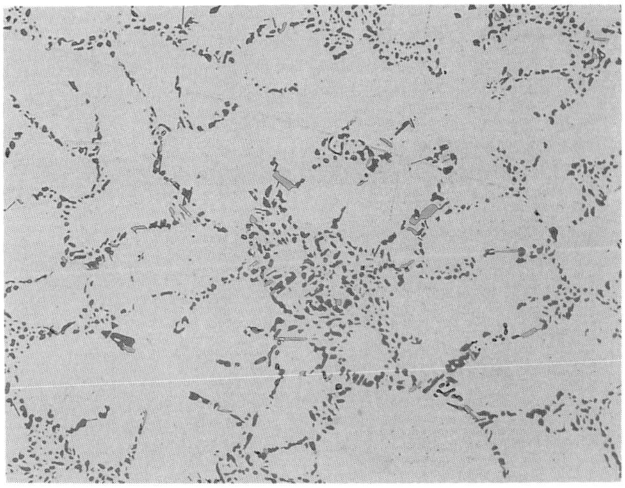

(16-28b) T6 temper

Fig. 16-28. Structure of Alloy A356.0 sand castings (250X). a) No heat treatment—shows modified eutectic Si. b) Solution heat treated—shows spheroidized eutectic Si.

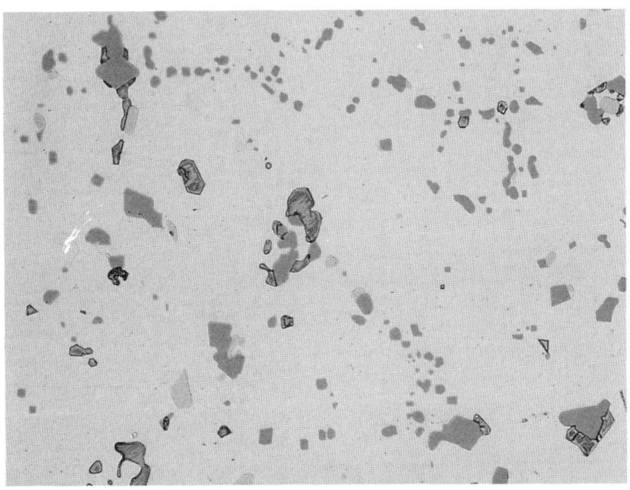

Fig. 16-29. Structure of an A356.0–T6 permanent mold casting overheated during solution heat treatment (250X).

Precipitation Heat Treatment

The precipitate particles that are responsible for the mechanical property changes derived from precipitation heat treating (either directly after casting or after solution heat treating) are of submicroscopic size for light microscopy and can be resolved only by electron microscopy. Consequently, it is not possible to distinguish between the light microstructures before and after precipitation heat treatment for many alloys.

With copper-containing alloys, grain contrast revealed by Keller's etch decreases as a result of precipitation heat treatment, and after the higher temperature treatments of the T7 stabilizing type precipitate particles may be detectable. These effects are qualitative and permit no judgment as to the adequacy of precipitation.

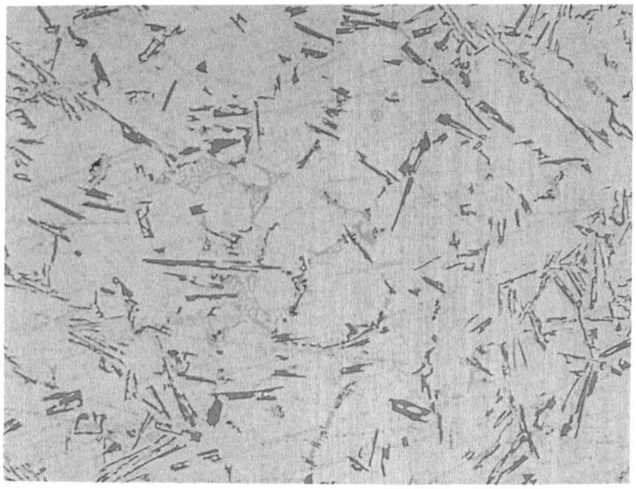

Fig. 16-30. Alloy 319.0–F sand casting showing unmodified eutectic Si (250X).

MODIFICATION OF Al-Si ALLOYS

Aluminum-silicon alloys with silicon contents from about 5% to that of the eutectic composition (12.6% Si) are frequently treated with metallic additions or by salt fluxes of alkaline earth metals, usually sodium or strontium, to modify the Si particles of the Al-Si eutectic. This is called "modification" and the microstructure is a "modified structure" (Figs. 16-30 and 16-31). For more information, see Chapter 3, Chemical and Physical Characteristics of Molten Aluminum.

Modification improves mechanical properties through microstructural refinement as well as increased soundness resulting from improved feeding characteristics. These benefits are most pronounced for castings or portions of castings that solidify at relatively slow rates because similar, though not identical, effects on structure and soundness derive from very high solidification rates without modification by chemical addition (Fig. 16-32).

Metallographic and/or fracture examinations are used frequently as a part of quality control procedures to ascertain whether appropriate modification is being achieved with practices employed in producing foundry alloy ingot and in the foundry operations.

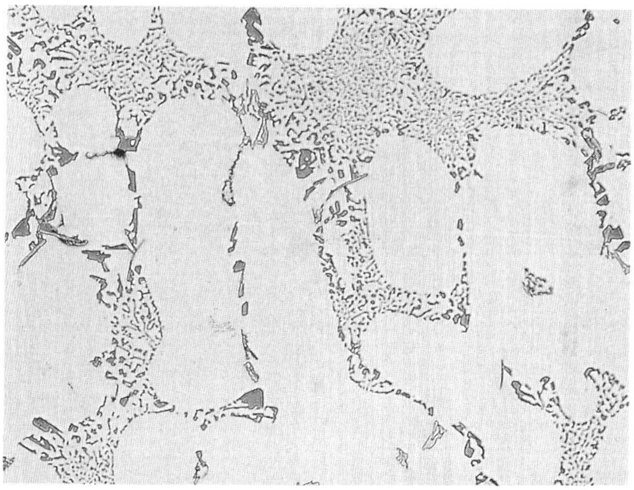

Fig. 16-31. Alloy 319.0–F sand casting showing modified eutectic Si (250X).

REFINEMENT OF PRIMARY Si IN HYPEREUTECTIC Al-Si ALLOYS

Marked microstructural changes in hypereutectic Al-Si alloys, such as 390.0, A390.0, 392.0 or 393.0, can be produced by treating the melt with additives containing a small amount of phosphorus (see Chapter 3 for more details). This treatment refines the primary Si particles (Figs. 16-33 and 16-34), reduces segregation and improves machining characteristics.

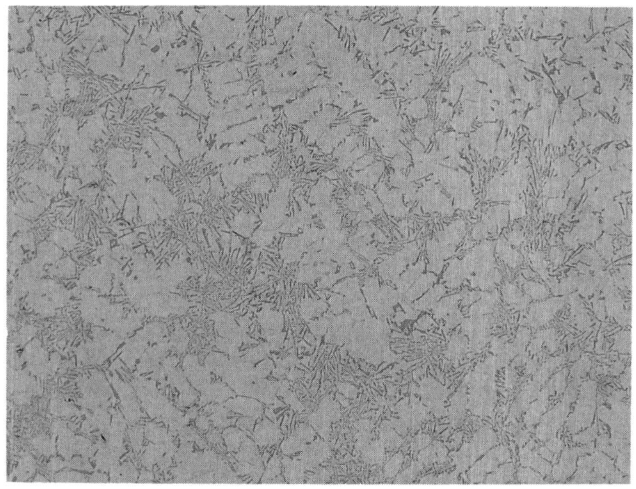

Fig. 16-32. The microstructure of an Alloy A360.0–F die casting. It exhibits a fine eutectic Si structure because of rapid solidification (250X).

OTHER Al-Si-Mg AND Al-Si-Cu-Mg ALLOYS

A number of other compositions of the 3XX group for production of sand, permanent mold, composite mold and premium quality castings have silicon contents intermediate to those discussed previously, with some differences in nominal copper and/or magnesium contents and allowable impurity limits. The basic microstructures of these alloys are similar to those discussed previously, although additional and more complex phases are formed in some cases. Figures 16-35 and 16-36 show alloys 308.0–F and 512.0–F.

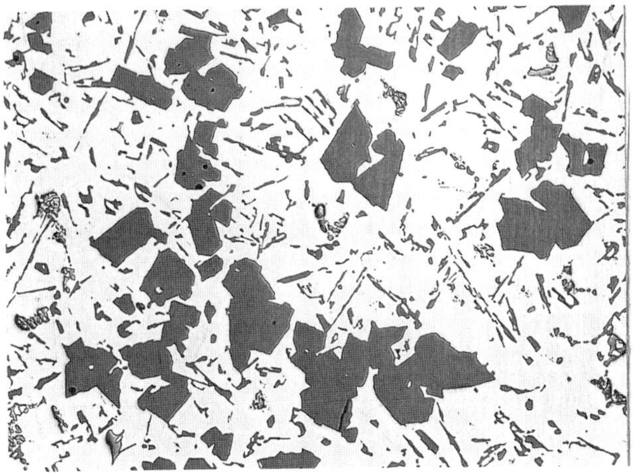

(16-33a) unrefined

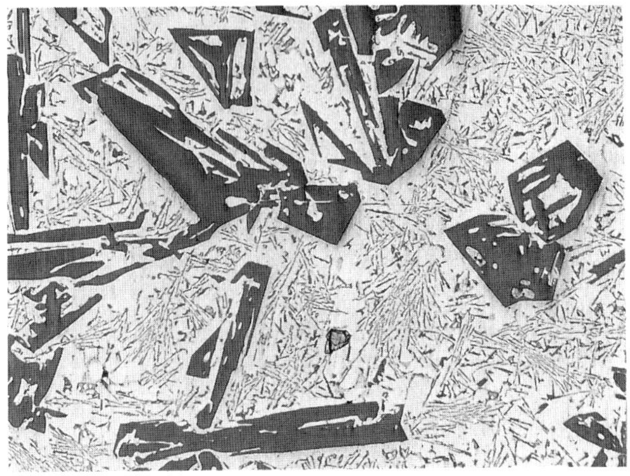

(16-34a) unrefined

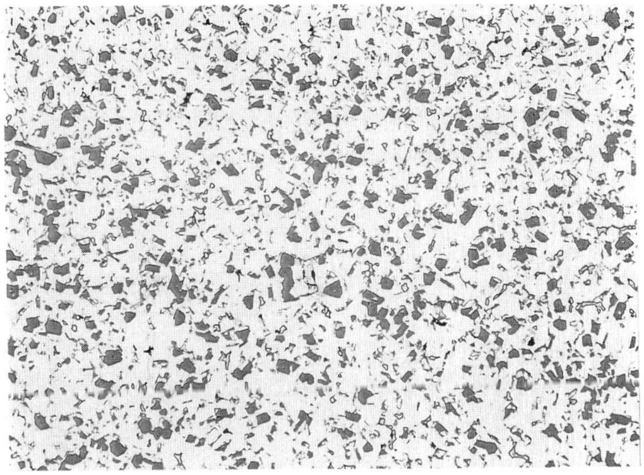

(16-33b) refined

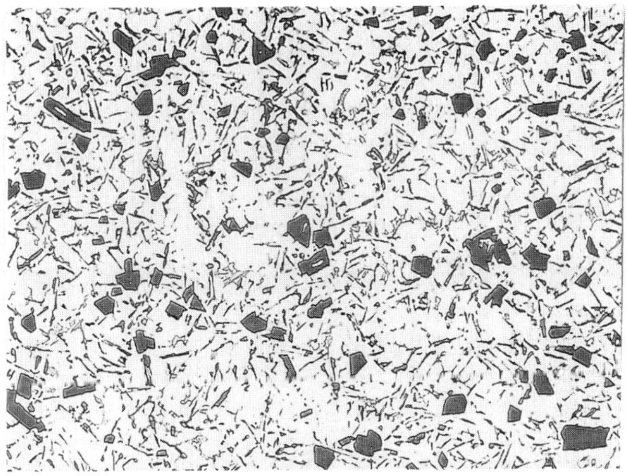

(16-34b) refined

Fig. 16-33. The structure of permanent mold cast B390.0 alloy in the a) unrefined and b) refined states (100X).

Fig. 16-34. The structure of sand cast B390.0 alloy in the a) unrefined and b) refined states (100X).

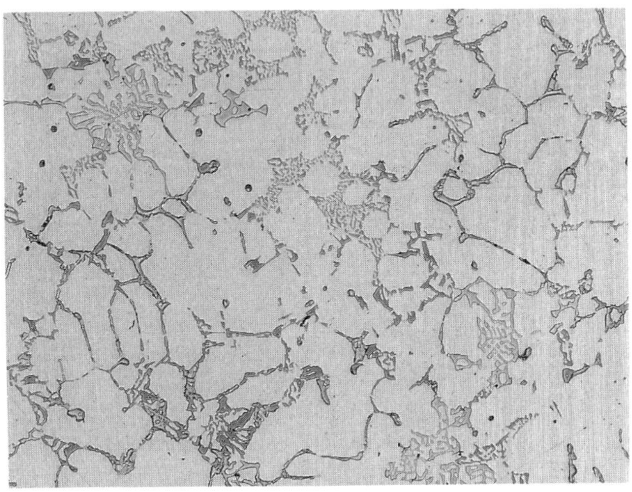

(16-35a) 100X

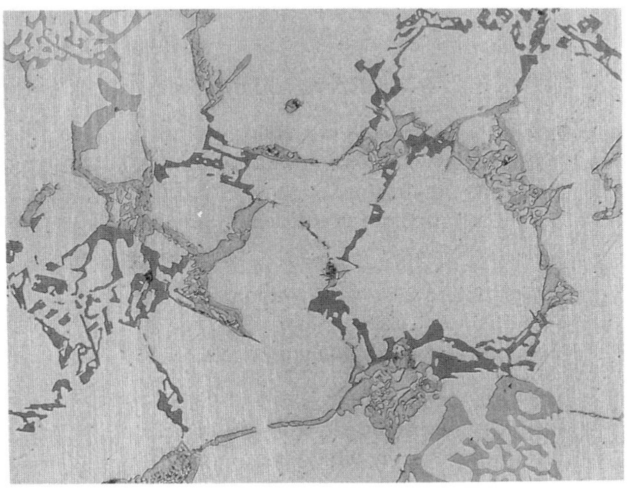

(16-35b) 250X

Fig. 16-35. These photomicrographs show the structure of sand cast Alloy 308.0–F.

DIE CASTINGS

The alloys used most widely for making die castings are those of the 3XX Al-Si-Mg and Al-Si-Cu types and 4XX Al-Si types that have basic constitution and structures analogous to those of similar compositions employed for other casting processes.

Because of the generally more rapid solidification rates of die castings, the microstructures are finer and the small size of the second-phase constituent particles generally increases difficulty of distinguishing between and identifying the phases (Fig. 16-37). Also, because the alloys usually have higher iron content, which is beneficial in minimizing die soldering, as well as increased amounts of other metallic impurities, the amounts and complexity of phases formed by these elements frequently are increased (Fig. 16-38).

Except for the Al-Mg alloys 515.0 (Fig. 16-39) and 518.0, which have essentially solid solution structures with modest contents of second-phase constituents, the diecasting alloys all contain 5% or more silicon and have hypoeutectic to

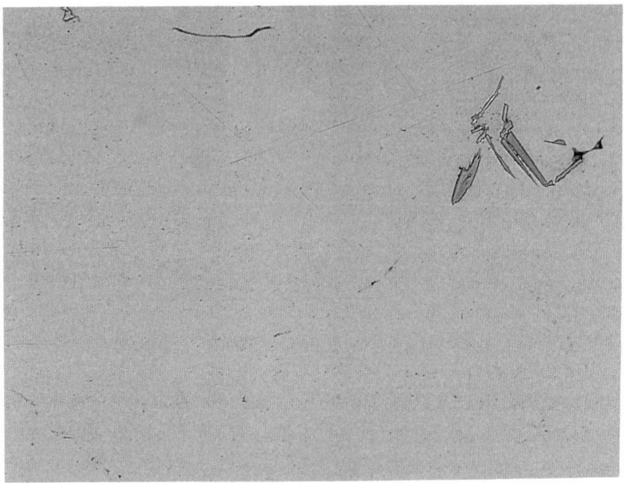

Fig. 16-36. Sand cast Alloy 512.0–F (250X).

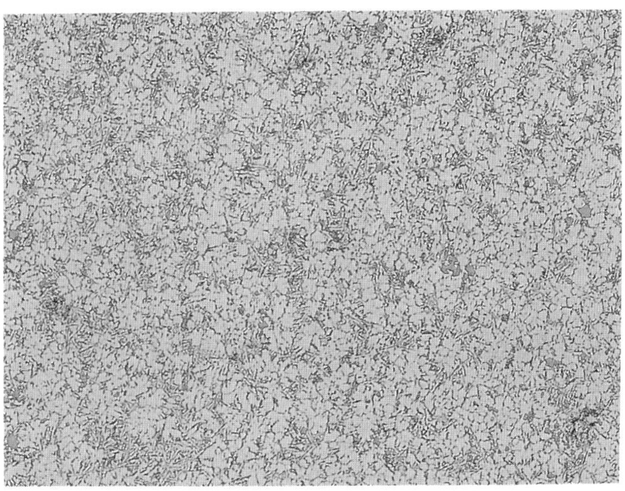

Fig. 16-37. General structure of diecast Alloy 383.0 (100X).

(16-38a) 100X

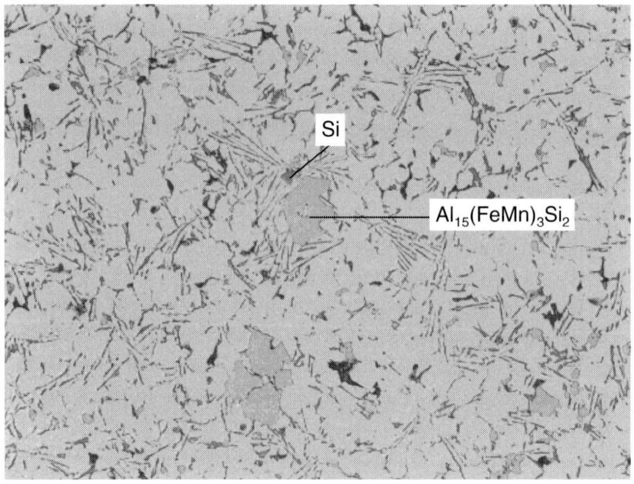

(16-38b) 500X

Fig. 16-38. General structure of diecast Alloy A380.0. Note the Fe phases in 16-38b.

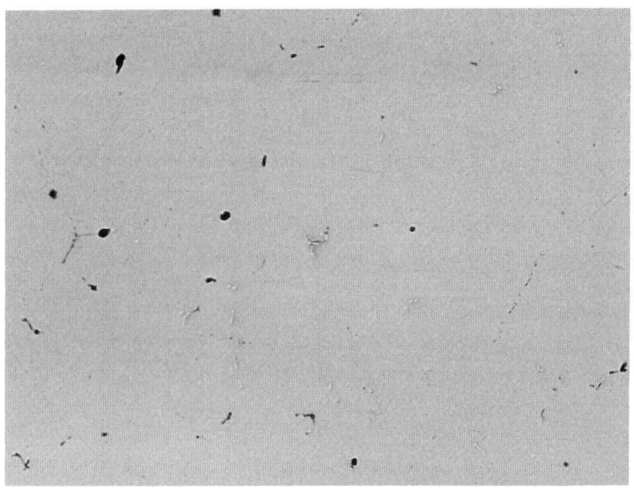

Fig. 16-39. Microstructure of an Alloy 515.0–F die casting (250X).

hypereutectic structures. At the eutectic composition, the structure is fine and relatively uniform, although areas differing in shape, size and spacing of Si particles may occur.

DISCONTINUITIES

A variety of types and sizes of discontinuities differing in severity with respect to their potential for functional impairment may be encountered in aluminum alloy castings. These usually can be classified and the probable cause or source identified by metallographic examination.

Voids or porosity can be caused by air entrapment (Fig. 16-40) (probably most frequent in die castings), by gas (hydrogen) evolution during solidification or by shrinkage. Voids from entrapped air are usually isolated and discrete and tend to be roughly spherical.

Gas porosity also may be spherical when present in relatively large quantities, but it usually is finer and more generally distributed in the casting. It frequently assumes an interdendritic shape because it forms after the alloy has started to solidify. Thus, it can be difficult to distinguish gas from shrinkage porosity (Fig. 16-41), which usually has an interdendritic appearance. [Note: shrinkage occurring only in isolated eutectic is not dendritic. It is round, just as in a 1XX alloy.]

The best criterion for distinguishing between the two forms is the distribution, which tends to be general for gas porosity (Fig. 16-42) while shrinkage porosity (Fig. 16-43) is concentrated in inadequately fed regions. Radiographic examination may be more efficient than extensive metallography for determining the porosity type because it provides more immediate and complete information concerning distribution.

Inclusions may occur as coarse, discrete particles; thin irregular films; or clusters of films of nonmetallic materials. Particulate inclusions usually are harder than the alloy and are apparent visually on as-polished metallographic specimens because they stand in relief and cause drag marks in polishing. Transparent or translucent nonmetallics normally appear black under vertical illumination, while under oblique

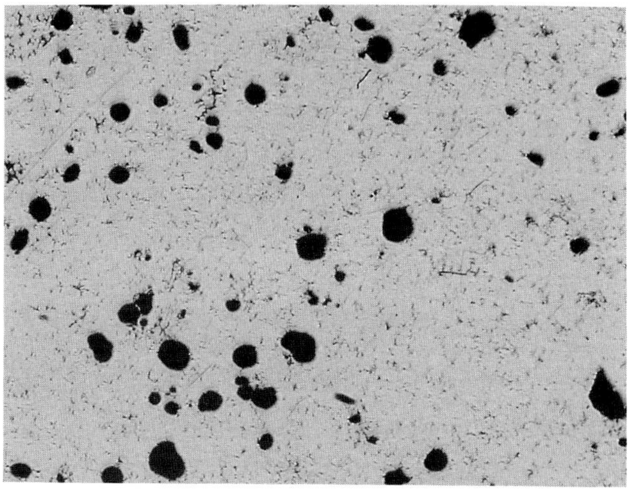

Fig. 16-41. Gross hydrogen porosity and microshrinkage can be seen in this Alloy 535.0–F sand casting (10X).

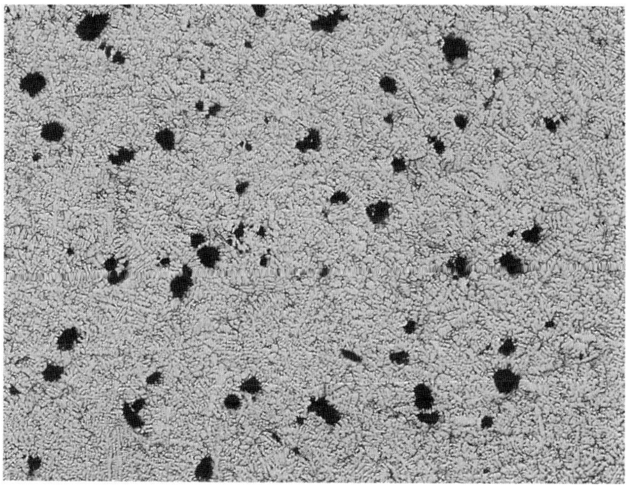

Fig. 16-42. This Alloy A356.0–F sand casting shows gross hydrogen porosity (0.6 cc/100g) (10X).

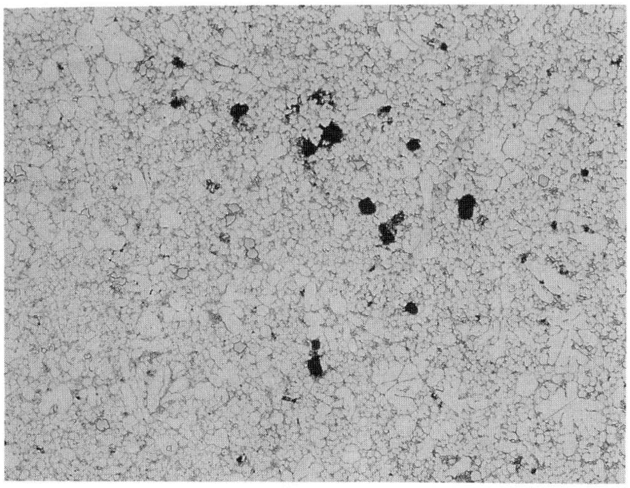

Fig. 16-40. This Alloy 443.0–F die casting shows air bubbles that were entrapped during turbulent die filling and then compressed to small size during pressurization up to 20,000 psi (100X).

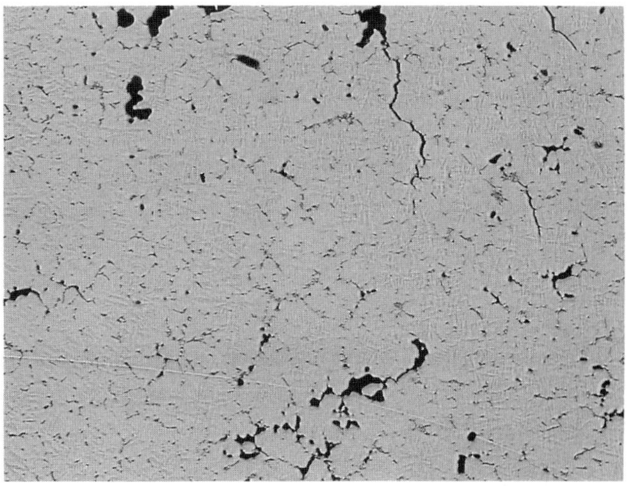

Fig. 16-43. This photograph of Alloy 771.0–F sand casting shows gross interdendritic/intergranular microshrinkage, which leads to hot tearing because of constrained solidification (10X).

illumination distinctions between some materials may be made by color or translucency differences. Al_4C, for example, appears in its natural brownish-yellow color and contrasts with brown-black for spinel ($MgAl_2O_4$) or black for corundum (Al_2O_3).

Cracks encountered during casting and before heat treating most frequently are of the hot cracking type. These are interdendritic or intergranular and sometimes partially filled with eutectic. Cracks also may result from mechanical damage or abnormal conditions during solution heat treatment or quenching.

Cracks formed under these conditions may appear similar to hot cracks, but they generally do not evidence eutectic refilling unless the casting was overheated. Also, features on one side of the crack tend to match those on the other. Discontinuities, such as misruns and cold shuts, generally do not require metallographic examination for identification.

Segregation may be detected in metallographic specimens by the presence of higher than normal concentrations of alloy constituents. It may be associated with hot cracks, shrinkage, porosity or liquation.

SURFACE COATINGS

Surface coatings that may be applied to castings—such as electrolytic oxide coatings, electroplated coatings or paint—are examined by mounting a cross section in plastic. Thickness measurements may be made by a direct reading optical microscope or by projecting an image of the coating and measuring the magnified image.

Electrolytic oxide (anodized) coatings (Fig. 16-44) are translucent in a dark field and characterized by the protrusion of alloy constituents into the coating, since the constituents either form little coating or the rate of formation is lower than that of the aluminum matrix.

Electro-deposited metal coatings are characterized by metallic appearance and frequently exhibit a columnar structure normal to the plated surface (Fig. 16-45). Characteristic color differences may allow identification of layers produced by the plating system. Paint coatings appear opaque, colored and lamellated under dark field illumination.

SERVICE-RELATED EFFECTS

Corrosion

Corrosion is not usually an important factor in the normal service life of most aluminum castings since the section thickness is great enough that the corrosion is of insufficient depth to cause structural failure or penetration. Corrosion, of course, may cause undesirable changes in appearance and is a factor in alloy selection for applications in which appearance or functional factors require the highest resistance.

Corrosion may be of different types characterized by the relationship to microstructural features. Pitting corrosion exhibits little location or rate differences directly related to the microstructure and results in simple, relatively smooth walled cavities. Localized intragranular corrosion may result from preferential attack of the Al solid solution dendrites or by galvanic effects between alloy constituents and the matrix

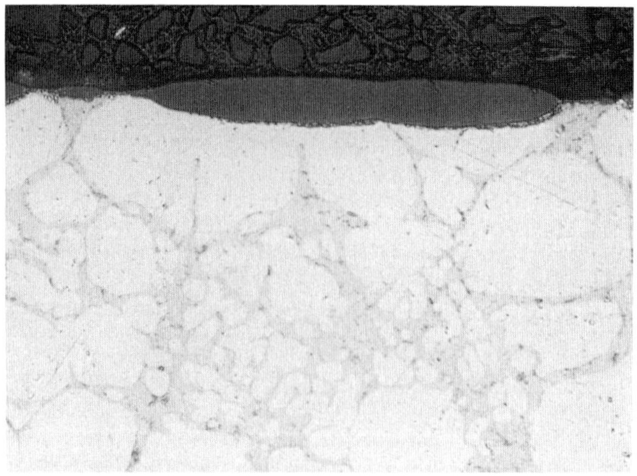

(16-44a) dark field illumination (standard)

(16-44b) bright field illumination

Fig. 16-44. Anodized surface coating on an A356.0–F permanent mold casting. Note that the primary aluminum phase readily anodizes, but the Al-Si eutectic does not. The photo in (a) is slightly out of focus to emphasize the anodized surface cross section (500X).

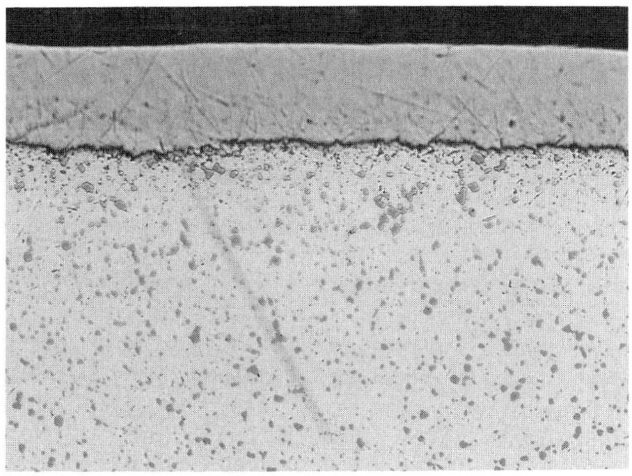

Fig. 16-45. Decorative chromium plating on an A356.0–T6 permanent mold casting (250X).

within the grains. Intergranular corrosion proceeds principally along grain boundaries and, if extensive, may form a network of corrosion that envelops the grains.

Intergranular corrosion and stress corrosion cracking are encountered infrequently and generally are restricted to the heat treated, high-strength 2XX alloys; improperly heat treated 520.0–T4; and alloys of the 7XX series. With good processing, good design, and careful consideration of the environmental conditions and stresses, difficulty with stress corrosion cracking can be avoided. Predominantly intergranular fractures in these susceptible alloys, when associated with a condition producing sustained high tensile stress, however, may raise suspicion of stress corrosion cracking.

Overloading

Castings are sometimes underdesigned, or the strength of the alloy/temper selected is inadequate. This is most likely to become apparent and be rectified in development stages or early service experience. This condition or the occurrence of abnormally high loads in service may result in tensile or bending failure.

Fractures from overloading and exceeding strength exhibit no preferential microstructural path. They are random or erratic, with portions through the aluminum matrix as well as along interdendritic networks and grain boundaries.

Fatigue

Fatigue fractures usually exhibit both fractographic and metallographic features distinguishing them from the fractures resulting from overload. The fracture surface may provide evidence of progressive development, usually from a distinguishable originating geometric condition or an included discontinuity. Early portions may be smoothed or faceted by relative movement of the mating surfaces and exhibit characteristic beach mark or clamshell curved progression lines concentric with the origin.

In metallographic section, the characteristics of the branching or the secondary cracks are frequently most informative because they are not disfigured by wear from relative motion. The fracture paths follow certain crystallographic planes with jogs to parallel planes giving a jagged or lightning-like appearance.

TYPICAL MICROSTRUCTURES OF ALUMINUM CASTING ALLOYS

Representative photomicrographs of the common aluminum casting alloys were illustrated in Figs. 16-1 through 16-39. These alloys are of nominal composition cast at commercial casting temperatures. The major constituents were indicated for the micrographs and were listed in Table 16-3.

Most photomicrographs were made at 250X not only to obtain a general view of the microstructure including the size, distribution and quantity of constituents and the size of aluminum solid solution dendrites, but also to identify the constituents. In practice, other magnifications also are used, ranging from 100X for general examination to 500X for identification of constituents.

Examination of these micrographs will enable metallographers who are not familiar with the structures of aluminum casting alloys to determine if the composition of an alloy specimen is approximately correct by a comparison of the type and quantity of the specimen's constituents with the corresponding micrograph. Thus, for example, the presence of larger needles, β(Al-Fe-Si), than those in the micrograph of alloy 443.0 (Fig. 16-17), may indicate excessive iron content.

ACKNOWLEDGMENTS

AFS would like to thank Richard Hughes and John Jorstad, Reynolds Metals Co., for providing the photomicrographs for this chapter, and John Horwath, AFS/Cast Metals Institute, for his assistance.

Chapter 17

POSTCASTING OPERATIONS

INTRODUCTION

Whether a casting is simple or complex, small or large, formed in molds of metal or sand, custom-made or the product of an automated process line, it is not a finished product when it is removed from the mold. The casting process is complete, but handling of the casting, from shakeout or ejection from the mold to the shipping or assembly department, always includes additional operations. These are secondary to the casting process but are of major significance to the total cost of the casting.

Postcasting (secondary) operations commonly performed on aluminum castings include trimming and cleanup, inspection and testing, finishing, machining, heat treating, impregnating and straightening. Not all of these secondary operations are performed on every aluminum casting; many castings are finished with only a trimming operation to remove gates and flash.

Most secondary operations—especially removal of mold sand, cores, flash and excess cast metal—normally are considered unwanted production steps because they contribute no real value to the finished product. Cleaning operations quite often are given insufficient attention or creative thought because they are the least profitable of all foundry processes.

Machining operations, when required, are necessary to compensate for the inherent limitations of the various casting processes. If a casting process could meet all the design requirements, machining of holes, details and surfaces would be unnecessary.

Postcasting operations do not lend themselves to a high degree of automation. They increase production costs and should be minimized in any way possible. Even in diecasting, a casting process that gives excellent dimensional accuracy and very good as-cast surface finishes, secondary operations often amount to 30–60% of raw casting costs.

The casting producer can reduce these costs by selecting the most effective and efficient equipment with which to perform postcasting operations. Often the cost of postcasting operations can be minimized by the designer of the casting. Mechanical finishing costs, for example, generally can be reduced substantially by the designer and manufacturing engineer who are aware that:

- Finish specifications and general casting texture and condition should be defined.
- To minimize polishing and finishing costs, parting lines should be placed on inconspicuous or readily accessible surfaces. If possible, these operations should be coordinated with planned machining in order to eliminate some of the secondary operations.
- Depths of recesses and heights of protrusions should be held to a minimum.
- Prominent surface designs that might be marred during postcasting operations should be avoided or at least designed with finishing methods in mind.

The Aluminum Association (AA) has published standards that are established for the individual requirements involved in producing a "usable, unmachined casting consistent with normal production practices; reproducibility; reasonable mold or pattern life and maintenance costs; and normal inspection practices, packing and shipping procedures." According to the introduction to these standards, special requirements may be added wherever desired or required, although additional costs may be involved. The principal standards dealing with postcasting operations are found in AA standards, Engineering Series AA-CS E1 through E21.

Heat treating is a crucial postcasting operation that has come to be controlled carefully in modern industry. Demands for aluminum castings with the high mechanical properties obtainable by thermal treatments make the process significant. Thermal treating is discussed in detail in Chapter 15.

CLEANING AND TRIMMING

After castings are taken from the mold, any core or mold refractory that adheres to the casting and the gates and risers must be removed. This removal process is a cleaning operation and varies for different castings. Table 17-1 lists most of the methods commonly used to clean aluminum castings from shakeout to preparation for welding.

Removing Mold Material

Adhering mold or core sand refractory usually is removed by wire brushing or surface blastcleaning. Air, water and mechanical blasting are the surface cleaning methods used. In air blasting, a high-pressure stream of air is used to blast silica sand, aluminum oxide, glass beads or shot (shotblasting) against casting surfaces. In water blastcleaning, high-pressure water is used as the medium of force. In mechanical blasting, various mechanical units are used to propel silica sand or shot against the casting by rapidly rotating paddles.

An example of a surface blastcleaner is shown in Fig. 17-1. This equipment uses a rotating hanger to expose all casting surfaces to the full effect of a centrifugally thrown abrasive. It also is used frequently for final cleaning to minimize texture differences between as-cast and ground surfaces.

Removing Core Material

The removal of sand or aggregate cores from aluminum castings frequently requires special equipment and added operations. Many binders used to make the core shapes are

Table 17-1.
Cleaning Methods Used for Aluminum Castings

Method	Purpose of Cleaning			
	Removing Refractories	**Removing Gates and Risers**	**Finish Cleaning**	**Preparation for Welding or Repair**
Wire Brushing	X			
Air Blastcleaning (sand, shot, glass)	X		X	
Mechanical Blastcleaning	X		X	
Water Blastcleaning	X		X	
Shearing		X		
Band Sawing		X		X
Hack Sawing		X		X
Grinding		X	X	X
Hand Filing			X	X
Rotary Filing			X	X
Polishing			X	
Brushing			X	
Buffing			X	
Tumbling (with or without media)			X	
Thermal Deburring, etc.			X	

relatively unaffected by thermal degradation at the pouring temperatures used for most aluminum casting production. It then becomes necessary to adopt special procedures to ensure complete removal of the core mass. Methods commonly used include:

- mechanical vibration and impact (Fig. 17-2).
- lengthened cleaning cycles in airless blastcleaners especially adapted to accommodate or recycle the molding and core aggregate.
- baking the casting and enclosed core in an oven to destroy the binder system. (This method must be used with some caution because the temperatures used for the operation can have a significant effect on the mechanical properties of the casting. If the operation is combined with a traditional solution heat treatment and quench, the effect of the sand mass on the rate of quench of the casting must be considered because lower mechanical properties could result. With many techniques, the sand is removed before quenching to avoid this problem.)

Fig. 17-1. Castings are loaded into a rotating hanger for blastcleaning.

Fig. 17-2. With a horizontal knockout machine, a casting is clamped by its gating or by a riser between the workhead and the bumper head (heavy castings can be supported by a table or hoist). Vibration of up to 1800 blows per minute breaks down the internal cores. [Photo courtesy of Herschal Products, Inc.]

- techniques that involve ultrasonics or spark discharge under water. (These methods are under commercial development but have not been widely adopted. They are receiving intensive study because of environmental considerations associated with current methods.)

Removing Gates and Risers

Trim presses are used to remove parting line flash, gates and other excess cast metal from die or permanent mold castings when production quantities are large enough to justify the design and construction of the special trim dies required. A trim press can be operated manually or by robot.

During manual operation (Fig. 17-3), an operator places a casting on the lower half of a trim die and begins the trim sequence by pulling two safety bars. A movable platen lowers, and the casting is trimmed. When the press returns to the fully open position, the operator can remove the casting.

In automatic operation (Fig. 17-4), a robot removes a casting from a diecasting (or permanent mold) machine and places it into the trim press. The robot signals the trim press to start the trim sequence. After the casting is trimmed and the press returns to the open position, a part unloader mechanism moves into position below the casting. The casting is ejected onto the unloader tray and removed.

Band saws are used extensively for removing and trimming gates, risers and fins from castings other than die castings. The blades should be rugged—preferably ¾ in. to 1 in. wide and 20 or 21 gauge thick—with 3–4 teeth per in. for faster cutting of aluminum. Skip-tooth blades cut faster and are less inclined to load than those with a regular tooth pattern; however, they do not give as smooth a cut.

Cutting speeds may be from 2500–4000 feet per minute (fpm) or higher. Cutting oil may be sprayed on the blade, but to avoid contamination of the casting, lubrication usually is supplied by occasionally touching the blade with a stick lubricant. For sawing larger castings, automatic hack saws are sometimes used.

Until the advent of the carbide-tipped saw blade, abrasive cutoff methods using reinforced wheels were used to cut aluminum alloys. Most of these alloys were of the medium-silicon and high-silicon alloy families, which previously required severe reductions in blade surface speed (e.g., to 100 fpm) to maintain a reasonable tool life.

Abrasive cutoff also is used in swing frame machines to cut off castings too heavy or awkward to be lifted onto the table of a regular cutoff saw. Typically, the wheels are the breakage-resistant, reinforced rubber-bonded type, 24 grit and Q grade.

Many medium and high-volume permanent mold and sand castings currently are being cut off with high-speed rotary saws that use carbide-tipped blades. Normally, the casting is held in place with hydraulic or pneumatic actuated fixtures,

Fig. 17-3. This 50-ton vertical trim press is loaded manually by an operator. [Photo courtesy of Prince Machine Corp.]

Fig. 17-4. This 50-ton horizontal trim press is loaded by a robot and unloaded automatically by a parts unloader. [Photo courtesy of Prince Machine Corp.]

and the parting line flash and risers are trimmed to tolerances of ±1/32 in. (or better) in a single high-speed pass. With proper design, no additional cleaning or stock removal operations are required on the casting surfaces. An added benefit from this type of cutoff is that the chips from the operation have a higher recovery potential because of their uniform size and the absence of fines and contamination.

Pneumatic chipping hammers are sometimes used to remove gates and risers in locations inaccessible to a saw or portable cutoff wheel. Torch burning also is used but not frequently because aluminum saws so easily that flame-cutting generally is not economical.

Finish Cleaning by Grinding and Filing

After the gates, risers and other unwanted cast protrusions are removed, aluminum castings usually are given one or more finish cleaning operations. These mechanical operations are performed to smooth off the gate and riser areas, to remove any remaining excess metal, and to improve surface finish or appearance.

Snagging and Rough Grinding

Abrasive wheels and belts have become important as foundry tools not only for rough removal of gates and risers, but also for basic finishing of aluminum castings. Stubs, fins, parting lines, other surplus material and rough spots are removed by abrasive wheels, belts or discs. This process is called snagging. Sharpness and the ability to cut free and clear are requisites for the abrasives used on aluminum.

The main types of foundry abrasive grinding products used for finish cleaning are floor or bench-mounted wheels, belts or discs, and swing frame and portable grinders. Most aluminum castings are so lightweight that they can be handled easily on floor or bench stands.

Floor and bench stand grinders are used when the castings are small enough to hold against the grinding wheel (Fig. 17-5). Grinding wheels generally vary from 8–30 in. in diameter and are resinoid-bonded, reinforced around 20 grit, 0 grade, and specially treated to retard loading of the wheel face. The wheels usually are made of aluminum oxide, silicon carbide, blended zirconia alumina or a combination of these. Portable grinders usually are used only on large castings.

For snagging large aluminum castings, belt swing frame grinders are used. The grit size of the coated, resin-bonded abrasive belt will vary from 36–80 and even finer, depending on the stock to be removed and the finish desired. To retard loading, the belt must be lubricated by using the appropriate grease stick or spray mist.

Grinding Wheels

Resinoid-bonded grinding wheels are preferred because they have less tendency to load with particles of the metal, particularly if these wheels are paraffin-treated. A typical wheel for floor stand or portable grinding is 24 grit, medium hardness, resinoid-bonded, paraffin-treated, and made from a blend of zirconia alumina and silicon carbide.

When choosing the most efficient snagging wheel, the following must be considered: 1) the shape, size and amount of metal to be removed; 2) the type, condition and horsepower of grinding equipment; peripheral speed and contact area of the wheel; and 3) the finish required.

The desired rate of stock removal and the required finish are important factors to consider when selecting the grain of the wheel. A coarse-grain wheel cuts faster, but a finer-grain wheel gives a better finish. The strongly bonded, open structure of snagging wheels gives the cutting grains an opportunity to function at their maximum cutting efficiency. These grains penetrate deep, giving snagging wheels their fast cutting action.

The high production levels that must be achieved and maintained to realize the low grinding costs necessary in today's competitive markets have been made possible primarily by the development and use of high-speed wheels. Silicon carbide resinoid-bonded wheels permit cool, fast grinding action without burn at peripheral speeds up to 12,500 sfm.

The portable, reinforced type 27 wheel has achieved an excellent, high-production record in snagging aluminum castings. This development has virtually eliminated grinding wheel breakage caused by strong pressure on the sides of the wheel. Now offered by several companies, the reinforced wheel is manufactured from laminated sheets of bonded abrasives and reinforcing fabric. It possesses great strength and durability, permitting high-speed operation with safety. Grinding with the reinforced wheel is done on its edge or on either side, simultaneously or singly.

Fig. 17-5. Aluminum castings that are small enough to handle can be held against a grinding wheel by hand.

In addition to cutting off excess metal, such as gates and risers, the reinforced type 27 wheel is rugged and rigid enough to double as a general-purpose tool for plowing off stubs, smoothing surfaces and doing any portable tool grinding work on castings. In one particularly versatile application, the wheel is mounted in a light, tubular swing frame, so the wheel and its immediate support can be swiveled 90 degrees in either direction and locked in place. Risers of any size and in either plane are removed readily with this equipment.

Considerable light snagging is handled with mounted abrasive points that are small wheels of assorted cylindrical, spherical and conical shapes. These abrasive points are made of much harder grains than the average grinding wheel or belt because their diameters are very small and the spindle speeds do not result in a very high peripheral speed. A typical specification for abrasive points to be used on light snagging of aluminum is size 36 grain aluminum oxide, T hardness, 4 structure, resinoid-bonded.

Abrasive-Coated Belts

Abrasive-coated belt grinding is an efficient method for the light snagging of castings, particularly die castings (Fig. 17-6). Advocates of the abrasive belt system claim a number of savings inherent in this method:

- improved production schedules;
- less downtime to change setup or belt;
- elimination, in many cases, of subsequent or intermediate grinding operations;
- less rework on the product;
- cooler working.

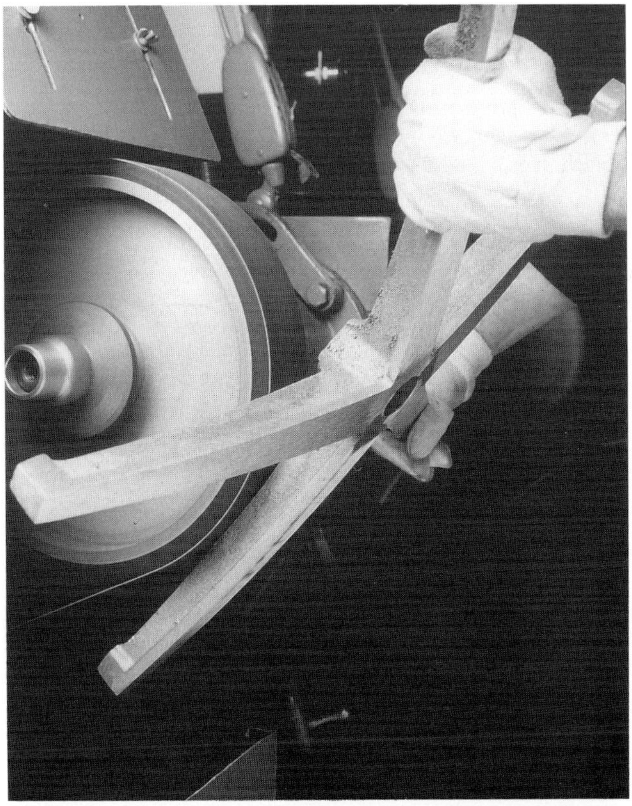

Fig. 17-6. An aluminum casting is ground using an abrasive-coated belt. [Photo courtesy of Norton Co.]

For this form of grinding, the endless, coated, abrasive belt is driven by a wheel or pulley fixed to a polishing lathe; an idler wheel is located at the other end. The abrasive belt has a much greater cutting area than the edge of a grinding wheel and, consequently, cuts cooler at high speeds.

Downtime for belt changing is almost negligible because the old belt is slipped off and the new one put on merely by moving the idler pulley forward slightly to relieve tension. There are no flanges, spindle nuts or guards to be removed.

As with grinding wheels, the abrasives used most commonly on coated grinding belts are silicon carbide, aluminum oxide and blended zirconia aluminas. For rough grinding and snagging operations, a belt with relatively coarse grit (60 to 36 grain size) is used. Cloth belts, 3 in. wide and coated with 36 grit, are used extensively for rough metal removal for aluminum and sand castings. The grain size numbers coincide with the mesh number of the grit grading screen—the number of holes per square inch of screening. The larger the grain size, the smaller the number.

There are two methods of coating abrasives to a backing—with a closed coat or with an open coat. Open coating, where about 50% of the surface is covered, is used for aluminum. The voids between the grains prevent loading.

Deburring and Finish Cleaning

Aluminum castings are being used more in the aerospace industry and in other applications where tolerances are critical and the finish must meet exact requirements. A high-speed die grinder, equipped with a nonwoven nylon abrasive that disc removes any light burrs and sharp corners without disturbing critical dimensions, often is used.

On less critical parts, where stock removal is heavier, small coated abrasive discs, available in various sizes and shapes and driven by high-speed portable power tools, make quick work of deburring and removing surface imperfections.

Rotary Files

Rotary files (carbide cutters or burrs) are used in finish cleaning of aluminum. A rotary file is a hardened steel grinding wheel. Rotational speeds of standard rotary files, which are usually 1/4 to 1 5/8 in. in diameter, vary from 2500–10,000 rpm. Both rotary and hand files are used frequently to clean areas that cannot be reached by the standard types of grinders.

Rotary files are preferred for light-duty cleaning operations, such as trimming light flash or veins in accessible areas. Although the initial cost of a rotary wheel is higher than that of a mounted wheel, rotary wheels resist loading and remove more metal in a shorter time. When used with stick lubricants or light oils, they have a long cutting life and can be resharpened for additional use.

Rotary wheels are most effective when operated against the direction of their rotation, or in an upcut manner. They should be worked freely to achieve maximum stock removal. In general, coarse cutters (used for the heaviest removal) operate at slightly lower speeds; fine cutters are used at higher speeds. In both cases, a liberal application of lubricant promotes the most efficient cutting conditions.

INSPECTION OF CASTINGS

Castings are inspected to ensure that they conform to customer requirements. Postcasting inspection usually falls into three categories: 1) preliminary or "hot" inspection; 2) intermediate inspection; and 3) final inspection.[1] Intermediate inspections may follow specific operations, such as hardness checks after heat treatment or visual checks after grinding.

Casting producers cannot afford to spend time and money cleaning defective castings; hence, a visual inspection station at the point where castings are removed from the mold is justified. Castings that have obvious defects can be sent for repair or be scrapped. Those castings that are visually acceptable are sent to the cleaning area. When a casting is clean enough to reveal surface detail, it should receive a closer visual inspection.

Radiographic and other critical nondestructive testing methods are applied as required by customer specifications, especially for castings designed for severe service performance. Casting customers sometimes require radiographic inspection as assurance of the casting's quality, but casting producers now often use radiographic examination and other critical tests (even when not required by the customer) as a control indicator to maintain the quality of their castings.

Many foundries are depending on statistical quality control methods to inspect castings. In a statistical sampling procedure, only a predetermined percentage of the castings is subjected to critical inspection. This method saves time and money and often is as adequate as checking every casting in the production run.

Inspection of die castings usually consists only of chemical analysis, dimensional checks, leak tests and visual examination to check for surface defects. The inspection of permanent mold, sand and other castings usually includes the following procedures:

1) Hot inspection to detect defective castings quickly so corrective process control steps may be taken immediately, before more defective ones are produced. This inspection also reveals scrap castings before the expenses of trimming and other processing are incurred.
2) Rough dimensional check and visual examination after cores, gates and risers have been removed. A more careful examination can be made at this stage. Again, the expense of further processing of defective castings can be avoided. Castings may be sand-, shot- or grit-blasted before making this inspection.
3) Foundry control radiographic and penetrant inspections.
4) Leak tests.
5) Final inspection after castings are completely cleaned, heat treated, straightened, machined or welded.

Before a workable quality control program can be set up for a given casting, suitable standards must be established. These standards will depend on the type of casting, the engineering specifications and, often, the casting's end use. For many applications, the requirements are fairly simple—the casting needs only to be reasonably free from foundry defects to be acceptable. In other applications, the requirements for the casting may be much more exacting.

Rigid quality standards are incorporated in various association and government specifications controlling the acceptance of aluminum alloy castings for military, federal and commercial use—particularly those used in aircraft and other highly stressed equipment. Usually, castings are divided into three groups, with different inspection standards for each.

The first class of standards includes highly-stressed castings, the failure of which would result in serious consequences, such as loss of life and property. Such castings usually require 100% x-ray and penetrant inspection.

The second class of castings are those that are moderately stressed. The failure of these castings would not involve loss of life or serious interruption in the operation of the equipment. Such castings may be produced under x-ray control in which a certain percentage of castings would be x-rayed after the quality had been established by initial runs in the foundry.

The third class of castings includes normal commercial castings under relatively low stress in nonstructural applications. Their failure would not result in serious consequences.

The purpose of any inspection program is to determine whether the castings produced meet the established standards of quality. As already indicated, the inspection method and the standards of quality depend on the requirements of the casting and its application.

Regardless of the casting methods used or the care that is exercised, it is nearly impossible to produce castings without some minor defects. The establishment of quality standards for any given casting and the decision of whether or not a given casting meets these standards requires careful judgment and interpretation.

Nondestructive Testing

Inspection techniques may be classified into two broad groups: nondestructive methods and destructive methods. Because the casting is destroyed in destructive methods, these processes can be used only on a relatively small percentage of the castings produced. This type of inspection includes the attainment of tensile property data on specimens cut from the castings, sectioning to determine soundness, performance of breakdown tests and simulated service tests.

Nondestructive testing methods include visual and dimensional inspections, fluorescent powder and dye penetrant tests, radiographic (x-ray) inspections, leak tests, ultrasonic testing, and mechanical and physical property tests from separately cast samples.

Visual Inspection

Visual inspection is the examination of castings for surface defects, molding errors and nonfilling. Casting defects such as sand holes, excessively rough surfaces, surface shrinkage, blowholes, misruns, cold shuts and surface dross usually can be detected visually. Optical magnifiers often are used to help detect defects in casting surfaces. In most instances, a visual inspection most readily can be made after the castings have been shot-, grit- or sand-blasted.

Dimensional Inspection
Dimensional accuracy of castings not only is an indicator of process control, but also is a requirement because customers will not accept the high cost of machining out-of-tolerance or inconsistent castings. Casting dimensions are checked with dial indicators, micrometers, vernier calipers and height gauges. These checks may be aided by the use of "go-no go" gauges, jigs and fixtures.

The introduction of the coordinate measuring machine (CMM) has greatly improved the speed and accuracy of dimensional inspections of castings (Fig. 17-7). Computerization of the CMM has eased repetitive tasks and has allowed for its use as a statistical tool. In addition, it significantly reduces the man-hours required for inspection.

Fig. 17-7. An aluminum cylinder head is held in a fixture for measurement by a coordinate measuring machine.

Fluorescent Powder and Dye Penetrant Inspection
Minute surface cracks, pores and discontinuities may be detected by immersing castings in a suspension of fluorescent powder in warmed penetrating oil. The suspension penetrates cracks and pores, and excess suspension is washed away in warm water. The casting is dusted, sprayed or chalked with a drying powder that draws the suspension from the cracks where it fluoresces and is visible under ultraviolet light.

Fluorescent powder inspection detects only surface cracks and flaws, but it does so more effectively and more economically than radiographic examination.

In the dye penetrant method, a colored dye (instead of fluorescent powder) is used in suspension in the penetrating oil. After cleaning, the colored dye in cracks is visible under normal light, but defects are not seen as readily as under the ultraviolet light in the fluorescent powder method.

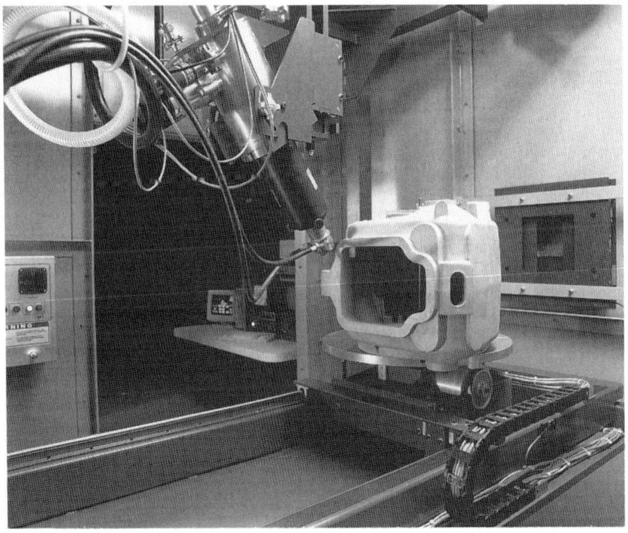

Fig. 17-8. A large aluminum casting is held by a manipulator inside a room-sized lead enclosure. The x-ray tube is shown poised above the casting. The programmable, four-post manipulator can handle castings as heavy as 150 lb. [Photo courtesy of IRT Corp.]

Radiographic Inspection
Internal shrinkage, gas porosity, internal dross films and particles of foreign inclusions that usually are not detectable by visual examination can be detected by radiographic inspection. Internal defects can be detected (within certain limitations) depending on the design of the casting; the size, shape and location of the defect; and the radiographic techniques used.

Film radiography involves exposing a casting to radiation from the x-ray tube. Part of the radiation is absorbed by the part,[2] and the remainder exposes the radiographic film. A 100 kv x-ray tube is adequate for x-raying aluminum castings up to two inches thick.

Correct radiographic techniques are the best nondestructive methods for determining the presence of internal defects, such as shrinkage and foreign inclusions. Experience, which can be gained through practice under initial metallurgical or qualified metal radiological supervision, is necessary to interpret the results of radiography.

Filmless or real-time radiographic inspection techniques (Fig. 17-8) allow a foundry to view an x-ray image on a video screen. This technique is being used effectively as a quality control tool as well as an inspection technique.

Advanced techniques—such as computerized axial tomography (CAT scanning), originally used in the medical field—are being used to develop three-dimensional computer imagery of castings and their internal soundness.[3] In the computed tomography (CT) technique,[4] images of a thin cross section of a casting are produced using a fan-shaped x-ray beam. As the casting is rotated to provide different viewing angles, a computer mathematically reconstructs an image of the cross section. Three-dimensional images are created by adding vertical part movement and image processing software (Fig. 17-9).

Leak Tests
A casting that must be pressure-tight or leakproof often is tested by sealing openings in the casting and pressurizing it with air, an inert gas or water.

Fig. 17-9. An operations technician analyzes a reconstructed cross-sectional image on a CT scan system. [Photo courtesy of Boeing Aerospace & Electronics.]

If hydrostatic pressure is used, leaks are indicated by the seeping of water through the casting wall. If air or gas pressure is used (the more sensitive leak testing methods), the pressurized casting is immersed in a tank of clear water. If bubbles appear, the casting is leaking.

Ultrasonic Testing

In this method, beams of high-frequency acoustic energy from a sending unit are introduced into the casting. The energy beam travels through the casting until it hits an interface or defect. This interface or defect reflects portions of the energy, which are collected in a receiving unit and displayed on a screen. This testing method can determine wall thickness as well as the position and size of defects.

Mechanical and Physical Property Tests

The physical properties of a casting can be determined by metallographic examination—photographs, macrographs and micrographs. The microscope can resolve many casting problems. Grain size, nonmetallic impurities, submicroscopic pinholes and the effects of thermal treatments can be observed by metallographic examination. Chemical analyses of aluminum castings can be made by the wet method or with an optical emission spectrometer.

Mechanical tests include:

- tensile tests to determine ultimate strength, yield strength and elongation. Both cast test bars and bars excised from castings are used;
- bend, notch, impact and transverse tests to check ductility and shock resistance;
- hardness tests, which are reasonably accurate measurements of the ductility and tensile strength of the cast metal;
- fatigue and fracture toughness tests to determine the durability of the material.

MACHINING ALUMINUM CASTINGS

Aluminum castings are machined on standard metalcutting machine tools, much the same as is cast iron, but at higher speeds and feeds. Machinability varies widely among the aluminum casting alloys and is greatly dependent on the composition and temper of the alloy and, to a lesser extent, on the casting process. All aluminum alloys can be machined readily when the proper tooling is used.

In general, aluminum alloys containing copper, zinc and magnesium machine most easily. Alloys containing silicon as the major alloying constituent are more difficult to machine and usually require carbide tools. Alloys in this hard-to-machine group are 336, 360, 380, 413, 443 and all of the hypereutectic Al-Si alloys.

The machinability ratings given for the aluminum casting alloys in Tables 17-2 to 17-4 are based on characteristics such as cutting ease, surface finish, chip form and especially tool wear.[5] For example, hypereutectic Al-Si alloys often are rated as having only fair to poor machinability because tool wear is poor, even though these alloys provide superior chip characteristics and excellent surface finish. These and similar ratings from other sources[6] are offered as useful comparisons and general guides and are based on the combined results of research and practical experience.

Tools and Cutting Speeds

For best results, aluminum castings should be machined with cutting tools specifically designed for aluminum. For occasional machining of aluminum castings, tools made for machining cast iron and steel may be used, but it is not advisable because the finish of machined surfaces usually is not as good and because machining time is increased.

In many instances, aluminum castings may be machined at the maximum speeds and feeds obtainable from the machine, but the following factors all influence speeds and feeds and must be given full consideration:

- the alloy and subsequent heat treatment of the casting;
- the configuration and rigidity of the casting and distribution of hold-down or clamping points;
- the feed and depth of cut;
- the cutting fluid, if used;
- the cutting tool material;
- the clearance and rake angles of the cutting edges;
- the shape and size of the tool, which affect its rigidity in the machine (e.g., a large-diameter flycutter compared to a 6-in. dia multitooth face mill; or an 8-in. long, 1-in. dia boring bar compared to a 4-in. long, 2-in. dia bar);
- the power of the machine; its range, speeds and feeds; its overall rigidity and condition;
- the anticipated (planned) cutting life of the tool before replacement or regrinding is necessary.

All cutters that are used for machining aluminum alloys should have cutting edges that are keen, smooth and free of burrs and grinding wheel marks. High-speed steel tools can be used, on occasion, for low-volume production runs, but when machining aluminum alloys on a continuous production basis, cemented-carbide tools are recommended.

Table 17-2.
Machinability Ratings for Aluminum Permanent Mold Castings

Alloy	Temper	Rating*
222.0	T551, T61, T65	Excellent
242.0	T571, T61	Very good
296.0	T4, T6, T7	Good
308.0	F	Good
319.0	F	Good
332.0	T5	Fair
333.0	F, T5, T6, T7	Good
336.0	T551, R65	Poor
355.0	T51, T6, T62, T71	Good
C355.0	T61	Good
356.0	T6, T7	Good
A356.0	T61	Good
357.0	T6, T61	Good
443.0	F	Fair
512.0	F	Excellent
513.0	F	Excellent
535.0	F	Excellent
711.0	F	Excellent
850.0	T5	Excellent
851.0	T5	Excellent
852.0	T5	Excellent

* Machinability rating is based on a range of all cutting operations.

Table 17-3.
Machinability Ratings for Aluminum Sand Castings

Alloy	Temper	Rating*
222.0	T	Excellent
242.0	T2, T571, T77	Very good
295.0	T4, T6, T62	Very good
308.0	F	Fair
319.0	F, T5, T6	Good
355.0	T51, T6, T7, T71	Good
C355.0	T61	Good
356.0	T51, T6, T7, T71	Good
A356.0	T61	Good
357.0	T6	Good
358.0	F	Good
443.0	F	Poor
511.0	F	Excellent
512.0	F	Very good
514.0	F	Excellent
520.0	T4	Excellent
535.0	F	Excellent
710.0	F	Excellent
713.0	F	Excellent

* Machinability rating is based on a range of all cutting operations.

Table 17-4.
Machinability Ratings for Aluminum Die Castings

Alloy	Rating*	Alloy	Rating*
360.0	Fair	A413.0	Fair
A360.0	Fair	443.0	Fair
380.0	Good	513.0	Excellent
A380.0	Good	515.0	Very good
384.0	Good		
413.0	Fair	518.0	Excellent

* Die castings generally are used in the as-cast (F) condition. Machinability rating is based on a range of all cutting operations, but machinability can be improved by reducing internal stresses in an intermediate thermal treating operation (see alloy tables in Chapter 4).

Cemented-carbide tools are preferred and are almost a necessity when machining any of the aluminum alloys that contain silicon as the major alloying constituent. Because of the better results, longer tool life and higher speeds possible, many machine shops have found it more economical to use carbide tools for machining all aluminum casting alloys.

In certain applications, diamonds (single crystal or polycrystalline) or diamond-tipped carbide inserts can be more advantageous than plain carbides. Diamond-tipped carbide has the hardness of a diamond and the strength of carbide. A diamond will have a cutting life of up to 30 times that of standard carbide and can be used for both roughing and finishing machining operations. Coated carbides (carbide tools coated by various vapor or ion methods with nitrides or even diamond) are becoming increasingly important substitutes for diamond when complex-shaped tools are required for cutting hypereutectic Al-Si alloys or MMCs.

Cutting speeds of 5000 surface ft/min (sfm) are not uncommon. If all factors influencing machinability approach optimum, this figure can be attained or exceeded. However, manufacturers of milling cutters specify 2500 sfm and faster for inserted carbide tooth cutters. This is considered a realistic figure for the general run of castings used by industry.

The feeds per tooth of milling cutters range from 0.010–0.025 in. The feed actually used depends on all the factors affecting machinability. Feeds in the range of 0.010–0.020 in. are used more frequently when milling cast machine parts. The higher feeds and speeds are used for rapid stock removal on relatively plain, rigid parts such as flat plates that have a continuous surface.

Milling cutters and other tools, such as boring bars, drills, taps and reamers designed for machining aluminum, have rake and clearance angles different from those used for machining cast iron. For example, teeth on milling cutters designed for aluminum generally have a relatively large positive radial and helical rake angle, while cutters designed for cast iron usually have rake angles ranging from only slightly positive to as much as seven degrees negative. The clearance angles specified for aluminum cutters are from 1.5–2 times as great as those used for milling cast iron. Radial primary clearance angles of 7–9 degrees and a side clearance of seven degrees are usually specified for aluminum cutters.

Many of the disposable carbide insert milling cutters in use can be used to mill different materials by changing the insert's geometry and carbide grade. When using a new cutter or insert, caution should be used; starting speeds and feeds should be conservative rather than aggressive.

Several sources of machining speed and feed data are available.[6] These publications give details of materials; machining process; and type of tooling including high-speed steel, uncoated and coated carbide, ceramics, and diamond (see Tables 17-5 through 17-12 in the appendix to this chapter).

Die castings milled with carbide cutting tools can be machined at speeds from 500–1800 sfm. Sand and permanent mold castings can be machined at higher speeds, such as 1300–2500 sfm. Because of possible sand inclusions, speeds used for machining sand castings generally are lower than those used for permanent mold castings.

Average drilling speeds usually range from 150–300 sfm. This relatively low speed range is due largely to the limitations of available spindle speeds. For example, a ⅜-in. dia drill operating at 1000 sfm would require a spindle speed of over 10,000 rpm, which is seldom possible.

Drills of high-speed steel are recommended, and for deep holes, polished flutes are preferable. Drill life when drilling more abrasive alloys, such as hypereutectic Al-Si or MMCs, often is improved by using drills with a parabolic flute configuration. Drills tipped with polycrystalline diamond also are useful with these alloys. Cutting fluids generally are used.

Tapping speeds are even lower than speeds used for drilling. For example, it would not be feasible to tap a ⅜-in. dia, 16 threads/in. hole, ⅝-in. deep at 200 sfm (2000 rpm). In this example, it would take only 0.30 sec for the tap to travel the ⅝ in. distance before it would have to be stopped and reversed. Chrome-finished, high-speed stem taps designed for aluminum are recommended, and cutting fluids are required.

REPAIRING CASTINGS

The basic purpose of all casting practices is to produce sound castings within established specifications. A number of factors in each method may determine whether or not a part can be cast without critical defects. Because of such factors, optimum qualities cannot be obtained in every unit cast, even though it is the desire of every casting producer to make the best possible product with the fewest defective castings.

Quality standards for a casting are determined in advance by the end use of the casting. Specifications are established, and the producer meets the requirements, improving, if possible, the overall qualities by its techniques or practices.

Castings that have defects often can be salvaged. Methods of reclaiming castings include straightening, welding, peening, impregnating with sealants and hot isostatic pressing (HIP). However, salvage of castings through secondary operations should not be relied upon to compensate for poor-quality work. It is far better to control casting procedures and part handling carefully to prevent defects than to depend on salvaging methods, which often are expensive in comparison with proper process control.

Straightening Castings

Straightening often is necessary because of stresses and strains set up in different sections of the casting. Warping and distortion may be attributed, in part, to handling during and after removal of the part from the mold.

The timing of handling is important because the aluminum alloy may not have had time to cool sufficiently before removal from the mold. If castings are handled carelessly from the shakeout racks or storage space while they retain heat, the chances for distortion increase. This cause of defects may easily be corrected or eliminated. Obviously, if more careful handling will eliminate distortion, then the extra effort should be taken.

It is not always possible to prevent undue distortion, even if care has been taken at shakeout and in storage or racking. Distortion may occur when heat treatable alloys are heat treated. If distortion occurs too frequently, changes should be made in the placement and distribution of the castings in the heat treating furnace. Improper air circulation within the furnace and around the castings may cause distortion. If the castings are solution heat treated, straightening should be done before aging.

Fortunately, many warped or bent aluminum castings can easily be straightened by mechanical or manual methods. Usually lead or rawhide mallets can be used to straighten castings that are properly supported and held in a fixed position. Hydraulic jacks and even hydraulic presses are used to straighten large castings. If cracks occur in straightening, the casting should be scrapped. If counter stresses are set up by the cold work, the casting may be weakened structurally. In all straightening operations, critical areas of the casting must be considered.

Peening

Surface irregularities and small holes and leaks often can be covered or closed by pneumatic, mechanical or manual peening with a suitable tool. To be permanently effective, the peening action should work the metal together below the surface, rather than merely smear the metal over the outside of the pores. However, a casting should not be peened so deeply that the section thickness is reduced to less than the minimum allowed by the specifications.

Successful leak-proofing of certain die castings has been accomplished by overall peening of the surface using shotblasting. The effect is only skin deep but is satisfactory for certain purposes if the peened areas are not machined after the shotblasting and if the operating pressures are low.

Foundry Welding

No matter how closely a casting process is controlled, deficiencies sometimes occur. These deficiencies may be sand holes, porosity, saw cuts, grinding errors or other similar defects. These small process defects in aluminum castings often are repaired by welding.

Several factors determine the feasibility of foundry welding of castings. The ultimate function of the part and the nature and accessibility of the defect have to be considered. In structural applications or high-stress areas, permission may have to be obtained from the customer before welding is permitted. Surface appearance can be affected, and anodized surfaces may be off-color. It has been shown that properly welded castings can be satisfactorily heat treated when the process is controlled properly.

Foundry welding is a difficult process that presents greater problems to the welder than does weld fabrication. Aluminum is a highly reactive metal and instantly oxidizes when exposed to air. This tenacious oxide layer made welding rare until the development of gas tungsten arc (GTA) and gas metal arc (GMA) welding.

Aluminum oxide melts at 3725F (2052C), more than three times the melting temperature of aluminum, and can lead to inclusions in the weld. The oxide surface is porous, which makes it trap moisture and other contaminants that can lead to weld porosity. Weld porosity is caused principally by hydrogen coming out of solution as the weld pool solidifies. Hydrogen contamination usually comes from oil or water.

Many factors influence the welding of castings,[5] including casting design, equipment used, welding process, shielding gas, welding environment, edge preparation, heat control, alloy properties and casting temper. The casting designer should give consideration to the following factors: 1) all surfaces should be accessible to the welder, finisher and inspector; and 2) a criteria for weld repair of defects should be established.

The selection of process and equipment is important to successful welding. Gas welding using oxyhydrogen or oxyacetylene flame, while feasible, is no longer widely used. Most repair welding operations in aluminum foundries are with GTA or GMA welding processes, with the GTA welding process used for the bulk of foundry repair operations. The GMA welding process should be used where there is good access and large volumes of metal are required. GMA welding has a deposition rate of 6+ lb/hr, while GTA welding has rates of 2–3 lb/hr.

For aluminum welding, equipment with 300 amp solid state power supply and both AC and DC capability is recommended. This would offer several welding process capabilities, such as: 1) pulsed GTA; 2) GTA DC electrode negative (DCEN); 3) GMA DC electrode positive (DCEP); and 4) AC asymmetrical waveform.

The pulsed GTA welding process controls distortion in thin-walled castings by minimizing heat input. The GTA DCEN provides a deep penetrating arc and is best suited for materials over ¼ in. thick. The GMA DCEP process provides cleaning action from the arc and produces excellent welds.

The AC GTA process with asymmetrical waveform is used widely because of the cleaning action of the reverse portion of the wave. Both argon and helium gases, either alone or mixed, are used as shielding gases for the arc welding processes. The effects of theses gases for the GTA welding of aluminum castings can be seen in Table 17-13.[7] Care must be taken to make sure that the shielding gases are not contaminated because this can cause porosity or inclusions in the

Table 17-13.
Shielding Gases for GTA Welding of Aluminum Alloys

Shielding Gas	Welding Characteristics
100% helium	Requires high gas flow rates.
	Provides good wetting action with DCEP GTA.
	Poor arc ignition characteristics.
	Produces higher arc voltage than argon for a given current.
	Produces deep, penetrating arc.
	Poor cleaning action.
100% argon	Requires moderate flow rates.
	Produces best cleaning action with AC GTA.
	Good arc ignition.
	Produces lower arc voltage than helium for a given current.
	Weld bead tends to broaden and exhibit less penetration than with helium.
90% argon - 10% helium	Has increased penetration over pure argon.
helium + 5–25% argon	Improves arc ignition.
	Improves arc stability.

weld. The environment in weld areas is important because drafts—even a moderate breeze—can cause loss of shielding gases and result in dirty welds.

The area to be welded must be prepared properly to receive the weld. The following should be observed to ensure a quality weld.

- The defect must be completely removed, and the surface of the groove must be sound metal.
- The groove must be fully accessible to the welder.
- The surface of the groove should be clean and bright with no evidence of smeared metal.
- Feather edges should be avoided.
- Diameter of holes drilled to remove a defect should be equal to the depth. Deeper holes should be beveled.
- When several defects are close together, a single groove should be used to connect the defects.

The use of preheating before welding to minimize distortion is not as effective as proper weld settings, interpass temperature control and weld sequencing. Preheating can be used to reduce stresses induced from solidification of the weld bead, especially in thin sections. Preheat temperatures in the range of 150–300F (66–149C) should be used. For welding castings, the choice of filler metal usually is limited to the same alloy chemistry as the casting.

The heat treatable alloys may be welded in any temper. As-cast (F) or annealed (O) tempers are preferred because alloys are more ductile in these tempers. Castings should be welded prior to solution heat treatment to achieve full mechanical properties in the weld area.

Table 17-14 is a troubleshooting list that can be used to analyze welding defects in order to reduce rewelding.[8]

Impregnation

Impregnation has been used for many years by foundries as a method to seal castings where there are problems with microporosity. The three types of microporosity generally associated with impregnation are shown in Fig. 17-10.[9] Impregnation is generally thought of as a repair procedure for castings, however, in processes such as diecasting and permanent mold casting, where porosity can be a problem, it is sometimes used as part of the process.

The purpose of impregnation is to fill the microporosity with a material that is chemically resistant and can withstand the conditions (such as temperature, pressure and stress) where the casting will be used. Pressure impregnation techniques that have been used in the past result in incomplete filling of the voids. This happens because air is trapped in the porosity, which, after the release of the pressure, expands and pushes the sealant out of the pores.

Vacuum impregnation techniques have overcome most of this problem by removing the air prior to introducing the sealant. There are two vacuum impregnation techniques in use: the wet-vacuum and dry-vacuum processes.

In the wet-vacuum process, castings are immersed in the sealant in a pressure vessel and subjected to a high vacuum to remove air. The vacuum is then released and the vessel sometimes pressurized, forcing sealant into the porosity.

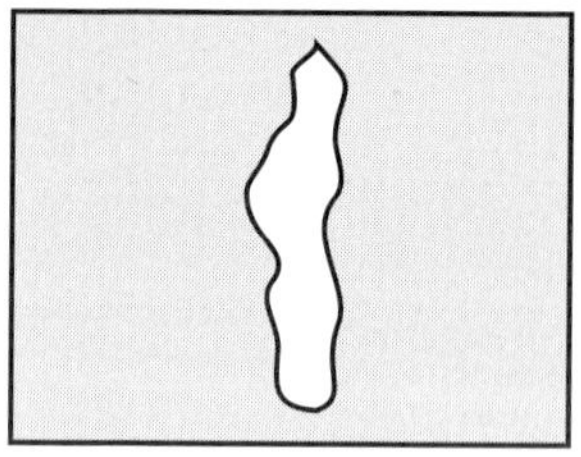

Enclosed porosity is not a problem unless it is uncovered by machining operations.

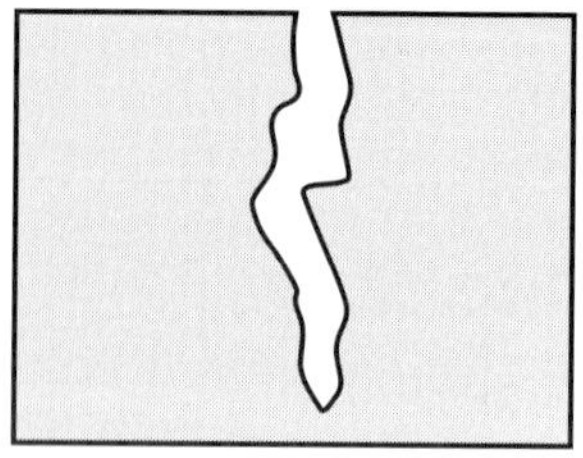

Blind porosity encourages internal corrosion, leading to "spotting out" of plating and "blow out" of paintwork.

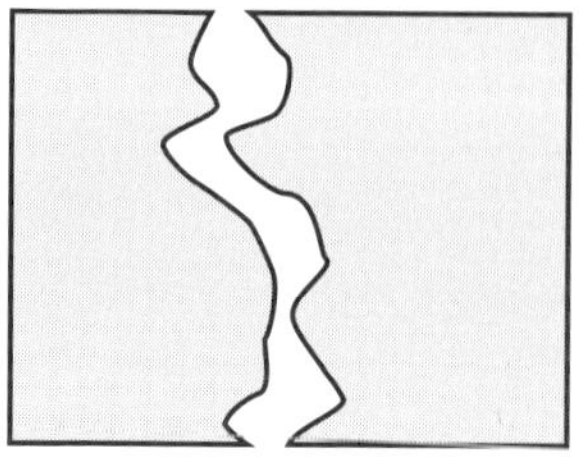

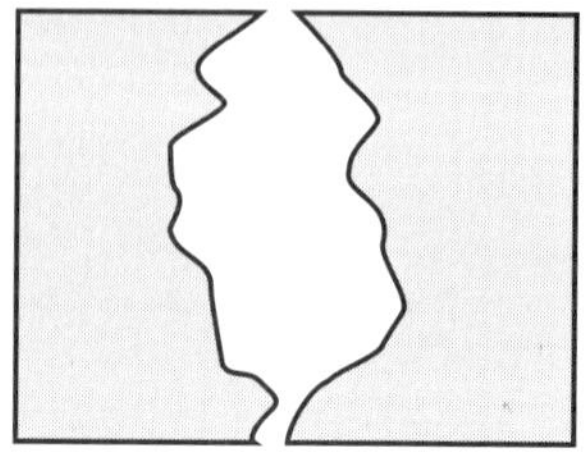

Through porosity allows gas or liquid to leak through the casting. The size of porosity need not be the same as what is seen on the surface. Porosity often is an irregular phenomenon within the casting.

Fig. 17-10. Three types of microporosity.

In the dry-vacuum process, castings are subjected to the vacuum before they are immersed in the sealant. After immersion, the vacuum is released to the atmosphere, or the vessel may be pressurized to allow sealant to enter the porosity (Fig. 17-11).

Castings should be thoroughly cleaned and degreased prior to impregnation. After the impregnation step, the castings

Fig. 17-11. Batch impregnation of aluminum castings. [Photo courtesy of Ultraseal America, Inc.]

Table 17-14.
Troubleshooting Aluminum Weld Defects

Defect	Possible Causes	Suggested Action
Tungsten inclusion	"Spitting" (older power supply).	Replace power supply.
	"Spitting" (new power supply).	Change electrode to 2% thor.
	Welding current too high.	Reduce current. Increase electrode diameter. Change electrode type.
	Damaged tungsten.	Replace electrode.
	Excessive electrode extension.	Reduce electrode extension.
	Inadequate gas shielding.	Increase gas flow. Reduce drafts from open doors, fans, vent tubes, etc.
Oxide inclusion	Insufficient cleaning of repair groove and surrounding weld areas.	Thoroughly wire brush groove before welding and after each pass, and wipe clean.
	Oxide layer on weld wire.	Abrade wire and wipe clean. Cast strips should be filed and wiped clean.
	Sharp edges on weld groove.	Break edges of edge penetration.
Porosity at fusion zone	Hydrogen in base metal.	Improve melt degassing. Reduce sodium additions. Weld with DCEP 100% He.
Round (in weld)	Inadequate shielding.	Increase gas flow. Eliminate drafts. Reduce electrode extension.
	Welding current too high.	Reduce welding current.
	Penetrant oils, lubricants.	Completely remove defect. Wipe surfaces clean with solvent. Eliminate lubricants from the weld area.
	Contaminated or no shielding gas.	Check gas lines for damage, loose connections. Replace gas bottles.
	Travel speed too high.	Reduce travel speed.
	Torch angle too flat.	Correct torch angle.
	Moisture on edge preparation.	Clean surfaces. Preheat.
	Condensation in torch head.	Eliminate water flow during downtime.
Undercutting	Welding current too high.	Reduce welding current.
	Arc too long.	Reduce the gap between the electrode and the part.
	Travel speed too high.	Gradually reduce travel speed until the defect is eliminated.
	Insufficient amount of filler metal.	Add more filler metal.
Lack of fusion	Welding current too low.	Increase welding current. Reduce filler wire additions.
	Travel speed too high.	Reduce travel speed.
	Poor joint preparation.	Make all surfaces accessible to arc.
	Improper torch angle. Weld puddle in front of arc.	Change torch angle.
Overlap	Slow travel speed.	Increase travel speed.
	Welding current too low.	Increase welding current.
	Too much filler metal.	Reduce filler wire addition.
	Incorrect torch angle.	Change torch angle.
Drop through	Slow travel speed.	Increase travel speed.
	Welding current too high.	Reduce welding current.
	Trying to close too wide a gap.	Build up edges of repair groove and reduce gap.
	Too much heat built up in the part.	Reduce interpass temperature.
Cracking crater	Improper breaking or arc.	Gradually reduce arc current.
Hot cracking	Weld bead too small (under bead).	Preheat base metal.
	Too high current.	Add more filler wire to increase weld size.
	Interpass temperature too high.	Reduce current. Cool between passes, sequence welds.
Cold (delayed) cracking	High joint restraint.	Preheat. Remove thin web between adjacent repair weld areas.

must be drained of excess sealant and washed to remove any sealant remaining on the surface. The next step is the curing of the sealant, which takes from 5 min to 24 hr under temperatures up to 250F (121C), depending on the sealant material used.

There are three types of sealants in use for impregnation:[9] 1) aqueous sodium silicate; 2) styrene polyesters; and 3) methacrylate sealants. Each has its own distinct advantages and disadvantages.

Impregnation is a cost-effective way of filling microvoids. In addition to preventing leakage and improving pressure tightness, this method eliminates bleedout and blemishes in plating and surface conversion coatings, white spotting of anodizing, and blistering of paint finishes due to air expelled on curing.

Hot Isostatic Pressing

Hot isostatic pressing (HIP) is a process that can be used to improve the internal soundness or integrity, increase the density, and improve the mechanical and fatigue properties of castings.[10] It has been recognized as a viable tool to the industry and has become part of the processing specifications for critical applications such as structural aerospace and automotive castings.

The HIP process (Fig. 17-12) requires that castings be placed in a pressure vessel and the temperature be raised to near the solidus, as in solution heat treatment. This is done in an inert atmosphere. The pressure in the vessel is increased to approximately 15,000 psi. The external pressure and temperature cause plastic deformation, which collapses the internal porosity.

The amount of hydrogen removed by outgassing during HIP is insignificant. Where high levels of hydrogen are present in the castings, it is possible that reheat treatment can cause reformation of objectionable levels of internal porosity. This is not the case at more normal hydrogen levels.

The HIP of castings has proved capable of substantially eliminating microporosity that formed as the result of hydrogen precipitation and internal shrinkage formed during solidification. The process makes possible the acceptance of castings that might otherwise be scrapped. This is especially true for castings that are subject to radiographic inspection. Some of the properties benefited by HIP are ductility, low and high cycle fatigue, tensile and yield strength, and high temperature creep strength.[11]

Fig. 17-12. Aluminum castings are subjected to HIP. [Photo courtesy of Howmet Turbine Components Corp.]

FINISHING ALUMINUM CASTINGS

Aluminum is an inherently attractive metal with excellent resistance to many common types of corrosion. For numerous uses, the as-cast surface of aluminum castings does not require any applied decorative or protective finishes.

However, in the constantly increasing list of aluminum casting applications in all types of industries, specified decorative effects, corrosion protection or special design functions often require surface finishes. Because aluminum readily receives finishes by mechanical, chemical and electrochemical methods, the applications for aluminum castings are multiplied considerably where appearance and utility are significant considerations in the end uses of the castings. A wide variety of finishes offers many advantages in commercial and industrial applications.

With a wide variety of methods employed by individual finishing sources, many users of castings rarely recognize the full range of finishes available—particularly from the standpoint of the operations involved, competitive cost analysis or pertinent inspection standards.

The average manufacturer regularly finishes at least three basic types of aluminum castings: die castings, permanent mold castings and sand castings for various applications. The granular and surface characteristics of each of these types call for entirely different operational sequences to obtain a satisfactory finish.

Surface Finish Measurement

It is important to recognize that the finish (smoothness) of cast surfaces is related directly to the casting process used and the craftsmanship of the foundry worker. A part made in sand usually has more surface roughness or irregularity than one made in any other mold. Diecast parts, generally, have the smoothest surfaces of all aluminum castings.

Sometimes in the manufacture of sand, permanent or semi-permanent mold castings it is necessary to produce castings with less than the smoothest surface possible. Thus, the designer should recognize the economic implications of surface texture when performing operations such as filling thin section thicknesses or removing gas or trapped air from the mold cavity.

The as-cast surface finishes of aluminum castings are usually specified as microinch roughness. This is an unsatisfactory method of measurement for cast surfaces because microinch roughness measures deviation of tool marks from a smooth condition on a machined surface.

The various instruments for measuring surface smoothness all are designed to work on machined surfaces and do not give accurate or reproducible results on as-cast surfaces. Furthermore, because the surface on a casting varies from place to place, the point on the casting surface where the roughness measurements should be taken varies from casting to casting.

One method for measuring surface finish is to use comparison standards as a visual match. Samples are made by the casting supplier, and when a satisfactory surface is produced, both supplier and buyer agree on it. But even this method can cause complications because the apparent surface varies with different amounts, kinds and angle of incident light.

A second method that has gained acceptance uses a cast surface comparator block (Fig. 17-13) distributed by the Aluminum Association and described in its Standard AA-CS-E18. Other methods used to compare or measure surface roughness include: optical inspection with a comparascope and using profilometers and electronic surface indicators.

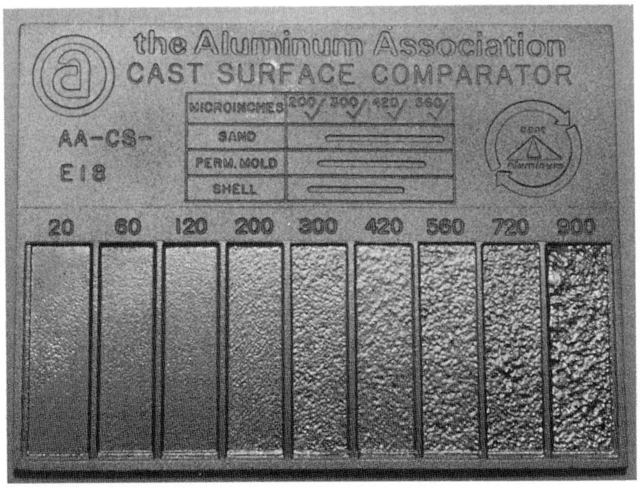

Fig. 17-13. An aluminum cast surface comparator, such as this one provided by the Aluminum Association, can be used to estimate the smoothness (rms) of as-cast aluminum surfaces by making visual and touch comparisons.

While significant emphasis often is given to surface finishes for specific types of castings, precision measurements are not necessary for all cast aluminum parts. However, it is important to note that surface measuring methods are currently used in many foundries to control product quality and to meet engineering requirements.

Establishing Finishing Specifications

Metal finishes are based on visual standards. They are subject to light refraction and light reflection. The luster obtained varies greatly with the different operations involved. This sometimes leads to visual interpretation that varies from desired results. The proper identification of finishing methods and interpretations of the surface textures they offer are the first steps to establishing a finishing specification.

The many proprietary designations that have been introduced for finishes on aluminum have created a great deal of confusion and have made it difficult for users to specify finishes in such a way that suppliers will know what to furnish. In an attempt to remedy this situation, the Aluminum Association, through its technical committee, has developed a system of designations broad enough to cover all types of finishes now being used and to meet anticipated future needs as new finishes are developed. Examples of this designation system are shown in Table 17-15.[1]

It is important to properly visualize and identify the type of finish required by preparing samples before the casting is released for finishing operations. Two people will not always agree on the finished appearance of a casting.

Finishes applied to aluminum castings can be chemical, electrochemical, mechanical or painted. Whichever process is used, finishes for aluminum castings can only be successful when proper techniques are used to ensure that the cast surfaces are sound and free from imperfections.

Integrity of cast surfaces is related principally to the casting process, the alloys, and the melting and pouring practices used. The control of alloy grain size and crystal structure is equally important. Cast surfaces are unsatisfactory if they contain gas holes, oxides, cracks, cold shuts, hard spots, cold folds, inclusions, discolorations, large differences in grain or crystal size, lubricants or corrosion products.

Finishing Characteristics of Alloys

Any aluminum casting alloy can be finished in one manner or another. Certain alloys exhibit excellent finishing characteristics while others do not. Chemical composition of the material is the controlling factor in finishing response.

It is important to note that, until recently, the casting alloys used in various finishing processes were materials designed for other purposes. Casting alloys now can be tailored specifically for almost any casting and finishing process.

Alloys with the best casting characteristics are those containing up to 12% silicon. However, these alloys exhibit poor finishing characteristics. Very little silicon is soluble in aluminum. The silicon imparts varying shades of dull gray upon treatment with some chemical and mechanical finishing processes. However, alloys containing as much as 7% silicon can be electroplated successfully.

Table 17-15.
Designations for Aluminum Finishes Developed by the Aluminum Association

Designation	Finish		Designation	Finish	
Mechanical finishes—M			**Chemical finishes*—C**		
M10	As fabricated	Unspecified	C10	Nonetched cleaned	Unspecified
M11		Specular finish as fabricated	C11		Degreased
M12		Nonspecular finish as fabricated	C12		Inhibited chemical cleaned
M1X		Other (to be specified)	C1X		Other (to be specified)
M20	Buffed	Unspecified			
M21		Smooth specular	C20	Etched	Unspecified
M22		Specular	C21		Fine matte
M2X		Other (to be specified)	C22		Medium matte
M30	Directional textured	Unspecified	C23		Coarse matte
M31		Fine satin	C2X		Other (to be specified)
M32		Medium satin			
M33		Coarse satin	C30	Brightened	Unspecified
M34		Hand rubbed	C31		Highly specular
M35		Brushed	C32		Diffuse bright
M3X		Other (to be specified)	C3X		Other (to be specified)
M40	Non-directional textured	Unspecified			
M41		Extra fine matte	C40	Chemical coatings	Unspecified
M42		Fine matte	C41		Acid chromate-fluoride
M43		Medium matte	C42		Acid chromate-fluoride-phosphate
M44		Coarse matte			
M45		Fine shotblast	C43		Alkaline chromate
M46		Medium shotblast	C44		Nonchromate
M47		Coarse shotblast	C45		Nonrinsed chromate
M4X		Other (to be specified)	C4X		Other (to be specified)
Anodic coatings—A					
A10	General	Unspecified	A31	Architectural class II** (10–18 μm or 0.4–0.7 mil)	Clear (natural)
A11		Preparation for other applied coatings	A32		Integral color
A12		Chromic acid coatings	A33		Impregnated color
A13		Hard, wear- and abrasion-resistant coatings	A34		Electrolytically deposited color
A1X		Other (to be specified)	A3X		Other (to be specified)
A21	Protective and decorative (less than 10μm or 0.4 mil)	Clear (natural)	A41	Architectural class I** (18 μm or 0.7 mil)	Clear (natural)
A22		Integral color	A42		Integral color
A23		Impregnated color	A43		Impregnated color
A24		Electrolytically deposited color	A44		Electrolytically deposited color
A2X		Other (to be specified)	A4X		Other (to be specified)
Resinous and other organic coatings *—R**			**Electroplated and other metal coatings***—E**		
R10	Unspecified		E10	Unspecified	
R1X	Other (to be specified)		E1X	Other (to be specified)	
Vitreous (porcelain and ceramic coatings*—V**			**Laminated coatings***—L**		
V10	Unspecified		L10	Unspecified	
V1X	Other (to be specified)		L1X	Other (to be specified)	

* Includes chemical conversion coatings, chemical or electrochemical brightening and cleaning treatments.
** Classification established in AA Standards for Anodically Coated Aluminum Alloys for Architectural Applications (Oct 1978).
*** These designations may be used until a more complete series of designations is developed for these coatings.

Aluminum alloys containing magnesium, zinc or copper respond well to almost any finishing method, but, unfortunately, they do not always exhibit the best casting characteristics. When silicon is added to these alloys, the casting characteristics are greatly improved; the amount of silicon required depends on the casting process used and the finishing technique used. It will be seen, therefore, that compromises in alloy composition are necessary in order to meet specific casting and finishing requirements.

Aluminum-silicon alloys can be cast successfully in any casting process. All other alloys have limitations. Aluminum-magnesium and aluminum-zinc alloys generally are cast in sand molds.

Aluminum-copper alloys are more versatile than aluminum-magnesium or aluminum-zinc alloys, but, again, if varying amounts of silicon are added to these alloys they can be cast using processes other than sand.

Aluminum-magnesium alloys, such as the 514 and 535 types, respond well to clear and colored anodizing. The same is true of such aluminum-zinc alloys as the 712, 713 and 771 types. Electroplated finishes are best on aluminum-copper alloys, such as 295; aluminum-copper-silicon alloys, such as the 319, 355 and 380 types; and aluminum-silicon alloys, such as 356 and 360.

Effects of Alloy Structure on Finishing

Grain size, crystal size and phase distribution are factors that affect casting surface soundness and uniformity of the physical makeup of the alloy structure. If grain size varies drastically, the casting surface will appear wavy or blotchy after polishing or plating. Because of different thermal conditions in a mold, the grain size will vary, but the deleterious effect of grain size differences on surface finishing is much less noticeable in fine-grained material. Grain refinement is covered in detail in Chapter 3.

Aluminum-silicon alloys have a special problem. Because a negligible amount of silicon is soluble in aluminum, the silicon forms a mixture with aluminum called the aluminum-silicon eutectic. In this mixture, the silicon crystals are coarse, hard, platelike particles that can grow relatively large. These large particles promote uneven distribution of the eutectic phase. The results can be unsound metal and a blotchy appearance after polishing.

The morphology of the silicon can be controlled by adding elements such as sodium or strontium. The result of such additions is to convert the eutectic mixture from coarse silicon flakes into a finely dispersed mixture. This treatment, called modification, is covered in more detail in Chapter 3. The appearance and quality of subsequent surface finishes are greatly enhanced by modifying aluminum-silicon alloys.

Mechanical Finishes

Aluminum casting polishing usually consists of a series of operations progressing from coarse to very fine wheels or belts, from 60 to 240 or 320 grit. This generally is followed by buffing using a compound to produce mirror finishes. Coloring produces a surface of high gloss, used either as a final polished finish or as a foundation for subsequent finishes.

An excellent burnished finish can be developed on an aluminum-silicon alloy by treating the surface of the part with hydrofluoric acid. This acid dissolves the silicon on the surface, leaving the soft aluminum matrix, which burnishes to a bright silver finish.

When cut buffing aluminum castings, high buffing pressures can open the surface structure giving the appearance of gas and/or shrinkage porosity.

Scratch brushing is a textured effect obtained by using rotating wire brushes made of stainless steel, brass or nickel-silver. The size of the wire used determines the coarseness of the texture. Satin finishes are produced in a manner similar to scratch brushing except that smaller and softer wheels with wires on the order of 0.002–0.005 in. dia are used.

Abrasive blastcleaning is an effective method of cleaning and preparing castings for subsequent finishes (Fig. 17-14). Both wet and dry methods are used. Various blast media (such as silica sand, aluminum oxide, glass beads or steel grit) can produce from coarse to fine matte finishes.

Fig. 17-14. A batch barrel tumbler is used to blastclean small castings. [Photo courtesy of Pangborn Corp.]

Silica sand should not be used to blastclean castings that subsequently will be anodized. The silica sand will imbed in the aluminum surface and react with the electrolyte to produce a milky gray appearance. Because these finishes are easily scratched or stained by fingerprints, they usually are anodized or coated with clear lacquer for protection.

Barrel or vibratory finishing is an inexpensive way to clean, deburr and burnish small castings, both raw and machined. The process smoothens sharp edges, imparts a matte finish and prepares the surface for operations, such as anodizing, plating or painting. A variety of finishes can be imparted by using different media and abrasives.

Surfaces of barrels and vibratory tubs should be lined to protect the castings from contamination by ferrous material. When wet methods are used, it is important to control the pH of the solution because aluminum is susceptible to etching by both acids and alkalis. All aluminum alloys can be finished by these methods.

Chemical Cleaning, Brightening and Etching

Common chemical cleaners are produced by treating castings with chemical solutions to provide etched or decreased surfaces. These cleaners are used to clean casting surfaces for subsequent coatings, for inhibiting surface corrosion or for securing uniform coloration.

Acid dipping creates a frosted appearance on aluminum, which is considered pleasing. Chemical finishes are decorative and resist corrosion but are not resistant to abrasion. There are several proprietary processes available.

Chemical brightening or *bright dipping* smoothens and brightens aluminum parts by making use of the solution potential of the aluminum in various baths. Chemical brighteners are usually acid solutions with oxidizing agents. Chemical brightening can be used to partially or completely replace buffing.

Chemical etching with acid or alkaline solutions, produces a matte finish and can be used as a final finish. It generally is used prior to other finishing, such as lacquering, anodizing or conversion coating. Chemical etching removes oxide films, slightly roughens the surface and minimizes color match differences when used prior to anodizing. Acid etching generally is used for high-silicon alloys, such as die castings.

Anodized Finishes

These finishes, of which clear anodizing is the most common, are formed by anodically oxidizing the casting in a suitable electrolyte to form an oxide coating. This coating can be given supplementary coloring or sealing treatments. For satisfactory color appearance, special high-purity alloys must be used. An extremely dense metal structure is required. Other anodic finishes that give an integral color, produced by varying alloy composition, bath salts, current and time, also have limited use on cast aluminum but are used more widely on certain wrought alloys.

Anodizing is performed to:

- increase corrosion resistance;
- increase paint adhesion;
- permit subsequent plating;
- improve decorative appearance;
- provide electrical insulation;
- increase emissivity;
- increase abrasion resistance;
- detect surface flaws.

Hard anodizing, which produces a heavier coating of oxide than regular anodizing, uses lower operating temperatures and higher current densities.

Chemical Conversion Coatings

Chemical conversion coatings are surface layers of oxide, phosphate and chromate compounds formed by chemical reaction with the aluminum surface. These coatings usually involve removal of 0.3–2.5 μm of the part surface. Chemical conversion coatings differ from anodize coatings in that they are formed by oxidation-reduction reactions, while anodic coatings are formed by electrochemical reactions.[1]

These coatings affect the surface properties of the part. They are used to retard corrosion under other finishes (such as oil, wax and organic), to improve adhesion of organic finishes, to slightly improve wear resistance, to aid in drawing and forming characteristics, and to provide corrosion resistance without changing electrical resistivity. Because of their electrical properties, chemical conversion coatings are used widely for electronic parts.

Painting

Aluminum castings may be finished in a wide variety of lacquers or enamels for decoration or protection. Zinc chromate primers are recommended for maximum corrosion resistance of the organic finish. The final finish coatings may be any one of a large number of synthetic resin organic materials that include oil base paints, varnishes, enamels and synthetic resin finishes.

Lacquer may be used for fast drying. The aluminum-pigmented lacquers are the most durable. Clear and colored epoxy lacquers may be used to obtain a transparent look. Common types of finishes frequently specified include clear organic, high gloss and wrinkle-type (diffused pattern).

Electroplated Finishes

Even though aluminum is more difficult to plate than other metals because of the quick-forming oxide film and because most plating metals are cathodic to aluminum, many aluminum parts are electroplated with chromium, nickel, cadmium, copper, tin, zinc, gold or silver.

Before electroplating, most aluminum is immersion coated in a zincate solution. Doing this provides a thin, continuous film of metallic zinc on the aluminum surface, which is a good base for subsequent electroplating.[1]

Casting surfaces that are to be electroplated require a first-class "hardware" finish. The surfaces must be absolutely smooth and free from all blemishes, such as flow marks, cold shuts, porosity and pits. Parting lines must be solid and polished flush.

Successful applications of electroplated finishes require optimum conditions in casting operations, trimming and secondary finishing techniques. Parts with excessive porosity should not be used.

ACKNOWLEDGMENTS

AFS would like to thank the following people for their help with this chapter: Stephen M. Anderson, Norton Co., for the abrasives section; Steve Gryglas, Acme Industries, Inc., for the machining section; and Wayne Rasmussen, AFS, for the sections on casting inspection, repair and finishing.

REFERENCES

1. *ASM Metals Handbook,* 9th edition, vol 5 (1982).
2. P.M. Bralower, "Nondestructive Testing, Part 1—The New Generation of Radiography," *Modern Casting,* pp 21–23 (July 1986).
3. P.M. Bralower, "Nondestructive Testing, Part 2—New Applications for Traditional Methods," *Modern Casting,* pp 25–27 (Aug 1986).
4. R.H. Bossi, J.L. Cline, G.E. Georgeson; "NDT: Computed Tomography of Castings," *Modern Casting,* pp 19–22 (Mar 1991).
5. *Machining Data Handbook,* Institute of Advanced Manufacturing Sciences, Cincinnati, Ohio.
6. *Machinery Handbook,* 24th edition, Society of Mfg. Engineers (SME), R. Green ed (1992).
7. W. Tuttle; "Understanding Aluminum Welding," *Welding Journal,* pp 43–46 (Feb 1991).
8. T.J. Bosworth; "Repair Welding of Aluminum Castings," *AFS Transactions,* vol 87, pp 621–628 (1983).
9. N.H. Hewett, "Advances in Chemicals and Processes for the Impregnation of Castings," *AFS Transactions,* vol 94, pp 859–864 (1990).
10. E.L. Rooy; "Improving Casting Properties and Integrity with Hot Isostatic Processing," *Modern Casting,* pp 18–20 (Dec 1983).
11. R. Widmer; "The Current Status of HIP Technology and Its Prospects for the 1990s," presented at the 4th Annual Conference on Isostatic Pressing (Nov 5–7, 1990).

CHAPTER 17 APPENDIX

MACHINING DATA TABLES

Tables 17-5 to 17-12 are excerpted from the third edition of the *Machining Data Handbook* (©1980) and are used for this publication with the permission of the Institute of Advanced Manufacturing Sciences, Inc.

Other sources, such as the 24th edition of the *Machinery Handbook* published by SME (edited by R. Green, © 1992), offer similar information.

Table 17-5.
Machining Speeds and Feeds for Turning with Single Point and Box Tools

Casting Method and Alloy	Bhn	Condition	Depth of Cut*	High-Speed Steel Tool Speed	High-Speed Steel Tool Feed	High-Speed Steel Tool Tool Material	Uncoated Carbide Tool Speed Brazed	Uncoated Carbide Tool Speed Index-able	Uncoated Carbide Tool Feed	Uncoated Carbide Tool Tool Material Grade
			in.	fpm	ipr	AISI	fpm	fpm	ipr	C
			mm	m/min	mm/r	ISO	m/min	m/min	mm/r	ISO
sand and permanent mold			0.040	1000	0.007	M2, M3	2000	Max.	0.010	C-3
201.0 332.0 514.0			0.150	900	0.015	M2, M3	1800	Max.	0.020	C-2
203.0 333.0 520.0			0.300	800	0.030	M2, M3	1200	Max.	0.040	C-2
208.0 336.0 535.0	40–100 500 kg	as cast	0.625	500	0.040	M2, M3	1000	Max.	0.080	C-2
213.0 354.0 705.0			1	305	0.18	S4, S5	610	Max.	0.25	K01, M10
222.0 355.0 707.0			4	275	0.40	S4, S5	550	Max.	0.50	K10, M10
224.0 C355.0 710.0			8	245	0.75	S4, S5	365	Max.	1.0	K10, M10
240.0 356.0 712.0			16	150	1.0	S4, S5	305	Max.	2.05	K20, M20
242.0 A356.0 713.0			0.040	800	0.007	M2, M3	1800	Max.	0.010	C-3
295.0 357.0 771.0			0.150	700	0.015	M2, M3	1400	Max.	0.020	C-2
296.0 359.0 850.0			0.300	600	0.030	M2, M3	1000	Max.	0.040	C-2
308.0 B443.0 851.0	70–125 500 kg	solution treated and aged	0.625	400	0.040	M2, M3	800	Max.	0.080	C-2
319.0 512.0 852.0			1	245	0.18	S4, S5	550	Max.	0.25	K01, M10
328.0 513.0			4	215	0.40	S4, S5	425	Max.	0.50	K10, M10
			8	185	0.75	S4, S5	305	Max.	1.0	K10, M10
			16	120	1.0	S4, S5	245	Max.	2.05	K20, M20
diecasting			0.040	1000	0.007	M2, M3	2000	Max.	0.010	C-3
360.0 380.0 C443.0	40–100 500 kg	as cast	0.150	900	0.015	M2, M3	1800	Max.	0.020	C-2
A360.0 A380.0 518.0			1	305	0.18	S4, S5	610	Max.	0.25	K01, M10
			4	275	0.40	S4, S5	550	Max.	0.50	K10, M10
			0.040	800	0.007	M2, M3	1800	Max.	0.010	C-3
	70–125 500 kg	solution treated and aged	0.150	700	0.015	M2, M3	1400	Max.	0.020	C-2
			1	245	0.18	S4, S5	550	Max.	0.25	K01, M10
			4	215	0.40	S4, S5	425	Max.	0.50	K10, M10
diecasting			0.040	800	0.007	M2, M3	1800	Max.	0.010	C-3
383.0 413.0 A413.0	40–100 500 kg	as cast	0.150	700	0.015	M2, M3	1600	Max.	0.020	C-2
A384.0			1	245	0.18	S4, S5	550	Max.	0.25	K01, M10
			4	215	0.40	S4, S5	490	Max.	0.50	K10, M10
			0.040	700	0.007	M2, M3	1600	Max.	0.010	C-3
	70–125 500 kg	solution treated and aged	0.150	600	0.015	M2, M3	1200	Max.	0.020	C-2
			1	215	0.18	S4, S5	490	Max.	0.25	K01, M10
			4	185	0.40	S4, S5	365	Max.	0.50	K10, M10
diecasting			0.040	150	0.007	M2, M3	525	550	0.010	C-2
390.0 392.0	40–100 500 kg	as cast	0.150	135	0.015	M2, M3	475	500	0.020	C-2
			1	46	0.18	S4, S5	160	170	0.25	K10, M10
			4	41	0.40	S4, S5	145	150	0.50	K10, M10
			0.040	140	0.007	M2, M3	450	475	0.010	C-2
	70–125 500 kg	solution treated and aged	0.150	125	0.015	M2, M3	425	450	0.020	C-2
			1	43	0.18	S4, S5	135	145	0.25	K10, M10
			4	38	0.40	S4, S5	130	135	0.50	K10, M10

Table 17-6.
Machining Speeds and Feeds for Turning with Diamond Tools

Casting Method and Alloy	Bhn	Condition	Depth of Cut in.	Depth of Cut mm	Speed fpm	Speed m/min	Feed ipr	Feed mm/r
sand and permanent mold 201.0 308.0 A356.0 705.0 203.0 319.0 357.0 707.0 208.0 328.0 359.0 710.0 213.0 332.0 B443.0 712.0 222.0 333.0 512.0 713.0 224.0 336.0 513.0 771.0 240.0 354.0 514.0 850.0 242.0 355.0 520.0 851.0 295.0 C355.0 535.0 852.0 296.0 356.0	40–100 500 kg	as cast	0.005–0.015	0.13–0.40	3000	915	0.003–0.006	0.075–0.15
			0.015–0.050	0.40–1.25	2500	760	0.006–0.012	0.15–0.30
			0.050–0.125	1.25–3.20	1500	460	0.012–0.020	0.30–0.50
	70–125 500 kg	solution treated and aged	0.005–0.015	0.13–0.40	2800	855	0.003–0.006	0.075–0.15
			0.015–0.050	0.40–1.25	2200	670	0.006–0.012	0.15–0.30
			0.050–0.125	1.25–3.20	1300	395	0.012–0.020	0.30–0.50
diecasting 360.0 380.0 C443.0 518.0 A360.0 A380.0	40–100 500 kg	as cast	0.005–0.015	0.13–0.40	3500	1065	0.003–0.006	0.075–0.15
			0.015–0.050	0.40–1.25	3000	915	0.006–0.012	0.15–0.30
			0.050–0.125	1.25–3.20	2500	760	0.012–0.020	0.30–0.50
diecasting 383.0 A384.0 413.0 A413.0	40–100 500 kg	as cast	0.005–0.015	0.13–0.40	3000	915	0.003–0.006	0.075–0.15
			0.015–0.050	0.40–1.25	2500	760	0.006–0.012	0.15–0.30
			0.050–0.125	1.25–3.20	2000	610	0.012–0.020	0.30–0.50
diecasting 390.0 392.0	40–100 500 kg	as cast	0.005–0.015	0.13–0.40	2600	790	0.003–0.006	0.075–0.15
			0.015–0.050	0.40–1.25	2000	610	0.006–0.012	0.15–0.30
			0.050–0.125	1.25–3.20	1400	425	0.012–0.020	0.30–0.50

Table 17-7.
Machining Speeds and Feeds for Boring with Diamond Tools

Casting Method and Alloy	Bhn	Condition	Depth of Cut in.	Depth of Cut mm	Speed fpm	Speed m/min	Feed ipr	Feed mm/r
sand and permanent mold 201.0 308.0 A356.0 705.0 203.0 319.0 357.0 707.0 208.0 328.0 359.0 710.0 213.0 332.0 B443.0 712.0 222.0 333.0 512.0 713.0 224.0 336.0 513.0 771.0 240.0 354.0 514.0 850.0 242.0 355.0 520.0 851.0 295.0 C355.0 535.0 852.0 296.0 356.0	40–100 500 kg	as cast	0.005–0.015	0.13–0.40	2500	760	0.003–0.006	0.075–0.15
			0.015–0.050	0.40–1.25	2000	610	0.006–0.012	0.15–0.30
			0.050–0.125	1.25–3.20	1000	305	0.012–0.020	0.30–0.50
	70–125 500 kg	solution treated and aged	0.005–0.015	0.13–0.40	2200	670	0.003–0.006	0.075–0.15
			0.015–0.050	0.40–1.25	1700	520	0.006–0.012	0.15–0.30
			0.050–0.125	1.25–3.20	700	215	0.012–0.020	0.30–0.50
diecasting 360.0 380.0 C443.0 518.0 A360.0 A380.0	40–100 500 kg	as cast	0.005–0.015	0.13–0.40	3000	915	0.003–0.006	0.075–0.15
			0.015–0.050	0.40–1.25	2500	760	0.006–0.012	0.15–0.30
			0.050–0.125	1.25–3.20	2000	610	0.012–0.020	0.30–0.50
diecasting 383.0 A384.0 413.0 A413.0	40–100 500 kg	as cast	0.005–0.015	0.13–0.40	2500	760	0.003–0.006	0.075–0.15
			0.015–0.050	0.40–1.25	2000	610	0.006–0.012	0.15–0.30
			0.050–0.125	1.25–3.20	1500	460	0.012–0.020	0.30–0.50
diecasting 390.0 392.0	40–100 500 kg	as cast	0.005–0.015	0.13–0.40	2000	610	0.003–0.006	0.075–0.15
			0.015–0.050	0.40–1.25	1500	460	0.006–0.012	0.15–0.30
			0.050–0.125	1.25–3.20	1000	305	0.012–0.020	0.30–0.50

Table 17-8.
Machining Speeds and Feeds for Boring

Casting Method and Alloy	Bhn	Condition	Depth of Cut*	High-Speed Steel Tool			Uncoated Carbide Tool			
							Speed			
			Depth of Cut*	Speed	Feed	Tool Material	Brazed	Index-able	Feed	Tool Material Grade
			in.	fpm	ipr	AISI	fpm	fpm	ipr	C
			mm	m/min	mm/r	ISO	m/min	m/min	mm/r	ISO
sand and permanent mold 201.0 332.0 514.0 203.0 333.0 520.0 208.0 336.0 535.0 213.0 354.0 705.0 222.0 355.0 707.0 224.0 C355.0 710.0 240.0 356.0 712.0 242.0 A356.0 713.0 295.0 357.0 771.0 296.0 359.0 850.0 308.0 B443.0 851.0 319.0 512.0 852.0 328.0 513.0	40–100 500 kg	as cast	0.010	1000	0.003	M2, M3	1800	Max.	0.006	C-3
			0.050	800	0.005	M2, M3	1200	Max.	0.008	C-2
			0.100	700	0.012	M2, M3	1000	Max.	0.015	C-2
			0.25	305	0.075	S4, S5	550	Max.	0.15	K01, M10
			1.25	245	0.13	S4, S5	365	Max.	0.20	K10, M10
			2.50	215	0.30	S4, S5	305	Max.	0.40	K20, M20
	70–125 500 kg	solution treated and aged	0.010	750	0.003	M2, M3	1800	Max.	0.006	C-3
			0.050	600	0.005	M2, M3	1200	Max.	0.008	C-2
			0.100	550	0.012	M2, M3	1000	Max.	0.015	C-2
			0.25	230	0.075	S4, S5	550	Max.	0.15	K01, M10
			1.25	185	0.13	S4, S5	365	Max.	0.20	K10, M10
			2.50	170	0.30	S4, S5	305	Max.	0.40	K20, M20
diecasting 360.0 380.0 C443.0 A360.0 A380.0 518.0	40–100 500 kg	as cast	0.010	1000	0.003	M2, M3	1800	Max.	0.006	C-3
			0.050	800	0.005	M2, M3	1200	Max.	0.008	C-2
			0.100	700	0.012	M2, M3	1000	Max.	0.015	C-2
			0.25	305	0.075	S4, S5	550	Max.	0.15	K01, M10
			1.25	245	0.13	S4, S5	365	Max.	0.20	K10, M10
			2.50	215	0.30	S4, S5	305	Max.	0.40	K20, M20
	70–125 500 kg	solution treated and aged	0.010	750	0.003	M2, M3	1600	Max.	0.006	C-3
			0.050	600	0.005	M2, M3	1400	Max.	0.008	C-2
			0.100	550	0.012	M2, M3	800	Max.	0.015	C-2
			0.25	230	0.075	S4, S5	490	Max.	0.15	K01, M10
			1.25	185	0.13	S4, S5	425	Max.	0.20	K10, M10
			2.50	170	0.30	S4, S5	245	Max.	0.40	K20, M20
diecasting 383.0 413.0 A413.0 A384.0	40–100 500 kg	as cast	0.010	750	0.003	M2, M3	1600	Max.	0.006	C-3
			0.050	600	0.005	M2, M3	1400	Max.	0.008	C-2
			0.100	550	0.012	M2, M3	800	Max.	0.015	C-2
			0.25	230	0.075	S4, S5	490	Max.	0.15	K01, M10
			1.25	185	0.13	S4, S5	425	Max.	0.20	K10, M10
			2.50	170	0.30	S4, S5	245	Max.	0.40	K20, M20
	70–125 500 kg	solution treated and aged	0.010	675	0.003	M2, M3	1350	Max.	0.006	C-3
			0.050	550	0.005	M2, M3	1100	Max.	0.008	C-2
			0.100	480	0.012	M2, M3	850	Max.	0.015	C-2
			0.25	205	0.075	S4, S5	410	Max.	0.15	K01, M10
			1.25	170	0.13	S4, S5	335	Max.	0.20	K10, M10
			2.50	145	0.30	S4, S5	260	Max.	0.40	K20, M20
diecasting 390.0 392.0	40–100 500 kg	as cast	0.010	150	0.003	M2, M3	410	Max.	0.006	C-3
			0.050	120	0.005	M2, M3	325	Max.	0.008	C-2
			0.100	110	0.012	M2, M3	300	Max.	0.012	C-2
			0.25	46	0.075	S4, S5	125	Max.	0.15	K01, M10
			1.25	37	0.13	S4, S5	100	Max.	0.20	K10, M10
			2.50	34	0.30	S4, S5	90	Max.	0.30	K20, M20
	70–125 500 kg	solution treated and aged	0.010	140	0.003	M2, M3	350	Max.	0.006	C-3
			0.050	110	0.005	M2, M3	280	Max.	0.008	C-2
			0.100	100	0.012	M2, M3	230	Max.	0.012	C-2
			0.25	43	0.075	S4, S5	105	Max.	0.15	K01, M10
			1.25	34	0.13	S4, S5	85	Max.	0.20	K10, M10
			2.50	30	0.30	S4, S5	70	Max.	0.30	K20, M20

Table 17-9.
Machining Speeds and Feeds for Face Milling

Casting Method and Alloy	Bhn	Condition	Depth of Cut*	High-Speed Steel Tool: Speed	High-Speed Steel Tool: Feed per Tooth	High-Speed Steel Tool: Tool Material	Uncoated Carbide Tool: Speed, Brazed	Uncoated Carbide Tool: Speed, Index-able	Uncoated Carbide Tool: Feed per Tooth	Uncoated Carbide Tool: Tool Material Grade
			in.	fpm	in.	AISI	fpm	fpm	in.	C
			mm	m/min	mm	ISO	m/min	m/min	mm	ISO
sand and permanent mold 201.0 332.0 514.0 203.0 333.0 520.0 208.0 336.0 535.0 213.0 354.0 705.0 222.0 355.0 707.0 224.0 C355.0 710.0 240.0 356.0 712.0 242.0 A356.0 713.0 295.0 357.0 771.0 296.0 359.0 850.0 308.0 B443.0 851.0 319.0 512.0 852.0 328.0 513.0	40–100 500 kg	as cast	0.040	1200	0.008	M2, M7	2000	2500	0.010	C-2
			0.150	800	0.012	M2, M7	1800	2000	0.015	C-2
			0.300	650	0.016	M2, M7	1200	1500	0.020	C-2
			1	365	0.20	S4, S2	610	760	0.25	K10, M20
			4	245	0.30	S4, S2	550	610	0.40	K20, M30
			8	200	0.40	S4, S2	365	460	0.50	K30, M40
	70–125 500 kg	solution treated and aged	0.040	1000	0.008	M2, M7	1800	2500	0.010	C-2
			0.150	700	0.012	M2, M7	1400	1800	0.015	C-2
			0.300	550	0.016	M2, M7	1000	1300	0.020	C-2
			1	305	0.20	S4, S2	550	760	0.25	K10, M20
			4	215	0.30	S4, S2	425	550	0.40	K20, M30
			8	170	0.40	S4, S2	305	395	0.50	K30, M40
diecasting 360.0 380.0 C443.0 A360.0 A380.0 518.0	40–100 500 kg	as cast	0.040	1200	0.008	M2, M7	1400	1800	0.005	C-2
			0.150	800	0.012	M2, M7	1150	1500	0.010	C-2
			1	365	0.20	S4, S2	425	550	0.13	K10, M20
			4	245	0.30	S4, S2	350	460	0.25	K10, M20
	70–125 500 kg	solution treated and aged	0.040	1000	0.008	M2, M7	1250	1600	0.005	C-2
			0.150	700	0.012	M2, M7	1000	1300	0.010	C-2
			1	305	0.20	S4, S2	380	490	0.13	K10, M20
			4	215	0.30	S4, S2	305	395	0.25	K10, M20
diecasting 383.0 413.0 A413.0 A384.0	40–100 500 kg	as cast	0.040	700	0.006	M2, M7	1150	1400	0.005	C-2
			0.150	600	0.010	M2, M7	900	1100	0.010	C-2
			1	215	0.15	S4, S2	350	425	0.13	K10, M20
			4	185	0.25	S4, S2	275	335	0.25	K10, M20
	70–125 500 kg	solution treated and aged	0.040	700	0.006	M2, M7	1150	1400	0.005	C-2
			0.150	600	0.010	M2, M7	900	1100	0.010	C-2
			1	215	0.15	S4, S2	350	425	0.13	K10, M20
			4	185	0.25	S4, S2	275	335	0.25	K10, M20
diecasting 390.0 392.0	40–100 500 kg	as cast	0.040	165	0.006	M2, M7	425	500	0.005	C-2
			0.150	140	0.009	M2, M7	370	450	0.010	C-2
			1	50	0.15	S4, S2	130	150	0.13	K10, M20
			4	43	0.23	S4, S2	115	135	0.25	K10, M20
	70–125 500 kg	solution treated and aged	0.040	155	0.006	M2, M7	375	475	0.005	C-2
			0.150	130	0.009	M2, M7	350	425	0.010	C-2
			1	43	0.18	S4, S2	135	145	0.25	K10, M20
			4	38	0.40	S4, S2	130	135	0.50	K10, M20

Table 17-10.
Machining Speeds and Feeds for Face Milling with Diamond Tools

Casting Method and Alloy	Bhn	Condition	Depth of Cut in.	Depth of Cut mm	Speed fpm	Speed m/min	Feed per Tooth in.	Feed per Tooth mm
sand and permanent mold 201.0 308.0 A356.0 705.0 203.0 319.0 357.0 707.0 208.0 328.0 359.0 710.0 213.0 332.0 B443.0 712.0 222.0 333.0 512.0 713.0 224.0 336.0 513.0 771.0 240.0 354.0 514.0 850.0 242.0 355.0 520.0 851.0 295.0 C355.0 535.0 852.0 296.0 356.0	40–100 500 kg	as cast	0.010–0.030	0.25–0.75	3300	1005	0.005	0.13
			0.030–0.050	0.75–1.25	2750	840	0.010	0.25
			0.050–0.100	1.25–2.50	1800	550	0.015	0.40
	70–125 500 kg	solution treated and aged	0.010–0.030	0.25–0.75	3000	915	0.005	0.13
			0.030–0.050	0.75–1.25	2500	760	0.010	0.25
			0.050–0.100	1.25–2.50	1800	550	0.015	0.40
diecasting 360.0 380.0 C443.0 518.0 A360.0 A380.0	40–100 500 kg	as cast	0.010–0.030	0.25–0.75	3600	1100	0.005	0.13
			0.030–0.050	0.75–1.25	3200	975	0.010	0.25
			0.050–0.100	1.25–2.50	2750	840	0.015	0.40
diecasting 383.0 A384.0 413.0 A413.0	40–100 500 kg	as cast	0.010–0.030	0.25–0.75	3300	1005	0.005	0.13
			0.030–0.050	0.75–1.25	2750	840	0.010	0.25
			0.050–0.100	1.25–2.50	1800	550	0.015	0.40
diecasting 390.0 392.0	40–100 500 kg	as cast	0.010–0.030	0.25–0.75	2800	855	0.005	0.13
			0.030–0.050	0.75–1.25	2200	670	0.010	0.25
			0.050–0.100	1.25–2.50	1550	470	0.015	0.40

Table 17-11.
Machining Speeds for Tapping

Casting Method and Alloy	Bhn	Condition	Threads/in.: 7 or less Pitch (mm): over 3	8–15 1.5–3	16–24 1–1.5	over 24 1 or less	HSS Tool Material
			Speed fpm / m/min				AISI / ISO
sand and permanent mold 201.0 242.0 333.0 A356.0 514.0 712.0 203.0 295.0 336.0 357.0 520.0 713.0 208.0 296.0 354.0 359.0 535.0 771.0 213.0 308.0 355.0 B443.0 705.0 850.0 222.0 319.0 C355.0 512.0 707.0 851.0 224.0 328.0 356.0 513.0 710.0 852.0 240.0 332.0	40–100 500 kg	as cast	50	85	110	115	M10, M7, M1
			15	26	34	35	S2, S3
	70–125 500 kg	solution treated and aged	40	65	85	90	M10, M7, M1
			12	20	26	27	S2, S3
diecasting 360.0 A360.0 380.0 A380.0 C443.0 518.0	40–100 500 kg	as cast	65	110	140	150	M10, M7, M1
			20	34	43	46	S2, S3
	70–125 500 kg	solution treated and aged	45	75	95	100	M10, M7, M1
			14	23	29	30	S2, S3
diecasting 383.0 A384.0 413.0 A413.0	40–100 500 kg	as cast	65	110	140	150	M10, M7, M1
			20	34	43	46	S2, S3
	70–125 500 kg	solution treated and aged	45	75	95	100	M10, M7, M1
			14	23	29	30	S2, S3
diecasting 390.0 392.0	40–100 500 kg	as cast	55	100	110	120	M10, M7, M1
			17	30	34	37	S2, S3
	70–125 500 kg	solution treated and aged	40	70	90	95	M10, M7, M1
			12	21	27	29	S2, S3

Table 17-12.
Machining Speeds and Feeds for Drilling

Casting Method and Alloy	Bhn	Condition	Speed	Nominal Hole Diameter								Tool Material Grade
				1/16 in.	1/8 in.	1/4 in.	1/2 in.	3/4 in.	1 in.	1-1/2 in.	2 in.	
				1.5 mm	3 mm	6 mm	12 mm	18 mm	25 mm	35 mm	50 mm	
			Speed	Feed								Grade
			fpm	ipr								AISI or C
	Bhn	Condition	m/min	mm/r								ISO
sand and permanent mold 201.0 308.0 A356.0 705.0 203.0 319.0 357.0 707.0 208.0 328.0 359.0 710.0 213.0 332.0 B443.0 712.0 222.0 333.0 512.0 713.0 224.0 336.0 513.0 771.0 240.0 354.0 514.0 850.0 242.0 355.0 520.0 851.0 295.0 C355.0 535.0 852.0 296.0 356.0	40–100 500 kg	as cast	140	0.001								M10, M7, M1
			350	—	0.003	0.007	0.012	0.016	0.019	0.025	0.030	
			43	0.025								S2, S3
			105	—	0.075	0.18	0.30	0.40	0.50	0.65	0.75	
	70–125 500 kg	solution treated and aged	140	0.001								M10, M7, M1
			275	—	0.003	0.007	0.012	0.016	0.019	0.025	0.030	
			43	0.025								S2, S3
			84	—	0.075	0.18	0.30	0.40	0.50	0.65	0.75	
diecasting 360.0 380.0 C443.0 518.0 A360.0 A380.0	40–100 500 kg	as cast	150	0.001								M10, M7, M1
			375	—	0.003	0.007	0.012	0.016	0.019	0.025	0.030	
			46	0.025								S2, S3
			115	—	0.075	0.18	0.30	0.40	0.50	0.65	0.75	
	70–125 500 kg	solution treated and aged	150	0.001								M10, M7, M1
			300	—	0.003	0.007	0.012	0.016	0.019	0.025	0.030	
			46	0.025								S2, S3
			90	—	0.075	0.18	0.30	0.40	0.50	0.65	0.75	
diecasting 383.0 A384.0 413.0 A413.0	40–100 500 kg	as cast	100	0.001								M10, M7, M1
			140	—	0.003	0.007	0.012	0.016	0.019	0.025	0.030	
			30	0.025								S2, S3
			43	—	0.075	0.18	0.30	0.40	0.50	0.65	0.75	
	70–125 500 kg	solution treated and aged	90	0.001								M10, M7, M1
			120	—	0.003	0.007	0.012	0.016	0.019	0.025	0.030	
			27	0.025								S2, S3
			37	—	0.075	0.18	0.30	0.40	0.50	0.65	0.75	
diecasting 390.0 392.0	40–100 500 kg	as cast	25	0.001								M10, M7, M1
			50	—	0.003	0.007	0.012	0.016	0.019	0.025	0.030	
			8	0.025								S2, S3
			15	—	0.075	0.18	0.30	0.40	0.50	0.65	0.75	
	70–125 500 kg	solution treated and aged	25	0.001								M10, M7, M1
			45	—	0.003	0.007	0.012	0.016	0.019	0.025	0.030	
			8	0.025								S2, S3
			14	—	0.075	0.18	0.30	0.40	0.50	0.65	0.75	

INDEX

A

Abrasive-coated belts 329
Acid demand value 216
Acurad process 164
AFS clay test 201
AFS grain fineness number 187–189, 216
Age hardening 289, 293–294
Alloy identification 311
Alloys—tables and characteristics
 100.0 56
 170.0 57
 201.0 and A201.0 57–58
 A206.0 59
 208.0 60
 222.0 60–61
 242.0 and A242.0 62–63
 295.0 64
 296.0 65
 308.0 66
 319.0, A319.0, B319.0 and 320.0 66–67
 332.0 68
 333.0 and A333.0 69
 336.0 70
 354.0 71
 355.0 72–73
 C355.0 74
 356.0 75–76
 A356.0 76–77
 357.0 78
 A357.0 79
 359.0 80
 360.0 and A360.0 81
 A380.0 and B380.0 82
 383.0 and 384.0 83
 A390.0 and B390.0 84
 413.0 and A413.0 85
 443.0, A443.0 and C443.0 86–87
 A444.0 87
 511.0 88
 512.0 89
 513.0 90
 514.0 91
 518.0 92
 520.0 93
 535.0 94–95
 710.0 95
 711.0 96
 712.0 97
 713.0 98
 771.0 and 772.0 99–100
 850.0 101
 851.0 102
 852.0 103
 853.0 104
α-aluminum 289, 291–293
Alumina 45–46—*See also Aluminum oxide*
Aluminum (pure)
 diecasting motor rotors 166
 melting point 5
 solidification characteristics 54
 standard pure ingot 48
Aluminum alloys
 adding alloying elements 6
  master alloys 6
 castability 1–3
 characteristics 1–2, 19–43, 46, 47
  electrical conductivity 46
  specific gravity 1
  weight 46
 constituent phases 292
 designation system 50–51
 determining composition 49
 effect of alloying elements 293
 gating systems for composites 279–280
 fabricating characteristics 52
 mechanical properties 52
 production 45–46
 selection 2–3, 47–48
 tables
  composition of alloys 56–104, 105–112
  density 50
  electrical conductivity 51
  elongation 32
  solidification range table 50
  specific gravity 51
  tensile strength 32
  thermal conductivity 51
  thermal expansion coefficient 51
  yield strength 32
  Young's modulus 51
 uses for 45
Aluminum Association
 alloy numbering system 50–51
 registration record (pink sheets) 105–112
Aluminum oxide—*See also Alumina*
 formation 5
 on molten metal surface 21
 skimming off ingot 48
Annealing 290
Anodizing
 definition of 52
 effects of grain refining on 30
 microstructural appearance 322
 surface finish 342
Antimony
 modification 35
Antioch process 241
Archimedes Principle 22
Argon
 degassing 24–25, 26
Artificial nucleation—*See Refinement*
As-cast finishes
 die castings 147
 EPC castings 141
 investment castings 140
 permanent mold castings 132
 plaster mold castings 136
 sand castings 133
As-cast microstructure 287, 311–313
Autoclave dewaxing 259–260
Automated functions
 diecasting 160–161
  fully automated machines 161
  robots 161
 EPC 8, 269
 green sand casting 8
 investment casting 261
 ladling 8, 269
  filtration systems 17
 molding and mold handling 186–187
 permanent mold casting
  pouring 178
  tilt-pour machines 179
 shell process 225–226
 spraying die lubricants 159
 trimming castings 327

B

Bakelite 305
Bauxite 45
Binary alloys
 microstructure 307
Blastcleaning 325
BMI bond 128–129
Boron
 for grain refining 30
  commercial hardeners 30
Bosses
 design considerations 121–123
Brazing 52
Brinell hardness
 definition of 52
Bubble traps 279

C

CAD—*See Computer-aided design*
Carbide tools 334
Carbide-tipped saws 327
Cast surface comparator 339
Cast-in inserts
 design considerations 118, 125
 die castings 128–129
 permanent mold castings 131–132
Castability 1–3
 definition of 52
Casting processes
 centrifugal casting 245–250
 diecasting 145–166
 expendable pattern casting 263–270
 investment casting 251–261
 permanent mold casting 167–183
 plaster mold casting 235–243
 selecting 47–48
 shell process 223–234
Centrifugal casting 245–250
 centrifuge casting 246, 248–249

molds and cores 249
rotating speeds 249
spray coatings 249
defects 249–250
designing castings 137–139, 248
cores 138
directional solidification 138
draft 138
fillets 137, 138
machining allowances 137
undercuts 138
foundry practices 248
section thickness 138
semicentrifugal casting 245–246
tolerances 137–138, 138–139
true centrifugal casting 245
types of machines
horizontal axis 248
vertical axis 246–248
Centrifuge casting 138–139, 246, 248–249
—*See also Centrifugal casting*
cores 249
molds 249
rotating speeds 249
section thickness 138
spray coatings 249
Ceramic filters 277–279
Channel induction furnaces 14–15
Chaplets 118
Charging 6
Chemical conversion coating 52
Chemically bonded sand molding and coremaking 203–221
additives 217
binder identification, classification 203
binder system development 203
coldbox binder systems 207–210
acrylic/SO_2 208
epoxy/acrylic/SO_2 208–209
furan/SO_2 208
phenolic/CO_2 209
phenolic/methyl formate 209
phenolic/urethane/amine 207–208
silicate/CO_2 209–210
heat-activated systems 206–207
core oils 206
hotbox systems 207
shell process 206
warmbox systems 207
mold and core coatings 219–220
nobake binder systems 210–215
alkyd oil/urethane 211
ester phenolic 214
furan/acid 214
phenolic urethane 211–214
phenolic/acid 214–215
phosphate/metal oxide 211
polyol urethane 214
silicate/ester 211
relative humidity 220–221
sand testing
acid demand value 216
AFS grain fineness number 216
gas evolution 215–216
loss on ignition 215
mold quality indicator 216
permeability 216
tensile strength 215
transverse strength 215
shakeout 221
Chilling in permanent molds 130, 170, 172
Chills 284–285
Chinese script 311
Chlorine
degassing 24–25, 26
effects on modification 34
effects on phosphorus refinement 37
Chokes 271
Clay
calcium bentonite 192
fireclay 192
sodium bentonite 192–193
Cleaning castings 325–330
Cleaning fluxes 26
CNC machines for making dies 155
Coatings
chemical conversion coatings 342
for centrifugal casting molds 249
for EPS patterns 267
testing 267
for permanent molds 177
for resin-bonded molds and cores 219–220
parting compounds for plaster molds 239
surface coatings on castings 322
Coldbox binder systems 207–210
acrylic/SO_2 208
epoxy/acrylic/SO_2 208–209
furan/SO_2 208
phenolic/CO_2 209
phenolic/methyl formate 209
phenolic/urethane/amine 207–208
silicate/CO_2 209–210
Cold-chamber diecasting 150
Columnar grain structure 305
Combination dies 128, 154
Compactibility testing 198
Compressive yield strength
definition of 52
Computed tomography 331
Computer-aided design 117
for making dies 155
Computer-aided engineering 117
Contraction
design to minimize stress of 113–114
during solidification 2, 52
in die castings 157, 165–166
in shell process castings 225
Cooling rate
effect on permanent mold alloys 170
Coordinate measuring machines 261
Coreless induction furnaces 15
Cores
core oils 206
design considerations 129–130
centrifugal casting 138
coring holes 125, 132–133, 133, 136
EPC 141
permanent mold casting 130
sand casting 134–135
sand cores 117–118, 133
shell process 224–225
in diecasting 153, 158–159
in permanent mold casting 174–176
shell cores 227–229
strainer cores 25, 274
using chaplets for support 118
Coring 307–308
effect of heat treatment on 287, 315, 316
Corrosion
microstructural 322–323
Corrosion resistance 52
Covering fluxes 26
Cracking—*See Hot tearing*
Crucible furnaces 9–10
melting in 6
Crucibles
filtration systems 16
life expectancy 10
melting/holding 16
selection 10
Cryogenic treatments 291
Cryolite
for aluminum production 46
Crystallography 292–293

D

Dalton cores 133
Deburring 329
Defects
in centrifugal castings 249–250
inclusions 321
in die castings 164–166
table of 166
in EPC castings 269–270
in permanent mold castings 181–183
table of 182–183
in shells and shell castings 232–234
table of 233–234
macroscopic examination 304
metallographic examination 321–322
repairing 334–338
welding (table of) 337
Degassing
ingot production 49
methods 22–24
types of gases used 24–25
using fluxes 26
effects on phosphorus refinement 37
Dendrite arm spacing
effect of solidification rate on 311
permanent mold castings 168
Density
for determining hydrogen content 22
table of 50
Designing castings 113–143
calculating casting weight 118–119
centrifugal castings 137–139
choosing a process 113
choosing an alloy 47–48
composite designs 116
computer-aided design 117
computer-aided engineering 117
die castings 127–129, 129–130
checklist 128
economic considerations 116–117
EPC castings 141–143
checklist 143
general suggestions 117, 119–127
bosses and pads 121–123
cored holes 125
fillets 120
hot spots 119–120
joining sections 120
raised letters 126
ribs 121, 123–124
sharp internal corners 119–120
uniform thickness 119
webs 123, 124
investment castings 139–140

modelmaking 117
permanent mold castings 129–130, 130–133
plaster mold castings 136–137, 236
recording changes to design 118
safety factors 116
sand castings 133–135
shell mold castings 135–136
using sand cores 117–118
Diecasting 2, 145–166
Acurad process 164
advantages and disadvantages 147
alloys used for 150
automated functions 160–161
casting ejection 163
casting size 147, 148
choosing a casting method 48
combination dies 128, 154
compared to
forgings 149
permanent mold castings 149, 167
plastic moldings 149
sand casting 149
screw machine products 3, 149
shell process 223–224
defects 164–166
remedies for (table) 166
designing castings 127–129, 129–130
cast-in inserts 128–129
checklist 128
draft 128
parting lines 127
threads 129
die height adjustment 162
die lubricants 159
die release agents 159
dies
building 154–157
cores 153, 158–159
designing 154–157
ejector pins 153, 158
guide pins 154
gating 153, 157–158
hydraulic fluids 152–153
injection speeds and pressures 162
machines
air-operated 152
automated 161
cold chamber 150
hot chamber 150
hydraulically operated 152–153
types and sizes 150
vertical 150–152
machining 159–160
speeds 334
microstructure 320–321
multiple-cavity dies 128, 154
pore-free process 164–165
process control 160
prototypes 127, 136, 236
robots 161
section thickness 147
shot monitoring 160
shot stroke control 163
shrinkage 157
single-cavity dies 128, 154
squeeze process 163–164
surface finishes 147
temperature control 147, 154, 165
tolerances 127–128, 147, 148
unit dies 154
uses for castings 146
venting 153, 158
Dimensional inspection of castings 331
Direct-fired reverberatory furnaces 11
Directional solidification 280
general design factors 114–115
in centrifugal casting 138
in permanent mold casting 172
Draft
centrifugal castings 138
die castings 128
EPC castings 142, 143
investment castings 140
permanent mold castings 131, 132
shell castings 135
Drilling 334
Dross
drossing-off fluxes 26–27
formation 19–20
removal from ingot 48
Dual-energy reverberatory furnaces 13–14
Ductility 293
Dye penetrant inspection 331
DYPUR® system 278

E

Economic considerations
alloy selection 3
casting design 116–117
diecasting 145–147
furnace selection 8
permanent mold casting 169
postcasting operations 325
Ejector pins
design considerations 130
diecasting 153, 158
permanent mold casting 176
shell process 225, 228, 230
Electric furnaces
electric radiant reverberatory 13
induction 14–15
resistance 15
Electrical conductivity (table of) 51
Electrical discharge machine
for designing dies 155
Electromagnetic pumps 8
Electroplating
definition of 52
microstructural appearance 322
surface finish 342
Elongation
definition of 52
table of 32
Endurance limit 52
Environmental effects of chlorine 25
EPA regulations 25
Equiaxed grain structure 305
Etching
macroscopic examination 304–305
microscopic examination 306
surface finish 341–342
Eutectic Al-Si alloys
solidification characteristics 54–55
Eutectic arrest temperature
modified alloys 40
unmodified alloys 42
Eutectic structure
effect of modification on 32–33
in permanent mold alloys 170
microstructure 307–308
Evaporative pattern casting—*See Expendable pattern casting*
Expandable polystyrene 263, 264–267
molding cycle 264–265
pattern assembly 266
pattern storage and handling 265
preexpansion 264
tooling 266–267
Expendable pattern casting 2, 263–270
alloys used 269
casting quality 269–270
casting size 141
defects
folds 269–270
inclusions 270
porosity 269
designing castings 141–143
checklist 143
draft 142, 143
ribs 141
flasks 267–268
gating 268
pattern coatings 267
testing 267
patterns 141–142, 264–267
assembly 266
EPS molding cycle 264–265
EPS preexpansion 264
storage and handling 265
pouring 268–269
automatic 8, 269
sand 267
section thickness 142
surface finishes 141
tolerances 142–143
tooling 266–267
pulls and pullbacks 267

F

Fabricating characteristics
definitions of terms 52
Fatigue, defects caused by 323
Feeding
designing for 115–116, 130
EPC 268
permanent mold casting 171–172
effects of grain refining on 30
feeding distance of risers 283–284
Fillets
design considerations 120
centrifugal casting 137, 138
investment casting 140
permanent mold casting 131, 132
shell process 136
Filtering metal
effect on hydrogen content 22
in gating systems 277–279
removing large inclusions 25
systems for melting equipment 16–17
Finishing
anodized finishes 342
characteristics 339–341
chemical conversion coatings 342
cleaning, brightening and finishing 341–342
designations 339–341

effect of microstructure on finish 341
electroplating 342
measuring surface finish 338–339
mechanical finishes 341
painting 342
Flatness tolerance
plaster molding 137
sand casting 133–134
Fluid flow principles 271–272
Fluidity
considerations in gating systems 272
definition of 52
of hypereutectic alloys 54
Fluorescent powder inspection 331
Fluorides
choosing a flux 28
cleaning fluxes 26
in aluminum production 45–46
Fluxing
cleaning fluxes 26
covering fluxes 26
degassing 24, 26
effects on phosphorus refinement 37
determining which to use 27–28
drossing-off fluxes 26–27
effects of chlorine on modification 34
grain refining 30–31
Foam plaster process 242–243
Fractures
caused by fatigue 323
caused by overloading 323
macroscopic examination 304
Front charging reverberatory furnaces 13
Furnaces
charging 6
electric 14–15
heat treating 290–291
induction 14–15
operating temperatures 1
reverberatory 10–14
direct-fired 11
dual-energy 13–14
electric radiant 13
front charging 13
sloping dry-hearth 11
stack melting 11–12
wet bath 12–13
selecting 9
types of 8–9

G

Gas evolution, resin-bonded sands 215–216
Gas metal arc welding 335–336
Gas tungsten arc welding 335–336
Gating
diecasting 153, 157–158
EPC 268
horizontal systems 272–276
nonpressurized systems 273–276
permanent mold casting 171
tilt-pouring 179–180, 277
plaster molding 239
pressurized systems 273
recommended ratios 274
requirements 272
vertical systems 276
Grain refinement
chemicals used 30
amounts added 31
commercial hardeners 30
effects on alloy properties 28–30
fluxes 30–31
methods 30
reasons for 305
titanium additions 30
Grain size
AFS grain fineness number 187–189, 216
average 305
effect of solidification rate on 311
for green sand casting 187–189
test procedure 31
thermal analysis 37–43
Grain structure 305
Graphite lance
for degassing 23
for injecting powder fluxes 24
Green compression strength test 199–200
Green sand casting 187–202
additives
cellulose 195–196
cereal 196
automatic pouring 8
clay characteristics 191–192
deformation 192
dry strength 192
flowability 192
green strength 192
hot strength 192
wet strength 192
clays
calcium bentonite 192
fireclay 192
sodium bentonite 192–193
mulling 196–197
naturally bonded sands 194
oil-bonded molding sands 193
reusing sand 197
sand properties
grain shape 190–191
grain size 187–189
permeability 191
screen distribution 189–190
surface area 190
sand segregation 191
sand testing 197–202
AFS clay test 201
compactibility 198
green compression strength 199–200
loss on ignition 201
methylene blue clay 198–199
moisture 198
mold hardness testing 202
mold quality indicator 202
permeability 200–201
25 micron clay test 201
synthetic sands 194–195
water in sand mixtures 193
Grinding
abrasive-coated belts 329
grinding wheels 328–329
snagging and rough grinding 328
Guinier-Preston (GP) zones 293–295

H

Hardeners 48
commercial 30
Heat reclamation systems 14
Heat treatment 47, 287–299
age hardening 289
annealing 290
benefits of 287
cryogenic treatments 291
effect of alloy and casting method 48
effect of grain refining on 30
effects on microstructure 313–318
effects on properties 287–290
example of 295–296
furnaces for 290–291
GP zones 293–295
hardening effects 294–295
heat treatable alloys 291
metallurgical aspects 291–293
non-heat treatable alloys 291
of tool steels 156
Orowan looping 293, 294–295
Ostwald ripening 294
precipitation hardening 293–294
effect on microstructure 318
quenching 289
quenchant and quench interval 291
reducing residual stresses 291
solution treatment 287–289
effect on microstructure 314–316
non-heat treatable alloys 290
strengthening aluminum 293–295
stress relief or stabilization 290
tempers
designations 47, 289–290
solution and age 289–290
solution treatment for non-heat treatable alloys 290
Holding furnaces for diecasting 147–148
Holding time
effects on molten aluminum 6
effects on phosphorus refinement 37
Holes
diecasting 129
EPC 142
investment casting 140
permanent mold casting 132–133
sand casting 133
shell molding 136
Horizontal gating systems 272–276
Hotbox binder systems 207
Hot chamber diecasting 150
Hot isostatic pressing 338
Hot shortness—*See Hot tearing*
Hot spots
designing to minimize 114–115, 119–120
in hypereutectic alloys 55
in sand castings 134
Hot tearing 19, 322
designing to minimize 114
effects of grain refining on 28
in 713.0 55
in die castings 165–166
in permanent mold castings 172–173
induced porosity to combat 27–28
resistance to 52
Hotbox binder systems 207
Hycon test—*See Initial bubble test*
Hydraulically operated diecasting machines 152
hydraulic fluids 152–153
Hydrogen
hot isostatic pressing 338
in molten aluminum 2, 5, 20–21

measuring concentration
density measurements 22
initial bubble test 22
quantitative reduced pressure test 22
recirculating gas test 22
reduced pressure test 21–22
porosity 20–25, 321
Hypereutectic alloys
foundry characteristics 53–56
general properties 53
microstructure 307–308
modification 35–37
Hypoeutectic alloys
microstructure 307–308
modification 32–35

I

Impregnation 336–338
Inclusions
effect on fluidity 272
effect on hydrogen porosity 22
filtration to remove 25, 277–279
microscopic appearance 321
Induction furnaces
channel 14–15
coreless 15
Inert gases for degassing 24–25, 26
Ingot
aluminum hardeners 48
degassing of 49
for rotors 48
modification and grain refinement 49
quality assurance 48–49
standard alloyed 48
standard pure 48
Initial bubble test 22
Inoculations
grain refining 28
refining using phosphorus 35–37
Inserts
cast-in
design considerations 118, 125
die castings 128–129
permanent mold castings 131–132
in permanent molds 174
Inspection of castings 330–332
Investment casting 251–261
advantages 139
alloys used 260
automation 261
ceramic shell process
control of ceramic slurries 258–259
mold drying 258
mold setting or gelling 258
stucco materials 257–258
cores 254
designing castings 139–140
draft 140
fillets 140
threads 140
dewaxing ceramic shell molds
autoclave dewaxing 259–260
flash firing 260
dewaxing solid molds 259
melting metals 260
mold firing 260
mold materials
colloidal silica binders 255
ethyl silicate binders 255
refractories 255–256
patternmaking
dimensional stability of patterns 253
materials used 252–253
pattern assembly 254–255
wax injection 253
postcasting
inspection 261
knockout 261
machining 261
pouring 260
prototypes 139
section thickness 140
solid mold process binders 256–257
surface finishes 140
tolerances 140
tooling costs 139–140

J–K

Jet melters—*See Stack melters*
Keller's etch 306
Knockout of core material 326

L

Ladling
automatic 8, 269
filtration systems 16–17
permanent mold casting 178, 181
Lamellar structure 40, 41
Latent heat of crystallization 54
Launder system 147–148
Law of Continuity 271–272
Leak testing of castings 331–332
Loss on ignition 201, 215
Lost foam process—*See Expendable pattern casting*
Lost wax process—*See Investment casting*
Low-pressure permanent molding 181
Lubricants
diecasting 159
producing shell molds 230

M

Machinability
definition of 52
effects of grain refining on 30
tool steels 156
Machining 332–334
allowances
centrifugal castings 137
design considerations 117, 130
die castings 157
EPC castings 143
investment casting 140
permanent molding 132
shell process 136
die castings 159–160
locating points 117, 126, 132
speeds and feeds
band saws 327
boring 344, 345
cutting 333–334
drilling 334, 348
face milling 346, 347
milling cutters 334
tapping 334, 347
turning 343, 344
Macrostructure examination 303–305
Magnesium
alloying with 6
effect on pure aluminum 166
Master alloys
adding alloying elements 6
grain refinement 30
strontium modification 34–35
Mechanical finishes 341
Mechanical properties
definitions of terms 52
effects of grain refining on 29
effects of modification on 32
general 46–47
tests for determining 332
Melting practices 5–6
permanent mold casting 177–178
Metal inserts (cast-in)
design considerations 118, 125
die castings 128–129
permanent mold castings 131–132
Metallography 303–323
alloy type identification 311
finding defects 321–322
macroscopic examination 303–305
microscopic examination 305–311
sampling 303
Methylene blue clay testing 198–199
Microstructure 303–323
as cast 287, 311–313
constituent phases 292
coring 307–308
crystal structures 292–293
effect of heat treatment on 313–318
effect of solidification rate on 311
effect on surface finish 341
examination of 305–311
grain refined structure 28
in die castings 320–321
lamellar 40, 41
modified 318
sodium 32–34, 40–43
strontium 34–35, 40–43
of hypereutectic alloys 53
control of 55
of tool steels 156
refined 318
second-phase constituents 309–311
unmodified 40
Milling cutters 334
Modification
effect on mechanical properties 32
effect on microstructure 318
eutectic alloys 32
hypereutectic alloys 35–37
hypoeutectic alloys 32–35
using antimony 35
using sodium 32–34, 40–43
using strontium 34–35, 40–43
Moisture testing 198
Mold hardness testing 202
Mold quality indicator 202, 216
Molds
automated handling 186–187
design for low cost 116–117

permanent 176–177
Molten aluminum
characteristics 19–43
filtration 16–17, 277–279
flow rate 271
fluidity 272
pumping 17–18, 49
electromagnetic pumps 8
transfer 46, 147–148, 177
velocity 271
Monitoring systems for diecasting 160
Movable cores
diecasting 158–159
Mulling 196–197
Multiple-cavity dies 128, 154

N

Nitriding 156–157
Nitrogen degassing 24–25, 26
effects on phosphorus refinement 37
Nobake binder systems 210–215
alkyd oil/urethane nobake 211
ester phenolic 214
furan/acid 214
phenolic urethane 211–214
phenolic/acid 214–215
phosphate/metal oxide 211
polyol urethane 214
silicate/ester 211
Non-heat treatable alloys 291
Nondestructive testing
dimensional inspection 331
fluorescent powder, dye penetrant 331
leak tests 331–332
radiographic (x-ray) inspection 331
ultrasonic inspection 332
visual inspection 330
Nonpressurized gating systems 273–276
gates 275–276
runners 274–275
vertical gating 276

O

Oil-bonded green sand molding 193
Operating temperatures 1
producing shell cores 228
producing shell molds 229
Optical emission spectrograph 49
Orowan looping 293, 294
OSHA regulations for chlorine 25
Ostwald ripening 294
Oxides
buildup removal 5
formation 5
in remelted scrap 178
removal by degassing 24
tendency to form 2

P

Pads
design considerations 121
sand castings 134
Painting casting surfaces 342
Parting lines
lack of on EPC castings 141
lack of on investment castings 139
on cores 118
on die castings 127
on sand castings 133
Patterns
design for low cost 116–117
EPC 141–142, 264–267
Peening 335
Permanent mold casting 167–183
advantages and disadvantages 168–169
alloys used for 170
applications 169
automated functions 178, 179
chilling 170, 172
choosing a casting method 48
compared to
diecasting 149, 167
sand casting 167
shell process 223–224
cores 174–176
defects 181–183
table of 182–183
designing castings 129–130, 130–133
cast-in inserts 131–132
cores 130, 132–133
draft 131, 132
factors that affect soundness 170–172
feeding 171–172
fillets 131, 132
gating 171
raised letters 132, 168
undercuts 131
venting 170–171
directional solidification 115, 172
economic considerations 169
fluxing 27–28
gravity method 2
hot cracking 172–173
low-pressure method 2, 181
melting practices 177–178
mold coating 177
mold materials 173–174
inserts 174
mold operation 176–177
pouring castings 178–181
quality control 181
section thickness 115, 130
solidification 172
step-by-step description 167
surface finishes 132
tilt-pouring 179–181, 277
gating 179–180, 277
mold temperatures 180
pouring time 180–181
venting 180
tolerances 130–131, 174
Permeability testing 200–201, 216
Phosphorus refinement 35–37
effects of sodium and strontium 37
Pinhole porosity—*See Hydrogen: porosity*
Plaster molding 235–243
characteristics (table) 235
cores 237, 243
description of process 236–238
designing castings 136–137, 236
drying the mold 239–240
gating 239
mold shrinkage 240–241
parting compounds 239
pattern material 238
pattern removal 239
permeable plaster methods
Antioch process 241
foamed plaster process 242–243
plasters 238–241
mixing 238–239
pouring 239
pressure cast process 241
prototypes for diecasting 127, 136, 236
section thickness 137, 235
shrinkage 235
surface finishes 136, 235
tolerances 136–137
cores 137
Plunger lubricants 159
Plunger speeds 162
Polishing rating 52
Pore-free process 164
Porosity
gas vs. shrinkage 321
hydrogen 20–25
in EPC castings 269
induced 27–28
voids in die castings 147
Postcasting
diecasting trimming, finishing 159–160
investment casting
inspection 261
knockout 261
machining 261
machining 332–334
removing core material 325–327
removing gates and risers 327
removing mold material 325
Pouring
automatic 269
EPC 268–269
ladle filtration systems 16–17
permanent mold casting 178–181
shell molds 232
Pouring basin 273
Pouring rate
effects on phosphorus refinement 36–37
preventing collapse in EPC 268
Precipitation hardening 293–294
effect on microstructure 318
Pressure tightness 52
Pressurized gating systems 273
Process control
diecasting 160
shell process 232
Progressive solidification 54, 114–115, 280
—*See also Directional solidification*
Prototypes 127, 136, 139, 236
Pumping molten aluminum 17–18, 49

Q

Quality control
die castings 164–166
ingot 48–49
inspection of castings 330–332
investment castings 261
permanent mold castings 181
repairing castings 334–338
shell process 232
Quantitative reduced pressure test 22

Quenching
during solution treatment 289
quenchant and quench interval 291

R

Radiographic inspection of castings 331
Raised letters 126, 132, 168
Recirculating gas test 22
Reclamation
EPC sand 267
resin-bonded sand 217–219
mechanical reclamation 217–219
thermal reclamation 219
wet reclamation 219
Recuperators 14
Reduced pressure test 21–22
Reduction cells 45–46
Refinement
effect on microstructure 318
using phosphorus 35–37, 55
effects of sodium and strontium 37
Refractories
in investment casting 255–256
reverberatory furnace linings 10
Regenerators 14
Relative humidity effect on sand binders 220–221
Release agents for diecasting 159
Repairing castings
hot isostatic pressing 338
impregnation 336–338
peening 335
straightening 334–335
Residual stress reduction 291
Resistance furnaces 16
Reuse of sand 197, 219
Reverberatory furnaces 10–14
direct-fired 11
dual-energy 13–14
electric radiant 13
front charging 13
sloping dry hearth 11
stack melting 11–12
wet bath 12–13
Ribs
design considerations 121, 123–124
Risering
feeding distance 283–284
permanent mold casting 171–172
riser aids 282–283
riser shape 281–282
riser size 282
Robots
used in diecasting 161
used in investment casting 261
Rotary files 329
Rotary impeller degassing 23
Runners
diecasting 157–158
in nonpressurized gating 274–275

S

Safety factors
design considerations 116
Sample preparation 303
etching 304–305, 306
mechanical 304–305, 306
Sand
angular grains 191
chromite 195
compounded grains 191
naturally bonded 194
olivine 195, 217
rounded grains 191
silica 194–195, 217
subangular grains 191
synthetic sand 194–195
zircon 195, 217
Sand casting 185–221
—See also Chemically bonded sand molding and coremaking; Green sand casting
alloys used for 186
automated molding 186–187
choosing a casting method 48
compared to
diecasting 149
permanent mold casting 167
shell process 223–224
designing castings
cores 133, 134–135
pads 134
parting lines 133
undercuts 134
flatness tolerance 133–134
fluxing 27–28
machining 117
sand testing 197–202
acid demand value 216
AFS clay test 201
compactibility 198
gas evolution 215–216
green compression strength 199–200
loss on ignition 201, 215
methylene blue clay 198–199
moisture 198
mold hardness testing 202
mold quality indicator 202, 216
permeability 200–201, 216
tensile strength 215
transverse strength 215
25 micron clay test 201
surface finishes 133
tolerances 133
Sawing 327–328
Scrap melting and charging 6
Screen distribution of sands 189–190, 216
Second-phase constituents 309–311
Secondary operations—*See Postcasting*
Section thickness
centrifugal castings 138
die castings 147
EPC castings 142
investment castings 140
permanent mold castings 130, 170
plaster mold castings 137
shell castings 135–136
Segregation 318, 322
Semicentrifugal casting 245–246
Shakeout
of hypereutectic alloys 55
of resin-bonded molds and cores 221
of shell molds 232
Shaw Process 155
Shearing strength 52
Shell process 223–234
advantages 223
automated functions 225–226
compared to other processes 223–224
coreboxes 224–225
defects 232–234
table of 233–234
designing castings
cored holes 136
draft 135
fillets 136
threads 136
disadvantages 223
ejector pins 225, 228, 230
equipment 225–226
materials
binders 227
fillers 227
sand 226–227
patterns 224–225
pouring 232
process control 232
producing cores 206, 227–229
operating temperatures 228
producing molds 229–232
assembly 230–231
blown molds 230
curing 230
heating the pattern 229–230
investment (dwell) time 230
lubricating the pattern 230
operating temperatures 229
sand coating methods 206
shakeout 232
shrinkage allowance 225
surface finishes 135
defects 232–233
testing 232
tolerances 135
tooling costs 135
Shot monitors 160
Shrinkage
design to minimize stress of 113–114
during solidification 2, 281
in die castings 157, 165–166
in shell process castings 225
tendency during solidification 52
Shrinkage porosity
effects of grain refining on 29
microstructural appearance 321
Sievert's Law 22
Silicon
alloying with 6
effect of heat treatment on 316
effect on fluidity 54
effect on permanent mold alloys 170
effect on pure aluminum 166
eutectic structure 307
in hypereutectic alloys 53
surface depletion 55
in phosphorus-refined alloys 35–36
structure in modified and unmodified alloys 32–35
Single-cavity dies 128, 154
Sloping dry-hearth reverberatory furnace 11
Sodium
effects on phosphorus refinement 37
modification 32–34, 40–43
Solidification 280–281
designing to minimize stress 113–114
directional 114–115
in permanent mold casting 172
isothermal 54–55

software to predict 117
Solidification range 280–281
table of 50
Solidification rate 280
effect on microstructure 311
effect on phosphorus refinement 36
Solution heat treatment 287–289
effect on microstructure 314–316
Specific gravity 1
table of 51
Spectrograph 49
Sprue base 274
Squeeze process 163–164
Stabilization treatment 290
Stack melters 11–12
Statistical process control 49
Straightening castings 334–335
Strainer cores 25
Straube-Pfeiffer test—*See Reduced pressure test*
Strength
designing for 114
increasing by heat treating 293–295
Stress
calculating 114
controlling residual stresses 291
heat treatment to relieve 290
minimizing during design 113–114
safety factors 116
Strontium
effects on phosphorus refinement 37
modification 34–35, 40–43
effect of solution treatment time 288
Surface coatings 322
Surface finish
die castings 147
EPC castings 141
investment castings 140
measuring 338–339
permanent mold castings 132
plaster mold castings 136
sand castings 133
shell mold castings 135
defects 232–233

T

Tapping 334
Temperature
control 7–10
diecasting 147, 154, 165
shell process 229
tilt-pour permanent molding 180
effect on fluidity 272
effect on green sand systems 197
effect on phosphorus refinement 36
effects on molten aluminum 6–7
measurement 7–10
Tempers—*See Heat treatment*
Tensile properties
at low temperatures 47, 73
Tensile strength
definition of 52
table of 32
testing in resin-bonded sand 215
Thermal analysis 37–43
Thermal conductivity
table of 51
Thermal expansion coefficient
of hypereutectic alloys 53
table of 51
Thermal treatment—*See Heat treatment*
Thermocouples 7–10
Threads
in die castings 129
in investment castings 140
in shell castings 136
Tilt-pour permanent mold casting 179–181
gating 277
Titanium
for grain refining 30
commercial hardeners 30
Tolerances
in centrifugal castings 137–138, 138–139
in die castings 127–128, 147, 148
in EPC castings 142–143
in investment castings 140
in permanent mold castings 130–131, 174
in plaster mold castings 136–137
cores 137
in sand castings 133
in shell process 135
Tooling
EPC 266–267
investment casting 139–140
shell process 135
Tool maintenance 6
Tool steel for making dies
heat treatment 156
machinability 156
nitriding 156–157
resistance to heat checking 155–156
resistance to softening 156
soundness 156
surface finish 156
toughness 156
Transverse strength testing 215
Trimming
of die castings 159–160
of hypereutectic alloys 55–56
trim presses 327
Turbulence
effects on phosphorus refinement 37
elimination in tilt-pouring 179
minimizing in diecasting 158
25 micron clay test 201

U

Ultimate strength—*See Tensile strength*
Ultrasonic inspection of castings 332
Undercuts
centrifugal casting 138
diecasting 147
investment casting 140
permanent mold casting 131
sand casting 134
Unit dies 154
Unmodified alloys
characteristics 32
effect of solution treatment time 288
eutectic arrest temperature 42
microstructure 40
structure 33
Uprunners 276

V

Vacuum impregnation 336
Venting
diecasting 153, 158
permanent mold casting 170–171
tilt-pouring 180
resin-bonded sand casting 220
Vertical diecasting machine 150–152
Vertical gating systems 276–277
Visual inspection of castings 330
Voids—*See Porosity*

W

Warmbox binder systems 207
Web gates 276
Weldability 52
Wet bath reverberatory furnaces 12–13

X

X-ray inspection of castings 331

Y

Yield strength
definition of 52
table of 32
Young's modulus (table of) 51

Z

Zirconium
effect on pure aluminum 166
for grain refining 30